AI Deep Learning in Image Processing

Image processing plays a crucial role in various fields, including digital multimedia, automated vision detection and inspection, and pattern recognition. *AI Deep Learning in Image Processing* aims to provide a comprehensive overview of the mechanisms and techniques involved, with a particular focus on the application of advanced AI deep-learning technologies in image processing.

The field of image processing has experienced unprecedented growth in recent years. Image processing presents the latest state-of-the-art developments alongside clear explanations of fundamental concepts and contemporary applications. By emphasizing essential principles, this book enables readers to not only implement algorithms and techniques with ease but also identify new challenges and explore innovative applications in the field.

Features:

- New and advanced AI deep-learning techniques for image processing as comparing against traditional image-processing methods
- Numerous practical examples and AI image-processing-related applications
- A more intuitive development and clear explanation to the complex technology
- Updated image-processing technology in medical, chemical, and ecological fields
- Extensive discussions of performance comparisons of various AI deep-learning image-processing methods

This book is designed for students, researchers, and professionals seeking to enhance their knowledge, gain practical insights, and explore the evolving role of image processing in modern technology.

AI Deep Learning in Image Processing

Frank Y. Shih

CRC Press
Taylor & Francis Group
Boca Raton London New York

CRC Press is an imprint of the
Taylor & Francis Group, an **informa** business

Designed cover image: Getty Images

First edition published 2026
by CRC Press
2385 NW Executive Center Drive, Suite 320, Boca Raton FL 33431

and by CRC Press
4 Park Square, Milton Park, Abingdon, Oxon, OX14 4RN

CRC Press is an imprint of Taylor & Francis Group, LLC

ISBN: 978-1-032-75530-4 (hbk)
ISBN: 978-1-032-75532-8 (pbk)
ISBN: 978-1-003-47439-5 (ebk)

DOI: 10.1201/9781003474395

Typeset in Times
by Apex CoVantage, LLC

To my beloved wife and children, whose endless love, support, and understanding have been my greatest source of strength.

To my dear parents, whose unwavering encouragement has guided me through every step of my journey.

And to all those whose kindness and help made the creation of this book possible – my heart is forever grateful.

Contents

PART II Fundamentals of AI Deep Learning

PART III Practical Applications

Preface

Image processing plays a vital role in today's technology-driven world, fueled by the widespread use of digital cameras, camcorders, multimedia devices, and the rapid expansion of the Internet. Its applications extend across diverse fields, including remote sensing and forensic data analysis. The field of image processing has experienced unprecedented growth in recent years. *Image Processing* presents the latest state-of-the-art developments alongside clear explanations of fundamental concepts and contemporary applications. By emphasizing essential principles, this book enables readers to not only implement algorithms and techniques with ease but also identify new challenges and explore innovative applications in the field.

This book presents a comprehensive range of both fundamental and advanced image-processing algorithms, alongside cutting-edge AI deep-learning techniques, to illustrate a cohesive framework. It offers valuable technical guidance and practical tools to help readers understand and implement the essential principles of image processing with confidence and clarity.

The topics covered in this book encompass a wide range of fundamental and advanced concepts in image processing. These include basic image transformations, enhancement and segmentation techniques, and multimedia signal processing through mathematical morphology. Key areas, such as distance transformation, shortest path planning, image representation and description, and feature extraction, are thoroughly explored.

The book also delves into advanced machine learning and artificial intelligence methods, including active learning, neural networks, support vector machines, fuzzy logic, and pattern recognition. Application-specific areas are highlighted as well, featuring practical use cases in drug toxicity classification, ecological data classification, medical image classification, image watermarking, and adversarial attack detection. Together, these topics provide readers with a comprehensive understanding of both theoretical foundations and real-world applications in the field.

This book is structured into three comprehensive parts to guide readers from foundational concepts to practical applications:

- **Part I** covers the fundamentals of image processing, providing the essential principles and techniques that form the basis of the field.
- **Part II** explores the core concepts of AI and deep learning, offering a solid understanding of the algorithms and architectures that drive modern advancements.
- **Part III** focuses on a variety of practical applications, demonstrating how the theories and methods discussed in the earlier sections are implemented in real-world scenarios.

This organized approach ensures a clear progression from foundational knowledge to advanced applications, making the content accessible and valuable to readers at all levels.

The principles of image processing presented in this book are illustrated with abundant graphs and practical examples, designed to simplify complex concepts and make both the theoretical foundations and real-world applications easily accessible to readers. This book aims to provide a comprehensive overview of various mechanisms and techniques in image processing. It is tailored for students, researchers, and professionals who seek to deepen their understanding, enhance their knowledge, and gain valuable insights into the field. By thoroughly exploring the essential principles of the underlying algorithms and architectures, readers will not only develop innovative ideas for implementing advanced techniques but also be inspired to identify and address emerging challenges in the discipline.

This book guides readers through the analysis of key principles and architectures, helping them leverage the author's innovative approaches to implement advanced algorithms and develop practical, detailed plans to foster their own ideas.

Covering both the historical evolution and state-of-the-art techniques in image processing and pattern recognition, the text is enriched with practical examples that bridge theory and real-world application. A clear, tutorial-style approach to complex technologies, complemented by intuitive tools, ensures that readers gain a thorough understanding of the subject matter.

An updated bibliography, along with numerous informative graphs and illustrations, further supports the learning process. The book also introduces several new algorithms, enabling readers to adapt and refine them to develop their own solution strategies.

Praised as an invaluable reference, this volume provides essential data and methodologies to master current theoretical advancements and practical applications – empowering readers to not only understand existing solutions but also create innovative ones of their own.

Acknowledgments

Portions of this book originated from earlier works, including conference papers, journal articles, and theses developed in collaboration with my students at the New Jersey Institute of Technology. As a result, some sections reflect a blend of my words and those of my students, whose contributions I deeply appreciate.

I am sincerely grateful to the following publishers for granting permission to reuse texts and figures from my previous publications: the Institute of Electrical and Electronics Engineers (IEEE), Elsevier, and World Scientific Publishing. Their generosity has been invaluable in bringing this book to fruition.

Frank Y. Shih
New Jersey Institute of Technology

Author

Frank Y. Shih, Ph.D., received his B.S. from National Cheng Kung University, Tainan, Taiwan, in 1980; M.S. from State University of New York, Stony Brook, U.S.A., in 1984; and Ph.D. from Purdue University, West Lafayette, Indiana, U.S.A., in 1987. He is a professor jointly appointed in the Department of Computer Science, the Department of Electrical and Computer Engineering, and the Department of Biomedical Engineering at New Jersey Institute of Technology, Newark, New Jersey. He currently serves as director of Artificial Intelligence and Computer Vision Laboratory.

Dr. Shih held a visiting professor position at Princeton University, Columbia University, National Taiwan University, National Institute of Informatics (Tokyo, Japan), Conservatoire National Des Arts Et Metiers (Paris, France), and Nanjing University of Information Science and Technology (China). He is an internationally renowned scholar and currently serves as editor-in-chief for the *International Journal of Pattern Recognition and Artificial Intelligence.* He was editor-in-chief for the *International Journal of Multimedia Intelligence and Security.* In addition, he is on the editorial board of 12 international journals. He has served as a steering member, session chair, and committee member for numerous professional conferences and workshops. He has received numerous grants from the National Science Foundation, the NIH, NASA, the Navy and Air Force, and industry companies Industry. He has won the Research Initiation Award from NSF, the Outstanding Teaching Award and the Board of Overseers Excellence in Research Award from NJIT, and the Best Paper Awards from journals and conferences.

Dr. Shih is internationally recognized as an expert in artificial intelligence and pattern recognition, deep learning, watermarking, steganography, and forensics. He has authored seven books, including *Digital Watermarking and Steganography, Image Processing and Mathematical Morphology, Image Processing and Pattern Recognition*, and *Multimedia Security: Watermarking, Steganography, and Forensics.* He has published over 160 journal papers, 110 conference papers, and 23 book chapters. His current research interests include artificial intelligence, deep learning, image processing, watermarking and steganography, digital forensics, pattern recognition, bioinformatics, biomedical engineering, fuzzy logic, and neural networks.

Part I

Fundamentals of Image Processing

1 Introduction

An image is a subset of a signal. A *signal* is a function that carries information, often revealing the behavior of a physical system or the characteristics of a phenomenon. From traffic lights to high-tech communication systems, signals are all around us. Take, for instance, the traffic signal. Using three universal colors – red, yellow, and green – it conveys clear and vital instructions: stop, prepare to move, or go. This simple example illustrates how signals use patterns to communicate information, regardless of the medium through which they're transmitted.

Although signals can be represented in many ways, in all cases the information is contained in a pattern of variations of some form, with that information typically is transmitted and received over a medium. Electrical signals, for example, involve variations in current or voltage and are essential in technologies like radios, radars, telephones, and televisions. Meanwhile, acoustic wave signals carry sound information, such as speech or music. When people talk about a "strong" or "weak" signal, they are often referring to the clarity and audibility of the sound. Other signals include temperature readings from a thermocouple or acidity levels from a pH meter. These demonstrate that signals can represent virtually any kind of data, making them a universal tool for conveying information.

A signal takes its patterns and variations, which can change over time or space. Mathematically, signals are represented as functions of one or more independent variables. These variables can either be continuous – changing smoothly over time – or discrete, occurring at specific intervals. *Continuous-time* signals are defined at a continuum of the time variable. *Discrete-time* signals are defined at discrete instants of time. *Digital* signals are those for which both time and amplitude are discrete. The continuous time and amplitude signals are called *analog* signals. When analog signals are converted into digital forms, they become compatible with computers and digital devices, unlocking the potential for advanced processing and analysis.

Signal processing is the process of extracting meaningful information from signals. *Digital signal processing* (DSP) is a field focused on representing signals as sequences of numbers or symbols and manipulating these sequences to uncover valuable insights. It was initiated in the 17th century and has become an important modern tool in the tremendously diverse fields of science and technology. The purpose of such processing is to estimate characteristic parameters of a signal or to transform a signal into a form that is more interpretable to humans.

DSP is a versatile field with numerous specialized areas, including: digital image processing, video processing, statistical signal processing, signal processing for communications, biomedical signal processing, audio and speech signal processing, sonar and radar signal processing, sensor array processing, spectral estimation, and so on. From enhancing medical imaging to enabling reliable communication, DSP continues to reshape how we interact with and interpret the world around us.

DOI: 10.1201/ 9781003474395-2

Human beings are equipped with a remarkable natural signal-processing system. The act of "seeing" occurs in the *human visual system* (HVS), while "hearing" takes place in the *auditory system*. These biological systems excel at interpreting signals from our environment, enabling a wide array of functions essential for survival and interaction. The human visual system plays a central role in activities like navigation, identification, verification, interpreting gestures, analyzing posture, and even psychological assessments. It allows us to decode visual cues for communication and understanding the world around us. Similarly, the human auditory system transforms sound waves into nerve impulses, enabling us to analyze auditory events, recognize sound sources, remember them, and perceive sequences of sounds. Together, these systems form a foundation for our ability to interact with and interpret our environment.

As the speed, capability, and affordability of modern signal-processing devices continue to advance, there has been a surge in efforts to create sophisticated, real-time automatic systems that mimic human abilities. These systems aim to replicate the incredible efficiency and adaptability of natural signal processing, further bridging the gap between human and machine intelligence. The rise of the digital revolution has transformed the way signals are used and processed. Digital signals now dominate, powering most household electronic devices and forming the backbone of modern technology. The Internet itself is a vast network of digital signals, as is the infrastructure of mobile phone communication. From smart home appliances to global connectivity, digital signal processing has become integral to daily life.

1.1 THE WORLD OF SIGNALS

The world around us is brimming with countless types of signals, each carrying its own unique physical meaning. However, the human body often lacks the ability to detect or interpret certain signals, leaving the information they carry out of reach. Far from being meaningless or insignificant, these elusive signals are precisely what scientists and researchers strive to understand. Each discovery brings us closer to harnessing their potential, improving our lives, and safeguarding our planet.

Understanding the signals of the natural world can help us create better living environments and even avert disasters. For instance, historical accounts suggest that animals can detect signals humans cannot. In 373 B.C., animals like rats, snakes, and weasels fled the Greek city of Helice days before a devastating earthquake struck. Similarly, many pet owners have reported unusual behavior in dogs and cats – such as barking, whining, or restlessness – before an earthquake occurs. These examples underscore the importance of learning from the world's hidden signals. By decoding them, we gain not only insight into the workings of nature but also the ability to predict and prepare for events that might otherwise catch us unawares.

The characteristics of a signal may be one of a broad range of shapes, amplitudes, time durations, and perhaps other physical properties. Based on the sampling of time axis, signals can be divided into continuous-time signals and discrete-time signals. Based on the sampling of time and amplitude axes, signals can be divided into analog signals and digital signals. If signals repeat in some period, they are called periodic signals; otherwise, aperiodic or nonperiodic signals. If each value of a signal is fixed by a mathematical function, it is called a deterministic signal; otherwise, a random signal has uncertainty about its behavior. In the category of dimensionality, signals are divided into one-dimensional (1D) signals, two-dimensional (2D) signals, three-dimensional (3D) signals, and multidimensional signals, which are further explained in the following sections.

1.1.1 One-Dimensional Signals

A one-dimensional (1D) signal is typically modeled as an ensemble of time waveforms, such as $x(t)$ or $f(t)$. This type of signal processing has a long and rich history, playing a vital role in various fields, including biomedical engineering (Chakraborty et al., 2021), acoustics (Li and Cox, 2019), sonar (Le, 2002), radar (Gamba, 2019), seismology (Mousa, 2020), speech communication, and many others. When we are using a telephone, our voice is converted to an electrical signal and through telecommunication systems circulates around Earth. The radio signals, which are propagated through free space and by radio receivers, are converted into sound. In speech transmission and recognition, one may wish to extract some characteristic parameters of the linguistic messages, representing the temporal and spectral behavior of acoustical speech input. Alternatively, one may wish to remove interference, such as noise, from the signal or to modify the signal to present it in a form from which is more easily interpreted by an expert.

1.1.2 Two-Dimensional Signals

Signal processing extends far beyond the realm of 1D signals. A 2D signal is a function of two independent variables, often represented as $f(x, y)$. In particular, the functional behavior of a 2D signal is frequently expressed as an intensity variation across the (x, y)-plane. Everyday scenes viewed by a human observer can be considered to be composed of illuminated objects. The light energy reflected from these objects can be considered to form a 2D intensity function, which is commonly referred to as an *image*.

As a result of numerous applications, not least as a consequence of cheap computer technology, image processing now influences almost all areas of our daily life: automated acquisition, processing and production of documents, industrial process automation, acquisition and automated analysis of medical images, enhancement and analysis of aerial photographs for detection of forest fires or crop damage, analysis of satellite weather photos, and enhancement of television transmission from lunar and deep-space probes.

1.1.3 Three-Dimensional Signals

Photographs of a still scene are the images which are functions of the (x, y)-plane. By adding a time variable, the 3D signals represent image sequences of a dynamic scene that are called *video signals*. Computer analysis of image sequences requires the development of internal representations for the entities in a depicted scene as well as for discernible changes in appearance and configuration of such entities. More fundamental approaches result from efforts to improve application-oriented solutions. Some illustrative examples are given as follows.

Image sequences obtained from satellite sensors are routinely analyzed in order to detect and monitor changes. Evaluation of image series recorded throughout the growth and harvest periods can result in more reliable cover type mapping as well as improved estimates of crop field. Very important is the determination of cloud displacement vector fields. These are used to estimate wind velocity distributions, which in turn are employed for weather prediction and meteorological modeling (Desportes et al., 2007).

Biomedical applications are concerned with the study of growth, transformation, and transport phenomena. Angiocardiography, blood circulation, and studies of metabolism are the primary areas of medical interests for the evaluation of temporal image sequences (Charalampidis et al., 2006). Architects who have to design pedestrian circulation areas would appreciate quantitative data about how pedestrians walk in halls and corridors. Efforts to extract such data from TV-frame sequences could be considered as behavioral studies. They might as well be assigned to a separate topic like object tracking (Qu and Schonfeld, 2007), which is of special concerns in cases of traffic monitoring (Zhou et al., 2007), target tracking, and visual feedback for automated navigation (Negahdaripour and Xun, 2002; Xu and Tso, 1999).

1.1.4 Multidimensional Signals

When a signal is represented in more than one dimension, it is often called a multidimensional signal. As discussed in previous sections, an image is a 2D signal, and a video is a 3D signal. A multidimensional signal is vector valued and may be a function of multiple relevant independent variables. One chooses the variable domain in which to process a signal by making an informed guess as to which domain best represents the essential characteristics of the signal. Multidimensional signal processing is an innovative field interested in developing technology which can capture and analyze information in more than one dimension. Some of its applications include 3D face modeling (Roy-Chowdhury et al., 2004), 3D object tracking (Wiles et al., 2001), and multidimensional signal filtering.

The need for a generally applicable artificial intelligence approach for optimal dimensionality selection in high-dimensional signal spaces is evident in problems involving vision since the dimensionality of the input data often exceeds 10^6. It is likely to fail if vision problems are handled by reducing the dimensionality by means of throwing away almost certain available information in a basically ad hoc manner. Therefore, designing a system capable of learning the relevant information extraction mechanisms is critical.

1.2 DIGITAL IMAGE PROCESSING

Images are produced by a variety of physical devices, including still and video cameras, scanners, X-ray devices, electron microscopes, radar, and ultrasound, and are used for a variety of purposes, including entertainment, medical, business, industrial, military, civil, security, and scientific. The interests in digital image processing stem from the improvement of pictorial information for human interpretation and the processing of scene data for autonomous machine perception.

Webster's Dictionary defines an *image* as "[a]n image is a representation, likeness or imitation of an object or thing, a vivid or graphic description, something introduced to represent something else." The word "picture" is a restricted type of image. *Webster's Dictionary* defines a *picture* as "[a] representation made by painting, drawing, or photography; a vivid, graphic, accurate description of an object or thing so as to suggest a mental image or give an accurate idea of the thing itself." In image processing, the word "picture" is sometimes equivalent to "image."

Digital image processing starts with one image and produces a modified version of that image. *Webster's Dictionary* defines *digital* as "[t]he calculation by numerical methods or discrete units," *digital image* as "[a] numerical representation of an object," *processing* as "[t]he act of subjecting something to a process," and *process* as "[a] series of actions or operations leading to a desired result." An example of a process is car wash, which changes an automobile from dirty to clean.

Digital image analysis is a process that takes a digital image into something other than a digital image, such as a set of measurement data or a decision. Image digitization is a process that converts a pictorial form to numerical data. A digital image is an image $f(x, y)$ that has been discretized in both spatial coordinates and brightness (intensity). The image is divided into small regions called *picture elements* or *pixels* (see Figure 1.1).

Image digitization includes image sampling (i.e., digitization of spatial coordinates (x,y)) and gray-level quantization (i.e., brightness amplitude digitization). An image is represented by a rectangular array of integers. The image sizes and the number of gray levels are usually integer powers of 2. The number at each pixel

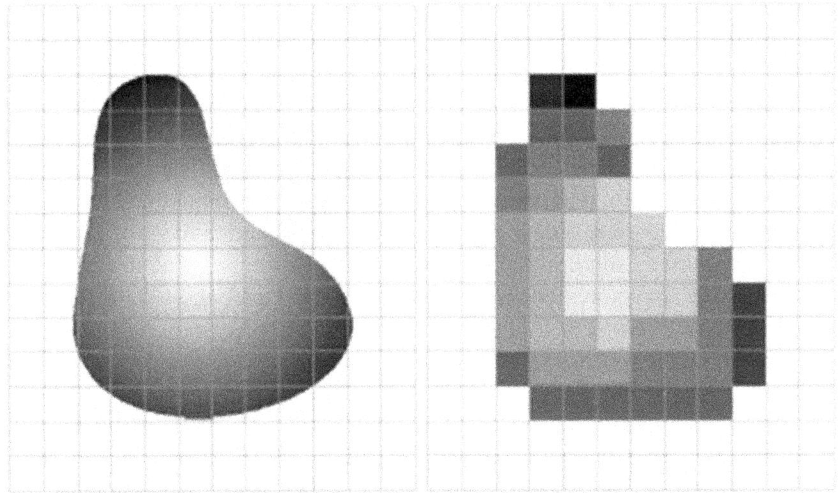

FIGURE 1.1 Image digitization.

Source: Gonzalez and Woods, 2008

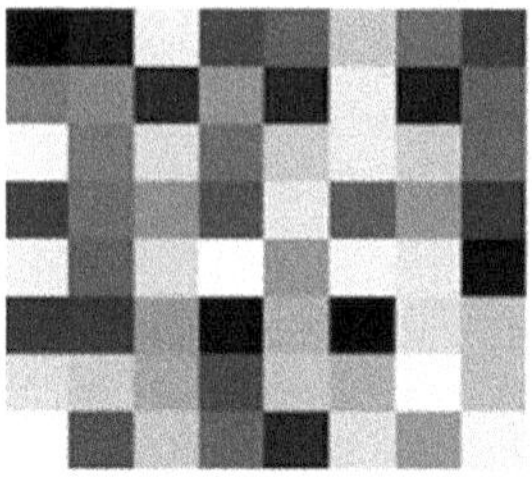

1	8	219	51	69	171	81	41
94	108	20	121	17	214	15	74
233	93	197	83	177	215	183	78
41	84	118	62	210	71	122	38
222	73	197	248	125	226	210	5
35	36	127	5	151	2	197	165
196	180	142	52	173	151	243	164
254	62	172	75	21	196	126	224

FIGURE 1.2 A digital image and its numerical representation.

represents the brightness or darkness (generally called the intensity) of the image at that point. For example, Figure 1.2 shows a digital image of size 8 × 8 with one byte (i.e., 8 bits = 256 gray levels) per pixel.

The quality of an image strongly depends upon the number of samples and gray levels; the more are these two, the better would be the quality of an image. But this will result in a high amount of storage space as well because the storage space for an image is the product of dimensions of an image and the number of bits required to store gray levels. At lower resolution an image can result in checkerboard effect or graininess. When an image of size 1024 × 1024 is reduced to 512 × 512, it may not show much deterioration, but when reduced to 256 × 256 and then rescaled back to 1024 × 1024 by duplication, it might show discernible graininess.

The visual quality of an image required depends upon its applications. To achieve the highest visual quality and at the same time the lowest memory requirement, we can perform fine sampling of an image in the neighborhood of sharp gray-level transitions and coarse sampling in the smooth areas of an image. This is known as sampling based on the characteristics of an image (Damera-Venkata et al., 2000). Another method, known as tapered quantization, can be used for the distribution of gray levels by computing the occurrence frequency of all allowed levels. Quantization level is finely spaced in the regions where gray levels occur frequently, but when gray levels occur rarely in other regions, the quantization level can be coarsely spaced. Images with large amounts of details can sometimes still enjoy a satisfactory appearance despite possessing a relatively small number of gray levels. This can be seen by examining isopreference curves using a set of subjective tests for images in the *Nk*-plane, where *N* is the number of samples and *k* is the number of gray levels (Huang, 1965).

In general, image-processing operations can be categorized into four types:

(1) Pixel operations – the output at a pixel depends only on the input at that pixel, independent of all other pixels in that image. Thresholding, a process of making the corresponding input pixels above a certain threshold level white, others black, is simply a pixel operation. Other examples include brightness addition/subtraction, contrast stretching, image inverting, log, and power-law.

(2) Local (neighborhood) operations – the output at a pixel depends on the input values in a neighborhood of that pixel. Some examples are edge detection, smoothing filters (e.g., the averaging filter and the median filter), and sharpening filters (e.g., the Laplacian filter and the gradient filter). This operation can be adaptive because results depend on the particular pixel values encountered in each image region.
(3) Geometric operations – the output at a pixel depends only on the input levels at some other pixels defined by geometric transformations. Geometric operations are different from global operations, such that the input is only from some specific pixels based on geometric transformation. They do not require the input from the entire pixels to make its transformation.
(4) Global operations – the output at a pixel depends on all the pixels in an image. It may be independent of the pixel values in an image, or it may reflect statistics calculated for all pixels, but not a local subset of pixels. A popular distance transformation of an image, which assigns to each object pixel the minimum distance from it to all the background pixels, belongs to a global operation. Other examples include histogram equalization/specification, image warping, Hough transform, and connected components.

Today, digital image processing touches nearly every aspect of our lives, with applications spanning countless fields. Its applications include:

(1) Remote Sensing:
   a. Earth Resource Tracking: Satellite imagery helps monitor and manage natural resources, offering insights into environmental changes, agricultural patterns, and deforestation.
   b. Solar Observations: Images of solar activity enable scientists to study solar flares, sunspots, and other phenomena that influence space weather.
   c. Geographical Mapping: High-resolution images provide detailed and accurate maps for urban planning, disaster management, and land use analysis. (See Figure 1.3.)

FIGURE 1.3 Remote sensing images for tracking earth climate and resources.

FIGURE 1.4 Space image applications.

   d. Space Exploration: Remote sensing extends beyond Earth, enabling the study of planetary surfaces, asteroid features, and celestial objects. (See Figure 1.4.)
(2) Image Transmission and Storage for Business:
   a. Broadcast Television: Enhancing the quality of video content and facilitating seamless distribution to global audiences.
   b. Teleconferencing: Supporting real-time communication with high-resolution visuals, enabling remote collaboration across the globe.
   c. Facsimile Transmission: Automating office processes by transmitting high-quality facsimile images for efficient documentation.
   d. Computer Networks: Enabling image sharing and storage over the Internet, from email attachments to cloud-based systems.
   e. Security Monitoring Systems: Powering surveillance systems to capture and analyze visual data for enhanced safety and crime prevention.
   f. Military Communications: Supporting secure and reliable image transmission for strategic operations and intelligence gathering.
(3) Medical Image Processing:
   a. X-Ray Imaging: Providing detailed views of bones and internal structures for fracture diagnosis, lung screening, and more.
   b. Cineangiography: Capturing moving images of blood flow in vessels, crucial for diagnosing cardiovascular conditions.
   c. Transaxial Tomography: Producing cross-sectional images of the body, as seen in CT scans, to identify abnormalities with precision.
   d. Nuclear Magnetic Resonance (NMR): Now widely known as MRI, this technique generates high-resolution images of soft tissues, enabling the detection of tumors and other diseases. (See Figure 1.5.)

(4) Radar, Sonar, and Acoustic Image Processing:
   a. Target Detection and Recognition: Radar and sonar systems process reflected signals to detect and identify objects, such as ships, aircraft, or underwater features, even in challenging conditions like fog, darkness, or murky waters.
   b. Aircraft Maneuvering: Radar image processing plays a crucial role in guiding aircraft, enabling precise navigation and collision avoidance, especially in areas with poor visibility. (See Figure 1.6.)

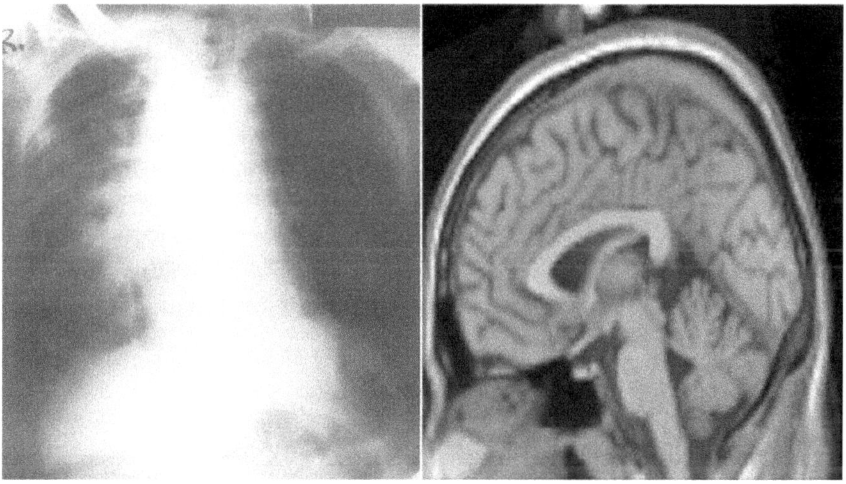

FIGURE 1.5 Medical imaging applications.

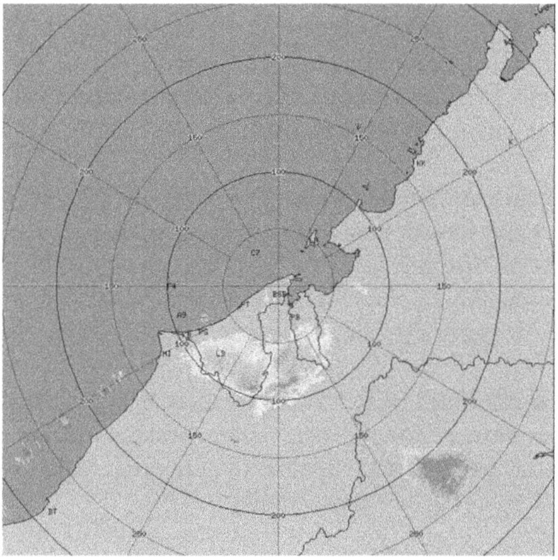

FIGURE 1.6 Radar imaging.

FIGURE 1.7 Robot and machine vision applications.

(5) Robot/machine Vision: Its applications include the recognition and description of objects or industrial parts within 3D scenes, allowing for precise quality control, sorting, and assembly in manufacturing environments. (See Figure 1.7.)

1.3 ELEMENTS OF AN IMAGE-PROCESSING SYSTEM

An image-processing system comprises several key elements, each contributing to the effective capture, storage, processing, and communication of visual data:

1. **Image Acquisition**:
   Image acquisition starts with a physical device sensitive to a specific band of the electromagnetic spectrum, such as a camera or scanner, which produces an electrical signal output. This signal is then converted into a digital form by a digitizer. Digital images can be obtained in two ways:
   - By converting analog images (e.g., 35 mm prints, slides, or reflective artwork) into digital format using a scanner.
   - By directly capturing objects or scenes into digital form using a digital camera or video-capturing device.
2. **Storage**:
   Proper storage is essential for managing image data during processing and for long-term preservation. Storage types include:
   - **Short-Term Storage**: Used during processing, often in computer memory or a specialized frame buffer board.
   - **Online Storage**: Provides relatively fast access for frequently used images.
   - **Archival Storage**: Designed for infrequent access, this storage type preserves images for the long term. Archival-quality materials are permanent, durable, and immune to tampering, deletion, viruses, and disasters, ensuring safe preservation.
3. **Processing**:
   Most image-processing tasks are carried out using software running on a host computer. These tasks include enhancing image quality, extracting

features, and manipulating the data for specific applications, such as object recognition or medical imaging.

4. **Communication**:
   Image data often needs to be transmitted over networks. Cable modem services typically offer higher bandwidth than DSL, translating to faster communication speeds. Cable technology can support speeds up to 30 Mbps, while most forms of DSL top out around 10 Mbps. This difference impacts the speed at which image data can be shared or accessed remotely.
5. **Display**:
   To view images, display devices are crucial. These can include:
   - **Monochrome and Color Monitors**: Commonly used for displaying images on personal computers and workstations.
   - **Random-Access Cathode Ray Tubes (CRTs)**: Historically popular for image display, though now largely replaced by newer technologies.
   - **Printing Devices**: Provide physical copies of digital images for various applications, from photography to technical documentation.

To illustrate the systematic procedures for an image-processing system, we give an example of human face identification (Adler and Schuckers, 2007). The problem domain is the faces of people. The objective is to associate the face with the identity of the person. The output is a person's unique identifier (e.g., social security number). The necessary procedures to achieve the goal of face identification could include:

(1) Image Acquisition: The face image could be acquired through a high-resolution still digital camera and compressed to an image file.
(2) Pre-Processing: The acquired image may be enhanced by improving contrast, sharpness, color, etc.
(3) Segmentation: The image may first be cropped to only the facial area. Then, the face may be segmented into eyes, mouth, nose, chin, etc.
(4) Representation and Description: In this step, each of the segmented areas may be characterized by statistical data; for examples, principal components analysis, texture, aspect ratios of eyes and nose, or the color of eyes.
(5) Matching Recognition and Interpretation: This step may involve using the characteristics derived in the previous step to match each individually segmented area based on specific recognition algorithms. For example, eyes may be processed to determine, based on its features, what class of eye it is. Then, all of these interpretations are used to create a composite description of the "ensemble," perhaps in the form of a feature vector for the subject.
(6) Knowledge Base: Finally, the feature vector may be fed to a knowledge base of all known subjects to associate it with one of the subjects in the database, thus returning perhaps the individual's social security number or perhaps a confidence score of the match.

REFERENCES

Adler, A. and Schuckers, M. E., "Comparing human and automatic face recognition performance," *IEEE Transactions on Systems, Man, and Cybernetics, Part B*, vol. 37, no. 5, pp. 1248–1255, Oct. 2007.

Chakraborty, C., Sinha, G. R., and Bajaj, V., *Biomedical Signal Processing for Healthcare Applications*, CRC Press, 2021.

Charalampidis, D., Pascotto, M., Kerut, E. K., and Lindner, J. R., "Anatomy and flow in normal and ischemic microvasculature based on a novel temporal fractal dimension analysis algorithm using contrast enhanced ultrasound," *IEEE Transactions on Medical Imaging*, vol. 25, no. 8, pp. 1079–1086, Aug. 2006.

Damera-Venkata, N., Kite, T. D., Geisler, W. S., Evans, B. L., and Bovik, A. C., "Image quality assessment based on a degradation model," *IEEE Transactions on Image Processing*, vol. 9, no. 4, pp. 636–650, Apr. 2000.

Desportes, C., Obligis, E., and Eymard, L., "On the wet tropospheric correction for altimetry in coastal regions," *IEEE Transactions on Geoscience and Remote Sensing*, vol. 45, no. 7, pp. 2139–2149, Jul. 2007.

Gamba, J., *Radar Signal Processing for Autonomous Driving*, Springer Nature Singapore, 2019.

Gonzalez, R. C. and Woods, R. E., *Digital Image Processing*, 3rd edition, Prentice Hall, 2008.

Huang, T. S., "PCM picture transmission," *IEEE Spectrum*, vol. 2, no. 12, pp. 57–63, Dec. 1965.

Le, C. F., *Principles of Radar and Sonar Signal Processing*, Artech House, 2002.

Li, F. F. and Cox T. J., *Digital Signal Processing in Audio and Acoustical Engineering*, CRC Press, 2019.

Mousa, W. A., *Advanced Digital Signal Processing of Seismic Data*, Cambridge University Press, 2020.

Negahdaripour, S. and Xun X., "Mosaic-based positioning and improved motion-estimation methods for automatic navigation of submersible vehicles," *IEEE Journal of Oceanic Engineering*, vol. 27, no. 1, pp. 79–99, Jan. 2002.

Qu, W. and Schonfeld, D., "Real-time decentralized articulated motion analysis and object tracking from videos," *IEEE Transactions on Image Processing*, vol. 16, no. 8, pp. 2129–2138, Aug. 2007.

Roy-Chowdhury, A. K., Chellappa, R., and Keaton, T., "Wide baseline image registration with application to 3-D face modeling," *IEEE Transactions on Multimedia*, vol. 6, no. 3, pp. 423–434, Jun. 2004.

Wiles, C. S., Maki, A., and Matsuda, N., "Hyperpatches for 3-D model acquisition and tracking," *IEEE Transactions on Pattern Analysis and Machine Intelligence*, vol. 23, no. 12, pp. 1391–1403, Dec. 2001.

Xu, W. L. and Tso, S. K., "Sensor-based fuzzy reactive navigation of a mobile robot through local target switching," *IEEE Transactions on Systems, Man, and Cybernetics, Part C*, vol. 29, no. 3, pp. 451–459, Aug. 1999.

Zhou, J., Gao, D., and Zhang, D., "Moving vehicle detection for automatic traffic monitoring," *IEEE Transactions on Vehicular Technology*, vol. 56, no. 1, pp. 51–59, Jan. 2007.

2 Image Enhancement

Whenever a picture is converted from one form to another, e.g., imaged, copied, scanned, transmitted, or displayed, the "quality" of the output picture may be lower than that of the input. In the absence of knowledge about how the given picture was actually degraded, it is difficult to predict in advance how effective a particular enhancement method will be. Image enhancement aims to improve human perception and interpretability of information in images or to provide more useful input for other automated image-processing techniques (Munteanu and Rosa, 2004; Kober, 2006; Panetta et al., 2008). In general, image enhancement techniques can be divided into three categories:

(1) Spatial domain methods, which manipulate directly on pixels in an image.
(2) Frequency domain methods, which operate on the Fourier transform or other frequency domains of an image.
(3) Combinational methods, which process an image in both spatial and frequency domains.

When conducting image enhancement, we need to keep in mind that there is no general rule for determining what the best image enhancement technique is. Consequently, the enhancement methods are applications specific and often developed empirically. When image enhancement techniques are used as pre-processing tools for other image-processing techniques, some quantitative measures can be used to determine which techniques are most appropriate.

In general, the process of image enhancement can involve in three types: point process, mask process, and global process. In a point process, each pixel is modified according to a particular equation depending on the input only at the same pixel, which is independent of other pixel values. The input may be one or more images. For example, the difference or product of two images can be taken point by point. In a mask process, each pixel is modified according to the values of the pixel's neighbors using convolution masks. For example, an average of the pixels can be taken in the neighborhood as a lowpass filter. In a global process, all the pixel values in the image (or sub-image) are taken into consideration. For example, histogram equalization remaps the histogram of the entire input pixels to a uniformly distributed histogram. Spatial-domain processing methods include all three types, but frequency-domain methods, by the nature of frequency transforms, are global processes. Of course, frequency-domain operations can become mask operations based only on a local neighborhood by performing the transform on small image blocks instead of the entire image.

In practice, it is very difficult to enhance an image that has been recorded in the presence of one or more sources of degradation in a single-pass filtering. Noise must be first eliminated and edges have to be preserved. Several passes may be needed to clarify the perceived subpatterns. It is similar to the activation of a human's visual

DOI: 10.1201/ 9781003474395-3

system. When we see a picture, our brain will request the visual system to focus on certain subpatterns which are the most interesting. Humans can identify object edges and regions with exceptional accuracy seemingly instantaneously, while conventional computers are not always nearly as accurate or as fast.

In this chapter we introduce the commonly used image enhancement techniques, including grayscale transformation, piecewise linear transformation, bit plane slicing, histogram equalization, histogram specification, enhancement by arithmetic operations, smoothing filter, and sharpening filter. We also introduce image blur types and image quality measures.

2.1 GRAYSCALE TRANSFORMATION

Grayscale transformation aims to change the gray levels of an entire image in a uniform way, or intends to modify the gray levels within a defined window by a mapping function. This transformation is usually expected to enhance the image contrast, so the details of an image can be more visible. The value of a pixel with coordinates (x, y) in the enhanced image $\hat{f}$ is the result of performing some operation on the value of f at (x, y). Thresholding is the simplest case to replace the intensity profile by a step function, jumping at a chosen threshold value. In this case any pixel with a gray level below the threshold in the input image receives 0 in the output image, and above or equal to the threshold receives 255.

Another simple operation, image negative, reverses the order of pixel intensities from black to white, so the intensity of output decreases as the intensity of input increases. It is a reversed image where the image that is usually black on a white background is reversed to be white on a black background. Let the input gray level r and the output gray level s be in the range $[0, L-1]$. The relationship between the input and output gray levels is given by $s = L-1-r$. Figure 2.1 shows the Lena image and its image negative. This image negative operation is equivalent to a

FIGURE 2.1 (a) The Lena image, (b) its image negative.

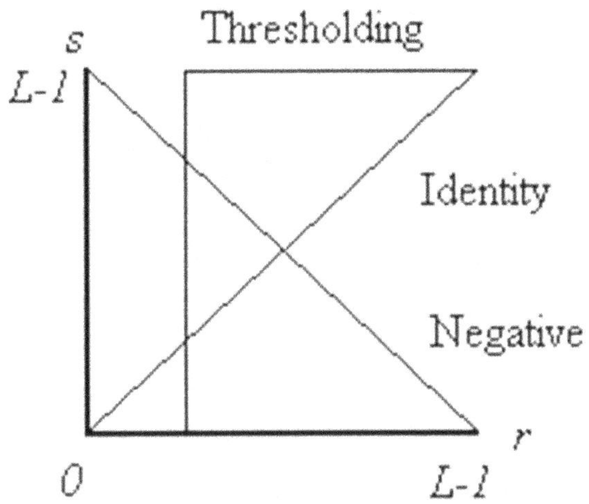

FIGURE 2.2 The mapping from an input image to an output image.

photographic negative. It is quite useful to show the details of small white or gray regions when they appear in a large dark background.

Some examples of mapping from the input gray level r to the output gray level s is illustrated in Figure 2.2, including identity, negative, and thresholding. The mapping function can use mathematical operations (e.g., logarithm, exponential, root, power, etc.) and any degree of polynomial functions. For example, the log mapping is given by $s = c\log(1+r)$, where c is a constant. A useful application of log mapping is to compress the large dynamic range of gray-level values (e.g., Fourier spectrum), so the brightest pixels will not dominate the display and the darker pixels will still be visible. Grayscale transformation can be performed using look-up tables and is often used for human perception purposes.

2.2 PIECEWISE LINEAR TRANSFORMATION

The piecewise linear transformation uses different linear functions to map different input gray-level intervals. A low-contrast image could occur due to poor illumination, lack of dynamic range in the sensor, or a wrong setting in a lens. Contrast stretching (or sometimes called *image normalization*) aims to improve the contrast in an image by stretching the narrow range of input intensity values to span a desired range of intensity values (usually the full range of gray values). It only applies a linear scaling function to the input. A general function for contrast stretching is illustrated in Figure 2.3. The locations of (r_1, s_1) and (r_2, s_2) control the shape of the transformation function. The constraints for this function are $r_1 \leq r_2$ and $s_1 \leq s_2$. The function is single valued and monotonically increasing, so the order of gray levels in the output is preserved.

If $r_1 = s_1$ and $r_2 = s_2$, it is a linear function indicating no changes in the output gray levels. If $r_1 = r_2$, $s_1 = 0$ and $s_2 = L-1$, it is a thresholding function and the

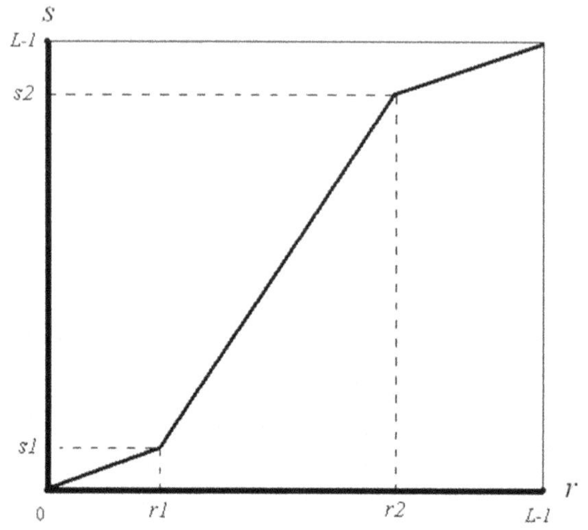

FIGURE 2.3 The mapping function of contrast stretching.

output is a binary image. If $r_1 < r_2$, $s_1 = 0$ and $s_2 = L-1$, it is a linear scaling. Let the input gray level r be mapped to the output gray level s in the full range of $[0, L-1]$. Let the minimum and maximum values of the input image be denoted as "min" and "max," respectively. The following equation is used to perform the linear scaling and round the floating point value to the closest integer for the output image to display as

$$s = \frac{L-1}{\max - \min}(r - \min) \tag{2.1}$$

Figure 2.4 shows an example of contrast stretching.

Example 2.1: Given $r_1 = 2$, $s_1 = 1$, $r_2 = 5$, $s_2 = 7$. Show the contrast-stretched output of the following image in the range [0,7].

2	2	3	3	3	3	3	2
2	2	2	2	3	2	3	2
2	0	4	4	5	5	3	3
3	0	4	4	5	5	5	5
2	4	4	4	4	5	7	3
2	2	4	5	6	5	6	2
2	3	4	4	2	5	6	2
2	3	2	2	2	3	5	3

Answer: The transform function is depicted by the following graph using the given (2,1) and (5,7) coordinates.

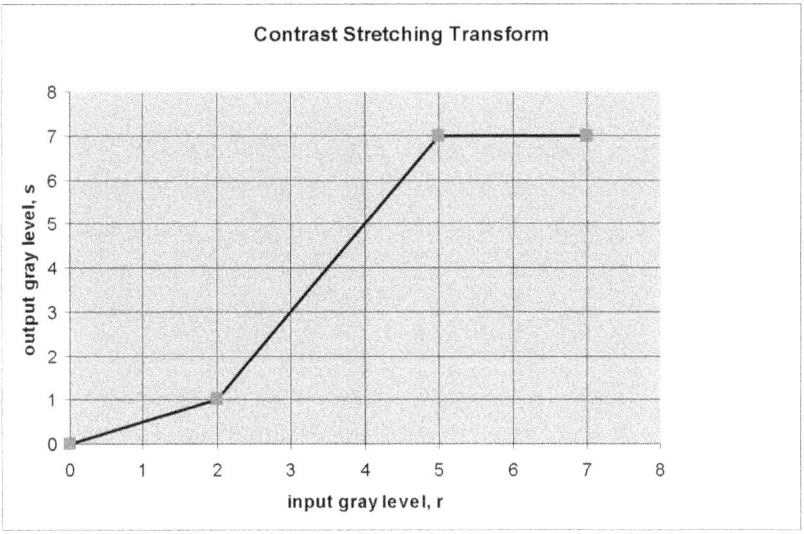

This results in the following intensity transformations from a given input original value to the contrast-stretched value.

FIGURE 2.4 (a) A low-contrast image, (b) the image after contrast stretching.

Original value	Contrast-stretched value
0	0
1	0
2	1
3	3
4	5
5	7
6	7
7	7

The following contrast-stretched image is obtained by making the replacements in the original image.

1	1	3	3	3	3	3	1
1	1	1	1	3	1	1	1
1	0	5	5	7	7	3	3
3	0	5	5	7	7	7	7
1	5	5	5	5	7	7	3
1	1	5	7	7	7	7	1
1	3	5	5	1	7	7	1
1	3	1	1	1	3	7	3

Another piecewise linear transformation uses the histogram of an input image to find the cluster of pixels that contain the most relevant information. The histogram is a diagram which shows the distribution of gray levels. In a digital image, the histogram shows at each gray level, the number of pixels in the image having that gray level. It has a particular inclination for most images to be multimodal. The determination of this cluster can become a non-trivial process. This application searches for the lowest valley between the background and the start of the cluster. For example, if a histogram is trimodal, the two valley points A and B around the clusters are selected. Let the output gray levels be in the range [0,255]. This grayscale transformation maps the input gray level r into the output gray level s and is given as

$$s = \begin{cases} \dfrac{255 - A + B}{2(B - A)}(r - A) + \dfrac{A}{2} & \text{if } A \leq r \leq B \\ \dfrac{r}{2} & \text{if } r < A \\ \dfrac{r + 255}{2} & \text{if } r > B \end{cases} \tag{2.2}$$

This transformation compresses the grayscale by a factor of 2 in the ranges $[0, A]$ and $[B, 255]$. The cluster is expanded by a factor of 2 over the range $[A, B]$. Note that the linear functions used can be replaced by any desired mathematical functions, such as quadratic and high-degree polynomials.

2.3 BIT PLANE SLICING

Since a pixel contains 8 bits in a 256-grayscale image, the bit plane slicing collects all the bits in the specific 8-bit plane of the image at hand. An 8-bit per pixel image can be sliced into eight bit-planes. Zero is the least significant bit (LSB) and 7 is the most significant bit (MSB). Each plane constitutes a collection of 0s and 1s from that plane where 0 translates to gray level 0 and 1 translates to gray level 255. The higher order bits contain more visually significant data, and the lower order bits contribute the more detailed information about the image. Figure 2.5 shows the bit plane slicing

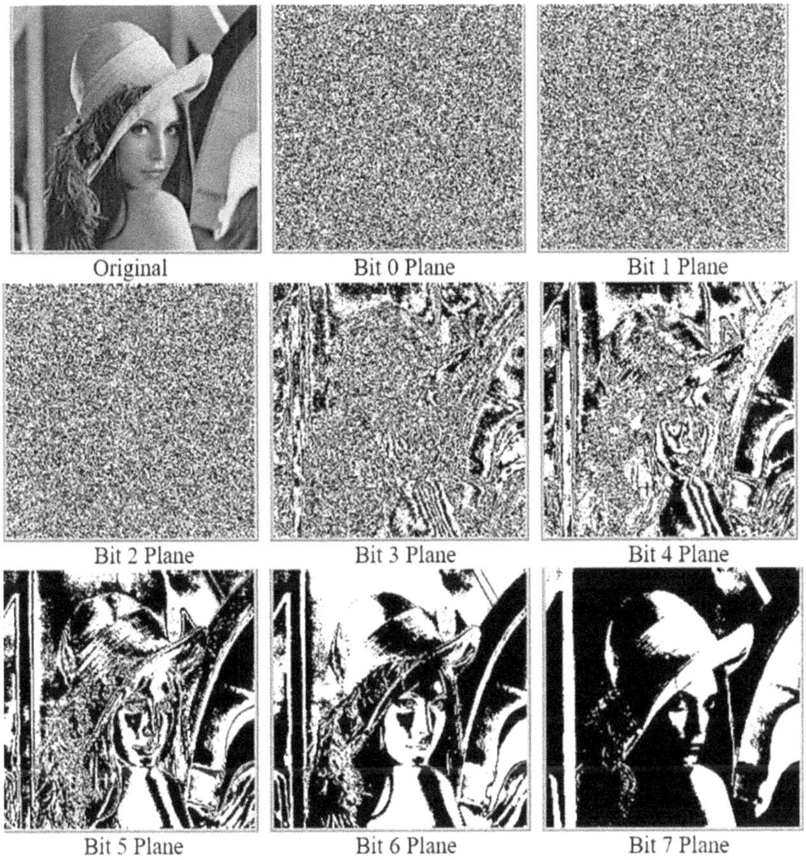

FIGURE 2.5 The Lena image and its bit planes 0 (least significant bit) through 7 (most significant bit).

of the Lena image. The images show that most image information is contained in the higher (i.e., more significant) bits, whereas the less significant bits contain some of the finer details and noise.

An application of bit-plane slicing is for data compression in image processing. Bit plane combining is the reverse process of the slicing. The planes are recombined in order to reconstruct the image. But it is not needed to take into consideration all the slice contributions. Especially, in the case where the data rate is important, some planes can be ignored until the changes in gray level have an acceptable impact on the image.

2.4 HISTOGRAM EQUALIZATION

Histogram equalization employs a monotonic, non-linear mapping which reassigns the intensity values of pixels in an input image, such that the output image contains a uniform distribution of intensities (i.e., a histogram that is constant for all brightness values). This corresponds to a brightness distribution where all values are equally probable. Unfortunately, we can only achieve the approximation of this uniform distribution for a digital image. This technique is often used in image comparison because it is effective in detail enhancement and in the correction of non-linear effects introduced by a digitizer or a display system. In general, since the histogram equalization causes the dynamic range of an image to be stretched, the density distribution of the resultant image is made flat so that the contrast of the image is enhanced. However, the histogram equalization method causes some problems. Since the contrast is enhanced by stretching the dynamic range, background noise is simultaneously increased by the equalization, and the image quality in a near-constant region may be degraded.

Consider an image pixel value $r \geq 0$ to be an element of random variable R with a continuous probability density function $p_R(r)$ and cumulative probability distribution $F_R(r) = P[R \leq r]$. Let the mapping function be $s = f(r)$ between the input and output images. To equalize the histogram of the output image, we let $p_S(s)$ be a constant. In particular, if the gray levels are assumed to be in the ranges between 0 and 1, then $p_S(s) = 1$ forms a uniform random variable S. The mapping function for histogram equalization

$$s = F_R(r) = \int_0^r p_R(r)dr \tag{2.3}$$

will be uniformly distributed over (0, 1).

To implement this transformation on digital images, let n denote the total number of pixels, n_G the total number of gray levels, and n_{r_j} the number of pixels in the input image with intensity value r_j. Let the input and output gray values be in the range of $[0, 1, \cdots, n_G - 1]$. Then, the histogram equalization transformation maps the input value r_k (where $k = 0, 1, \cdots, n_G - 1$) to the output value s_k as

$$s_k = T(r_k) = (n_G - 1)\sum_{j=0}^{k} \frac{n_{r_j}}{n} = \frac{(n_G - 1)}{n}\sum_{j=0}^{k} n_{r_j} \tag{2.4}$$

Note that the resulting floating-point value will be rounded to its closest integer as the output value. Figure 2.6 shows a low-contrast input image and its resulting image after histogram equalization.

Example 2.2: Suppose that a digital image is subjected to histogram equalization. Show that a second pass of histogram equalization will produce exactly the same result as the first pass.

Answer: Assume that a first pass of histogram equalization transforms the input value r_k to s_k and then a second pass transforms s_k to v_k. The second pass is conducted according to the following transformation:

$$v_k = T(s_k) = \frac{(n_G - 1)}{n}\sum_{j=0}^{k} n_{s_j}$$

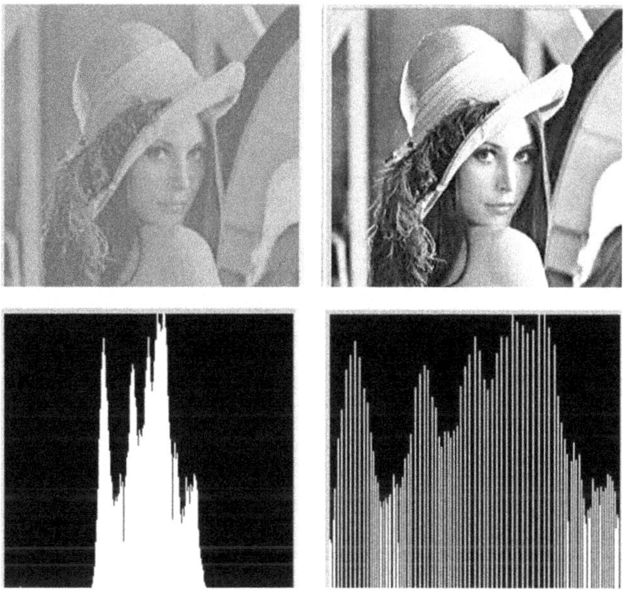

FIGURE 2.6 Column (a) shows a low-contrast image and its histogram, and column (b) shows the equalized image and its histogram.

Since every pixel (and no others) with value r_k is mapped to value s_k, it follows $n_{s_k} = n_{r_k}$. Therefore,

$$v_k = T(s_k) = \frac{(n_G - 1)}{n} \sum_{j=0}^{k} n_{r_j} = s_k$$

which shows that a second pass of histogram equalization would yield the same result as the first pass. Note that negligible round-off errors are assumed in this case.

Example 2.3: Apply histogram equalization on the input image of 8 × 8. Let the input and output gray levels be in the range of [0,7].

1	1	5	5	0	0	1	0
1	1	2	2	0	1	0	1
1	7	6	6	5	5	0	0
0	7	6	7	5	5	5	5
4	7	6	7	3	5	7	0
1	1	4	1	6	5	6	1
2	2	4	1	1	5	1	1
1	2	2	0	0	0	0	5

Answer: In this case, $n_G = 8$ and $n = 64$. Using equation (2.4), it yields

$$s_k = \frac{7}{64}\sum_{j=0}^{k} n_{r_j}$$

r_k	n_{r_j}	$\sum_{j=0}^{k} n_{r_j}$	s_k
0	13	13	1
1	17	30	3
2	6	36	4
3	1	37	4
4	3	40	4
5	12	52	6
6	6	58	6
7	6	64	7

Therefore, the output image is

3	3	6	6	1	1	3	1
3	3	4	4	1	3	1	3
3	7	6	6	6	6	1	1
1	7	6	7	6	6	6	6
4	7	6	7	4	6	7	1
3	3	4	3	6	6	6	3
4	4	4	3	3	6	3	3
3	4	4	1	1	1	1	6

Histogram equalization has been extensively applied. Menotti et al. (2007) developed multi-histogram equalization methods for contrast enhancement and brightness preserving. Ibrahim and Kong (2007) proposed the method of brightness preserving dynamic histogram equalization (BPDHE), which can produce the output image with the mean intensity almost equal to the mean intensity of the input. Agaian et al. (2007) presented transform coefficient histogram-based image enhancement algorithms using contrast entropy. Wang and Ward (2007) developed fast image/video contrast enhancement based on weighted thresholded histogram equalization.

2.5 HISTOGRAM SPECIFICATION

Histogram equalization intends to map any input image into an output image with the uniformly distributed histogram. Sometimes, a particularly distributed histogram of output images is desired in specific applications. Therefore, histogram specification is used to convert an image so that it has a particular histogram of output images as specified (Sun et al., 2005; Coltuc et al., 2006; Wan and Shi, 2007). Let x and y denote the gray levels of the input and output images, respectively. The probability

$p_x(x)$ from the input image can be computed, and the specified probability $p_z(z)$, that the output image is desired to have, is designed for a particular application. In this method, the histogram equalization is first conducted as

$$y = f(x) = \int_0^x p_x(u)du \tag{2.5}$$

When the gray levels of the desired image z are available, it can be also equalized as

$$y' = g(z) = \int_0^z p_z(u)du \tag{2.6}$$

The inverse of the transform is $z = g^{-1}(y')$. Since the images y and y' have the same equalized histogram, they are the same image and the overall transform from the given image x to the desired image z can be obtained by $z = g^{-1}(y) = g^{-1}(f(x)) = h(x)$, where $h(x) = g^{-1}(f(x))$ is the overall transform, and both f and g can be found from the corresponding histograms of the given image x and the desired image z, respectively.

Example 2.4: Apply histogram specification on the following image. Let the input and output gray levels be in the range of [0, 7]. Assume that the expected grayscale specification is {0: 5%, 1: 5%, 2: 10%, 3: 10%, 4: 25%, 5: 5%, 6: 25%, 7: 15%}. Show the output image.

1	1	0	0	0	0	0	1
1	1	1	1	0	1	0	1
1	3	4	4	5	5	0	0
0	3	4	4	5	5	5	5
2	4	4	4	3	5	7	0
1	1	4	5	6	5	6	1
1	0	4	4	1	5	6	1
1	0	1	0	0	0	5	0

Answer: From the input image:

x	p_x	$\int_0^x p_x(u)du$
0	17/64	17/64 = 0.2656
1	18/64	35/64 = 0.5469
2	1/64	36/64 = 0.5625
3	3/64	39/64 = 0.6094
4	10/64	49/64 = 0.7656
5	11/64	60/64 = 0.9375
6	3/64	63/64 = 0.9844
7	1/64	64/64 = 1.0000

From the expected specification:

z	p_z	$\int_0^z p_z(u)du$
0	0.05	0.05
1	0.05	0.10
2	0.10	0.20
3	0.10	0.30
4	0.25	0.55
5	0.05	0.60
6	0.25	0.85
7	0.15	1.00

Mapping from the input gray level to the output gray level of the expected specification yields:

x	mapping	z
0	$0.2656 \approx 0.30$	3
1	$0.5469 \approx 0.55$	4
2	$0.5625 \approx 0.55$	4
3	$0.6094 \approx 0.60$	5
4	$0.7656 \approx 0.85$	6
5	$0.9375 \approx 1.00$	7
6	$0.9844 \approx 1.00$	7
7	$1.0000 \approx 1.00$	7

Therefore, the output image after histogram specification is:

4	4	3	3	3	3	3	4
4	4	4	4	3	4	3	4
4	5	6	6	7	7	3	3
3	5	6	6	7	7	7	7
4	6	6	6	5	7	7	3
4	4	6	7	7	7	7	4
4	3	6	6	4	7	7	4
4	3	4	3	3	3	7	3

2.6 ENHANCEMENT BY ARITHMETIC OPERATIONS

Arithmetic operations (e.g., subtraction, addition, multiplication, division, mean) are often used to combine and transform two or more images into a new image which can better display or highlight certain features in the scene. It is also possible to just use a single image as input and perform arithmetic operations on all the pixels to modify brightness and enhance contrast.

The image subtraction operator takes the difference of two input images. It usually uses the absolute difference between pixel values rather than the straightforward signed output. Image subtraction can be used to detect changes in a series of images of the same scene or recognize a moving object.

Image averaging works if the noise in the image pixels and the associated noise are not correlated and the noise has a zero-averaging value. These conditions are necessary because the image-averaging method relies on the summing of N different noisy images. If the noise did not average out to 0, then artifacts of the noise would appear in the averaged image. The mathematical representation of this method is described in the following.

Let the captured K images be $g_i(x,y), i=1,2,\cdots,K$. The averaged image, denoted as $\bar{g}(x,y)$, can be computed by

$$\bar{g}(x,y)=\frac{1}{K}\sum_{i=1}^{K}g_i(x,y). \tag{2.7}$$

The captured image, $g_i(x,y)$, is assumed to be formed by the addition of noise $\eta_i(x,y)$ to an original image $f(x,y)$ as

$$g_i(x,y)=f(x,y)+\eta_i(x,y). \tag{2.8}$$

At every pair of coordinates (x,y), the noise is uncorrelated and has zero average value. If an image $\bar{g}(x,y)$ is formed by averaging K different noisy image as in equation (2.7), then it follows that

$$E\{\bar{g}(x,y)\}=f(x,y) \text{ and } \sigma^2_{\bar{g}(x,y)}=\frac{1}{K}\sigma^2_{\eta(x,y)}, \tag{2.9}$$

where $E\{\bar{g}(x,y)\}$ is the expected value of $\bar{g}$, $\sigma^2_{\bar{g}(x,y)}$ and $\sigma^2_{\eta(x,y)}$ are the variances of $\bar{g}$ and η, all at coordinates (x,y). The standard deviation at any point in the average image is

$$\sigma_{\bar{g}(x,y)}=\frac{1}{\sqrt{K}}\sigma_{\eta(x,y)}. \tag{2.10}$$

As K increases, the variability of the pixel values at each location (x,y) decreases. Because $E\{\bar{g}(x,y)\}=f(x,y)$, this means that $\bar{g}(x,y)$ approaches $f(x,y)$ as the number of noisy images used in the averaging process increases. This method can result in improved image quality because the noise is effectively cancelled if enough images are averaged together since a largely distributed set of images should bring the noise average to 0 if it is truly uncorrelated.

Similarly, frame averaging provides a way to average multiple video frames to create a more stable image. This module can be used to eliminate pixel vibrations or high-frequency image changes. It works by adding each frame into a moving average of frames. This effectively creates the same effect of averaging many frames without the significant memory and time that averaging hundreds of frames would take.

2.7 SMOOTHING FILTER

A simple mean smoothing filter or operation intends to replace each pixel value in an input image by the mean (or average) value of its neighbors, including itself. This has an effect of eliminating pixel values which are unrepresentative of their surroundings. Like other convolution filters, it is based around a kernel, which represents the shape and size of the neighborhood to be sampled in calculation. Often a 3 × 3 square kernel is used, as shown in the following although larger kernels (e.g., a 5 × 5 square) can be used for more severe smoothing. Figure 2.7 shows an example of applying the 3 × 3 averaging kernel.

$$\frac{1}{9}\begin{bmatrix} 1 & 1 & 1 \\ 1 & 1 & 1 \\ 1 & 1 & 1 \end{bmatrix}$$

Another method of image smoothing convolves an input image by the Gaussian filter. The Gaussian filter, representing the shape of a Gaussian (bell-shaped) hump, will screen noise with the high spatial frequencies and will produce a smoothing effect. The 1D Gaussian filter function is

FIGURE 2.7 (a) The Lena image, (b) the resulting image after a 3 × 3 averaging filter.

$$G(x) = \frac{1}{\sqrt{2\pi}\sigma} e^{-x^2/2\sigma^2} \tag{2.11}$$

where σ is the standard deviation of the distribution. The distribution of mean zero and $\sigma = 1$ is shown in Figure 2.8.

In two dimensions, an isotropic (i.e., circularly symmetric) Gaussian filter function is

$$G(x, y) = \frac{1}{\sqrt{2\pi\sigma^2}} e^{-(x^2+y^2)/2\sigma^2} \tag{2.12}$$

The distribution of mean (0,0) and $\sigma = 1$ is shown in Figure 2.9.

In image processing, a discrete representation of the Gaussian function is required for conducting the convolution. In theory, the Gaussian distribution is non-zero

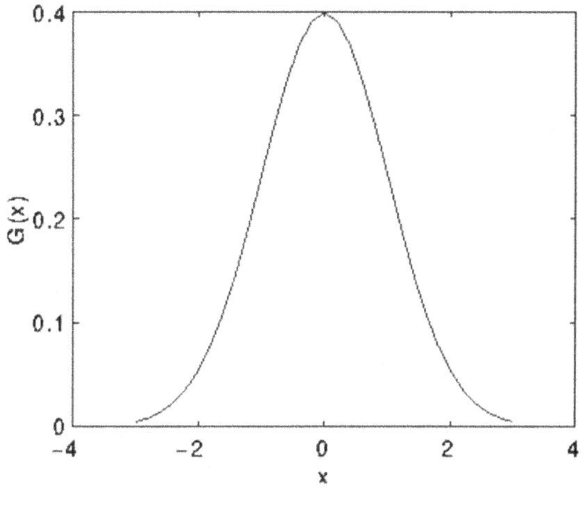

FIGURE 2.8 The 1D Gaussian filter function with mean zero and $\sigma = 1$.

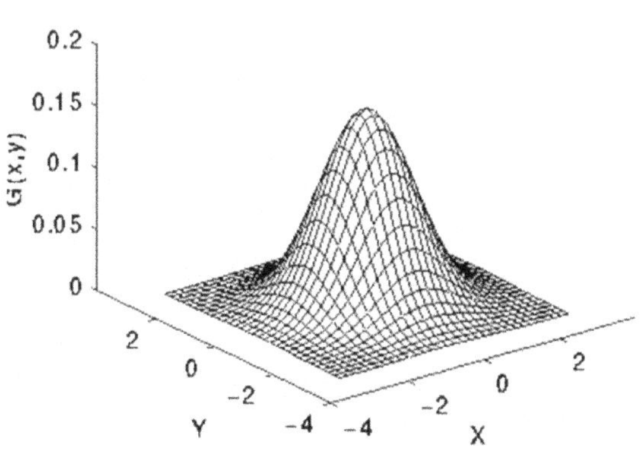

FIGURE 2.9 The 2D Gaussian filter function with mean (0,0) and $\sigma = 1$.

$$\frac{1}{16}\begin{bmatrix} 1 & 2 & 1 \\ 2 & 4 & 2 \\ 1 & 2 & 1 \end{bmatrix} \qquad \frac{1}{273}\begin{bmatrix} 1 & 4 & 7 & 4 & 1 \\ 4 & 16 & 26 & 16 & 4 \\ 7 & 26 & 41 & 26 & 7 \\ 4 & 16 & 26 & 16 & 4 \\ 1 & 4 & 7 & 4 & 1 \end{bmatrix}$$

FIGURE 2.10 Discrete Gaussian low-pass convolution filters (a) 3 × 3 and (b) 5 × 5.

everywhere, meaning an infinitely large convolution kernel. In practice, it is effectively zero more than about three standard deviations from the mean, so the kernels beyond this point can be truncated. Figure 2.10 shows integer-valued convolution kernels of 3 × 3 and 5 × 5 that approximate a Gaussian filter function with $\sigma = 1$. Note that a constant scaling factor is multiplied to ensure that the output gray levels are in the same range as the input gray levels.

Example 2.5: Suppose a low-pass spatial filter is formed by averaging the four immediate neighbors of a point (x,y) but excluding the point itself. Find the equivalent filter $H(u,v)$ in the frequency domain.

Answer: The spatial average is

$$g(x,y) = \frac{1}{4}\left[f(x,y+1) + f(x+1,y) + f(x-1,y) + f(x,y-1)\right]$$

Perform Fourier transform to obtain

$$G(u,v) = \frac{1}{4}\left[e^{j2\pi v/N} + e^{j2\pi u/M} + e^{-j2\pi u/M} + e^{-j2\pi v/N}\right]F(u,v) = H(u,v)F(u,v),$$

where $H(u,v) = \frac{1}{2}\left[\cos(2\pi u / M) + \cos(2\pi v / N)\right]$ is the filter transfer function in the frequency domain.

Another smoothing filter, called a *median filter*, is used to reduce noise in an image, somewhat like a mean filter. However, it performs better than a mean filter in the sense of preserving useful detail in the image. It is especially effective for removing impulse noise, which is characterized by bright and/or dark high-frequency features appearing randomly over the image. Statistically, impulse noise falls well outside the peak of the distribution of any given pixel neighborhood, so the median is well suited to learn where impulse noise is not present, and hence to remove it by exclusion. The median of a distribution is the value for which larger and smaller values are equally probable. To calculate the median of a list of sample values, we sort them in a descending or ascending order, and then select the central value. An example of applying a median filter on an added salt-and-pepper noisy Lena image is shown in Figure 2.11.

FIGURE. 2.11 (a) A noisy image, (b) the resulting image after median filtering.

A median filter is a kind of nonlinear filters and a subset of so-called *order-statistic filters*. A maximum, minimum or any rank in the descending or ascending order, instead of the median value, can be picked up in the neighborhood around a pixel. Note that the maximum filter corresponds to the morphological dilation and the minimum filter corresponds to the morphological erosion, which will be discussed in Chapter 4.

2.8 SHARPENING FILTER

Sharpening filter is used to enhance the edges of objects and adjust the contrast of object and background transitions. They are sometimes used as edge detectors by combining with thresholding. Sharpening or high-pass filter allows high-frequency components to pass and delete the low-frequency components. For a kernel to be a high-pass filter, the coefficients near the center must be set positive and in the outer periphery must be set negative. Sharpening filter can be categorized into four types: high-pass filter, Laplacian of Gaussian filter, high-boost filter, and derivative filter.

A high-pass filter, opposite of a low-pass filter, is a filter that passes high-frequency components but attenuates (or deletes) the components whose frequency is lower than the cutoff frequency. A simple 3×3 high-pass filter is given as

$$\begin{bmatrix} -1 & -1 & -1 \\ -1 & 8 & -1 \\ -1 & -1 & -1 \end{bmatrix}.$$

An example of applying the high-pass filter on the Lena image is shown in Figure 2.12.

Another sharpening filter is the derivative (or gradient) operator. For an image function $f(x, y)$, the gradient vector is defined as $\nabla f \equiv (\partial f / \partial x)\vec{i} + (\partial f / \partial y)\vec{j}$,

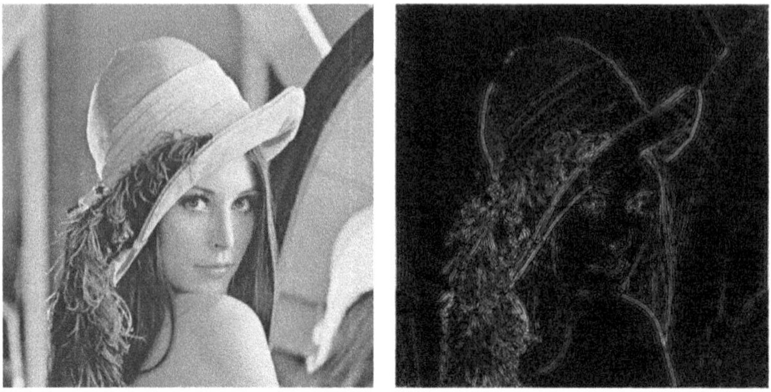

FIGURE 2.12 Applying the high-pass filter on the Lena image.

where $\vec{i}$ and $\vec{j}$ are respectively the unit vectors along x- and y-axes. Let $f_x \equiv \partial f / \partial x$ and $f_y \equiv \partial f / \partial y$. The magnitude of ∇f is $\sqrt{f_x^2 + f_y^2}$, and its direction is $\theta = \tan^{-1}(f_y / f_x)$. For digital images, the derivatives of the gradient operator are the two differences: $f_x = f(x+1, y) - f(x, y)$ and $f_y = f(x, y+1) - f(x, y)$. Both can be obtained by convolution with the following kernels:

$$\begin{bmatrix} -1 & 1 \\ 0 & 0 \end{bmatrix}, \begin{bmatrix} -1 & 0 \\ 1 & 0 \end{bmatrix}.$$

If we consider the diagonal directions, the two kernels are

$$\begin{bmatrix} 0 & 1 \\ -1 & 0 \end{bmatrix}, \begin{bmatrix} 1 & 0 \\ 0 & -1 \end{bmatrix},$$

which are called *Roberts cross gradient operators*. If considering kernels of size 3×3 and first finding the averages of one direction and then finding the difference of these averages in the other direction, we have

$$\begin{bmatrix} -1 & 0 & 1 \\ -1 & 0 & 1 \\ -1 & 0 & 1 \end{bmatrix}, \begin{bmatrix} -1 & -1 & -1 \\ 0 & 0 & 0 \\ 1 & 1 & 1 \end{bmatrix},$$

which are called *Prewittt operators*. If considering a double weight in the horizontal and vertical neighbors, we have

$$\begin{bmatrix} -1 & 0 & 1 \\ -2 & 0 & 2 \\ -1 & 0 & 1 \end{bmatrix}, \begin{bmatrix} -1 & -2 & -1 \\ 0 & 0 & 0 \\ 1 & 2 & 1 \end{bmatrix},$$

which are called *Sobel operators*.

The Laplacian of Gaussian filter (LoG) is the combination of the Laplacian and Gaussian filters where its characteristic is determined by the σ parameter and the kernel size, as shown in the mathematical expression of the kernel:

$$\mathrm{LoG}(i,j) = -\frac{1}{\pi\sigma^4}\left(1 - \frac{i^2 + j^2}{2\sigma^2}\right) e^{-(i^2+j^2)/2\sigma^2} \tag{2.13}$$

The 3D graphics of the LoG filter is shown in Figure 2.13.

A discrete 9 × 9 kernel that approximates this function (for a Gaussian σ = 1.4) is shown in Figure 2.14. An example of applying this kernel on the Lena image is shown in Figure 2.15.

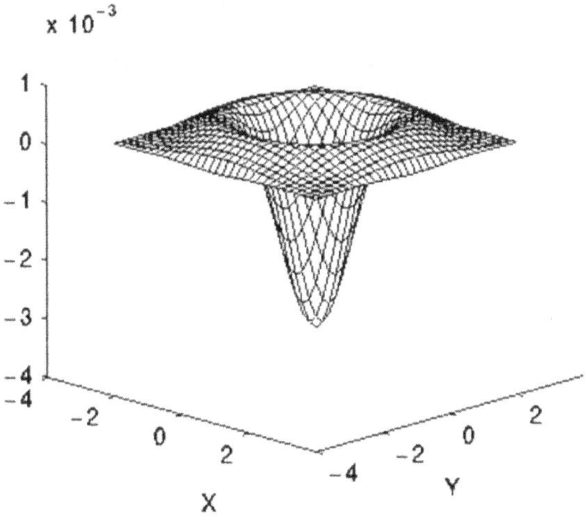

FIGURE 2.13 The Laplacian of Gaussian filter.

0	1	1	2	2	2	1	1	0
1	2	4	5	5	5	4	2	1
1	4	5	3	0	3	5	4	1
2	5	3	−12	−24	−12	3	5	2
2	5	0	−24	−40	−24	0	5	2
2	5	3	−12	−24	−12	3	5	2
1	4	5	3	0	3	5	4	1
1	2	4	5	5	5	4	2	1
0	1	1	2	2	2	1	1	0

FIGURE 2.14 The discrete approximation to LoG function with Gaussian σ = 1.4.

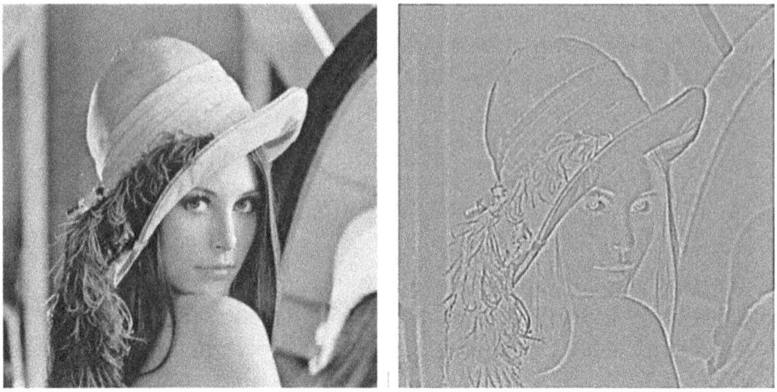

FIGURE 2.15 (a) The Lena image, (b) the result after applying a discrete 9 × 9 LoG filter.

The high-boost filter is used to emphasize high-frequency components representing the image details without eliminating low-frequency components. It multiplies the original image by an amplification factor A as

$$\begin{aligned} \text{High boost} &= A \times \text{Original} - \text{Lowpass} \\ &= (A-1) \times \text{Original} + (\text{Original} - \text{Lowpass}) \\ &= (A-1) \times \text{Original} + \text{Highpass} \end{aligned} \tag{2.14}$$

Note that if $A = 1$, it becomes the standard high-pass filter. For a 3 × 3 kernel,

$$\begin{bmatrix} 0 & 0 & 0 \\ 0 & A & 0 \\ 0 & 0 & 0 \end{bmatrix} - \frac{1}{9}\begin{bmatrix} 1 & 1 & 1 \\ 1 & 1 & 1 \\ 1 & 1 & 1 \end{bmatrix} = \frac{1}{9}\begin{bmatrix} -1 & -1 & -1 \\ -1 & 9A-1 & -1 \\ -1 & -1 & -1 \end{bmatrix} \tag{2.15}$$

An example of applying the high-boost filter with $A = 1.5$ on the Lena image is shown in Figure 2.16.

2.9 IMAGE BLUR TYPES AND QUALITY MEASURES

Three types of blur usually occurred in image processing are motion, Gaussian, and compression blurs (Molina et al., 2006). Motion blur may be due to object movement when a camera shutter remains open for an extended period of time and the object motion within this interval is visible in a single snapshot. Motion blur may also be caused by camera movement, in which the cameraman shifts the camera along objects. A 2D motion blur is shown in Figure 2.17(b). Gaussian blur is made by a soft lens to spread out the light on the focal plane, rather than all going toward a spot. It produces a smoothing effect by removing image details or noises. It can be simulated by mixing each pixel with a Gaussian density function, as shown in Figure 2.17(c). Compression blur is triggered by the loss of high-frequency components in JPEG compression. An example of JPEG compression ratio 1:160 to simulate the compression blur is shown in Figure 2.17(d).

FIGURE 2.16 Applying the high-boost filter with $A = 1.5$ on the Lena image.

FIGURE 2.17 (a) Original image, (b) motion blur at orientation 45° and magnitude 40, (c) Gaussian blur with a 17 × 17 window, (d) compression blur with compression ratio of 1:160.

Image quality is characterized by a close interaction between the display and human visual perception. There are many *image quality measures* (IQMs) proposed to evaluate imaging systems and image coding/processing techniques (Sanchez-Marin et al., 2001; Shnayderman et al., 2006). With the understanding of these measures, it is hoped that subjective decisions can be kept to a minimum. Therefore, computational image quality analysis aims at finding an accurate and meaningful model to evaluate image quality. The computational model takes an image as the input and returns a quality number as output. If the input takes one parameter, it is the blurred image. If the input takes two parameters, they are the original image and the blurred image.

Let $f(x_i)$ denote the IQM score of an image under the degree of blur x_i. The IQMs used for measuring image blurs must satisfy the monotonically increasing or decreasing property. That is, if $x_{i+1} > x_i$, then $f(x_{i+1}) - f(x_i) > 0$ for monotonically increasing or $f(x_{i+1}) - f(x_i) < 0$ for monotonically decreasing. The sensitivity of IQMs is defined as the score of the aggregate relative distance as

$$\sum_i \frac{f(x_{i+1}) - f(x_i)}{f(x_i)}. \tag{2.16}$$

Nine IQMs, which are grouped into three categories based on pixel distance, correlation, and mean square error, are introduced here. Let $F(j,k)$ denote the pixel value of row j and column k in a reference image of size $M \times N$, and $\hat{F}(j,k)$ denote the pixel value in a testing image.

Category I: IQMs based on pixel distance

1. AD (Average Distance)

$$AD = \sum_{j=1}^{M}\sum_{k=1}^{N} |F(j,k) - \hat{F}(j,k)| / MN \tag{2.17}$$

2. L2D (L2 Euclidean Distance)

$$L2D = \frac{1}{MN}\left(\sum_{j=1}^{M}\sum_{k=1}^{N}(F(j,k) - \hat{F}(j,k))^2\right)^{1/2} \tag{2.18}$$

Category II: IQMs based on correlation

3. SC (Structure Content)

$$SC = \sum_{j=1}^{M}\sum_{k=1}^{N} F(j,k)^2 / \sum_{j=1}^{M}\sum_{k=1}^{N} \hat{F}(j,k)^2 \tag{2.19}$$

4. IF (Image Fidelity)

$$IF = 1 - \left(\sum_{j=1}^{M}\sum_{k=1}^{N}(F(j,k) - \hat{F}(j,k))^2 / \sum_{j=1}^{M}\sum_{k=1}^{N} F(j,k)^2\right) \tag{2.20}$$

5. NK (N Cross-Correlation)

$$NK = \sum_{j=1}^{M}\sum_{k=1}^{N} F(j,k)\hat{F}(j,k) / \sum_{j=1}^{M}\sum_{k=1}^{N} F(j,k)^2 \tag{2.21}$$

Category III: IQMs based on mean square error

6. NMSE (Normal Mean Square Error)

$$NMSE = \sum_{j=1}^{M}\sum_{k=1}^{N}(F(j,k) - \hat{F}(j,k))^2 / \sum_{j=1}^{M}\sum_{k=1}^{N}F(j,k)^2 \tag{2.22}$$

7. LMSE (Least Mean Square Error)

$$LMSE = \sum_{j=1}^{M}\sum_{k=1}^{N}(F(j,k) - \hat{F}(j,k))^2 / \sum_{j=1}^{M}\sum_{k=1}^{N}O(F(j,k))^2, \tag{2.23}$$

where $O(F(j,k)) = F(j+1,k) + F(j-1,k) + F(j,k+1) + F(j,k-1) - 4F(j,k)$

8. PMSE (Peak Mean Square Error)

$$PMSE = \frac{1}{MN}\sum_{j=1}^{M}\sum_{k=1}^{N}\left[F(j,k) - \hat{F}(j,k)\right]^2 \Big/ \left\{\max_{j,k}\left[F(j,k)\right]\right\}^2 \tag{2.24}$$

9. PSNR (Peak Signal-to-Noise Ratio)

$$PSNR = 20 \times \log_{10}\left\{255 \Big/ \left[\sum_{j=1}^{M}\sum_{k-1}^{N}\left[F(j,k) - \hat{F}(j,k)\right]^2\right]^{1/2}\right\} \tag{2.25}$$

An automatic method to measure the degree of similarity with respect to the human vision system (HVS) is described as follows:

(1) Perform *Discrete Cosine Transform* (DCT) on the original image $C(x,y)$ and the blurred image $\hat{C}(x,y)$ to obtain $D(u,v)$ and $\hat{D}(u,v)$, respectively.
(2) Convolve $D(u,v)$ and $\hat{D}(u,v)$ with the following band-pass filter H to obtain $E(u,v)$ and $\hat{E}(u,v)$, respectively.

$$H(p) = \begin{cases} 0.05e^{p^{0.554}}, & \text{if } p < 7 \\ e^{-9|\log_{10} p - \log_{10} 9|^{2.3}}, & \text{if } p \geq 7 \end{cases}, \text{ where } p = \sqrt{u^2 + v^2}. \tag{2.26}$$

(3) Perform inverse DCT of $E(u,v)$ and $\hat{E}(u,v)$ respectively to obtain $F(x,y)$ and $\hat{F}(x,y)$.
(4) Calculate the Euclidean distance between $F(x,y)$ and $\hat{F}(x,y)$ using equation (2.18).

REFERENCES

Agaian, S. S., Silver, B., and Panetta, K. A., "Transform coefficient histogram-based image enhancement algorithms using contrast entropy," *IEEE Transactions on Image Processing*, vol. 16, no. 3, pp. 741–758, Mar. 2007.

Coltuc, D., Bolon, P., and Chassery, J.-M., "Exact histogram specification," *IEEE Transactions on Image Processing*, vol. 15, no. 5, pp. 1143–1152, May 2006.

Ibrahim, H. and Kong, N. S. P., "Brightness preserving dynamic histogram equalization for image contrast enhancement," *IEEE Transactions on Consumer Electronics*, vol. 53, no. 4, pp. 1752–1758, Nov. 2007.

Kober, V., "Robust and efficient algorithm of image enhancement," *IEEE Transactions on Consumer Electronics*, vol. 52, no. 2, pp. 655–659, May 2006.

Menotti, D., Najman, L., Facon, J., and de Araujo, A. A., "Multi-histogram equalization methods for contrast enhancement and brightness preserving," *IEEE Transactions on Consumer Electronics*, vol. 53, no. 3, pp. 1186–1194, Aug. 2007.

Molina, R., Mateos, J., and Katsaggelos, A. K., "Blind deconvolution using a variational approach to parameter, image, and blur estimation," *IEEE Transactions on Image Processing*, vol. 15, no. 12, pp. 3715–3727, Dec. 2006.

Munteanu, C. and Rosa, A., "Gray-scale image enhancement as an automatic process driven by evolution," *IEEE Transactions on Systems, Man, and Cybernetics, Part B*, vol. 34, no. 2, pp. 1292–1298, Apr. 2004.

Panetta, K. A., Wharton, E. J., and Agaian, S. S., "Human visual system-based image enhancement and logarithmic contrast measure," *IEEE Transactions on Systems, Man, and Cybernetics, Part B*, vol. 38, no. 1, pp. 174–188, Feb. 2008.

Sanchez-Marin, F. J., Srinivas, Y., Jabri, K. N., and Wilson, D. L., "Quantitative image quality analysis of a nonlinear spatio-temporal filter," *IEEE Transactions on Image Processing*, vol. 10, no. 2, pp. 288–295, Feb. 2001.

Shnayderman, A., Gusev, A., and Eskicioglu, A. M., "An SVD-based grayscale image quality measure for local and global assessment," *IEEE Transactions on Image Processing*, vol. 15, no. 2, pp. 422–429, Feb. 2006.

Sun, C., Ruan, S., Shie, M., and Pai, T., "Dnamic contrast enhancement based on histogram specification," *IEEE Transactions on Consumer Electronics*, vol. 51, no. 4, pp. 1300–1305, Nov. 2005.

Wan, Y. and Shi, D., "Joint exact histogram specification and image enhancement through the wavelet transform," *IEEE Transactions on Image Processing*, vol. 16, no. 9, pp. 2245–2250, Sep. 2007.

Wang, Q. and Ward, R. K., "Fast image/video contrast enhancement based on weighted thresholded histogram equalization," *IEEE Transactions on Consumer Electronics*, vol. 53, no. 2, pp. 757–764, May 2007.

3 Mathematical Morphology

Image-processing techniques have been tremendously developed during the past five decades, and among them, mathematical morphology has been continuously receiving a great deal of attention. It is because mathematical morphology provides quantitative description of geometric structure and shape as well as mathematical description of algebra, topology, probability, and integral geometry. Mathematical morphology has been proved to be extremely useful in many image processing and analysis applications.

According to *Webster's Dictionary*, the word *morphology* refers to any scientific study of form and structure. This term has been widely used in biology, linguistics, and geography. In image processing, a well-known general approach is provided by *mathematical morphology*, where the images being analyzed are considered as sets of points and the set theory is applied on the morphological operations. This approach is based upon logical relations between pixels, rather than arithmetic relations, and can extract geometric features by choosing a suitable structuring shape as a probe.

According to Haralick et al. (1987), "*As the identification and decomposition of objects, object features, object surface defects, and assembly defects correlate directly with shape, mathematical morphology clearly has an essential structural role to play in machine vision.*" Also, Shih and Mitchell (1989) state, "*Mathematical morphology has been becoming increasingly important in image processing and analysis applications for object recognition and defect inspection.*" Numerous morphological architectures and algorithms have been developed during last decades. The history of morphological image processing has followed a series of developments in the areas of mathematics and computer architecture.

From the mathematics aspect, the early work includes Minkowski (1903), Dineen (1955), Kirsch (1957), Preston (1961), Moore (1968), and Golay (1969). It was formalized at the Centre de Morphologie Mathematique on the campus of Paris School of Mines at Fontainebleau, France, in the mid 1970s for studying geometric and milling properties of ores. Matheron (1975) wrote a book on mathematical morphology, titled *Random Sets and Integral Geometry*. Two volumes containing sophisticated mathematical theory were written by Serra (1982, 1988). Other books describing fundamental applications can be referred to (Giardina and Dougherty, 1988; Dougherty, 1993; Goutsias and Heijmans, 2000; Shih, 2009). Haralick et al. (1987) presented a tutorial providing a quick understanding for a beginner. Shih and Mitchell (1989, 1991) created threshold decomposition of grayscale morphology into binary morphology, which provides a new insight in exploring grayscale morphological processing.

From the computer architecture aspect, morphological operations are the essence of the cellular logic machines, such as Golay logic processor (Golay, 1969), Diff3 (Graham and Norgren, 1980), PICAP (Kruse, 1977), Leitz texture analysis system (TAS) (Klein and Serra, 1977), CLIP processor arrays (Duff, 1979), cytocomputers (Lougheed and McCubbrey, 1980; Sternberg, 1980), and Delft image processor (DIP) (Gerritsen and Aardema, 1981).

DOI: 10.1201/ 9781003474395-4

Mathematical morphology can extract image shape features, such as edges, fillets, holes, corners, wedges, and cracks, by operating with various shaped structuring elements (Maragos and Schafer, 1987b). In industrial vision applications, mathematical morphology can be used to implement fast object recognition, image enhancement, segmentation, and defect inspection.

In this chapter, we will introduce binary morphology, opening and closing, hit-or-miss transform, grayscale morphology, basic morphological algorithms, and several variations of morphological filters, including alternating sequential filters, recursive morphological filters, soft morphological filters, order-statistic soft morphological filters, recursive soft morphological filters, recursive order-statistic soft morphological filters, regulated morphological filters, and fuzzy morphological filters.

3.1 BINARY MORPHOLOGY

Mathematical morphology involves geometric analysis of shapes and textures in images. An image can be represented by a set of pixels. Morphological operators work with two images. The image being processed is referred to as the *active image*, and the other image, being a kernel, is referred to as the *structuring element*. Each structuring element has a designed shape, which can be thought of as a probe or a filter of the active image. The active image can be modified by probing it with various structuring elements. The elementary operations in mathematical morphology are dilation and erosion, which can be combined in sequence to produce other operations, such as opening and closing.

Let E^N denote the set of all points $p = (x_1, x_2, \cdots, x_N)$ in N-dimensional Euclidean space. A binary image in E^2 is a silhouette, a set representing foreground regions. A binary image in E^3 is a solid, a set representing the surface and interior of objects. The definitions and properties of dilation and erosion from a tutorial on mathematical morphology by Haralick et al. (1987) are adopted, which are slightly different from Matheron's (1975) and Serra's (1982) definitions.

3.1.1 Binary Dilation

Binary dilation combines two sets using vector addition of set elements. It was first introduced by Minkowski and is named *Minkowski addition*. Let A and B denote two sets in E^N with elements a and b, respectively, where $a = (a_1, a_2, \cdots, a_N)$ and $b = (b_1, b_2, \cdots, b_N)$ being N-tuples of element coordinates. The binary dilation of A by B is the set of all possible vector sums of pairs of elements, one coming from A and the other from B. Since A and B are both binary, the morphological operators applied on the two sets are called *binary morphology*.

Definition 3.1: Let a set $A \subset E^N$ and an element $b \in E^N$. The *translation* of A by b, denoted by $(A)_b$, is defined as

$$(A)_b = \{c \in E^N \mid c = a + b \text{ for some } a \in A\} \tag{3.1}$$

Example 3.1: Let the coordinates (r, c) in A denote (row number, column number).

$$A = \{(0,2), (1,1), (1,2), (2,0), (2,2), (3,1)\}$$
$$b = (0,1)$$
$$(A)_b = \{(0,3),\ (1,2),\ (1,3),\ (2,1),\ (2,3),\ (3,2)\}$$

	0	1	2	3
0	0	0	1	0
1	0	1	1	0
2	1	0	1	0
3	0	1	0	0

A

	0	1	2	3
0	0	0	0	1
1	0	0	1	1
2	0	1	0	1
3	0	0	1	0

$(A)_{(0,1)}$

Definition 3.2: Let $A, B \subset E^N$. The *binary dilation* of A by B, denoted by $A \oplus_b B$, is defined as

$$A \oplus_b B = \left\{ c \in E^N \mid c = a + b \text{ for some } a \in A \text{ and } b \in B \right\}. \tag{3.2}$$

The subscript of dilation b indicates binary. Equivalently, we may write

$$A \oplus_b B = \bigcup_{b \in B} (A)_b = \bigcup_{a \in A} (B)_a. \tag{3.3}$$

The representation, $A \oplus_b B = \bigcup_{b \in B} (A)_b$, states that the dilation of A by B can be implemented by delaying the raster scan of A by the amounts corresponding to the points in B and then ORing the delayed raster scans. That is,

$$\left(A \oplus_b B\right)_{(i,j)} = \underset{m,n}{\text{OR}} \left[[\text{AND}\left(B_{(m,n)}, A_{(i-m,j-n)}\right) \right]. \tag{3.4}$$

Example 3.2:

$$A = \{(0,1),\ (1,1),\ (2,1),\ (3,1)\}$$
$$B = \{(0,0),\ (0,2)\}$$
$$A \oplus_b B = \{(0,1),\ (1,1),\ (2,1),\ (3,1),\ (0,3),\ (1,3),\ (2,3),\ (3,3)\}$$

	0	1	2	3
0	0	1	0	0
1	0	1	0	0
2	0	1	0	0
3	0	1	0	0

A

	0	1	2
0	1	0	1

B

	0	1	2	3
0	0	1	0	1
1	0	1	0	1
2	0	1	0	1
3	0	1	0	1

$A \oplus_b B$

In dilation, the roles of sets *A* and *B* are symmetric. Dilation has a local interpretation. The dilation $A \oplus_b B$ is the locus of all centers *c*, such that the translation $(B)_c$ (by placing the origin of B at *c*) hits the set *A*. That is, if we think of each point, $a \in A$, as a seed from which grows a flower $(B)_a$, then the union of all the flowers is the dilation of *A* by B. Note that the dilation by a disk structuring element corresponds to the isotropic expansion algorithm popular in image processing. Dilation by a small square (3×3) is an eight-neighborhood operation easily implemented by adjacently connected array architectures and is the one known by the name "fill," "expand," or "grow." Other properties of dilation are the following:

Properties of Dilation:

(1) If B contains the origin, i.e., $0 \in B$, then $A \oplus_b B \supseteq A$
(2) $A \oplus_b B = B \oplus_b A$ (Commutative)
(3) $(A \oplus_b B) \oplus_b C = A \oplus_b (B \oplus_b C)$ (Associative)
(4) $(A)_x \oplus_b B = (A \oplus_b B)_x$ (Translation Invariance)
(5) $(A)_x \oplus_b (B)_{-x} = A \oplus_b B$
(6) If $A \subseteq B$, then $A \oplus_b C \subseteq B \oplus_b C$ (Increasing)
(7) $(A \cap B) \oplus_b C \subseteq (A \oplus_b C) \cap (B \oplus_b C)$
(8) $(A \cup B) \oplus_b C \subseteq (A \oplus_b C) \cap (B \oplus_b C)$ (Distributive)

3.1.2 Binary Erosion

Erosion is the morphological dual to dilation. It combines two sets using vector subtraction of set elements. If *A* and *B* denote two sets in E^N with elements *a* and *b* respectively, then the *binary erosion* of *A* by *B* is the set of all elements *x*, for which $x + b \in A$ for every $b \in B$.

Definition 3.3: The *binary erosion* of *A* by *B*, denoted by $A \ominus_b B$, is defined as

$$A \ominus_b B = \left\{ x \in E^N \mid x + b \in A \text{ for every } b \in B \right\}. \tag{3.5}$$

Equivalently, we may write

$$A \ominus_b B = \bigcap_{b \in B} (A)_{-b}. \tag{3.6}$$

Note that the binary erosion defined here is slightly different from *Minkowski subtraction*, which is $\bigcap_{b\in B}(A)_b$. The previous equation indicates that the implementation of binary erosion is similar to dilation except changing the OR function to an AND function and using the image translated by the negated points of B. That is,

$$(A \ominus_b B)_{(i,j)} = \underset{m,n}{\text{AND}}\left\{\text{OR}\left[\text{AND}\left(B_{(m,n)}, A_{(i+m,j+n)}\right), \overline{B}_{(m,n)}\right]\right\}, \tag{3.7}$$

where $\overline{B}$ denotes an inverse function. Equivalently, we may simplify:

$$(A \ominus_b B)_{(i,j)} = \underset{m,n}{\text{AND}}\left[\text{OR}\left(A_{(i+m,j+n)}, \overline{B}_{(m,n)}\right)\right] \tag{3.8}$$

Erosion $A \ominus_b B$ can be interpreted as the locus of all centers c, such that the translation $(B)_c$ is entirely contained within the set A.

$$A \ominus_b B = \left\{x \in E^N \mid (B)_x \subseteq A\right\}. \tag{3.9}$$

Example 3.3:

$$A = \{(0,1),(1,1),(2,1),(2,2),(3,0)\}$$
$$B = \{(0,0),(0,1)\}$$
$$A \ominus_b B = \{(2,1)\}$$

	0	1	2	3
0	0	1	0	0
1	0	1	0	0
2	0	1	1	0
3	1	0	0	0

A

	0	1	2
0	1	1	0

B

	0	1	2	3
0	0	0	0	0
1	0	0	0	0
2	0	1	0	0
3	0	0	0	0

$A \ominus_b B$

Erosion does not possess the commutative property. Some erosion equivalent terms are "shrink" and "reduce." It should be noted that the erosion by B given here is defined as in Haralick et al. (1987). In Serra (1982), this definition is the erosion by $\hat{B}$ (the dual structuring element). Other properties of erosion are as follows:

Properties of Erosion:

(1) If B contains the origin, i.e., $0 \in B$, then $A \ominus_b B \subseteq A$
(2) $(A \ominus_b B) \ominus_b C = A \ominus_b (B \oplus_b C)$

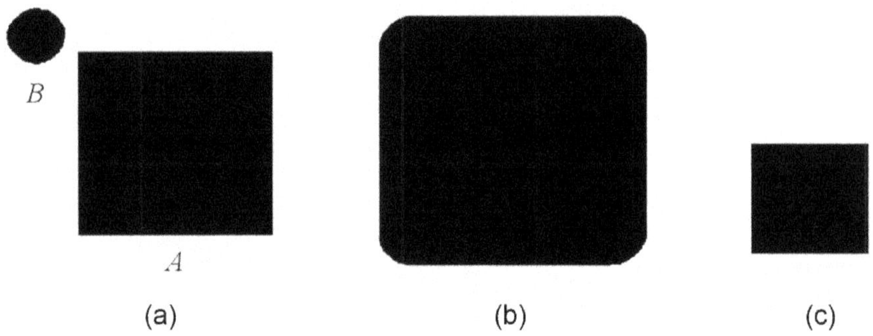

FIGURE 3.1 An example of binary morphological operations. (a) A and B, (b) $A \oplus_b B$, (c) $A \ominus_b B$.

(3) $A \oplus_b (B \ominus_b C) \subseteq (A \oplus_b B) \ominus_b C$
(4) $A_x \ominus_b B = (A \ominus_b B)_x$ (Translation Invariance)
(5) If $A \subseteq B$, then $A \ominus_b C \subseteq B \ominus_b C$ (Increasing)
(6) $(A \cap B) \ominus_b C = (A \ominus_b C) \cap (B \ominus_b C)$ (Distributive)
(7) $(A \cup B) \ominus_b C \supseteq (A \ominus_b C) \cup (B \ominus_b C)$
(8) $A \ominus_b (B \cup C) = (A \ominus_b B) \cap (A \ominus_b C)$
(9) $A \ominus_b (B \cap C) \supseteq (A \ominus_b B) \cup (A \ominus_b C)$
(10) Erosion Dilation Duality:

Let $A^c = \{x \in E^N \mid x \notin A\}$ and $\hat{B} = \{x \mid \text{ for some } b \in B,\ x = -b\}$. Then $(A \ominus_b B)^c = A^c \oplus_b \hat{B}$.

It should be noted that although dilation and erosion are dual, this does not imply that the equality can hold for any morphological cancellation. For example, if $C = A \oplus_b B$, then eroding both sides of the expression by B results in $C \ominus_b B = (A \oplus_b B) \ominus_b B \neq A$. Instead of equality, a containment relationship does hold: $(A \oplus B) \ominus B \supseteq \mathrm{A}$. Note that if the dilation and erosion are clearly used in the binary image A and binary structuring element B, then the binary b subscript will be skipped. An example of binary dilation and erosion is shown in Figure 3.1.

3.2 OPENING AND CLOSING

In practical applications, dilation and erosion pairs are combined in sequence, either the dilation of an image followed by the erosion of the dilated result, or vice versa. In either case, the result of iteratively applying dilations and erosions is an elimination of specific image details whose sizes are smaller than the structuring element without the global geometric distortion of unsuppressed features. The properties were first explored by Matheron (1975) and Serra (1982). Both of their definitions for opening and closing are identical to the ones given here, but their formulas appear different because they use the symbol $\ominus$ to mean Minkowski subtraction rather than erosion.

Definition 3.4: The *opening* of image *A* by structuring element *B*, denoted by $A \circ B$, is defined as

$$A \circ B = (A \ominus B) \oplus B. \tag{3.10}$$

Definition 3.5: The *closing* of image *A* by structuring element *B*, denoted by $A \bullet B$, is defined as

$$A \bullet B = (A \ominus B) \oplus B. \tag{3.11}$$

Note that the symbols $\oplus$ and $\ominus$ can represent either binary or grayscale dilation and erosion. Equivalently, we may write

$$A \bullet B = \bigcup_{B_y \subseteq A} B_y. \tag{3.12}$$

The opening and closing can be interpreted as follows: The opening will remove all of the pixels in the regions that are too small to contain the probe. The opposite sequence, closing, will fill in holes and concavities smaller than the probe. Such filters can be used to suppress object features or discriminate against objects based on their shape or size distribution. As an example, if a disk-shaped structuring element of diameter *h* is used, the opening of an image is equivalent to a *low-pass filter*. The opening residue is a *high-pass filter*. The difference of the openings of an image with two nonequal diameters is a *band-pass filter*. The symbolic expressions are

$$\text{LOW-PASS} = A \circ B^h$$

$$\text{HIGH-PASS} = A - \left(A \circ B^h\right)$$

$$\text{BAND-PASS} = \left(A \circ B^{h_1}\right) - \left(A \circ B^{h_2}\right), \text{ where diameters of } B,\ h_1 < h_2$$

Properties of Opening and Closing:

(1) $(A \circ B) \subseteq A \subseteq (A \bullet B)$

(2) $A \oplus B = (A \oplus B) \circ B = (A \bullet B) \oplus B$

(3) $A \ominus B = (A \ominus B) \bullet B = (A \circ B) \ominus B$

(4) If $X \subseteq Y$, then $X \circ B \subseteq Y \circ B$ (Increasing)

(5) If $X \subseteq Y$, then $X \bullet B \subseteq Y \bullet B$ (Increasing)

(6) $(A \circ B) \circ B = A \circ B$ (Idempotent)

(7) $(A \bullet B) \bullet B = A \bullet B$ (Idempotent)

(8) Opening-Closing Duality:

$$(A \bullet B)^c = A^c \circ \hat{B}$$

Example 3.4: Let A and B be shown in the following. Draw the diagrams of $A \circ B$ and $A \bullet B$.

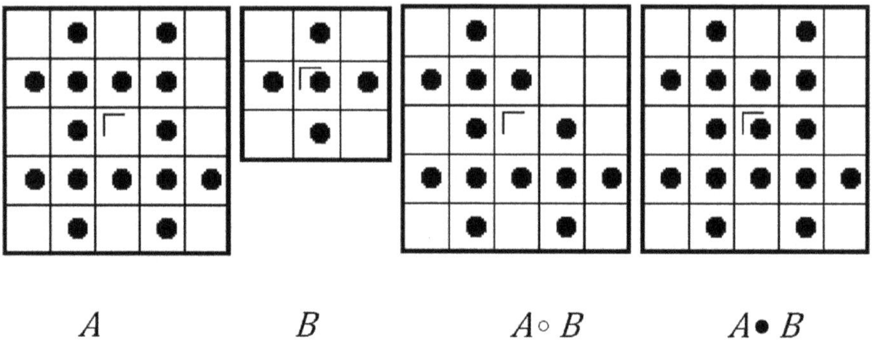

3.3 HIT-OR-MISS TRANSFORM

The hit-or-miss transformation on a binary image A is defined as follows: The structuring element B is a pair of binary images B_1 and B_2, where $B_1 \subseteq A$ and $B_2 \subseteq \bar{A}$ (complement of A). The *hit-or-miss transformation* of A by (B_1, B_2), denoted by $A \circledast (B_1, B_2)$, is defined as

$$A \circledast (B_1, B_2) = (A \ominus B_1) \cap (\bar{A} \ominus B_2) \tag{3.13}$$

Note that erosion is a special case of hit-or-miss transform, where B_2 is an empty set. Because $\bar{A} \ominus B_2 = \overline{A \oplus \hat{B}_2}$, we have

$$A \circledast (B_1, B_2) = (A \ominus B_1) \cap \left(\overline{A \oplus \hat{B}_2} \right). \tag{3.14}$$

It is equivalent to set difference:

$$A \circledast (B_1, B_2) = (A \ominus B_1) - \left(A \oplus \hat{B}_2 \right). \tag{3.15}$$

The morphological hit-or-miss transform is a natural tool for shape recognition. The hit-or-miss transform (HMT) is a powerful morphological tool for the processing of binary images. There have been several attempts to generalize it to grayscale images, based on the grayscale erosion (Bloomberg and Maragos, 1990; Naegel et al., 2007a, 2007b). Khosravi and Schafer (1996) proposed a class of rank-order-based template matching criteria using grayscale hit-or-miss transform, that are multiplier-free and independent of the dc variations of the image. Raducanu and Grana (2000) proposed the hit-or-miss transform based on level sets, called the level set hit-or-miss transform (LSHMT), to obtain a translation invariant recognition tool, with some robustness regarding small deformations and variations of illumination. It is applied to face localization on grayscale images based on a set of face patterns.

Example 3.5: Let A be

0	0	0	0	0	0	0	0
0	0	1	0	1	0	0	0
0	1	0	1	1	1	1	0
0	0	1	0	1	0	0	0
0	0	0	0	1	1	1	0
0	1	1	0	1	0	1	0
0	1	1	0	0	0	1	0
0	0	0	0	0	0	0	0

Let B_1 and B_2 be

B_1 :

0	1	0
0	1	1
0	1	0

B_2 :

1	0	1
1	0	0
1	0	1

The $A \circledast (B_1, B_2)$ is

0	0	0	0	0	0	0	0
0	0	0	0	0	0	0	0
0	0	0	0	0	0	0	0
0	0	0	0	0	0	0	0
0	0	0	0	1	0	0	0
0	0	0	0	0	0	0	0
0	0	0	0	0	0	0	0
0	0	0	0	0	0	0	0

3.4 GRAYSCALE MORPHOLOGY

Mathematical morphology represents image objects as sets in a Euclidean space. In morphological analysis, the set is the primary notion and a function is viewed as a particular case of a set (e.g., an N-dimensional, multi-valued function can be viewed as a set in $(N + 1)$-dimensional space). In this viewpoint then, any function- or set-processing system is viewed as a set mapping (transformation) from one class of sets into another class of sets. The extensions of the morphological transformations from binary to grayscale processing by Serra (1982) and Sternberg (1986) introduce a natural morphological generalization of the dilation and erosion operations.

3.4.1. Grayscale Dilation and Erosion

We begin with the definitions of the top surface of a set and the umbra of a surface. Given a set *A* in E^N, the top surface of *A* is a function defined on the projection of *A* onto its first (*N*-1) coordinates, and the highest value of the *N*-tuple is the function value (or grayvalue). We assume that for every $x \in F, \{y \mid (x,y) \in A\}$ is topologically closed, and also the sets *E* and *K* are finite.

Definition 3.6: Let $A \subseteq E^N$ and $F = \{x \in E^{N-1} \mid \text{for some } y \in E, (x,y) \in A\}$.
The *top* or *top surface* of *A*, denoted by $T[A]: F \to E$, is defined as

$$T[A](x) = \max\{y \mid (x,y) \in A\}. \tag{3.16}$$

Definition 3.7: Let $F \subseteq E^{N-1}$ and $f: F \to E$. The *umbra* of *f*, denoted by $U[f]$, $U[f] \subseteq F \times E$, is defined as

$$U[f] = \{(x,y) \in F \times E \mid y \leq f(x)\}. \tag{3.17}$$

Example 3.6: Let *f* be $\{2,1,0,2\}$. Show the umbra $U[f]$.

	0	1	2	3
0	2	1	0	2

f

	0	1	2	3
3	0	0	0	0
2	1	0	0	1
1	1	1	0	1
0	1	1	1	1
.	1	1	1	1
.	1	1	1	1
.	1	1	1	1
$-\infty$	1	1	1	1

$U[f]$

The rows below the zeroth row will not be shown in the following examples whenever they are all ones. A function is the top of its umbra. The umbra can also be illustrated by a graphic in R^2, as done by Serra (1982, ch. XII) (for example, (1982, Fig. XII.1)).

Definition 3.8: Let $F, K \subseteq E^{N-1}$ and $f: F \to E$ and $k: K \to E$. The *gray-scale dilation* of f by k, denoted by $f \oplus_g k$, is defined as

$$f \oplus_g k = T\left[U[f] \oplus_b U[k]\right]. \tag{3.18}$$

An alternate definition for $f \oplus_g k$ is

$$U[f] \oplus_b U[k] = U\left[f \oplus_g k\right]. \tag{3.19}$$

Example 3.7: Let f and k be $\{2,1,0,2\}$ and $\{2,1,0\}$, respectively. Show $U[f] \oplus_b U[k]$.

	0	1	2
0	2	1	0

k

	0	1	2
4	0	0	0
3	0	0	0
2	1	0	0
1	1	1	0
0	1	1	1

$U[k]$

	0	1	2	3
4	1	0	0	1
3	1	1	0	1
2	1	1	1	1
1	1	1	1	1
0	1	1	1	1

$U[f] \oplus_b U[k]$

Definition 3.9: Let F, $K \subseteq E^{N-1}$ and $f: F \to E$ and $k: K \to E$. The *gray-scale erosion* of f by k, denoted by $f \ominus_g k$, is defined as

$$f \ominus_g k = T\left[U[f] \ominus_b U[k]\right]. \tag{3.20}$$

An alternate definition for $f \ominus_g k$ is

$$U[f] \ominus_b U[k] = U\left[f \ominus_g k\right]. \tag{3.21}$$

Example 3.8: Let f and k be $\{2,1,0,2\}$ and $\{2,1,0\}$, respectively. Show $U[f] \ominus_b U[k]$.

k:

	0	1	2
0	2	1	0

$U[k]$:

	0	1	2
4	0	0	0
3	0	0	0
2	1	0	0
1	1	1	0
0	1	1	1

$U[f] \ominus_b U[k]$:

	0	1	2	3
1	0	0	0	0
0	1	0	0	1
-1	1	1	0	1
-2	1	1	1	1
-3	1	1	1	1

The following propositions give an alternative method of computing grayscale dilation and erosion (for a proof, see (Haralick et al., 1987)). Let F and K be the domains of the grayscale image $f(x,y)$ and the grayscale structuring element $k(m,n)$, respectively. When f or k is grayscale, the morphological operators applied on the two functions are called *grayscale morphology*.

Proposition 3.1: Let $f(x,y): F \rightarrow E$ and $k(m,n): K \rightarrow E$. Then $(f \oplus_g k)(x,y): F \oplus_g K \rightarrow E$ can be computed by

$$(f \oplus_g k)(x,y) = \max\{f(x-m, y-n) + k(m,n)\}, \tag{3.22}$$

for all $(m,n) \in K$ and $(x-m, y-n) \in F$.

Example 3.9: Compute $f \oplus_g k$.

f:

	0	1	2	3
0	2	1	0	2

k:

	0	1	2
0	2	1	0

$f \oplus_g k$:

	0	1	2	3
0	4	3	2	4

Example 3.10: Perform the grayscale dilation of f and k as shown here. Use the upper-left pixel of k as the origin (as circled). Note: do not consider the computation on the pixels which are located outside the boundary of f.

f:

2	3	0	4
1	3	5	7

k:

(−1)	1
2	3

Answer: The grayscale dilation is equivalent to reflecting k about the origin, pointwise adding it to f and taking the maximum. The reflected k is therefore:

$\hat{k}$:

3	2
1	(−1)

where the origin is now located at the lower-right pixel, and the values produced can be represented by

2 + (−1)	2 + 1, 3 + (−1)	3 + 1, 0 + (−1)	0+1, 4+(−1)
2 + 2, 1 + (−1)	2 + 3, 3 + 2, 1 + 1, 3 + (−1)	3 + 3, 0 + 2, 3 + 1, 5+(−1)	0 + 3, 4 + 2, 5 + 1, 7+(−1)

where the maximum is selected among those values. The result of grayscale dilation is

1	3	4	3
4	5	6	6

Proposition 3.2: Let $f(x,y): F \to E$ and $k(m,n): K \to E$. Then $(f \ominus_g k)(x,y): F \ominus_g K \to E$ can be computed by

$$(f \ominus_g k)(x,y) = \min\{f(x+m, y+n) - k(m,n)\}, \tag{3.23}$$

for all $(m,n) \in K$ and $(x+m, y+n) \in F$.

Example 3.11: Compute $f \ominus_g k$.

f:

	0	1	2	3
0	2	1	0	2

k:

	0	1	2
0	2	1	0

$f \ominus_g k$:

	0	1	2	3
0	0	-1	-2	0

Example 3.12: Perform the grayscale erosion of f and k, which are shown in the following. Use the upper-left pixel of k as the origin (as circled), and do not consider the computation on the pixels which are outside the boundary of f.

f:

2	3	0	4
1	3	5	7

k:

(−1)	1
2	3

Answer: The grayscale erosion is equivalent to pointwise subtracting k from f and taking the minimum. In this case the values can be represented by

2 − (−1), 3 − 1, 1 − 2, 3 − 3	3 − (−1), 0 − 1, 3 − 2, 5−3	0 − (−1), 4 − 1, 5 − 2, 7 − 3	4 − (−1), 7 − 2
1 − (−1), 3 − 1	3 − (−1), 5 − 1	5 − (−1), 7 − 1	7 − (−1)

Taking the minimum gives the result of grayscale erosion as

−1	−1	1	5
2	4	6	8

Properties of Umbra and Top Surface:

(1) $T[U[f]]=f$

(2) $U[T[U[f]]]=U[f]$

(3) $A\subseteq U[T[A]]$

(4) $U\left[f\oplus_g k\right]=U[f]\oplus_b U[k]$

(5) $U\left[f\ominus_g k\right]=U[f]\ominus_b U[k]$

Properties of Dilation:

(1) $f\oplus_g k=k\oplus_g f$

(2) $\left(f\oplus_g k_1\right)\oplus_g k_2=f\oplus_g\left(k_1\oplus_g k_2\right)$

(3) If $\vec{P}_c$ is a constant vector, then $\left(f+\vec{P}_c\right)\oplus_g k=\left(f\oplus_g k\right)+\vec{P}_c$

(4) $\left(f+\vec{P}_c\right)\oplus_g\left(k-\vec{P}_c\right)=f\oplus_g k$

(5) If $f_1\le f_2$, then $f_1\oplus_g k\le f_2\oplus_g k$

Properties of Erosion:

(1) $\left(f\ominus_g k_1\right)\ominus_g k_2=f\ominus_g\left(k_1\oplus_g k_2\right)$

(2) If $\vec{P}_c$ is a constant vector, then $\left(f+\vec{P}_c\right)\ominus_g k=\left(f\ominus_g k\right)+\vec{P}_c$

(3) $\left(f+\vec{P}_c\right)\ominus_g\left(k-\vec{P}_c\right)=f\ominus_g k$

(4) If $f_1\le f_2$, then $f_1\ominus_g k\le f_2\ominus_g k$

Note that the grayscale dilation and erosion are similar to the convolution operator, except that addition/subtraction is substituted for multiplication, and maximum/minimum is substituted for summation. Unlike convolution, morphological operations are, however, highly non-linear. Let a structuring element be a 3 × 3 square with all values 0. An example of grayscale dilation, erosion, opening, and closing is shown in Figure 3.2.

3.4.2 Grayscale Dilation Erosion Duality Theorem

Let $f : F \to E$ and $k : K \to E$. Let $x \in (F \oplus K) \cap (F \ominus \hat{K})$ be given. Then

$$-\left(f \oplus_g k\right) = (-f) \ominus_g \hat{k}, \text{ where } \hat{k}(x) = k(-x) \text{ denotes the reflection of } k. \quad (3.24)$$

Hence, the grayscale erosion can be obtained by computing a grayscale dilation as follows:

$$f \ominus_g k = -\left[(-f) \oplus_g \hat{k}\right]. \quad (3.25)$$

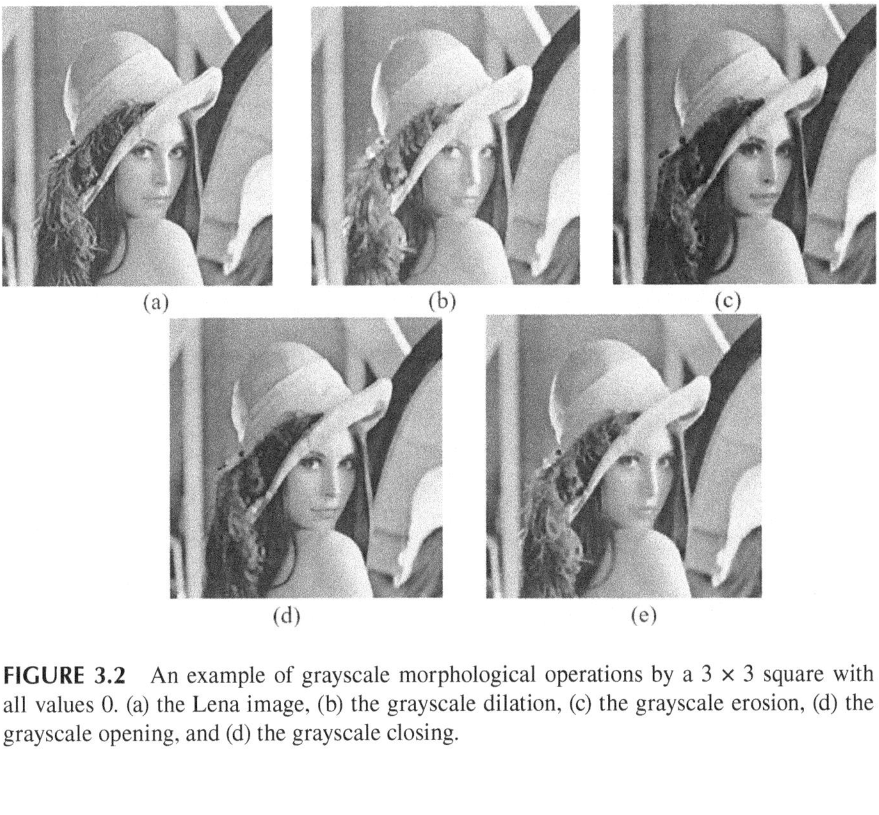

FIGURE 3.2 An example of grayscale morphological operations by a 3 × 3 square with all values 0. (a) the Lena image, (b) the grayscale dilation, (c) the grayscale erosion, (d) the grayscale opening, and (d) the grayscale closing.

[Proof]:

According to equation (3.24), $(-f') \ominus_g \hat{k}' = -(f' \oplus_g k')$.

Let $f' = -f$. We get $f \ominus_g \widehat{k'} = -((-f) \oplus_g k')$.

Let $k' = \hat{k}$. Because $\hat{\hat{k}}(x) = \hat{k}(-x) = k(x)$, we get $f \ominus_g k = -\left((-f) \oplus_g \hat{k}\right)$.

3.5 BASIC MORPHOLOGICAL ALGORITHMS

In image processing and analysis, it is important to extract object features, describe shapes, and recognize patterns. Such tasks often refer to geometric concept, such as size, shape, and orientation. Mathematical morphology takes the concept from set theory, geometry, and topology to analyze geometric structures in an image. Most essential image-processing algorithms can be represented in the form of morphological operations.

The operations desired for image processing, described in morphological notations, can be easily implemented in a parallel processor designed for cellular transformations. Each pixel can be thought of as a cell in a given state. If we define a neighborhood relationship and a pixel transition function, the application of the transition function in parallel across the space will cause the configuration of pixel states forming the image to be modified or transformed into new configurations. The implementation of these transforms by neighborhood processing stages allows the parallel processing of images in real-time with conceptually high-level operations. This allows the development of extremely powerful image-processing algorithms in a short time.

In this section, we introduce several basic morphological algorithms, including boundary extraction, region filling, connected components extraction, convex hull, thinning, thickening, skeletonization, pruning, and morphological edge operator.

3.5.1 Boundary Extraction

When an input image is grayscale, one can perform the segmentation process by thresholding the image. This involves the selection of a gray-level threshold between 0 and 255 via a histogram. Any pixels greater than this threshold are assigned 255 and otherwise 0. This produces a simple binary image in preparation for boundary extraction, in which 255 is replaced by 1 as the foreground pixels.

In a binary image, an object set is a connected component having the pixels of value 1. The boundary pixels are those object pixels whose 8-neighbors have at least one with value 0. *Boundary extraction* of a set A is first eroding A by a structuring element B and then taking the set difference between A and its erosion. The structuring element must be isotropic; ideally, it is a circle, but in digital image processing, a 3 × 3 matrix of 1s is often used. That is, the boundary of a set A is obtained by

$$\partial A = A - (A \ominus B). \tag{3.26}$$

For example, let A be

0	0	0	0	0	0	0	0	0
0	**1**	**1**	**1**	0	0	0	0	0
0	**1**	**1**	**1**	0	0	0	0	0
0	**1**	**1**	**1**	0	**1**	**1**	**1**	0
0	**1**	**1**	**1**	0	**1**	**1**	**1**	0
0	**1**	**1**	**1**	**1**	**1**	**1**	**1**	0
0	**1**	**1**	**1**	**1**	**1**	**1**	**1**	0
0	**1**	**1**	**1**	**1**	**1**	**1**	**1**	0
0	0	0	0	0	0	0	0	0

The boundary of A is extracted as

0	0	0	0	0	0	0	0	0
0	**1**	**1**	**1**	0	0	0	0	0
0	**1**	0	**1**	0	0	0	0	0
0	**1**	0	**1**	0	**1**	**1**	**1**	0
0	**1**	0	**1**	0	**1**	0	**1**	0
0	**1**	0	**1**	**1**	**1**	0	**1**	0
0	**1**	0	0	0	0	0	**1**	0
0	**1**	**1**	**1**	**1**	**1**	**1**	**1**	0
0	0	0	0	0	0	0	0	0

Note that the size of structuring element determines the thickness of object contour. For instance, a 3×3 structuring element will generate the thickness of 1 and a 5×5 structuring element will generate the thickness of 3.

3.5.2 Region Filling

Region filling intends to fill value 1 into the entire object region. It is given a binary image set A that contains all boundary pixels labeled 1 and nonboundary pixels labeled 0. Region filling starts by assigning 1 to a pixel p inside the object boundary, and grows the seed by performing iterative dilations under a limited condition restricted by A^c. If the restriction is not placed, the growing process would flood the entire image area. Let the initial step $X_0 = p$ and B be a cross-structuring element as shown in Figure 3.3. The iterative process is conducted as in the kth step (Gonzalez and Woods, 2007):

$$X_k = \left(X_{k-1} \oplus B\right) \cap A^c, k = 1, 2, 3, \cdots \tag{3.27}$$

The iteration terminates at step k if $X_k = X_{k-1}$. Therefore, X_{k-1} is the interior region of the set A. The union of X_{k-1} and A will yield the filled region and the

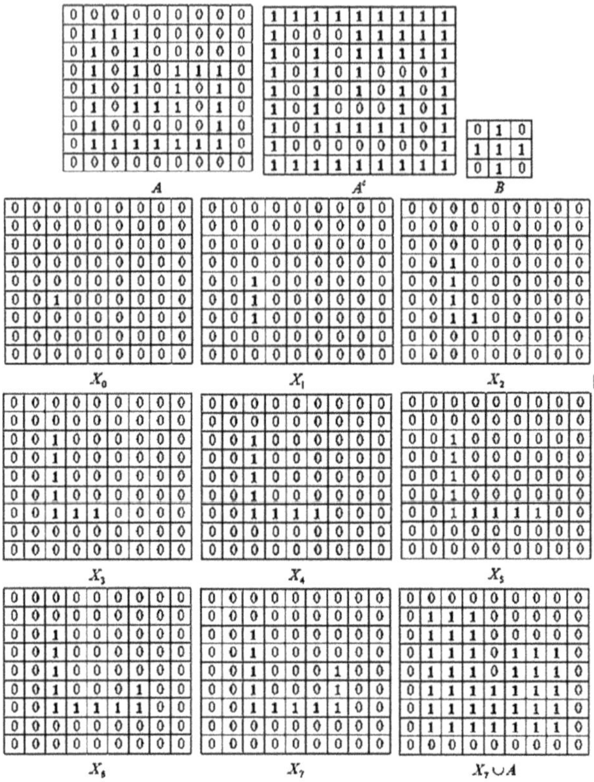

FIGURE 3.3 An example of region filling using mathematical morphology.

boundary. An example of region filling is shown in Figure 3.3, where the region is entirely painted after seven iterations.

3.5.3 Extraction of Connected Components

Similar to region filling, the extraction of connected components applies the iterative process. Assume that a binary image set A contains several connected components, in which all object pixels are labeled 1 and background pixels are labeled 0. One can extract a desired connected component by picking up a pixel in the component and growing the seed by performing iterative dilations under a limited condition restricted by A. Let the initial step $X_0 = p$ and B be a 3×3 structuring element of 1s. The iterative process is conducted as in the kth step

$$X_k = \left(X_{k-1} \oplus B\right) \cap A, k = 1, 2, 3, \cdots \tag{3.28}$$

The iteration terminates at step k if $X_k = X_{k-1}$. Therefore, X_{k-1} is the extracted connected component. An example of connected component extraction is shown in Figure 3.4.

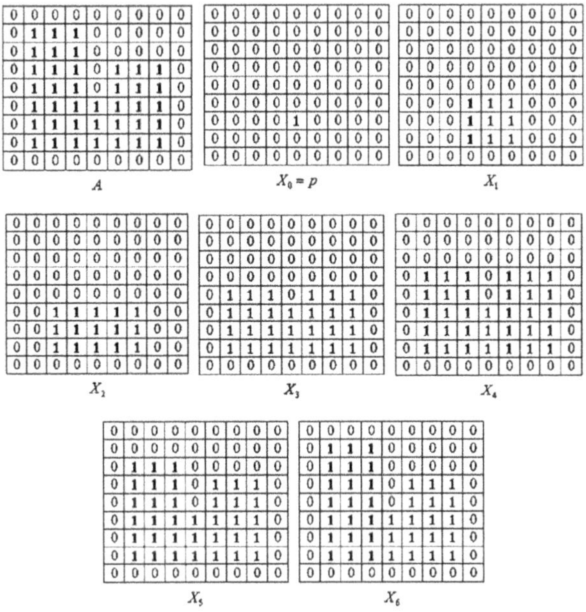

FIGURE 3.4 An example of connected component extraction by mathematical morphology.

3.5.4 Convex Hull

A *convex* set is the one for which the straight line joining any two points in the set consists of points that are also in the set. The *convex hull* of a set A is defined as the smallest convex set that contains A. One can visualize the convex hull of an object as using an elastic band surrounding the object, so the elastic band crosses over concavity parts and follows the convex contours of the object. It results in a convex set informing the object's shape without concavities. In digital images, we use an approximated 45° convex hull set, which considers its contours in orientations of multiples of 45°.

The convex hull can be easily obtained by hit-or-miss operations with four 45°-rotated structuring elements. Let A be a binary image set and let B^i, i = 1, 2, 3, 4 represent four structuring elements as

$$B^1 = \begin{bmatrix} 1 & \times & \times \\ 1 & 0 & \times \\ 1 & \times & \times \end{bmatrix}, B^2 = \begin{bmatrix} 1 & 1 & 1 \\ \times & 0 & \times \\ \times & \times & \times \end{bmatrix}, B^3 = \begin{bmatrix} \times & \times & 1 \\ \times & 0 & 1 \\ \times & \times & 1 \end{bmatrix}, B^4 = \begin{bmatrix} \times & \times & \times \\ \times & 0 & \times \\ 1 & 1 & 1 \end{bmatrix}, \quad (3.29)$$

where the symbol × denotes "don't care." Let the initial step be $X_0^i = A$. The iterative process is conducted as in the kth step

$$X_k^i = \left(X \circledast B^i\right) \cup A, i = 1,2,3,4 \text{ and } k = 1,2,3,\cdots \quad (3.30)$$

If $X_k^i = X_{k-1}^i$, then it converges. Let $D^i = X_{conv}^i$. Finally, the convex hull of A is

$$C(A) = \bigcup_{i=1}^{4} D^i \tag{3.31}$$

An example of convex hull by mathematical morphology is shown in Figure 3.5. Note that in this example, we obtain $D^1 = X_1^1$, $D^2 = X_1^2$, $D^3 = X_1^3$, and $D^4 = X_4^4$.

3.5.5 Thinning

Thinning is similar to erosion, but it does not cause disappearance of object components. It intends to reduce objects to the thickness of one pixel, generating a minimally connected axis that is equidistant from the object edges. Thinning is a fundamental early processing step in representing the structural shape of a pattern as a graph. It can be applied in industrial parts inspection, fingerprint recognition, optical character recognition, and biomedical diagnosis.

Digital skeletons, generated by thinning algorithms, are often used to represent objects in a binary image for shape analysis and classification. Jang and Chin (1990) presented a precise definition of digital skeletons and a mathematical framework for the analysis of a class of thinning algorithms based on morphological set transformation. The thinning of a binary image set can be represented in terms of hit-or-miss operation as in Serra (1982) and Meyer (1988):

$$A \otimes B = A - (A \circledast B) = A \cap (A \odot B)^c. \tag{3.32}$$

0	0	0	0	0	0	0	0	0
0	0	0	0	0	0	0	0	0
0	0	1	1	0	0	0	0	0
0	1	1	1	0	1	0	0	0
0	1	1	1	0	1	1	0	0
0	1	1	1	1	1	1	1	0
0	0	1	1	1	1	1	0	0
0	0	0	0	1	1	0	0	0
0	0	0	0	1	0	0	0	0

$X_0^i = A$

0	0	0	0	0	0	0	0	0
0	0	0	0	0	0	0	0	0
0	0	1	1	0	0	0	0	0
0	1	1	1	1	1	0	0	0
0	1	1	1	1	1	1	0	0
0	1	1	1	1	1	1	1	0
0	0	1	1	1	1	1	0	0
0	0	0	0	1	1	0	0	0
0	0	0	0	1	0	0	0	0

X_1^1

0	0	0	0	0	0	0	0	0
0	0	0	0	0	0	0	0	0
0	0	1	1	0	0	0	0	0
0	1	1	1	0	1	0	0	0
0	1	1	1	0	1	1	0	0
0	1	1	1	1	1	1	1	0
0	0	1	1	1	1	1	0	0
0	0	0	1	1	1	0	0	0
0	0	0	0	1	0	0	0	0

X_1^2

0	0	0	0	0	0	0	0	0
0	0	0	0	0	0	0	0	0
0	0	1	1	0	0	0	0	0
0	1	1	1	0	1	0	0	0
1	1	1	1	1	1	1	0	0
0	1	1	1	1	1	1	1	0
0	0	1	1	1	1	1	0	0
0	0	0	1	1	1	0	0	0
0	0	0	0	1	0	0	0	0

X_1^3

0	0	0	0	0	0	0	0	0
0	0	0	1	0	0	0	0	0
0	0	1	1	1	0	0	0	0
0	1	1	1	1	1	0	0	0
0	1	1	1	1	1	1	0	0
0	1	1	1	1	1	1	1	0
0	0	1	1	1	1	1	0	0
0	0	0	0	1	1	0	0	0
0	0	0	0	1	0	0	0	0

X_4^4

0	0	0	0	0	0	0	0	0
0	0	0	1	0	0	0	0	0
0	0	1	1	1	0	0	0	0
0	1	1	1	1	1	0	0	0
1	1	1	1	1	1	1	0	0
0	1	1	1	1	1	1	1	0
0	0	1	1	1	1	1	0	0
0	0	0	1	1	1	0	0	0
0	0	0	0	1	0	0	0	0

$C(A)$

FIGURE 3.5 An example of convex hull by mathematical morphology.

This can be thought of as a search-and-delete process. The operation $A \circledast B$ locates all occurrences of B in A, and the set subtraction removes those pixels which have been located from A. For thinning A symmetrically, we use a sequence of eight structuring elements:

$$\{B\} = \{B^1, B^2, \cdots, B^8\}, \tag{3.33}$$

where B^i is a rotated version of B^{i-1}. In digital images, this sequence contains eight structuring elements as

$$B^1 = \begin{bmatrix} 0 & 0 & 0 \\ \times & 1 & \times \\ 1 & 1 & 1 \end{bmatrix}, B^2 = \begin{bmatrix} \times & 0 & 0 \\ 1 & 1 & 0 \\ 1 & 1 & \times \end{bmatrix}, B^3 = \begin{bmatrix} 1 & \times & 0 \\ 1 & 1 & 0 \\ 1 & \times & 0 \end{bmatrix}, B^4 = \begin{bmatrix} 1 & 1 & \times \\ 1 & 1 & 0 \\ \times & 0 & 0 \end{bmatrix},$$
$$B^5 = \begin{bmatrix} 1 & 1 & 1 \\ \times & 1 & \times \\ 0 & 0 & 0 \end{bmatrix}, B^6 = \begin{bmatrix} \times & 1 & 1 \\ 0 & 1 & 1 \\ 0 & 0 & \times \end{bmatrix}, B^7 = \begin{bmatrix} 0 & \times & 1 \\ 0 & 1 & 1 \\ 0 & \times & 1 \end{bmatrix}, B^8 = \begin{bmatrix} 0 & 0 & \times \\ 0 & 1 & 1 \\ \times & 1 & 1 \end{bmatrix}. \tag{3.34}$$

Therefore, thinning a sequence of eight structuring elements is represented as

$$A \otimes \{B\} = \left(\cdots\left(\left(A \otimes B^1\right) \otimes B^2\right)\cdots\right) \otimes B^8 \tag{3.35}$$

The thinning process removes pixels from the outside edges of an object. The structuring elements are designed to find those edge pixels whose removal will not change the object's connectivity. After thinning for the first pass with these eight structuring elements is completed, the entire process for the second pass is repeated until no further changes occur. An example of thinning is shown in Figure 3.6, where the convergence is achieved in two passes. Note that the skeleton after this thinning process often contains undesirable short spurs as small irregularities in the object boundary. These spurs can be removed by a process called *pruning*, which will be introduced later.

3.5.6 Thickening

Thickening is similar to dilation, but it does not cause merging of disconnected objects. Thickening is the morphological dual to thinning. It is used to grow some concavities in an object and can be represented in terms of hit-or-miss operation and union as

$$A \odot B = A \cup (A \circledast B). \tag{3.36}$$

The thickened image contains the original set plus some filled-in pixels determined by the hit-or-miss operation. The sequence of structuring elements used in thickening is the same as that in thinning. The sequence of thickening operations can be represented as

$$A \odot \{B\} = (\cdots((A \odot B^1) \odot B^2)\cdots) \odot B^n. \tag{3.37}$$

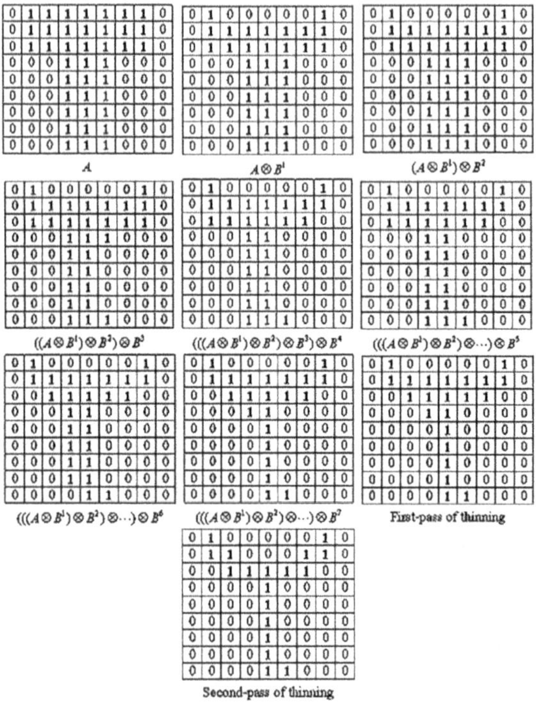

FIGURE 3.6 An example of morphological thinning.

3.5.7 Skeletonization

Skeletonization is similar to thinning, but it explores more structure information of an object. The skeleton emphasizes certain properties of images; for instance, curvatures of the contour correspond to topological properties of the skeleton. The concept of skeleton was first proposed by Blum (1967). *Skeleton*, *medial axis*, or *symmetrical axis* has been extensively used for characterizing objects satisfactorily using the structures that are composed of line or arc patterns. This has the advantages of reducing the memory space required for storage of the essential structural information and simplifying the data structure required in pattern analysis. Applications include the representation and recognition of handwritten characters, fingerprint ridge patterns, biological cell structures, circuit diagrams, engineering drawings, robot path planning, and the like. Note that the original shape can be reconstructed using the skeleton points and their distances to shape boundary. On the other hand, one cannot reconstruct the original shape from a thinned image obtained by a thinning algorithm.

Let us visualize a connected object region as a field of grass and place a fire starting from its contour. Assume that this fire burning spreads uniformly in all directions. The skeleton is where waves collide with each other in a frontal or circular manner. It receives the name of medial axis because the pixels are located at midpoints or along local symmetrical axes of the region. The skeleton of an object

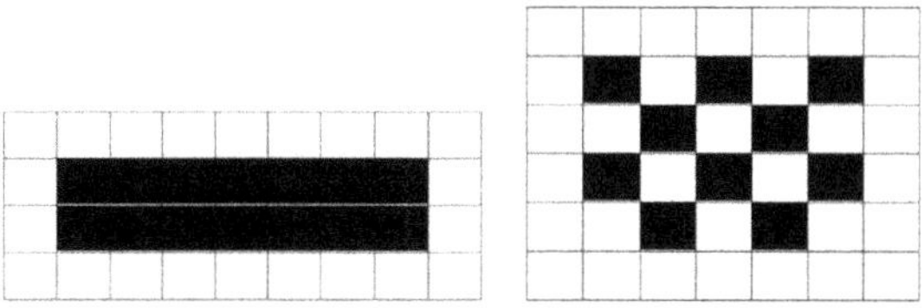

FIGURE 3.7 Two cases of non-realizable skeletonization.

pattern is a line representation of the object. It must preserve the topology of the object, have one-pixel thick, and locate in the middle of the object. However, in digital images some objects are not always realizable for skeletonization, as shown in Figure 3.7. In case (a), it is impossible to generate one-pixel thick skeleton from a two-pixel thick line. In case (b), it is impossible to produce the skeleton from an alternative-pixel pattern in order to preserve the topology.

In this section we introduce the skeletonization of an object by morphological erosion and opening (Serra, 1982). A study on the use of morphological set operations to represent and encode a discrete binary image by parts of its skeleton can be found in Maragos and Schafer (1986). The morphological skeleton was shown to unify many skeletonization algorithms, and fast algorithms for skeleton decomposition and reconstruction were developed. Fast homotopy-preserving skeletons using mathematical morphology was proposed to preserve homotopy (Ji and Piper, 1992). Another approach to obtaining the skeleton by Euclidean distance transformation and maximum value tracking can be referred to (Shih and Pu, 1995a; Shih and Wu, 2004).

Let A denote a binary image set containing object pixels of 1s and background pixels of 0s. Let B be a 3 × 3 structuring element of 1s. The skeleton of A is obtained by

$$S(A) = \bigcup_{k=0}^{K} S_k(A), \tag{3.38}$$

where $S_k(A) = (A \ominus kB) - [(A \ominus kB)^{\circ} B]$. Note that $(A \ominus kB)$ denotes k successive erosions of $(A \ominus B)$, and K is the final step before A is eroded to be an empty set. An example of morphological skeleton is shown in Figure 3.8. After obtaining the skeleton, one can easily reconstruct the original set A from these skeleton subsets using successive dilations as

$$A = \bigcup_{k=0}^{K} (S_k(A) \oplus kB) \tag{3.39}$$

3.5.8 Pruning

The skeleton of a pattern after thinning usually appears extra short noisy branches, which need to be cleaned up in the post-processing. This process is called *pruning*.

0	1	1	1	1	1	1	1	0
0	1	1	1	1	1	1	1	0
0	1	1	1	1	1	1	1	0
0	0	0	1	1	1	0	0	0
0	0	0	1	1	1	0	0	0
0	0	0	1	1	1	0	0	0
0	0	0	1	1	1	0	0	0
0	0	0	1	1	1	0	0	0
0	0	0	1	1	1	0	0	0

A

0	0	0	0	0	0	0	0	0
0	0	1	1	1	1	1	0	0
0	0	0	0	1	0	0	0	0
0	0	0	0	1	0	0	0	0
0	0	0	0	1	0	0	0	0
0	0	0	0	1	0	0	0	0
0	0	0	0	1	0	0	0	0
0	0	0	0	1	0	0	0	0
0	0	0	0	0	0	0	0	0

$A \ominus B$

0	0	0	0	0	0	0	0	0
0	0	0	0	0	0	0	0	0
0	0	0	0	0	0	0	0	0
0	0	0	0	0	0	0	0	0
0	0	0	0	0	0	0	0	0
0	0	0	0	0	0	0	0	0
0	0	0	0	0	0	0	0	0
0	0	0	0	0	0	0	0	0
0	0	0	0	0	0	0	0	0

$(A \ominus B) \circ B$

0	0	0	0	0	0	0	0	0
0	0	1	1	1	1	1	0	0
0	0	0	0	1	0	0	0	0
0	0	0	0	1	0	0	0	0
0	0	0	0	1	0	0	0	0
0	0	0	0	1	0	0	0	0
0	0	0	0	1	0	0	0	0
0	0	0	0	1	0	0	0	0
0	0	0	0	0	0	0	0	0

$S_1(A) = (A \ominus B) - (A \ominus B) \circ B$

0	0	0	0	0	0	0	0	0
0	0	0	0	0	0	0	0	0
0	0	0	0	0	0	0	0	0
0	0	0	0	0	0	0	0	0
0	0	0	0	0	0	0	0	0
0	0	0	0	0	0	0	0	0
0	0	0	0	0	0	0	0	0
0	0	0	0	0	0	0	0	0
0	0	0	0	0	0	0	0	0

$S_2(A)$

0	0	0	0	0	0	0	0	0
0	0	1	1	1	1	1	0	0
0	0	0	0	1	0	0	0	0
0	0	0	0	1	0	0	0	0
0	0	0	0	1	0	0	0	0
0	0	0	0	1	0	0	0	0
0	0	0	0	1	0	0	0	0
0	0	0	0	1	0	0	0	0
0	0	0	0	0	0	0	0	0

$S(A)$

FIGURE 3.8 An example of morphological skeleton.

For instance, in automated recognition of hand-printed characters, the skeleton is often characterized by "spurs" caused during erosion by nonuniformities in the strokes composing the characters. The short noisy branch can be located by finding end points, which have only one neighboring pixel in the set, and moving along the path until touching another branch within a very limited number of pixels. The length of a noisy branch is related to object characteristics and image size; it is often given as at most three pixels. A branch junction is considered as the pixel having at least three neighboring pixels. The pruning process could be complicated if an object has a complex shape structure. A detailed pruning process can be referred to (Serra, 1982, chapter 11).

The pruning is represented in terms of thinning as $X = A \otimes \{B\}$, where $\{B\} = \{B^1, B^2, \cdots, B^8\}$ is given here:

$$B^1 = \begin{bmatrix} \times & 0 & 0 \\ 1 & 1 & 0 \\ \times & 0 & 0 \end{bmatrix}, B^2 = \begin{bmatrix} x & 1 & \times \\ 0 & 1 & 0 \\ 0 & 0 & 0 \end{bmatrix}, B^3 = \begin{bmatrix} 0 & 0 & \times \\ 0 & 1 & 1 \\ 0 & 0 & \times \end{bmatrix}, B^4 = \begin{bmatrix} 0 & 0 & 0 \\ 0 & 1 & 0 \\ \times & 1 & \times \end{bmatrix},$$
$$B^5 = \begin{bmatrix} 1 & 0 & 0 \\ 0 & 1 & 0 \\ 0 & 0 & 0 \end{bmatrix}, B^6 = \begin{bmatrix} 0 & 0 & 1 \\ 0 & 1 & 0 \\ 0 & 0 & 0 \end{bmatrix}, B^7 = \begin{bmatrix} 0 & 0 & 0 \\ 0 & 1 & 0 \\ 0 & 0 & 1 \end{bmatrix}, B^8 = \begin{bmatrix} 0 & 0 & 0 \\ 0 & 1 & 0 \\ 1 & 0 & 0 \end{bmatrix}. \tag{3.40}$$

An example of pruning a character "b" after two passes is shown in Figure 3.9.

0	1	0	0	1	1	0	0	0
0	0	1	1	0	0	0	0	0
0	0	0	1	0	0	0	0	0
0	0	0	1	0	0	0	0	0
0	0	0	1	0	1	1	0	0
0	0	0	1	1	0	0	1	0
0	0	0	1	0	0	0	1	0
0	0	0	1	1	0	0	1	0
0	0	0	1	0	1	1	0	1

A

0	0	0	0	0	0	0	0	0
0	0	0	1	0	0	0	0	0
0	0	0	1	0	0	0	0	0
0	0	0	1	0	0	0	0	0
0	0	0	1	0	1	1	0	0
0	0	0	1	1	0	0	1	0
0	0	0	1	0	0	0	1	0
0	0	0	1	1	0	0	1	0
0	0	0	1	0	1	1	0	0

X

FIGURE 3.9 An example of pruning a character "b" by mathematical morphology.

FIGURE 3.10 (a) Lena image (b) the grayscale gradient.

3.5.9 Morphological Edge Operator

Edge operators based on grayscale morphological operations were introduced in (Lee et al., 1987). These operators can be efficiently implemented in real-time machine vision systems, which have special hardware architecture for grayscale morphological operations. The simplest morphological edge detectors are the dilation residue and erosion residue operators. Another *grayscale gradient* is defined as the difference between the dilated result and the eroded result. Figure 3.10 shows an example of grayscale gradient. A blur-minimum morphological edge operator is also introduced.

3.5.9.1 The Simple Morphological Edge Operators

A simple method of performing grayscale edge detection in a morphology-based vision system is to take the difference between an image and its erosion by a small structuring element. The difference image is the image of edge strength. We can then select an appropriate threshold value to threshold the edge strength image into a binary edge image.

Let f denote a grayscale image, and let k denote a structuring element as the following, where × means "don't care."

$k =$

×	0	×
0	0	0
×	0	×

The erosion of f by k is given by $e(x,y) = \min[f(x+i, y+j) - k(x,y)]$, for $(i,j) \in N_4(x,y)$. Since k contains 4-connected neighbors of position (0, 0), the erosion residue edge operator produces the edge strength image G_e as

$$G_e(x,y) = f(x,y) - e(x,y) = \max[f(x,y) - f(i,j)], \text{ for } (i,j) \in N_4(x,y), \quad (3.41)$$

where $N_4(x,y)$ is the set of 4-connected neighbors of pixel (x, y).

A natural nonmorphological variation of this operator takes the summation instead of maximization. This is the familiar *linear digital Laplacian operator*, $\nabla^2 f(x,y)$, which is the digital convolution of *f*(*x*,*y*) with the kernel

0	−1	0
−1	4	−1
0	−1	0

It is also possible to increase the neighborhood size of the morphological edge operator by increasing the size of the structuring element used for the erosion. For example, we can have an 8-connected neighborhood edge operator by changing the structuring element to the following:

$k =$

0	0	0
0	0	0
0	0	0

The erosion residue edge operator produces the edge strength image G_e defined by

$$G_e(x,y) = \max[f(x,y) - f(i,j)], \text{ for } (i,j) \in N_8(x,y), \quad (3.42)$$

where $N_8(x,y)$ is the set of 8-connected neighbors of pixel (x,y).

The corresponding linear Laplacian operator, which has 8-connected neighborhood support, can be implemented as the digital convolution of $f(x, y)$ with

−1	−1	−1
−1	8	−1
−1	−1	−1

The performance of the edge operator can be easily evaluated. This operator performs perfectly on ideal step edge patterns. However, it is noise-sensitive and position-biased. To correct this bias and give both inside and outside boundaries of the checkers their corresponding edge strengths, the erosion residue morphological edge detector can be used in conjunction with dilation residue operator. The dilation residue operator takes the difference between a dilated image and its original image. The dilation residue edge strength image is

$$G_d(x,y) = \max[f(i,j) - f(x,y)], \text{ for } (i,j) \in N_8(x,y) \quad (3.43)$$

A position unbiased edge operator can be obtained by a combination of the operators $G_e(x,y)$ and $G_d(x,y)$ using the pixel-wise minimum, maximum, or summation (Lee et al., 1987). For example, considering summation we have the edge operator strength: $E(x,y) = G_e(x,y) + G_d(x,y)$. The summation and maximum versions perform perfectly on ideal step edge patterns, but they are also noise-sensitive. The minimum version is noise-insensitive but has no response when applied to a single noise point. Unfortunately, it is unable to detect ideal step edge patterns. This motivates a new edge operator that first performs a blur operation to convert all the ideal step edges into ideal ramp edges and then applies the minimum version of edge operator on them.

3.5.9.2 Blur-Minimum Morphological Edge Operator

The *blur-minimum morphological edge operator* is defined by

$$I_{\text{edge-strength}} = \min\{\tilde{I} - Erosion(\tilde{I}), Dilation(\tilde{I}) - \tilde{I}\}, \quad (3.44)$$

where $\tilde{I} = \text{Blur}(I)$, which is the blurred image after a blurring operation (Lee et al., 1987). An example of the blur-minimum morphological edge operator is shown in Figure 3.11. The blur-minimum morphological edge operator is noise-insensitive. For the ideal step edge, it produces a result which has nonzero edge strength on both the edge pixels. However, due to the effect of blurring, the edge strength assigned to the edge pixels is one third of the edge contrast. For ideal ramp edges of larger spatial extent, it will assign nonzero edge strength to more than one pixel. However, the true edge pixel is usually given edge strength higher than its neighbors.

The performance of the blur-minimum morphological edge operator may be poor in the presence of noise. To overcome this problem, Feehs and Arce (1987) proposed a twofold modification, which uses a so-called *alpha-trimmed mean* (ATM) filter to

FIGURE 3.11 (a) The original image, (b) the edge image by the blur-minimum morphological edge operator.

smooth noise models other than Gaussian distribution. They defined the morphological edge operator as

$$I_{\text{atm}} = \min\{[(f_{\text{atm}} \circ k) - (f_{\text{atm}} \ominus k)], [(f_{\text{atm}} \oplus k) - (f_{\text{atm}} \bullet k)]\}, \tag{3.45}$$

where f_{atm} is the image f smoothed by an alpha-trimmed mean filter. Since opening and closing are smoothing and noise-suppressing operations, this result will be less sensitive to noise. Furthermore, Moran (1990) developed a local neighborhood operator to sharpen edges using a selective erosion followed by a selective dilation.

3.6 MORPHOLOGICAL FILTERS

Image filters are specifically designed to remove unwanted image components and/or enhance wanted ones. Mathematical morphology, applying set-theoretic concept, provides an approach to digital image processing and analysis based on the geometric shape of objects. Using an a priori determined structuring element, features in an image can be extracted, suppressed, or preserved by applying morphological operators. *Morphological filters*, using the basic operations such as dilation, erosion, opening, and closing, are suited for many purposes including digitization, enhancement, compression, restoration, segmentation, and description.

Linear filtering techniques have serious limitations in dealing with the images that have been created or processed by a system exhibiting some degree of nonlinearity. A nonlinear filter produces the output which is not a linear function of its input. Nonlinear filtering techniques have shown clear superiority over linear filters in image processing. For example, an image often contains meaningful high-frequency components, such as edges and fine details. A linear lowpass filter would blur sharp edges thus producing unacceptable results, so nonlinear filters must be used.

Morphological filters belong to nonlinear operations. Morphological dilation is used to smooth small dark regions. If all the values in the structuring element are positive, the output image tends to be brighter than the input. Dark elements are reduced or eliminated depending on how their shapes and sizes relate to the structuring element used. Morphological erosion is used to smooth small light regions as opposed to dilation. Morphological opening removes bright objects which are small in size and breaks narrow connections between two bright objects. Morphological closing preserves small objects which are brighter than background and connects bright objects with small gaps in between. A morphological top-hat operation, i.e., the difference between the original image and the opened image, is used to enhance low-contrast, high-frequency details within an image.

In this section, we will introduce alternating sequential filters, recursive morphological filters, soft morphological filters, order-statistic soft morphological filters, recursive soft morphological filters, recursive order-statistic soft morphological filters, regulated morphological filters, and fuzzy morphological filters. Another type of morphological filters called *general sweep morphology*, which allows simultaneous rotation and scaling of the structuring element when operating with an image, was proposed by Shih and Gaddipati (2003). Its geometric modeling and representation can be referred to (Shih and Gaddipati, 2005).

3.6.1 Alternating Sequential Filters

Alternating sequential filters (ASFs) in mathematical morphology are a combination of iterative morphological filters with increasing sizes of structuring elements. They offer a hierarchical structure for extracting geometric characteristics of objects. In addition, they provide less distortion in feature extraction than those filters which directly process the images with the largest structuring element (Serra, 1982). The alternating filter is composed of morphological openings and closings whose primitive morphological operations are dilation and erosion.

The class of alternating filters (AFs) has been demonstrated to be useful in image analysis applications. Sternberg (1986) introduced a new class of morphological filters called *alternating sequential filters* (ASFs), which consist of iterative operations of openings and closings with structuring elements of increasing sizes. Schonfeld and Goutsias (1991) showed that ASFs are the best in preserving crucial structures of binary images in the least differentiation sense. Pei et al. (1997) proposed an efficient class of alternating sequential filters in mathematical morphology to reduce the computational complexity in the conventional ASFs about a half.

The ASFs have been successfully used in a variety of applications, such as remote sensing and medical imaging (Destival, 1986; Preteux et al., 1985). Morales et al. (1995) presented a relationship between alternating sequential filters and the morphological sampling theorem (MST) developed by Haralick et al. (1989). The motivation is to take advantage of the computational efficiency offered by the MST to implement morphological operations.

Let X denote a binary image and let B denote a binary structuring element. The *alternating filter* (AF) is defined as an opening followed by a closing or a closing followed by an opening, and is represented as

$$AF_B(X) = (X \circ B) \bullet B \tag{3.46}$$

or

$$AF_B(X) = (X \bullet B) \circ B \tag{3.47}$$

Another type of AF is defined as

$$AF_B(X) = ((X \circ B) \bullet B) \circ B \tag{3.48}$$

or

$$AF_B(X) = ((X \bullet B) \circ B) \bullet B \tag{3.49}$$

An alternating sequential filter (ASF) is an iterative application of $AF_B(X)$ with increasing sizes of structuring elements, denoted as

$$ASF(X) = AF_{B_N}(X) AF_{B_{N-1}}(X) \cdots AF_{B_1}(X), \tag{3.50}$$

where N is an integer and $B_N, B_{N-1}, \cdots, B_1$ are structuring elements with decreasing sizes. The B_N is constructed by

$$B_N = B_{N-1} \oplus B_1, \text{ for } N \geq 2. \tag{3.51}$$

The ASF offers a method of extracting image features hierarchically. The features can be divided into different layers according to their corresponding structuring element sizes. The features, such as size distribution (Giardina and Dougherty, 1988), of each layer can be used in many applications – for example, feature classification and recognition.

3.6.2 Recursive Morphological Filters

Recursive morphological filters can be used to avoid the iterative processing in the distance transform and skeletonization (Shih and Mitchell, 1992; Shih et al., 1995). The intent of the recursive morphological operations is to feed back the output at the current scanning pixel to overwrite its corresponding input pixel to be considered into computation at the following scanning pixels. Therefore, each output value obtained depends on all the updated input values at positions preceding the current location. The resulting output image in recursive mathematical morphology inherently varies with different image scanning sequences.

Let $N(O)$ be the set of the pixel O plus the neighbors prior to it in a scanning sequence of the picture within the domain of a defined structuring element. Suppose that a set A in the Euclidean N-space $\left(E^N\right)$ is given. Let F and K be $\{x \in E^{N-1} \mid \text{for some } y \in E, (x, y) \in A\}$, and let the domain of the grayscale image be f and of the grayscale structuring element be k. The *recursive dilation* of f by k, which is denoted by $f \circledast k$, is defined as follows:

$$(f \circledast k)(x, y) = \max\{[(f \circledast k)(x-m, y-n)] + k(m, n)\} \tag{3.52}$$

for all $(m,n) \in K$ and $(x-m, y-n) \in (N \cap F)$. The *recursive erosion* of f by k, which is denoted by $f \circleddash k$, is defined as follows:

$$(f \circleddash k)(x, y) = \min\{[f \circleddash k)(x+m, y+n)] - k(m, n)\} \tag{3.53}$$

for all $(m,n) \in K$ and $(x+m, y+n) \in (N \cap F)$.

The scanning order will affect the output image in recursive morphology. Starting with a corner pixel, the scanning sequences in a 2D digital image are in general divided into the following:

(i) "LT" denotes left-to-right and top-to-bottom. (Note that this is a usual television raster scan.)
(ii) "RB" denotes right-to-left and bottom-to-top.
(iii) "LB" denotes left-to-right and bottom-to-top.
(iv) "RT" denotes right-to-left and top-to-bottom.

Assume that a 3 × 3 structuring element k is denoted by

$$k = \begin{bmatrix} A_1 & A_2 & A_3 \\ A_4 & A_5 & A_6 \\ A_7 & A_8 & A_9 \end{bmatrix}.$$

Because they are associated with different scanning directions, not all of the nine elements $A_1, \cdots, A_9$ are used in computation. The redefined structuring element in recursive mathematical morphology must satisfy the following criterion: Wherever a pixel is being dealt with, all its neighbors defined within the structuring element k must have already been visited before by the specified scanning sequence. Thus, the k is redefined as

$$k_{LT} = \begin{bmatrix} A_1 & A_2 & A_3 \\ A_4 & A_5 & \times \\ \times & \times & \times \end{bmatrix}, k_{RB} = \begin{bmatrix} \times & \times & \times \\ \times & A_5 & A_6 \\ A_7 & A_8 & A_9 \end{bmatrix}, k_{LB} = \begin{bmatrix} \times & \times & \times \\ A_4 & A_5 & \times \\ A_7 & A_8 & A_9 \end{bmatrix}, k_{RT} = \begin{bmatrix} A_1 & A_2 & A_3 \\ \times & A_5 & A_6 \\ \times & \times & \times \end{bmatrix} \tag{3.54}$$

where × means "don't care." A complete scanning in the image space must contain the dual directions, that is, the first one being any of the four scans *LT*, *RB*, *LB*, or *RT* working with its associated structuring element and the second one being the opposite scanning sequence of the first scan and its associated structuring element. For example, the opposite scanning of *LT* is *RB*.

The advantage of recursive mathematical morphology can be illustrated by applying distance transform. The distance transform is to convert a binary image which consists of object (foreground) and non-object (background) pixels into an image in which every object pixel has a value corresponding to the minimum distance from the background. Three types of distance measures in digital image processing are usually used: Euclidean, city-block, and chessboard. The city-block and chessboard distances are our concern since their structuring elements can be decomposed into iterative dilations of small structuring components. The distance computation is in principle a global operation. By applying the well-developed decomposition properties of mathematical morphology, we can significantly reduce the tremendous cost of global operations to that of small neighborhood operations, which is suited for parallel pipelined computers.

The iteration in the distance transform results in a bottleneck in real-time implementation. To avoid iteration, the recursive morphological operation is used. Only two scannings of the image are required: one is left-to-right top-to-bottom, and the other is right-to-left bottom-to-top. Thus, the distance transform becomes independent of the object size.

By setting the origin point as *P*, we can represent all of the points by their distances from *P* (so that *P* is represented by 0). The values of the structuring element are selected to be negative of the associated distance measures because the erosion is the minimum selection after the subtraction is applied. Let *f* denote a binary image which consists of two classes, object pixels (foreground) and nonobject pixels (background). Let the object pixels be the value "$+\infty$" (or any number larger than a half of the object's diameter) and the nonobject pixels be the value "0." Let k_1 and k_2 be the structuring elements associated with the *LT* and *RB* scannings, respectively; that is, in the chessboard distance

$$k_1 = \begin{bmatrix} -1 & -1 & -1 \\ -1 & 0 & \times \\ \times & \times & \times \end{bmatrix}, \quad k_2 = \begin{bmatrix} \times & \times & \times \\ \times & 0 & -1 \\ -1 & -1 & -1 \end{bmatrix}. \tag{3.55}$$

In the city-block distance

$$k_1 = \begin{bmatrix} -2 & -1 & -2 \\ -1 & 0 & \times \\ \times & \times & \times \end{bmatrix}, \quad k_2 = \begin{bmatrix} \times & \times & \times \\ \times & 0 & -1 \\ -2 & -1 & -2 \end{bmatrix}. \tag{3.56}$$

Distance Transform using Recursive Morphology:

(1) $d_1 = f \circleddash k_1$
(2) $d_2 = d_1 \circleddash k_2$

The distance transform algorithm has widely used applications in image analysis. One of the applications is to compute a shape factor that is a measure of the compactness of the shape based on the ratio of the total number of object pixels to the summation of all distance measures (Danielsson, 1978). Another is to obtain the medial axis (or skeleton) that can be used for features extraction.

Example 3.13: Apply the two-step distance transformation using recursive morphology to calculate the chessboard distance transform on the following image.

0	0	0	0	0	0	0	0	0	0	0	0	0
0	7	7	7	7	7	0	0	0	7	7	7	0
0	7	7	7	7	7	0	0	0	7	7	7	0
0	7	7	7	7	7	0	0	0	7	7	7	0
0	7	7	7	7	7	7	7	7	7	7	7	0
0	7	7	7	7	7	7	7	7	7	7	7	0
0	7	7	7	7	7	7	7	7	7	7	7	0
0	7	7	7	7	7	7	7	7	7	7	7	0
0	7	7	7	7	7	7	7	7	7	7	7	0
0	0	0	0	0	0	0	0	0	0	0	0	0

Answer: First, we apply $d_1 = f \ominus k_1$, where $k_1 = \begin{bmatrix} -1 & -1 & -1 \\ -1 & 0 & \times \\ \times & \times & \times \end{bmatrix}$, and obtain the following:

0	0	0	0	0	0	0	0	0	0	0	0	0
0	1	1	1	1	1	0	0	0	1	1	1	0
0	1	2	2	2	1	0	0	0	1	2	1	0
0	1	2	3	2	1	0	0	0	1	2	1	0
0	1	2	3	2	1	1	1	1	1	2	1	0
0	1	2	3	2	2	2	2	2	2	2	1	0
0	1	2	3	3	3	3	3	3	3	2	1	0
0	1	2	3	4	4	4	4	4	3	2	1	0
0	1	2	3	4	5	5	5	4	3	2	1	0
0	0	0	0	0	0	0	0	0	0	0	0	0

Second, we apply $d_2 = d_1 \ominus k_2$, where $k_2 = \begin{bmatrix} \times & \times & \times \\ \times & 0 & -1 \\ -1 & -1 & -1 \end{bmatrix}$, and obtain the chess-board distance transform as

0	0	0	0	0	0	0	0	0	0	0	0	0
0	1	1	1	1	1	0	0	0	1	1	1	0
0	1	2	2	2	1	0	0	0	1	2	1	0
0	1	2	3	2	1	0	0	0	1	2	1	0
0	1	2	3	2	1	1	1	1	1	2	1	0
0	1	2	3	2	2	2	2	2	2	2	1	0
0	1	2	3	3	3	3	3	3	3	2	1	0
0	1	2	2	2	2	2	2	2	2	2	1	0
0	1	1	1	1	1	1	1	1	1	1	1	0
0	0	0	0	0	0	0	0	0	0	0	0	0

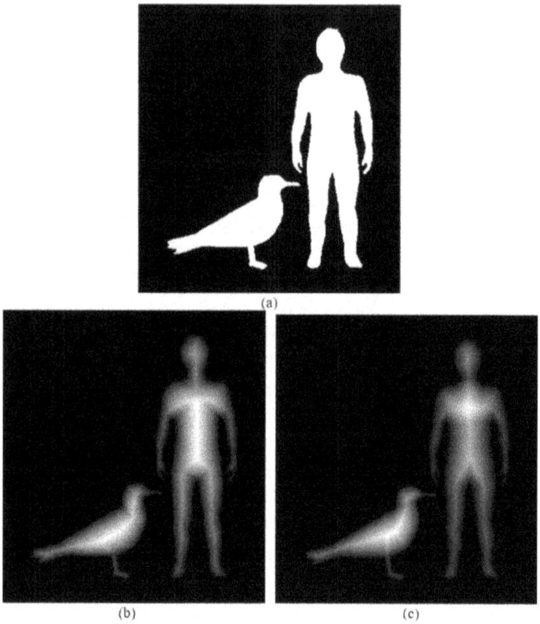

FIGURE 3.12 (a) A binary image of human and bird figures, (b) its chessboard distance transformation, (c) its city-block distance transformation.

Figure 3.12 shows a binary image of human and bird figures and its chessboard and city-block distance transformations. Note that after distance transformations, the distances are linearly rescaled to be in the range of [0, 255] for clear display.

Note that the recursive morphological filters defined here are different from the commonly used recursive filters in signal processing, which reuse one or more of the outputs as an input and can efficiently achieve a long impulse response, without having to perform a long convolution (Smith, 1997). Other recursive algorithms may use different definitions. For example, the recursive algorithm based on observation of the basis matrix and block basis matrix of grayscale structuring elements was used to avoid redundant steps in computing overlapping local maximum or minimum operations (Ko et al., 1996). Another recursive two-pass algorithm that runs at a constant time for simultaneously obtaining binary dilations and erosions with all possible length line structuring elements can be referred to (Nadadur and Haralick, 2000).

3.6.3 Soft Morphological Filters

Soft morphological filters adopt the concept of weighted order-statistic filters and morphological filters. The primary difference from traditional morphological filters is that the maximum and minimum operations are replaced by general weighted order statistics and the "soft" boundary is added to the structuring element. In general, soft morphological filters are less sensitive to additive noise and to small variations in object shape, and can preserve most of the desirable properties of traditional morphological filters (Koskinen et al., 1991; Kuosmanen and Astola, 1995; Shih and Pu, 1995b; Shih and Puttagunta, 1995). Hamid et al. (2003) presented a technique for the optimization of multidimensional grayscale soft morphological filters for applications in automatic film archive restoration, specific to the problem of film dirt removal.

The basic idea of soft morphological operations is that the structuring element B is split into two subsets: the *core* subset A (i.e., $A \subseteq B$) and the *soft boundary* subset $B \setminus A$, where $\setminus$ denotes the *set difference*. Soft morphological dilation (erosion) of a function with respect to these two finite sets A and B using the order index k is a function whose value at location x is obtained by sorting in a descendent (ascendent) order the total $\text{Card}(B \setminus A) + k \times \text{Card}(A)$ values of the input function, including the elements within $(B \setminus A)_x$ and repetition k times of the elements within $(A)_x$, and then selecting the kth order from the sorted list. Let $\{k \lozenge f(a)\}$ denote the repetition k times of $f(a)$, which means $\{k \lozenge f(a)\} = \{f(a), f(a), \cdots, f(a)\}$ (k times). The soft morphological operations of function-by-set are defined as follows, where f is a grayscale image and A and B are flat structuring elements.

Definition 3.10: The *soft morphological dilation* of f by $[B, A, k]$ is defined as

$$(f \oplus [B, A, k])(x) = k\text{th largest of}\left(\left\{k \lozenge f(a) \mid a \in A_x\right\} \cup \left\{f(b) \mid b \in (B \setminus A)_x\right\}\right). \quad (3.57)$$

Definition 3.11: The *soft morphological erosion* of f by $[B, A, k]$ is defined as

$$(f \ominus [B, A, k])(x) = k\text{th smallest of}\left(\left\{k \lozenge f(a) \mid a \in A_x\right\} \cup \left\{f(b) \mid b \in (B \setminus A)_x\right\}\right). \quad (3.58)$$

The soft morphological operations of set-by-set can be simplified as in the following definition, where $n = \text{Card}(A)$ and $N = \text{Card}(B)$.

Definition 3.12: The *soft morphological dilation* of a set X by $[B, A, k]$ is defined as

$$X \oplus [B, A, k] = \left\{x \mid k \times \text{Card}\left(X \cap A_x\right) + \text{Card}\left(X \cap (B \setminus A)_x\right) \geq k\right\}. \quad (3.59)$$

Definition 3.13: The *soft morphological erosion* of a set X by $[B, A, k]$ is defined as

$$X \ominus [B, A, k] = \left\{x \mid k \times \text{Card}\left(X \cap A_x\right) + \text{Card}\left(X \cap (B \setminus A)_x\right) \geq N + (k-1) \times n - k + 1\right\}. \quad (3.60)$$

The logic-gate implementation of soft morphological dilation and erosion is illustrated in Figure 3.13, where the parallel counter counts the number of 1s of the input signal and the comparator outputs 1 if the input is greater than or equal to the index k.

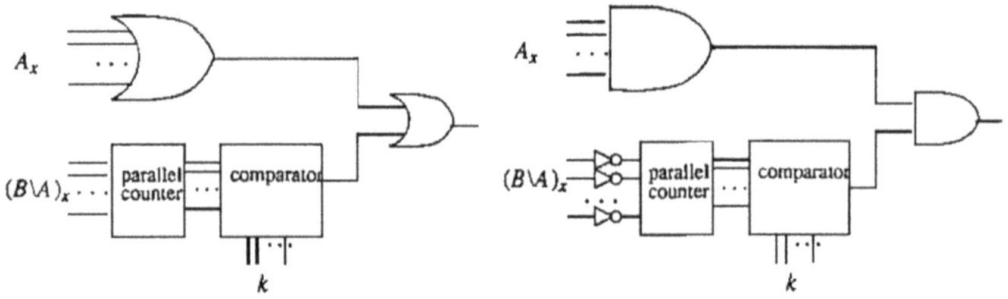

FIGURE 3.13 The logic-gate implementation of soft morphological operations (a) dilation and (b) erosion.

The cross-section $X_t(f)$ of f at level t is the set obtained by thresholding f at level t:

$$X_t(f) = \{x \mid f(x) \geq t\}, \text{ where } -\infty < t < \infty. \tag{3.61}$$

Theorem 3.1: The soft morphological operations of function-by-set commute with thresholding. That is, for any t, we have

$$X_t(f \oplus [B, A, k]) = X_t(f) \oplus [B, A, k]. \tag{3.62}$$

This theorem can be easily derived from Shih and Mitchell (1989) and Shih and Pu (1995b). The implementation and analysis of soft morphological operations of function-by-set can be achieved by using the set-by-set operations, which are much easier to deal with because they are involved in only counting the number of pixels instead of sorting numbers. Consider the thresholding of a binary image:

$$f_a(a) = \begin{cases} 1 & \text{if } f(a) \geq a \\ 0 & \text{otherwise} \end{cases}, \tag{3.63}$$

where $0 \leq a \leq L$ and L is the largest value in f. It is simple to show that f can be reconstructed from its thresholded binary images as

$$f(x) = \sum_{a=1}^{L} f_a(x) = \max\{a \mid f_a(x) = 1\}. \tag{3.64}$$

A transformation Ψ is said to possess *threshold-linear superposition* if it satisfies

$$\Psi(f) = \sum_{a=1}^{L} \Psi(f_a) \tag{3.65}$$

Such a transformation Ψ can be realized by decomposing f into all its binary images f_a s, processing each thresholded images by Ψ, and creating the output $\Psi(f)$ by summing up the processed f_a s. The soft morphological operations of function-by-set can be easily shown to obey the threshold-linear superposition.

Theorem 3.2: The soft morphological operations of functions by sets obey the threshold-linear superposition. That is,

$$f \oplus [B,A,k] = \sum_{a=1}^{L} \left(f_a \oplus [B,A,k] \right). \tag{3.66}$$

This theorem is very useful in the sense that the grayscale input image is thresholded at each gray level a to get f_a, followed by the soft morphological operations of sets by sets for f_a, and then summing up all f_a s, which is exactly the same as the soft morphological operations of functions by sets. Considering A_1, A_2, and B in Figure 3.14, an example of illustrating the threshold decomposition is given in Figure 3.15.

The grayscale soft morphological operation can be decomposed into binary soft morphological operation (Pu and Shih, 1995). A new hardware structure for implementation of soft morphological filters can be referred to (Gasteratos et al., 1997). Let the reflected sets of A and B be $\hat{A}(x) = A(-x)$ and $\hat{B}(x) = B(-x)$, respectively.

Definition 3.14: The *soft morphological closing* of f by $[B, A, k]$ is defined as

$$f \bullet [B,A,k] = (f \oplus [B,A,k]) \ominus [\hat{B},\hat{A},k]. \tag{3.67}$$

Definition 3.15: The *soft morphological opening* of f by $[B, A, k]$ is defined as

$$f \circ [B,A,k] = (f \ominus [B,A,k]) \oplus [\hat{B},\hat{A},k]. \tag{3.68}$$

It is interesting to note that if $k = 1$, the soft morphological operations are equivalent to standard morphological operations. If $k > \text{Card}(B \setminus A)$, the soft morphological operations are reduced to consider only the hard set A (i.e., the soft boundary set $(B\setminus A)$ will actually not affect the result). Therefore, in order to preserve the nature of soft morphological operations, the constraint that $k \le \min\{\text{Card}(B)/2, \text{Card}(B \setminus A)\}$ is used.

In Figure 3.16, (a) shows an original image, and (b) shows the same image but with added Gaussian noise being zero mean and standard deviation 20. Figure 3.16(c) shows the result of applying a standard morphological closing with a structuring

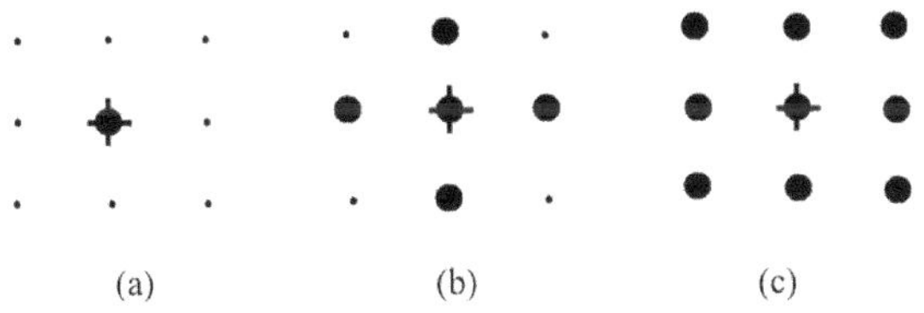

FIGURE 3.14 Two core sets (a) A_1, (b) A_2, and (c) the structuring element B.

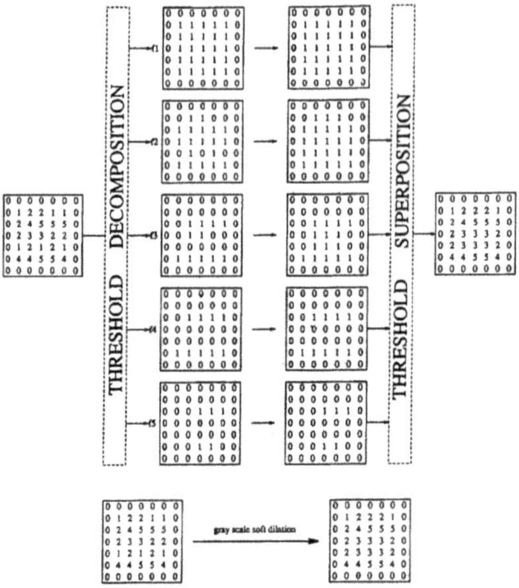

FIGURE 3.15 The soft morphological dilation obeys threshold superposition with A_1 and B as shown in Figure 3.14, and k is equal to 4.

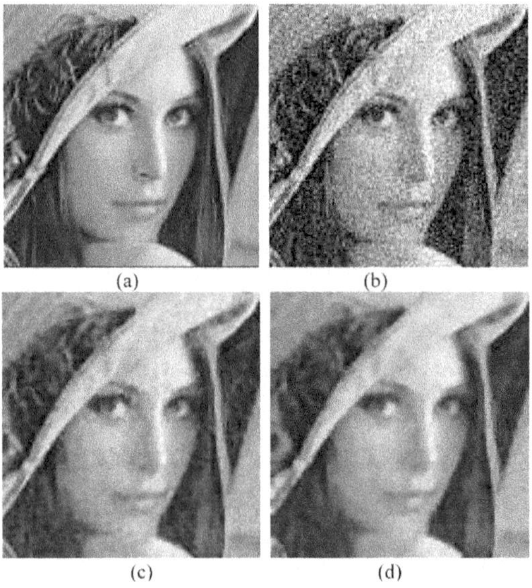

FIGURE 3.16 (a) An original image, (b) image corrupted by Gaussian noise, (c) the result of applying a standard morphological closing, (d) the result of applying a soft morphological closing.

element of size 3 × 3, and (d) shows the result of applying a soft morphological closing with $B = \langle(0,-1),(-1,0),(0,0),(1,0),(0,1)\rangle$, $A = \langle(0,0)\rangle$, and $k = 3$. The mean-square signal-to-noise ratios for Figures 3.16(c) and (d) are 39.78 and 55.42, respectively. One can observe that the result of soft morphological filtering produces a smoother image and a higher signal-to-noise ratio.

Example 3.14: Perform the soft morphological dilation $f \oplus [B, A, 3]$ and the soft morphological erosion $f \ominus [B, A, 3]$. Use the center (underlined) of A and B as the origin.

$$f: \begin{bmatrix} 0 & 0 & 0 & 1 & 0 \\ 0 & 5 & 5 & 5 & 1 \\ 4 & 5 & 1 & 5 & 0 \\ 3 & 6 & 5 & 6 & 0 \\ 0 & 0 & 0 & 0 & 1 \end{bmatrix}, A = \begin{bmatrix} 0 & 0 & 0 \\ 0 & \underline{1} & 0 \\ 0 & 0 & 0 \end{bmatrix}, \quad B = \begin{bmatrix} 1 & 1 & 1 \\ 1 & 1 & 1 \\ 1 & 1 & 1 \end{bmatrix}$$

Answer: The soft morphological dilation $f \oplus [B, A, 3]$ repeats the central element of the 3 × 3 window three times and others once, and then arranges them in an ascending order. Taking out the third largest element we obtain the following result:

$$\begin{matrix} 0 & 0 & 5 & 1 & 1 \\ 4 & 5 & 5 & 5 & 1 \\ 5 & 5 & 5 & 5 & 5 \\ 4 & 6 & 5 & 6 & 1 \\ 0 & 3 & 5 & 1 & 1 \end{matrix}$$

The soft morphological erosion $f \ominus [B, A, 3]$ repeats the central element of the 3 × 3 window three times and others once, and then arranges in an ascending order. Taking out the third smallest element, we obtain the following result:

$$\begin{matrix} 0 & 0 & 0 & 1 & 0 \\ 0 & 0 & 1 & 0 & 1 \\ 4 & 3 & 1 & 1 & 0 \\ 3 & 0 & 0 & 0 & 0 \\ 0 & 0 & 0 & 0 & 1 \end{matrix}$$

3.6.4 Order-Statistic Soft Morphological (OSSM) Filters

The disadvantage of soft morphological filters is that the repetition number is set up to be the rank order. This constraint limits the flexibility and applicability of the filters in noise removal and edge preservation. In this section we present the order-statistic soft morphological (OSSM) filters which provide a free order selection parameter. We also develop the properties and explore the restoration performance when a signal is corrupted by variant noises.

Many nonlinear image-processing techniques have their root in statistics. Their applications in images, image sequences, and color images with order statistics (David, 1981) and their closely related morphological filters have by far been very successful (Maragos and Schafer, 1987a, 1987b; Stevenson and Arce, 1987; Soille, 2002). The median filter was first introduced in statistics for smoothing economic time series, and was then used in image processing applications to remove noises with retaining sharp edges. The stacking property unifies all subclasses of stack filters, such as ranked order, median, weighted median, and weighted order-statistic filters.

Let f be a function defined on m-dimensional discrete Euclidean space Z^m. Let W be a window of a finite subset in Z^m which contains N points, where $N = \text{Card}(W)$, the *cardinality* of W. The kth order statistics (OS^k) of a function $f(x)$ with respect to the window W is a function whose value at location x is obtained by sorting in a descendent order the N values of $f(x)$ inside the window W whose origin is shifted to location x and picking up the kth number from the sorted list, where k ranges from 1 to N. Let W_x denote the translation of the origin of the set W to location x. The kth OS is represented by

$$OS^k(f,W)(x) = k\text{th largest of}\left\{f(a) \mid a \in W_x\right\}. \tag{3.69}$$

The order-statistic soft morphological (OSSM) operations, like in the soft morphology, adopt order statistics, and use the split subsets, core set A and boundary set $B \backslash A$, as the structuring element. The obvious difference is that a free order selection parameter is introduced to increase flexibility and efficiency of the operations. That is, we select the lth order, rather than the kth order, from the sorted list of the total "$\text{Card}(B \setminus A) + k \times \text{Card}(A)$" values of the input function. The definitions of function by sets are given as follows, where f is a grayscale image and B is a flat structuring element.

Definition 3.16: The *order-statistic soft morphological dilation* of f by $[B, A, k, l]$ is defined as

$$(f \oplus [B,A,k,l])(x) = l\text{th largest of}\left(\left\{k \Diamond f(a) \mid a \in A_x\right\} \cup \left\{f(b) \mid b \in (B \setminus A)_x\right\}\right). \tag{3.70}$$

Definition 3.17: The *order-statistic soft morphological erosion* of f by $[B, A, k, l]$ is defined as

$$(f \ominus [B,A,k,l])(x) = l\text{th smallest of}\left(\left\{k \Diamond f(a) \mid a \in A_x\right\} \cup \left\{f(b) \mid b \in (B \setminus A)_x\right\}\right). \tag{3.71}$$

One of the advantages of using the OSSM operators over standard morphological operators can be observed in the following example. Assume that we have a flat signal f corrupted by impulsive noise as shown in Figure 3.17(a). In order to remove the impulsive noise, a standard morphological opening and a closing are applied to the distorted signal. Note that the positive impulse is removed by the opening operation

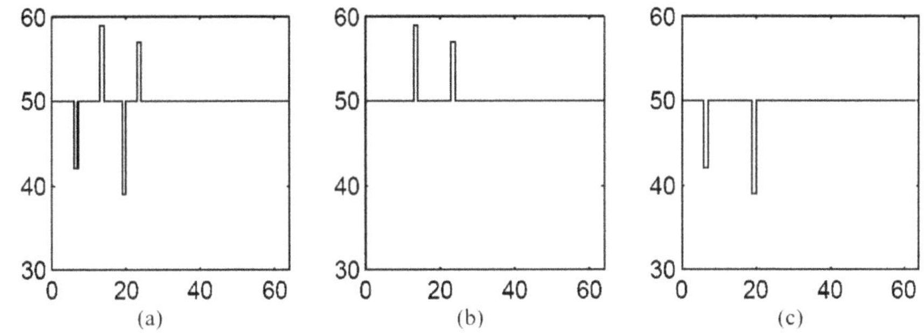

FIGURE 3.17 (a) A flat signal corrupted by positive and negative impulsive noises, and the filtered results by using (b) a soft dilation filter and (c) a soft erosion filter.

and the negative impulse by closing. On the other hand, by applying a single soft opening and closing, the impulse noise can be removed. It is due to that the soft dilation can remove one part (positive or negative) of the noise and the soft erosion can remove the other. Figures 3.17(b) and (c) respectively show the results of a soft dilation and a soft erosion by $[(1,1,\underline{1},1,1),(1),3]$, where $B=(1,1,\underline{1},1,1)$ with underlined one indicating the origin, $A=(\underline{1})$, and $k=3$. Therefore, four standard morphological operations or two soft morphological operations are needed to eliminate the impulse noise. The same goal of filtering can be achieved by using an OSSM dilation by $[(1,1,\underline{1},1,1),(1),3,4]$. This leads to a significant reduction in computational complexity near a half amount.

Another important characteristic of the OSSM filters is the edge-preserving ability. Figures 3.18(a) and (b) respectively show a signal with upward and downward step edges. The edges of applying a standard morphological dilation or an erosion to the signal are shifted and are preserved if an opening or a closing is used. However, if an OSSM dilation by $[(1,1,\underline{1},1,1),(1),2,3]$ is used, the location of edges is preserved.

3.6.5 Recursive Soft Morphological Filters

The structuring element in morphological filters can be regarded as a template which is translated to each pixel location in an image. These filters can be implemented in parallel because each pixel's value in the transformed image is only a function of its neighboring pixels in the given image. Also, the sequence in which the pixels are processed is completely irrelevant. Thus, these parallel image operations can be applied to each pixel simultaneously if a suitable parallel architecture is available. Parallel image transformations are also referred to as *non-recursive transformations*.

In contrast to non-recursive transformations, a class of *recursive transformations* is also widely used in signal and image processing, for example, sequential block labeling, predictive coding, and adaptive dithering (Serra, 1982). The main distinction between these two classes of transformations is that in the recursive

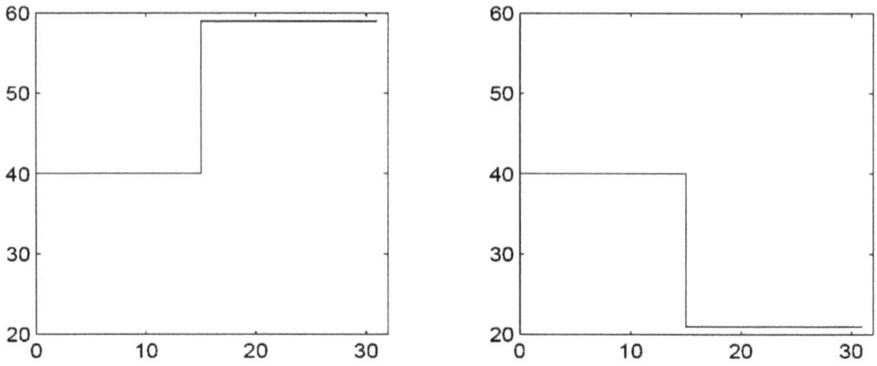

FIGURE 3.18 A signal with (a) upward step edge and (b) downward step edge.

transformations, the pixel's value of the transformed image depends upon the pixel's values of both the input image and the transformed image itself. Due to this reason, some partial order has to be imposed on the underlying image domain so that the transformed image can be computed recursively according to this imposed partial order. In other words, a pixel's value of the transformed image may not be processed until all the pixels preceding it have been processed.

Recursive filters are the filters which use previously filtered outputs as their inputs. Let x_i and y_i denote the input and output values at location i, respectively, where $i = \{0, 1, \cdots, N-1\}$. Let the domain of the structuring element be $\langle -L, \cdots, -1, 0, 1, \cdots, R \rangle$, where L is the left margin and R is the right margin. Hence, the structuring element has the size of $L + R + 1$. Start up and end effects are accounted for by appending L samples to the beginning and R samples to the end of the signal sequence. The L-appended samples are given the value of the first signal sample; similarly, the R-appended samples receive the value of the last sample of the signal.

Definition 3.18: The recursive counterpart of a nonrecursive filter Ψ given by

$$y_i = \Psi\left(x_{i-L}, \cdots, x_{i-1}, x_i, x_{i+1}, \cdots, x_{i+R}\right) \tag{3.72}$$

is defined as

$$y_i = \Psi\left(y_{i-L}, \cdots, y_{i-1}, x_i, x_{i+1}, \cdots, x_{i+R}\right) \tag{3.73}$$

by assuming that the values of $y_{i-L}, \cdots, y_{i-1}$ are already given.

Example 3.15: Let $B = \langle -1, 0, 1 \rangle, A = \langle 0 \rangle$, and $k = 2$. Let the input signal $f = \{472968547\}$. We have

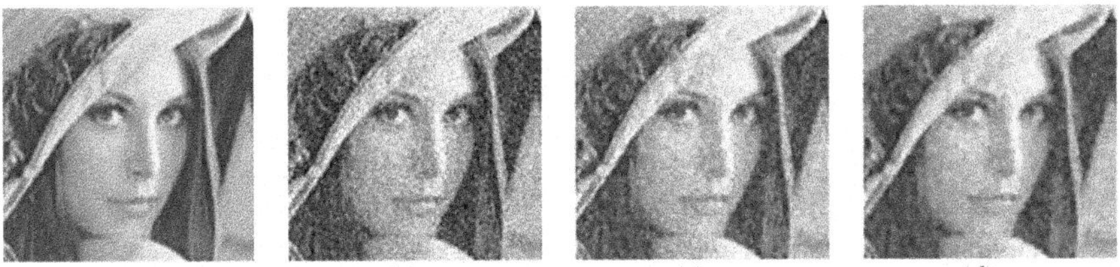

(a) (b) (c) (d)

FIGURE 3.19 (a) An original image, (b) image corrupted by Gaussian noise, (c) result of a recursive soft morphological dilation, (d) followed by a recursive soft morphological erosion (i.e., closing).

$$f \oplus_r [B, A, 2] = \{477988557\},$$

$$f \ominus_r [B, A, 2] = \{442666547\},$$

where "$\oplus_r$" and "$\ominus_r$" denote the recursive soft morphological dilation and erosion, respectively.

In Figure 3.19, (a) shows a 125 × 125 Lena image, and (b) shows the same image but with added Gaussian noise having zero mean and standard deviation 20. The mean-square signal-to-noise ratio (*SNR*) is used. Figure 3.19(b) has *SNR* = 40.83. Figure 3.19(c) shows the result of applying a recursive soft morphological dilation with $B = \langle(0,-1),(-1,0),(0,0),(1,0),(0,1)\rangle$, $A = \langle(0,0)\rangle$, and $k = 3$, where *SNR* = 63.65. Figure 3.19(d) shows the result of applying a recursive soft morphological erosion on Figure 3.19(c) (i.e., closing) with the same structuring elements and rank order, where *SNR* = 101.45.

In Figure 3.20, (a) shows a 512 × 432 grayscale image, and (b) shows the same image but with added Gaussian noise having zero mean and standard deviation 20. Figure 3.20(b) has *SNR* = 35.54. Figure 3.20(c) shows the result of applying a standard morphological closing with a structuring element of size 3 × 3, where *SNR* = 25.15. Figure 3.20(d) shows the result of applying a recursive soft morphological closing with $B = \langle(0,-1),(-1,0),(0,0),(1,0),(0,1)\rangle$, $A = \langle(0,0)\rangle$, and $k = 3$, where *SNR* = 69.41.

3.6.6 Recursive Order-Statistic Soft Morphological (ROSSM) Filters

A class of *recursive order-statistic soft morphological* (ROSSM) filters is introduced and their properties are described in this section. It has been shown through experimental results that the ROSSM filters, compared to the order-statistic soft morphological filters or other nonlinear filters, have better outcomes in signal reconstruction (Pei et al., 1998).

To define the recursive OSSM filters, let Φ denote an ROSSM operation and *mBnA* be the symmetric structuring element sets with *m* pixels in *B* and *n* pixels in *A*. For example, the structuring element $[-2,-1,\underline{0},1,2]$ is denoted as 5*B*1*A* while

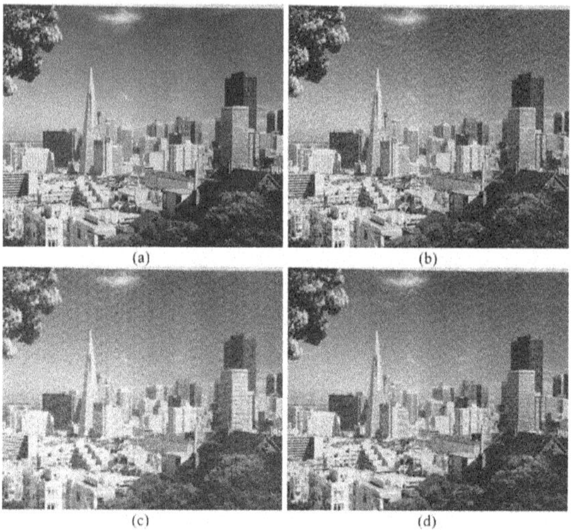

FIGURE 3.20 (a) An original image, (b) image corrupted by Gaussian noise, (c) result of a standard morphological closing, (d) result of a recursive soft morphological closing.

$[-2,\underline{-1,0,1},2]$ is denoted as $5B3A$, where the center of the underlined denotes the center of symmetry. Then, applying a recursive order-statistic soft morphological filter on input f at index i can be written as

$$y_i = \Phi\left(y_{i-L},\cdots,y_{i-1},y_i,y_{i+1},\cdots,y_{i+R}\right) \tag{3.74}$$

where $y_{i-L},\cdots,y_{i-1}$ are already obtained by $\Phi\left(f_{i-L}\right),\cdots,\Phi\left(f_{i-1}\right)$.

The two-dimensional ROSSM operation on an image F with a structuring element S is represented as

$$Y_{ij} = \Phi\left(Y_p, F_{ij}, F_q\right) \tag{3.75}$$

where $S = (S_B, S_A)$ is a two-dimensional structuring element with core set S_A and boundary set $S_B \setminus S_A$ (Note that $S_A \subseteq S_B$), and p and q are domain index sets that precede and succeed the index (i, j) in a scanning order (see Figure 3.21).

An example is given here to show the difference between the OSSM and ROSSM filters.

Example 3.16: Let the structuring element be $5B1A$. Let $k = 2$ and $l = 3$. Assume that the input sequence be $f = \{74,14,28,65,55,8,61,20,17,65,26,94\}$ and the values that are outside the region of f are set to be boundary values of f. Then the output sequences obtained from the OSSM and the ROSSM dilations are

$$\text{OSSM}(f,B,A,k,l) = \{74,65,55,55,55,55,55,20,26,65,65,94\}$$

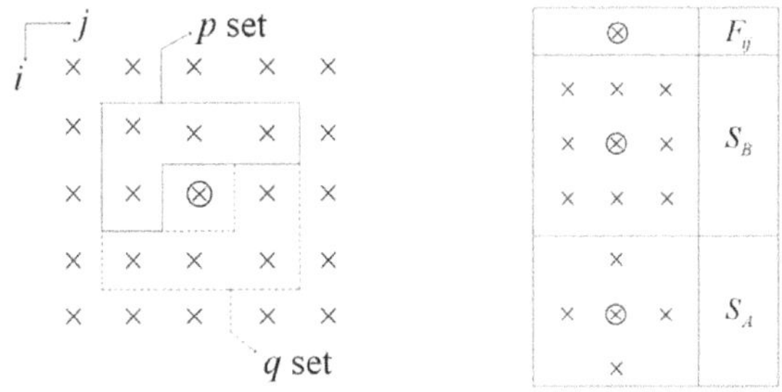

FIGURE 3.21 Two-dimensional structuring element and the preceding (succeeding) domain index sets when using the rater-scan order.

and

$$\text{ROSSM}(f, B, A, k, l) = \{74, 65, 65, 65, 61, 61, 61, 61, 61, 65, 65, 94\}$$

respectively. Figures 3.22(a) and (b) show a 256 × 256 image and its disturbed version with additive Gaussian noise having zero mean and standard deviation 20. The reconstructed images of applying the OSSM and ROSSM filters with $S_B = [(-1,0),(0,-1),(0,0),(1,0),(0,1)]$, $S_A = [(0,0)]$, $k = 3$, and $l = 4$ are shown in Figures 3.22(c) and (d). The mean square signal-to-noise ratio, defined as

$$SNR_{ms} = \frac{\Sigma\left(y_i'\right)^2}{\Sigma\left(y_i' - y_i\right)^2} \tag{3.75}$$

with y and y' denoting the reconstructed and original signal respectively, is used to measure the fidelity of the reconstructed signals. The SNR_{ms} values of Figures 3.22(c) and (d) are 83.3 and 91.6, respectively. It is observed that the visual perception effect of Figure 3.22(d) is slightly better than Figure 3.22(c).

An important feature of the OSSM filter is the duality property. This relationship property also holds on the ROSSM operators. That is, choosing l in an ROSSM dilation is the same as choosing $T - l + 1$ in an ROSSM erosion, where T represents the total number in the order list. If $\tilde{\oplus}_R$ and $\tilde{\Theta}_R$ denote the ROSSM dilation and erosion respectively, the duality relationship can be represented as $f \,\tilde{\oplus}_R\, [B, A, k, l] = f \,\tilde{\Theta}_R\, [B, A, k, T - l + 1]$. However, if the structuring element is obtained by taking the intersection (union) of two or more structuring elements, this duality property may be unavailable.

3.6.7 Regulated Morphological Filters

Agam and Dinstein (1999) developed regulated morphological operators and showed how the fitting property can be adapted for analyzing the map and line-drawing images. Since regulated morphology inherits many properties of the traditional

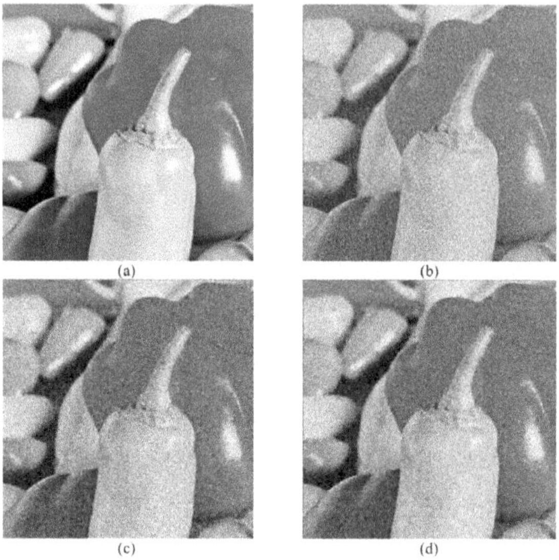

FIGURE 3.22 (a) original image, (b) noisy image disturbed by zero mean Gaussian noise ($\sigma = 20$), (c) reconstructed image obtained from the OSSMF, (d) reconstructed image obtained from the ROSSMF.

standard morphology, it is feasible to apply in image processing and optimize its strictness parameters. Tian et al. (2002) extended the regulated morphological operators by adjusting the weights in the structuring element.

Regulated morphological operations adopt a strictness parameter to control their sensitivity with respect to noise and small intrusions or protrusions on object boundary, and thereby prevent excessive dilation or erosion. The *regulated dilation* of a set A by a structuring element set B with a strictness parameter s is defined by (Agam and Dinstein, 1999):

$$A \oplus^{s} B = \left\{ x \mid \#\left(A \cap (\hat{B})_x \right) \geq s, s \in [1, \min(\#A, \#B)] \right\}, \tag{3.77}$$

where the symbol # denotes the cardinality of a set.

The properties of regulated morphological dilation are introduced here. Their proofs can be referred to (Agam and Dinstein, 1999).

Property 3.1: The regulated dilation is decreasing with respect to the strictness s.

$$A \oplus^{s1} B \subseteq A \oplus^{s2} B \Leftrightarrow s1 \geq s2 \tag{3.78}$$

Property 3.2: A regulated dilation generates a subset of a standard dilation.

$$A \oplus^{s} B \subseteq A \oplus B \tag{3.79}$$

Property 3.3: The regulated dilation is commutative, increasing with respect to the first and the second arguments, and translation invariant.

$$A \oplus^s B = B \oplus^s A \tag{3.80}$$

$$A \subseteq B \Rightarrow A \oplus^s D \subseteq B \oplus^s D \tag{3.81}$$

$$B \subseteq C \Rightarrow A \oplus^s B \subseteq A \oplus^s C \tag{3.82}$$

$$(A)_x \oplus^s B = (A \oplus^s B)_x \tag{3.83}$$

$$A \oplus^s (B)_x = (A \oplus^s B)_x \tag{3.84}$$

Property 3.4: The regulated dilation of a union (intersection) of sets is bigger (smaller than) or equal to the union (intersection) of the regulated dilation of the individual sets.

$$(A \cup B) \oplus^s D \supseteq (A \oplus^s D) \cup (B \oplus^s D) \tag{3.85}$$

$$(A \cap B) \oplus^s D \subseteq (A \oplus^s D) \cap (B \oplus^s D) \tag{3.86}$$

The *regulated erosion* of a set A by a structuring element set B with a strictness parameter s is defined by (Agam and Dinstein, 1999)

$$A \ominus^s B = \left\{ \mathrm{x} \mid \#\left(A^c \cap (B)_x\right) < s, s \in [1, \#B] \right\}. \tag{3.87}$$

The properties of regulated morphological erosion are introduced in the following. Their proofs can be referred to (Agam and Dinstein, 1999).

Property 3.5: The regulated erosion is increasing with respect to the strictness s.

$$A \ominus^{s1} B \subseteq A \ominus^{s2} B \Leftrightarrow s1 \le s2 \tag{3.88}$$

Property 3.6: A regulated erosion results in a superset of a standard dilation.

$$A \ominus^s B \supseteq A \ominus B \tag{3.89}$$

Property 3.7: The regulated erosion is increasing with respect to the first argument, decreasing with respect to the second argument, and translation invariant.

$$A \subseteq B \Rightarrow A \ominus^s D \subseteq B \ominus^s D \tag{3.90}$$

$$B \subseteq C \Rightarrow A \ominus^s B \supseteq A \ominus^s C \tag{3.91}$$

$$(A)_x \ominus^s B = \left(A \ominus^s B\right)_x \tag{3.92}$$

$$A \ominus^s (B)_x = \left(A \ominus^s B\right)_{-x} \tag{3.93}$$

Property 3.8: The regulated erosion of a union (intersection) of sets is bigger (smaller) or equal to the union (intersection) of the regulated dilation of the individual sets.

$$(A \cup B) \ominus^s D \supseteq \left(A \ominus^s D\right) \cup \left(B \ominus^s D\right) \tag{3.94}$$

$$(A \cap B) \ominus^s D \subseteq \left(A \ominus^s D\right) \cap \left(B \ominus^s D\right) \tag{3.95}$$

Property 3.9: The regulated dilation and erosion are dual in the same sense that exists for the standard dilation and erosion.

$$A \ominus^s B = \left(A^c \oplus^s \hat{B}\right)^c \tag{3.96}$$

In order to maintain the same properties as standard morphological operators, the *regulated closing* of a set A by a structuring element set B with a strictness parameter s is defined by

$$A \bullet^s B = \left(\left(A \dot{\underline{\oplus}}^s B\right) \ominus B\right) \cup A, \tag{3.97}$$

where $A \underline{\oplus}^s B = \left(A \oplus^s B\right) \cup A$ is defined as the *extensive regulated dilation*. The *regulated opening* of a set A by a structuring element set B with a strictness parameter s is defined by

$$A \circ^5 B = \left(\left(A \underline{\ominus}^s B\right) \oplus B\right) \cap A, \tag{3.98}$$

where $A \underline{\ominus}^s B = \left(A \ominus^s B\right) \cap A$ is defined as the *anti-extensive regulated erosion.*

By adjusting the strictness parameter, we can alleviate the noise sensitivity problem and small intrusions or protrusions on the object boundary. An application of using regulated morphological filters is for corner detection. The regulated morphological corner detector is described as follows (Shih et al., 2005):

Step 1: $A_1 = \left(A \oplus^s B\right) \ominus^s B$. Corner strength: $C_1 = |A - A_1|$.

Step 2: $A_2 = \left(A \ominus^s B\right) \oplus^s B$. Corner strength: $C_2 = |A - A_2|$.

Step 3: Corner detector $= C_1 \cup C_2$.

The original image is performed using a regulated dilation by a 5 × 5 circular structuring element with a strictness s, and then followed by a regulated erosion by the same structuring element with the same strictness. The corner strength C_1 is computed by finding the absolute value of the difference between the original and resulting images. This step is to extract concave corners. By reversing the order such that the regulated erosion is applied first and then followed by the regulated dilation, the convex corners can be extracted. Finally, both types of corners are combined. Figure 3.23 shows the corner detection for an airplane image using the regulated morphological corner detector at strictness = 2, 3, and 4.

3.6.8 Fuzzy Morphological Filters

Fuzzy set theory has found a promising field of applications in the domain of digital image processing since fuzziness is an intrinsic property of image. The fuzzy logic has been developed in order to capture the uncertainties associated with human cognitive processes such as in thinking, reasoning, perception, etc. Images are modeled as fuzzy subsets of the Euclidean plane or Cartesian grid, and the morphological operators are defined in terms of a fuzzy index function. Zadeh (1965) introduced fuzzy sets, which contain elementary operations such as intersection, union, complementation, and inclusion. The fuzzy set theoretical operations are defined as follows. Let the characteristic function of a crisp set A be denoted as $\mu_A : U \to \{0,1\}$, and it is defined as

$$\chi_A(x) = \begin{cases} 1 & \text{if } x \in A \\ 0 & \text{otherwise} \end{cases} \tag{3.99}$$

where A is any finite set. However, the object classes generally encountered in the real world are not so "precisely" or "crisply" defined. In most cases, several ambiguities arise in the determination of whether a particular element belongs to a set or not.

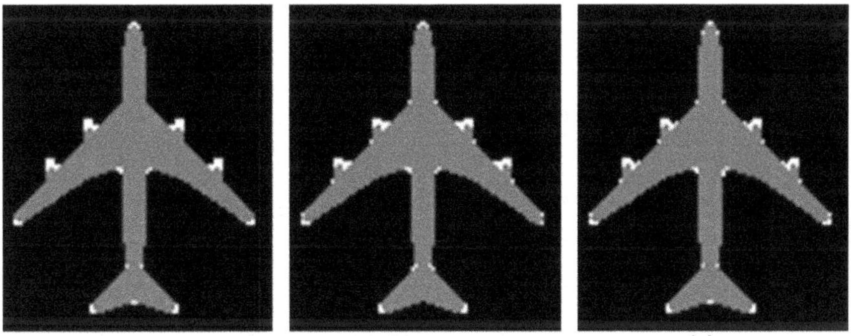

(a) (b) (c)

FIGURE 3.23 (a)–(c) Corner detection for an airplane image using the regulated morphological corner detector at strictness = 2, 3, and 4, respectively.

A good example mentioned by Zadeh is a class of animals. This class clearly includes dogs, cats, tigers, etc. and excludes rocks, plants, houses, etc. However, an ambiguity arises in the context of objects such as bacteria and starfish with respect to the class of animals.

The membership function of a fuzzy set A, denoted as $\mu_A : U \rightarrow [0,1]$, is defined in such a way that $\mu_A(x)$ denotes the degree to which x belongs to A. Note that in the crisp set, the membership function has the value either 1 or 0, but in the fuzzy set, the member function has the value in the range of 0 to 1. The higher the value of $\mu_A(x)$, the more x belongs to A, and conversely, the smaller the value of $\mu_A(x)$, the less likelihood of x being in the set A. For further details see Zadeh (1965, 1977).

The *union*, *intersection*, *difference*, and *complement* operations on crisp as well as fuzzy sets can be defined in terms of their characteristic/membership functions as

$$\mu_{A\cup B}(x) = \max\left[\mu_A(x), \mu_B(x)\right] \tag{3.100}$$

$$\mu_{A\cap B}(x) = \min\left[\mu_A(x), \mu_B(x)\right] \tag{3.101}$$

$$\mu_{A|B}(x) = \min\left[\mu_A(x), 1-\mu_B(x)\right] \tag{3.102}$$

$$\mu_A^c(x) = 1-\mu_A(x) \tag{3.103}$$

The subset relation can be expressed as

$$A \subseteq B \Leftrightarrow \mu_B = \max\left[\mu_A, \mu_B\right] \Leftrightarrow \mu_A = \min\left[\mu_A, \mu_B\right] \tag{3.104}$$

The *support* of a set A, denoted as $S(A)$, is a crisp set of those elements of U which belong to A with some certainty:

$$S(A) = \left\{x \mid \mu_A(x) > 0\right\} \tag{3.105}$$

The *translation* of a set A by a vector $v \in U$, denoted by $\Im(A;v)$, is defined as

$$\mu_{\Im(A;v)}(x) = \mu_A(x-v) \tag{3.106}$$

The *reflection* of a set A, denoted by $\hat{A}$, is defined as

$$\mu_{\hat{A}}(x) = \mu_A(-x) \tag{3.107}$$

The *scalar addition* of a fuzzy set A and a constant α, denoted as $A \dagger \alpha$, is defined as

$$\mu_{A\dagger\alpha}(x) = \min(1, \max[0, \mu_A(x)+\alpha]) \tag{3.108}$$

Giles (1976) proposed fuzzy operations, *bold union*, $X \Delta Y$ of two sets X and Y as

$$\mu_{X\Delta Y}(z) = \min[1, \mu_X(z), \mu_Y(z)] \tag{3.109}$$

and *bold intersection* $X\nabla Y$ as

$$\mu_{X\nabla Y}(z) = \max[0, \mu_X(z), \mu_Y(z) - 1] \tag{3.110}$$

If X and Y are crisp sets, then $X\Delta Y \equiv X \cup Y$ and $X\nabla Y \equiv X \cap Y$.

An *index* function I: $2^U \times 2^U \rightarrow \{0,1\}$ is defined as

$$I(A,B) = \begin{cases} 1 & \text{if } A \subseteq B \\ 0 & \text{otherwise} \end{cases} \tag{3.111}$$

The previous equation can be rewritten so as to express the index function directly in terms of characteristic functions as

$$\begin{aligned} I(A,B) &= \inf_{x\in A} \mu_B(x) \\ &= \min[\inf_{x\in A} \mu_B(x), \ \inf_{x\notin A} 1] \\ &= \inf_{x\in U} \mu_{A^c \Delta B}(x) \end{aligned} \tag{3.112}$$

The last relation follows because for crisp sets the bold union has the following properties:

$$x \in A \Rightarrow \mu_{A^c \Delta B}(x) = \mu_B(x)$$

$$x \notin A \Rightarrow \mu_{A^c \Delta B}(x) = 1.$$

The index function can be generalized so that $I(A, B)$ gives the degree to which A is a subset of B. The formulation in equation (3.112) suffices for this purpose. The properties of morphological operations will be induced by the properties of the index function.

Consider any two fuzzy subsets $A, B \subset U$; index function $I(A, B)$ for different values of set B, and set A essentially fixed and $A \neq \phi$. The properties (axioms) for the index function are as follows:

$$I(A,B) \in [0,1].$$

If A and B are crisp sets, then $I(A,B) \in \{0,1\}$.

$$A \subseteq B \Leftrightarrow I(A,B) = 1.$$

If $B \subseteq C$, then $I(A,B) \leq I(A,C)$.

If $B \subseteq C$, then $I(C,A) \leq I(B,A)$.

Invariant under translation, complement, and reflection:

(a) $I(A,B) = I(\Im(A;v), \Im(B;v))$.
(b) $I(A,B) = I(A^c, B^c)$.
(c) $I(A,B) = I(\hat{A}, \hat{B})$.

If B and C are subsets of A, then so is $B \cup C$. The converse also holds:

$$(B \subseteq A) \wedge (C \subseteq A) \Leftrightarrow (B \cup C) \subseteq A; \text{i.e.}, I(B \cup C, A) = \min[I(B,A), I(C,A)] \cdot$$

Similarly, $(A \subseteq B) \wedge (A \subseteq C) \Leftrightarrow A \subseteq (B \cup C); \text{i.e.}, I(A, B \cup C) = \min[I(A,B), I(A,C)]$.

Gasteratos et al. (1998) proposed a framework which extends the concepts of soft mathematical morphology into fuzzy sets. They studied its compatibility with binary soft mathematical morphology as well as the algebraic properties of fuzzy soft operations. The fuzzy morphological operations, dilation, erosion, opening, and closing (Sinha and Dougherty, 1992, 1995) are defined in terms of index function as follows. The fuzzy erosion of a set A by another set B, denoted as $A \ominus B$, is defined by

$$\mu_{A \ominus B}(x) = I(\Im(B;x), A). \tag{3.113}$$

The fuzzy dilation of a set A by another set B, denoted as $A \oplus B$, is defined by

$$\mu_{A \oplus B}(x) = \mu_{(A^c \ominus \hat{B})^c}(x) = 1 - \mu_{A^c \ominus \hat{B}}(x)$$

$$= I^c(\Im(\hat{B};x), A^c) = \sup_{z \in U} \max[0, \mu_{\Im(\hat{B};x)}(z) + \mu_A(z) - 1] \tag{3.114}$$

The fuzzy opening of a set A by another set B, denoted as $A \circ B$, is defined by

$$A \circ B = (A \ominus B) \oplus B. \tag{3.115}$$

The fuzzy closing of a set A by another set B, denoted as $A \bullet B$, is defined by

$$A \bullet B = (A \oplus B) \ominus B. \tag{3.116}$$

Sinha and Dougherty (1992) did not just introduce the modeling of grayscale images and the fuzzy morphological operations by simply replacing ordinary set theoretic operations by their fuzzy counterparts. Instead, they developed an intrinsically fuzzy approach to mathematical morphology. A new fitting paradigm is used by employing an index for set inclusion to measure the degree to which one image is beneath another image for erosion. Fuzzy morphological operations are defined based on this approach.

In binary morphology, erosion of image A by structuring element B is defined by

$$A \ominus B = \bigcap_{x \in -B} T(A;x) \tag{3.117}$$

where $T(A; x)$ is translation of A by a vector. Dilation of image A by structuring element B can be defined by

$$A \oplus B = \bigcup_{x \in B} T(A; x) \tag{3.118}$$

The definition of fuzzy morphology requires an index function for set inclusion:

$$\begin{aligned} I(A,B) &= \inf_{x \in A} \mu_B(x) A \\ &= \min\left[\inf_{x \in A} \mu_B(x), \inf_{x \notin A} 1\right] = \inf_{x \in U} \mu_{A^c \Delta B}(x) \end{aligned} \tag{3.119}$$

The *fuzzy complement* of the set inclusion indicator is then defined as

$$I^c(A,B) = 1 - I(A,B) = \sup_{x \in U} \mu_{A \nabla B^c}(x) \tag{3.120}$$

Based on equations (3.119) and (3.120), the *fuzzy erosion* is defined as

$$\begin{aligned} \mu_{E(A,B)}(x) &= I(T(B;x), A) \\ &= \inf_{x \in U} \ \min\left[1,\ 1 + \mu_A(x) - \mu_T(B;x)(x)\right] \end{aligned} \tag{3.121}$$

and the *fuzzy dilation* is defined as

$$\begin{aligned} \mu_{D(A,B)}(x) &= 1 - \mu_{E(A,B)}(x) \\ &= \sup_{x \in U} \ \max\left[0,\ \mu_T(-B;x)(x) + \mu_A(x) - 1\right] \cdot \end{aligned} \tag{3.122}$$

The property of being dual operations between fuzzy erosion and fuzzy dilation is preserved.

$$\mu_{D(A,B)}(x) = \mu_{E(A^c, -B)^c}(x) \tag{3.123}$$

The fuzzy morphological operations are illustrated in the following examples. The area of interest in a given image is represented by a rectangle, and the membership values within this rectangular region are specified in a matrix format. The membership values outside this region are assumed to be fixed as a constant. This value is specified as the superscript of the matrix, and the coordinates of the upper-left element of the matrix are indicated by the subscript.

Example 3.17:

$$A = \begin{bmatrix} 0.2 & 1.0 & 0.8 & 0.1 \\ 0.3 & 0.9 & 0.9 & 0.2 \\ 0.1 & 0.9 & 1.0 & 0.3 \end{bmatrix}_{(0,0)}^{0.0} \quad \text{and } B = [0.8 \quad 0.9]_{(0,0)}^{0.0}$$

Since the structuring element fits entirely under the image when it translated by $(1,-1)$, $A \ominus B(1,-1) = 1.0$. Similarly, $A \ominus B(1,-2) = 1.0$. For vector (1, 0), even though $\Im(B;1,0) \not\subset A$, the subset relationship almost holds:

$$A \cap \Im(B;1,0) = (0.8 \quad 0.8)_{(1,0)} \approx \Im(B;1,0).$$

Therefore, a relatively high value of $A \ominus B(1,0)$ is expected.

$$I[\Im(B;1,0), A] = \min\{\min[1, 1+1-0.8], \min[1, 1+0.8-0.9]\} = 0.9.$$

Thus, $A \ominus B(1,0) = 0.9$. Proceeding along this way, the eroded image is obtained as

$$A \ominus B = \begin{bmatrix} 0.2 & 0.4 & 0.9 & 0.2 \\ 0.2 & 0.5 & 1.0 & 0.3 \\ 0.2 & 0.3 & 1.0 & 0.4 \end{bmatrix}^{0.1}_{(-1,0)}.$$

Note that the coordinates of the upper-left element are now (−1, 0) instead of (0, 0). By appropriately thresholding the image at a value between 0.5 and 0.9, we obtain

$$A \ominus B = \begin{bmatrix} 0.9 \\ 1.0 \\ 1.0 \end{bmatrix}^{0.0}_{(1,0)}.$$

Example 3.18:

$$A = \begin{bmatrix} 0.7 \\ 0.9 \\ 0.8 \end{bmatrix}^{0.0}_{(0,0)} \quad \text{and } B = [0.8 \quad 0.9]^{0.0}_{(0,0)}$$

We obtain the reflected set: $\hat{B} = [0.9 \quad 0.8]^{0.0}_{(-1,0)}$.

$$I^c(\Im(\hat{B};x), A^c) = \begin{cases} 0.5 & \text{if } x = (0,0) \\ 0.6 & \text{if } x = (1,0),(0,-2) \\ 0.7 & \text{if } x = (0,-1),(1,-2) \\ 0.8 & \text{if } x = (1,-1) \\ 0.0 & \text{otherwise} \end{cases}$$

Hence, the fuzzy dilation is

$$A \oplus B = \begin{bmatrix} 0.5 & 0.6 \\ 0.7 & 0.8 \\ 0.6 & 0.7 \end{bmatrix}^{0.0}_{(0,0)}$$

Example 3.19: Consider the image from Example 3.17.

$$P \equiv A \ominus B = \begin{bmatrix} 0.2 & 0.4 & 0.9 & 0.2 \\ 0.2 & 0.5 & 1.0 & 0.3 \\ 0.2 & 0.3 & 1.0 & 0.4 \end{bmatrix}_{(-1,0)}^{0.1}$$

Therefore,

$$P^c \ominus \hat{B} = \begin{bmatrix} 0.8 & 0.3 & 0.2 & 0.9 \\ 0.7 & 0.2 & 0.1 & 0.8 \\ 0.9 & 0.2 & 0.1 & 0.7 \end{bmatrix}_{(-1,0)}^{1.0}$$

Hence,

$$A^\circ B = \left(P^c \ominus \hat{B}\right)^c = \begin{bmatrix} 0.2 & 0.7 & 0.8 & 0.1 \\ 0.3 & 0.8 & 0.9 & 0.2 \\ 0.1 & 0.8 & 0.9 & 0.3 \end{bmatrix}_{(-1,0)}^{0.0}$$

By appropriately thresholding the image at a value between 0.3 and 0.7, we obtain

$$A \circ B = \begin{bmatrix} 0.7 & 0.8 \\ 0.8 & 0.9 \\ 0.8 & 0.9 \end{bmatrix}_{(0,0)}^{0.0}$$

Some applications of fuzzy morphological filters to image processing are given in the following. Grobert et al. (1996) applied fuzzy morphology to detect infected regions of a leaf. It was shown that the fuzzy morphological operation achieves a compromise between the deletion of isolated bright pixels and the connection of scattered regions of bright pixels. Wirth and Nikitenko (2005) applied fuzzy morphology to contrast enhancement. Yang et al. (2006) proposed an edge detection algorithm based on adaptive fuzzy morphological neural network. The gradient of the fuzzy morphology utilizes a set of structuring elements to detect the edge strength with a view to decrease the spurious edge and suppressed the noise.

Chatzis and Pitas (2000) proposed the generalized fuzzy mathematical morphology (GFMM) based on a novel definition of the fuzzy inclusion indicator (FII). FII is a fuzzy set used as a measure of the inclusion of a fuzzy set into another. The GFMM provides a very powerful and flexible tool for morphological operations. It can be applied on the skeletonization and shape decomposition of two-dimensional and three-dimensional objects. Strauss and Comby (2007) presented variable structuring element based fuzzy morphological operations for single-viewpoint omnidirectional images.

REFERENCES

Agam, G. and Dinstein, I., "Regulated morphological operations," *Pattern Recognition,* vol. 32, no. 6, pp. 947–971, Jun. 1999.

Bloomberg, D. and Maragos, P., "Generalized hit-miss operations," *Image Algebra and Morphological Image Processing 1998, Proceedings*, P. Gader, Ed., vol. 1350, San Diego, CA, pp. 116–128, SPIE, 1990.

Blum, H., "A transformation for extracting new descriptors of shape," *Models for the Perception of Speech and Visual Forms*, Proceedings of Meeting held Nov. 1964, W. Dunn, Ed., pp. 362–380, MIT Press, 1967.

Chatzis, V. and Pitas, I., "A generalized fuzzy mathematical morphology and its application in robust 2-D and 3-D object representation," *IEEE Transactions on Image Processing*, vol. 9, no. 10, pp. 1798–1810, Oct. 2000.

Danielsson, P. E., "A new shape factor," *Computer Graphics and Image Processing*, vol. 292, 1978.

David, H., *Order Statistics*, Wiley, 1981.

Destival, I., "Mathematical morphology applied to remote sensing," *Acta Astronautica*, vol. 13, no. 6–7, pp. 371–385, Jun.-Jul. 1986.

Dineen, G. P., "Programming pattern recognition," *Proceedings of the Western Joint Computer Conference*, Los Angeles, CA, pp. 94–100, Mar. 1955.

Dougherty, E. R., *Mathematical Morphology in Image Processing*, CRC Press, 1993.

Duff, M. J. B., "Parallel processors for digital image processing," *Advances in Digital Image Processing*, P. Stucki, Ed., pp. 265–276, Plenum, 1979.

Feehs, R. J. and Arce, G. R., "Multidimensional morphological edge detection," *Proceedings of the SPIE Visual Communications and Image Processing II*, vol. 845, pp. 285–292, 1987.

Gasteratos, A., Andreadis, I., and Tsalides, P., "A new hardware structure for implementation of soft morphological filters," *Computer Analysis of Images and Patterns, Lecture Notes in Computer Science*, vol. 1296, pp. 488–494, Springer, 1997.

Gasteratos, A., Andreadis, I., and Tsalides, P., "Fuzzy soft mathematical morphology," *IEE Proceedings of Vision, Image and Signal Processing*, vol. 145, no. 1, pp. 41–49, Feb. 1998.

Gerritsen, F. A. and Aardema, L. G., "Design and use of DIP-1: A fast flexible and dynamically microprogrammable image processor," *Pattern Recognition*, vol. 14, no. 1–6, pp. 319–330, 1981.

Giardina, C. and Dougherty, E., *Morphological Methods in Image and Signal Processing*, Prentice-Hall, 1988.

Giles, R., "Lukasiewicz logic and fuzzy theory," *International Journal of Man-Machine Studies*, vol. 8, no. 3, pp. 313–327, 1976.

Golay, M. J. E., "Hexagonal parallel pattern transformations," *IEEE Transactions on Computers*, vol. 18, no. 8, pp. 733–740, Aug. 1969.

Gonzalez, R. and Woods, R., *Digital Image Processing*, 3rd edition, Prentice Hall, 2007.

Goutsias, J. and Heijmans, H. J., *Mathematical Morphology*, IOS Press, 2000.

Graham, D. and Norgren, P. E., "The Diff3 analyzer: A parallel/serial Golay image processor," *Real Time Medical Image Processing*, P. Onoe and Rosenfeld, Eds., pp. 163–182, Plenum, 1980.

Grobert, S., Koppen, M., and Nickolay, B. A., "New approach to fuzzy morphology based on fuzzy integral and its application in image processing," *Proceedings of International Conference on Pattern Recognition*, vol. 2, pp. 625–630, Aug. 1996.

Hamid, M. S., Harvey, N. R., and Marshall, S., "Genetic algorithm optimization of multidimensional grayscale soft morphological filters with applications in film archive restoration," *IEEE Transactions on Circuits and Systems for Video Technology*, vol. 13, no. 5, pp. 406–416, May 2003.

Haralick, R., Zhuang, X., Lin, C., and Lee, J., "The digital morphological sampling theorem," *IEEE Transactions on Acoustics, Speech, and Signal Processing*, vol. 37, no. 12, pp. 2067–2090, Dec. 1989.

Haralick, R. M., Sternberg, S. R., and Zhuang, X., "Image analysis using mathematical morphology," *IEEE Transactions on Pattern Analysis and Machine Intelligence*, vol. 9, no. 7, pp. 532–550, Jul. 1987.

Jang, B. and Chin, R. T., "Analysis of thinning algorithms using mathematical morphology," *IEEE Transactions on Pattern Analysis and Machine Intelligence*, vol. 12, no. 6, pp. 541–551, Jun. 1990.

Ji, L. and Piper J., "Fast homotopy – preserving skeletons using mathematical morphology," *IEEE Transactions on Pattern Analysis and Machine Intelligence*, vol. 14, no. 6, pp. 653–664, Jun. 1992.

Khosravi, M. and Schafer, R.W., "Template matching based on a grayscale hit-or-miss transform," *IEEE Transactions on Image Processing*, vol. 5, no. 6, pp. 1060–1066, Jun. 1996.

Kirsch, R. A., "Experiments in processing life motion with a digital computer," *Proceedings of the Eastern Joint Computer Conference*, pp. 221–229, 1957.

Klein, J. C. and Serra, J., "The texture analyzer," *Journal of Microscopy*, vol. 95, no. 2, pp. 349–356, Apr. 1977.

Ko, S.-J., Morales, A., and Lee, K.-H., "Fast recursive algorithms for morphological operators based on the basis matrix representation," *IEEE Transactions on Image Processing*, vol. 5, no. 6, pp. 1073–1077, Jun. 1996.

Koskinen, L., Astola, J. T., and Neuvo, Y. A., "Soft morphological filters," *Proceedings of the SPIE, vol. 1568, Image Algebra and Morphological Image Processing II*, P. D. Gader and E. R. Dougherty, Eds., pp. 262–270, Jul. 1991.

Kruse, B., "Design and implementation of a picture processor," Science and Technology Dissertations, no. 13, University of Linkoeping, 1977.

Kuosmanen, P. and Astola, J., "Soft morphological filtering," *Journal of Mathematical Imaging and Vision*, vol. 5, no. 3, pp. 231–262, Sep. 1995.

Lee, J. S., Haralick, R. M., and Shapiro, L. G., "Morphologic edge detection," *IEEE Transactions on Robotics and Automation*, vol. 3, no. 4, pp. 142–156, Apr. 1987.

Lougheed, R. M. and McCubbrey, D. L., "The cytocomputer: A practical pipelined image processor," *Proceedings of the IEEE Annual Symposium on Computer Architecture*, La Baule, pp. 271–278, 1980.

Maragos, P. and Schafer, R., "Morphological filters – Part I: Their set-theoretic analysis and relations to linear shift-invariant filters," *IEEE Transactions on Acoustics, Speech, and Signal Processing*, vol. 35, no. 8, pp. 1153–1169, Aug. 1987a.

Maragos, P. and Schafer, R., "Morphological filters – Part II: Their relations to median, order-statistics, and stack filters," *IEEE Transactions on Acoustics, Speech, and Signal Processing*, vol. 35, no. 8, pp. 1170–1184, Aug. 1987b.

Maragos, P. A. and Schafer, R. W., "Morphological skeleton representation and coding of binary images," *IEEE Transactions on Acoustics, Speech, and Signal Processing*, vol. 34, no. 5, pp. 1228–1244, Oct. 1986.

Matheron, G., *Random Sets and Integral Geometry*, John Wiley & Sons Inc., 1975.

Meyer, F., "Skeletons in digital spaces," *Image Analysis and Mathematical Morphology*, J. Serra, Ed., vol. 2, pp. 257–296, Academic Press, 1988.

Minkowski, H., "Volumen und oberflache," *Mathematical Annals*, vol. 57, pp. 447–495, 1903.

Moore, G. A., "Automatic sensing and computer process for the quantitative analysis of micrographs and equivalued subjects," *Pictorial Pattern Recognition*, pp. 275–326, Thompson Book Co., 1968.

Morales, A., Acharya, R., and Ko, S.-J., "Morphological pyramids with alternating sequential filters," *IEEE Transactions on Image Processing*, vol. 4, no. 7, pp. 965–977, Jul. 1995.

Moran, C. J., "A morphological transformation for sharpening edges of features before segmentation," *Compute Vision, Graphics and Image Processing*, vol. 49, no. 1, pp. 85–94, Jan. 1990.

Nadadur, D. and Haralick, R. M., "Recursive binary dilation and erosion using digital line structuring elements in arbitrary orientations," *IEEE Transactions on Image Processing*, vol. 9, no. 5, pp. 749–759, May 2000.

Naegel, B., Passat, N., and Ronse, C., "Grey-level hit-or-miss transforms-part I: Unified theory," *Pattern Recognition*, vol. 40, no. 2, pp. 635–647, Feb. 2007a.

Naegel, B., Passat, N., and Ronse, C., "Grey-level hit-or-miss transforms-part II: Application to angiographic image processing," *Pattern Recognition*, vol. 40, no. 2, pp. 648–658, Feb. 2007b.

Pei, S., Lai, C., and Shih, F. Y., "An efficient class of alternating sequential filters in morphology," *Graphical Models and Image Processing*, vol. 59, no. 2, pp. 109–116, Mar. 1997.

Pei, S., Lai, C., and Shih, F. Y., "Recursive order-statistic soft morphological filters," *IEE Proceedings Vision, Image and Signal Processing*, vol. 145, no. 5, pp. 333–342, Oct. 1998.

Preston, K. Jr., "Machine techniques for automatic identification of binucleate lymphocyte," *Proceedings of the International Conference on Medical Electronics*, 1961.

Preteux, F., Laval-Jeantet, A. M., Roger, B., and Laval-Jeantet, M. H., "New prospects in C.T. image processing via mathematical morphology," *European Journal of Radiology*, vol. 5, no. 4, pp. 313–317, Nov. 1985.

Pu, C. C. and Shih, F. Y., "Threshold decomposition of gray-scale soft morphology into binary soft morphology," *Graphical Model and Image Processing*, vol. 57, no. 6, pp. 522–526, Nov. 1995.

Raducanu, B. and Grana, M., "A grayscale hit-or-miss transform based on level sets," *Proceedings of the IEEE International Conference on Image Processing*, vol. 2, Vancouver, BC, pp. 931–933, Sep. 2000.

Schonfeld, D. and Goutsias, J., "Optimal morphological pattern restoration from noisy binary image," *IEEE Transactions on Pattern Analysis and Machine Intelligence*, vol. 13, no. 1, pp. 14–29, Jan. 1991.

Serra, J., *Image Analysis and Mathematical Morphology*, Academic Press, 1982.

Serra, J., *Image Analysis and Mathematical Morphology, vol 2: Theoretical Advances*, Academic, 1988.

Shih, F. Y., *Image Processing and Mathematical Morphology: Fundamentals and Applications*, CRC Press, 2009.

Shih, F. Y., Chuang, C., and Gaddipati, V., "A modified regulated morphological corner detector," *Pattern Recognition Letters*, vol. 26, no. 7, pp. 931–937, Jun. 2005.

Shih, F. Y. and Gaddipati, V., "General sweep mathematical morphology," *Pattern Recognition*, vol. 36, no. 7, pp. 1489–1500, Jul. 2003.

Shih, F. Y. and Gaddipati, V., "Geometric modeling and representation based on sweep mathematical morphology," *Information Sciences*, vol. 171, no. 3, pp. 213–231, Mar. 2005.

Shih, F. Y., King, C. T., and Pu, C. C., "Pipeline architectures for recursive morphological operations," *IEEE Transactions on Image Processing*, vol. 4, no. 1, pp. 11–18, Jan. 1995.

Shih, F. Y. and Mitchell, O. R., "Threshold decomposition of grayscale morphology into binary morphology," *IEEE Transactions on Pattern Analysis and Machine Intelligence*, vol. 11, no. 1, pp. 31–42, Jan. 1989.

Shih, F. Y. and Mitchell, O. R., "Decomposition of gray scale morphological structuring elements," *Pattern Recognition*, vol. 24, no. 3, pp. 195–203, Mar. 1991.

Shih, F. Y. and Mitchell, O. R., "A mathematical morphology approach to Euclidean distance transformation," *IEEE Transactions on Image Processing*, vol. 1, no. 2, pp. 197–204, Apr. 1992.

Shih, F. Y. and Pu, C. C., "A skeletonization algorithm by maxima tracking on Euclidean distance transform," *Pattern Recognition*, vol. 28, no. 3, pp. 331–341, Mar. 1995a.

Shih, F. Y. and Pu, C. C., "Analysis of the properties of soft morphological filtering using threshold decomposition," *IEEE Transactions on Signal Processing*, vol. 43, no. 2, pp. 539–544, Feb. 1995b.

Shih, F. Y. and Puttagunta, P., "Recursive soft morphological filters," *IEEE Transactions on Image Processing*, vol. 4, no. 7, pp. 1027–1032, Jul. 1995.

Shih, F. Y. and Wu, Y., "The efficient algorithms for achieving Euclidean distance transformation," *IEEE Transactions on Image Processing*, vol. 13, no. 8, pp. 1078–1091, Aug. 2004.

Sinha, D. and Dougherty, E. R., "Fuzzy mathematical morphology," *Journal of Visual Communication and Image Representation*, vol. 3, no. 3, pp. 286–302, Sep. 1992.

Sinha, D. and Dougherty, E. R, "A general axiomatic theory of intrinsically fuzzy mathematical morphologies," *IEEE Transactions on Fuzzy Systems*, vol. 3, no. 4, pp. 389–403, Nov. 1995.

Smith, S. W., *The Scientist and Engineer's Guide to Digital Signal Processing*, California Technical Publishing, 1997.

Soille, P., "On morphological operators based on rank filters," *Pattern Recognition*, vol. 35, no. 2, pp. 527–535, Feb. 2002.

Sternberg, S. R., "Grayscale morphology," *Computer Vision, Graphics, and Image Processing*, vol. 35, no. 3, pp. 333–355, Sep. 1986.

Sternberg, S. R., "Language and architecture for parallel image processing," *Proceedings of the International Conference on Pattern Recognition in Practice*, Gelsema and Kanal, Eds., Amsterdam, p. 35, 1980.

Stevenson, R. L. and Arce, G. R., "Morphological filters: Statistics and further syntactic properties," *IEEE Transactions on Circuits and Systems*, vol. 34, no. 11, pp. 1292–1305, Nov. 1987.

Strauss, O. and Comby, F., "Variable structuring element based fuzzy morphological operations for single viewpoint omnidirectional images," *Pattern Recognition*, vol. 40, no. 12, pp. 3578–3596, Dec. 2007.

Tian, X.-H., Li, Q.-H., and Yan, S.-W., "Regulated morphological operations with weighted structuring element," *Proceedings of the International Conference on Machine Learning and Cybernetics*, Beijing, pp. 768–771, Nov. 2002.

Wirth, M. A. and Nikitenko, D., "Applications of fuzzy morphology to contrast enhancement," *Proceedings of the Annual Meeting of the North American Fuzzy Information Processing Society*, Ann Arbor, MI, pp. 355–360, Jun. 2005.

Yang, G.-Q., Guo, Y.-Y., and Jiang, L.-H., "Edge detection based on adaptive fuzzy morphological neural network," *Proceedings of the International Conference on Machine Learning and Cybernetics*, Heibei, pp. 3725–3728, Aug. 2006.

Zadeh, L. A., "Fuzzy sets," *Information Control*, vol. 8, no. 3, pp. 338–353, Jun. 1965.

Zadeh, L. A., "Theory of fuzzy sets," *Encyclopedia of Computer Science and Technology*, J. Belzer, A. Holzman, and A. Kent, Eds., Dekker, 1977.

4 Image Segmentation

Image segmentation, a process of pixel classification, aims to extract or segment objects or regions from the background. It is a critical pre-processing step to the success of image recognition (Pachowicz, 1994), image compression (Belloulata and Konrad, 2002), image visualization (Hartmann and Galloway, 2000), and image retrieval (Chen and Wang, 2002). Pal and Pal (1993) provided a review on various segmentation techniques. It should be noted that there is no single standard approach to segmentation. Many different types of scene parts can serve as the segments on which descriptions are based, and there are many different ways in which one can attempt to extract these parts from the image. Selection of an appropriate segmentation technique depends on the type of images and applications.

The level of segmentation or subdivision relies on the problem domain being dealt with. For example, in the optical character recognition (OCR), the text is separated from the document image, and further partitioned into columns, lines, words, and connected components. In building character sub-images, one is often confronted with touching or broken characters that occur in degraded documents (such as fax, scan, photocopy, etc.). It is still challenging to develop techniques for properly segmenting words into their characters.

There are primarily four types of segmentation techniques: thresholding, boundary-based, region-based, and hybrid techniques. Thresholding is based on the assumption that clusters in the histogram correspond to either background or objects of interest that can be extracted by separating these histogram clusters. In addition to thresholding, many image segmentation algorithms are based on two basic properties of the pixel intensities in relation to their local neighborhood: discontinuity and similarity. Methods based on pixel discontinuity are called boundary-based or edge extraction methods, whereas methods based on pixel similarity are called region-based methods. Boundary-based methods assume that the pixel properties, such as intensity, color, and texture (Khotanzad and Chen, 1989; Panjwani and Healey, 1995), should change abruptly between different regions. Region-based methods assume that neighboring pixels within the same region should have similar values (e.g., intensity, color, texture).

It is well known that such segmentation techniques – based on boundary or region information alone – often fail to produce accurate segmentation results. Hence, there has been a tendency toward hybrid segmentation algorithms which take advantage of the complementary nature of such information. Hybrid methods combine boundary detection and region growing together to achieve better segmentation (Pavlidis and Liow, 1990; Haris et al., 1998; Fan et al., 2001). Note that both results should achieve the foreground and background segmentation coherently.

The evaluation of segmentation algorithms has been mostly subjective; therefore, people usually judge an algorithm's effectiveness based on intuition and the

DOI: 10.1201/9781003474395-5

results from several segmented images. Cardoso and Corte-Real (2005) proposed a generic framework for segmentation evaluation using symmetric and asymmetric distance metric alternatives to meet the specificities of a wide class of applications. Unnikrishnan et al. (2007) demonstrated how a proposed measure of similarity, the normalized probabilistic rand (NPR) index, can be used to perform a quantitative comparison between image segmentation algorithms using a hand-labeled set of ground-truth segmentations.

In this chapter, we discuss a number of image segmentation techniques. They include thresholding, component labeling, locating object contours by the snake model, edge detection, linking edges by adaptive mathematical morphology, automatic seeded region growing, and top-down region dividing. Other variations of image segmentation algorithms can be referred to hidden Markov Model (Pyun et al., 2007; Wu and Chung, 2007; Rivera et al., 2007), Bayesian methods (Wong and Chung, 2005), genetic algorithm and artificial neural network (Awad et al., 2007), and multicue partial differential equation (Sofou and Maragos, 2008).

4.1 THRESHOLDING

Thresholding provides an easy and convenient way to perform image segmentation based on the different intensities or colors in the foreground and background regions of an image (Saha and Udupa, 2001). The automatic selection of optimum thresholds has remained a challenge over decades. Not all images can be segmented successfully into foreground and background using simple thresholding. Its validity relies on the distribution of the intensity histogram. If the intensity distribution of foreground objects is quite distinct from the intensity distribution of background, it will be clear to apply thresholding for image segmentation. In this case, we expect to see distinct peaks in the histogram corresponding to foreground objects, such that threshold values can be picked to isolate those peaks accordingly. If such a peak does not exist, it is unlikely that simple thresholding can achieve a good segmentation.

There are several methods of choosing the threshold value λ. For example, the universal thresholding by Donoho et al. (1996) sets

$$\lambda = \frac{\sigma\sqrt{2\log n}}{\sqrt{n}} \tag{4.1}$$

where σ is the standard deviation of the wavelet coefficients and n is the total size of samples. Other possibility is quantile thresholding where λ is statistically set to replace a percentage of the coefficients with the smallest magnitude to zero. Another adaptive method of automatically choosing the best threshold λ consists of four steps: (1) Choose an initial estimate λ. (2) Calculate the two mean values μ_1 and μ_2 within the two groups of pixels after thresholding at λ. (3) Calculate the new threshold value $\lambda = (1/2)(\mu_1 + \mu_2)$. (4) If the new threshold value has a little change (i.e., smaller than a predefined constant), then the threshold selection is done; otherwise, go back to step 2.

Example 4.1: Apply the adaptive threshold selection method on the following image.

1	1	1	1	0	0	1	0
1	1	0	1	1	1	1	1
1	2	8	8	8	9	1	0
0	1	8	9	9	9	1	0
0	1	8	8	9	8	1	0
1	1	8	9	9	8	1	1
0	1	1	1	1	1	0	0
1	0	0	0	0	0	0	0

Answer: We first compute the average gray level of the image, which is 2.5781, as the initial threshold λ_0. We segment the image by λ_0 into two groups of pixels and calculate the mean values μ_1 and μ_2. The new threshold value $\lambda_1 = 0.5(\mu_1 + \mu_2) = 4.5312$. Then the image is segmented by λ_1 into two groups of pixels and calculate the new mean values μ_1 and μ_2. The new threshold value $\lambda_2 = 0.5(\mu_1 + \mu_2) = 4.5312$. Since $\lambda_1 = \lambda_2$, the best threshold value is 4.5312. Note that if a random number is chosen as the initial threshold, it would take more iterations to converge to the best threshold value.

An image can be regarded as a fuzzy subset of a plane. A fuzzy entropy measuring the blur in an image is a functional which increases when the sharpness of its argument image decreases. Zenzo et al. (1998) generalized and extended the relation "sharper than" between fuzzy sets in view of implementing the properties of a relation "sharper than" between images. They showed that there are infinitely many implementations of this relation into an ordering between fuzzy sets (equivalently, images). Relying upon these orderings, classes of fuzzy entropies are constructed which are useful for image thresholding by cost minimization.

Saha and Udupa (2001) took advantages of both intensity-based class uncertainty (a histogram-based property) and region homogeneity (an image morphology-based property) to achieve optimal thresholding. A scale-based formulation is used for region homogeneity computation. At any threshold, intensity-based class uncertainty is computed by fitting a Gaussian to the intensity distribution of each of the two regions segmented at that threshold. The theory of the optimum thresholding method is based on the postulate that objects manifest themselves with fuzzy boundaries in any digital image acquired by an imaging device. The optimal threshold value is selected at which pixels having high class uncertainty accumulate mostly around object boundaries.

Tao et al. (2008) used a normalized graph-cut measure as thresholding principle to distinguish an object from the background. The weight matrices used in evaluating the graph cuts are based on the gray levels of the image, rather than the commonly used image pixels. For most images, the number of gray levels is much smaller than the number of pixels. Therefore, their algorithm requires much smaller storage space and lower computational complexity than other image segmentation algorithms based on graph cuts.

4.2 OBJECT (COMPONENT) LABELING

It is very possible to have more than one object in a scene. All the objects must be individually extracted for the purpose of establishing the object model base. The object-labeling technique is used so that the array representation of these objects is a multivalued picture, in which the points of each component all have a unique nonzero label and the points of background are all zeros. This technique only requires two raster scans. To label 4-connected components, only the upper and left neighbors are checked. If the 8-connectedness is used, the two upper diagonal neighbors are also included.

Athanasiadis et al. (2007) presented a framework for simultaneous image segmentation and object labeling leading to automatic image annotation. Xu et al. (2003) proposed object segmentation and labeling by learning from examples. They used color and two-dimensional shape matching to automatically group those low-level segments into objects based on their similarity to a set of example object templates presented by the user. A hierarchical content tree data structure is used for each database image to store matching combinations of low-level regions as objects. The learning phase refers to labeling of combinations of low-level regions that have resulted in successful color and/or 2D shape matches with the example templates. Moga and Gabbouj (1997) developed parallel image component labeling using watershed transformation.

Let the value of object points be 1 and of background points be 0. Assume that the 8-connectedness is adopted. If all the four previously scanned neighbors of the point P are 0s, then P is assigned a new label. If one of the four neighbors is 1, then P gets the same label as that neighbor. If two or more of them are 1s, then P gets any one of their labels, and the equivalence table is established by marking the different labels together for later adjustment. Equivalence processing consists in merging the equivalent pair into the same class; i.e., a unique label is assigned to each class. Finally, a second scan is performed to replace each label by the representative of its class. Each component has now been uniquely labeled. After these processes, each individual model object can be fetched by its label.

Example 4.2: Perform object labeling on the following figure using 4-connectedness for foreground.

				1	1	1			1	1		1	1
		1	1	1		1	1	1	1				1
1	1			1	1	1		1	1	1	1	1	

Answer: The two-step procedure is conducted as follows:

					a	a	a			b	b		c	c	
(1)			d	d	a		a	a	a	a				c	(Mark d = a, b = a)
	e	e			a	a	a		a	a	a	a	a		
					a	a	a			a	a		c	c	
(2)			a	a	a		a	a	a	a				c	
	e	e			a	a	a		a	a	a	a	a		

He et al. (2008) presented an efficient run-based two-scan algorithm for labeling connected components in a binary image. Unlike the label-equivalence-based algorithm using label equivalences between provisional labels, their algorithm resolves label equivalences between provisional label sets. At any time, all provisional labels that are assigned to a connected component are combined in a set, and the smallest label is used as the representative label. The corresponding relation of a provisional label and its representative label is recorded in a table. Whenever different connected components are found to be connected, all provisional label sets concerned with these connected components are merged together, and the smallest provisional label is taken as the representative label. When the first scan is finished, all the provisional labels that were assigned to each connected component in the given image will have a unique representative label. During the second scan, each provisional label is replaced by its representative label.

4.3 LOCATING OBJECT CONTOURS BY THE SNAKE MODEL

The snake model is a variation model in image segmentation. It moves the snake points according to the minimization of the sum of internal and external energies integrated along the length of the snake to approach edges. In the original snake formulation of Kass et al. (1987), the best snake position was defined as the solution of a variational problem requiring the minimization of the sum of internal and external energies integrated along the length of the snake. The corresponding Euler equations, which give the necessary conditions for this minimizer, comprise a force balance equation. The snake model has provided a number of applications in object segmentation, stereo matching, motion tracking, etc. In image processing, the snake model defines a snake as an energy-minimizing spline guided by external constraint forces and influenced by such forces that pull it toward image features such as lines and edges. It is a kind of the active contour model in the way that it locks on nearby edges, localizing them accurately.

There are two parts in the energy function of the snake model. The first part reflects geometric properties of the contour, and the second part utilizes the external force field to drive the snake. The first part serves to impose a piecewise smoothness constraint, and the second part is responsible for putting the snake near the local minimum of energy. The traditional snakes suffer from a great disadvantage that when an object resides in a complex background, the strong edges may not be the object edges of interest. Therefore, researchers have proposed statistical and variational methods to enrich the energy function and extend its flexibility. They calculated the difference between the target object and the background using statistical analysis, but these limitations come from the priori knowledge requirements, such as independent probability models and template models. Unfortunately, such priori knowledge is usually unavailable unless the captured images are under very constrained settings. A problem with the snake model is that a user needs to place the initial snake points sufficiently close to the feature of interest.

Li et al. (2005) described a variational formulation for geometric active contours that forces the level set function to be close to a signed distance function. They used the level set function to allow flexible initialization. Shen and Davatzikos (2000) proposed an adaptive-focus deformable model using statistical and geometric information. Michailovich et al. (2007) proposed a generalization of the segmentation methods, in which active contours maximize the difference between a finite number of empirical moments. Melonakos et al. (2008) proposed an image segmentation technique based on augmenting the conformal (or geodesic) active contour framework with directional information.

Most snake models do not discern between positive and negative step edges. Cheng and Foo (2006) proposed dynamic directional gradient vector flow to uses this information for better performance. Recently the proposed Sobolev active contours (Sundaramoorthi et al., 2007) introduced a new paradigm for minimizing energies defined on curves by changing the traditional cost of perturbing a curve and thereby redefining gradients associated to these energies. Sobolev active contours evolve more globally and are less attracted to certain intermediate local minima than traditional active contours, and it is based on a well-structured Riemannian metric, which is important for shape analysis and shape priors. Sundaramoorthi et al. (2008) analyzed Sobolev active contours using scale-space analysis in order to understand their evolution across different scales. Chen et al. (2006) presented an integrated approach to extend the geodesic and edgeless active contours approaches for texture segmentation. Xie and Mirmehdi (2008) proposed an active contour model using an external force field that is based on magnetostatics and hypothesized magnetic interactions between the active contour and object boundaries.

4.3.1 The Traditional Snake Model

A snake is a controlled continuity spline that moves and localizes onto a specified contour under the influence of the objective function. Let a snake be a parametric curve: $v(s)=[x(s),y(s)]$, where parameter $s\in[0,1]$. It moves around the image spatial domain to minimize the objective energy function as defined by

$$E_{snake}(v)=\sum_{i=1}^{n}\left[\alpha\times E_{cont}(v_i)+\beta\times E_{curv}(v_i)+\gamma\times E_{image}(v_i)\right], \tag{4.2}$$

where α, β, and γ are weighting coefficients that control the snake's tension, rigidity, and attraction, respectively. The first and second terms are correspondingly the first- and second-order continuity constraints. The third term measures the edge strength (i.e., the image force).

The continuity force E_{cont} encouraging even spacing of points can be calculated as

$$E_{cont}\left[v_i\right] = \frac{\left|\bar{d} - \left|v_i - v_{i-1}\right|\right|}{\max_j \left\{\left|\bar{d} - \left|v_i(j) - v_{i-1}\right|\right|\right\}}, \tag{4.3}$$

where $\{v_i(j) \mid j = 1, 2, \ldots, m\}$ denotes the snake point v_i's m neighbors, and $\bar{d}$ denotes the average length of all the pairs of adjacent points on the snake contour as given by

$$\bar{d} = \frac{\sum_{i=1}^{n} \left|v_i - v_{i-1}\right|}{n}, \tag{4.4}$$

where $v_0 = v_n$. This term tends to keep the distances between each pair of adjacent vertices equal.

The energy of the second-order continuity E_{curv} is represented by

$$E_{curv}[v_i] = \frac{\left|v_{i-1} - 2v_i + v_{i+1}\right|}{\max\left\{\left|v_{i-1} - 2v_i + v_{i+1}\right|\right\}}. \tag{4.5}$$

The numerator can be rearranged as

$$v_{i-1} - 2v_i + v_{i+1} = \left(v_{i+1} - v_i\right) - \left(v_i - v_{i-1}\right). \tag{4.6}$$

If the ith vertex is pushed toward the midpoint of two adjacent vertices, E_{curv} is minimized; i.e., the shape of the contour will remain C^2 continuity.

The third term, image energy E_{image}, is derived from the image so that it takes on its smaller values at the features of interest, such as boundaries. It considers the gradient (denoted as *grad*) magnitude, leading the active contour toward step edges. It is normalized to measure the relative magnitude as

$$E_{image}[v_i] = \frac{\min\left\{\left|grad\right|\right\} - \left|grad_{v_i}\right|}{\max\left\{\left|grad\right|\right\} - \min\left\{\left|grad\right|\right\}}, \tag{4.7}$$

where min and max denote the minimum and maximum gradients in the v_i's local m-neighborhood, respectively. Note that because the numerator in equation (4.7) is always negative, E_{image} can be minimized for locating the largest gradient, which is the edge. In general, the traditional snake model can locate object contours in a simple background. If the background becomes complex, it will fail since the complex background will generate noisy edges to compete with the object edges for attracting the snake.

4.3.2 The Improved Snake Model

The following two issues are observed when the snake model fails to locate object contours in complex backgrounds. One is the gray-level sensitivity; i.e., the more abrupt change the gray levels have (e.g., noises), the larger impact on the energy function the snake makes. The other is that the snake mistakenly locates the edges belonging to the background details due to their closeness to the snake point. Figure 4.1 shows a disk object in a complex background. If the snake points are initialized outside the disk, the snake cannot locate the disk contour accurately due to the disturbances from the background grids. The idea is to push the mean intensity of the polygon enclosed by the snake contour to be as close as to the mean intensity of the target object. The smaller intensity difference between the polygon and the object, the closer the snake approaches the object contour. Therefore, a new energy term, called the *regional similarity energy* (RSE), is established for calculating the gray-level differences to be added into the overall energy (Shih and Zhang, 2004).

Supposedly, the intensities of the target disk are homogeneous while its background is complex. As illustrated in Figure 4.2, let the snake contour in the current iteration be C, which contains a point i. An area of neighborhood is defined and the neighbors of i are searched for the new point in the next iteration, say i', which produces the minimum area as compared to the other neighbors. Such a contour C' is regarded as the base area and the mean intensity of its enclosed area is regarded as the base mean intensity M_i.

FIGURE 4.1 A disk object resides in a complex background.

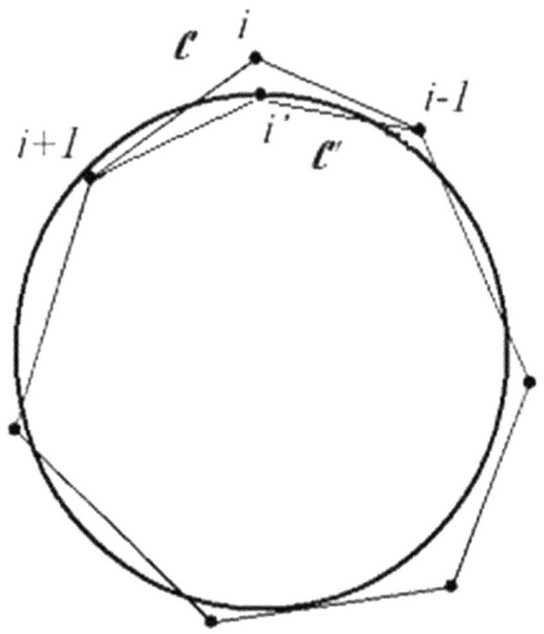

FIGURE 4.2 The regional similarity energy. See context for explanation.

Let p denote a point in the neighborhood of i. The RSE of point p is calculated as

$$E_{RSE}(i,p) = |m(p) - M_i|, \tag{4.8}$$

where $m(p)$ denotes the mean intensity in the neighborhood of point p. Therefore, the RSE of the snake contour is

$$E_{RSE}[v(s)] = \sum_i E_{RSE}(i,p) \tag{4.9}$$

For convex and concave regions, the snake point which could be located outside, on, or inside the object contour is analyzed. As shown in Figure 4.3, there are six cases and the left-side image in each case displays a convex or concave region being overlaid with a snake contour. The right-side image shows the RSE distribution, where the brightness indicates the energy value. The square box implies the region of neighborhood. Since the mean intensity of the snake contour is desired to be the closest to that of the object contour, a minimum RSE is favored. Cases (a)–(c) are for convex regions. In case (a), the RSE will pull the snake point toward the boundary. In case (b), part of the square box inside the object boundary (i.e., black region) will keep the snake point the same; however, the other part outside the object boundary (i.e., white region) will pull the snake point toward the boundary. In case (c), the RSE will have no effect since the square box is entirely inside the object boundary. Similarly, cases (d)–(f) are for concave regions.

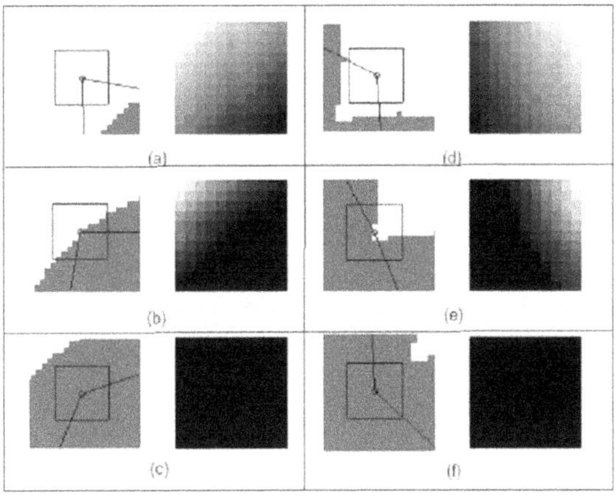

FIGURE 4.3 Cases (a)–(c) show outside, on, and inside the boundary for convex regions; cases (d)–(f) show outside, on, and inside the boundary for concave regions, respectively.

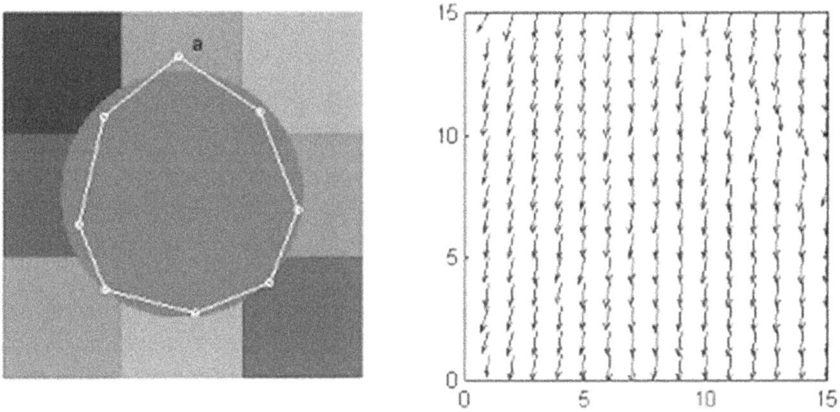

FIGURE 4.4 The energy flow of the snake point a within a 15×15 area.

Figure 4.4(a) shows an image of 300×300 pixels, and (b) illustrates the RSE for point a within a 15×15 area. It is observed that the energy flow pushes the snake point a downward to the disk contour.

4.3.3 The Gravitation External Force Field and the Greedy Algorithm

In this section, the gravitation external force field is introduced and the greedy algorithm (Ji and Yan, 2002) is used for the active contour. The concept of gravitation external force field is taken from physics. Two objects attract each other by a force, which is proportional to their mass product and inversely proportional to the distance

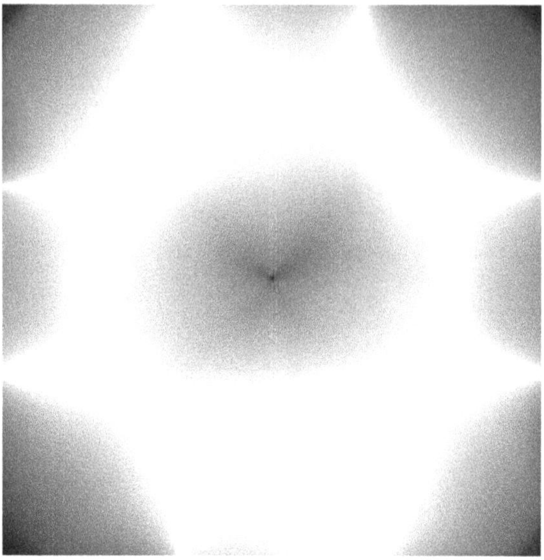

FIGURE 4.5 The effect of the gravitation force field on Figure 4.1.

between their mass centers. Based on this concept, an external energy field, called the *gravitation energy field* (GEF), is developed as given by

$$E_{gravitation} = \int \frac{g(\vec{r})}{\|\vec{r}\|}\vec{r}d\vec{r}, \tag{4.10}$$

where $\vec{r}$ is a position vector and $g(\vec{r})$ is a first-order derivative. Note that the edge pixels have the local maxima in the first-order derivative. The attractive force enables the snake points to move toward the object. Figure 4.5 illustrates the effect of the gravitation force field on Figure 4.1. Note that the brightness represents the energy value. With the gravitation energy field, the active contour can be dragged toward the object even if the snake points are far away. Therefore, the total energy function becomes

$$E_{snake} = aE_{cont}[v(s)] + \beta E_{curv}[v(s)] + \gamma E_{image}[v(s)] + \mu E_{gravitation}[v(s)] + \delta E_{RSE}[v(s)]. \tag{4.11}$$

These weighting coefficients can be adjusted according to the user's application. For simplicity, $a = \beta = \gamma = \mu = \delta = 1$.

A set of snake points are initialized, and then a loop is entered. At each loop, every snake point is temporally moved to its neighbor within a defined neighborhood, and the energy is calculated. The one with the least energy is chosen to be the new location. When all the snake points are processed, the next loop is repeated until all the snake points are stabilized or a given number of iterations are reached. The greedy algorithm is described as follows:

```
while n < iterations
  for j = 1 to numPoints  /* Check every snake point. */
    for k = 1 to numNeighbors
      E_snake = α * E_cont[v_k(s) ] + β * E_curv[v_k(s) ] + γ * E_image[v_k(s) ];
    end;
    Move this point to the new location that has the least
    energy;
  end;    /* main loop*/
end;    /* iteration loop*/
```

4.3.4 Experimental Results

Both the improved and the traditional snake models are applied on the added salt-and-pepper noise of Figure 4.1. The results are shown in Figure 4.6. It is observed that the improved model can locate the disk contour, but the traditional model fails. The improved model is suitable for random noise or fixed pattern noise. The banding noise is highly camera-dependent, and is the one which is introduced by the camera when it reads data from the digital sensor. The improved model may not perform well in an image with such noise.

Figure 4.7 illustrates that the proposed model can work successfully for a concave object, but the traditional model fails. Figure 4.7(d) shows that an interpolation between adjacent snake points can be applied for smooth appearance of the object contour. Figure 4.8 shows more results for convex and concave objects in a variety of complex backgrounds using the proposed model. It is observed that some snake point is attracted by the background grid using the traditional model; however, it can locate the object boundary using the proposed model.

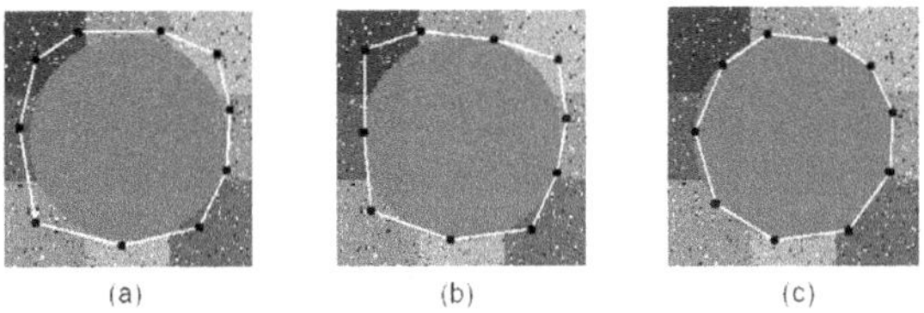

FIGURE 4.6 (a) The initial snake points, (b) the result by the traditional snake model, (c) the result by the improved snake model.

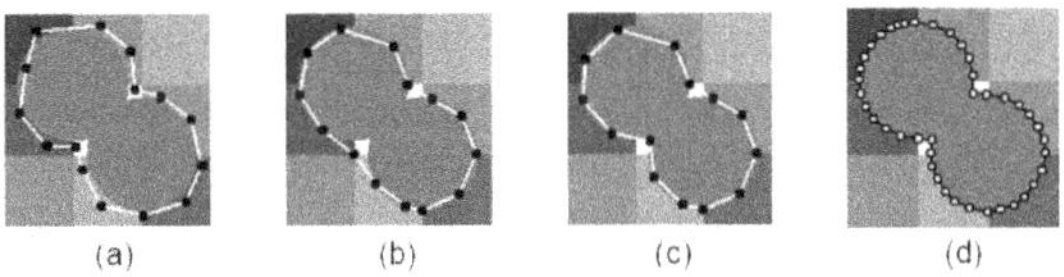

FIGURE 4.7 (a) The initial snake points, (b) the result by the traditional snake model, (c) the result by the improved snake model, (d) the interpolation appearance.

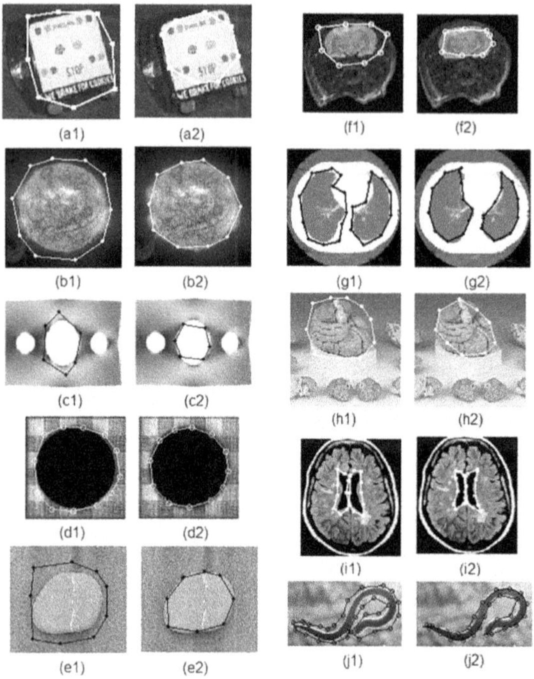

FIGURE 4.8 More examples for convex and concave objects in complex backgrounds. The first column shows the initial snake points and the second column shows the resulting snake points using the improved model.

The snake model expects the resulting contour to be stabilized on the boundary. If a snake point is inside the object, the force field can pull it out toward the boundary. In Figure 4.8(c2), due to the display limitation and in order to clearly augment the snake points, the snake contour points may look deviating from the true boundary, but in fact they are close to the boundary. The size of neighborhood affects the smoothness of snake contour. In the experiments, a 7×7 neighborhood and 50 as the maximum number of iterations are used. The improved snake model works more effective in an image with convex or concave objects under a variety of complex backgrounds.

4.4 EDGE OPERATORS

Image segmentation is the fundamental step in image analysis to partitioning an image into individual regions of interest. Gray-level edge detection is one of image segmentation techniques. The output of edge operators should form continuous boundaries of the regions of interest to be further processed. Many different edge operators have already been proposed (Davis, 1974).

Law et al. (1996) characterized the problem of detecting edges in images as a fuzzy reasoning problem. The edge detection problem is divided into three stages: filtering, detection, and tracing. Images are filtered by applying fuzzy reasoning based on local pixel characteristics to control the degree of Gaussian smoothing. Ahmad and Choi (1999) developed an edge detection method based on local threshold and Boolean function. Meer and Georgescu (2001) proposed edge detection with embedded confidence. Konishi et al. (2003) formulated edge detection as statistical inference. Bao et al. (2005) analyzed the technique of scale multiplication in the framework of Canny edge detection (Canny, 1986). A scale multiplication function is defined as the product of the responses of the detection filter at two scales. Edge maps are constructed as the local maxima by thresholding the scale multiplication results. The detection and localization criteria of the scale multiplication are derived. Wu et al. (2007a) proposed a fast multilevel fuzzy edge detection (FMFED) algorithm to realize the fast and accurate detection of the edges from the blurry images.

The derivative filter emphasizes the edges (i.e., high-frequency components) of objects. As an image is a 2D function, it is needed to define the direction in which the derivative is taken. Edges could be in the horizontal, vertical, or arbitrary directions. Let h_x, h_y, and h_θ respectively denote horizontal, vertical, and arbitrary angle derivative filters. Their relationship can be represented as

$$[h_\theta] = \cos\theta \cdot [h_x] + \sin\theta \cdot [h_y] \tag{4.12}$$

For a function $f(x, y)$, the gradient filter generates a vector derivative $\nabla \mathbf{f}$ as the *gradient*:

$$\nabla \mathbf{f} = \begin{bmatrix} \partial f / \partial x \\ \partial f / \partial y \end{bmatrix} = \frac{\partial f}{\partial x}\mathbf{u}_x + \frac{\partial f}{\partial y}\mathbf{u}_y = (h_x * f)\mathbf{u}_x + (h_y * f)\mathbf{u}_y \tag{4.13}$$

where "$*$" denotes the convolution operation, and $\mathbf{u_x}$ and $\mathbf{u}_y$ are unit vectors in the horizontal and vertical directions, respectively. The magnitude of the gradient vector is

$$|\nabla \mathbf{f}| = \left[\left(\frac{\partial f}{\partial x}\right)^2 + \left(\frac{\partial f}{\partial y}\right)^2\right]^{1/2} = \sqrt{(h_x * f)^2 + (h_y * f)^2}, \tag{4.14}$$

and the direction of the gradient vector is

$$\theta(\nabla \mathbf{f}) = \tan^{-1}\left(\frac{\partial f / \partial y}{\partial f / \partial x}\right) = \tan^{-1}\left(\frac{h_y * f}{h_x * f}\right). \tag{4.15}$$

For computational simplicity, the gradient magnitude is approximately taken as

$$|\nabla \mathbf{f}| = |h_x * f| + |h_y * f| \tag{4.16}$$

Basic derivative filters are $h_x = \begin{bmatrix} -1 & 1 \end{bmatrix}$ and $h_y = \begin{bmatrix} 1 \\ -1 \end{bmatrix}$. The center of the filter is at the middle (i.e., the grid) of two horizontal or vertical neighboring pixels. In order to use a pixel (instead of a grid) as the center, the filter is extended to 3 pixels as

$$h_x = \begin{bmatrix} -1 & 0 & 1 \end{bmatrix} \text{ and } h_y = \begin{bmatrix} 1 \\ 0 \\ -1 \end{bmatrix}.$$

Roberts cross-gradient operator uses the 2D case of 2 pixels as

$$\text{Roberts: } h_x = \begin{bmatrix} 1 & 0 \\ 0 & -1 \end{bmatrix} \text{ and } h_y = \begin{bmatrix} 0 & 1 \\ -1 & 0 \end{bmatrix}.$$

Prewitt gradient operator uses the 2D case of 3 pixels as

$$\text{Prewitt: } h_x = \begin{bmatrix} -1 & 0 & 1 \\ -1 & 0 & 1 \\ -1 & 0 & 1 \end{bmatrix} \text{ and } h_y = \begin{bmatrix} 1 & 1 & 1 \\ 0 & 0 & 0 \\ -1 & -1 & -1 \end{bmatrix}.$$

Sobel gradient operator slightly modifies the Prewitt gradient operator by putting more weight on the pixels closer to the center as

$$\text{Sobel: } h_x = \begin{bmatrix} -1 & 0 & 1 \\ -2 & 0 & 2 \\ -1 & 0 & 1 \end{bmatrix} \text{ and } h_y = \begin{bmatrix} 1 & 2 & 1 \\ 0 & 0 & 0 \\ -1 & -2 & -1 \end{bmatrix}.$$

Another popularly used edge operation, called Canny edge operator (Canny, 1986), used the calculus of variations to find the function which optimizes a given functional, described by the sum of four exponential terms in horizontal, vertical, and diagonal edges.

Example 4.3: Apply two masks of Sobel operator on the following image and combine the two results using the so-called "squares and square roots." Show the output edge image after rounding real numbers to integers. After that, how can we separate the edge pixels and non-edge pixels? What is the rough shape of the object in this image?

1	1	1	1	0	0	1	0
1	1	0	1	1	1	1	1
1	2	8	8	8	9	1	0
0	1	8	9	9	9	1	0
0	1	8	8	9	8	1	0
1	1	8	9	9	8	1	1
0	1	1	1	1	1	0	0
1	0	0	0	0	0	0	0

Answer: Let the parts of the masks falling outside the image are zeroed out. After taking the square root of the sum of the squares of the two results convolving with h_x and h_y, we obtain the output image as

4	3	2	3	4	4	4	4
5	10	23	29	32	27	13	4
6	22	31	32	32	33	28	4
6	31	29	3	1	31	35	4
4	31	30	4	1	32	32	4
4	24	31	29	30	34	25	3
3	13	27	35	35	28	14	3
1	3	4	4	4	3	1	0

We can separate edge pixels from non-edge pixels by thresholding. Using histogram thresholding by selecting a valley between two significant peaks, we can obtain the binary edge image as follows. The rough shape of the object in this image is a square.

0	0	0	0	0	0	0	0
0	1	1	1	1	1	1	0
0	1	1	1	1	1	1	0
0	1	1	0	0	1	1	0
0	1	1	0	0	1	1	0
0	1	1	1	1	1	1	0
0	1	1	1	1	1	1	0
1	0	0	0	0	0	0	0

Figure 4.9 shows the Lena image and the resulting images of applying the Roberts, Prewitt, and Sobel operators, respectively. It should be noted that both h_x and h_y are separable. That is, they can be decomposed into 1D cases. For example, in the Prewitt gradient operator,

$$\text{Prewitt: } h_x = \begin{bmatrix} 1 \\ 1 \\ 1 \end{bmatrix} \cdot \begin{bmatrix} -1 & 0 & 1 \end{bmatrix} \text{ and } h_y = \begin{bmatrix} 1 \\ 0 \\ -1 \end{bmatrix} \cdot \begin{bmatrix} 1 & 1 & 1 \end{bmatrix}.$$

Higher-order derivatives of functions of two variables can be also adopted in image derivative filters. Laplacian second derivative filter plays an important role. Unlike the first derivative filters, the Laplacian is isotropic; i.e., it gives the same response to edges in any directions. However, the disadvantages are the loss of edge direction (which may be useful in some applications) and the edge responses being not closely related to the edge magnitudes.

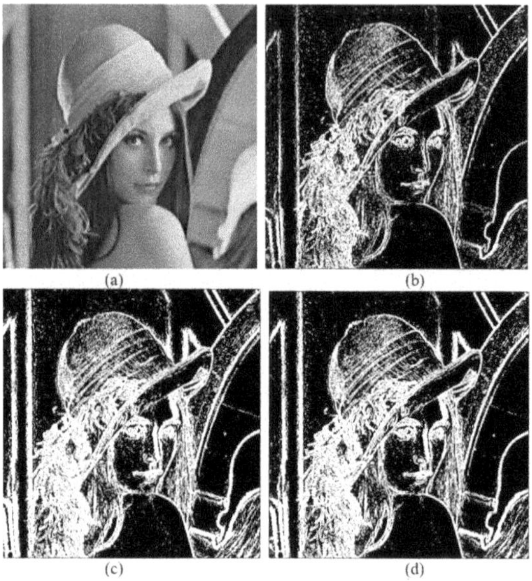

FIGURE 4.9 (a) The Lena image and the resulting images after (b) Roberts, (c) Prewitt, and (d) Sobel filters.

Let a 3 × 3 window of an input image *f* be represented as

$f(x-1, y+1)$	$f(x, y+1)$	$f(x+1, y+1)$
$f(x-1, y)$	$f(x, y)$	$f(x+1, y)$
$f(x-1, y-1)$	$f(x, y-1)$	$f(x+1, y-1)$

The Laplacian is defined as

$$\nabla^2 f = \frac{\partial^2 f}{\partial x^2} + \frac{\partial^2 f}{\partial y^2} \tag{4.17}$$

since

$$\begin{aligned} \frac{\partial^2 f}{\partial x^2} &= [f(x+1, y) - f(x, y)] - [f(x, y) - f(x-1, y)] \\ &= f(x+1, y) + f(x-1, y) - 2f(x, y). \end{aligned} \tag{4.18}$$

Similary,

$$\frac{\partial^2 f}{\partial y^2} = f(x, y+1) + f(x, y-1) - 2f(x, y). \tag{4.19}$$

Therefore,

$$\nabla^2 f = [f(x+1,y)+f(x-1,y)+f(x,y+1)+f(x,y-1)]-4f(x,y). \quad (4.20)$$

The kernel of the Laplacian second derivative filter h is denoted as

$$h = \begin{bmatrix} 0 & 1 & 0 \\ 1 & -4 & 1 \\ 0 & 1 & 0 \end{bmatrix}$$

Example 4.4: Apply the Laplacian second-derivatives filter on the following image. Show the output edge image after rounding real numbers to integers. After that, how can the edge pixels and non-edge pixels can be separated? What is the rough shape of the object in this image?

1	1	1	1	0	0	1	0
1	1	0	1	1	1	1	1
1	2	8	8	8	9	1	0
0	1	8	9	9	9	1	0
0	1	8	8	9	8	1	0
1	1	8	9	9	8	1	1
0	1	1	1	1	1	0	0
1	0	0	0	0	0	0	0

Answer: The Laplacian second-derivatives filter mask used is as follows:

0	1	0
1	–4	1
0	1	0

The output is

–2	–1	–2	–2	2	2	–3	2
–1	0	11	6	6	7	0	–3
–1	3	–14	–6	–5	–17	7	2
2	7	–6	–3	–1	–9	7	1
2	6	–7	3	2	–5	6	2
–3	7	–13	–10	–9	–13	6	–3
3	–2	6	7	7	5	2	1
–4	2	1	1	1	1	0	0

The negative and positive values are rescaled to be from 0 to 255. Sometimes one encounters the absolute value being used for this purpose, but this really is not

(a) (b)

FIGURE 4.10 (a) The Lena image, (b) the resulting image of the Laplacian filter.

correct because it produces double lines of nearly equal magnitude, which can be confusing. After performing thresholding on the image to obtain the edge image, it is observed that the object is a square.

Figure 4.10 shows an example of applying the Laplacian second derivative filter. Sometimes, the kernel of the Laplacian second derivative filter h is extended to include diagonal neighbors as

$$h = \begin{bmatrix} 1 & 1 & 1 \\ 1 & -8 & 1 \\ 1 & 1 & 1 \end{bmatrix}$$

4.5 EDGE LINKING BY ADAPTIVE MATHEMATICAL MORPHOLOGY

In edge or line linking, two assumptions are usually made: (1) True edge and line points in a scene follow some continuity patterns, whereas the noisy pixels do not follow any such continuity. (2) The strengths of the true edge or line pixels are greater than those of the noisy pixels. The "strength" of pixels is defined differently for different applications.

Nevatia (1976) presented an algorithm to link edge points by fitting straight lines. Linking direction is based on the direction of edge elements within a defined angular interval. However, portions of small curved edge segments may be neglected. Nalwa and Pauchon (1987) presented an approach based upon local information. The defined *edgels* (i.e., short, linear edge elements, each characterized by a direction and a position) and curved segments are linked based solely on proximity and relative orientation. No global factor has been taken into consideration, and there is little concern about noise reduction.

Liu et al. (1988) proposed an edge-linking algorithm to fill gaps between edge segments. The filling operation is performed in an iterative manner rather than a single step. They first connect so-called *tip ends* (i.e., the set of edge pixels which are very likely to be the knots of some contour containing the one being considered) of two segments with a line segment and then try to modify the resulting segment by straight-line fitting. In each iteration, they define the dynamic threshold, and noises are removed gradually. However, it is difficult to get accurate tip ends. Another method (Russ, 1992) locates all end points of broken edges and uses a relaxation method to link them up such that line direction is maintained. Lines are not allowed to cross, and the points which are closer are matched first. However, this method may fail if unmatched end points or noises are present.

Hajjar and Chen (1999) presented a real-time algorithm and its VLSI implementation for edge linking. The linking process is based on the break points' directions and the weak level points. Stahl and Wang (2007, 2008) proposed an edge-grouping method to detect perceptually salient structures in noisy images by combining boundary and region information. Papari and Petkov (2008) introduced the adaptive pseudo-dilation, which uses context dependent structuring elements in order to identify long curvilinear structure in the edge map.

4.5.1 The Adaptive Mathematical Morphology

Traditional morphological operators process an image using a structuring element of fixed shape and size over the entire image pixels. In adaptive morphology, rotation and scaling factors are incorporated. The structuring elements are adjusted according to local properties of an image. In order to introduce adaptive mathematical morphology, the following terminologies need to be first introduced.

Assume that the sets in consideration are always connected and bounded. Let the boundary ∂B of a set B be the set of points, all of whose neighborhoods intersect both B and its complement B^c. If a set B is connected and has no holes, it is called *simply connected*. If it is connected but has holes, it is called *multiply connected*. The concept of ∂B is referred to as the continuous boundary of a structuring element in Euclidean plane and only the simply connected structuring element is considered. Therefore, the transformations performed on the structuring element B are replaced by the transformations on its boundary ∂B in Euclidean plane and followed by a positive filling operator. The positive filling $[\partial B]_+$ of a set B is defined as the set of points that are inside ∂B. The adaptive morphological dilation and erosion can be defined by slightly modifying the traditional morphological operations (Shih and Cheng, 2004).

Definition 4.1: Let A and B be subsets of E^N. The *adaptive morphological dilation* of A by a structuring element B is defined by

$$A \hat{\oplus} B = \{c \in E^N \mid c = a + \hat{b}, \text{ for some } a \in A \text{ and } \hat{b} \in [(R(a)S(a)\partial B]_+\}, \tag{4.21}$$

where S and R are the scaling and rotation matrices, respectively.

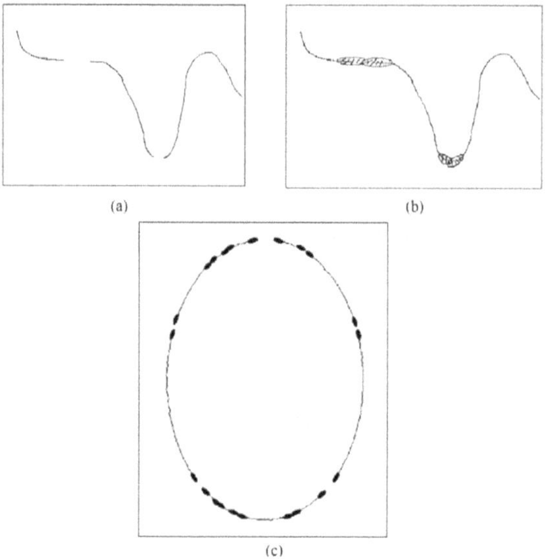

FIGURE 4.11 (a) Input signal with gaps, (b) adaptive dilation using elliptical structuring elements, and (c) the result of an adaptive dilation operation. Note that the white intensity of edge pixels and the black intensity of background pixels are inverted for clear display purposes.

Definition 4.2: Let A and B be subsets of E^N. The *adaptive morphological erosion* of A by a structuring element B is defined by

$$A \hat{\Theta} B = \left\{ c \in E^N \mid c + \hat{b} \in A, \text{for some } a \in A \text{ and every } \hat{b} \in [(R(a)S(a)\partial B]_+ \right\}. \quad (4.22)$$

The broken edge segments can be linked gradually using the adaptive morphological dilation as shown in Figure 4.11. Figure 4.11(a) is the input signal with gaps and (b) shows an adaptive dilation at the endpoints. The reason for choosing the elliptical structuring element is that by using appropriate major and minor axes, all kinds of curves can be linked smoothly. Figure 4.11(c) shows the result of an adaptive dilation operation. Note that the white intensity of edge pixels and the black intensity of background pixels are inverted for clear display purposes.

4.5.2 The Adaptive Morphological Edge-Linking Algorithm

Removing noisy edge segments, checking endpoints, applying adaptive dilation, thinning, and pruning constitute a complete edge-linking algorithm. If large gaps occur, this algorithm needs to be applied several times until no more gap exists or a predefined number of iterations has been reached.

Step 1: Removing noisy edge segments

If the length of one edge segment is shorter than a threshold value, then this segment is removed. A reasonable threshold value 3 is used.

Step 2: Detecting all the endpoints

An *endpoint* is defined as any pixel having only one 8-connected neighbor. All the edge points are checked to extract the entire endpoint set.

Step 3: Applying adaptive dilation at each endpoint

From each endpoint, a range of pixels along the edge are chosen. From the properties of this set of pixels, the rotation angle and size of the elliptical structuring element are obtained. In the program, an increasingly sized range of pixels for each iteration is used. It is referred to as an adjustable parameter s denoting the size of the range of pixels from the endpoint. Two pixels are taken from this range of pixels, the left-most side $p_1(x_1, y_1)$ and the right-most side $p_2(x_2, y_2)$. The equation $slope = (y_2 - y_1)/(x_2 - x_1)$ is used to compute the slope. The rotation angle of the elliptical structuring element can be obtained by using the equation $\theta = \tan^{-1}(slope)$. The size of the elliptical structuring element is adjusted according to the number of the pixels in the set, the rotation angle θ, and the distance between p_1 and p_2. In the program, the elliptical structuring element with fixed $b = 3$ and a changing from 5 to 7 are used. These values are based on the experimental results. For a big gap, by using $a = 7$ the gap can be linked gradually in several iterations. For a small gap, using $a = 5$ is better than using 7. If $a = 3$ or 4 is used for a small gap, the structuring element will be almost a circle, and the thinning algorithm does not work well.

An ellipse can be represented as

$$\frac{x^2}{a^2} + \frac{y^2}{b^2} = 1, \tag{4.23}$$

where a and b denote the major and minor axes, respectively. If the center of the ellipse is shifted from the origin to (x_0, y_0), the equation becomes

$$\frac{(x - x_0)^2}{a^2} + \frac{(y - y_0)^2}{b^2} = 1. \tag{4.24}$$

If the center is at (x_0, y_0) and the rotation angle is $\theta(-\pi/2 \le \theta \le \pi/2)$, the equation becomes

$$\frac{\left((x - x_0)\cos\theta - (y - y_0)\sin\theta\right)^2}{a^2} + \frac{\left((x - x_0)\sin\theta - (y - y_0)\cos\theta\right)^2}{b^2} = 1. \tag{4.25}$$

Because the image is discrete, the rounded values are used in defining the elliptical structuring element. At each endpoint, an adaptive dilation is performed using the elliptical structuring element. Therefore, the broken edge segment will be extended along the slope direction by the shape of ellipse.

Step 4: Thinning

After applying the adaptive dilation at each endpoint, the edge segments are extended in the direction of local slope. Because the elliptical structuring element is used, the edge segments grow a little fat. Morphological thinning is used to obtain the edge with one pixel of width.

Step 5: Branch pruning

The adaptively dilated edge segments after thinning may contain noisy short branches. These short branches must be pruned away. The resulting skeletons after thinning have width of one pixel. A *root point* is defined as a pixel having at least three pixels in 8-connected neighbors. From each endpoint, it is traced back along the existing edge. If the length of this branch is shorter than a threshold value after it reaches a root point, the branch is pruned.

Step 6: Decision

The program terminates when no endpoint exists or when a predefined number of iterations has been reached. In the program, 8 iterations are used as the limit.

4.5.3 Experimental Results

Figure 4.12(a) shows an original elliptical edge, and (b) shows its randomly discontinuous edge. The edge-linking algorithm is experimented on Figure 4.12(b).

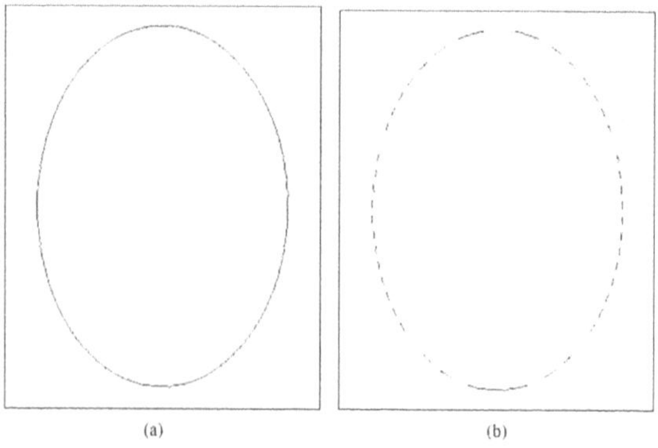

FIGURE 4.12 (a) Original elliptical edge, and (b) its randomly discontinuous edge.

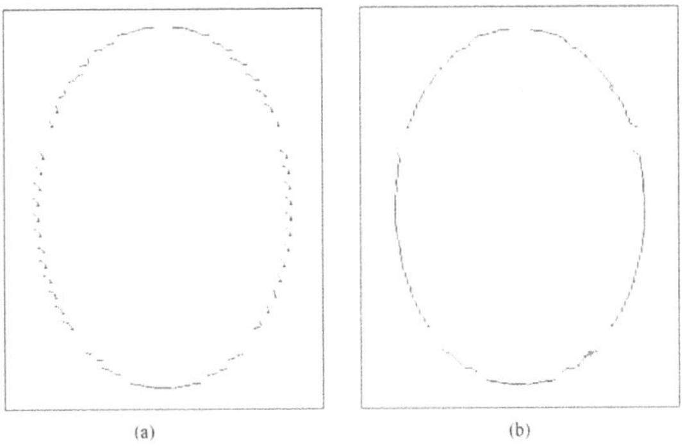

FIGURE 4.13 Using circular structuring elements in 5 iterations with (a) $r = 3$ and (b) $r = 5$.

4.5.3.1 Using Circular Structuring Elements

Figures 4.13 shows the results of using circular structuring elements with $r = 3$ and $r = 5$, respectively, in five iterations. As comparing with the original ellipse in Figure 4.12(a), it is known if the gap is larger than the radius of the structuring element, it is difficult to link smoothly. However, if a very big circular structuring element is used, the edge will look hollow and protuberant. Besides, it can obscure the details of the edge.

4.5.3.2 Using a Fixed-Sized Elliptical Structuring Element and a Fixed Range of Pixels to Measure the Slope

In this experiment, an elliptical structuring element with a fixed minor axis $b = 3$ and a fixed major axis $a = 5$, 7, and 9, respectively, are used. A fixed range of pixels is used to measure the local slope for each endpoint. That is, the same range of pixels counting from the endpoint in each iteration is used. The parameter s denotes the range of pixels from endpoint.

Figure 4.14 shows the result of using an elliptical structuring element with $a = 5$, $b = 3$, and $s = 7$. The result has some big gaps because $a = 5$ is too small. Figure 4.15 shows the result of using $a = 7$, $b = 3$, and $s = 9$. Figure 4.16 shows the result of using $a = 7$, $b = 3$, and $s = 11$. There is not much difference between Figures 4.15 and 4.16. Compared with the original ellipse in Figure 4.12(a), Figures 4.15 and 4.16 have a few shortcomings, but they are much better than Figure 4.14. Figure 4.17 shows the result of using $a = 9$, $b = 3$, and $s = 11$. Therefore, using $a = 7$ or $a = 9$, a reasonably good result is obtained except for a few shifted edge pixels.

4.5.3.3 Using a Fixed-Sized Elliptical Structuring Element for Every Endpoint but Using Adjustable-Sized Range of Pixels to Measure Local Slope in Each Iteration

Figure 4.18 shows the result of using $a = 5$, $b = 3$, and an adjustable s, and Figure 4.19 shows the result of using $a = 7$, $b = 3$, and an adjustable s. Compared with Figures 4.16

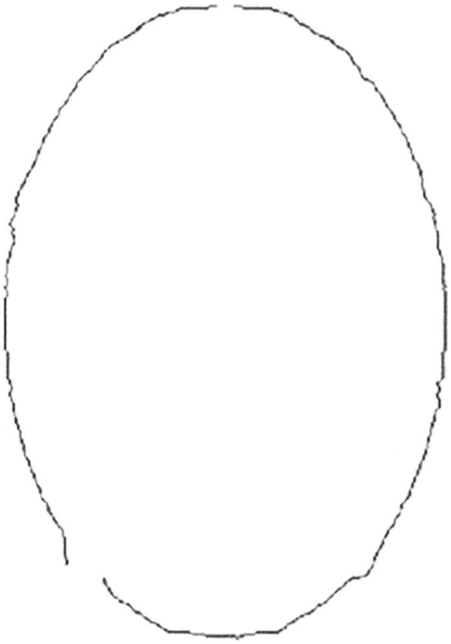

FIGURE 4.14 An elliptical structuring with $a = 5$, $b = 3$, and $s = 7$.

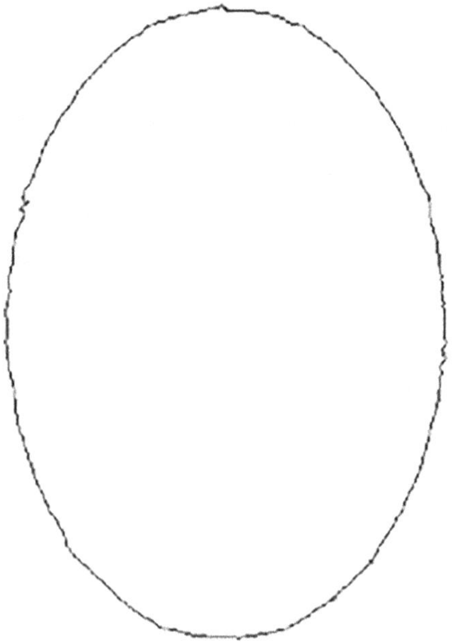

FIGURE 4.15 An elliptical structuring element with $a = 7$, $b = 3$, and $s = 9$.

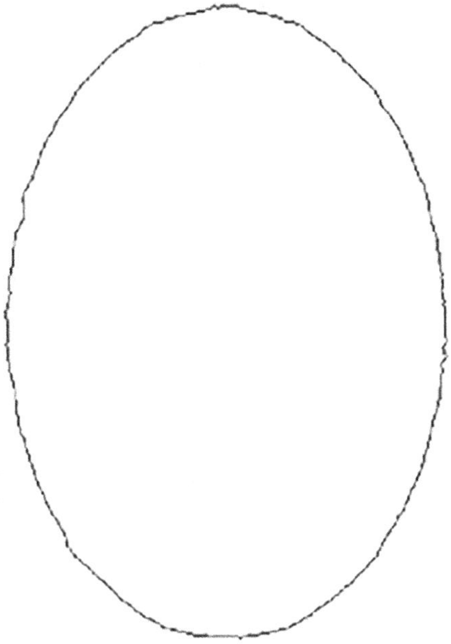

FIGURE 4.16 An elliptical structuring element with $a = 7$, $b = 3$, and $s = 11$.

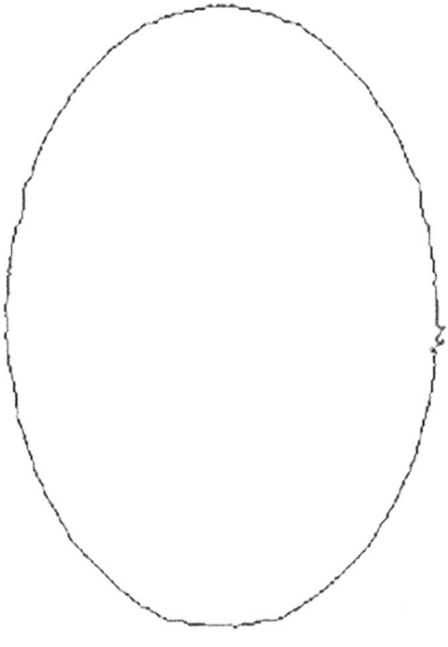

FIGURE 4.17 An elliptical structuring element with $a = 9$, $b = 3$, and $s = 11$.

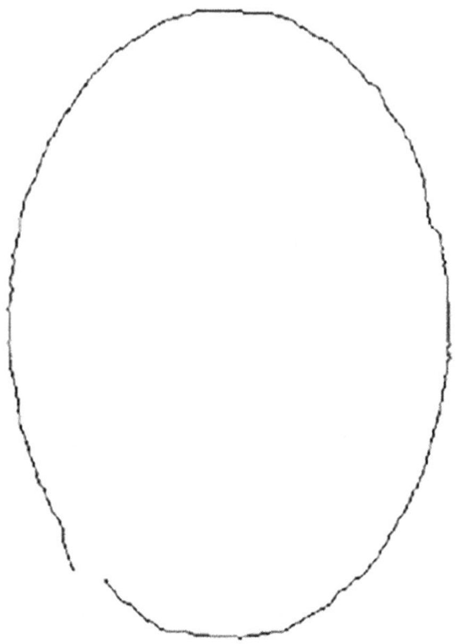

FIGURE 4.18 Using an elliptical structuring element with $a = 5$, $b = 3$, and an adjustable s.

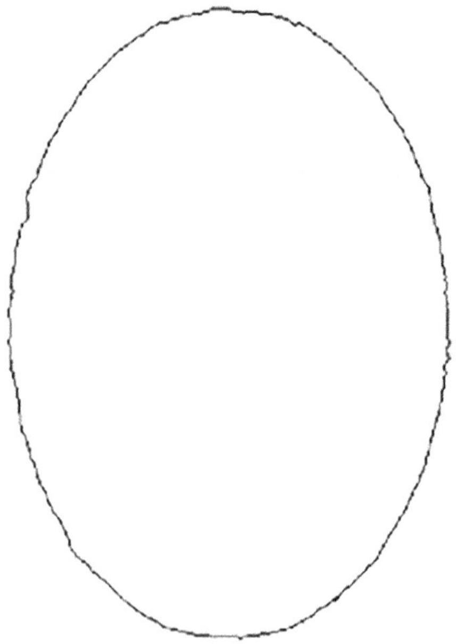

FIGURE 4.19 Using an elliptical structuring element with $a = 7$, $b = 3$, and an adjustable s.

and 4.17, Figures 4.18 and 4.19 are better in terms of elliptical smoothness. This is true because after each iteration, the range of pixels used to measure local slope is increased and more information from the original edge is taken into account.

4.5.3.4 Using Adjustable-Sized Elliptical Structuring Elements for Every Endpoint and Using Adjustable-Sized Range of Pixels to Measure Local Slope in Each Iteration (the Adaptive Morphological Edge-Linking Algorithm)

In this experiment, the adjustable-sized elliptical structuring elements with a changing from 5 to 7, $b = 3$, and an adjustable s are used.

Figure 4.20 shows the result of using the adaptive morphological edge-linking algorithm. Compared with Figures 4.16 and 4.19, Figure 4.20 seems about the same. However, the adaptive method has the advantage of adjusting a and s automatically, while the parameters in Figures 4.16 and 4.19 are fixed, i.e., they work well in a certain case, but may not work well in other cases.

Figure 4.21(a) shows the elliptical shape with added uniform noise. Figure 4.21(b) shows the shape after removing the noise. Figure 4.21(c) shows the result of using the adaptive morphological edge-linking algorithm. Compared with Figure 4.20, Figure 4.21(b) has several shortcomings. This is because after removing the added noise, the shape is changed at some places.

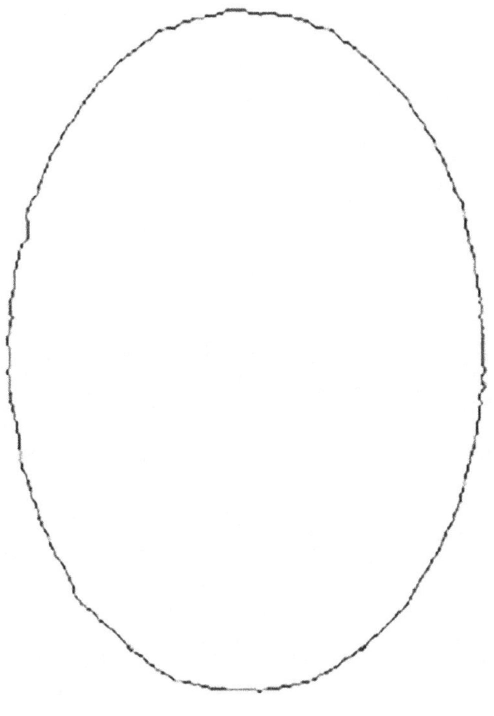

FIGURE 4.20 Using the adaptive morphological edge-linking algorithm.

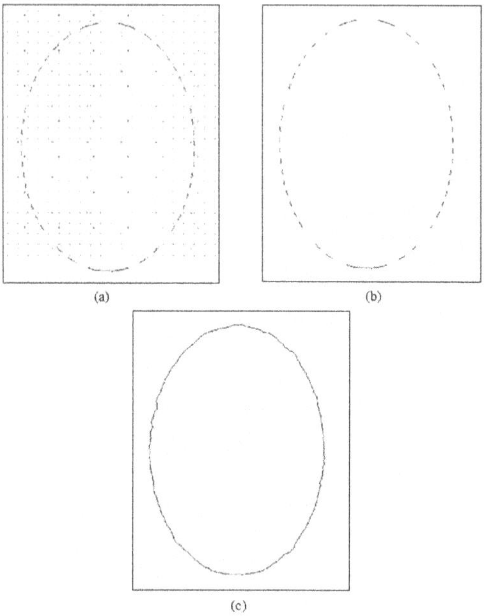

FIGURE 4.21 (a) The elliptical shape with added uniform noise, (b) the shape after removing noise, and (c) using the adaptive morphological edge-linking algorithm.

4.6 AUTOMATIC SEEDED REGION GROWING

Seeded region growing (SRG) is one of hybrid methods proposed by Adams and Bischof (1994). It starts with assigned seeds, and grow regions by merging a pixel into its nearest neighboring seed region. Mehnert and Jackway (1997) pointed out that SRG has two inherent pixel order dependencies that cause different resulting segments. The first order dependency occurs whenever several pixels have the same difference measure to their neighboring regions. The second order dependency occurs when one pixel has the same difference measure to several regions. They used parallel processing and re-examination to eliminate the order dependencies.

Fan et al. (2001) presented an automatic color image segmentation algorithm by integrating color-edge extraction and seeded region growing on the *YUV* color space. Edges in *Y*, *U*, and *V* are detected by an isotropic edge detector, and the three components are combined to obtain edges. The centroids between adjacent edge regions are taken as the initial seeds. The disadvantage is that their seeds are over-generated. Tao et al. (2007) incorporated the advantages of the mean shift segmentation and the normalized cut partitioning methods to achieve low computational complexity and real-time color image segmentation.

4.6.1 Overview of the Automatic Seeded Region Growing Algorithm

Figure 4.22 presents the overview of the automatic seeded region growing algorithm (Shih and Cheng, 2005). First, the color image is converted from *RGB* to YC_bC_r color

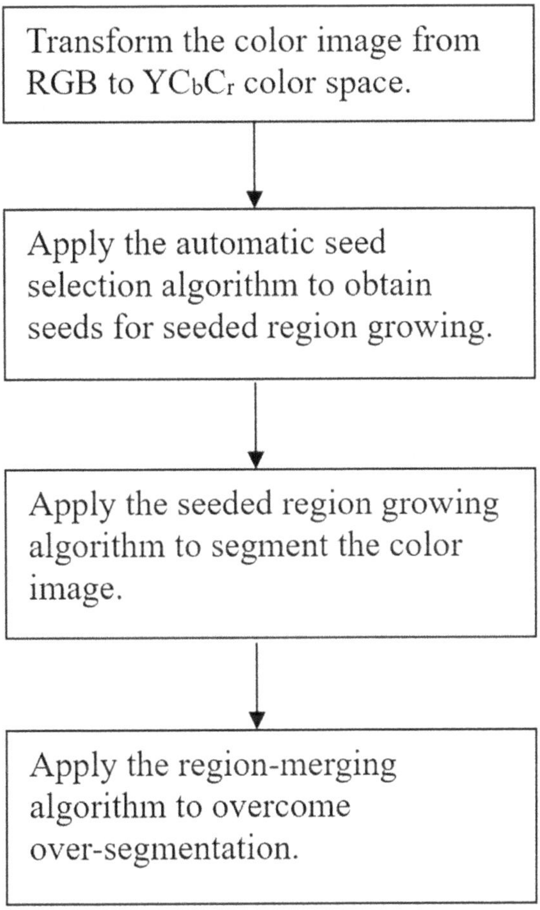

FIGURE 4.22 Outline of the automatic seeded region growing algorithm.

space. Second, automatic seed selection is applied to obtain initial seeds. Third, the seeded region growing algorithm is used to segment the image into regions, where a region corresponds to a seed. Fourthly, the region-merging algorithm is applied to merge similar regions, and small regions are merged into their nearest neighboring regions.

A captured color image is stored in *RGB* components. The *RGB* model is suitable for color display, but is not good for color analysis because of its high correlation among *R*, *G*, and *B* components. Besides, the distance in *RGB* color space does not represent the perceptual difference in a uniform scale. In image processing and analysis, the *RGB* color space is often converted into other color spaces. Cheng et al. (2001) compared several color spaces, including *RGB*, *YIQ*, *YUV*, normalized *RGB*, *HIS*, *CIE* $L^*a^*b^*$, and *CIE* $L^*u^*v^*$ for color image segmentation purposes. Every color space has its advantages and disadvantages. Garcia and Tziritas (1999) noticed that the intensity value Y has little influence on the distribution in the C_bC_r plane and the sample skin colors form a small and very compact cluster in the C_bC_r plane. The YC_bC_r color space has been widely chosen for skin color segmentation (Hsu et al., 2002). The reasons are as follows: (1) YC_bC_r color space is used in video compression standards (e.g., MPEG and JPEG); (2) the color difference of human perception is related to the Euclidean distance of the YC_bC_r color space; (3) the intensity and chromatic components can be easily and independently controlled.

The YC_bC_r color space is a scaled and offset version of YUV color space, where Y, U, and V denote luminance, color, and saturation, respectively. The C_b (C_r, respectively) is the difference between the blue (red, respectively) component and a reference value (Chai and Bouzerdoum, 2000). The transformation from RGB to YC_bC_r can be performed using the following equation, where R, G, and B are in the range of [0, 1], Y is [16, 235], and C_b and C_r are [16, 240].

$$\begin{bmatrix} Y \\ C_b \\ C_r \end{bmatrix} = \begin{bmatrix} 65.481 & 128.553 & 24.966 \\ -39.797 & -74.203 & 112 \\ 112 & -93.786 & -18.214 \end{bmatrix} \times \begin{bmatrix} R \\ G \\ B \end{bmatrix} + \begin{bmatrix} 16 \\ 128 \\ 128 \end{bmatrix}$$

4.6.2 The Method for Automatic Seed Selection

For automatic seed selection, the following three criteria must be satisfied. First, the seed pixel must have high similarity to its neighbors. Second, for an expected region, at least one seed must be generated in order to produce this region. Third, seeds for different regions must be disconnected.

The similarity of a pixel to its neighbors can be computed as follows. Considering a 3×3 neighborhood, the standard deviations of Y, C_b and C_r components are calculated using

$$\sigma_x = \sqrt{\frac{1}{9}\sum_{i=1}^{9}(x_i - \bar{x})^2}, \tag{5.27}$$

where x can be Y, C_b, or C_r, and the mean value $\bar{x} = \frac{1}{9}\sum_{i=1}^{9} x_i$. The total standard deviation is

$$\sigma = \sigma_Y + \sigma_{C_b} + \sigma_{C_r} \tag{5.28}$$

The standard deviation is normalized to [0, 1] by

$$\sigma_N = \sigma / \sigma_{\max} \tag{5.29}$$

where $\sigma_{\max}$ is the maximum of the standard deviation in the image. The similarity of a pixel to its neighbors is defined as

$$H = 1 - \sigma_N. \tag{5.30}$$

From the similarity, the first condition for the seed pixel candidate is defined as follows:

Condition 1: A seed pixel must have the similarity higher than a threshold value.

Second, the relative Euclidean distances (in terms of YC_bC_r) of a pixel to its 8 neighbors is calculated as

$$d_i = \frac{\sqrt{(Y - Y_i)^2 + (C_b - C_{b_i})^2 + (C_r - C_{r_i})^2}}{\sqrt{Y^2 + C_b^2 + C_r^2}} \qquad i = 1, 2, \cdots, 8. \tag{5.31}$$

From experiments, the performance of using relative Euclidean distance is better than using normal Euclidean distance. For each pixel, the maximum distance to its neighbors is calculated as

$$d_{\max} = \max_{i=1}^{8}(d_i). \tag{5.32}$$

From the maximum distance, the second condition for the seed pixel candidate is defined here.

Condition 2: A seed pixel must have the maximum relative Euclidean distance to its eight neighbors which is less than a threshold value.

A pixel is classified as a seed pixel if it satisfies the two conditions. In order to choose the threshold automatically in condition 1, Otsu's method (Otsu, 1979) is used. The threshold is determined by choosing the value that maximizes the discrimination criterion σ_B^2 / σ_w^2, where σ_B^2 is the between-class variance and σ_w^2 is the within-class variance. In condition 2, the value 0.05 is selected as the threshold based on our experiments.

Each connected component of seed pixels is taken as one seed. Therefore, the seeds generated can be one pixel or one region with several pixels. Condition 1 checks whether the seed pixel has high similarity to its neighbors. Condition 2 makes sure that the seed pixel is not on the boundary of two regions. It is possible that for one desired region, several seeds are detected to split it into several regions. The over-segmented regions can be merged later in the region-merging step. Figure 4.23(a) shows a color image, and (b) shows the detected seeds marked in red color. Note that the connected seed pixels are considered as one seed.

Regarding the threshold value for the relative Euclidean distance, if a higher value is used, a smaller number of pixels will be classified as seeds and some objects may be missed; oppositely, a higher number of pixels will be classified as seeds and different regions may be connected. For example, in Figure 4.24, if the threshold is 0.04, one flower is missed in *c*, that is because there is no seed pixel on the flower as shown in *b*. On the other hand, if the threshold is 0.08, then Figure 4.24(f) shows that some part of the boat is merged with the water since these seed pixels are connected as shown in *e*. Note that due to variant background colors appearing in the figures, different colors are used to display object boundaries.

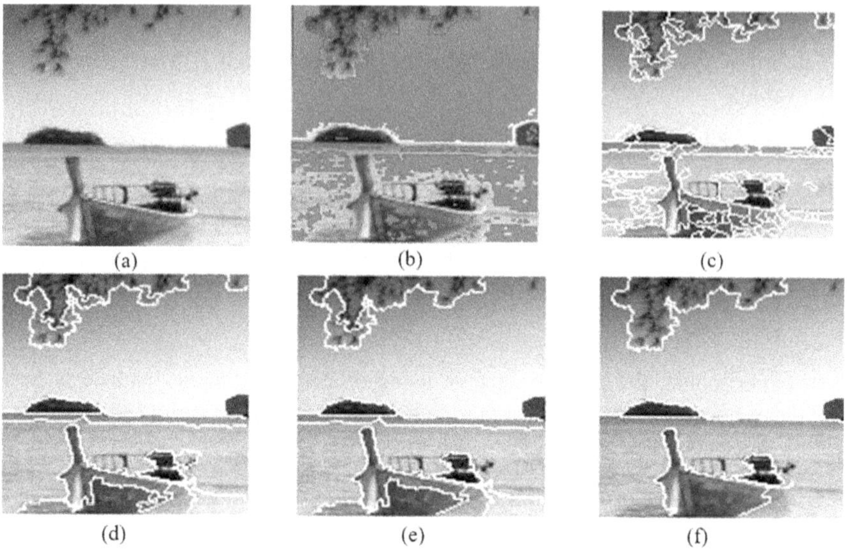

FIGURE 4.23 (a) Original color image, (b) the detected seeds are shown in red color, (c) seeded region growing result, (d) the result of merging adjacent regions with relative Euclidean distance less than 0.1, (e) the result of merging small regions with size less than 1/150 of the image, and (f) final segmented result.

FIGURE 4.24 Examples of setting threshold for relative Euclidean distance. (a) A color image, (b) the red pixels are seed pixels generated using threshold 0.04, (c) the segmented result, (d) a color image with boat, water, and sky, (e) the red pixels are the seed pixels generated using threshold 0.08, and (f) the segmented result.

4.6.3 The Segmentation Algorithm

Let $A_1, A_2, \ldots, A_i$ denote initial seeds and S_i denote the region corresponding to A_i. The mean of all seed pixels in S_i in terms of Y, C_b and C_r components is denoted as $\left(\bar{Y}, \bar{C}_b, \bar{C}_r\right)$. The segmentation algorithm is described as follows:

(1) Perform automatic seed selection.
(2) Assign a label to each seed region.
(3) Record neighbors of all regions in a sorted list T in a decreasing order of distances.
(4) While T is not empty, remove the first point p and check its 4-neighbors. If all labeled neighbors of p have a same label, set p to this label. If the labeled neighbors of p have different labels, calculate the distances between p and all neighboring regions and classify p to the nearest region. Then, update the mean of this region and add 4-neighbors of p, which are neither classified yet nor in T, to T in a decreasing order of distances.
(5) Perform region merging.

Note that in step 3, T denotes the set of pixels that are unclassified and are neighbors of at least one of the region S_i. The relative Euclidean distance d_i between the pixel i and its adjacent region is calculate by

$$d_i = \frac{\sqrt{\left(Y_i - \bar{Y}\right)^2 + \left(C_{b_i} - \bar{C}_b\right)^2 + \left(C_{r_i} - \bar{C}_r\right)^2}}{\sqrt{Y_i^2 + C_{b_i}^2 + C_{r_i}^2}}, \tag{5.33}$$

where $(\bar{Y}, \bar{C}_b, \bar{C}_r)$ are the mean values of Y, C_b, and C_r components in that region. In step 4, the pixel p with the minimum distance value is extracted. If several pixels have the same minimum value, the pixel corresponding to the neighboring region having the largest size is chosen. If p has the same distance to several neighboring regions, it is classified to the largest region. Figure 4.23(c) shows the result of the algorithm, where boundaries of regions are marked in white color.

Step 4 is explained graphically in Figure 4.25. Figure 4.25(a) is a small window of Figure 4.23(b), where the red pixels are the seed pixels and the green pixels are the 4-neighbor pixels of the seed regions. Therefore, the green pixels are actually the pixels stored in the sorted list T. In Figure 4.25(b), the white pixel is the one among all the green pixels that has the minimum distance to its adjacent regions. Therefore, the white pixel will be connected to its neighboring red seed region. In Figure 4.25(c), the black pixels are the three 4-neigbor pixels of the white pixel that

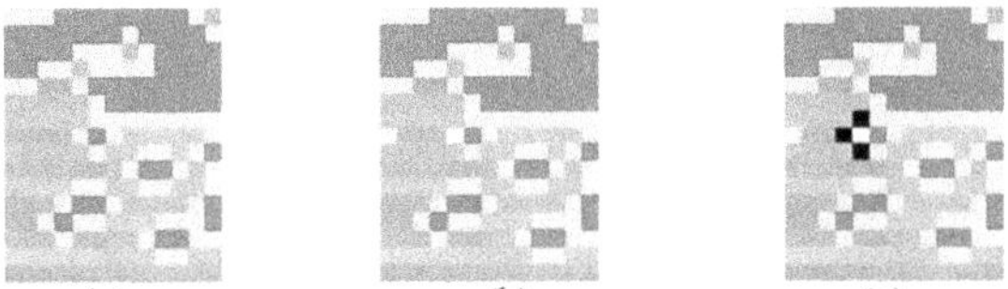

(a) (b) (c)

FIGURE 4.25 Step 4 of region growing. (a) A small window of Figure 4.23(b), where the red pixels are the seeds and the green pixels are the pixels in the sorted list T, (b) the white pixel is the pixel with the minimum distance to the seed regions, and (c) the white pixel is connected to the neighboring red region and the black pixels are added to T.

are added to T. Note that these three black pixels are neither classified nor previously stored in T.

It is possible that several seeds are generated to split a region into several small ones. To overcome the over-segmentation problem, region merging is applied. Two criteria are used: one is the similarity and the other is the size. If the mean color difference between two neighboring regions is less than a threshold value, the two regions are merged. Unfortunately, the result of region merging depends on the order in which regions are examined. The two regions having the smallest distance value among others are first checked. In each iteration, if this value is less than a threshold, the two regions are merged, and the mean of the new region and the distances between the new region and its neighboring regions are re-calculated. The process is repeated until no region has the distance less than the threshold. The color difference between two adjacent region R_i and R_j is defined as the relative Euclidean distance

$$d\left(R_i, R_j\right) = \frac{\sqrt{\left(\bar{Y}_i - \bar{Y}_j\right)^2 + \left(\bar{C}_{b_i} - \bar{C}_{b_j}\right)^2 + \left(\bar{C}_{r_i} - \bar{C}_{r_j}\right)^2}}{\min\left(\sqrt{\bar{Y}_i^2 + \bar{C}_{b_i}^2 + \bar{C}_{r_i}^2}, \sqrt{\bar{Y}_j^2 + \bar{C}_{b_j}^2 + \bar{C}_{r_j}^2}\right)}. \tag{5.345}$$

Based on the experiments, 0.1 is selected as the threshold for color similarity measurement.

Next, the size of regions is checked. If the number of pixels in a region is smaller than a threshold, the region is merged into its neighboring region with the smallest color difference. This procedure is repeated until no region has size less than the threshold. Based on the experiments, 1/150 of the total number of pixels is selected in an image as the threshold. Figure 4.23(d) shows the result of merging all similar neighboring regions, and (e) shows the result of merging small regions.

Since variations of image complexity could be large, some has a high number of regions and some has small-sized regions. In order to achieve better segmentation results, the merging by controlling the size of regions and the color difference between regions is conducted. In the region merging, the relative Euclidean difference threshold is 0.1 for initial merging. If this threshold is too high, some desired region may be merged with other regions; oppositely, there will be too many regions. The size of region threshold is set as 1/150 of the image size. By using too high threshold, some important objects may be merged to other regions, while too low threshold may lead to over-segmentation. In the further merging step, the threshold value 0.2 is chosen for final merging because all the regions with large difference from their surroundings and with the size larger than 1/150 of the image must be kept. It is observed that two regions are obviously different if the relative Euclidean distance is over 0.2. The size threshold is set as 1/10 of the image. All the regions with difference higher than 0.1 and size larger than 1/10 of the image must be kept. If this is not used, the sky and water will be merged as shown in Figure 4.23(f). If the number of regions in an image is known, this information will help in improving the segmentation result; however, it is usually unknown. Therefore, in the improved algorithm the final result is obtained by controlling the size of regions and the difference between regions.

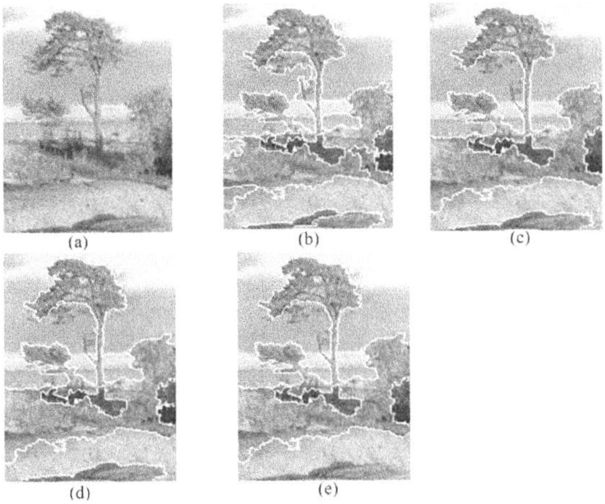

FIGURE 4.26 Example of choosing the relative Euclidean distance threshold for further region merging. (a) Original image, (b) the result after merging using 0.1 as the distance threshold value and 1/150 of the image size as the size threshold, (c) further merging using the relative Euclidean distance threshold 0.2, (d) further merging using threshold 0.15, and (e) further merging using threshold 0.25.

The relative Euclidean distance is used as the merging condition because it outperforms the fixed Euclidean distance. Figure 4.23(d) shows that using 0.1 as the threshold value for initial merging can merge the most similar regions, while the water and sky can be separated. If 0.15 is used, then the sky and water are merged. Figure 4.26(b) shows the segmented result of using 0.1 as the distance threshold and 1/150 of the image size as the size threshold. Figure 4.26(c) shows the result of using threshold 0.2 for further merging, which produces a reasonably good result. Figure 4.26(d) shows the result of using threshold 0.15 for further merging. The two small regions on the right should be merged. Figure 4.26(e) shows the result of using threshold 0.25 for further merging. It is observed that the two different regions at the bottom are merged.

4.6.4 Experimental Results and Discussions

The proposed segmentation algorithm is conducted on 150 color nature scene images randomly collected from the Internet. Figure 4.27 illustrates some of the proposed results. In Figure 4.28, the comparison with the JSEG algorithm developed by Deng and Manjunath (2001) is shown. Figures 4.28(a_3), (b_3), and (c_3) are their segmented results on Figures (a_1), (b_1), and (c_1), and Figures 4.28(a_2), (b_2), and (c_2) are the proposed results, respectively. Compared to Figure 4.28(a_2), Figure 4.25(a_3)'s nose of baboon is separated into several regions and its background is over-segmented. Their JSEG algorithm is downloaded from the website and Figures 4.28(d_3) and (e_3) are their segmented results.

The results of the segmentation algorithm are less noisy as compared to the results in Deng and Manjunath (2001) and Fan et al. (2001). This is due to the fact that the

FIGURE 4.27 Some results of the color image segmentation algorithm.

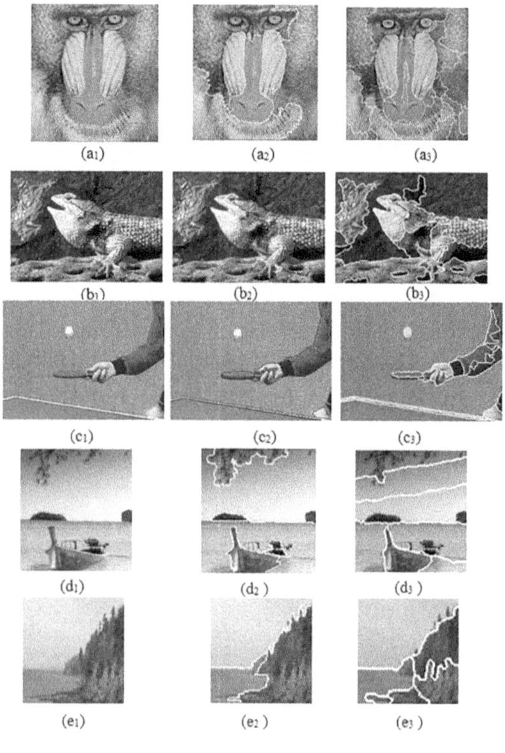

FIGURE 4.28 The first column shows the original images, the second column shows our segmented results, and the third column shows the results of JSEG algorithm.

merging algorithm uses the dynamic control over the size of regions and the difference between regions. However, the merging step in Deng and Manjunath (2001) and Fan et al. (2001) uses a fixed threshold value. In Figure 4.28(d_3), two objects are missed in Fan et al. (2001). The reason is that either there is no seed generated in these objects or the merging threshold is too high so that they are merged to the surroundings. Compared to Figure 4.28(e_3), (e_2) produces more detailed and accurate boundaries. The reason is that in Fan et al. (2001), color quantization is applied to form the class-map so that some details of color information may be lost.

The comparison with the segmentation algorithm by Fan et al. (2001) is shown in Figure 4.29. Figures 4.29(a_3), (b_3), (c_3), and (d_3) are their respectively segmented results on Figures (a_1), (b_1), (c_1), and (d_1), and Figures 4.29(a_2), (b_2), (c_2), and (d_2) are the proposed results. By observing Figure 4.29(a_3), their result is over-segmented and the hair is inappropriately merged to background. By comparing Figures 4.29(b_2), (c_2), and (d_2) with (b_3), (c_3), and (d_3), respectively, their results are over-segmented.

There are mainly two disadvantages in the proposed algorithm. First, although using the fixed threshold values can produce reasonably good results as shown in Figures 4.27–4.30, it may not generate the best results for all the images. From Figures 4.28(b2), (c2), and (d2) and Figure 4.29(a2), some small objects are missed. Some applications require more detailed information. Users can specify these threshold values according to their applications. For example, if the threshold values are adjusted for the size of objects and color differences, more detailed information

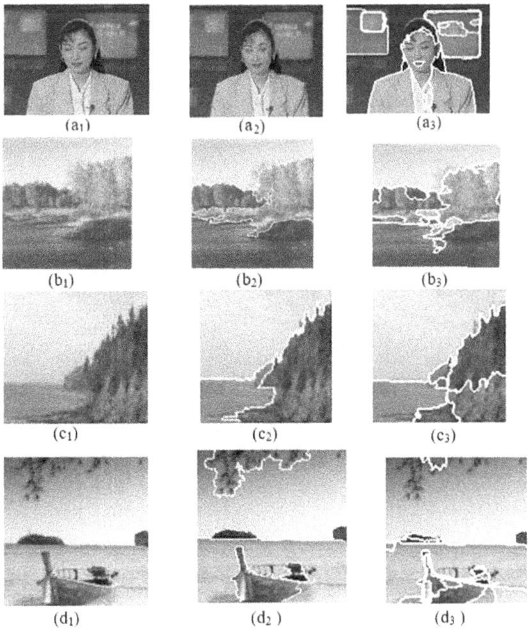

FIGURE 4.29 The first column on the left side shows the original images, the second column shows our segmented results, and the third column shows the results of Fan's algorithm.

FIGURE 4.30 The segmented results with more details by adjusting the size of the minimum objects and the threshold value for the relative Euclidean distance.

FIGURE 4.31 (a) An image with highly colored texture, (b) our segmented result showing the tree is merged to the flower region, and (c) the results of Fan et al. (2001).

can be segmented as shown in Figure 4.30. Second, when an image is highly color textured (i.e., there are numerous tiny objects with mixed colors), the proposed algorithm may fail to obtain satisfied results because the mean value could not represent the property of the region well. Figure 4.31 shows an example that the proposed algorithm performs worse than the method in Fan et al. (2001). Figure 4.31(b) shows that the tree is merged to the flower.

The time complexity for the segmentation algorithm consists of three components: seed selection, region growing, and region merging. In automatic seed selection, calculating the standard deviation and maximum distance for each pixel takes $O(n)$, where n is the total number of pixels in an image. In region growing, each unclassified pixel is inserted into the sorted list exactly once. Checking neighboring regions and calculating distances can be done in constant time. Putting a pixel into the sorted list requires $\log(n)$. Therefore, it takes $O(n\log(n))$ for region growing.

In region merging, calculating the differences between regions takes $O(m^2)$, where m is the number of regions. To calculate sizes for all the regions, it takes $O(n)$. Usually, m is much less than n. To merge two regions, it is needed to label the pixels in the two regions, calculate the mean for the new merged region, and calculate the distances between this and other regions. Therefore, it takes $O(n+m) \approx O(n)$ to merge two regions. The complexity for region merging is $O\left(m^2+n+mn\right) \approx O(mn)$. Therefore, the total time complexity for the proposed algorithm is $O(n\$\$+n\log(n)+mn) \approx O((m+\log(n))n)$.

4.7 A TOP-DOWN REGION DIVIDING APPROACH

Histogram-based and region-based segmentation approaches have been widely used in image segmentation. Difficulties arise when we use these techniques, such as the selection of a proper threshold value for the histogram-based technique and the over-segmentation followed by the time-consuming merge processing for the region-based technique. To provide efficient algorithms that not only produce better segmentation results but also maintain low computational complexity, a novel top-down region dividing based approach is developed for image segmentation, which combines the advantages of both histogram-based and region-based approaches.

4.7.1 Introduction

The histogram-based (or feature-based) segmentation technique produces a binary image based on the threshold value (Tobias and Seara, 2002). The intensities of object and background pixels tend to cluster into two sets in the histogram. The histogram will be bimodal by a threshold value, which is selected from the valley between the two sets. The optimal segmentation intends to find the threshold value that minimizes the misclassification. If the threshold is too high, many object pixels will be lost and object contour will be severely destroyed. Although the complexity of histogram-based technique is low, the threshold selection is difficult, especially when the histogram is multi-modal. Furthermore, the histogram-based technique only considers the feature image (histogram) without checking the spatial relationship among connected pixels.

Watershed-based segmentation, a region-based approach, uses a bottom-up strategy that segments an image into several small regions, followed by a merge procedure. The immersion-based system (Vincent and Soille, 1991) and the drainage rainfall system (Chibanga et al., 2001) are two approaches for performing watershed transformation. It considers an image as a topographic surface and the image intensity as the altitude. The drop of water will progressively fill up the ascending catchment basins from the minima of lowest altitude (lowest intensity) of the surface. Each pixel will flow along a descending path to a local minimum. When the altitude of water gradually increases, two catchment basins will reach at some points, called *watershed points*. A collection of watershed pixels on the contour is defined as the *watershed line*. Since the watershed algorithm is highly sensitive to the local minimum, it usually results in over-segmentation. In other words, there are overcrowded regions segmented in an image. Furthermore, in order to merge similarly smaller connected regions, the *region adjacency graph* (RAG) is used for region growing. Although the watershed-based image segmentation provides better results than the histogram-based approach, its computational complexity is high.

In order to provide more efficient algorithms that not only obtain better results, but also maintain low complexity, a novel top-down region dividing (TDRD) based approach is presented to iteratively divide sub-regions if the size of a sub-region is larger than a predefined threshold or the homogeneity of a sub-region is larger than a predefined threshold.

4.7.2 Overview of the TDRD-Based Image Segmentation

4.7.2.1 Problem Motivation

The histogram-based image segmentation method, although its complexity is low, does not consider the spatial relationship of neighborhood and may fail in some cases. Figure 4.32(a) shows an image containing dark and bright parts on the left and right areas, respectively. In each part, gray values gradually increase from left to right. For example, the pixels on the leftmost column of left part are 32, and are increased by 1 on each column to the right, until the pixels on the rightmost column of left part are 96. Similarly, the pixels on the leftmost and the rightmost columns of right part are 160 and 224, respectively. In Figure 4.32 (b), we reverse a small rectangular area on each part by swapping the pixels on the corresponding right and left columns within the rectangle. Figure 4.32(c) shows the histogram of the original image. Since we only exchange the positions of pixels, the histogram remains the same. After thresholding, the same result as shown in Figure 4.32(d) is obtained for both Figures (a) and (b).

The watershed-based image segmentation method, although offering better performance than the histogram-based method in general, may fail in some cases. Figure 4.33(a) shows an image composed of five rectangular layers with different gray values. From the property of watershed, the central lowest intensity box is considered as the catchment basin, and after watershed-based image segmentation, we obtain Figure 4.33(b), which only show one object in existence. It is because the watershed approach could destroy the spatial structure of an image.

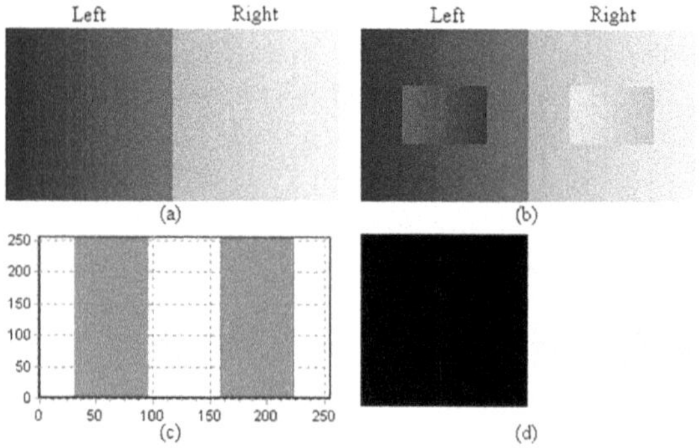

FIGURE 4.32 An example shows the problem of histogram-based image segmentation.

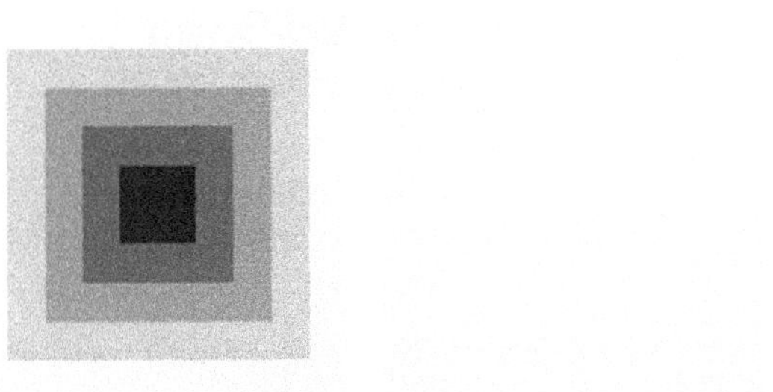

(a) (b)

FIGURE 4.33 A counterexample shows the problem of watershed-based image segmentation.

4.7.2.2 The Top-Down Region Dividing Based Image Segmentation

In order to perform image segmentation in arbitrary cases successfully and reduce computational complexity, the top-down region dividing (TDRD) based method is proposed. Its concept is opposite to the watershed algorithm. The watershed algorithm finds local minimum areas as the catchment basins and segments each catchment basin. Then, the region growing procedure is repeated to merge connected similar catchment basins. It is a bottom-up approach that first segments an image as detailed as possible and then merges the segments from small to large regions. The TDRD-based segmentation consists of two main steps: region dividing and sub-region evaluation procedures, to iteratively segment an input region into several sub-regions by evaluating both feature and spatial properties. In the region dividing procedure, the previously developed image simplification algorithm (Wu et al., 2006a) is used to segment an input region. First, two histograms of each input region are used to determine suspicious intensities. Second, suspicious pixels are obtained using the suspicious intensities. A pixel will be considered as suspicious if its intensity is suspicious. Third, the final intensity of a suspicious pixel can be derived by analyzing the fuzziness of the suspicious pixel and its 8-connected neighbors. In the sub-regions evaluation procedure, the size and homogeneity of each divided sub-region will be evaluated to determine whether the sub-region has to be continuously divided. The TDRD-based image segmentation algorithm is presented here and its flowchart is shown in Figure 4.34.

TDRD-Based Image Segmentation Algorithm:

1. Consider an input image as a big region and add it into a dividing list.
2. Region dividing procedure: Segment each region in the dividing list into several sub-regions.
3. Sub-region evaluation procedure: On each divided obtained sub-region, determine whether a sub-region needs to be continuously divided. If it does, add the sub-region into the dividing list.
4. Repeat steps 2 and 3 until the dividing list is empty.
5. Output the segmentation result by collecting all the sub-regions.

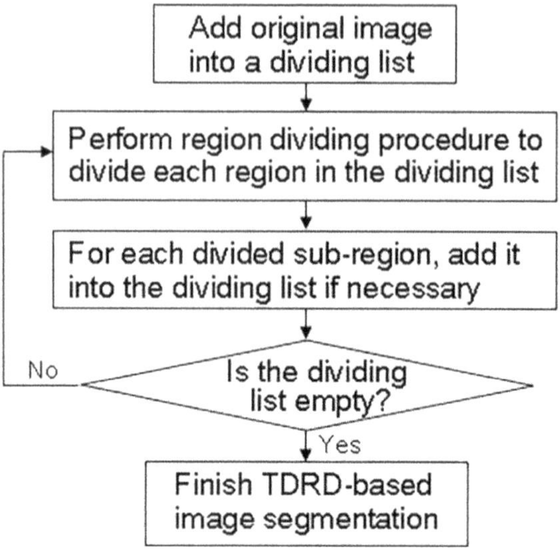

FIGURE 4.34 The flowchart of our TDRD-based image segmentation.

4.7.3 The Region Dividing and Sub-Region Examination Strategies

4.7.3.1 Region Dividing Procedure

The region dividing procedure combines the advantages from histogram-based and region-based segmentation methods. It includes three steps: (a) suspicious intensities determination, (b) suspicious pixels determination, and (c) final intensity determination.

4.7.3.1.1 Suspicious Intensities Determination

The suspicious intensities are determined by comparing the histograms of two transformed images using histogram equalization. As mentioned earlier, the selection of an appropriate threshold value is challenging in the histogram-based approach. Many researchers have tried to solve this problem. However, it is really difficult to find a general rule for all cases to determine the threshold value for optimal segmentation because this method lacks of the relationship between a pixel and its neighbors. To solve this problem, we propose a new idea by avoiding the selection of such a threshold value on the histogram. That is, we do not intend to find a general rule for determining the threshold value; instead, we determine the suspicious intensities by comparing histograms of two transformed image using histogram equalization (HE) as shown in Figure 4.35, and determine the final intensity of suspicious pixels by checking their local spatial information.

Figure 4.35(a) shows the thresholding of the histogram after HE into binary. The input grayscale image of 256 colors will become a binary image of 2 colors. Let $Num(m)$ denote the total number of pixels of gray level m in an image. In an ideal case, we obtain the intensity, say i, such that $S_1 = S_2$ where $S_1 = \sum_{m=0}^{i} Num(m)$

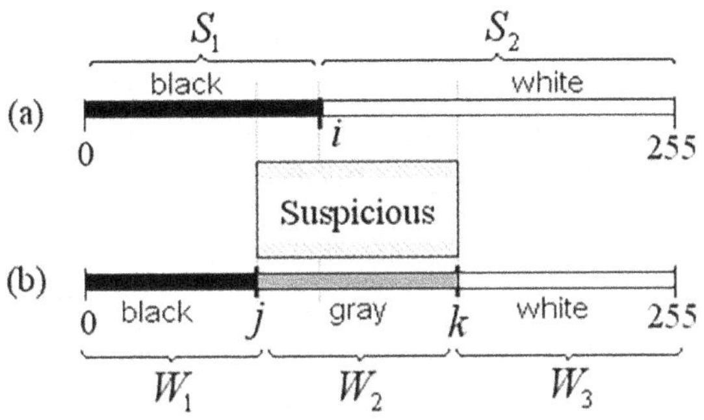

FIGURE 4.35 The idea of finding suspicious intensities by the DHE.

and $S_2 = \sum_{m=i+1}^{255} Num(m)$. Similarly, we perform another histogram graph of a translated image after HE, in which the intensity range is changed from [0,255] to [0,2] (i.e., 3 colors) as shown in Figure 4.35(b). Also, we can find two intensities, say j and k, such that $W_1 = W_2 = W_3$ where $W_1 = \sum_{m=0}^{j} Num(m)$, $W_2 = \sum_{m=j+1}^{k} Num(m)$ and $W_1 = \sum_{m=k+1}^{255} Num(m)$.

Let *OriGray* and *NewGray* indicate the original and the translated intensities, respectively. In Figure 4.35(a), $NewGray = 0$ (black) if $OriGray \leq i$, and $NewGray = 1$ (white) for the others. Similarly, in Figure 4.35(b), $NewGray = 0$ (black) if $OriGray \leq j$, $NewGray = 1$ (gray) if $j < OriGray \leq k$, and $NewGray = 2$ (white) if $OriGray > k$. The *suspicious intensity* is defined as the one whose new intensities are different between two histograms of an image. For example, the suspicious intensities in Figure 4.35 posses the following two properties: *NewGray* is black in Figure 4.35(a) but gray in Figure 4.35(b), and *NewGray* is white in Figure 4.35(a) but gray in Figure 4.35(b).

4.7.3.1.2 Suspicious Pixels Determination

A pixel is considered as *a suspicious pixel* if its intensity is one of the suspicious intensities. Figure 4.36 shows an example of determining suspicious pixels based on the suspicious intensities. Figure 4.36(a) is the original image. Let HE_i indicate a transformed image after HE, meaning that there are only *i* colors remaining. Figures 4.36(b) and (c) are obtained by translating Figure 4.36(a) into HE_2 and HE_3 after performing histogram equalization, respectively. Figure 4.36(d) is obtained by checking the pixel intensity in the corresponding location between Figures 4.36(b) and (c), where white and gray pixels are used to indicate suspicious pixels. Note that the white pixels in Figure 4.36(d) mean that the intensity is white in Figure 4.36(b) but gray in Figure 4.36(c), and the gray pixels in Figure 4.36(d) mean that the intensity is black in Figure 4.36(b) but gray in Figure 4.36(c).

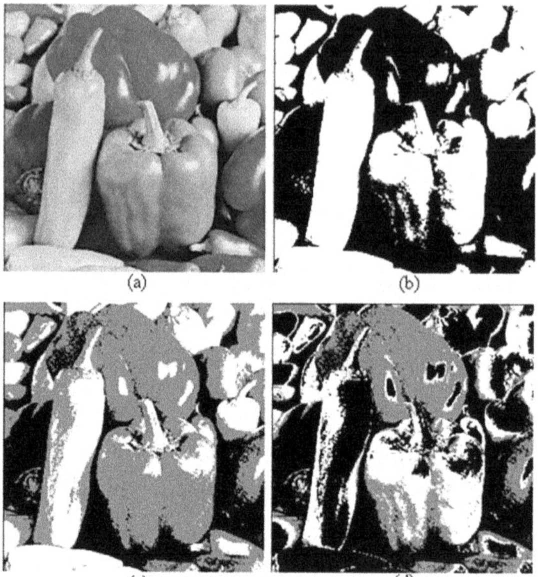

FIGURE 4.36 An example of determining the suspicious pixels.

4.7.3.1.3 Final Intensity Determination

As aforementioned, it is difficult to determine the proper threshold in the histogram-based approach. In order to avoid the difficulty, a range of suspicious intensities are first determined based on the histogram analysis. Once the suspicious pixels are derived, the local spatial information is used to determine the final intensity of suspicious pixels. We analyze the pixels and their neighbors to determine the final intensity. The weighted fuzzy mean (WFM) filter introduced by Lee et al. (1997) works well in image enhancement, especially for removing additive impulse noise. Therefore, we develop our final intensity determination (FID) based on the WFM filter as follows.

In the first step, three intensities *i*, *j*, and *k* are derived. For each of them, a membership function is generated as shown in Figure 4.37. Let FM_i, FM_j, and FM_k denote the fuzzy subsets for each intensity, respectively. The fuzzy subsets used are the *L*-*R* type fuzzy member (Zimmermann, 1991) as

$$f(x)=\begin{cases} L\ (\dfrac{m-x}{\alpha}) & \text{for } x \leq m \\ R\ (\dfrac{m-x}{\beta}) & \text{for } x \geq m \end{cases} \tag{4.35}$$

where the functions $L(y)=R(y)=\max(0,1-y), f(x)$ is represented by a triple $[m,a,\beta]$, and α and β are used to determine the slopes for the values respectively smaller than and larger than *m*. Note that once *i*, *j*, and *k* are obtained,

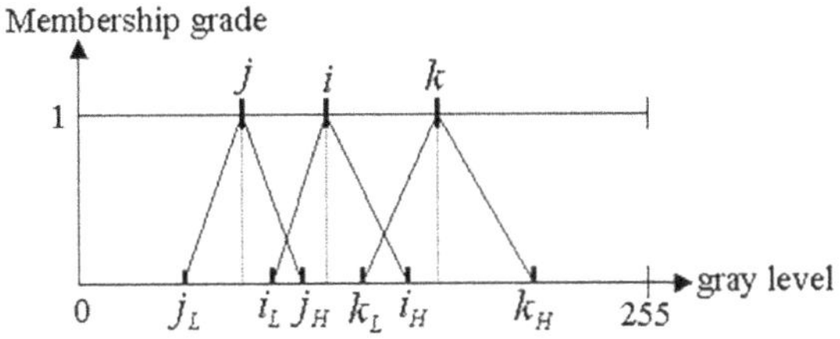

FIGURE 4.37 The membership functions of intensities *i*, *j*, and *k*.

the three fuzzy subsets can be derived by the three triples $[m_i, \alpha_i, \beta_i]$, $[m_j, \alpha_j, \beta_j]$, and $[m_k, \alpha_k, \beta_k]$ using the following formulas, where $0.5 \le c_1 \le 1$.

$$m_i = i,\ a_i = c_1(i-j), \beta_i = c_1(k-i),$$
$$m_j = j,\ a_j = c_1(i-j), \beta_j = c_1(i-j),$$
$$m_k = k,\ a_k = c_1(k-i), \beta_k = c_1(k-i).$$

Let $p(n,m)$ denote a suspicious pixel located at (n,m) of an image. Three fuzziness evaluations, $Fuzzy_i(n,m)$, $Fuzzy_j(n,m)$, and $Fuzzy_k(n,m)$, for their corresponding intensities, *i*, *j* and *k*, are calculated for determining the proper intensity.

Fuzziness evaluation for Intensity *i*:

$$\text{If } \sum_{u=-1}^{1} \sum_{v=-1}^{1} \sum_{i} FM_i(\text{Ori}(n+u, m+v)) \neq 0,$$

$$\text{then } \text{Fuzzy}_i(n,m) = \frac{\sum_{u=-1}^{1} \sum_{v=-1}^{1} FM_i(\text{Ori}(n+u, m+v)) * \text{Ori}(n+u, m+v)}{\sum_{u=-1}^{1} \sum_{v=-1}^{1} FM_i(\text{Ori}(n+u, m+v))},$$

else $\text{Fuzzy}_i(n,m) = 0$. where $Ori(n,m)$ is the intensity of pixel *p* in the original image. We obtain the other two fuzziness evaluations, $Fuzzy_j(n,m)$, and $Fuzzy_k(n,m)$, similarly.

In order to determine the proper intensity of a suspicious pixel, we use *fuzzy estimator* (FE) (Lee et al., 1997) to evaluate the maximum-likelihood. Let FM_{FE} be the fuzzy set of fuzzy estimator, which is a fuzzy interval *I* of *L-R* type as shown in Figure 4.38 and denoted by $I = [FE_L, FE_H, a, \beta]_{LR}$, where FE_L and FE_H are the low and high gray levels, and α and β are used to determine the slopes for the values smaller than FE_L and larger than FE_H, respectively.

$$f_{LR\_I}(x) = \begin{cases} L(\frac{FE_L - x}{\alpha}) & \text{for } x \le FE_L \\ 1 & \text{for } FE_L \le x \le FE_H \\ R(\frac{x - FE_H}{\beta}) & \text{for } x \ge FE_H \end{cases} \tag{4.36}$$

Note that FE_L and FE_R in Figure 4.38 can be respectively calculated by $FE_L = j - c_2(i - j)$ and $FE_H = k + c_2(k - i)$, where $0 \le c_2 \le 0.5$.

The fuzzy estimator of a suspicious pixel can be calculated by

$$Fuzzy_{FE}(n,m) = \frac{\sum_{u=-1}^{1}\sum_{v=-1}^{1} f_{LR\_I}(Ori(n+u,m+v)) \times Ori(n+u,m+v)}{\sum_{u=-1}^{1}\sum_{v=-1}^{1} f_{LR\_I}(Ori(n+u,m+v)).}$$

Let "*New*" be the adjusted image after performing the final intensity determination (FID) on an original region. The FID algorithm is presented here, and Figure 4.39 shows its result where the whole image is considered as an original region. Figure 4.39(a) is obtained by performing FID on Figure 4.36(d), and Figure 4.39(b) is the result of Figure 4.39(a) after edge detection. Note that Figure 4.39(a) is also

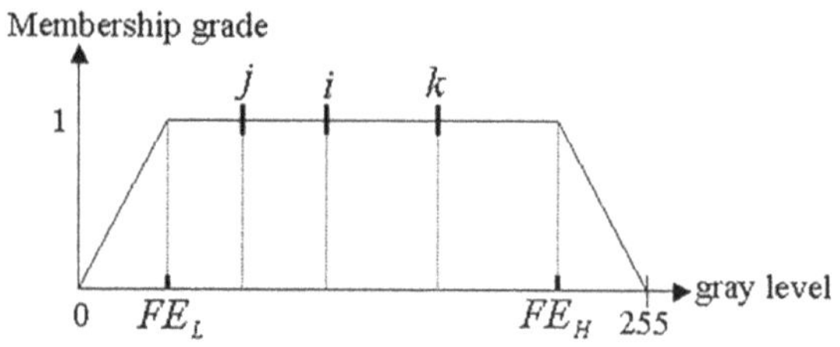

FIGURE 4.38 The membership functions of fuzzy interval.

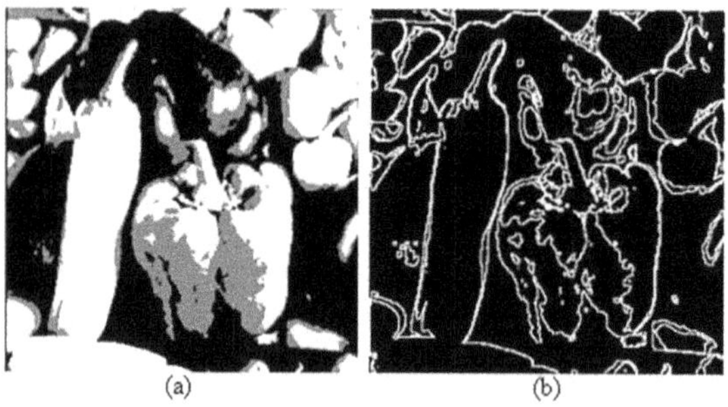

FIGURE 4.39 (a) The result of performing FID on Figure 4.36(d), (b) the result after edge detection.

the result after performing the TDRD-based image segmentation, in which all sub-regions derived only require at most one dividing procedure.

The Final Intensity Determination Algorithm:

1. Setup *L-R* type fuzzy subsets: FM_i, FM_j, FM_k, and FM_{FE} using arrays. For example,

   $FM_i = array\_fm_i[h]$, where $0 \le h \le 255$ and $0 \le array\_fm_i[h] \le 1$.
2. For each suspicious pixel (*n*,*m*), calculate $Fuzzy_i(n,m)$, $Fuzzy_j(n,m)$, $Fuzzy_k(n,m)$ and $Fuzzy_{FE}(n,m)$.
3. If $(|Fuzzy_j(n,m) - Fuzzy_{FM}(n,m)| < |Fuzzy_i(n,m) - Fuzzy_{FM}(n,m)|)$ and $(|Fuzzy_j(n,m) - Fuzzy_{FM}(n,m)| < |Fuzzy_k(n,m) - Fuzzy_{FM}(n,m)|)$ then $Fin(n,m) = 0$

   else If $(|Fuzzy_i(n,m) - Fuzzy_{FM}(n,m)| \le |Fuzzy_j(n,m) - Fuzzy_{FM}(n,m)|)$ and $(|Fuzzy_i(n,m) - Fuzzy_{FM}(n,m)| \le |Fuzzy_k(n,m) - Fuzzy_{FM}(n,m)|)$ then $Fin(n,m) = 2$

   else $New(n,m) = 1$.
4. Repeat steps 1 to 3 to determine the final intensity for all suspicious pixels.

4.7.3.2 Sub-Region Examination Strategy

In the top-down dividing image segmentation, two factors of size and homogeneity are used to determine whether a sub-region needs to be further segmented. The size factor is the total number of pixels in the sub-region, and the homogeneity factor is its standard deviation. The sub-region is continuously segmented if its size is larger than a pre-defined size threshold and its homogeneity value is larger than a pre-defined standard deviation threshold. Otherwise, the sub-region is not considered for further segmentation. Note that the criteria are varied with the purpose of using the TDRD-based segmentation algorithm. For example, for an application to the breast boundary segmentation, only one decomposing iteration is enough since we do not need to consider the detailed patterns of the breast. If the TDRD-based image segmentation is merely used to reduce the gray levels of an image such as the texture feature extraction in the application of mass detection, more decomposing iterations would be required.

4.7.4 Experimental Results

We perform our algorithm using two dividing iterations on an input image in Figure 4.36(a). After obtaining Figure 4.39(a) which performs one dividing iteration, we continuously segment all the sub-regions to achieve two dividing iterations for an input image. The results of performing the second iteration of our TDRD-based image simplification are shown in Figure 4.40. Since there are three classes in white, gray, and black in Figure 4.39(a), the results of the second iteration are shown here for each class.

(1) Figures 4.40 (a1)–(a4): The processing on the "black" class in Figure 4.39(a).
(2) Figures 4.40 (b1)–(b4): The processing on the "gray" class in Figure 4.39(a).
(3) Figures 4.40 (c1)–(c4): The processing on the "white" class in Figure 4.39(a).

FIGURE 4.40 An example of performing the second iteration of the TDRD-based image segmentation on each class of Figure 4.39(a).

Figures 4.40(a1) and (a2) are the HE_2 and HE_3 of the "black." Figure 4.40(a3) shows the suspicious pixels represented as "light gray" and "dark gray." Note that the white pixels in Figure 4.40 indicate "ignored pixels," which are not considered during the second iteration of the TDRD-based image segmentation. Figure 4.40(a4) shows the result after performing the FID procedure on the suspicious pixels of Figure 4.40(a3). Similarly, Figures (4.40(b1), (b2), (b3), and (b4)) and (4.40(c1), (c2), (c3), and (c4)) correspond to (HE_2 , HE_3 , the suspicious-pixel image, and the final result) after performing the second iteration of the TDRD-based segmentation respectively on the "gray" and "white" areas of Figure 4.39(a). Figure 4.41 shows the final result of collecting all sub-regions in Figures 4.40(a4), (b4), and (c4).

Figure 4.42 shows the results when we fix the homogeneity threshold (Th_H) but change the size threshold (Th_S). Figure 4.42(a) shows four original images: vegetable, Lena, camera man, and baboon. Figures 4.42(b), (c) and (d) show the results after applying the TDRD-based image segmentation with Th_S = 500, 250, and 50, respectively. The Th_H is fixed to be 10. The number of connected component (NCC) and the within-class standard deviation (WCSD) in each condition are listed

FIGURE 4.41 The final result of performing the second iteration of the TDRD-based image segmentation.

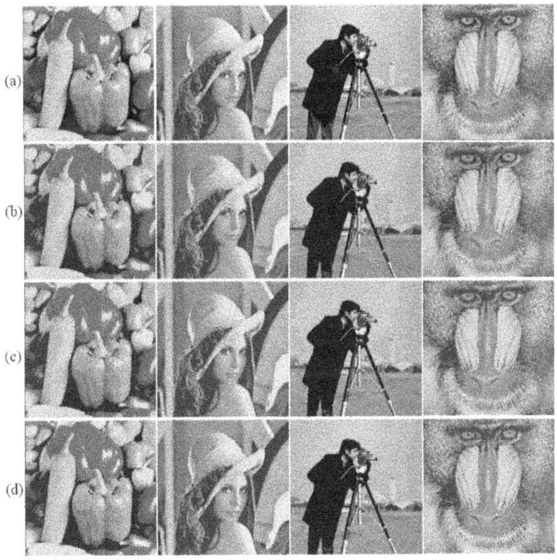

FIGURE 4.42 An example of performing the TDRD algorithm in different Th_S when Th_H is fixed.

in Table 4.1. Note that the WCSD is defined here and is used to represent the homogeneity of all the segmented sub-regions.

$$WCSD = \sqrt{\sum_{i=1}^{m} \left(\frac{N_i \sigma_i^2}{N_{\text{Total}}} \right)}, \tag{4.37}$$

TABLE 4.1
The Number of Connected Component (NCC) and the Within-Class Standard Deviation (WCSD) of Performing the TDRD Algorithm in Conditions (I) Th_H = 10, Th_S = 500, (II) Th_H = 10, Th_S = 250, and (III) Th_H = 10, Th_S = 50

		Vegetable	Lena	Camera Man	Baboon
(I)	NCC	1556	2294	1407	8876
	WCSD	11.84	8.82	11.69	12.57
(II)	NCC	2055	2226	1519	9296
	WCSD	10.92	8.62	11.47	12.34
(III)	NCC	3555	3401	2396	11476
	WCSD	9.14	7.91	10.10	11.38

FIGURE 4.43 An example of performing the TDRD algorithm in different Th_H when Th_S is fixed.

where m is the total number of sub-regions in the segmented image, N_i is the number of pixels in the i-th sub-region, and $\sigma_i^2 = \sum_{k=1}^{N_i} (I_i(k) - \overline{I_i})$ in which $\overline{I_i}$ and $I_i(k)$ are the average intensity and the intensity of the k-th pixels in i-th sub-region, respectively. The WCSD measures the average homogeneity of all sub-regions in a given image. The ideal value of WCSD is 0, which indicates that each sub-region has the same pixel intensity. In practice, the WCSD is subject to the range of pixel intensities. Here all images are normalized by the pixel intensities ranging from 0 to 255. We estimate that the good WCSD range is from 0 to 12, and it would be bad if the WCSD value is larger than 24.

Figure 4.43 shows the results when we fix the size threshold (Th_S) but change the homogeneity threshold (Th_H). Figures 4.43(a), (b) and (c) show the results after applying the TDRD-based image segmentation with Th_H= 40, 25, and 10, respectively. The Th_S is fixed to be 100. The NCC and the WCSD in each condition are presented in Table 4.2.

We compare the performance of the TDRD-based image segmentation against the watershed segmentation provided in Matlab and the histogram-based image segmentation presented by Tobias and Seara (2002). Since the watershed algorithm contains the problem of over-segmentation, the thresholds (Th_S and Th_H) are intentionally set to be as low as possible, such that the NCC derived in the TDRD approach is close to the one derived in watershed. Figure 4.44 shows the segmentation results of the

TABLE 4.2
The Number of Connected Component (NCC) and the Within-Class Standard Deviation (WCSD) of Performing the TDRD Algorithm in Conditions (I) Th_S = 100, Th_H = 40, (II) Th_S = 100, Th_H = 25, and (III) Th_S = 100, Th_H = 10

		Vegetable	Lena	Camera Man	Baboon
(I)	NCC	230	267	365	1586
	WCSD	22.95	19.12	16.97	21.17
(II)	NCC	753	267	694	3305
	WCSD	14.59	19.12	13.84	17.64
(III)	NCC	3198	2958	2213	10290
	WCSD	8.89	7.50	10.44	11.85

FIGURE 4.44 The results of the watershed and the TDRD-based image segmentations.

watershed and our TDRD-based approaches. Figures 4.44(a) and (b) show the results of watershed segmentation without and with displaying the edges, respectively. Similarly, Figures 4.44(c) and (d) show the results of our TDRD-based segmentation without and with displaying the edges, respectively. The thresholds, (Th_S, Th_H), are (100,8), (100,8), (80,2), and (30,16) for images "Vegetable," "Lena," "Camera Man," "Baboon," respectively. The detailed results are listed in Table 4.3. Note that the watershed segmentation will somehow destroy the spatial structure of an image since the strong edges will be ignored if there is only one catchment basin in the area as shown in Figure 4.33. In Figure 4.44(a), it shows again the drawback of watershed segmentation. Our TDRD-based segmentation will avoid the spatial-structure destruction problem, such that the WCSD derived in our algorithm is much lower than the one from the watershed segmentation.

The histogram-based image segmentation by Tobias and Seara (2002) segments a given image into a binary image. In order to compare our TDRD-based and their histogram-based algorithm, only one decomposing iteration is performed. Table 4.4 shows the comparison.

TABLE 4.3
The Number of Connected Component (NCC) and the Within-Class Standard Deviation (WCSD) of Performing the Watershed and the TDRD Segmentations in Four Images

		Vegetable	Lena	Camera Man	Baboon
Watershed	NCC	3686	3744	4287	5663
	WCSD	28.51	24.80	21.70	23.39
TDRD	NCC	3782	3443	4341	5625
	WCSD	8.54	6.98	9.66	14.60
Original Image	NCC	1	1	1	1
	SD	54.06	47.95	62.35	42.57

TABLE 4.4
The Number of Connected Component (NCC) and the Within-Class Standard Deviation (WCSD) of Performing the Histogram (Tobias and Seara, 2002) and the TDRD Segmentations in Four Images

		Vegetable	Lena	Camera Man	Baboon
Histogram	NCC	130	210	213	850
	WCSD	27.05	24.12	29.40	23.34
TDRD	NCC	228	300	307	1217
	WCSD	22.92	19.12	16.97	21.17

In order to conduct numerical analysis, the 1,000 images used in the SIMPLIcity paper (Wang et al., 2001) were downloaded for testing our TDRD-based algorithm. Tables 4.5 and 4.6 show the numerical results when performing the watershed and the TDRD-based algorithm, in which two parameter sets parameter_1 (Th_S = 150, Th_H = 15) and parameter_2 (Th_S = 300, Th_H = 30) are applied. There are 81%, 40%, and 20% of images having NCC values that are smaller than or equal to 2,000 for the watershed, TDRD-parameter_1, and TDRD-parameter_2, respectively, and 55%, 91%, and 67% of images having WCSD values that are smaller than or equal to 15 for the watershed, TDRD-parameter_1, and TDRD-parameter_2, respectively. The results show that the both NCC and WCSD values in the TDRD-based algorithm are smaller than the watershed approach.

The complexity of the TDRD-based algorithm relies on the total number of pixels (*Num_Image*), the number of decomposing iterations (*Num_Iteration*), and the required calculations (e.g., the important intensities and their corresponding fuzzy subsets) in each iteration. By using the array structure, the required calculation of each connected component can be completed within a certain number of image scans. The complexity of calculating the required information of all connected components and determining the final intensity of suspicious pixels is O(*Num_Image*). Thererfore, the overall computation complexity of the TDRD algorithm is O(*Num_Image* × *Num_Iteration*).

TABLE 4.5
The NCC Values When Performing the Watershed, and the TDRD-Based Algorithm in Two Parameter Sets to the 1,000 Images Used in the SIMPLIcity Paper (Wang et al., 2001)

	<1,000	1,000	2,000	3,000	4,000	5,000	6,000	>6,000
Watershed	0	65	135	185	203	174	150	88
TDRD(150,15)	32	212	155	120	192	190	46	53
TDRD(300,30)	191	425	189	112	56	17	5	5

TABLE 4.6
The WCSD Values When Performing the Watershed, and the TDRD-Based Algorithm in Two Parameter Sets to the 1,000 Images Used in the SIMPLIcity Paper (Wang et al., 2001)

	5	10	15	20	25	>25
Watershed	4	72	473	326	82	43
TDRD(150,15)	26	690	196	48	30	10
TDRD(300,30)	10	244	416	280	40	10

4.7.5 Potential Applications in Medical Image Analysis

Medical image analysis is important since it provides assistance for medical doctors to find out the diseases inside the body without the surgery procedure (Duncan and Ayache, 2000). The TDRD-based image segmentation provides useful applications due to the properties of medical images. Generally, medical images contain three major regions: background, soft tissue, and object. We present the ideas of two potential applications, breast boundary segmentation and lung segmentation. Further detailed implementation is still ongoing under investigation.

4.7.5.1 Breast Boundary Segmentation

Mammography is a useful tool for detecting breast cancer (Zuckerman, 1987). It has been shown that CAD can provide assistance to radiologists to increase the cancer detection rate (Chan et al., 1990). Automated detection of breast boundary is the fundamental step in the mammographic CAD system (Wu et al., 2006b). A breast boundary detection algorithm was presented in (Wu et al., 2007b) based on multiple thresholding. However, it can be improved by incorporating more related information. Figure 4.45 shows an example of applying the TDRD approach in breast boundary segmentation. Figure 4.45(a) is a mammogram image obtained from the mammography database in http://marathon.csee.usf.edu/Mammography/Database.html. Figure 4.45(b) is the segmented result after the first iteration of the TDRD-based approach. The image is segmented into three regions marked in dark-gray, gray, and light-gray indicating background, soft tissue, and breast areas, respectively. Most breast boundary detection algorithms (Wu et al., 2006b) obtain an initial breast area by analyzing the histogram followed by a refinement procedure to determine the final breast boundary. However, they may fail if the initial breast boundary obtained from the thresholding approach is incorrect, such that the refinement procedure is led to the wrong direction. After edge detection, we obtain two outer and inner breast boundaries, as shown in Figure 4.45(c), in which the outer is close to the true boundary and the inner can evaluate the correctness of the outer boundary using the slopes of two boundaries in the corresponding positions.

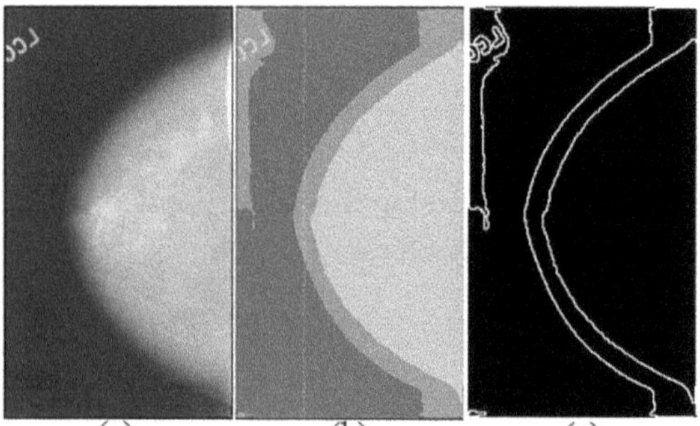

(a) (b) (c)

FIGURE 4.45 An example of applying the TDRD approach in breast boundary segmentation.

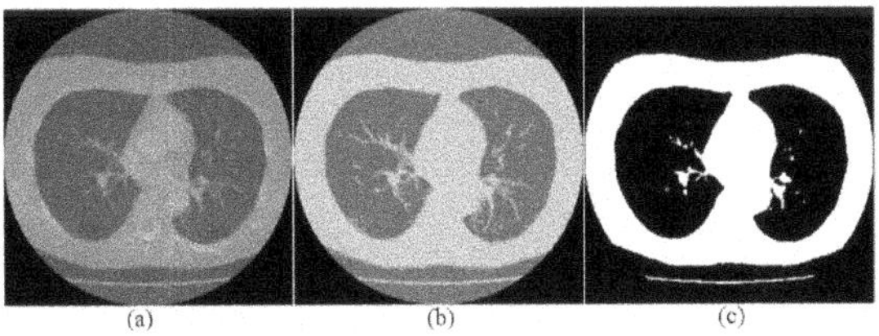

FIGURE 4.46 An example of applying the TDRD approach in lung segmentation.

4.7.5.2 Lung Segmentation

Lung segmentation plays an important role in the CAD systems. Appropriate segmentation helps lung nodule detection (Ge et al., 2005) and alignment of lung nodules from multiple chest CT scans (Shi et al., 2007). One lung cancer CAD system uses the k-means clustering as the initial segmentation. Its disadvantage is the high computational cost. Since the proposed segmentation method is computationally efficient, it can be used to substitute for the k-means clustering. Figure 4.46(a) shows a typical chest CT image from the LIDC database (see http://imaging.cancer.gov/programsandresources/InformationSystems/LIDC). The left and right lungs with low intensities are well separated by the mediastinum in the middle. Using the new method with one dividing iteration yields Figure 4.46(b), where the three major regions are marked in dark-gray, gray, and light-gray indicating the lung, vessel, and body areas, respectively. The profiles of two lungs are delineated roughly as shown in Figure 4.46(c) by a simple thresholding to keep the light-gray area. Furthermore, it is possible to build a 3D vessel tree by considering the gray area in Figure 4.46(b), such that the false positive can be reduced.

REFERENCES

Adams, R. and Bischof, L., "Seeded region growing," *IEEE Transactions on Pattern Analysis and Machine Intelligence*, vol. 16, no. 6, pp. 641–647, Jun. 1994.

Ahmad, M. B. and Choi, T., "Local threshold and Boolean function based edge detection," *IEEE Transactions on Consumer Electronics*, vol. 45, no. 3, pp. 674–679, Aug. 1999.

Athanasiadis, T., Mylonas, P., Avrithis, Y., and Kollias, S., "Semantic image segmentation and object labeling," *IEEE Transactions on Circuits and Systems for Video Technology*, vol. 17, no. 3, pp. 298–312, Mar. 2007.

Awad, M., Chehdi, K., Nasri, A., "Multicomponent image segmentation using a genetic algorithm and artificial neural network," *IEEE Transactions on Geoscience and Remote Sensing Letters*, vol. 4, no. 4, pp. 571–575, Oct. 2007.

Bao, P., Zhang, L., and Wu, X., "Canny edge detection enhancement by scale multiplication," *IEEE Transactions on Pattern Analysis and Machine Intelligence*, vol. 27, no. 9, pp. 1485–1490, Sep. 2005.

Belloulata, K. and Konrad, J., "Fractal image compression with region-based functionality," *IEEE Transactions on Image Processing*, vol. 11, no. 4, pp. 351–362, Apr. 2002.

Canny, J., "A computational approach to edge detection," *IEEE Transactions on Pattern Analysis and Machine Intelligence*, vol. 8, no. 6, pp. 679–698, Nov. 1986.

Cardoso, J. S. and Corte-Real, L., "Toward a generic evaluation of image segmentation," *IEEE Transactions on Image Processing*, vol. 14, no. 11, pp. 1773–1782, Nov. 2005.

Chai, D. and Bouzerdoum, A., "A Bayesian approach to skin color classification in YCbCr color space," *Proceedings of the IEEE TENCON 2000*, Kualu Lumpur, Malaysia, pp. 421–424, 2000.

Chan, H. P., Doi, K., Vyborny, C. J., Schmidt, R. A., Metz, C. E., Lam, K. L., Ogura, T., Wu, Y., and Macmahon, H., "Improvement in radiologists' detection of clustered microcalcifications on mammograms: The potential of computer-aided diagnosis," *Investigative Radiology*, vol. 25, pp. 1102–1110, 1990.

Chen, S., Sochen, N. A., and Zeevi, Y. Y., "Integrated active contours for texture segmentation," *IEEE Transactions on Image Processing*, vol. 15, no. 6, pp. 1633–1646, Jun. 2006.

Chen, Y. and Wang, J. Z., "A region-based fuzzy feature matching approach to content-based image retrieval," *IEEE Transactions on Pattern Analysis and Machine Intelligence*, vol. 24, no. 9, pp. 1252–1267, Sep. 2002.

Cheng, H. D., Jiang, X. H., Sun, Y., and Wang, J., "Color image segmentation: Advance and prospects," *Pattern Recognition*, vol. 34, no. 12, pp. 2259–2281, Dec. 2001.

Cheng, J. and Foo, S., "Dynamic directional gradient vector flow for snakes," *IEEE Transactions on Image Processing*, vol. 15, no. 6, pp. 1563–1571, Jun. 2006.

Chibanga, R., Berlamont, J., and Vandewalle, J., "Artificial neural networks in hydrological watershed modeling: Surface flow contribution from the ungaged parts of a catchment," *Proceedings of the 13th IEEE International Conference on Tools with Artificial Intelligence*, Dallas, TX, pp. 367–374, Nov. 2001.

Davis, L. S., "A survey of edge detection technique," *Computer Graphics and Image Processing*, vol. 4, pp. 248–270, 1974.

Deng, Y. and Manjunath, B. S., "Unsupervised segmentation of color-texture regions in images and video," *IEEE Transactions on Pattern Analysis and Machine Intelligence*, vol. 23, no. 8, pp. 800–810, Aug. 2001.

Donoho, D., Johnstone, I., Kerkyacharian, G., and Picard, D., "Density estimation by wavelet thresholding," *Annals of Statistics,* vol. 24, no. 2, pp. 508–539, 1996.

Duncan, J. S. and Ayache, N., "Medical image analysis: Progress over two decades and the challenges ahead," *IEEE Transactions on Pattern Analysis and Machine Intelligence*, vol. 22, no. 1, pp. 85–106, Jan. 2000.

Fan, J., Yau, D. K., Elmagarmid, A. K., and Aref, W. G., "Automatic image segmentation by integrating color-edge extraction and seeded region growing," *IEEE Transactions on Image Processing*, vol. 10, no. 10, pp. 1454–1466, Oct. 2001.

Garcia, C. and Tziritas, G., "Face detection using quantized skin color regions merging and wavelet packet analysis," *IEEE Transactions on Multimedia*, vol. 1, no. 3, pp. 264–277, Sep. 1999.

Ge, Z., Sahiner, B., Chan, H. P., Hadjiiski, L. M., Cascade, P. N., Bogot, N., Kazerooni, E. A., Wei, J., and Zhou, C., "Computer aided detection of lung nodules: False positive reduction using a 3D gradient field method and 3D ellipsoid fitting," *Medical Physics*, vol. 32, no. 8, pp. 2443–2454, Aug. 2005.

Hajjar, A. and Chen, T., "A VLSI architecture for real-time edge linking," *IEEE Transactions on Pattern Analysis and Machine Intelligence*, vol. 21, no. 1, pp. 89–94, Jan. 1999.

Haris, K., Efstratiadis, S. N., Maglaveras, N., and Katsaggelos, A. K., "Hybrid image segmentation using watersheds and fast region merging," *IEEE Transactions on Image Processing*, vol. 7, no. 12, pp. 1684–1699, Dec. 1998.

Hartmann, S. L. and Galloway, R. L., "Depth-buffer targeting for spatially accurate 3-D visualization of medical images," *IEEE Transactions on Medical Imaging*, vol. 19, no. 10, pp. 1024–1031, Oct. 2000.

He, L., Chao, Y., and Suzuki, K., "A run-based two-scan labeling algorithm," *IEEE Transactions on Image Processing*, vol. 17, no. 5, pp. 749–756, May 2008.

Hsu, R.-L., Abdel-Mottaleb, M., and Jain, A. K., "Face detection in color image," *IEEE Transactions on Pattern Analysis and Machine Intelligence*, vol. 24, no. 5, pp. 696–706, May 2002.

Ji, L. and Yan, H., "Attractable snakes based on the greedy algorithm for contour extraction," *Pattern Recognition*, vol. 35, no. 4, pp. 791–806, Apr. 2002.

Kass, M., Witkin, A., and Terzopoulos, D., "Snakes: Active contour models," *International Journal of Computer Vision*, vol. 1, no. 4, pp. 321–331, 1987.

Khotanzad, A. and Chen, J. Y., "Unsupervised segmentation of textured images by edge detection in multidimensional feature," *IEEE Transactions on Pattern Analysis and Machine Intelligence*, vol. 11, no. 4, pp. 414–421, Apr. 1989.

Konishi, S., Yuille, A. L., Coughlan, J. M., and Zhu, S., "Statistical edge detection: Learning and evaluating edge cues," *IEEE Transactions on Pattern Analysis and Machine Intelligence*, vol. 25, no. 1, pp. 57–74, Jan. 2003.

Law, T., Itoh, H., and Seki, H., "Image filtering, edge detection, and edge tracing using fuzzy reasoning," *IEEE Transactions on Pattern Analysis and Machine Intelligence*, vol. 18, no. 5, pp. 481–491, May 1996.

Lee, C.-S., Kuo, Y.-H., and Yu, P.-T., "Weighted fuzzy mean filters for image processing," *Fuzzy Sets and Systems*, vol. 89, no. 2, pp. 157–180, 1997.

Li, C., Xu, C., Gui, C., and Fox, M. D., "Level set evolution without re-initialization: A new variational formulation," *Proceedings of the IEEE Conference on Computer Vision and Pattern Recognition*, San Diego, CA, vol. 1, pp. 430–436, Jun. 2005.

Liu, S.-M., Lin, W.-C., and Liang, C.-C., "An iterative edge linking algorithm with noise removal capability," *Proceedings of the 9th International Conference on Pattern Recognition,* Rome, pp. 1120–1122, Nov. 1988.

Meer, P. and Georgescu, B., "Edge detection with embedded confidence," *IEEE Transactions on Pattern Analysis and Machine Intelligence*, vol. 23, no. 12, pp. 1351–1365, Dec. 2001.

Mehnert, A. and Jackway, P., "An improved seeded region growing algorithm," *Pattern Recognition Letters*, vol. 18, no. 10, pp. 1065–1071, Oct. 1997.

Melonakos, J., Pichon, E., Angenent, S., and Tannenbaum, A., "Finsler active contours," *IEEE Transactions on Pattern Analysis and Machine Intelligence*, vol. 30, no. 3, pp. 412–423, Mar. 2008.

Michailovich, O., Rathi, Y., and Tannenbaum, A., "Image segmentation using active contours driven by the Bhattacharyya gradient flow," *IEEE Transactions on Image Processing*, vol. 16, no. 11, pp. 2787–2801, Nov. 2007.

Moga, A. N. and Gabbouj, M., "Parallel image component labelling with watershed transformation," *IEEE Transactions on Pattern Analysis and Machine Intelligence*, vol. 19, no. 5, pp. 441–450, May 1997.

Nalwa, V. S. and Pauchon, E., "Edgel aggregation and edge description," *Computer Vision, Graphics, and Image Processing*, vol. 40, no. 1, pp. 79–94, Oct. 1987.

Nevatia, R., "Locating objects boundaries in textured environments," *IEEE Transactions on Computers*, vol. 25, no. 11, pp. 1170–1175, Nov. 1976.

Otsu, N., "A threshold selection method from gray-level histogram," *IEEE Transactions on Systems, Man, and Cybernetics*, vol. 9, no. 1, pp. 62–66, Jan. 1979.

Pachowicz, P. W., "Semi-autonomous evolution of object models for adaptive object recognition," *IEEE Transactions on System Man and Cybernetics*, vol. 24, no. 8, pp. 1191–1207, Aug. 1994.

Pal, N. R. and Pal, S. K., "A review on image segmentation techniques," *Pattern Recognition*, vol. 26, no. 9, pp. 1277–1294, Sep. 1993.

Panjwani, D. K. and Healey, G., "Markov random field models for unsupervised segmentation of textured color images," *IEEE Transactions on Pattern Analysis and Machine Intelligence*, vol. 17, no. 10, pp. 939–954, Oct. 1995.

Papari, G. and Petkov, N., "Adaptive pseudo dilation for Gestalt edge grouping and contour detection," *IEEE Transactions on Image Processing*, vol. 17, no. 10, pp. 1950–1962, Oct. 2008.

Pavlidis, T. and Liow, Y. T., "Integrating region growing and edge detection," *IEEE Transactions on Pattern Analysis and Machine Intelligence*, vol. 12, no. 3, pp. 225–233, Mar. 1990.

Pyun, K. P., Lim, J., Won, C., and Gray, R. M., "Image segmentation using hidden Markov Gauss mixture models," *IEEE Transactions on Image Processing*, vol. 16, no. 7, pp. 1902–1911, Jul. 2007.

Rivera, M., Ocegueda, O., and Marroquin, J. L., "Entropy-controlled quadratic Markov measure field models for efficient image segmentation," *IEEE Transactions on Image Processing*, vol. 16, no. 12, pp. 3047–3057, Dec. 2007.

Russ, J. C., *The Image Processing Handbook*, CRC Press, 1992.

Saha, P. K. and Udupa, J. K., "Optimum image thresholding via class uncertainty and region homogeneity," *IEEE Transactions on Pattern Analysis and Machine Intelligence*, vol. 23, no. 7, pp. 689–706, Jul. 2001.

Shen, D. and Davatzikos, C., "An adaptive-focus deformable model using statistical and geometric information," *IEEE Transactions on Pattern Analysis and Machine Intelligence*, vol. 22, no. 8, pp. 906–913, Aug. 2000.

Shi, J., Sahiner, B., Chan, H.-P., Zhou, L. H. C., Cascade, P. N., Bogot, N., Kazerooni, E. A., Wu, Y.-T., and Wei, J., "Pulmonary nodule registration in serial CT scans based on rib anatomy and nodule template matching," *Medical Physics*, vol. 34, no. 4, pp. 1336–1347, Apr. 2007.

Shih, F. Y. and Cheng, S., "Adaptive mathematical morphology for edge linking," *Information Sciences*, vol. 167, nos. 1–4, pp. 9–21, Dec. 2004.

Shih, F. Y. and Cheng, S., "Automatic seeded region growing for color image segmentation," *Image and Vision Computing*, vol. 23, no. 10, pp. 877–886, Sep. 2005.

Shih, F. Y. and Zhang, K., "Efficient contour detection based on improved snake model," *Pattern Recognition and Artificial Intelligence*, vol. 18, no. 2, pp. 197–209, Mar. 2004.

Sofou, A. and Maragos, P., "Generalized flooding and multicue PDE-based image segmentation," *IEEE Transactions on Image Processing*, vol. 17, no. 3, pp. 364–376, Mar. 2008.

Stahl, J. S. and Wang, S., "Edge grouping combining boundary and region information," *IEEE Transactions on Image Processing*, vol. 16, no. 10, pp. 2590–2606, Oct. 2007.

Stahl, J. S. and Wang, S., "Globally optimal grouping for symmetric closed boundaries by combining boundary and region information," *IEEE Transactions on Pattern Analysis and Machine Intelligence*, vol. 30, no. 3, pp. 395–411, Mar. 2008.

Sundaramoorthi, G., Yezzi, A., and Mennucci, A. C., "Sobolev active contours," *International Journal of Computer Vision*, vol. 73, no. 3, pp. 345–366, Jul. 2007.

Sundaramoorthi, G., Yezzi, A., and Mennucci, A. C., "Coarse-to-fine segmentation and tracking using Sobolev active contours," *IEEE Transactions on Pattern Analysis and Machine Intelligence*, vol. 30, no. 5, pp. 851–864, May 2008.

Tao, W., Jin, H., and Zhang, Y., "Color image segmentation based on mean shift and normalized cuts," *IEEE Transactions on Systems, Man, and Cybernetics, Part B*, vol. 37, no. 5, pp. 1382–1389, Oct. 2007.

Tao, W., Jin, H., Zhang, Y., Liu, L., and Wang, D., "Image thresholding using graph cuts," *IEEE Transactions on Systems, Man, and Cybernetics, Part A*, vol. 38, no. 5, pp. 1181–1195, Sep. 2008.

Tobias, O. J. and Seara, R., "Image segmentation by histogram thresholding using fuzzy sets," *IEEE Transactions on Image Processing*, vol. 11, no. 12, pp. 1457–1465, Dec. 2002.

Unnikrishnan, R., Pantofaru, C., and Hebert, M., "Toward objective evaluation of image segmentation algorithms," *IEEE Transactions on Pattern Analysis and Machine Intelligence*, vol. 29, no. 6, pp. 929–944, Jun. 2007.

Vincent, L. and Soille, P., "Watersheds in digital spaces: An efficient algorithm based on immersion simulations," *IEEE Transactions on Pattern Analysis and Machine Intelligence*, vol. 13, no. 6, pp. 583–598, Jun. 1991.

Wang, J. Z., Li, J., and Wiederhold, G., "SIMPLIcity: Semantics-sensitive Integrated Matching for Picture LIbraries," *IEEE Transactions on Pattern Analysis and Machine Intelligence*, vol. 23, no. 9, pp. 947–963, Sep. 2001.

Wong, W. and Chung, A., "Bayesian image segmentation using local iso-intensity structural orientation," *IEEE Transactions on Image Processing*, vol. 14, no. 10, pp. 1512–1523, Oct. 2005.

Wu, J. and Chung, A., "A segmentation model using compound Markov random fields based on a boundary model," *IEEE Transactions on Image Processing*, vol. 16, no. 1, pp. 241–252, Jan. 2007.

Wu, J., Yin, Z. P., and Xiong, Y., "The fast multilevel fuzzy edge detection of blurry images," *IEEE Transactions on Signal Processing Letters*, vol. 14, no. 5, pp. 344–347, May 2007a.

Wu, Y.-T., Hadjiiski, L. M., Wei, J., Zhou, C., Sahiner, B., and Chan, H.-P., "Computer-aided detection of breast masses on mammograms: Bilateral analysis for false positive reduction," *Proceedings of the 2006 SPIE Medical Imaging*, San Diego, CA, Feb. 2006b.

Wu, Y.-T., Shih, F. Y., and Wu, Y.-T., "Differential histogram equalization based image simplification," *Proceedings of the Sixth IEEE International Conference Electro/ Information Technology*, East Lansing, MI, pp. 344–349, May 2006a.

Wu, Y.-T., Zhou, C., Hadjiiski, L. M., Shi, J., Paramagul, C., Sahiner, B., and Chan, H.-P., "A dynamic multiple thresholding method for automated breast boundary detection in digitized mammograms," *Proceedings of the 2007 SPIE Medical Imaging*, San Diego, CA, Feb. 2007b.

Xie, X. and Mirmehdi, M., "MAC: Magnetostatic active contour model," *IEEE Transactions on Pattern Analysis and Machine Intelligence*, vol. 30, no. 4, pp. 632–646, Apr. 2008.

Xu, Y., Saber, E., and Tekalp, A. M., "Object segmentation and labeling by learning from examples," *IEEE Transactions on Image Processing*, vol. 12, no. 6, pp. 627–638, Jun. 2003.

Zenzo, D. S., Cinque, L., and Levialdi, S., "Image thresholding using fuzzy entropies," *IEEE Transactions on Systems, Man, and Cybernetics, Part B*, vol. 28, no. 1, pp. 15–23, Feb. 1998.

Zimmermann, H. J., *Fuzzy Set Theory and Its Applications*, Kluwer Academic, 1991.

Zuckerman, H. C., "The role of mammography in the diagnosis of breast cancer," *Breast cancer, diagnosis and treatment*, I. M. Ariel and J. B. Cleary, Eds., pp. 152–172, McGraw-Hill, 1987.

5 Image Representation and Description

Representation and manipulation of digital images are two important tasks in image processing, pattern recognition, pictorial database, computer graphics, geographic information systems, and other related applications. Image representation and description is critical for successful detection and recognition of objects in a scene. After an image has been segmented into object and background regions, one intends to represent and describe them into characteristic features for computer processing of pattern recognition or into quantitative codes for efficient storage of image compression. There are basically two types of characteristics: external and internal. An external representation focuses on object shape characteristics such as boundary. An internal representation emphasizes object region characteristics such as texture, topology, and skeleton.

This chapter is organized as follows. Run-length coding is introduced in Section 5.1. Binary tree and quadtree representation are described in Section 5.2. In Section 5.3, we present different contour representation schemes, such as chain code, crack code, and mid-crack code. Skeletonization (or medial axis transformation) by thinning is presented in Section 5.4, and by maxima tracking on distance transform is presented in Section 5.5. Finally, in Section 5.6, we present object representation and tolerance by mathematical morphology.

5.1 RUN-LENGTH CODING

Pictures transmitted over communication channels and stored on disks or tapes in computer systems generally contain significant data redundancy. The technique or mechanism of image compression intends to represent pictures by as few bits as possible. The storage requirement for a digitized binary image, represented directly as a 2D $N \times N$ array of points, means that such a representation scheme requires N^2 bits. Different techniques have been used to reduce this storage requirement by eliminating some of the data correlation or redundancy that exists in an image at the expense of processing time, without the loss of any information and consequent introduction of misrepresentation in the recovered image. Many lossless image coding schemes have been proposed. In particular, run-length coding, chain coding, crack coding, rectangular coding, quadtree representation, and medial axis transformation (MAT) are among those schemes that have been considered for the use in image data compression.

Run-length coding is one of the most practical and simplest methods for binary images (Golomb, 1966). Each row of a picture consists of a sequence of maximal runs of points, such that the points in each run all have the same value. Thus, the row is completely determined, as far as a binary image is concerned, by specifying

DOI: 10.1201/ 9781003474395-6

the value of the first run in the row and the lengths of all runs. The value of each run does not have to be specified because these values must alternate in 0 and 1. For an $N \times N$ picture, if the average number of runs in each row is R, the total number of bits required for the run-length representation is $N(1 + R\log N)$. High efficiency is obtained for images with the small number of runs.

Capon (1959) presented a probabilistic model for run-length coding of pictures. Tanaka and Leon-Garcia (1982) proposed an efficient method for the run-length coding of binary memoryless sources when the symbol probabilities are unknown. Montani and Scopigno (1990) developed the "sticks" representation scheme data model based on a 3D extension of the run-length encoding methods. Bunke and Csirik (1995) proposed an improved algorithm for computing the edit distance of run-length coded strings. Berghorn et al. (2001) presented fast variable run-length coding for embedded progressive wavelet-based image compression.

Example 5.1: Represent the following figure using the run-length representation.

1	1	1	1	0	0	0	0
1	1	1	1	0	0	0	0
1	1	1	1	1	1	0	0
1	1	1	1	1	1	0	0
0	0	1	1	1	1	0	0
0	0	1	1	1	1	0	0
0	0	0	0	0	0	0	0
0	0	0	0	0	0	0	0

Answer: The run-length representation is:

1	4	4	
1	4	4	
1	6	2	
1	6	2	
0	2	4	2
0	2	4	2
0	8		
0	8		

5.2 BINARY TREE AND QUADTREE

In computer science, a binary tree is a tree data structure where each node contains at most two children. We first introduce a binary tree to represent a row of an image. Let the row length be $n = 2^k$. If the row all has one value, it is labeled by the value; otherwise, we add two descendants to the root node representing two halves. If the half has one constant value, it is labeled with that value; otherwise, we give two descendants corresponding two halves of its half. The division is stopped until it reaches a single pixel. Note that at level h in the tree, the nodes represent pieces of length 2^{k-h}.

Wang et al. (2003) proposed a region-based binary tree representation incorporating with adaptive processing of data structures for image classification. Robinson (1997) presented efficient general-purpose image compression with binary tree predictive coding, which uses a noncausal, shape-adaptive predictor to decompose an image into a binary tree of prediction errors and zero blocks. Robinson et al. (2000) presented video compression with binary tree recursive motion estimation and binary tree residue coding.

Example 5.2: Draw the binary tree for the following row:

10010110111111110100101111110110

Answer:

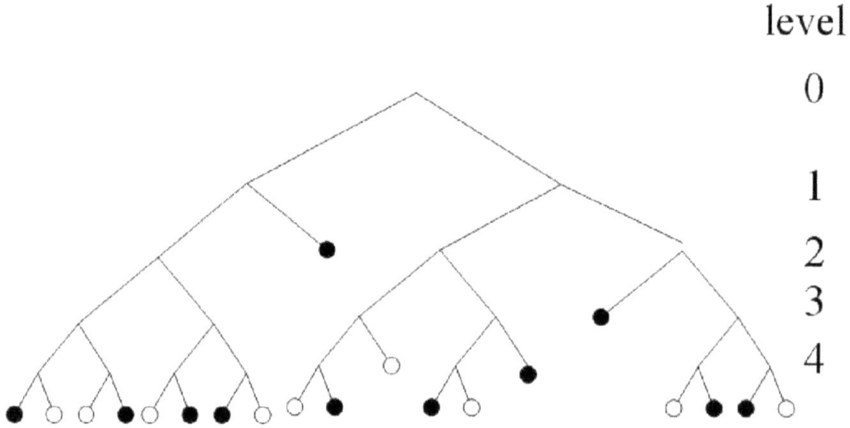

Quadtree is one of the compact hierarchical data structures in representing a binary image (Hunter and Steiglitz, 1979). Let the image size be $n = 2^k \times 2^k$. It is constructed by successively subdividing an image into four equal-size sub-images in the NW (northwest), NE (northeast), SW (southwest), and SE (southeast) quadrants. A homogeneously colored quadrant is represented by a leaf node in the tree. Otherwise, the quadrant is represented by an internal node and further divided into four sub-quadrants until each sub-quadrant has the same color. The leaf node with a black (white) color is called the black (white) node, and the internal node is called the gray node. Note that at level h in the tree, the nodes represent blocks of size $2^{k-h} \times 2^{k-h}$. An example of quadtree representation is shown in Figure 5.1.

Quadtree representation has been used for a variety of image processing and image compression algorithms. Samet (1980) presented an algorithm for smoothing images represented by quadtrees. Shneier (1981) proposed an algorithm for computing

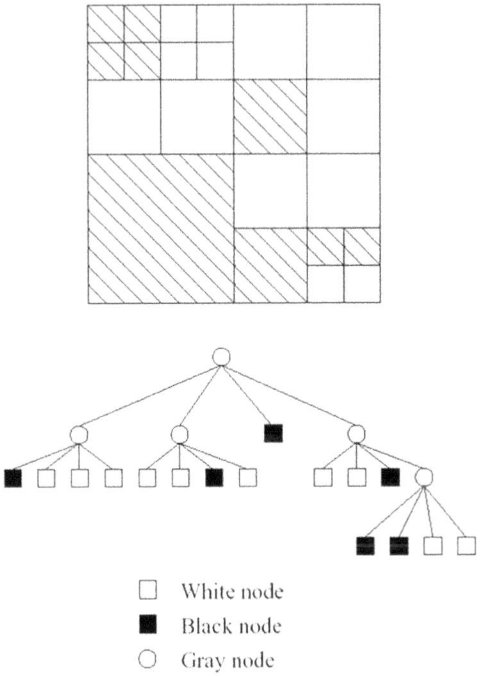

FIGURE 5.1 An image and its quadtree representation.

geometric properties of binary images represented by quadtrees. Sullivan and Baker (1994) described the theory needed to construct quadtree data structures of images and videos that optimally allocate rate, given a set of quantizers. Shusterman and Feder (1994) presented image compression via improved quadtree decomposition algorithms. Knipe and Li (1998) proposed improvements over the reconstruction method described by Shusterman and Feder (1994). Laferte et al. (2000) presented discrete Markov image modeling and inference on the quadtree. Wong et al. (2006) presented thinning algorithms based on quadtree and octree representations.

Example 5.3: Draw the quadtree representation for the following image:

00000000
00001000
11110011
11111000
00011111
00111100
00111100
00111111

Answer: In the following figure, we only draw the white and black nodes. White nodes denote "0" and black nodes denote "1."

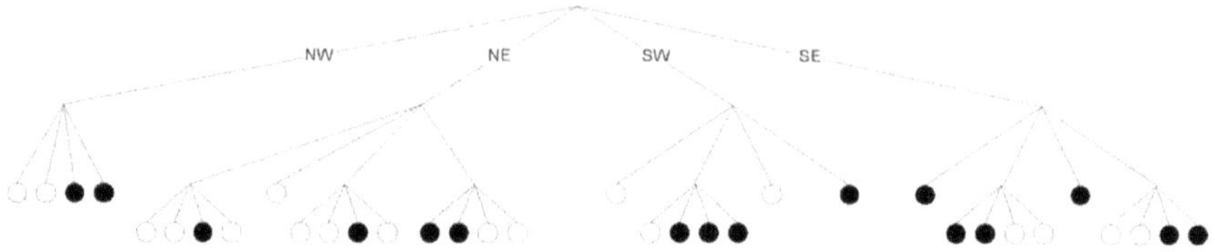

This tree is called *quadtree* since its nonleaf nodes all have degree 4. The storage requirement for the quadtree is proportional to number of nodes. The quadtree is not data redundant. However, it is shift-variant; a simple position shift will cause a big variation in the quadtree representation. Therefore, it is difficult to match whether two images are congruent.

There are two widely used representations of quadtrees. A pointer-based quadtree uses the standard tree representation. A linear quadtree can be either a preorder traversal of the nodes of a quadtree or the sorted sequence of the quadtree's leaves. The quadtree's data structure has been applied to the representation of maps in the geographic information systems (GIS) successfully. A fundamental operation intends to overlay two maps with union, intersection, and difference operators. Some research areas are parallel computational models of the quadtree algorithm.

In the *quadtrees representation* case, maximal blocks can be of any size and at any position. They are analogous to the runs in the 1D case. If the picture has all one value, the root node is labeled with that value and stop. Otherwise, four descendants are added to the root node, representing the four quadrants of the picture. The process is then repeated for each of these new nodes and so on. If the number of nodes in the tree is T, then $2T$ bits are required to represent the image. This scheme is conceptually clear and well structured, but it turns out that the compression efficiency of the scheme is not, in general, very high. Another disadvantage of the quadtree representation is that it is shift-variant. When a picture is shifted, it may generate a very different quadtree.

5.3 CONTOUR REPRESENTATION

In building the object model base, an object or multiple objects of a grayscale image are thresholded into a binary image, and object (or component) labeling is used to assign each object a unique nonzero label and reset the background pixels to zeros. Then a border operator is done by extracting all the boundary pixels in an object. Chain code, crack code, or mid-crack code can be used for contour representation.

There are other approaches for contour representation in the literature. Meer et al. (1990) proposed a hierarchical approach toward fast parallel processing of chain-codable contours. Saint-Marc et al. (1993) presented representation of edge contours by approximating B-splines and showed that such a representation facilitates the extraction of symmetries between contours. Hasan and Karam (2000) developed morphological reversible contour representation, in which a binary image is represented by a set of non-overlapping multilevel contours and a residual image. Ghosh and Petkov (2005) proposed shape descriptors to incomplete contour representations.

5.3.1 Chain Code and Crack Code

The contour representation of a binary image is determined by specifying a starting point and a sequence of moves around the borders of each region. Often-used methods of contour tracing are based on the chain code or crack code concept (Freeman, 1974; Rosenfeld and Kak, 1982; Wu, 1982). The chain code moves along a sequence of the center of border points, while the crack code moves along a sequence of "cracks" between two adjacent border points. Typically, they are based on the 4- or 8-connectivity of the segments, where the direction of each segment is encoded by a numbering scheme, such as 3-bit numbers $\{i \mid i = 0, 1, \cdots, 7\}$ denoting an angle of $45i°$ counter-clockwise from the positive x-axis for a chain code as shown in Figure 5.2, or 2-bit numbers $\{i \mid i = 0, 1, 2, 3\}$ denoting an angle of $90i°$ for a crack code. The elementary idea of the chain or crack coding algorithm is to trace the border-pixels or cracks and sequentially generate codes by considering the neighborhood adjacency relationship.

Kaneko and Okudaira (1985) proposed an efficient encoding scheme for arbitrary curves based on the chain code representation. Zingaretti et al. (1998) presented a fast single-pass algorithm to convert a multivalued image from a raster-based representation into chain codes. Madhvanath et al. (1999) presented chain code contour processing for handwritten word recognition. Shih and Wong (1999) proposed a one-pass algorithm for local symmetry of contours from chain codes. Bribiesca (2000) developed a chain code for representing 3D curves. Liu and Zalik (2005) proposed an efficient chain code with Huffman coding.

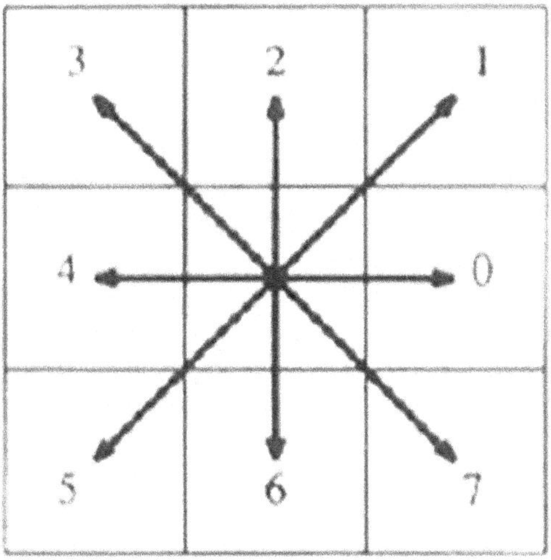

FIGURE 5.2 The chain code representation.

Example 5.4: What are the chain codes for the following binary region by considering 8-connectedness? Note that use the counterclockwise approach.

0000000
0011111
0111110
0011110
0001000

Answer: The chain codes are given as follows. Let the initial pixel be the upper-left pixel.

Starting pixel

0	0	0	0	0	0	0
0	0	1	1	1	1	1
0	1	1	1	1	1	0
0	0	1	1	1	1	0
0	0	0	1	0	0	0

Chain codes: (2, 3), 5, 7, 7, 1, 0, 2, 1, 4, 4, 4, 4. The crack codes are given as follows.

Example 5.5: What are the crack codes for the following binary region by considering 8-connectedness? Note that use the counterclockwise approach.

0000000
0011110
0111000
0010110
0000000

Answer: The crack code tracking is expressed as follows:

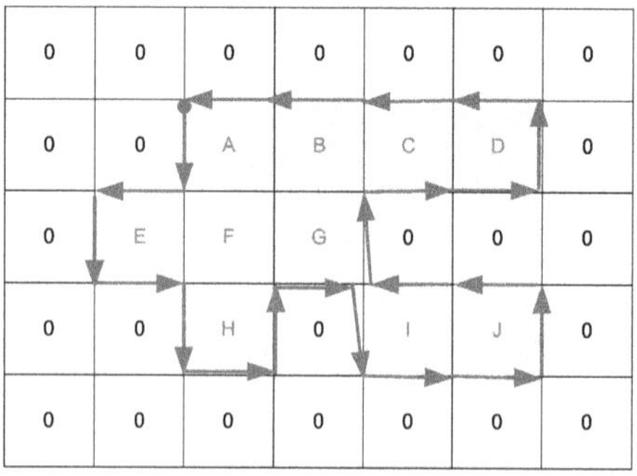

Therefore, the crack codes are: A(2,3) 3 2 3 0 3 0 1 0 3 0 0 1 2 2 1 0 0 1 2 2 2 2.

5.3.2 Difference Chain Code

For any given chain code, a difference chain code can be constructed, e.g., curvature, whose values represent the successive changes in direction. In the raster scan of an image, the first border point encountered is the starting point. The chain code is initialized as 0. According to the current chain code, it is checked whether the next point in the chain-code direction is a border point or not. If it is, this point is appended to border linked list and the code is recorded. If it is not, the chain code is subtracted by 1 repeatedly until the next neighboring border point is found. Repeat the procedures until all the border points have been already visited.

If the difference of successive unit slopes, i.e., chain code, is used, a multiple of a 45° curvature will be always obtained. To obtain more continuous range of the curvature values, a smoothing procedure can be used. The *k-slope* of a chain code at a point P is defined as the average of the unit slopes at a sequence of k points centered at P. The k-curvature of a point P is defined as the difference between its left and right k-slopes. The first direction is determined by a point before the current point, and the second direction is the one after the current point. The curvature is computed from the first direction to the second direction counterclockwise. That is, if the direction changes counterclockwise, the curvature is positive and negative otherwise.

Example 5.6: Give the chain code and the difference chain code of the following image.

1	1	1
1	1	1
1	1	1

Answer: The chain code is: 66002244. The difference chain code is: 02020202.

Example 5.7: Compute the 1-curvature and 3-curvature of the following image.

0	0	0	0	0	0	1	1	1
0	0	0	1	1	1	0	0	0
1	1	1	0	0	0	0	0	0

Answer: The 1-slope is: 0° or 45°, and the 1-curvature is 0° or 45°. The 3-slope is $\tan^{-1}(1/3)$, and the 3-curvature is 0°.

5.3.3 Shape Signature

After the chain code has been extracted, the information pertaining to the slope and the curvature of boundary curves can be easily obtained. One nature characteristic which satisfies the conditions of a suitable shape signature is the curvature of a curve. It is well known that there is a one-to-one correspondence between a regular curve and its curvature function. The curvature is the rate of change θ, the angle between the tangent vector to the curve and the horizontal axis, with respect to arc length s.

Let κ (s) be the curvature function, then

$$\kappa(s) = \frac{d\theta}{ds}, \quad \text{where } \theta = \tan^{-1}\left(\frac{dy/ds}{dx/ds}\right) \tag{5.1}$$

or

$$|\kappa(s)|^2 = \left[\frac{d^2x}{ds^2}\right]^2 + \left[\frac{d^2y}{ds^2}\right]^2. \tag{5.2}$$

One way to obtain the discrete approximation of the curvature is to draw the so-called *arc-length-versus-turning-angle graph* of the curve (Kalvin et al., 1986) and to sample this graph at equally spaced (ΔS) points, and at each point S_i $(i = 1, 2, \cdots, n)$ to compute the difference such as

$$\Delta\theta(S_i) = \theta(S_i + \Delta S) - \theta(S_i) \tag{5.3}$$

5.3.4 The Mid-Crack Code

The chain code and crack code can be viewed as a connected sequence of straight line segments with specified lengths and directions. An obvious disadvantage of the chain code is observed when it is used to compute the area and perimeter of an object. As seen in Figure 5.3, the inside chain code appears to underestimate the area and perimeter, whereas the outside chain code overestimates them. The disadvantages in the crack code are that many more codes are generated and the perimeter is greatly overestimated. The mid-crack code (Dunkelberger and Mitchell, 1985), located in between, should make a more accurate computation of the geometric features.

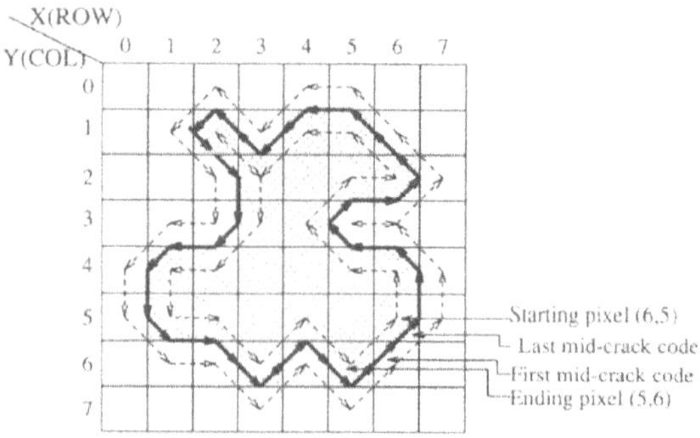

FIGURE 5.3 Silhouette with the inside and outside chain coded contours (dashed lines) and the mid-crack coded contour (solid line).

A method of generating the chain code by using the run-length coding was presented in (Kim et al., 1988). But, the COD-S and ABS-S tables must be produced in the first run-length step before the chain code generation and linking phases start. A useful concept for dealing with the inclusion relationship between boundaries was also suggested. In addition, a raster-scan chain code introducing the max-point and min-point concept, especially a linking concept of relation links, was proposed in (Cederberg, 1979). Another single-pass algorithm (Chakravarty, 1981) for generating the chain code adopts the use of a step-by-step concept, such as chain link generation, a link segment data structure, and a junction of links.

The mid-crack code is a variation of and an improvement on the traditional tracing methods between the chain code and the crack code. In contrast to Freeman chain code, which moves along the center of pixels, the mid-crack code moves along the edge mid-point of a pixel, producing codes of links. For the horizontal and vertical moves, the length of a move is 1, and for diagonal moves, it is $\sqrt{2}/2$. If the crack is located in between two adjacent object pixels in the vertical direction, it is said to be on a vertical crack. Similarly, if the crack is located in between two adjacent object pixels in the horizontal direction, it is said to be on a horizontal crack.

Figure 5.4 shows the mid-crack code on the vertical and horizontal cracks. There are two restrictions on the moves in the mid-crack code. If the move is from the vertical crack, the codes 0 and 4 are not allowed. Similarly, the codes 2 and 6 are not allowed in the moves from the horizontal crack. The experimental verification of the mid-crack code in the area and perimeter computations is shown in (Koplowitz, 1981; Saghri and Freeman, 1981), where the mean perimeter error value is –0.074% and the mean area error is –0.006%. Therefore, the mid-crack code is a desirable alternative method in contour tracing with its benefit in accuracy.

It is possible to invert a closed mid-crack code sequence into its corresponding silhouette. An internal boundary of the 8-connectedness object (or foreground) being traced counterclockwise can be reconstructed by treating it as the 4-connectedness background being traced clockwise. A disadvantage of the mid-crack code is that it is always longer than the Freeman chain code. An algorithm of conversion between a mid-crack code sequence and a chain code sequence is described in (Dunkelberger and Mitchell, 1985) to complement the defects as a compression process.

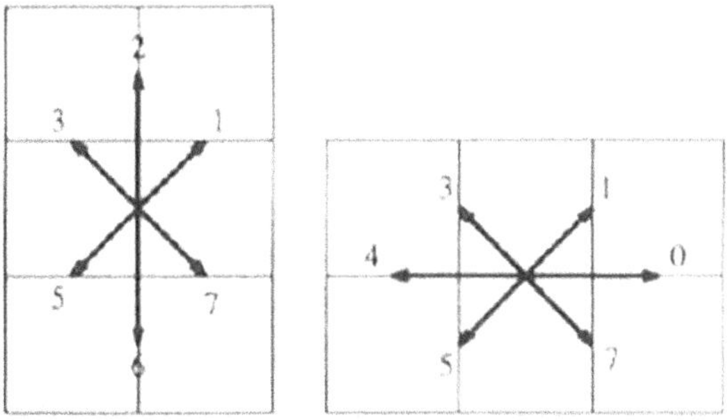

FIGURE 5.4 (a) The mid-crack code on the vertical crack and (b) the mid-crack code on the horizontal crack.

A fast algorithm for extracting the mid-crack code in a single-pass raster-scan fashion and its extension to parallel implementation are presented. The developed algorithm needs only one phase to generate all the code sequences even though a binary image is composed of several objects. Two types of boundaries, external and internal, are also considered, such that the inclusion relationship among region boundaries can be easily determined using the same algorithm.

An efficient algorithm to obtain the mid-crack codes can be referred to (Shih and Wong, 1992). In the contour description, the border points are represented in the form of a string of eight moving-directional codes. The mid-crack code description can be seen as movement along the mid-points of border cracks which produce codes of links. The codes from 0 to 7 are assigned to indicate the eight moving directions of 0° to 315° in intervals of 45°. For the horizontal or vertical move, the length of a move is 1, and for the diagonal move, it is $\sqrt{2}/2$. If the crack is located in between two adjacent pixels in the vertical direction, it is said to be on a *vertical crack*. Similarly, if the crack is located in between two adjacent pixels in the horizontal direction, it is said to be on a *horizontal crack*. There are two restrictions on the moves in the mid-crack code as illustrated in Figure 5.4. If the move is from a vertical crack, the codes 0 and 4 are not allowed. Similarly, the codes 2 and 6 are not allowed in the moves from a horizontal crack.

The tracking rule of 8-connectedness counterclockwise is used. The description form of all m contours in a binary image can be expressed as follows:

$$
\begin{array}{l}
sx_{00}y_{00}d_{00}d_{01}\cdots\cdots\cdots\cdots d_{0n_0}e \\
sx_{10}y_{10}d_{10}d_{11}\cdots\cdots\cdots\cdots d_{1n_1}e \\
\cdots\cdots\cdots\cdots\cdots\cdots\cdots\cdots \\
\cdots\cdots\cdots d_{i(j-1)}d_{ij}\cdots\cdots\cdots \\
\cdots\cdots\cdots\cdots\cdots\cdots\cdots\cdots \\
sx_{m0}y_{m0}d_{m0}d_{m1}\cdots\cdots\cdots\cdots d_{mn}\ e\#
\end{array}
$$

where # is the end symbol of the description; s and e are the starting and ending symbols for each contour, respectively; x and y are the coordinates of the starting point; and d is the directional code. The subscript of d_{ij} indicates that d is the code in the ith contour and the jth move.

Each contour could be either an in-contour or an out-contour. In mid-crack code description, since the starting and ending pixels in each contour must be adjacent to each other, the opened curve occurred in the chain-code description should not exist here. Besides, there are four cracks surrounding a pixel. Therefore, the problems in the chain code description of no-code for an isolated point and redundantly tracing a same pixel twice for a multi-way junction pixel will not again happen here.

A simple example of the mid-crack code description for a single-object image is illustrated in Figure 5.3. The resulting mid-crack codes are:

$$s(6,5)11771177076545677533554333101343 21e\#$$

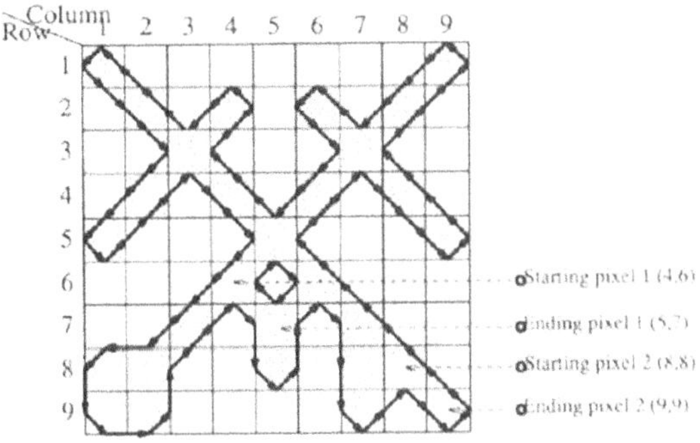

FIGURE 5.5 An image contains an object with a hole.

Referring to Figure 5.3, the starting pixel (e.g., (6, 5)) is defined as the pixel which is pointed by the link of the first mid-crack code located at the ending pixel (e.g., (5, 6)) in a raster scan. As mentioned, the counterclockwise tracing indicated by the arrow directions is adopted. Since the mid-crack codes are generated based on pixel-by-pixel according to a look-up table and then are concatenated using the code-linking from the head to the tail of arrows, it is in nature that the generated sequences of the mid-crack code description are in a clockwise direction.

According to their algorithm, two results are shown in the following, which are considered using 8-connectedness for the object regions. In Figure 5.5, an image containing an object with a hole is shown. The result is shown as follows:

The total number of pixels = 29.

(1) Starting pixel = (4,6).
Codes: 3571
Perimeter = 5.66, Area = 0.5, Boundary type: Internal
(2) Starting pixel = (8,8).
Codes: 331771176671217671112107654555557771111755557777533335531
1333555775
33555531111333317777111333333
Perimeter = 70.26, Area = 29, Boundary type: External

In Figure 5.6, an image with multiple objects is shown. The result is shown as follows:

The total number of pixels = 43.

(1) Starting pixel = (9,2).
Codes: 2107665432
Perimeter = 8.83, Area = 5.5, Boundary type: External
(2) Starting pixel = (3,3).
Codes: 217653
Perimeter = 4.83, Area = 1.5, Boundary type: External

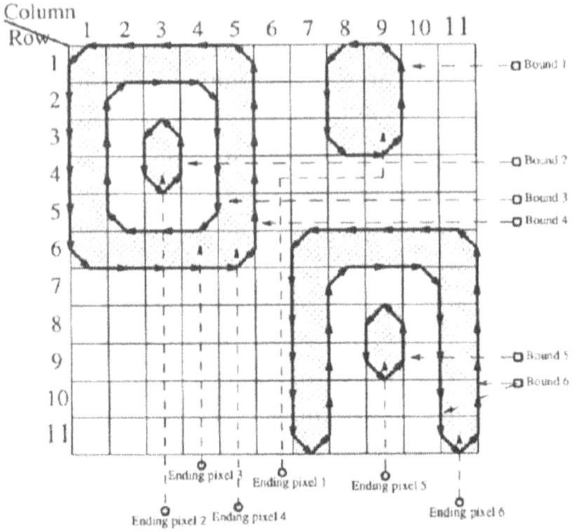

FIGURE 5.6 A five-object image with external and internal boundaries.

(3) Starting pixel = (3,6).
Codes: 45666700122234
Perimeter = 12.83, Area = 11.5, Boundary type: Internal
(4) Starting pixel = (5,5).
Codes: 2100007666665444432222
Perimeter = 20.83, Area = 18, Boundary type: External
(5) Starting pixel = (9,8).
Codes: 217653
Perimeter = 4.83, Area = 1.5, Boundary type: External
(6) Starting pixel = (11,10).
Codes: 21766667001222217666665444432222
Perimeter = 25.66, Area = 14.5, Boundary type: External

When the contours of multiple objects in an image are extracted, one of the most common problems is to fill the region inside each contour. A region consists of a group of adjacent, connected pixels. The task of filling primitives can be divided into two parts: the decision of which pixels to fill and the easier decision of with what value to fill them. If the image is binary, the same value is assigned to each pixel lying on a scan line running from the left edge to the right edge; i.e., fill each span from x_{min} to x_{max}. Spans exploit a primitive's *spatial coherence*: the fact that primitives often do not change from pixel to pixel within a span or from scan-line to scan-line. The coherence is obtained in general by looking for only those pixels at which changes occur. There are several algorithms performing the region-filling based on chain codes (Cai, 1988; Chakravarty, 1981) by the technique of parity checking (Ackland and Weste, 1981; Pavlidis, 1979) or a method of seed growing (Pavlidis, 1979; Shani, 1980). The technique of parity checking is often used because only a single scan is required instead of the iterations required in the seed-growing method.

However, it is possible to produce an incorrect count of the number of intersections if points from two or more sides are mapped on the identical pixel. Besides, a problem will arise if the test line is tangent to the contour.

A method of region filling by using discrete Green's theorem was presented by Tang and Lien (1988). They reduced the region filling problem to the inclusion problem and the intersection problem. The theorem is used to solve the inclusion problem if the spatial connectivity fails to make a decision. However, the decision causes an extra long computational time. A parity checking using the chain-coded contour (Cai, 1988; Chang and Leu, 1990) needs to deal with the characteristics of chain codes, such as those cases of opened contours, and isolated pixel, end-point pixels, internal line segment pixels, and multi-way pixels (Ali and Burge, 1988), that makes the region filling problem complicated.

In *chain coding*, each chain is composed of line segments connecting adjacent pixels in a horizontal, vertical, or diagonal direction. A closed boundary is chain coded from an arbitrary starting point. The initial border point coordinate is stored and the border around a region is then followed back to the first point, and it is repeated for each separate region. A region may be defined as a set of all connected points of one gray level, and a border of a region as a set of region's points that are adjacent to the points of different gray level (in the binary image case: adjacent to the points of the region's complement). In moving from point to point around a border, a point is moved to one of its eight neighbors. The eight neighbors are labeled from 0 to 7 starting from the right-hand side and then tracking counterclockwise. Thus, each move is defined by an octal digit. It means that three bits are needed to indicate where the next border point is, starting from any particular border point. If the total number of regions in an image is K, and if the sum of the lengths of the borders for each region in the image is L, then the total storage requirement is $2K\log N + 3L$ bits. It is possible to improve chain coding technique by constructing *difference chain code*, whose values represent the successive changes in direction, e.g., let 0 represent no turn, +1/−1 represent 45° right/left turns, +2/−2 and +3/−3 similarly represent 90° and 135° right/left turns, respectively, and 4 represent a 180° turn. Thus, the difference chain code, like the chain code, has eight possible values and requires three bits per move. However, the values are no longer equally likely, e.g., 0 and +1/−1 should appear often, while 4 rarely happens. Hence, the variable-length coding technique can be used. High efficiency is obtained for images with short total border length and a small number of regions.

5.4 SKELETONIZATION BY THINNING

Thinning is a fundamental early processing step in pattern analysis such as industrial parts inspection (Shih and Mitchell, 1988), fingerprint recognition (Moayer and Fu, 1976), optical character recognition (Ogawa and Taniguchi, 1982), and biomedical diagnosis (Lu and Wang, 1986). The result of thinning a binary image is called the *skeleton*. One advantage of thinning is the reduction of memory space required for storage of the essential structural information presented in a pattern. Moreover, it simplifies the data structure required in pattern analysis.

Although there is no general agreement on the exact mathematical definition of thinness, a reasonable compromise has been reached that the generated skeleton must be essential to preserve the object's topology and to represent the pattern's shape informatively. The resulting skeleton must be also a single pixel in width and lie approximately along the medial axes of the object regions. Many thinning algorithms are available in the literature. Different algorithms produce slightly different skeletons. Usually, thinning algorithms are divided into two groups: sequential and parallel. A more general classification is: sequential, parallel, hybrid, and others (Govindham and Shivaprasad, 1987). It is generally believed that a sequential algorithm will be faster than a parallel algorithm when they are both implemented on a serial computer (Rosenfeld and Pfaltz, 1966). However, the parallel thinning algorithms are best suited for parallel computers to gain efficiency.

In the parallel thinning approach, the new value of a given pixel at the *n*th pass depends on its neighbors over their neighborhood in the (n–1)th pass; therefore, all the pixels can be transformed simultaneously. In order to have a more efficient algorithm, the smaller neighborhood is considered. The 3 × 3 neighborhood is particularly desirable but has difficulty preserving the connectivity of the original shape. Thus, to solve this problem, partially serializing the parallel algorithms by the subiteration or subfield approaches are mostly used. Subiteration approaches employ different thinning operators to the entire image in different subiterations. Subfield approaches partition the image space in some manner and a parallel operator is applied to different partitions of the image in different iterations.

In order to maintain the completely parallel processing, one branch of thinning is to use thinning operators on supports over the neighborhood larger than 3 × 3 and accomplish the fully parallel characteristics that means that the same thinning operators are applied over the entire image at each iteration. An 11-pixel neighborhood thinning operator was developed as a fully parallel thinning algorithm by Guo and Hall (1989, 1992) that claims to be the optimally small support. This algorithm is very efficient because no subiteration is involved but its resulting skeleton is highly affected by boundary noise. No matter how fast the algorithm is, the characteristics and consistency of the resulting skeletons are the first concern; moreover, the boundary noise fluctuation must not be neglected.

Ahmed and Ward (2002) presented a rule-based rotation-invariant thinning algorithm to produce a single-pixel wide skeleton from a binary image. Rockett (2005) showed examples where Ahmed and Ward's algorithm fails on two-pixel wide lines and proposed a modified method which corrects this shortcoming based on graph connectivity. Ji et al. (2007) presented a coarse-to-fine binary-image-thinning algorithm by proposing a template-based pulse-coupled neural-network model.

So far in the existing literature, no precise standard evaluation of thinning algorithms has been established. However, it has been generally accepted that a good skeleton must possess the properties, such as topology, shape, connectivity, and sensitivity to boundary noise. In other words, a good thinning algorithm should have the following characteristics:

(1) The resulting skeleton must be topologically equivalent to the object.
(2) It must run along the medial axes of the object regions.
(3) It must be a single pixel thick or as thin as possible.

(4) It should preserve connectivity for both foreground and background.
(5) It should be noise-insensitive to small protrusions and indentations in the boundaries.
(6) It should prevent extensive erosion and not be completely deleted.
(7) It should require as few iterations as possible.
(8) It should delete pixels symmetrically to avoid bias in certain directions.

5.4.1 The Iterative Thinning Algorithm

In this section, we introduce the iterative thinning algorithm based on shrinking which deletes the object border pixels at each iteration, with consideration of such a deletion will not locally disconnect their neighborhoods (Rosenfeld and Kak, 1982). Let S denote the set of object pixels. If we delete all border points from S, then S will vanish completely. For example, in the following case:

11111111
11111111

Instead, we delete only border points that lie on a given side of S at each iteration. To ensure the obtained skeleton as close to "middle," we use opposite sides alternately, e.g., north, south, east, and west. We provide the conditions under which a border pixel can be deleted without affecting the object connectivity. The border pixel P is called *simple* if the set of 8-neighbors of P that lie in S has exactly one component that is adjacent to P. In other words, deleting a simple pixel from S does not change the connectedness properties of either S or $\overline{S}$; i.e., $S-\{P\}$ has the same component as S. For example, the following case is 8-simple, but not 4-simple because the 3 × 3 neighborhood contains 1 component originally and the deletion of P will cause this component to be disconnected into 2 components if considering 4-connectedness.

$$\begin{bmatrix} 0 & 1 & 1 \\ 1 & P & 0 \\ 1 & 1 & 0 \end{bmatrix}$$

According to the definition of simple point, the isolated point is not simple and the end point is simple. The iterative thinning algorithm is given here:

(1) Delete all border points from a given side of S if they are simple and not end points.
(2) Do this successively from the north, south, east, west, north, south, . . . sides of S until no further changes occur from a complete four-side of north, south, east, and west.

Example 5.8: Perform the iterative thinning algorithm on the following image using 8-connectedness.

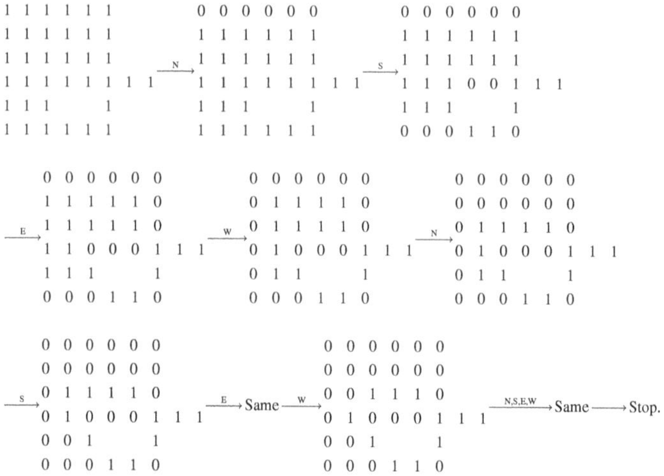

5.4.2 The Fully Parallel Thinning Algorithm

Most of existing thinning algorithms do not have the ability to handle boundary noise. The lack of this capability will lead to the generation of spurious branches and the unpredictable accuracy of the resulting skeleton. This defect will not be avoided if only local 3 × 3, 3 × 4, or 11-pixel neighborhood is considered in the thinning algorithm. Such noise effects can be eliminated by a pre-processing step to smooth out the noisy boundary of the original shape before the application of the thinning or by a post-processing step to delete the short branches which are shorter than a predefined length. Any of these additional steps is undesirable in some applications when real time constraints are required.

Chin et al. (1987) developed an alternative which applies an additional set of noise cleaning templates when a set of thinning templates is used and the templates at each iteration are restored to suppress the growth of spurious branch. However, the number of iterations must be predetermined. Yu and Tsai (1990) developed a thinning algorithm for grayscale images by applying the relaxation technique. A 9 × 9 neighborhood is considered in the computation to keep line straightness and to avoid noise affection. Nevertheless, the complicated formula computation is time-consuming.

5.4.2.1 Definition of a Safe Point

A pixel in a binary image can be either black or white. Usually, black pixels correspond to foreground and white pixels correspond to background. In a 3 × 3 neighborhood, if all eight neighbors are considered while a pixel is processed, it is called the *8-connectivity* for foreground. Similarly, if only four neighbors (north, south, west, and east) are used, it is called the *4-connectivity* for foreground. To avoid connectivity paradoxes, the 8-connectivity for foreground and 4-connectivity for background are used. Considering the eight neighbors, there exist totally 256 permutation. All the permutation can be analytically divided into 69 groups as listed in Figure 5.7. The number in parenthesis denotes the group number from 1 to 69. The number of cases corresponds to the number of different permutations to be generated by rotating the pattern by 90°, 180°, and 270°.

The definitions of border point, end point, break point, and excessive erosion point are given as follows:

- A *border point* is an object point which has at least one of its 4-neighbors belonging to background.
- An *end point* is a border point which has only one 8-adjacent border point.
- A *break point* is a border point and its deletion will cause the loss of connectivity.
- An *excessive erosion point* is a border point and its deletion will cause the loss of the object's shape.

A *safe point* is defined as one of the end point, break point, or excessive erosion point but not the noise point. Safe points are preserved in the current iteration. Non-safe points can be deleted in two ways: no condition or condition-checking. The safe-point status is effective in the present iteration and might be changed into non-safe in the next iteration. The thinning algorithm is defined in terms of the deletion process which transforms certain 1s (i.e., object pixels) to 0s (i.e., background pixels).

Referring to Figure 5.7, since the central points in cases (47), (56), (63), (65), (68), and (69) are not border points, totally 16 cases can not be deleted; otherwise, a hole will be produced. Usually, the central points in cases (1) and (2) are defined as end points, and in cases (3), (4), (13), and (66) are defined as excessive erosion points. In most thinning algorithms the end points and excessive erosion points are not eliminated in order to avoid a totally deleted event and preserve the object's shape. However, an observation is made if they are all preserved, the unnecessary branches would arise. In cases (2), (3), (4), and (13) by the proposed algorithm, the central points are not always safe points and will be deleted under certain deletability conditions. Because in cases (6)–(10), (14)–(18), (20), (22)–(27), (29), (34)–(39), (43), (45), (46), (48), (49), (51), (52), (57), (58), and (67), they are break points, totally 127 cases are safe points. The break points should be preserved in the resulting skeleton to ensure the connectivity.

The remaining 92 cases may be divided into three types: preserved, deleted, and conditionally deleted. Because in cases (5), (19), (40)–(42), (50), (53), (54), (61), and (62), the deletion of the central point will cause the shape's distortion, totally 38 cases are preserved. Because the central points in cases (30), (31), and (44) are deletable without the effectness of distortion and connectivity, there are totally 12 cases to be deleted immediately. Since the parallel thinning algorithm is used, if all the central points are deleted in cases (11), (12), (21), (28), (32), (33), (55), (59), (60), and (64) at the same time, the loss of connectivity and shape will be encountered. Therefore, they are conditionally deletable and some neighbors in the 5 × 5 neighborhood are checked. Overall, 256 cases have been explored.

The definition of boundary noise which depends on the object size can not be clearly given only over the 3 × 3 neighborhood of border points. The size of the neighborhood to be considered to determine the boundary noise increases with the size of an object. The noise can be classified into two types: *existent* and *post-existent.*

FIGURE 5.7 (a) Cases (1)–(35) of 69 groups from 256 permutation of 3 × 3 neighborhood.

FIGURE 5.7 (b) Cases (36)–(69) of 69 groups from 256 permutation of 3 × 3 neighborhood.

Existent noise has been existed before the image is processed. For example, cases (2), (3), (4), and (13) of Figure 5.7 may be an existent noise on a specific condition in a large object. Post-existent noise is produced after the image has been processed. For example, cases (11), (12), (21), (32), and (33) will be a post-existent noise on a specific condition in a large object after the first iteration of thinning. In the experiments and observation, the boundary noise can be eliminated in a reasonable and consistent way if a 5 × 5 neighborhood is considered.

5.4.2.2 Safe-Point Table

A 3 × 3 weighted template shown in Figure 5.8(a) is used to indicate the variations of eight neighboring pixels. The correlation operation is performed by using the template over the 3 × 3 neighborhood to obtain the index value for table look-up. Note that the object pixel is represented as "1" and the background pixel as "0." An example of the index calculation is shown in Figure 5.8(b).

The safe-point look-up table is shown in Table 5.1, which contains two items: *index* and *flag number*. The flag number is determined by looking at the 3 × 3 template

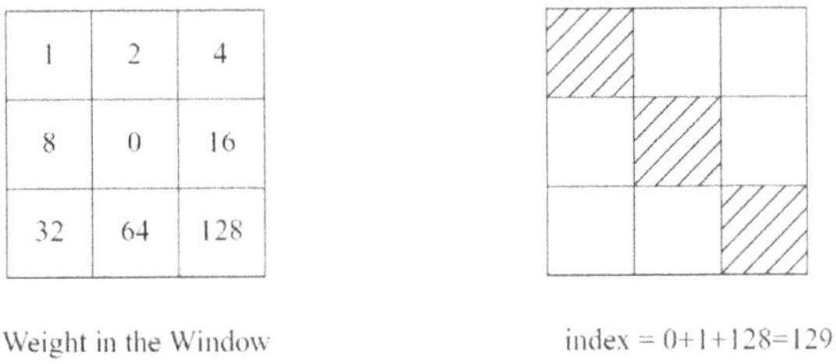

FIGURE 5.8 (a) A 3 × 3 template incorporated with different weights at different neighbors and (b) an example of the index calculation.

TABLE 5.1
The Safe-Point Look-Up Table

Index	Flag Number	Index	Flag Number	Index	Flag Number
1	2	75	2	210	2
3	2	84	1	212	1
4	2	86	2	214	1
6	2	96	2	215	1
7	2	104	2	216	2
9	2	105	1	224	2
11	1	106	2	232	1
14	2	107	2	233	1
15	1	111	2	235	2
19	2	112	1	239	2
20	2	116	1	240	1
22	1	120	2	244	1
23	1	128	2	246	1

Continued

TABLE 5.1
continued

Index	Flag Number	Index	Flag Number	Index	Flag Number
27	2	144	2	247	1
30	2	146	2	248	2
31	1	148	2	249	2
32	2	150	1	252	2
40	2	151	1	253	2
41	2	159	1		
42	2	191	1		
43	1	192	2		
47	1	200	1		
63	1	201	1		
73	1	208	1		

whether the central object point is deletable or conditionally deletable. After the index is calculated and the specific conditions are checked, the object point can be determined whether to be deleted or preserved.

Remark: (1) The flag number of erased points is 1 and of conditionally erased points is 2. (2) Any index number which is not listed here belongs to the safe point.

5.4.2.3 Deletability Conditions

As aforementioned, the 3 × 3 neighborhood is not sufficient to describe the noise. There is an apparent tradeoff to be considered here in order to use the 5 × 5 neighborhood. The strategy of safe-point checking conditions is given as follows: The index of the 3 × 3 neighbors is computed first. If it equals to any index number of the conditionally erasable points, some neighbors of the 5 × 5 neighborhood are necessarily checked. As illustrated in Figure 5.9, for example, the existent noise, three upper-left corner pixels of the 5 × 5 neighbors in case (2.1) are checked since the central pixel is possible to be a noise if all three pixels are object pixels. Any case except case (28) which is not mentioned in Figure 5.9 is a non-safe point.

In cases (3), (4), and (13), the conditions not only involve Boolean operation "AND" but also "OR." Referring to Figure 5.9, let P_{a1}, P_{a2}, P_{a3} denote the neighbors with label "1," P_{b1} and P_{b2} denote the neighbors with label "2," and P_{c1}, P_{c2}, and P_{c3} denote the neighbors with label "0." The deletability Boolean operation can be represented as

$$(P_{a1} \wedge P_{a2} \wedge P_{a3}) \wedge (P_{b1} \vee P_{b2}) \wedge (\bar{P}_{c1} \wedge \bar{P}_{c2} \wedge \bar{P}_{c3}).$$

For example, in case (3.1), if the neighbor denoted by "1" is an object pixel and the left side neighbor denoted by "2" is an object pixel, then the condition is true and the central pixel can be deleted. The conditions in cases (32) and (33) are similar, that is assumed to be a post-existent noise. In case (32.1) if three right side neighbors denoted by "0" belong to background and two neighbors denoted by "1" are object pixels, then the condition is true and the central pixel can be deleted.

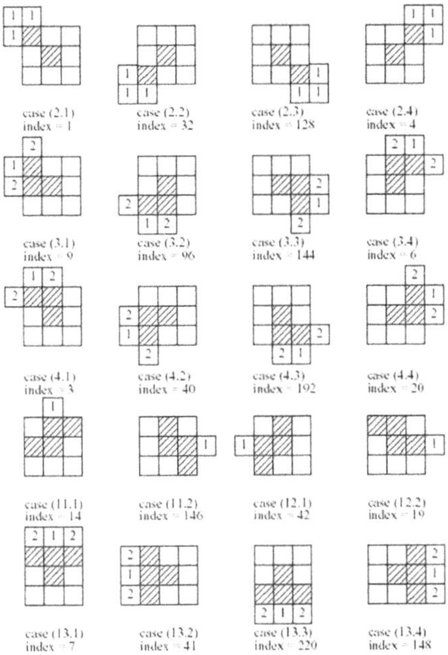

FIGURE 5.9 (a) Deletion-checking conditions of cases (2), (3), (4), (11), (12), and (13).

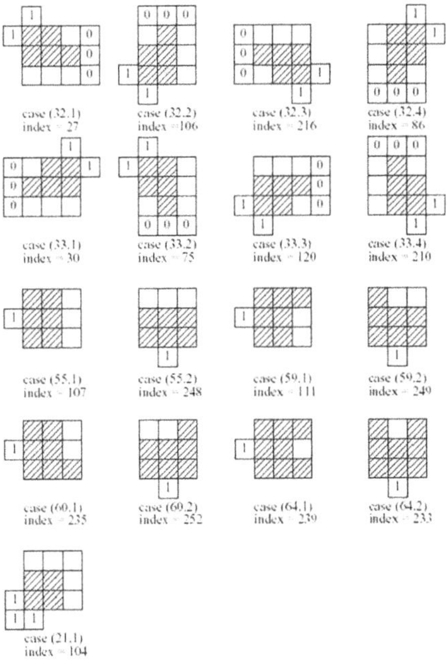

FIGURE 5.9 (b) Deletion-checking conditions of cases (21), (32), (33), (55), (59), (60), and (64).

In two subcases of cases (11), (12), (55), (59), (60), and (64), only one more neighbor needs to be checked in order to preserve the connectivity, and the excessive erosion incident and the other two subcases are deletable immediately. In order to preserve the consistent connectivity of diagonal lines as in case (28), the two subcases with the index numbers 46 and 147 are safe points and the other two are non-safe points, which can be deleted without condition-checking. Three more neighbors in the subcase (21.1), as illustrated in the bottom of Figure 5.9(b), need to be checked, and the other three subcases of case (21) are erasable without condition-checking. The reason is to preserve the consistent connectivity and avoid the excessive erosion. All the deletability conditions have been discussed and the critical concerns are not to conflict with each other during parallel processing and to fulfill the requirements of a good thinning algorithm.

5.4.2.4 The Fully Parallel Thinning Algorithm

Simply applying the safe-point definition and deletability conditions in parallel on the entire image, the thinning result will not be as good as expected. The reason is that the noise definition does not contain all the possibilities that will produce noise after the first few iterations. In the experiments, two additional "expansions" are performed. This expansion step will prevent the early generation of the safe points by adding an object pixel into the intermediate result. The expansion operation performs as follows: If the neighborhood satisfies the conditions shown in Figure 5.10 (note: the index number is the same as defined before, and the pixel denoted as "0" belongs to background), then the central pixel will be turned on. The fully parallel thinning algorithm is described like so (Shih and Wong, 1994, 1995):

(1) The index number is calculated for each object pixel.
(2) If the index number is one of the conditions for expansion, turn on the central pixel.
(3) Repeat step 1.
(4) Use the index number and look-up table to find the flag number.
(5) Check the conditions if it is a conditionally deletable case; otherwise, preserve or delete the pixel accordingly.
(6) If the deletability conditions are satisfied, erase the object pixel; otherwise, preserve it.
(7) Perform steps 1 and 2 once again.
(8) Recursively perform steps 3–5 until no further deletions occur.

5.4.2.5 Experimental Results and Discussion

In most of the existing thinning algorithms, a break point will be a safe point no matter what the circumstance is. It may cause trouble if the existent boundary noise is adjacent to any break point. Except when at least a whole 5 × 5 neighborhood is considered in the thinning algorithm, this problem cannot be solved. The results of the input images in Figures 5.11(a)–(d) are shown in Figures 5.11(e)–(h), respectively.

Three 37 × 31 input images shown in Figures 5.12(a), (b), (c) describe a noisy rectangle and their corresponding results are shown in Figures 5.12(d), (e), (f). The results using the proposed algorithm have the consistency that the similar input shapes will

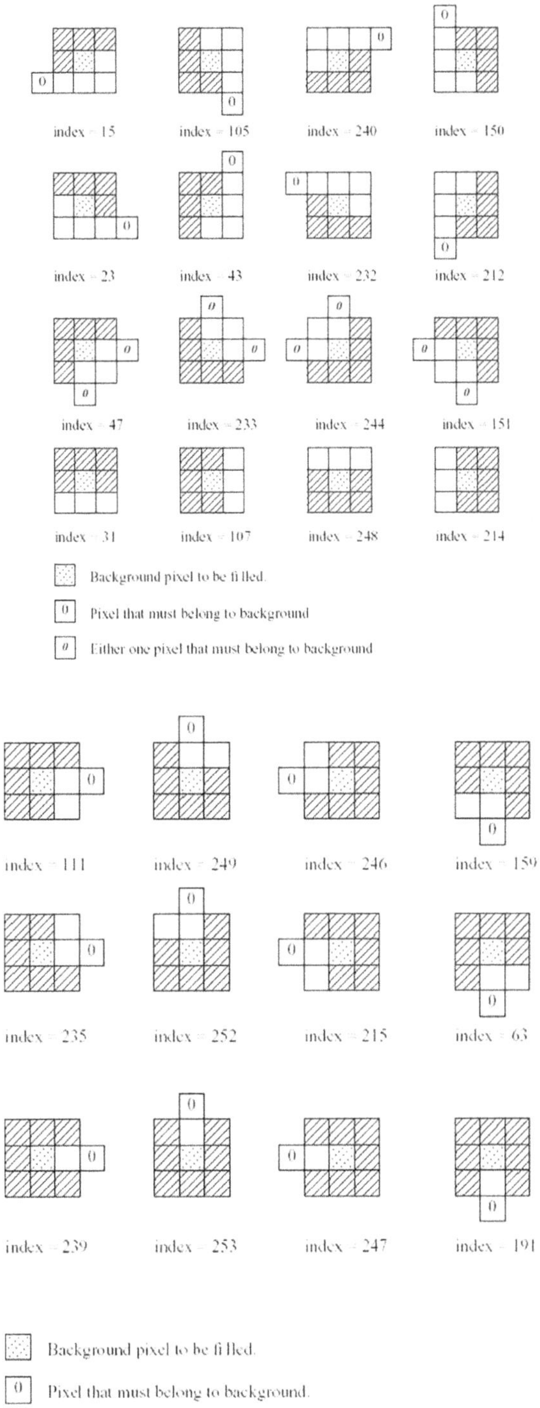

FIGURE 5.10 A set of conditions which are defined to have the expansion operation.

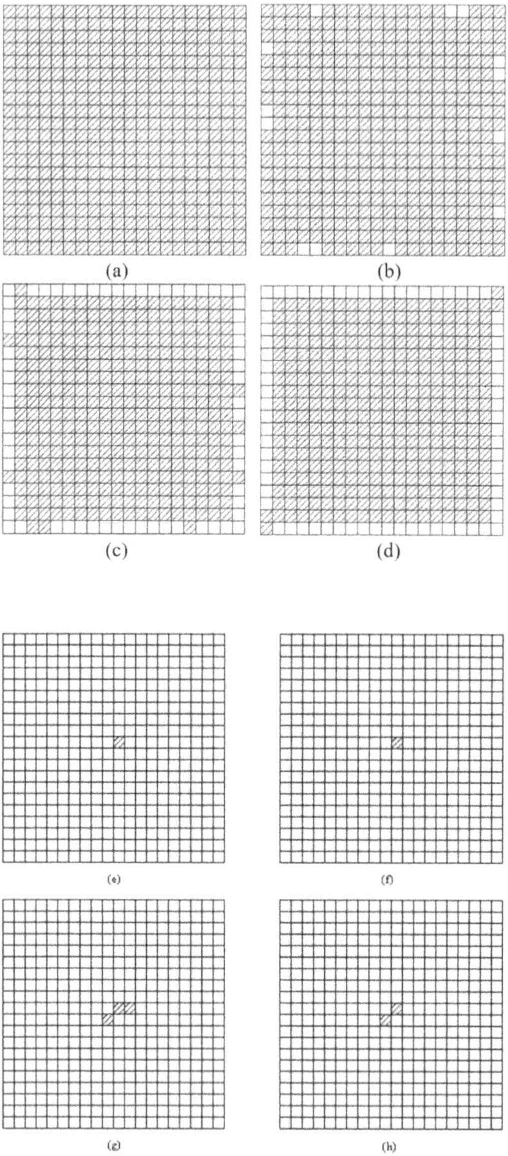

FIGURE 5.11 (e)–(h). The results of (a)–(d) by the thinning algorithm.

produce the similar output skeletons. The resulting skeleton of a wrench using the proposed algorithm is shown in Figure 5.13(a) and using Guo and Hall's algorithm (1992) is shown in Figure 5.13(b) in which lots of small branches are produced. The images of an English character "e" together with 30° and 45°, rotations are shown in Figure 5.14. A set of Chinese characters are also tested as shown in Figure 5.15.

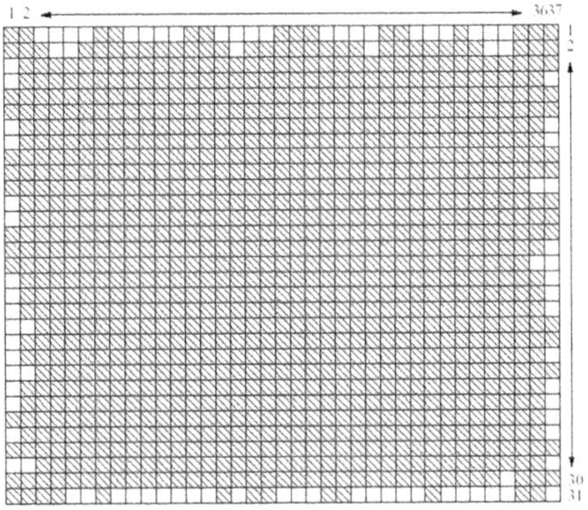

FIGURE 5.12 (a) A 37 × 31 input image describes a noisy rectangle.

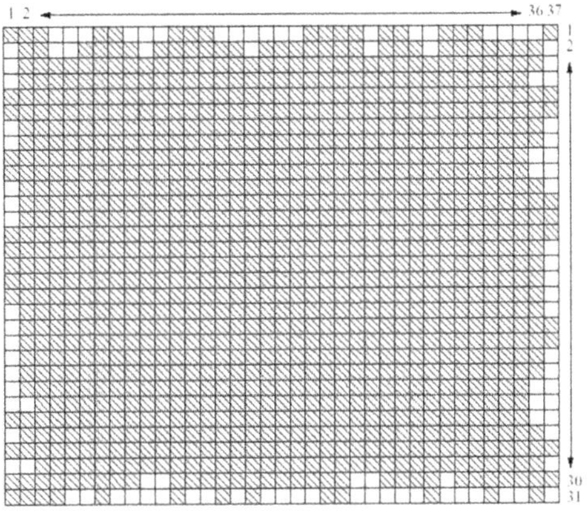

FIGURE 5.12 (b) A 37 × 31 input image describes a noisy rectangle.

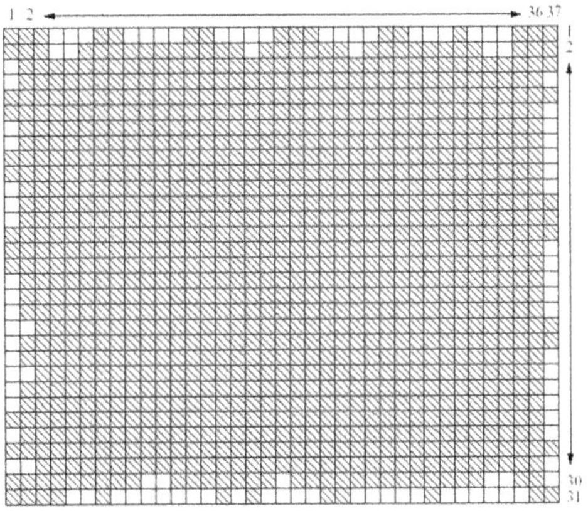

FIGURE 5.12 (c) A 37 × 31 input image describes a noisy rectangle.

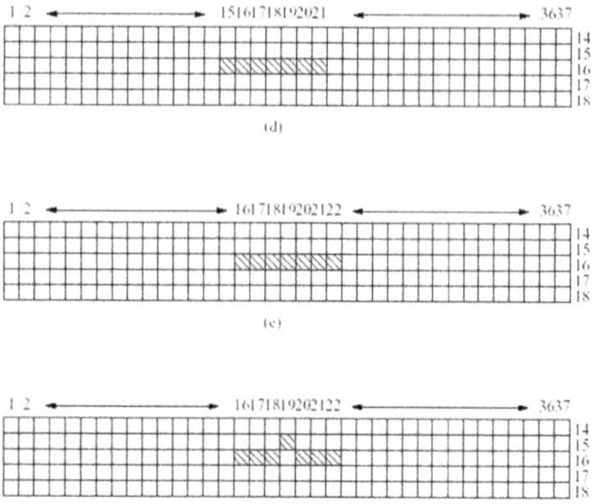

FIGURE 5.12 (d)–(f) The corresponding results of Figures 5.12(a)–(c).

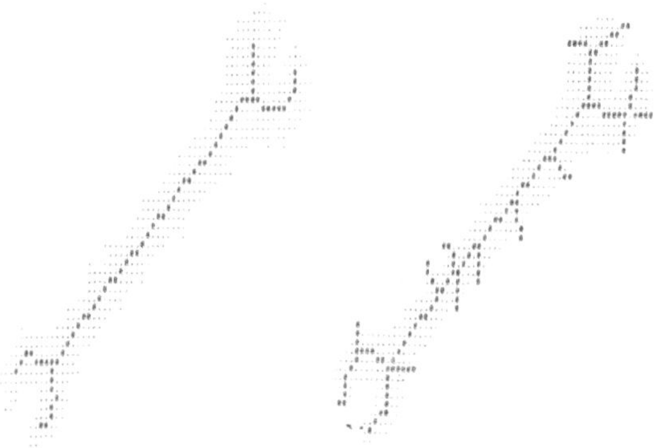

FIGURE 5.13 The resulting skeletons of a wrench (a) using the proposed algorithm (b) using Guo and Hall's algorithm (1992).

FIGURE 5.14 The resulting skeletons of (a) English character "e" and rotated by (b) 30° (c) 45°.

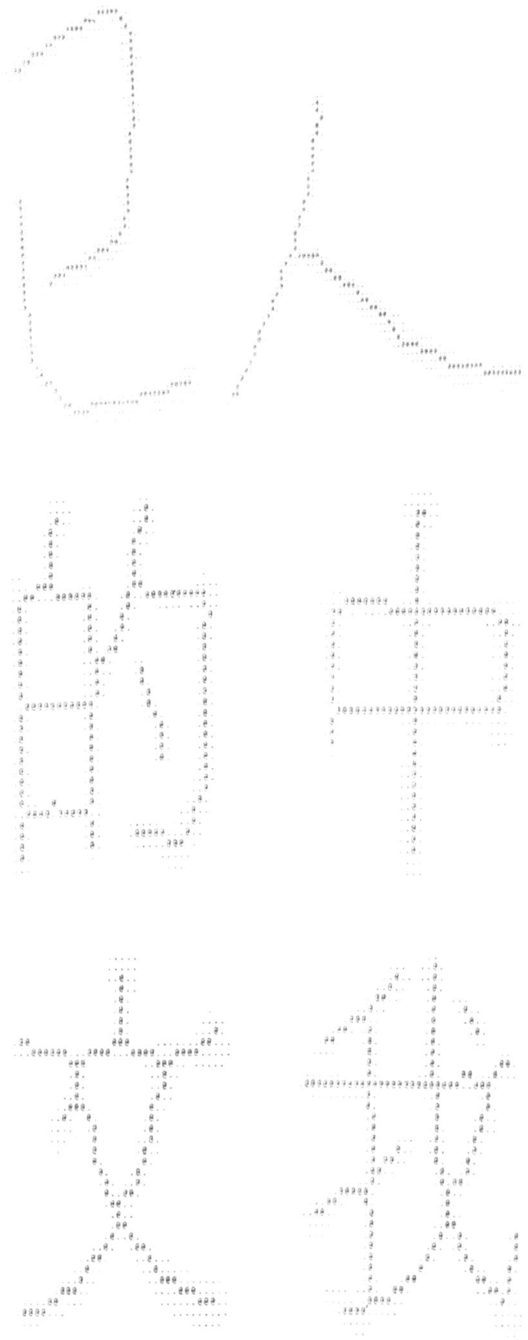

FIGURE 5.15 The resulting skeletons of a set of Chinese characters.

5.5 MEDIAL AXIS TRANSFORMATION

The set of centers and radii (values) of the maximal blocks is called the *medial axis* (or *symmetric axis*) *transformation*, abbreviated as MAT or SAT. Intuitively, if a block S_P is maximal and is contained in the constant-value region S, it must touch the border of S in at least two places; otherwise, we could find a neighbor Q of P that was further away than P from the border of S, and then S_Q would contain S_P. Let the output of the distance transformation be d. A distance transformation converts a binary image, which consists of object (foreground) and non-object (background) pixels, into an image where every object pixel has a value corresponding to the minimum distance to the background. Consider the slope variation in both X and Y axes for d. The union of the summit pixels of the slope in both X and Y directions will be the skeleton. From the distance transformation, one can see that distance change between two neighboring pixels is not more than 1. Hence, the four structuring elements are chosen as follows:

$$h_1 = \begin{bmatrix} 0 \\ 0 \\ \varepsilon \end{bmatrix}, h_2 = \begin{bmatrix} \varepsilon \\ 0 \\ 0 \end{bmatrix}, h_3 = \begin{bmatrix} 0 & 0 & \varepsilon \end{bmatrix}, h_4 = \begin{bmatrix} \varepsilon & 0 & 0 \end{bmatrix},$$

where ε denotes a small positive number less than 1.

MAT Algorithm:

(1) Raster scan the distance image d. Do $d \oplus_g h_1, d \oplus_g h_2, d \oplus_g h_3$, and $d \oplus_g h_4$ at each pixel.

(2) If any one of four resulting values is equal to the original grayvalue at this pixel, then the grayvalue is kept; otherwise, let the grayvalue of this pixel be zero. After all pixels are processed, the output is the skeleton with the grayvalue indicating the distance.

An example of the Euclidean distance transformation and the medial axis transformation is shown in Figure 5.16.

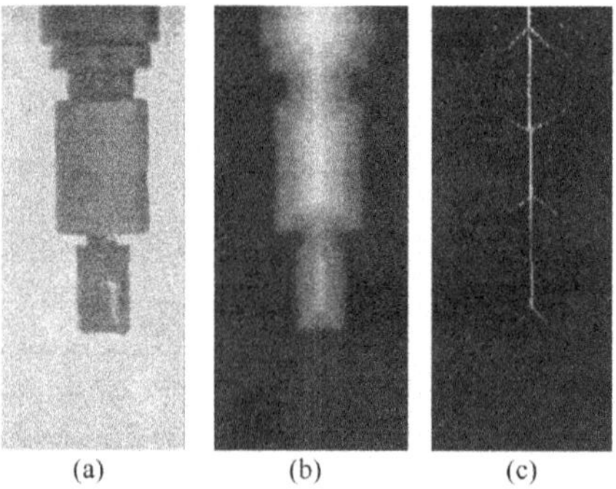

FIGURE 5.16 An example of Euclidean distance transformation and skeletonization. (a) Tool picture, (b) distance transform, (c) skeleton (medial axis).

The Distance Transform from MAT:

Let m denote the grayscale MAT. Using the distance structuring element k, the distance transform reconstruction from the MAT is achieved by the following algorithm:

Reconstruction Algorithm:

The grayscale MAT is dilated by the distance structuring element; i.e., $m \oplus_g k$.

This reconstruction algorithm is quite simple. The result is exactly the distance transform. After thresholding at 1, the original binary image can be recovered.

In *MAT representation*, with each pixel P of an image, the set of upright squares of odd side length centered at P is associated. The largest such a square that is contained in the image and has a constant value is called *a maximal block*. It is easy to see that if the set of centers P and the radii $r(P)$ of the maximal blocks are specified, the image is completely determined. For a binary $N \times N$ image, $2\log N$ bits are needed to specify the x- and y-coordinates of each block center, and $\log(N/2)$ bits to specify the radius. Thus, if there are B blocks in MAT, the total number of bits required to specify is $B(3\log N - 1)$. High efficiency is obtained for images consisting of only a few constant-valued regions which have simple shapes.

A *uniform-region representation* of binary images is usually constructed by introducing a set of basic shapes whose combinational operations construct the image. Although there are many possible basic shapes, the rectangle is one of the most appropriate elementary shapes for digitized binary images. In *rectangular coding*, each region in an image is decomposed into a set of rectangles, and the rectangles are encoded to represent the image in an efficient way. Rectangular coding requires four parameters (the x- and y-coordinates of one vertex, width and height of a rectangle). For an $N \times N$ picture, if the average number of rectangles is R, the total number of bits required for representation is $4R\log N$, i.e., $2\log N$ for each vertex and $2\log N$ for width and height of the maximal rectangle. An isolated point is considered to be a rectangle of zero width and height. High efficiency is obtained for images that can be decomposed into a small number of rectangles.

Since the MAT algorithm produces a skeleton which may be disconnected into several short branches for an object, we present a simple and efficient algorithm using the maxima tracking approach on Euclidean distance transform to detect skeleton points. The advantages of the skeleton obtained are (1) connectivity preservation, (2) single-pixel in width, and (3) its locations as close as to the most symmetrical axes.

5.5.1 Thick Skeleton Generation

A grayscale image is first segmented into a binary image, which is then applied by the Euclidean distance transformation algorithm. An example of the resulting Euclidean distance transformation is shown in Figure 5.17. For any pixel P, we associate a set of disks or circles with various radii centered at P. Let C_P be the largest disk which is completely contained in the object region, and let r_P be the radius

-	-	-	-	-	-	-	-	-	-	-	-	-	-	-	-	-	-	-	-	-	-	-	-	-	-	-	-	-
-	$\underline{10}$	$\underline{10}$	10	10	10	10	10	10	10	10	10	10	10	10	10	10	10	10	10	10	10	10	10	10	$\underline{10}$	$\underline{10}$	$\underline{10}$	-
-	$\underline{10}$	$\underline{20}$	$\underline{20}$	20	20	20	20	20	20	20	20	20	20	20	20	20	20	20	20	20	20	20	20	$\underline{20}$	$\underline{20}$	$\underline{10}$	-	-
-	10	$\underline{20}$	$\underline{30}$	$\underline{30}$	$\underline{30}$	$\underline{30}$	$\underline{30}$	$\underline{30}$	$\underline{30}$	$\underline{30}$	$\underline{30}$	$\underline{30}$	30	30	30	30	30	30	30	30	30	$\underline{30}$	$\underline{30}$	$\underline{22}$	$\underline{14}$	$\underline{10}$	-	-
-	10	20	$\underline{30}$	$\underline{40}$	$\underline{36}$	28	22	20	20	22	28	$\underline{36}$	$\underline{40}$	$\underline{40}$	40	40	40	40	40	$\underline{40}$	$\underline{40}$	$\underline{36}$	$\underline{28}$	20	10	-	-	-
-	10	20	30	$\underline{36}$	28	22	14	10	10	14	22	28	36	$\underline{45}$	$\underline{50}$	50	50	50	$\underline{50}$	$\underline{45}$	36	28	$\underline{22}$	$\underline{14}$	$\underline{10}$	-	-	-
-	10	20	$\underline{30}$	$\underline{32}$	22	14	10	-	-	10	14	22	32	41	$\underline{51}$	$\underline{60}$	$\underline{60}$	$\underline{60}$	$\underline{51}$	41	32	22	14	10	-	-	-	-
-	10	20	$\underline{30}$	$\underline{30}$	20	10	-	-	-	-	10	20	30	40	50	$\underline{60}$	$\underline{70}$	$\underline{60}$	50	40	30	20	10	-	-	-	-	-
-	10	20	$\underline{30}$	$\underline{30}$	20	10	-	-	-	-	10	20	30	40	50	$\underline{60}$	$\underline{70}$	$\underline{61}$	51	41	32	22	14	10	10	$\underline{10}$	$\underline{10}$	-
-	10	20	$\underline{30}$	$\underline{32}$	22	14	10	-	-	10	14	22	32	41	$\underline{51}$	$\underline{60}$	$\underline{60}$	$\underline{60}$	$\underline{54}$	45	36	28	$\underline{22}$	$\underline{20}$	$\underline{20}$	$\underline{20}$	$\underline{10}$	-
-	10	20	30	$\underline{36}$	28	22	14	10	10	14	22	28	36	$\underline{45}$	$\underline{50}$	50	50	50	$\underline{50}$	$\underline{45}$	36	28	$\underline{22}$	$\underline{20}$	$\underline{20}$	$\underline{20}$	$\underline{10}$	-
-	10	20	$\underline{30}$	$\underline{40}$	$\underline{36}$	28	22	20	20	22	28	$\underline{36}$	$\underline{40}$	$\underline{40}$	40	40	40	40	40	$\underline{40}$	$\underline{32}$	22	14	10	10	$\underline{10}$	$\underline{10}$	-
-	10	$\underline{20}$	$\underline{30}$	$\underline{30}$	$\underline{30}$	$\underline{30}$	$\underline{30}$	$\underline{30}$	$\underline{30}$	$\underline{30}$	$\underline{30}$	$\underline{30}$	30	30	30	30	30	30	30	30	$\underline{30}$	$\underline{20}$	10	-	-	-	-	-
-	$\underline{10}$	$\underline{20}$	$\underline{20}$	20	20	20	20	20	20	20	20	20	20	20	20	20	20	20	20	20	$\underline{20}$	$\underline{20}$	$\underline{10}$	-	-	-	-	-
-	$\underline{10}$	$\underline{10}$	10	10	10	10	10	10	10	10	10	10	10	10	10	10	10	10	10	10	10	$\underline{10}$	$\underline{10}$	-	-	-	-	-
-	-	-	-	-	-	-	-	-	-	-	-	-	-	-	-	-	-	-	-	-	-	-	-	-	-	-	-	-

FIGURE 5.17 The Euclidean distance function with the ridge points underlined. For expressional simplicity, we use ten times of the Euclidean distance in rounded numbers.

of C_P. There may exist another pixel Q such that C_Q contains C_P. If no such a Q exists, C_P is called a *maximal disk*. The set of the centers and radii (or Euclidean distance values) of these maximal inscribed disks is called the skeleton.

5.5.1.1 The Skeleton from Distance Function

The distance function of pixel P in the object region S is defined as the smallest distance of P from all the background pixels $\bar{S}$. That is,

$$D(P) = \min_{Q \in \bar{S}} \{d(P,Q)\}, \tag{5.4}$$

where $d(P, Q)$ is the Euclidean distance between P and Q. If we visualize the distance values as the altitude on a surface, then the "ridges" of the surface constitute the skeleton in which a tangent vector cannot be uniquely defined. It can be easily derived that if and only if a point P belongs to the skeleton of an object region S, the maximal disk with the radius $D(P)$ hits the contour of S at least two places. Let us define the set

$$A(P) = \{Q \mid d(P,Q) = D(P), Q \in \bar{S}\}. \tag{5.5}$$

If the set $A(P)$ contains more than one element, then P is a skeleton point.

5.5.1.2 Detection of Ridge Points

Considering two neighbors of P in any of the horizontal, vertical, 45°-diagonal, and 135°-diagonal directions, if the altitude of P is higher than one neighbor and not lower than the other, then P is called the *ridge point*. Figure 5.17 shows the Euclidean distance transform with ridge points underlined. Note that the resulting skeleton on the right-hand side is disconnected from the major skeleton.

5.5.1.3 Trivial Uphill Generation

The trivial uphill of a point P is the set of all the neighbors with a higher altitude. Figure 5.18 shows the resulting skeleton if we add the uphill of ridge points and continuously add the uphill of the new skeleton points until no further uphill can be

.	.	.	.	.	.	.	.	.	.	.	.	.	.	.	.	.	.	.	.	.	.	.	.	.	.	.	.	.
.	**10**	**10**	10	10	10	10	10	10	10	10	10	10	10	10	10	10	10	10	10	10	10	10	10	10	**10**	**10**	**10**	.
.	**10**	**20**	**20**	20	20	20	20	20	20	20	20	20	20	20	20	20	20	20	20	20	20	20	20	**20**	**20**	**10**	.	.
.	10	**20**	**30**	**30**	**30**	**30**	**30**	**30**	**30**	**30**	**30**	**30**	30	30	30	30	30	30	30	30	30	**30**	**30**	**22**	**14**	**10**	.	.
.	10	20	**30**	**40**	**36**	28	22	20	20	22	28	**36**	**40**	**40**	40	40	40	40	40	**40**	**40**	**36**	**28**	**20**	10	.	.	.
.	10	20	30	**36**	28	22	14	10	10	14	22	28	36	**45**	**50**	50	50	50	**50**	**45**	**36**	**28**	**22**	**14**	**10**	.	.	.
.	10	20	**30**	**32**	22	14	10	.	.	10	14	22	32	41	**51**	**60**	**60**	**60**	**51**	**41**	**32**	22	14	10	.	.	.	.
.	10	20	**30**	**30**	20	10	.	.	.	.	10	20	30	40	50	**60**	**70**	**60**	**50**	**40**	30	20	10	.	.	.	.	.
.	10	20	**30**	**30**	20	10	.	.	.	.	10	20	30	40	50	**60**	**70**	**61**	**51**	**41**	**32**	22	14	10	10	**10**	**10**	.
.	10	20	**30**	**32**	22	14	10	.	.	10	14	22	32	41	**51**	**60**	**60**	**60**	**54**	**45**	**36**	**28**	**22**	**20**	**20**	**20**	**10**	.
.	10	20	30	**36**	28	22	14	10	10	14	22	28	36	**45**	**50**	50	50	50	**50**	**45**	**36**	**28**	**22**	**20**	**20**	**20**	**10**	.
.	10	20	**30**	**40**	**36**	28	22	20	20	22	28	**36**	**40**	**40**	40	40	40	40	40	**40**	**32**	**22**	14	10	10	**10**	**10**	.
.	10	**20**	**30**	**30**	**30**	**30**	**30**	**30**	**30**	**30**	**30**	**30**	30	30	30	30	30	30	30	30	**30**	**20**	10	.	.	.	.	.
.	**10**	**20**	**20**	20	20	20	20	20	20	20	20	20	20	20	20	20	20	20	20	20	**20**	**20**	**10**	.	.	.	.	.
.	**10**	**10**	10	10	10	10	10	10	10	10	10	10	10	10	10	10	10	10	10	10	10	**10**	**10**	.	.	.	.	.
.	.	.	.	.	.	.	.	.	.	.	.	.	.	.	.	.	.	.	.	.	.	.	.	.	.	.	.	.

FIGURE 5.18 The ridges points and their trivial uphill generation are underlined. It is indeed connected but too thick.

generated. Finally, a connected skeleton can be obtained, but a thick one. In order to make the skeleton to be single-pixel wide, we need get rid of the loose definition of ridge points and take into account the directional-neighborhood.

5.5.2 Basic Definitions

Prior to the discussion of our algorithm, the basic definitions of base point, apex point, directional-uphill generation, and directional-downhill generation are introduced in the following sections.

5.5.2.1 Base Point

The base point is defined as a corner point which has the distance value 1 and is surrounded by a majority of background points. It belongs to one of the following three configurations as well as their variations up to eight 45°-rotations, respectively:

$$\begin{matrix} 1 & 0 & 0 \\ 1 & \underline{1} & 0 \\ 0 & 0 & 0 \end{matrix}, \quad \begin{matrix} 1 & 1 & 0 \\ 1 & \underline{1} & 0 \\ 0 & 0 & 0 \end{matrix}, \text{ and } \begin{matrix} 1 & 1 & 1 \\ 1 & \underline{1} & 0 \\ 0 & 0 & 0 \end{matrix},$$

where the center pixels underlined represent the 45°, 90°, and 135° corner points, respectively. In a digital image, only the above three degrees are considered in a local 3 × 3 window.

If all the three configurations are considered as base points, then more nonsignificant short skeletal branches are produced. In other words, an approach based on the skeletal branches originating from all corner points in a 3 × 3 window would lead to unmanageable complexity in the skeletal structure. Therefore, an appropriate shape-informative skeleton should reach a compromise among the representativity of the connected object structure, the required reconstructivity, and the cost of deleting nonsignificant branches.

We consider sharper convexities as more significant. In detail, if the amplitude of the angle formed by two intersecting wavefronts of the fireline is viewed as the

-	-	-	-	-	-	-	-	-	-	-	-	-	-	-	-	-	-	-	-	-	-	-	-	-	-	-	-	-
-	B	10	10	10	10	10	10	10	10	10	10	10	10	10	10	10	10	10	10	10	10	10	10	10	10	10	B	-
-	10	20	20	20	20	20	20	20	20	20	20	20	20	20	20	20	20	20	20	20	20	20	20	20	20	10	-	-
-	10	20	30	30	30	30	$\underline{30}$	$\underline{30}$	$\underline{30}$	$\underline{30}$	30	30	30	30	30	30	30	30	30	30	30	30	30	22	14	10	-	-
-	10	20	30	$\underline{40}$	36	28	22	20	20	22	28	36	40	40	40	40	40	40	40	40	40	36	28	20	10	-	-	-
-	10	20	30	36	28	22	14	10	10	14	22	28	36	45	50	50	50	50	50	45	36	28	22	14	10	-	-	-
-	10	20	30	32	22	14	10	-	-	10	14	22	32	41	51	60	60	60	51	41	32	22	14	10	-	-	-	-
-	10	20	30	30	20	10	-	-	-	-	10	20	30	40	50	60	$\underline{70}$	60	50	40	30	20	10	-	-	-	-	-
-	10	20	30	30	20	10	-	-	-	-	10	20	30	40	50	60	$\underline{70}$	61	51	41	32	22	14	10	10	10	B	-
-	10	20	30	32	22	14	10	-	-	10	14	22	32	41	51	60	60	60	54	45	36	28	22	20	$\underline{20}$	$\underline{20}$	10	-
-	10	20	30	36	28	22	14	10	10	14	22	28	36	45	50	50	50	50	50	45	36	28	22	20	$\underline{20}$	$\underline{20}$	10	-
-	10	20	30	$\underline{40}$	36	28	22	20	20	22	28	36	40	40	40	40	40	40	40	40	32	22	14	10	10	10	B	-
-	10	20	30	30	30	30	$\underline{30}$	$\underline{30}$	$\underline{30}$	$\underline{30}$	30	30	30	30	30	30	30	30	30	30	30	20	10	-	-	-	-	-
-	10	20	20	20	20	20	20	20	20	20	20	20	20	20	20	20	20	20	20	20	20	20	10	-	-	-	-	-
-	B	10	10	10	10	10	10	10	10	10	10	10	10	10	10	10	10	10	10	10	10	10	B	-	-	-	-	-
-	-	-	-	-	-	-	-	-	-	-	-	-	-	-	-	-	-	-	-	-	-	-	-	-	-	-	-	-

FIGURE 5.19 The Euclidean distance function with the base points labeled "B" and the apex points underlined.

parameter characterizing the sharpness, the values smaller than or equal to 90° are identified as suitable convexities. More strictly, only 45°-convexities are detected as base points. By using these points as the source to grow up the skeleton, the remaining procedures are acquired to preserve the skeletal connectedness.

5.5.2.2 Apex Point

The apex point is the pixel being the local maximum in its 3×3 neighborhood. Note that the local-maximum pixels only construct a small portion of a disconnected skeleton. They occur as 45°-corner points or interior elements which have the highest altitude locally. The 45°-corner points also serve as base points. The base points and apex points are considered as sources or elementary cells of the skeleton which grows up emitting from them. Figure 5.19 shows the Euclidean distance function with the base points labeled "B" and the apex points underlined.

5.5.2.3 Directional-Uphill Generation

The set of points $\{P_i\}$ is called the *directional-neighborhood* of P, denoted by D_P, if they are in the 8-neighborhood of P and located within ±45° slope changes from the current medial axis orientation of P. For example, using the 8-neighbors labeled $P_1, P_2, \cdots,$ and P_8 counterclockwise from the positive x-axis of P, if P_7 and P are the skeleton points, the points $P_2, P_3,$ and P_4 are the directional-neighbors of P; i.e., $D_P = \{P_2, P_3, P_4\}$. Several cases are illustrated in the following.

$$\begin{bmatrix} P_4 & P_3 & P_2 \\ \bullet & P & \bullet \\ \bullet & P_7 & \bullet \end{bmatrix}, \begin{bmatrix} \bullet & \bullet & P_2 \\ P_5 & P & P_1 \\ \bullet & \bullet & P_8 \end{bmatrix}, \begin{bmatrix} P_4 & P_3 & \bullet \\ P_5 & P & \bullet \\ \bullet & \bullet & P_8 \end{bmatrix}, \text{ and } \begin{bmatrix} \bullet & P_3 & P_2 \\ \bullet & P & P_1 \\ P_6 & \bullet & \bullet \end{bmatrix}$$

Note that the set of directional neighbors always contains three elements. The directional-uphill generation adds the point, which is the maximum of P, and its directional-neighborhood. That is to say,

$$P^U_{\text{next}} = \max_{P_i \in D_P \cup \{P\}} \{P_i\}. \tag{5.6}$$

-	-	-	-	-	-	-	-	-	-	-	-	-	-	-	-	-	-	-	-	-	-	-	-	-	-	-	-	-
-	10	10	10	10	10	10	10	10	10	10	10	10	10	10	10	10	10	10	10	10	10	10	10	10	10	10	10	-
-	10	20	20	20	20	20	20	20	20	20	20	20	20	20	20	20	20	20	20	20	20	20	20	20	20	10	-	-
-	10	20	30	30	30	30	30	30	30	30	30	30	30	30	30	30	30	30	30	30	30	30	30	22	14	10	-	-
-	10	20	30	40	36	28	22	20	20	22	28	36	40	40	40	40	40	40	40	40	40	36	28	20	10	-	-	-
-	10	20	30	36	28	22	14	10	10	14	22	28	36	45	50	50	50	50	50	45	36	28	22	14	10	-	-	-
-	10	20	30	32	22	14	10	-	-	10	14	22	32	41	51	60	60	60	51	41	32	22	14	10	-	-	-	-
-	10	20	30	30	20	10	-	-	-	-	10	20	30	40	50	60	70	60	50	40	30	20	10	-	-	-	-	-
-	10	20	30	30	20	10	-	-	-	-	10	20	30	40	50	60	70	61	51	41	32	22	14	10	10	10	10	-
-	10	20	30	32	22	14	10	-	-	10	14	22	32	41	51	60	60	60	54	45	36	28	22	20	20	20	10	-
-	10	20	30	36	28	22	14	10	10	14	22	28	36	45	50	50	50	50	50	45	36	28	22	20	20	20	10	-
-	10	20	30	40	36	28	22	20	20	22	28	36	40	40	40	40	40	40	40	40	32	22	14	10	10	10	10	-
-	10	20	30	30	30	30	30	30	30	30	30	30	30	30	30	30	30	30	30	30	30	20	10	-	-	-	-	-
-	10	20	20	20	20	20	20	20	20	20	20	20	20	20	20	20	20	20	20	20	20	20	10	-	-	-	-	-
-	10	10	10	10	10	10	10	10	10	10	10	10	10	10	10	10	10	10	10	10	10	10	10	-	-	-	-	-
-	-	-	-	-	-	-	-	-	-	-	-	-	-	-	-	-	-	-	-	-	-	-	-	-	-	-	-	-

FIGURE 5.20 The results of directional-uphill generation of Figure 5.19.

Figure 5.20 shows the result of directional-uphill generation of Figure 5.19.

5.5.2.4 Directional-Downhill Generation

From Figure 5.20, we observe that there should exist a vertical path connecting two apex points of value 40 underlined on the left-hand side, in which the altitude changes are not always increasing; instead, they are a mixture of decreasing and increasing values. The directional-downhill generation, which is similar to the directional-uphill generation except the maxima tracking of the set excluding the central point P, is used to produce this type of skeletal branch. That is to say,

$$P^{D}_{\text{next}} = \max_{P_i \in D_P} \{P_i\}. \tag{5.7}$$

The directional-downhill generation is initialized from the apex points which cannot be further tracked by the directional-uphill generation. Hence, the altitude of the next directional-downhill should be lower in the beginning. However, the tracking procedure is continued without taking into account the comparison of neighbors with P. The next directional-downhill altitude could be lower or even higher until a skeleton point appears in the directional neighborhood. Figure 5.21 shows the result of directional-downhill generation of Figure 5.20. The skeleton is now connected and possesses single-pixel in width except for two pixels having the same local maximum of Euclidean distance.

5.5.3 The Skeletonization Algorithm and Connectivity Properties

The skeletonization algorithm traces the skeleton points by choosing the local maxima on the Euclidean distance transform and takes into consideration of the least slope changes in medial axes. The algorithm is described as follows:

(1) The base points and apex points are detected as the initial skeleton points.
(2) Starting with these skeleton points, the directional-uphill generation in equation (8.6) is used to add more skeleton points, and continuously add the directional-uphill of the new skeleton points until no further point can be generated. The points which cannot be further tracked are marked.

.	.	.	.	.	.	.	.	.	.	.	.	.	.	.	.	.	.	.	.	.	.	.	.	.	.	.	.	.
.	**10**	10	10	10	10	10	10	10	10	10	10	10	10	10	10	10	10	10	10	10	10	10	10	10	10	**10**	**10**	.
.	10	**20**	20	20	20	20	20	20	20	20	20	20	20	20	20	20	20	20	20	20	20	20	20	20	**20**	10	.	.
.	10	20	**30**	30	30	**30**	**30**	**30**	**30**	**30**	**30**	30	30	30	30	30	30	30	30	30	30	30	**30**	**22**	14	10	.	.
.	10	20	30	**40**	**36**	28	22	20	20	22	28	**36**	**40**	40	40	40	40	40	40	40	**40**	**36**	28	20	10	.	.	.
.	10	20	30	**36**	28	22	14	10	10	14	22	28	36	**45**	50	50	50	50	50	**45**	36	28	22	14	10	.	.	.
.	10	20	30	**32**	22	14	10	.	.	10	14	22	32	41	**51**	60	60	60	**51**	41	32	22	14	10	.	.	.	.
.	10	20	30	**30**	20	10	.	.	.	.	10	20	30	40	50	**60**	**70**	**60**	50	40	30	20	10	.	.	.	.	.
.	10	20	30	**30**	20	10	.	.	.	.	10	20	30	40	50	**60**	**70**	61	51	41	32	22	14	10	10	10	**10**	.
.	10	20	30	**32**	22	14	10	.	.	10	14	22	32	41	**51**	60	60	**60**	**54**	**45**	**36**	**28**	**22**	**20**	**20**	**20**	10	.
.	10	20	30	**36**	28	22	14	10	10	14	22	28	36	**45**	50	50	50	50	**50**	45	36	28	22	20	**20**	**20**	10	.
.	10	20	30	**40**	**36**	28	22	20	20	22	28	**36**	**40**	40	40	40	40	40	40	**40**	32	22	14	10	10	10	**10**	.
.	10	20	**30**	30	30	**30**	**30**	**30**	**30**	**30**	**30**	30	30	30	30	30	30	30	30	30	**30**	20	10	.	.	.	.	.
.	10	**20**	20	20	20	20	20	20	20	20	20	20	20	20	20	20	20	20	20	20	20	**20**	10	.	.	.	.	.
.	**10**	10	10	10	10	10	10	10	10	10	10	10	10	10	10	10	10	10	10	10	10	10	**10**	.	.	.	.	.
.	.	.	.	.	.	.	.	.	.	.	.	.	.	.	.	.	.	.	.	.	.	.	.	.	.	.	.	.

FIGURE 5.21 The results of directional-downhill generation of Figure 5.20.

-	-	-	-	-	-	-	-	-	-	-	-	-	-	-	-	-	-	-	-	-	-	-	-	-	-	-	-	-
-	M	O	O	O	O	O	O	O	O	O	O	O	O	O	O	O	O	O	O	O	O	O	O	O	O	M	M	-
-	O	M	O	O	O	O	O	O	O	O	O	O	O	O	O	O	O	O	O	O	O	O	O	O	M	O	-	-
-	O	O	M	O	O	M	M	M	M	M	M	O	O	O	O	O	O	O	O	O	O	O	M	M	O	O	-	-
-	O	O	O	M	M	O	O	O	O	O	O	M	M	O	O	O	O	O	O	O	M	M	O	O	O	-	-	-
-	O	O	O	M	O	O	O	O	O	O	O	O	O	M	O	O	O	O	O	M	O	O	O	O	O	-	-	-
-	O	O	O	M	O	O	O	-	-	O	O	O	O	O	M	O	O	O	M	O	O	O	O	O	-	-	-	-
-	O	O	O	M	O	O	-	-	-	-	O	O	O	O	O	M	M	M	O	O	O	O	O	-	-	-	-	-
-	O	O	O	M	O	O	-	-	-	-	O	O	O	O	O	M	M	O	O	O	O	O	O	O	O	O	M	-
-	O	O	O	M	O	O	O	-	-	O	O	O	O	O	M	O	O	M	M	M	M	M	M	M	M	M	O	-
-	O	O	O	M	O	O	O	O	O	O	O	O	O	M	O	O	O	O	M	O	O	O	O	O	M	M	O	-
-	O	O	O	M	M	O	O	O	O	O	O	M	M	O	O	O	O	O	O	M	O	O	O	O	O	O	M	-
-	O	O	M	O	O	M	M	M	M	M	M	O	O	O	O	O	O	O	O	O	M	O	O	-	-	-	-	-
-	O	M	O	O	O	O	O	O	O	O	O	O	O	O	O	O	O	O	O	O	O	M	O	-	-	-	-	-
-	M	O	O	O	O	O	O	O	O	O	O	O	O	O	O	O	O	O	O	O	O	O	M	-	-	-	-	-
-	-	-	-	-	-	-	-	-	-	-	-	-	-	-	-	-	-	-	-	-	-	-	-	-	-	-	-	-

FIGURE 5.22 The resulting skeleton on a rectangle with holes and notches.

(3) Starting with the marked points, the directional downhill generation in equation (8.7) is used to complete the skeleton tracking.

Figure 5.22 illustrates the skeleton of a rectangle with holes and notches obtained by the skeletonization algorithm, where "M" indicates the skeleton pixel.

In the experiment, the 8-connectedness is applied for foreground and the 4-connectedness for background. The skeletonization algorithm possesses the following three connectivity properties:

(C1) After skeletonization, an originally connected object will not be separated into two or more sub-objects.

(C2) After skeletonization, a connected object will not disappear at all.

(C3) After skeletonization, the originally disconnected background components are not 4-connected.

The proofs of the three properties are given here.

Proof of C1. *Induction hypothesis:* for any m ($m < n$) marked apex points of an object, the skeleton obtained will preserve the property C1, where n could be any number.

The base case is $m = 1$. The algorithm starts tracking from each of the base and the apex points, and it will stop when the current point is connected with another skeleton point or touches with the point marked. That means every sub-skeleton branch starting from the base point will connect with some skeleton point or meet with another branch at the marked apex point.

When $m = 2$, it is regarded as two sub-objects, each sub-object contains a marked apex point. As discussed, each sub-object is 8-connected. The algorithm will trace from each marked-apex and stop when it is connected with some skeleton points contained in the other skeleton subset. The reason that it would not connect with a skeleton point which belongs to the same skeleton subset is the directional-neighbors that we use. Using directional-neighbors will enforce the tracking direction to go toward the region which has not been tracked up to now. Finally, it will lead tracking toward another skeleton subset.

As the induction hypothesis claim, when $m = n$, the skeleton obtained is 8-connected. Now consider the case when $m = n + 1$. Those $n + 1$ marked-apex points can be regarded as two sub-objects: One contains n marked-apex points and the other contains only a marked-apex (i.e., the $(n + 1)$th point). As discussed, the one with n marked-apex points is 8-connected, and the other with a marked-apex is also 8-connected. Tracking from the $(n + 1)$th marked-apex, it will lead the tracking to go toward the skeleton subset with n marked-apex points. That means the whole object is 8-connected.

Proof of C2. Tracking starts from each base and apex points. The algorithm marks base points as the skeleton points. For any size of an object, there must be at least one base point or at least one apex point, and the skeleton obtained must have at least one point. For the ideal case of a circle, there is no base point but one apex point is present, which is the local maximum representing the center of the circle. That is, after skeletonization, an 8-connected object will not disappear at all, and the skeleton contains at least a pixel.

Proof of C3. According to the definition of the apex points, there exists at least one apex in the object between any two background components. As the algorithm is performed, it will trace from each apex point which is not 8-connected to any skeleton point. After tracking, there will be one skeleton branch that will make these two background components disconnected. Therefore, the skeleton obtained will not allow the originally disconnected background components to be 4-connected.

The algorithm including three procedures and two functions is described in the following pseudo-codes.

```
Procedure SKEPIK
  /* trace starting from each base and apex points */
  for i, j in 1 ··· N, 1 ··· N loop
    if BASE_APEX (p(i, j)) then
      Uphill-Generation (p(i, j), p(i, j));
```

```
end if
end loop
/* trace from each marked apex point in the procedure
  UpHill-Generation */
while (marked _ pixel)
  DownHill-Generation (marked _ apex, marked _ apex);
end while
end SKEPIK
function APEX (p): Boolean
/* apex point is the local maximum point */
  if (p is local maxima) then
    return TRUE
  else
    return FALSE;
  end APEX
function BASE _ APEX (p): Boolean
/* Base point is the point with distance 1 and has 4
  or more zeros in its 8-neighborhood */
  if (distance of p = 1) then
    find 8-neighbors of p;
  if ((number of 8-neighbors with distance 0) > 4)
  then
    return TRUE;
  else
    if APEX (p) then
      return TRUE;
    else
      return FALSE;
  end if
end if
  end if
end BASE _ APEX
procedure Uphill-Generation (current-skeleton-pixel,
  previous-skeleton-pixel)
  if (number of maximum in 8-neighbors > 1 and
    distance of maximum ≥ distance of current-skele-
    ton-pixel) then
      maximum-pixel = the maximum of the
      directional-neighbors;
      UpHill-Generation (maximum-pixel,
      current-skeleton-pixel);
else
  if (number of maximum in 8-neighbors = 1 and
    distance of maximum > distance of current-skele-
    ton-pixel) then
      maximum-pixel = the maximum;
      UpHill-Generation (maximum-pixel,
      current-skeleton-pixel);
  else
    mark-apex; /* mark current-skeleton-pixel for
    later processing */
```

```
    end if
    end if
  end UpHill-Generation
  procedure DownHill-Generation (current-skeleton-
    pixel, previous-
    skeleton-pixel)
    maximum-pixel = the maximum of the
    directional-neighbors;
    if (maximum-pixel is not a skeleton point) then
      DownHill-Generation (maximum-pixel,
      current-skeleton-pixel);
    end if
  end DownHill-Generation
```

5.5.4 A Modified Algorithm

The modified maxima tracking algorithm will extract the skeleton by eliminating non-significant short skeletal branches, which touch the object boundary at corners. It is different from the previous algorithm that detects the base and apex points as initial skeleton points. Instead, the maxima tracking starts from the apex points only. The algorithm will recursively repeat the same procedure which selects the maxima in the directional-neighborhood as the next skeleton point until another apex points are reached.

The modified maxima tracking algorithm is given as follows (Shih and Pu, 1995):

(1) The apex points are detected as the initial skeleton points.
(2) Starting with each apex point, we use the directional-uphill generation to generate the skeleton points. Recursively repeat this procedure until an apex point is reached, and then the apex point is marked.
(3) Starting with these marked apex points, the directional-downhill generation is used to track the new skeleton points.

Figure 5.23 illustrates the resulting skeleton by using the modified algorithm, where there is a closed curve indicating an inside hole.

Figure 5.24 illustrates that the character "e" and its rotations by 30°, 45°, and 90° using the modified algorithm will produce the identical rotated skeleton provided the digitization error is disregarded.

```
- - - - - - - - - - - - - - - - - - - - - - - - - - - - -
- O O O O O O O O O O O O O O O O O O O O O O O O O M M -
- O O O O O O O O O O O O O O O O O O O O O O O O M O - -
- O O O O O M M M M M M O O O O O O O O O O O M M O O - -
- O O O M M O O O O O O M M O O O O O O O M M O O O - - -
- O O O M O O O O O O O O O M O O O O O M O O O O O - - -
- O O O M O O O - - O O O O O M O O O M O O O O O - - - -
- O O O M O O - - - - O O O O O M M M O O O O O - - - - -
- O O O M O O - - - - O O O O O M M O O O O O O O O O O -
- O O O M O O O - - O O O O O M O O M M M M M M M M M O -
- O O O M O O O O O O O O O M O O O O O O O O O O M M O -
- O O O M M O O O O O O M M O O O O O O O O O O O O O O -
- O O O O O M M M M M M O O O O O O O O O O O O - - - - -
- O O O O O O O O O O O O O O O O O O O O O O O - - - - -
- O O O O O O O O O O O O O O O O O O O O O O O - - - - -
- - - - - - - - - - - - - - - - - - - - - - - - - - - - -
```

FIGURE 5.23 The resulting skeleton using the modified algorithm on a rectangle with holes and notches.

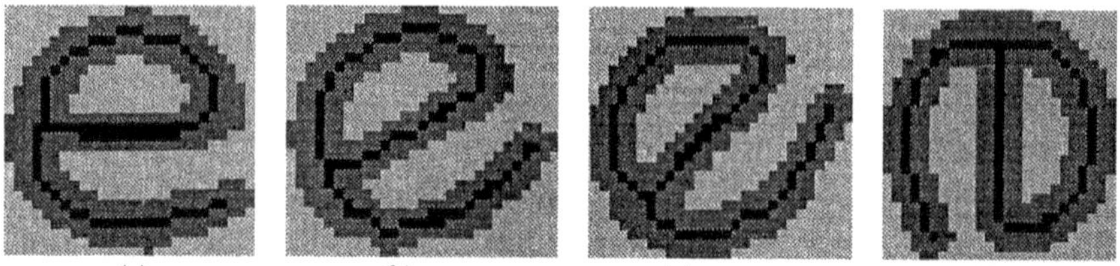

(a) (b) (c) (d)

FIGURE 5.24 An example of the modified skeleton for a set of the character "e" in various rotaions. (a) 0°, (b) 30°, (c) 45°, and (d) 90°.

5.6 OBJECT REPRESENTATION AND TOLERANCE

It has been well recognized that solid modeling is the foundation for CAD/CAM integration. The ultimate goal of planning for the automated manufacturing inspections and robotic assembly is to be able to generate a complete process plan automatically, starting from a CAD representation of the mechanical components. The representation must not only possess the nominal (or ideal) geometric shapes, but also reason the geometric inaccuracies (or tolerances) into the locations and shapes of solid objects. Most of the existing solid modeling systems (in the market) have the very efficient graphical interactive input front-end. Unfortunately, they are not satisfactory for such a fully integrated environment in the following two aspects:

(1) Information sufficiency problem: The existing solid models do not support sufficient information for design and manufacturing which require the specification of dimension, tolerance, surface finishing, data structure, and machining features.
(2) Input data redundancy problem: The input data of the existing solid models are inadequate or more than the data required to specify a part.

Although many solid models have been proposed (Requicha, 1980), only boundary representation and CSG representation are popularly used as the internal database (Requicha and Voelcker, 1983). Boundary representation consists of two kinds of information – topological information and geometric information – which represent the vertex coordinates, surface equations, and the connectivity among faces, edges, and vertices. There are several advantages in boundary representation: large domain, unambiguity, unique, and explicit representation of faces, edges, and vertices. There are also several disadvantages: verbose data structure, difficult to create, difficult to check validity, and variational information unavailability.

The idea of CSG representation is to construct a complex part by hierarchically combining simple primitives using Boolean set operations. There are several advantages in CSG representation: large domain, unambiguity, easy to check validity, and easy to create. There are also several disadvantages: non-unique, difficult to edit graphically, input data redundancy, and variational information unavailability.

The mathematical framework used for modeling solids is mathematical morphology (Shih, 1991; Shih and Gaddipati, 2003, 2005). Adopting mathematical morphology as a tool, the theoretical research aims at studying the representation schemes for the dimension and tolerance of the geometric structure. This section is divided into three parts. The first part defines the representation framework for characterizing dimension and tolerance of solid objects. The second part then adopts the framework to represent several illustrated 2D and 3D objects. The third part describes the added tolerance information to control the quality of the parts and the interchangeability of the parts among assemblies. With the help of variational information, it is known how to manufacture, setup and inspect to ensure the products within the required tolerance range.

5.6.1 Representation Framework: Formal Languages and Mathematical Morphology

There are rules that associate measured entities with features. Measured central planes, axes, and thinned components are associated with symmetric features. The representation framework is formalized as follows.

Let E^N denote the set of all points in the N-dimensional Euclidean space and $p = (x_1, x_2, \cdots, x_N)$ represent a point in E^N. In what follows, any object is a subset of E^N. The formal model will be a context-free grammar, G, consisting of a 4-tuple (Fu, 1982; Ghosh, 1988):

$$G = (V_N, V_T, P, S),$$

where V_N is a set of non-terminal symbols, such as complicated shapes; V_T is a set of terminal symbols which contains two sets: one is the decomposed primitive shapes, such as lines and circles, and the other is the shape operators; P is a finite set of rewrite rules or productions denoted by $A \to \beta$, where $A \in V_N$ and β is a string over $V_N \cup V_T$; and S is the start symbol which is the solid object. The operators include morphological dilation and erosion, set union and intersection, and set subtraction.

Note that a production of such a form allows the non-terminal A to be replaced by the string β independent of the context in which the A appears. The grammar G is context-free, since for each production in P, the left part is a single non-terminal and the right part is a non-empty string of terminals and non-terminals. The languages generated by context-free grammars are called context-free languages. The object representation task may be reviewed as the task of converting a solid shape into a sentence in the language, while object recognition is the task of "parsing" a sentence.

There is no general solution for the primitive selection problem at this time. This determination will be largely influenced by the nature of the data, the specific application in question, and the technology available for implementing the system. The following requirements usually serve as a guide for selecting the shape primitives.

(1) The primitives should serve as basic shape elements to provide a compact but adequate description of the object shape in terms of the specified structural relations (e.g., the concatenation relation).
(2) The primitives should be easily extractable by the existing non-syntactic (e.g., decision-theoretic) methods since they are considered to be simple and compact shapes and their structural information is not important.

5.6.2 Dimensional Attributes

Dimension and tolerance are the main components of the variational information. Dimensions are the control parameters that a designer uses to specify the shape of a part without redundancy, and tolerances are essentially associated with the dimensions to specify how accurate the part should be made. For a detailed description of dimension and tolerance refer to American National Standards Institute (1982).

5.6.2.1 The 2D Attributes

The commonly used 2D attributes are rectangle, parallelogram, triangle, rhomb, circle, and trapezoid. These can be represented easily in the morphological way and some of them are illustrated in Figure 5.25. These expressions are not unique, but the evaluation depends on the simplest combination and least computational complexity. The common method is to decompose the attributes into smaller components and apply morphological dilation to "grow up" these components. Most geometric units can be decomposed into thinned small subunits (skeleton) with a certain size structuring element using recursive dilation. In most of the cases, the N-dimensional solid object with the morphological operation can be decomposed into 1D lines.

The formal expressions of Figure 5.25 are written as follows:

For Figure 5.25(a): Rectangle $\rightarrow \vec{a} \oplus \vec{b}$

For Figure 5.25(b): Parallelogram $\rightarrow \vec{a} \oplus \vec{b}$

For Figure 5.25(c): Triangle $\rightarrow (\vec{a} \oplus \vec{b}) \cap (\vec{a} \oplus \vec{c})$

For Figure 5.25(d): Circle $\rightarrow$ (...(unit-circle $\oplus$ unit-circle) $\oplus$...) $\oplus$ unit-circle = (unit-circle))$^{\oplus R}$, where the "unit-circle" is a circle with the radius one. Note that the sweep representation scheme becomes a special case for the model. This same analogy can be easily extended to three dimensions.

5.6.2.2 The 3D Attributes

The basic geometric entities in manufacturing are features. Simple surface features are subsets of an object's boundary that lie in a single surface, which typically is a plane, cylinder, cone, sphere, or torus. Composite surface features are aggregates of simple ones. The three-dimensional attributes, which are similar to 2D attributes except addition of one more dimension, can also apply the similar decomposition

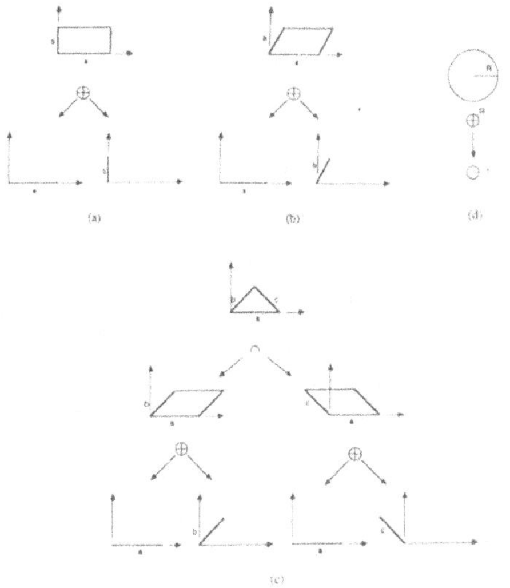

FIGURE 5.25 The decomposition of 2D attributes. (a) Rectangle, (b) parallelogram, (c) triangle, (d) circle.

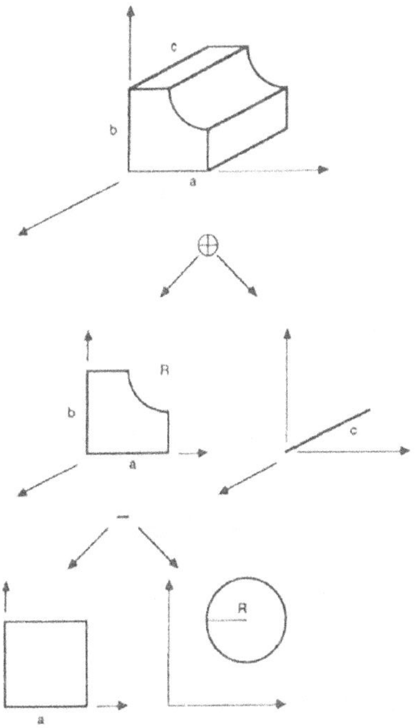

FIGURE 5.26 The decomposition of three-dimensional attributes. The right-handed slide shows the decomposition subunits that construct the attributes through the morphological dilation.

method. A representation scheme for a 3D shape model is described in Figure 5.26, and its formal expression is

$$\text{shape} \rightarrow [(\vec{a} \oplus \vec{b}) - (\text{unit} - \text{circle}^{\oplus R})] \oplus \vec{c}$$

5.6.2.3 Tolerancing Expression

Tolerances constrain an object's features to lie within regions of space called *tolerance zones*. Tolerance zones in Rossignac and Requicha (1986) were constructed by expanding the nominal feature to obtain the region bounded by the outer closed curve, shrinking the nominal feature to obtain the region bounded by the inner curve, and then subtracting the two resulting regions. This procedure is equivalent to the morphological dilation of the offset inner contour with a tolerance-radius disked structuring element. Figure 5.27(a) shows an annular tolerance zone that corresponds to a circular hole, and Figure 5.27(b) shows a tolerance zone for an elongated slot. Both could be constructed by dilating the nominal contour with a tolerance-radius disked structuring element whose representation becomes simpler.

Mathematical rules for constructing tolerance zones depend on the types of tolerances, but are independent of the specific geometry of the features and of feature representation methods. The resulting tolerance zones, however, depend on features' geometrics. The tolerance zone for testing the size of a round hole is an annular region lying between two circles with the specified maximal and minimal diameters; the zone corresponding to a form constraint for the hole is also an annulus, defined by two concentric circles whose diameters must differ by a specified amount but are otherwise arbitrary.

The mathematical morphology does support the conventional limit (±) tolerances on "dimensions" that appear in many engineering drawings. The positive deviation is equivalent to the dilated result and the negative deviation is equivalent to the eroded result. The industrial parts adding tolerance information can be expressed using dilation with a circle (see Figure 5.28).

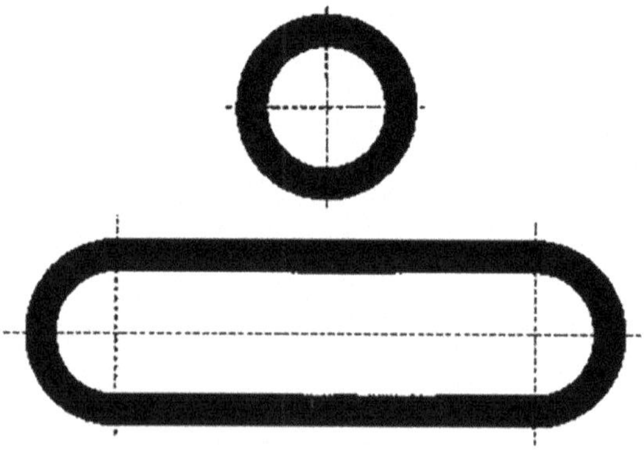

FIGURE 5.27 Tolerance zones. (a) An annular tolerance zone that corresponds to a circular hole, (b) a tolerance zone for an elongated slot.

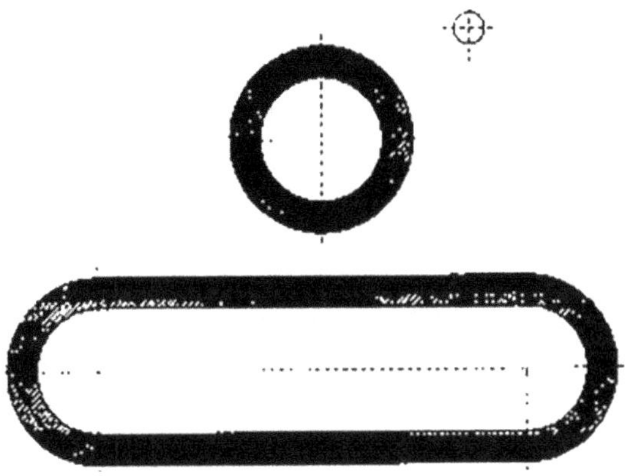

FIGURE 5.28 An example of adding tolerance by morphological dilation.

REFERENCES

Ackland, B. D. and Weste, N., "The edge flag algorithm – a fill method for raster scan display," *IEEE Transactions on Computers*, vol. 30, no. 1, pp. 41–48, Jan. 1981.

Ahmed, M. and Ward, R., "A rotation invariant rule-based thinning algorithm for character recognition," *IEEE Transactions on Pattern Analysis and Machine Intelligence*, vol. 24, no. 12, pp. 1672–1678, Dec. 2002.

Ali, S. M. and Burge, R. E., "A new algorithm for extracting the interior of bounded regions based on chain coding," *Computer Vision, Graphics, and Image Processing*, vol. 43, no. 2, pp. 256–264, Aug. 1988.

American National Standards Institute (ANSI), "Dimensioning and Tolerancing," ANSI Standard Y14.5M, ASME, 1982.

Berghorn, W., Boskamp, T., Lang, M., and Peitgen, H.-O., "Fast variable run-length coding for embedded progressive wavelet-based image compression," *IEEE Transactions on Image Processing*, vol. 10, no. 12, pp. 1781–1790, Dec. 2001.

Bribiesca, E., "A chain code for representing 3D curves," *Pattern Recognition*, vol. 33, no. 5, pp. 755–765, May 2000.

Bunke, H. and Csirik, J., "An improved algorithm for computing the edit distance of run-length coded strings," *Information Processing Letters*, vol. 54, no. 2, pp. 93–96, Apr. 1995.

Cai, Z., "Restoration of binary images using contour direction chain codes description," *Computer Vision, Graphics, and Image Processing*, vol. 41, no. 1, pp. 101–106, Jan. 1988.

Capon, J., "A probabilistic model for run-length coding of pictures," *IRE Transactions on Information Theory*, vol. 5, no. 4, pp. 157–163, Dec. 1959.

Cederberg, R. L. T., "Chain-link coding and segmentation for raster scan devices," *Computer Graphics and Image Processing*, vol. 10, no. 3, pp. 224–234, Jul. 1979.

Chakravarty, I., "A single-pass, chain generating algorithm for region boundaries," *Computer Graphics and Image Processing*, vol. 15, no. 2, pp. 182–193, Feb. 1981.

Chang, L.-W. and Leu, K.-L., "A fast algorithm for the restoration of images based on chain codes description and its applications," *Computer Graphics and Image Processing*, vol. 50, no. 3, pp. 296–307, Jun. 1990.

Chin, R. T., Wan, H.-K., Stover, D. L., and Iverson, R. D., "A one-pass thinning algorithm and its parallel implementation," *Computer Vision, Graphics, and Image Processing*, vol. 40, no. 1, pp. 30–40, Oct. 1987.

Dunkelberger, K. A. and Mitchell, O. R., "Contour tracing for precision measurement," *Proceedings of IEEE International Conference on Robotics and Automation*, St. Louis, MO, vol. 2, pp. 22–27, Mar. 1985.

Freeman, H., "Computer processing of line drawing images," *ACM Computing Surveys*, vol. 6, no. 1, pp. 57–97, Mar. 1974.

Fu, K. S., *Syntactic Pattern Recognition and Applications*, Prentice-Hall, 1982.

Ghosh, A. and Petkov, N., "Robustness of shape descriptors to incomplete contour representations," *IEEE Transactions on Pattern Analysis and Machine Intelligence*, vol. 27, no. 11, pp. 1793–1804, Nov. 2005.

Ghosh, P. K., "A mathematical model for shape description using Minkowski operators," *Computer Vision, Graphics, and Image Processing*, vol. 44, no. 3, pp. 239–269, Dec. 1988.

Golomb, S. W., "Run length encodings," *IEEE Transactions on Information Theory*, vol. 12, no. 3, pp. 399–401, Jul. 1966.

Govindham, V. K. and Shivaprasad, A. P., "A pattern adaptive thinning algorithm," *Pattern Recognition*, vol. 20, no. 6, pp. 623–637, Nov. 1987.

Guo, Z. and Hall, R. W., "Parallel thinning with two-subiteration algorithms," *Communications of the ACM*, vol. 32, no. 3, pp. 359–373, Mar. 1989.

Guo, Z. and Hall, R. W., "Fast fully parallel thinning algorithms," *Computer Vision, Graphics, and Image Processing: Image Understanding*, vol. 55, no. 3, pp. 317–328, May 1992.

Hasan, Y. M. Y. and Karam, L. J., "Morphological reversible contour representation," *IEEE Transactions on Pattern Analysis and Machine Intelligence*, vol. 22, no. 3, pp. 227–240, Mar. 2000.

Hunter, G. M. and Steiglitz, K., "Operations on images using quad trees," *IEEE Transactions on Pattern Analysis and Machine Intelligence*, vol. 1, no. 2, pp. 145–154, Apr. 1979.

Ji, L., Yi, Z., Shang, L., and Pu, X., "Binary fingerprint image thinning using template-based PCNNs," *IEEE Transactions on Systems, Man, and Cybernetics, Part B*, vol. 37, no. 5, pp. 1407–1413, Oct. 2007.

Kalvin, A., Schonberg, E., Schwartz, J. T., and Sharir, M., "Two dimensional model based boundary matching using footprints," *International Journal Robotics Research*, vol. 5, no. 4, pp. 38–55, 1986.

Kaneko, T. and Okudaira, M., "Encoding of arbitrary curves based on the chain code representation," *IEEE Transactions on Communications*, vol. 33, no. 7, pp. 697–707, Jul. 1985.

Kim, S.-D., Lee, J.-H., and Kim, J.-K., "A new chain-coding algorithm for binary images using run-length codes," *Computer Vision, Graphics, and Image Processing*, vol. 41, no. 1, pp. 114–128, Jan. 1988.

Knipe, J. and Li, X., "On the reconstruction of quadtree data," *IEEE Transactions on Image Processing*, vol. 7, no. 12, pp. 1653–1660, Dec. 1998.

Koplowitz, J., "On the performance of chain codes for quantization of line drawings," *IEEE Transactions on Pattern Analysis and Machine Intelligence*, vol. 3, no. 2, pp. 180–185, Mar. 1981.

Laferte, J.-M., Perez, P., and Heitz, F., "Discrete Markov image modeling and inference on the quadtree," *IEEE Transactions on Image Processing*, vol. 9, no. 3, pp. 390–404, Mar. 2000.

Liu, Y. K. and Zalik, B., "An efficient chain code with Huffman coding," *Pattern Recognition*, vol. 38, no. 4, pp. 553–557, Apr. 2005.

Lu, H. E. and Wang, P. S. P., "A comment on 'A fast parallel algorithm for thinning digital patterns'," *Communications of the ACM*, vol. 29, no. 3, pp. 239–242, Mar. 1986.

Madhvanath, S., Kleinberg, E., and Govindaraju, V., "Holistic verification of handwritten phrases," *IEEE Transactions on Pattern Analysis and Machine Intelligence*, vol. 21, no. 12, pp. 1344–1356, Dec. 1999.

Meer, P., Sher, C. A., and Rosenfeld, A., "The chain pyramid: Hierarchical contour processing," *IEEE Transactions on Pattern Analysis and Machine Intelligence*, vol. 12, no. 4, pp. 363–376, Apr. 1990.

Moayer, B. and Fu, K. S., "A tree system approach for fingerprint pattern recognition," *IEEE Transactions on Computers*, vol. 25, no. 3, pp. 262–274, Mar. 1976.

Montani, C. and Scopigno, R., "Rendering volumetric data using STICKS representation scheme," *ACM SIGGRAPH Computer Graphics*, vol. 24, no. 5, pp. 87–93, Nov. 1990.

Ogawa, H. and Taniguchi, K., "Thinning and stroke segmentation for handwritten Chinese character recognition," *Pattern Recognition*, vol. 15, no. 4, pp. 299–308, 1982.

Pavlidis, T., "Filling algorithms for raster graphics," *Computer Graphics & Image Processing*, vol. 10, no. 2, pp. 126–141, 1979.

Requicha, A. A., "Representations for rigid solids: Theory, method, and systems," *Computing Surveys*, vol. 12, no. 4, pp. 437–464, Dec. 1980.

Requicha, A. A. and Voelcker, H. B., "Solid modelling: Current status and research direction," *IEEE Computer Graphics and Applications*, vol. 3, no. 7, pp. 25–37, Oct. 1983.

Robinson, J. A., "Efficient general-purpose image compression with binary tree predictive coding," *IEEE Transactions on Image Processing*, vol. 6, no. 4, pp. 601–608, Apr. 1997.

Robinson, J. A., Druet, A., and Gosset, N., "Video compression with binary tree recursive motion estimation and binary tree residue coding," *IEEE Transactions on Image Processing*, vol. 9, no. 7, pp. 1288–1292, Jul. 2000.

Rockett, P. I., "An improved rotation-invariant thinning algorithm," *IEEE Transactions on Image Processing*, vol. 27, no. 10, pp. 1671–1674, Oct. 2005.

Rosenfeld, A. and Kak, A. C., *Digital Picture Processing*, vol. 2, Academic Press, 1982.

Rosenfeld, A. and Pfaltz, J. L., "Sequential operations in digital picture processing," *Journal of the ACM*, vol. 13, no. 4, pp. 471–494, Oct. 1966.

Rossignac, J. R. and Requicha, A. A. G., "Offsetting operations in solid modelling," *Computer Aided Geometric Design*, vol. 3, no. 2, pp. 129–148, Aug. 1986.

Saghri, J. A. and Freeman, H., "Analysis of the precision of generalized chain codes for the representation of planar curves," *IEEE Transactions on Pattern Analysis and Machine Intelligence*, vol. 3, no. 5, pp. 533–539, Sep. 1981.

Saint-Marc, P., Rom, H., and Medioni, G., "B-spline contour representation and symmetry detection," *IEEE Transactions on Pattern Analysis and Machine Intelligence*, vol. 15, no. 11, pp. 1191–1197, Nov. 1993.

Samet, H., "Region representation: Quadtrees from binary arrays," *Computer Graphics and Image Processing*, vol. 13, no. 1, pp. 88–93, May 1980.

Shani, U., "Filling regions in binary raster images," *SIGGRAPH Computer Graphics*, vol. 14, no. 3, pp. 321–327, Jul. 1980.

Shih, F. Y., "Object representation and recognition using mathematical morphology model," *International Journal of Systems Integration*, vol. 1, no. 2, pp. 235–256, Aug. 1991.

Shih, F. Y. and Gaddipati, V., "General sweep mathematical morphology," *Pattern Recognition*, vol. 36, no. 7, pp. 1489–1500, Jul. 2003.

Shih, F. Y. and Gaddipati, V., "Geometric modeling and representation based on sweep mathematical morphology," *Information Sciences*, vol. 171, no. 3, pp. 213–231, Mar. 2005.

Shih, F. Y. and Mitchell, O. R., "Industrial parts recognition and inspection by image morphology," *Proceedings of IEEE International Conference on Robotics and Automation*, Philadelphia, PA, vol. 3, pp. 1764–1766, Apr. 1988.

Shih, F. Y. and Pu, C. C., "A skeletonization algorithm by maxima tracking on Euclidean distance transform," *Pattern Recognition*, vol. 28, no. 3, pp. 331–341, Mar. 1995.

Shih, F. Y. and Wong, W., "A new single-pass algorithm for extracting the mid-crack codes of multiple regions," *Journal of Visual Communication and Image Representation*, vol. 3, no. 3, pp. 217–224, Sep. 1992.

Shih, F. Y. and Wong, W., "Fully parallel thinning with tolerance to boundary noise," *Pattern Recognition*, vol. 27, no. 12, pp. 1677–1695, Dec. 1994.

Shih, F. Y. and Wong, W., "A new safe-point thinning algorithm based on the mid-crack code tracing," *IEEE Transactions on Systems, Man, and Cybernetics*, vol. 25, no. 2, pp. 370–378, Feb. 1995.

Shih, F. Y. and Wong, W., "A one-pass algorithm for local symmetry of contours from chain codes," *Pattern Recognition*, vol. 32, no. 7, pp. 1203–1210, Jul. 1999.

Shneier, M., "Calculations of geometric properties using quadtrees," *Computer Graphics and Image Processing*, vol. 16, no. 3, pp. 296–302, Jul. 1981.

Shusterman, E. and Feder, M., "Image compression via improved quadtree decomposition algorithms," *IEEE Transactions on Image Processing*, vol. 3, no. 2, pp. 207–215, Mar. 1994.

Sullivan, G. J. and Baker, R. L., "Efficient quadtree coding of images and video," *IEEE Transactions on Image Processing*, vol. 3, no. 3, pp. 327–331, May 1994.

Tanaka, H. and Leon-Garcia, A., "Efficient run-length encodings," *IEEE Transactions on Information Theory*, vol. 28, no. 6, pp. 880–890, Nov. 1982.

Tang, G. Y. and Lien, B., "Region filling with the use of the discrete Green theorem," *Computer Graphics and Image Processing*, vol. 42, no 3, pp. 297–305, Jun. 1988.

Wang, Z., Feng, D., and Chi, Z., "Region-based binary tree representation for image classification," *Proceedings of International Conference on Neural Networks and Signal Processing*, Nanjing, vol. 1, pp. 232–235, Dec. 2003.

Wong, W., Shih, F. Y., and Su, T., "Thinning algorithms based on quadtree and octree representations," *Information Sciences*, vol. 176, no. 10, pp. 1379–1394, May 2006.

Wu, L. D., "On the chain code of a line," *IEEE Transactions on Pattern Analysis and Machine Intelligence*, vol. 4, no. 3, pp. 347–353, May 1982.

Yu, S.-S. and Tsai, W.-H., "A new thinning algorithm for gray-scale images by the relaxation technique," *Pattern Recognition*, vol. 23, no. 10, pp. 1067–1076, Oct. 1990.

Zingaretti, P., Gasparroni, M., and Vecci, L., "Fast chain coding of region boundaries," *IEEE Transactions on Pattern Analysis and Machine Intelligence*, vol. 20, no. 4, pp. 407–415, Apr. 1998.

6 Feature Extraction

To recognize or classify an object in an image, one must first extract some features out of the image, and then use these features inside a pattern classifier to obtain the final class. Feature extraction (or detection) aims to locate significant feature regions on images depending on their intrinsic characteristics and applications. These regions can be defined in global or local neighborhood and distinguished by shapes, textures, sizes, intensities, statistical properties, etc. Local feature extraction methods are divided into intensity-based and structure-based. Intensity-based methods analyze local intensity patterns to find regions that satisfy desired uniqueness or stability criteria. Structure-based methods detect image structures such as edges, lines, corners, circles, ellipses, etc. Feature extraction tends to identify the characteristic features that can form a good representation of the object so as to discriminate across the object category with tolerance of variations.

In general, local feature-based region extractors show their advantages of robustness to object occlusion or image deformation. They can detect locally distributed image features and provide a distinctive, compact representation scheme. Using all the pixel values can be redundant and noisy, whereas feature descriptors extract, filter, and summarize the most distinctive features and build compact feature vectors. These significant descriptors can facilitate the pattern classifiers with their compactness and discrimination.

This chapter presents selected often-used feature extraction topics. It is organized as follows. Fourier descriptor and moment invariants are introduced in Section 6.1. Shape number and hierarchical features are described in Section 6.2. Corner detection is presented in Section 6.3. Hough transform is presented in Section 6.4. We describe principal component analysis in Section 6.5 and linear discriminate analysis in Section 6.6. In Section 6.7, we discuss feature reduction in input and feature spaces.

6.1 FOURIER DESCRIPTOR AND MOMENT INVARIANTS

Shape is one of the most critical image features because shape is discriminative to human perception. Human beings tend to view a scene as composed of individual objects, which can be best described by their shapes. Pattern-matching techniques measure shape similarity using pattern transformation and calculation of their degrees of resemblance. The similarity of two patterns is measured as the distance between their feature vectors. Many shape representation methods (or shape descriptors) have been proposed in the literature, which can be summarized into two categories: region-based versus contour-based. In region-based methods, all the pixels within a shape are taken into consideration to obtain the shape feature. An example is the moment descriptors. In contour-based methods, only shape boundary

DOI: 10.1201/ 9781003474395-7

information is used. For examples, shape descriptors, shape signatures, and spectral descriptors belong to this class.

Spectral descriptors include Fourier descriptor (FD) and wavelet descriptor (WD). Wavelet descriptor has the advantage of describing shape features in both spatial and frequency domains (Chuang and Kuo, 1996; Hung, 2000). However, it requires intensive computation in shape matching because of not being rotation invariant. In FD, global shape features are attained by the first few low-frequency components, while high-frequency components describe finer features of the shape. There have been many FD methods applied in shape analysis, character recognition, shape coding, shape classification, and shape retrieval. We will focus on the Fourier descriptor.

Persoon and Fu (1977) proposed using Fourier descriptor as the shape representation. Rafiei and Mendelzon (2002) proposed an indexing technique for the fast retrieval of objects in 2D images based on similarity between their boundary shapes using Fourier descriptors. The shape signature is defined by a 1D continuous function $f(t)$. The Fourier transform of $f(t)$ is given in equation (6.1). Because discrete images are concerned, the discrete Fourier transform (DFT) is used. These were introduced in Section 2.2. For convenience, we briefly summarize them as follows. Let L be a closed polygonal line that consists of N vertices. Let also $[x(t), y(t)]$, $t = 0,1,\cdots,N-1$ be the Cartesian coordinates of each vertex. We consider the complex signal $f(t)$, where $f(t) = x(t) + i\,y(t)$, $t = 0,1,\cdots,N-1$, by randomly selecting a vertex as a starting point and tracing the vertices in a clockwise or counterclockwise direction. The DFT of $f(t)$ is given by equation (6.2), where the value of $u \in \{0,1,\cdots,N-1\}$. From Euler's equation, i.e., $e^{-j2\pi ut/N} = \cos(2\pi ut/N) - j\sin(2\pi ut/N)$, equation (6.3) can be obtained. The coefficient $F(u)$ is called *Fourier descriptor* (FD).

$$F(u) = \int_{-\infty}^{\infty} f(t)e^{-j2\pi ut}\,dt \tag{6.1}$$

$$F(u) = \frac{1}{N}\sum_{t=0}^{N-1} f(t)e^{-j2\pi ut/N} \tag{6.2}$$

$$F(u) = \frac{1}{N}\sum_{t=0}^{N-1} f(t)(\cos(2\pi ut/N) - j\sin(2\pi ut/N)) \tag{6.3}$$

Several shape signatures have been proposed to derive Fourier descriptor. Zhang and Lu (2002) proved empirically that the Fourier descriptor derived from the centroid distance (*CeFD*) can achieve significantly higher performance than other Fourier descriptors, such as the area FD, curvature FD, psi FD, position (complex) FD, affine FD, and chord-length FD. The centroid distance function $f(t)$ is defined as the distance of the boundary points to the centroid (x_c, y_c) as in equation (6.4). Only a half of Fourier descriptors $F(u)$ are used to index the corresponding shape because $f(t)$ is a set of real values. Therefore, equation (6.5) is obtained. The Fourier descriptors can be normalized to be independent of geometric translation, scaling, and rotation. A set of invariant descriptors *CeFD* can be obtained by equation (6.6). The first few terms of the invariant moments, like the first few terms of a Fourier series, capture the more general shape properties while the later terms capture finer details. Since $F(0)$ is the largest coefficient, it can be used as the normalization factor.

$$f(t) = \sqrt{(x(t) - x_c)^2 + (y(t) - y_c)^2} \tag{6.4}$$

$$|F(u)| = \frac{1}{N}\sqrt{\left(\sum_{t=0}^{N-1} f(t)\cos(2\pi ut / N)\right)^2 + \left(\sum_{t=0}^{N-1} f(t)\sin(2\pi ut / N)\right)^2} \tag{6.5}$$

$$CeFD = \frac{|F(u)|}{|F(0)|}, \text{ where } u = 1, 2, \ldots, N/2 \tag{6.6}$$

Two-dimensional moment invariants for planar geometric figures have been investigated. Hu (1962) introduced a set of moment invariants using nonlinear combination based on normalized central moments. The computation of these invariants is required to engage all the pixels in the shape. Chen (1993) proposed an improved moment invariants technique based on boundary pixels to speed up the computation. In addition, Chen's normalized moment invariants were designed to be invariant to scaling, translation, and rotation (Sun et al., 2003). The definitions of Hu's and Chen's moment invariants are briefly introduced here.

Let $f(x,y)$ be a continuous shape-based image function. The $(p,q)th$ central moment of $f(x,y)$ is defined as

$$\mu_{pq} = \int_{-\infty}^{\infty}\int_{-\infty}^{\infty} (x - x_c)^p (y - y_c)^q f(x, y)dxdy \tag{6.7}$$

where (x_c, y_c) are the coordinates of the centroid of the shape. The normalized $(p,q)th$ central moment is defined as

$$\eta_{pq} = \frac{\mu_{pq}}{\mu_{00}^{\gamma}}, \text{ where } \gamma = \frac{p+q+2}{2} \text{ and } p, q = 2, 3, \ldots \tag{6.8}$$

Based on these moments, Hu (1962) derived the following seven moment invariants, Φ_i, where $i = 1, 2, \ldots, 7$.

$$\left.\begin{aligned}
&\Phi_1 = \eta_{20} + \eta_{02} \\
&\Phi_2 = (\eta_{20} - \eta_{02})^2 + 4\eta_{11}^2 \\
&\Phi_3 = (\eta_{30} - 3\eta_{12})^2 + (\eta_{03} - 3\eta_{21})^2 \\
&\Phi_4 = (\eta_{30} + \eta_{12})^2 + (\eta_{03} + \eta_{21})^2 \\
&\Phi_5 = (\eta_{30} - 3\eta_{12})(\eta_{30} + \eta_{12})\left\{(\eta_{30} + \eta_{12})^2 - 3(\eta_{21} + \eta_{03})^2\right\} + \\
&\qquad (3\eta_{21} - \eta_{03})(\eta_{21} + \eta_{03})\left\{3(\eta_{30} + \eta_{12})^2 - (\eta_{21} + \eta_{03})^2\right\} \\
&\Phi_6 = (\eta_{20} - \eta_{02})\left\{(\eta_{30} + \eta_{12})^2 - (\eta_{21} + \eta_{03})^2\right\} + 4\eta_{11}(\eta_{30} + \eta_{12})(\eta_{21} + \eta_{03}) \\
&\Phi_7 = (3\eta_{21} - \eta_{03})(\eta_{30} + \eta_{12})\left\{(\eta_{30} + \eta_{12})^2 - 3(\eta_{21} + \eta_{03})^2\right\} + \\
&\qquad (3\eta_{21} - \eta_{30})(\eta_{21} + \eta_{03})\left\{3(\eta_{30} + \eta_{12})^2 - (\eta_{21} + \eta_{03})^2\right\}
\end{aligned}\right\} \quad (6.9)$$

Chen (1993) improved the central moments by applying the moment computation only on the boundary. Chen's $(p,q)th$ central moment is defined as

$$\mu_{pq} = \int_S (x - x_c)^p (y - y_c)^q \, ds \quad (6.10)$$

For a digital shape-based image, the discrete form is described as

$$\mu_{pq} = \sum_{(x,y)\in S} (x - \bar{x})^p (y - \bar{y})^q \quad (6.11)$$

Chen's normalized $(p,q)th$ central moment is defined as

$$\eta_{pq} = \frac{\mu_{pq}}{\mu_{00}^{\gamma}}, \text{ where } \gamma = p + q + 1 \text{ and } p, q = 2, 3, \ldots \quad (6.12)$$

Another kind of shape descriptor, called *shell descriptor*, is introduced here. Assume that Γ is a counterclockwise, simple, closed contour with parametric representation $(x(t), y(t)) = z(t)$, where t is the observation sequence and $0 \le t \le T$ (T is the perimeter of the contour). Let $z(t_{ref})$ be a reference observation and $t_{ref} \neq 0$. The observations can be represented in a polar form by adopting $z(t_{ref})$ and $z(0)$ as the reference line. Denote $\theta(t)$ as the angular direction of $z(t)$ and $\iota(t)$ as the normalized spatial displacement of the observation. In the polar form the observation contour Γ can be represented as $\{z(t) = \iota(t)e^{j\theta(t)}\}$, where $j = \sqrt{-1}$. The parameters $[\iota, \theta]$ are invariant to translation and rotation. To make the representation invariant to scaling, the length of the observations is normalized by the length of the reference line.

Let $\iota(0) = 0$ and $\theta(0) = 0$ be the initial values for the starting point $z(0)$. Let $\phi(r,s)$ be the angle between two lines: $\overline{z(0)z(r)}$ and $\overline{z(0)z(s)}$, as shown in

Figure 6.1. The angle $\phi(r,s)$ is called a *sector* if $\iota(r)=\iota(s)=\iota$. Thus, a sector can be represented as $(\iota,\theta(r),\phi(r,s))$ in addition to $\phi(r,s)$. A shell is a collection of all sectors. To simplify and generalize the symbol, $(\iota_l,\theta_{lk},\phi_{lk})$ is used to denote the k-th sector of the l-th shell. Thus, a contour Γ can be represented as

$$\{(t_l,\theta_{lk},\phi_{lk}),0\le t_l\le 1). \tag{6.13}$$

The 3D parametric space is then mapped onto a 2D plane (t,λ), where $t=\iota_l$ and $\lambda=\{x\,|\,\theta_{lk}\le x\le\theta_{lk}+\phi_{lk}\}$. A mapping of a uniform polar angular grid onto a uniform (t,λ) grid is illustrated in Figure 6.2. The values ι_l,θ_{lk}, and ϕ_{lk} which are the shell descriptors, are named *sector coefficients*.

All simple closed curves with the starting point are mapped onto the rings of shell descriptors in such a way that all curves of the identical shape and the starting point go into the same the descriptors. The shell-space with descriptor parameters (t,λ) of Figure 6.1 is shown in Figure 6.3.

Many different descriptors have been proposed in the literature. It is unclear which descriptors are more appropriate and how their performance depends on the interest region detector. Mikolajczyk and Schmid (2005) compared the

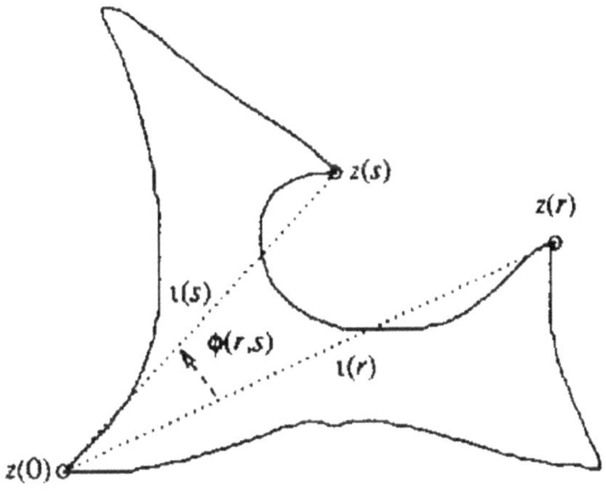

FIGURE 6.1 Sector notations.

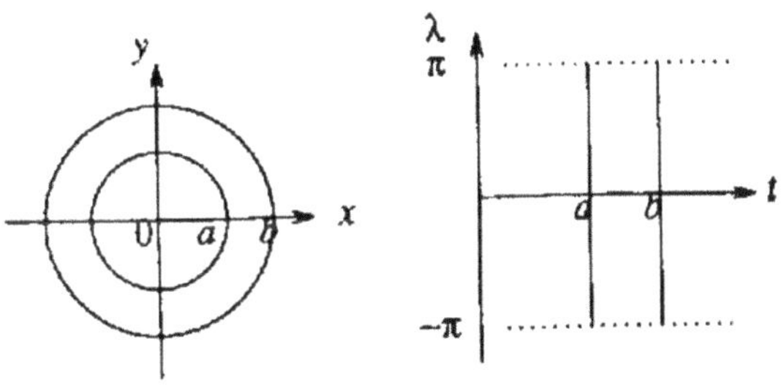

FIGURE 6.2 A mapping of a uniform polar angular grid onto a uniform (t,λ) grid.

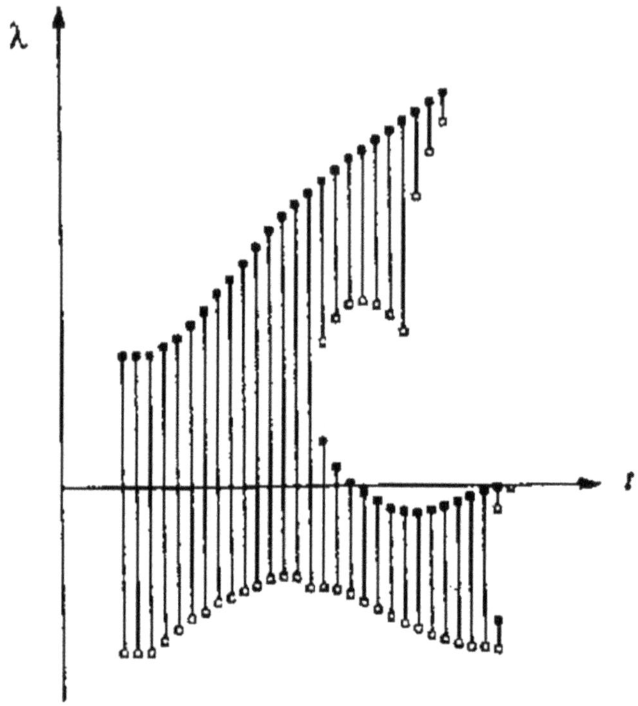

FIGURE 6.3 The shell descriptor of the shape shown in Figure 6.1.

performance of descriptors computed for local interest regions. Kiranyaz et al. (2008) proposed a generic shape descriptor, which can be extracted from the major object edges automatically and used for the multimedia content-based retrieval in multimedia databases. Hung (2000) presented a generalized uniqueness property inherent in the 1D discrete periodized wavelet transformation. Bober (2001) proposed techniques and tools for shape representation and matching, developed in the context of MPEG-7 standardization. Manjunath et al. (2001) proposed a histogram of the main edge directions (vertical, horizontal, and two diagonals) within fixed size blocks. It is an efficient texture descriptor for the images with heavy textural presence. Shen and Ip (1998) presented discriminative wavelet shape descriptors for recognition of 2D patterns. Zhang and Lu (2002) presented a comparative study of Fourier descriptors for shape representation and retrieval. Wong et al. (2007) proposed shape-based image retrieval using Fourier descriptor and support vector machine.

Many different moment invariants have been proposed in the literature. Yap et al. (2003) proposed a new set of orthogonal moments based on the discrete classical Krawtchouk polynomials. The Krawtchouk polynomials (Krawtchouk, 1929) are scaled to ensure numerical stability, thus creating a set of weighted

Krawtchouk polynomials. Dudani et al. (1977) presented aircraft identification by moment invariants. Belkasim (1991) presented a comparative study and new results of pattern recognition with moment invariants. Teh and Chin (1998) presented image analysis by the method of moments. Pawlak (1992) discussed the reconstruction aspect of moment descriptors. Liao and Pawlak (1996) presented image analysis by moments. Schlemmer et al. (2007) presented an approach for analyzing 2D flow field data based on the idea of invariant moments. Flusser and Suk (2006) proposed a set of moment invariants with respect to rotation, translation, and scaling suitable for recognition of objects having N-fold rotation symmetry. Mamistvalov (1998) presented n-dimensional moment invariants and conceptual mathematical theory of recognition n-dimensional solids. Rothe et al. (1996) developed a method of normalization to derive many sets of invariants. Flusser et al. (2003) presented the construction of combined blur and rotation moment invariants in arbitrary number of dimensions.

6.2 SHAPE NUMBER AND HIERARCHICAL FEATURES

The shape number of objects is designed to allow interpreting shape information into a number, which is supposed to be independent of their location, size, and orientation. A set of hierarchical shape features includes a shape number, several layers of significant point radii (i.e., Euclidean distances to the boundary), and their x- and y-coordinates. Each local maximum point of the distance transform is used as a significant point, which has the major contribution in representing features.

6.2.1 Shape Number

The shape number is computed from the total distance and the total number of all object points (Danielsson, 1978). It is a measure of shape compactness. Let N be the total number of the pixels belonging to the object, and let X_i be the distance of pixel i to the nearest pixel outside the object. The shape number is mathematically expressed as follows:

$$\text{Shape Number} = \frac{N^3}{9\pi(\sum_{i=1}^{N} X_i)^2} \tag{6.14}$$

Note that the constant is chosen such that the shape number of an ideal circle is 1. Hence, the larger the shape number is, the less compact the object shape looks. A 13 × 13 discrete circle is shown in Figure 6.4, where the total number of object pixels is 89. Its city-block distance transform is shown in Figure 6.5. It is the fact that we have labeled all border pixels with the distance value $x = 1$ (actually, it is 0.5).

0	0	0	0	0	0	0	0	0	0	0	0	0
0	0	0	0	0	1	1	1	0	0	0	0	0
0	0	0	1	1	1	1	1	1	1	0	0	0
0	0	1	1	1	1	1	1	1	1	1	0	0
0	0	1	1	1	1	1	1	1	1	1	0	0
0	1	1	1	1	1	1	1	1	1	1	1	0
0	1	1	1	1	1	1	1	1	1	1	1	0
0	1	1	1	1	1	1	1	1	1	1	1	0
0	0	1	1	1	1	1	1	1	1	1	0	0
0	0	1	1	1	1	1	1	1	1	1	0	0
0	0	0	1	1	1	1	1	1	1	0	0	0
0	0	0	0	0	1	1	1	0	0	0	0	0
0	0	0	0	0	0	0	0	0	0	0	0	0

FIGURE 6.4 A 13 × 13 discrete circle.

For precise measurement, each distance value is subtracted by 0.5. The summation of total distance values is $218 - 0.5 \times 89 = 173.5$. Therefore, the shape number is 0.8283. Note that the shape number is not equal to one because of the digitization and the city-block distance used. If we select Euclidean distance, it will increase the computational precision. Figure 6.6 shows the Euclidean distance transform of Figure 6.4. In order to make the distance summation more accurate, it would be appropriate to decrease every x value by the average error $\sqrt{0.5}/2$. The summation of total distances is $188.973 - \sqrt{0.5}/2 \times 89 = 157.507$. Hence, the shape number is 1.005, which gives a more accurate measure.

One example of using the shape number to extract circle feature is described here. We first use the distance transformation algorithm to obtain the distance information. According to the calculated shape number, we conclude whether a region is a circle or not. If it is a circle, the maximum distance is the radius, and the location of that value is the center. If there is more than one point, we estimate the average of those points as the center.

Circle Detection Algorithm

(1) A gray-level image is thresholded into a binary using a suitable thresholding algorithm.
(2) Do a distance transformation algorithm.

0	0	0	0	0	0	0	0	0	0	0	0	0
0	0	0	0	0	1	1	1	0	0	0	0	0
0	0	0	1	1	2	2	2	1	1	0	0	0
0	0	1	2	2	3	3	3	2	2	1	0	0
0	0	1	2	3	4	4	4	3	2	1	0	0
0	1	2	3	4	5	5	5	4	3	2	1	0
0	1	2	3	4	5	6	5	4	3	2	1	0
0	1	2	3	4	5	5	5	4	3	2	1	0
0	0	1	2	3	4	4	4	3	2	1	0	0
0	0	1	2	2	3	3	3	2	2	1	0	0
0	0	0	1	1	2	2	2	1	1	0	0	0
0	0	0	0	0	1	1	1	0	0	0	0	0
0	0	0	0	0	0	0	0	0	0	0	0	0

FIGURE 6.5 The city-block distance transform of Figure 6.4.

(3) Use the computed shape number to determine whether each region is a circle or not.
(4) If the region is a circle, detect the maximum distance value in the region and its coordinates.
(5) If there is more than one point with the maximum distance value, then select the average of the coordinates as the center and add 0.5 (approximated by a half of pixel unit-length) to the maximum distance as the radius.

6.2.2 Significant Points Radius and Coordinates

Significant points are defined to be the local maxima of 8-neighborhood in the distance transform. If the Euclidean distance is used, then the circle drawn with the significant point as the center and the distance as the radius, is the maximal inscribed circle in the local region.

Let d be the distance transform. The significant points extraction algorithm is:

(1) Execute $d \oplus_g k$. Here k is a 3×3 structuring element with all zeroes.
(2) If the object pixel value in the output of step 1 is equal to that in d, then keep the value; otherwise, let this point be zero.
(3) If the nonzero output points of step 2 are connected, then represent those points by a single average point.

0	0	0	0	0	0	0	0	0	0	0	0	0
0	0	0	0	0	1	1	1	0	0	0	0	0
0	0	0	1	1	$\sqrt{2}$	2	$\sqrt{2}$	1	1	0	0	0
0	0	1	$\sqrt{2}$	2	$\sqrt{5}$	$\sqrt{8}$	$\sqrt{5}$	2	$\sqrt{2}$	1	0	0
0	0	1	2	$\sqrt{8}$	$\sqrt{10}$	$\sqrt{13}$	$\sqrt{10}$	$\sqrt{8}$	2	1	0	0
0	1	$\sqrt{2}$	$\sqrt{5}$	$\sqrt{10}$	$\sqrt{17}$	$\sqrt{20}$	$\sqrt{17}$	$\sqrt{10}$	$\sqrt{5}$	$\sqrt{2}$	1	0
0	1	2	$\sqrt{8}$	$\sqrt{13}$	$\sqrt{20}$	$\sqrt{29}$	$\sqrt{20}$	$\sqrt{13}$	$\sqrt{8}$	2	1	0
0	1	$\sqrt{2}$	$\sqrt{5}$	$\sqrt{10}$	$\sqrt{17}$	$\sqrt{20}$	$\sqrt{17}$	$\sqrt{10}$	$\sqrt{5}$	$\sqrt{2}$	1	0
0	0	1	2	$\sqrt{8}$	$\sqrt{10}$	$\sqrt{13}$	$\sqrt{10}$	$\sqrt{8}$	2	1	0	0
0	0	1	$\sqrt{2}$	2	$\sqrt{5}$	$\sqrt{8}$	$\sqrt{5}$	2	$\sqrt{2}$	1	0	0
0	0	0	1	1	$\sqrt{2}$	2	$\sqrt{2}$	1	1	0	0	0
0	0	0	0	0	1	1	1	0	0	0	0	0
0	0	0	0	0	0	0	0	0	0	0	0	0

FIGURE 6.6 The Euclidean distance transform of Figure 6.4.

(4) List the values of radii and coordinates of all the nonzero points according to the decreasing order of the radius values.

Figure 6.7 shows a rectangular block with a hole and a slot and the extracted significant points.

Let s denote the shape number, and let r_i, x_i, y_i denote the radius and x, y coordinates in the decreasing order of radius, respectively. We can establish the shape database registered in the format: $(s, r_1, x_1, y_1, r_2, x_2, y_2, \cdots, r_n, x_n, y_n)$. As comparing an unknown object with the registered database and compute the error value using equation (6.15), if the error is within the tolerance range, we classify the object to be identical as the registered object. Otherwise, we classify it as a different object. Let s', r_i', x_i', y_i' denote the shape number, radius, and x, y coordinates for the unknown object. The error is computed as

$$\text{Error} = \frac{1}{2}\left(\frac{|s' - s|}{s} + \frac{|r_1' - r_1| + \cdots + |r_n' - r_n|}{r_1 + \cdots + r_n}\right) \tag{6.15}$$

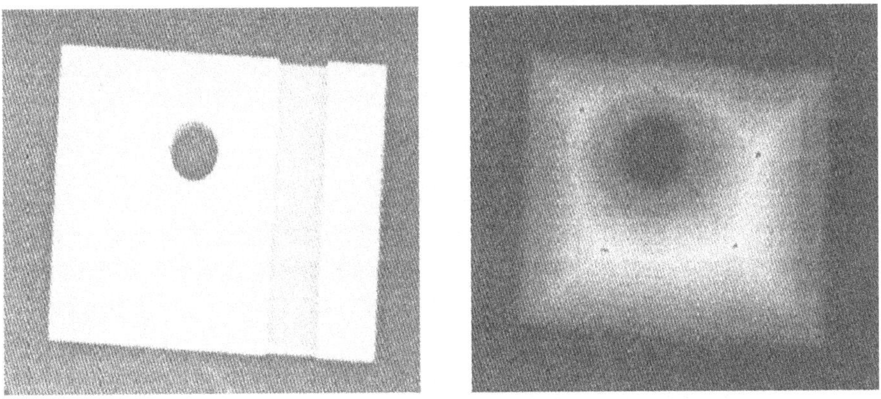

(a) (b)

FIGURE 6.7 (a) An industrial part with a hole and a slot, (b) the result of significant points extraction.

6.2.3 Localization by Hierarchical Morphological Band-Pass Filter

The problem considered in this section is how to quickly locate an object which was known and stored in the database. The morphological opening with an image will remove all of the pixels in regions which are too small to contain the probe. The opposite sequence (closing) will fill in holes and concavities which are smaller than the probe. Such filters can be used to suppress spatial features or discriminate against objects based on their size distribution. As an example, if a disk-shaped structuring element with radius h is used, then the morphological opening is equivalent to a low-pass filter. The morphological opening residue is a high-pass filter. The difference of two morphological openings of an image with two non-equal radii is a band-pass filter. It can be expressed as follows.

$$\text{LOW-PASS} = A \circ B^h,$$

$$\text{HIGH-PASS} = A - \left(A \circ B^h\right),$$

$$\text{BAND-PASS} = \left(A \circ B^{h_1}\right) - \left(A \circ B^{h_2}\right), \text{ where: radii of } B, h_1 < h_2.$$

If there are several unknown objects on the workbench and we want to pick up the one specified in the library database, the following algorithm is a simple and rapid approach.

The Localization Algorithm:

In order to locate the position of objects, we change from the opening to the erosion operation. Let A denote a binary image which contains several objects, and r_i denote the significant point radii in the library shape database.

(1) $\text{Do}\left(A \ominus_b B^{r_1-1}\right)-\left(A \ominus_b B^{r_1+1}\right)$.
(2) If the output of step 1 is empty, it means that the library object is not on the workbench. If the output has at least two nonzero points spaced more than a certain interval, it means that at least two objects match the library object possibly. Hence, we need to do erosion with the second radius. $\text{Do}\left(A \ominus_b B^{r_2-1}\right)-\left(A \ominus_b B^{r_2+1}\right)$. Apply this procedure recursively until only one point remains. This is the location of the object that matches the library object.

6.3 CORNER DETECTION

Corners are frequently employed in pattern recognition. They are the pixels whose slope changes abruptly, i.e., the absolute curvature is high. Given an image, a typical approach to detect corners involves segmenting the object from background, representing the object boundary by chain codes, and searching for significant turns in the boundary. Local measurement such as non-maxima suppression is used to obtain a binary corner map.

The algorithms of corner detection can be categorized into boundary-based and gray-level approaches. The boundary-based approach detects corners based on the boundary information of objects. Tsai et al. (1999) proposed a method using eigenvalues of the covariance matrix of data points on a curve segment. The gray-level approach directly works on gray-level images using the corner template matching or the gradients at edge pixels. Lee and Bien (1996) developed a real-time gray-level corner detector using fuzzy logic. Singh and Shneier (1990) proposed a fusion method by combining template-based and gradient-based techniques. Zheng et al. (1999) proposed a gradient-direction corner detector. Kitchen and Rosenfeld (1982) demonstrated that gray-level schemes perform better than boundary-based techniques.

Gao et al. (2007) presented two novel corner detection methods for gray-level images based on log-Gabor wavelet transform. Arnow and Bovik (2007) casted the problem of corner detection as a corner search process, and developed principles of foveated visual search and automated fixation selection to accomplish the corner search. Ando (2000) presented image field categorization and edge/corner detection from gradient covariance. Basak and Mahata (2000) proposed a connectionist model for corner detection in binary and gray images. Rockett (2003) described a generic methodology for evaluating the labeling performance of feature detectors, and proposed a method for generating a test set and applied the methodology to the performance assessment of corner detectors. Paler et al. (1984) presented local ordered gray levels as an aid to corner detection. Harris and Stephens (1988) proposed a combined corner and edge detector.

Corners can be extracted using morphological operations. The basic idea is described here, and is further explored by regulated morphological operations. A suitable-sized disk structuring element is selected. An image is opened with the disk structuring element and the result is subtracted from the original. The opening residue area varies with the angle of the corner. Then two structuring elements are selected: One is small enough to pass the residue area, and the other is too large. The opening residue is eroded with these two structuring elements and the two outputs are compared. This is equivalent to a shape band-pass filter. Different size structuring elements are used depending on the corner angles to be located. As the angle decreases, the opening residue area increases, hence the structuring element area increases. The largest area structuring element is first tested. If nothing is obtained, then the area of the structuring element is decreased until something is obtained. Thus, the angle information can be obtained.

Corner Detection Algorithm:

(1) A binary image is dilated with a suitable size of the disk structuring element.
(2) The output of step 1 is opened with a large size of the disk structuring element.
(3) Subtract the output of step 2 from the output of step 1.
(4) The output of step 3 is eroded with two disk structuring elements. The size of one structuring element is used in step 1 and the other size is a little bit larger.
(5) Subtraction of two eroded outputs of step 4. This is a band-pass filter. The remaining points are the corner points.

The size is determined by the largest area which can fit inside the opening residue corner (output of step 3). This is to estimate the location of the corner more accurately by correcting the bias introduced by the erosion in step 4. Figure 6.8 illustrates the corner detection algorithm which was applied on a rectangular block with two holes.

In the corner detection based on morphological operations, Laganiere (1998) proposed a method using the *asymmetrical closing* by four different structuring elements including plus, lozenge, cross, and square. Lin et al. (1998) proposed a morphological corner detector to find convex and concave corner points using integer computation. However, their disadvantages include the following: (1) It is difficult to choose a suitable structuring element. (2) As the size of the structuring element increases, the computational cost increases. Zhang and Zhao (1995) developed a morphological corner detector, but it can only detect convex corners and the result is sensitive to the structuring element size.

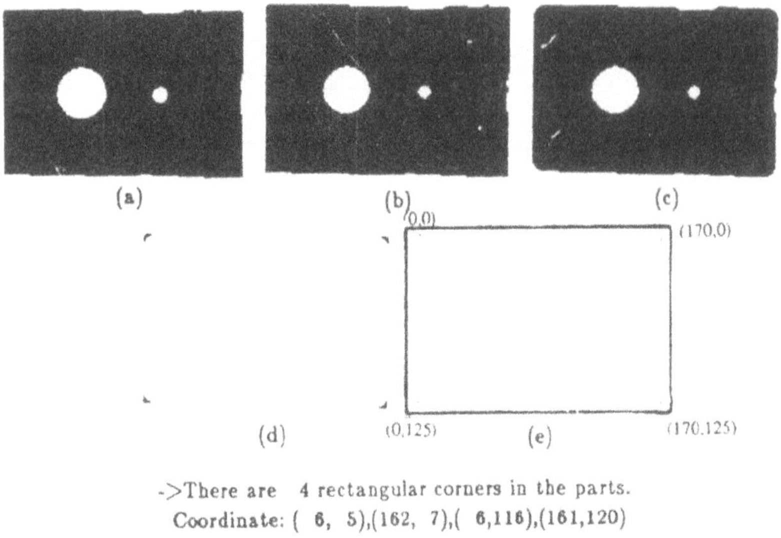

FIGURE 6.8 An example of a corner detector. Use morphology to detect the 90° corners. For clear display, the dark and white regions are reversed. (a) Original image, (b) output of step 1, (c) output of step 2, (d) output of step 3, (e) output of step 5.

6.3.1 Asymmetrical Closing for Corner Detection

Laganiere (1998) proposed corner detection using the *asymmetrical closing*, which is defined as a dilation of an image by a structuring element followed by an erosion by another structuring element. The idea is to make the dilation and erosion complementary and correspond to variant types of corners. Two structuring elements, cross (+) and lozenge (◊), are used. Let the asymmetrical closing of an image A by structuring elements + and ◊ be denoted by

$$A^c{}_{+.\Diamond} = (A \oplus +) \ominus \Diamond. \tag{6.16}$$

The corner strength is computed by

$$C+(A) = \left| A - A^c{}_{+.\Diamond} \right|. \tag{6.17}$$

For different types of corners, another corner strength (which is a 45°-rotated version) is computed as

$$C_\times(A) = \left| A - A^c{}_{\times.\Box} \right|. \tag{6.18}$$

By combining the four structuring elements in Figure 6.9, the corner detector is represented as

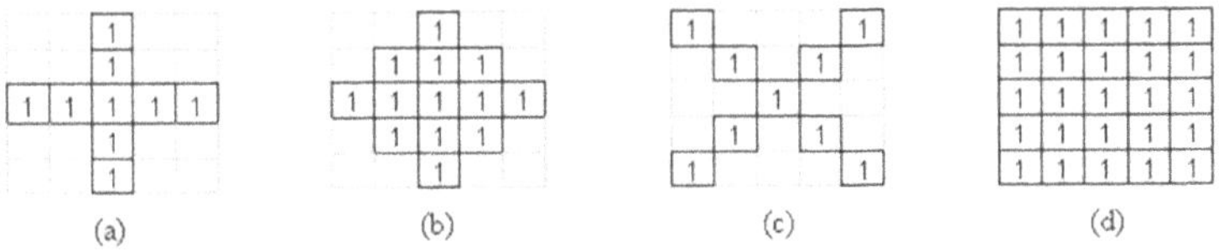

FIGURE 6.9 The four structuring elements: $(\text{a}) + (\text{b}) \Diamond (\text{c}) \times (\text{d}) \Box$.

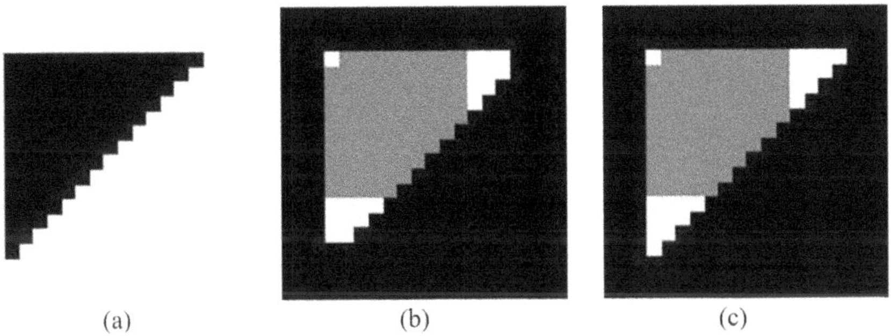

FIGURE 6.10 (a) A triangle image, (b) Laganiere's detector, (c) the modified detector.

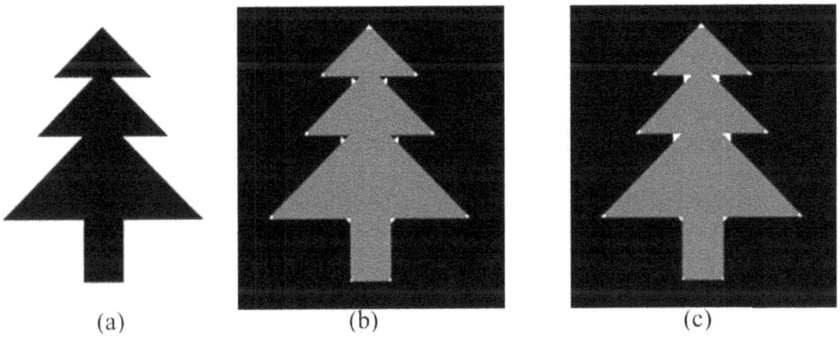

FIGURE 6.11 (a) A tree image, (b) Laganiere's detector, (c) the modified detector.

$$C_{+.\times}(A) = \left| A^c_{+.\Diamond} - A^c_{\times.\Box} \right|. \tag{6.19}$$

The disadvantages of the above corner detector are that it may miss obtuse-angle corners and remove sharp-angle corner pixels. Therefore, a modified corner detector uses: $C_{+.\times}(A) = |A - A^c{}_{+.\Diamond}| \cup |A - A^c{}_{\times.\Box}|$ (Shih et al., 2005). Two examples of comparing Laganiere's detector and the modified detector are shown in Figures 6.10 and 6.11. It is observed that the modified detector can locate the corner pixels more accurately.

6.3.2 Regulated Morphology for Corner Detection

Agam and Dinstein (1999) developed regulated morphological operators and showed how the fitting property can be adapted for analyzing the map and line-drawing images. Since regulated morphology inherits many properties of ordinary morphology, it is feasible to apply in image processing and optimize its strictness parameters using some criteria. Tian et al. (2002) extended the regulated morphological operators by adjusting the weights in the structuring element.

Note that the morphological hit-and-miss transform (Zhang and Zhao, 1995) can also be used to extract corners. For example, four different structuring elements can be designed as consisting of four types of right-angle corners and perform the hit-and-miss transform with each structuring element. The four results are combined to obtain all the corners. However, the corners extracted are only the locations of all right-angle convex corners in four directions.

The *regulated dilation* of a set A by a structuring element set B with a strictness parameter s is defined by

$$A \oplus^{s} B = \left\{x \middle| \#\left(A \cap (\hat{B})_x\right) \geq s, \quad s \in [1, \min(\# A, \# B)]\right\}, \tag{6.20}$$

where the symbol # denotes the cardinality of a set. The *regulated erosion* of a set A by a structuring element set B with a strictness parameter s is defined by

$$A \ominus^{s} B = \left\{\mathrm{x} \middle| \#\left(A^{c} \cap (B)_x\right) < s, \quad s \in [1, \# B]\right\}. \tag{6.21}$$

In order to maintain some properties the same as ordinary morphological operators, the *regulated closing* of a set A by a structuring element set B with a strictness parameter s is defined by

$$A \bullet^{s} B = \left(\left(A \underline{\oplus}^{s} B\right) \ominus B\right) \cup A, \tag{6.22}$$

where $A \underline{\oplus}^{s} B = \left(A \oplus^{s} B\right) \cup A$ is defined as the *extensive regulated dilation.* The *regulated opening* of a set A by a structuring element set B with a strictness parameter s is defined by

$$A \circ^{s} B = \left(\left(A \underline{\ominus}^{s} B\right) \oplus B\right) \cap A, \tag{6.23}$$

where $A \underline{\ominus}^{s} B = \left(A \ominus^{s} B\right) \cap A$ is defined as the *anti-extensive regulated erosion.*

By adjusting the strictness parameter, the noise sensitivity problem and small intrusions or protrusions on the object boundary can be alleviated. For corner detection, regulated openings or closings are not used because the resulting image will stay the same as the original image after the strictness parameter exceeds a certain value. The modified regulated morphological corner detector is described as follows:

Step 1: $A_1 = \left(A \oplus^s B\right) \ominus^s B$. Corner strength: $C_1 = |A - A_1|$.
Step 2: $A_2 = \left(A \ominus^s B\right) \oplus^s B$. Corner strength: $C_2 = |A - A_2|$.
Step 3: Corner detector $= C_1 \cup C_2$.

The original image is performed by a regulated dilation by a 5 × 5 circlular structuring element with a strictness s, and then followed by a regulated erosion by the same structuring element with the same strictness. The corner strength C_1 is computed by finding the absolute value of the difference between the original and resulting images. This step is to extract concave corners. By reversing the order such that the regulated erosion is applied first, and then followed by the regulated dilation, the convex corners can be extracted. Finally, both types of corners are combined.

6.3.3 Experimental Results

The comparisons of results by Laganiere's method and by the modified regulated morphology are shown in Figures 6.12–6.14. From Figures 6.12 and 6.13, the

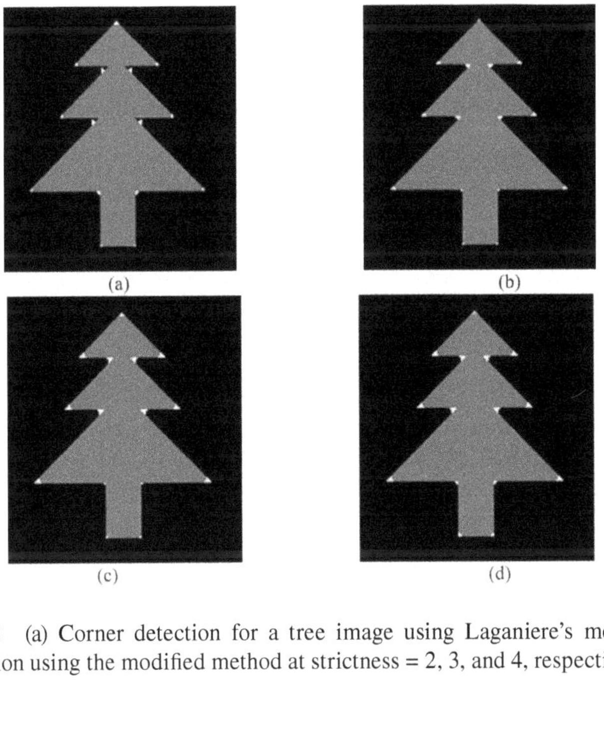

FIGURE 6.12 (a) Corner detection for a tree image using Laganiere's method. (b)–(d) Corner detection using the modified method at strictness = 2, 3, and 4, respectively.

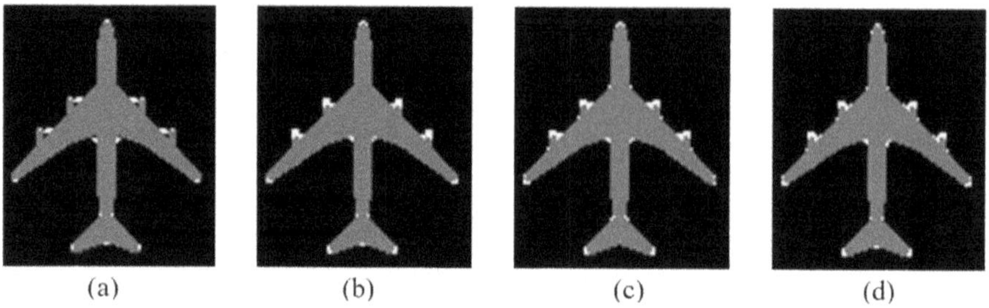

FIGURE 6.13 (a) Corner detection for an airplane image using Laganiere's method. (b)–(d) Corner detection using the modified method at strictness = 2, 3, and 4, respectively.

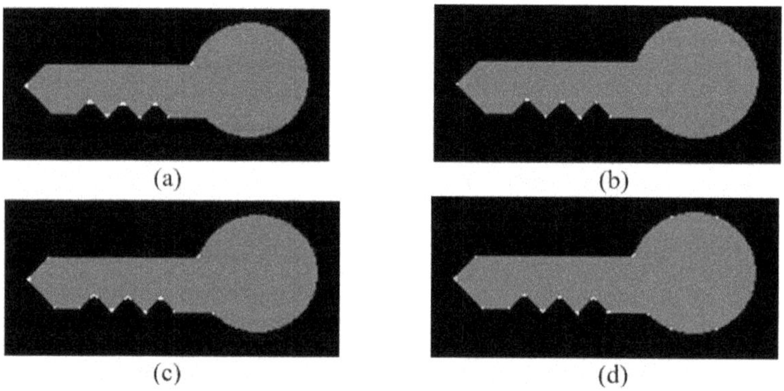

FIGURE 6.14 (a) Corner detection for a key image using Laganiere's method. (b)– (d) Corner detection using the modified method at strictness = 2, 3, and 4, respectively.

modified method can detect corners more accurately. From Figure 6.14, the modified method can detect corners more completely; for example, Laganiere's method can only detect 8 corners, but the modified method can detect all 12 corners at striness of 3. Implemented on the Patinum IV 2.0 Ghz PC, it takes 2.28 seconds to process Figure 6.14 by Laganiere's method; however, it spends 2.01 seconds by the proposed method.

A square is rotated by 30°, 45°, and 60° as illustrated in Figure 6.15. Experimental results show that the modified method can detect corners correctly. Table 6.1 summarizes the results of corner detection using Laganiere's and the modified methods with different values of strictness. From Table 6.1 and Figures 6.12–6.14, strictness of 3 is a better choice because it produces more accurate corners than strictness of 2 and generates less false-positive corners than strictness of 4.

In the experiments, a 5×5 circuluar structuring element is adopted. A 3×3 structuring element is applied on Figure 6.14, and the results are illustrated in Figure 6.16. It is observed that the 3×3 structuring element is worse than the 5×5 structuring element because some corners are missed.

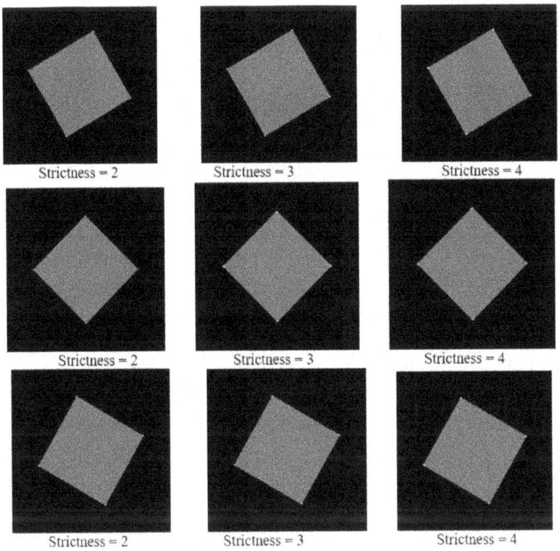

FIGURE 6.15 The corner detection results in different angles with strictness = 2, 3, and 4. The top-, middle-, and bottom-row images are rotated by 30°, 45°, and 60°, respectively.

TABLE 6.1
Number of Correctly Detected Corners, False Positives, and False Negatives

Test images	**Figure 8.12**	**Figure 8.13**	**Figure 8.14**
Total number of comers	15	24	12
Laganiere's method			
Correctly detected corners	15	18	8
False positives	0	0	0
False negatives	0	6	4
Our method at strictness = 2			
Correctly detected corners	15	18	7
False positives	0	0	0
False negatives	0	6	5
Our method at strictness = 3			
Correctly detected corners	15	23	12
False positives	0	0	0
False negatives	0	1	0
Our method at strictness = 4			
Correctly detected corners	15	24	12
False positives	0	2	4
False negatives	0	0	0

FIGURE 6.16 The detected corners using a 3 × 3 structuring element at strictness = 2, 3, and 4, respectively.

6.4 HOUGH TRANSFORM

Hough transform is a technique to identifying specific shapes in an image. It converts all the points in a curve into a single location in another parametric space by coordinate transformation. This method intends to map global features into local features. This concept can also apply to detect a circle, an ellipse, or other geometric shapes.

Considering detecting straight lines, let all the lines passing through a point (x', y') be represented as $y' = mx' + c$, where m is the line slope and c is the intercept with y-axis. The line equation can be rearranged as $c = -x'm + y'$. A point in an image of (x, y) domain corresponds to a line in the parametric space of (m, c) domain. Therefore, the intersection of all lines in the (m, c) parametric space indicates the slope and intercept of those collinear points (i.e., the line) in the image. Since the intersections are determined approximately by partitioning the parameter space into accumulator cells and then using whichever cell collects the greatest number of intersections as a solution, they are not always exact. Besides, the slope of vertical lines goes to infinity. In practice, the (ρ, θ) parametric space is used instead of (m, c) by the equation $x\cos\theta + y\sin\theta = \rho$, where ρ is the normal distance to the line from the origin and θ is the angle of this normal. Figure 6.17 shows the transformation from the (x, y) domain to the (ρ, θ) parametric space.

Hough transform to identifying a line:

(1) Let the parameters (ρ_{max}, ρ_{min}) and $(\theta_{max}, \theta_{min})$ represent the maximum and minimum distances from the origin and the maximum and minimum angles of the line, respectively.
(2) Subdivide the parametric space into accumulator cells.
(3) Initialize the accumulator cells to be all zeros.
(4) For every point of interest (x, y), incrementθ and solve for the correspondingρ using the normal line equation: $x\cos\theta_j + y\sin\theta_j = \rho_i$, to find the line intersect at (ρ_i, θ_j). The cell determined by (ρ_i, θ_j) is associated with the $A(i, j)$ of the accumulator cells, which would be incremented by one when the equation is solved.

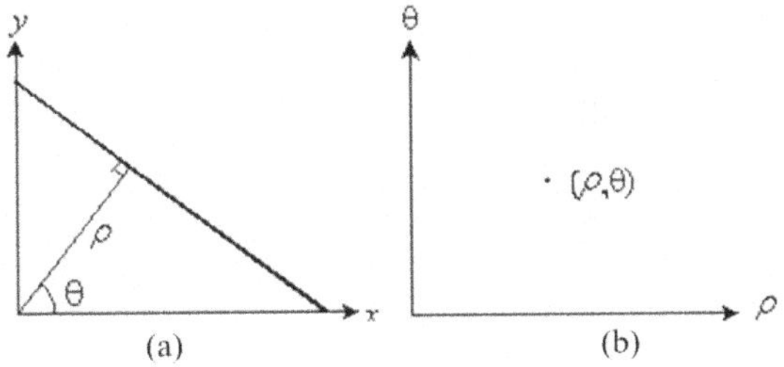

FIGURE 6.17 The transformation from the (x, y) domain to the (ρ, θ) parametric space.

To identify other geometric shapes, such as circles, arcs, squares, and triangles, using Hough transform, we extend the dimensions of the accumulator cells to a number of variables that need to be solved for the geometric shape equation. The Hough transform, when generalized, can be used to detect any geometric shape of the form g(**v**, **c**) = 0, where **v** and **c** are vectors of coordinates and coefficients. The procedure is similar to line detection except that all the curves through a single point will be represented by a surface in a parametric space that is higher dimensional. An example of Hough transform to detecting an image of several lines is shown in Figure 6.18.

Example 6.1: Apply the Hough transform to the following image to detect lines.

0	0	0	0	0	0	0	1
0	0	0	0	0	0	1	0
0	0	0	0	0	1	0	0
0	0	0	0	1	0	0	0
0	0	0	1	0	0	0	0
0	0	1	0	0	0	0	0
0	1	0	0	0	0	0	0
1	0	0	0	0	0	0	0

Answer: Find (m,c) for $c = y - mx$, we table the value of c as

(x,y)	$m = -3$	$m = -1$	$m = 1$	$m = 3$
(0, 0)	0	0	0	0
(1, 1)	4	2	0	−2
(2, 2)	8	4	0	−4
(3, 3)	12	6	0	−6
(4, 4)	16	8	0	−8
(5, 5)	20	10	0	−10
(6, 6)	24	12	0	−12
(7, 7)	28	14	0	−14

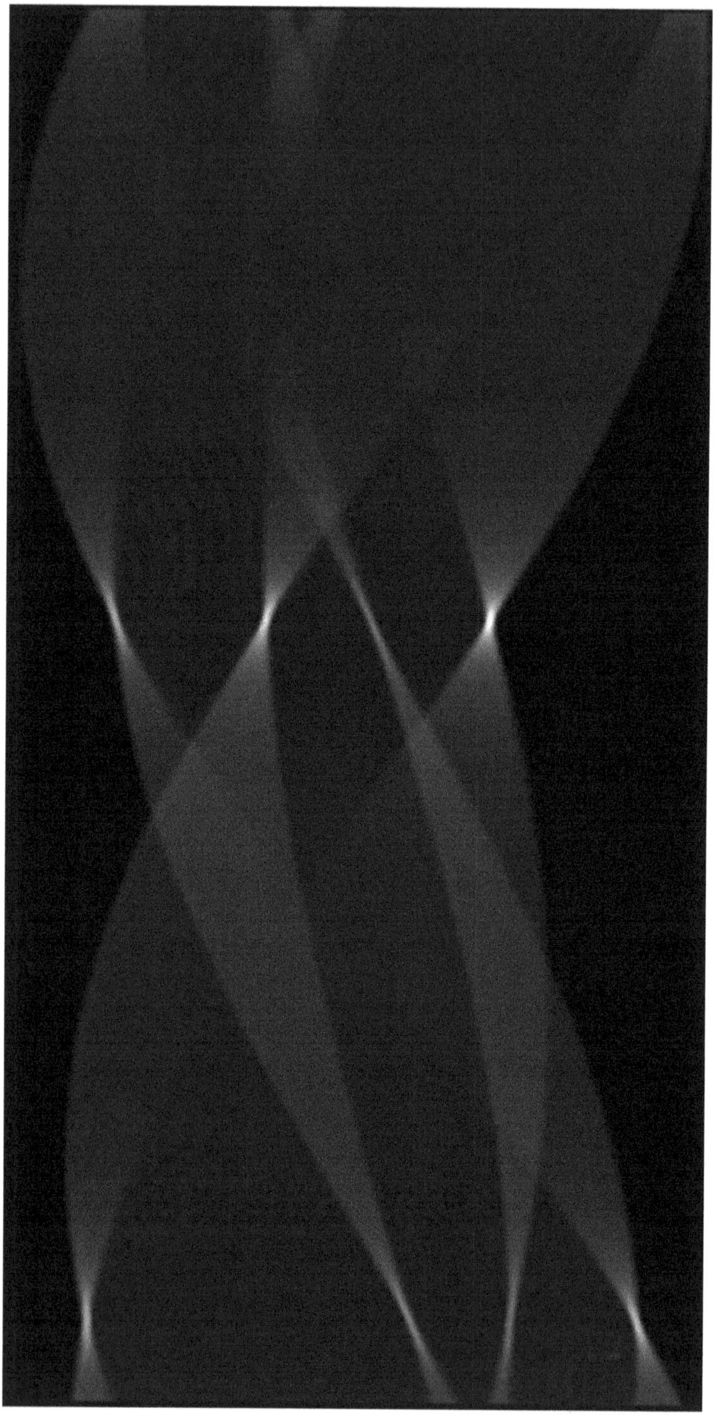

FIGURE 6.18 An example of Hough space with several lines.

The following graph shows the projection to (m,c) space, where A denotes the slope m and B denotes the intercept c.

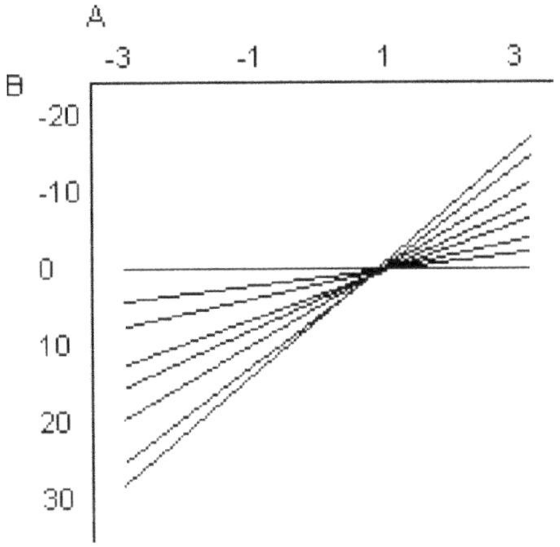

Since all lines cross $(m,c) = (1,0)$ and the accumulator has the peak value of 8, it indicates that there is a line with slope 1 and intercept 0 passing through the eight points in the image.

Koshimizu et al. (1989) presented global feature extraction using efficient Hough transform. Bonci et al. (2005) explained how to associate a rigorous probability value to the main straight line features extracted from a digital image, and proposed a Bayesian approach to the Hough transform. Aggarwal and Karl (2006) posed the problem of detecting straight lines in grayscale images as an inverse problem and proposed a formulation based on use of the inverse Radon operator, which relates the parameters determining the location and orientation of the lines in the image to the noisy input image. Satzoda et al. (2008) proposed a method called additive Hough transform (AHT) based on parallel processing to accelerate Hough transform computation.

6.5 PRINCIPAL COMPONENT ANALYSIS (PCA)

PCA is known as the best data representation in the least-square sense for classical recognition (Stork et al., 2000). It is commonly used to reduce the dimensionality of images and retain most of information. The central idea behind PCA is to find an orthonormal set of axes pointing in the direction of maximum covariance in the data. It is often used in representing facial images. The idea is to find the orthonormal basis vectors, or the eigenvectors, of the covariance matrix of a set of images, with each image treated as a single point in a high-dimensional space. It is assumed that the facial images form a connected sub-region in the image space. The eigenvectors map the most significant variations between faces and are preferred over other correlation techniques that assume every pixel in an image is of equal importance (Kosugi, 1995).

Since each image contributes to each of the eigenvectors which resemble ghostlike faces when displayed, it is referred to as *holon* (Cottrell and Fleming, 1990) or *eigenfaces* (Turk and Pentland, 1991), and the new coordinates system is referred to as the *face space*. Individual images can be projected onto the face space and represented exactly as weighted combinations of the eigenface components. The resulting vector of weights that describe each face can be used in data compression and face classification. Data compression relies on the fact that the eigenfaces are ordered, with each one accounting for a different amount of variation among the faces. Compression is achieved by reconstructing images using only those few eigenfaces that account for the most variability (Sirovich and Kirby, 1987). It results in dramatic reduction of dimensionality. Classification is performed by projecting a new image onto the face space and comparing the resulting weight vector with the weight vectors of a given class.

According to statistics, a dataset can be characterized by its mean value and its standard deviation, which is a measure of how spread out the data is. The principal components of a set of images can be derived directly as follows. Let $\mathbf{I}(x, y)$ be a two-dimensional array of intensity values of size $N \times N$. Let the set of facial images be $\Gamma_1, \Gamma_2, \Gamma_3, \ldots \Gamma_M$. The average face of the set is

$$\boldsymbol{\Psi} = \frac{1}{M} \sum_{k=1}^{M} \boldsymbol{\Gamma}_k \tag{6.24}$$

The distance from a given facial image Γ_k to the mean image of the set is

$$\Phi_k = \Gamma_k - \Psi \tag{6.25}$$

PCA seeks the set of M orthonormal vectors, υ_k, and their associated eigenvalues, λ_k, which best describes the distribution of the image points. The vectors υ_k and scalars λ_k are the eigenvectors and eigenvalues, respectively, of the covariance matrix

$$\mathbf{C} = \frac{1}{M} \sum_{k=1}^{M} \Phi_k \Phi_k^T = \mathbf{A}\mathbf{A}^T, \tag{6.26}$$

where the matrix $\mathbf{A} = [\Phi_1, \Phi_2, \ldots \Phi_M]$ (Turk and Pentland, 1991).

The size of $\mathbf{C}$ is N^2 by N^2 which for typical image sizes is an intractable task. However, since typically $M < N^2$, that is, the number of images is less than the dimension, there will only be $N - 1$ non-zero eigenvectors. Thus, the N^2 eigenvectors can be solved, in this case, by first solving for the eigenvectors of an $M \times M$ matrix, followed by taking the appropriate linear combinations of the data points $\mathbf{\Phi}$.

PCA is closely associated with the **S**ingular **V**alue **D**ecomposition of a data matrix and can be decomposed as

$$\mathbf{\Phi} = \mathbf{U}\mathbf{S}\mathbf{V}^T. \tag{6.27}$$

where $\mathbf{S}$ is a diagonal matrix whose diagonal elements are the singular values, or eigenvalues, of $\mathbf{\Phi}$, and $\mathbf{U}$ and $\mathbf{V}$ are unary matrices. The columns of $\mathbf{U}$ are the eigenvectors of $\mathbf{\Phi}\mathbf{\Phi}^T$, and are referred to as *eigenfaces*. The columns of $\mathbf{V}$ are the eigenvectors $\mathbf{\Phi}^T\mathbf{\Phi}$ and are not used in this analysis.

Faces can be classified by projecting a new face Γ onto the face space as follows:

$$\omega_k = \mathbf{u}_k^T \left(\Gamma_k - \Psi\right) \tag{6.28}$$

for $k = 1,...M'$ eigenvectors, with $M' \ll M$, if reduced dimensionality is desired. The weights form a vector $\Omega_k^T = \left[\omega_1, \omega_2, ...\omega_{M'}\right]$, which contains the projections onto each eigenvector. Classification is performed by calculating the distance of Ω_k from Ω, where Ω represents the average weight vector defining some class.

The applications of PCA include image compression, data mining, face detection (Sung and Poggio, 1998), face recognition, socio-economic analysis, protein dynamics, and especially the areas requiring the classification of the high-dimensional data (Jolliffe, 1986).

Examples of eigenfaces are shown in Figure 6.19. Individual images can be projected onto the face space and represented exactly as weighted combinations of the eigenface components (see Figure 6.20).

There are many PCA algorithms developed in the literature. Chatterjee et al. (2000) derived and discussed adaptive algorithms for PCA that are shown to converge faster than the traditional PCA algorithms. Weingessel and Hornik (2000) used a general framework to describe those PCA algorithms which are based on Hebbian learning. Martinez and Kak (2001) concluded that when the training dataset is small, PCA can outperform LDA and, also, that PCA is less sensitive to different training datasets by showing actual results on a face database. Kim et al. (2001) illustrated the potential of kernel PCA for texture classification. Kim et al. (2002) adopted the kernel PCA as a mechanism for extracting facial features. Liu (2004) presented a

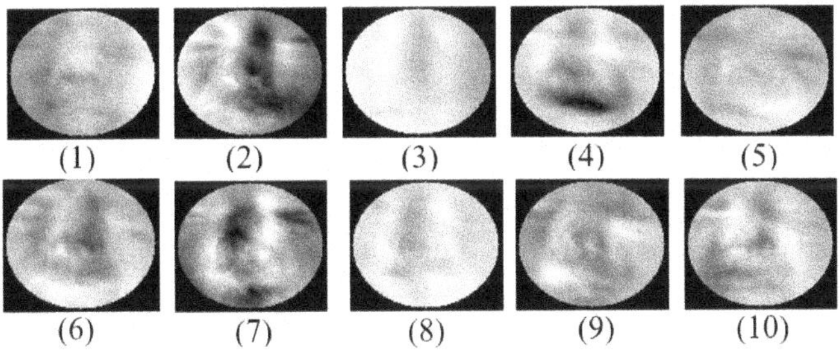

FIGURE 6.19 The first ten eigenfaces of 204 neonate images. The eigenfaces are ordered by magnitude of the corresponding eigenvalue.

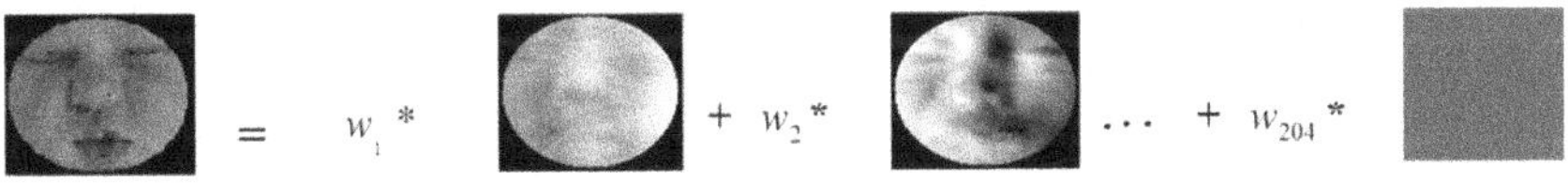

FIGURE 6.20 Illustration of the linear combination of eigenfaces. The face to the left can be represented as a weighted combination of eigenfaces.

Gabor-based kernel PCA method by integrating the Gabor wavelet representation of face images and the kernel PCA method for face recognition. Xie and Lam (2006) proposed a Gabor-based kernel PCA with doubly nonlinear mapping for human face recognition.

Nishino et al. (2005) presented a framework for applying PCA to visual data which takes advantage of the spatio-temporal correlation and localized frequency variations that are typically found in such data. Zhao et al. (2006) proposed a novel incremental PCA and its application for face recognition. Vaswani and Chellappa (2006) presented a classification algorithm, called principal component null space analysis (PCNSA), which is designed for classification problems like object recognition where different classes have unequal and nonwhite noise covariance matrices. Pang et al. (2008) incorporated both 2DPCA-based image decomposition and fast numerical calculations based on Haarlike bases to develop binary 2DPCA (B-2DPCA) for fast training and testing procedures.

6.6 LINEAR DISCRIMINATE ANALYSIS (LDA)

LDA is used to perform the feature extraction task and to classify samples of unknown classes based on training samples with known classes (Martinez and Kak, 2001). It provides a linear transformation of k-dimensional samples (or feature vectors in our experiments) into an m-dimensional space ($m < k$) so that samples belonging to the same class are close together, but samples from different classes are far apart from each other. This method maximizes the ratio of between-class variance to within-class variance in any dataset; thereby the theoretical maximum separation in the linear sense will be guaranteed. Since LDA seeks directions that are efficient for discrimination (Zhao et al., 1998), it is the optimal classifier for distinguishing classes that are Gaussian distribution and have equal covariance matrices.

After the PCA representation, each image is represented by k values as

$$y_i = (y_{i1}, y_{i2}, \ldots y_{ik})^T. \tag{6.29}$$

LDA seeks a transformation matrix that in some sense maximizes the ratio of the between-scatter matrix to the within-scatter matrix. The within-scatter matrix is defined as

$$S_w = \sum_{j=1}^{K} \sum_{i=1}^{N_j} \left(y_i^j - \mu_j\right)\left(y_i^j - \mu_j\right)^T, \tag{6.30}$$

where y_i^j is the ith sample of class j, μ_j is the mean of class j, K is the number of classes, and N_j is the number of samples in class j. The between-scatter matrix is defined as

$$S_b = \frac{1}{K} \sum_{i=1}^{K} (\mu_i - \mu)(\mu_i - \mu)^T, \tag{6.31}$$

where μ is the mean of all classes.

Our goal is to maximize the between-class measure while to minimize the within-class measure (i.e., $\max\left(\frac{\det|S_b|}{\det|S_w|}\right)$). To achieve this, we can find a transformation matrix

$$W_{opt} = \arg\max_W \left|\frac{W^T S_b W}{W^T S_w W}\right| = [w_1 w_2 \ldots w_m] \tag{6.32}$$

where $\{w_i\}$ is the set of generalized eigen vectors of S_b and S_w. We project y_i on W to find the LDA representation:

$$z_i = W^{-1}(y_i - \mu). \tag{6.33}$$

Once the transformation is found, the classification problem is simply a matter of finding the class whose transformed mean is closest to the transformed testing image.

PCA and LDA are two important feature extraction methods and have been widely applied in a variety of areas. A limitation of PCA and LDA is that when dealing with image data, the image matrices must be first transformed into vectors, which are usually of very high dimensionality. This causes expensive computational cost and sometimes the singularity problem. Li and Yuan (2005) introduced a new approach for image feature extraction and representation called *2D-LDA*. The difference between LDA and 2D-LDA is that in LDA, we use the image vector to compute the between-class and within-class scatter matrices, but in 2D-LDA, we use the original image matrix to compute the two matrices. They claimed that the 2D-LDA can achieve better results than other feature extraction methods, such as PCA, LDA, and 2D-PCA (Yang et al., 2004). The idea of 2D-LDA is described in the following.

Suppose that we have M training samples belonging to L classes ($L_1, L_2, \ldots, L_L$). The training samples in each class are denoted as N_i ($i = 1, 2, \ldots, L$). The size of each training image $\mathbf{A}_j$ ($j = 1, 2, \ldots, M$) is $m \times n$. Our purpose is to find a good projection vector, $\mathbf{x}$, such that when $\mathbf{A}_j$ is projected onto $\mathbf{x}$, we can obtain the projected feature vector, $\mathbf{y}_j$ of the image $\mathbf{A}_j$.

$$\mathbf{y}_j = \mathbf{A}_j\mathbf{x} \qquad j = 1, 2, \ldots, M \tag{6.34}$$

Similar to LDA, we can find the between-class scatter matrix, $\mathbf{TS}_B$, and the within-class scatter matrix, $\mathbf{TS}_W$, of the projected feature vectors by using training images. The criterion is to project the images onto a subspace that maximizes the between-class scatter and minimizes the within-class scatter of the projected data. Since the total scatter of the projected samples can be represented by the trace of the covariance matrix of the projected feature vectors, the Fisher linear projection criterion can be described as

$$\mathbf{J}(\mathbf{x}) = \frac{tr(\mathbf{TS_B})}{tr(\mathbf{TS_W})} = \frac{\mathbf{x}^T\mathbf{S_B}\mathbf{x}}{\mathbf{x}^T\mathbf{S_B}\mathbf{x}} \tag{6.35}$$

where $\mathbf{S}_\mathrm{B} = \sum_{i=1}^{L} N_i \left(\overline{\mathbf{A}}_i - \overline{\mathbf{A}}\right)^T \left(\overline{\mathbf{A}}_i - \overline{\mathbf{A}}\right)$ and $\mathbf{S}_\mathrm{W} = \sum_{i=1}^{L} \sum_{\mathbf{A}_k \in L_i} \left(\mathbf{A}_k - \overline{\mathbf{A}}_i\right)^T \left(\mathbf{A}_k - \overline{\mathbf{A}}_i\right)$.

The optimal projection, $\mathbf{x}_{\mathrm{opt}}$, can be decided when the criterion is maximized. That is, $\mathbf{x}_{\mathrm{opt}} = \arg\max_x \mathbf{J}(\mathbf{x})$. We can find out the solution by solving the generalized eigenvalue problem. Therefore, if we choose d optimal projection axes $\mathbf{x}_1$, $\mathbf{x}_2, \ldots, \mathbf{x}_d$ corresponding to the first d largest eigenvalues, then we can extract feature for a given image $\mathbf{I}$. The following equation can be used for feature extraction.

$$\mathbf{y} = \mathbf{I}\left[\mathbf{x}_1, \mathbf{x}_2, \ldots, \mathbf{x}_d\right] \tag{6.36}$$

Since $\mathbf{I}$ is an $m \times n$ matrix and $\left[\mathbf{x}_1, \mathbf{x}_2, \ldots, \mathbf{x}_d\right]$ is an $n \times d$ matrix, we can form an $m \times d$ matrix, $\mathbf{y}$, to represent the original image, $\mathbf{I}$. We use the 2D-LDA representation of the original image as the input feeding into the classifier in the next stage.

Ye and Li (2005) proposed a two-stage LDA method, which aims to overcome the singularity problems of classical LDA, while achieving efficiency and scalability simultaneously. Wang et al. (2006) showed that the matrices based 2D algorithms are equivalent to special cases of image-block-based feature extraction, i.e., partition each image into several blocks and perform standard PCA or LDA on the aggregate of all image blocks.

6.7 FEATURE REDUCTION IN INPUT AND FEATURE SPACES

Feature extraction and reduction are two main issues in feature selection, which is important for many supervised learning problems. The goal of image feature extraction is to achieve a better classification rate by extracting new features to represent objects from raw pixel data. The goal of feature reduction is to select a subset of features while preserving or improving the classification rate of a classifier. Feature reduction can speed up the classification process by keeping the most important class-relevant features.

6.7.1 Feature Reduction in the Input Space

Principal component analysis (PCA) is widely used in image representation for dimensionality reduction. To obtain m principal components, we need to multiply a transformation matrix of dimensions $m \times N$ with the input pattern $\mathbf{x}$ of dimensions $N \times 1$. Totally, there are $m \times N$ multiplications and $m \times N$ summations. This is computation costly. In this section, we propose a method of selecting a subset of features to save time and maintain competitive performance as the PCA representation.

A straightforward method for features reduction is using Fisher's criterion to choose a subset of features that have large between-class and small within-class variances. For example, in face and non-face classification, the within-class variance in dimension i is calculated as

$$\sigma_i^2 = \frac{\sum_{j=1}^{l} \left(g_{j,i} - m_i\right)^2}{l-1}, \text{ for } i = 1, \ldots, N, \tag{6.37}$$

where l is the number of samples, $g_{j,i}$ is the ith dimensional gray value of sample j, and m_i is the mean value of ith dimension. For between-class measurement, we use the Fisher's score as

$$S_i = \left| \frac{m_{i,\ face} - m_{i,\ nonface}}{\sigma^2_{i,\ face} + \sigma^2_{i,\ nonface}} \right|. \tag{6.38}$$

By selecting the features with highest Fisher's scores, we can retain the most discriminative features between face and non-face classes.

Alternatively, we propose a feature reduction method for the 2nd-degree support vector machine based on a decision function. For the 2nd-degree polynomial SVM with kernel $K(\mathbf{x},\mathbf{y}) = (1+\mathbf{x}\cdot\mathbf{y})^2$ the decision function for a given pattern $\mathbf{x}$ is defined as

$$f(\mathbf{x}) = \sum_{i=1}^{s} \alpha_i y_i \left(1+\mathbf{x}_i \cdot \mathbf{x}\right)^2 + b$$

$$= \sum_{i=1}^{s} \alpha_i y_i \left(1 + x_{i,1}x_1 + x_{i,2}x_2 + \ldots + x_{i,k}x_k + \ldots + x_{i,N}x_N\right)^2 + b, \tag{6.39}$$

where s is the number of support vectors, $\mathbf{x}_i$ is the i-th support vector, and $x_{i,k}$ and x_k are respectively the k-th dimension for the support vector $\mathbf{x}_i$ and the pattern $\mathbf{x}$. The component related to the k-th dimension (where $k = 1, 2, \ldots, N$) is

$$f(\mathbf{x},k) = \sum_{i=1}^{s} \alpha_i y_i \left[2x_k x_{i,k}\left(1 + x_{i,1}x_1 + \ldots + x_{i,k-1}x_{k-1} + x_{i,k+1}x_{k+1} + \ldots + x_{i,N}x_N\right) + x_k^2 x_{i,k}^2\right]$$

$$= \sum_{i=1}^{s} \alpha_i y_i \left[2x_k x_{i,k}\left(1 + x_{i,1}x_1 + \ldots + x_{i,N}x_N\right) - x_k^2 x_{i,k}^2\right]. \tag{6.40}$$

In order to select a subset of m features from the original N features, we use the largest m decision values. For the kth dimension, we need to estimate its contribution as

$$F(k) = \int_V f(\mathbf{x},k)\,dP(\mathbf{x}). \tag{6.41}$$

In fact, we can not obtain $F(k)$ because the probability distribution function $P(\mathbf{x})$ is unknown. We approximate $F(k)$ with a sum over the support vectors

$$F(k) = \sum_{i=1}^{s}\sum_{j=1}^{s} \left| \alpha_j y_j \left[2x_{i,k}x_{j,k}\left(1 + x_{j,1}x_{i,1} + \ldots + x_{j,N}x_{i,N}\right) - x_{i,k}^2 x_{j,k}^2\right]\right|. \tag{6.42}$$

We adopt a face image database from the Center for Biological and Computational Learning at Massachusetts Institute of Technology (MIT), which contains 2,429 face training samples, 472 face testing samples and 23,573 non-face testing samples. We collected 15,228 non-face training samples randomly from the images that do not contain faces. The size of all these samples is 19 × 19. A 2nd-degree polynomial SVM with kernel $K(\mathbf{x},\mathbf{y}) = (1+\mathbf{x}\cdot\mathbf{y})^2$ is used in our experiments.

In order to remove background pixels, a mask is applied to extract only the face. Prior to classification, we perform image normalization and histogram equalization. In normalization, we use zero mean and one variance. In Figure 6.21, (a) shows a face image, (b) shows the mask, and (c) and (d) are the images after normalization and histogram equalization, respectively. To calculate PCA values, we test the two methods: one is to combine face and non-face training samples, and the other is to use face training samples only.

Figure 6.22 shows the *receiver operating characteristic* (ROC) curves of using different numbers of principal components. The pre-processing is normalization. The ROC curve is defined as shifting the SVM hyperplane by changing the threshold value *b*. We perform face classification on the testing set and calculate the false positive and the detection rates. The horizontal axis shows the false positive rate over 23,573 non-face testing samples. The vertical axis shows the detection rate over 472 face testing samples. The PCA values are calculated using only the face training samples. We observe that using 60 or 100 PCA values can achieve almost the same performance as using all 283 gray values. Figure 6.23 shows the performance of using only positive training samples that is better than using the combination of

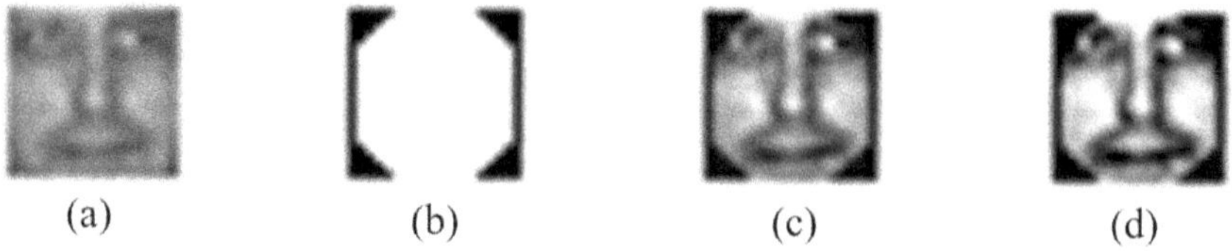

FIGURE 6.21 (a) Original face image, (b) the mask, (c) normalized image, and (d) histogram equalized image.

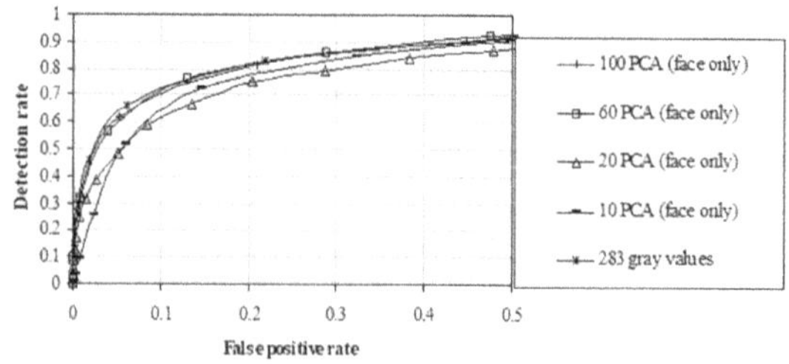

FIGURE 6.22 ROC curves for different numbers of PCA values. The pre-processing method is normalization and only face training samples are used to calculate PCA.

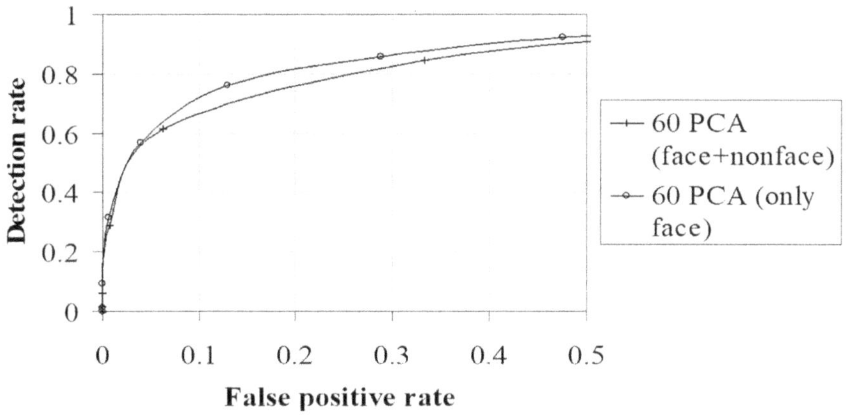

FIGURE 6.23 ROC curves for two methods to get PCA features. One method uses both face and non-face training samples, and the other uses only face training samples.

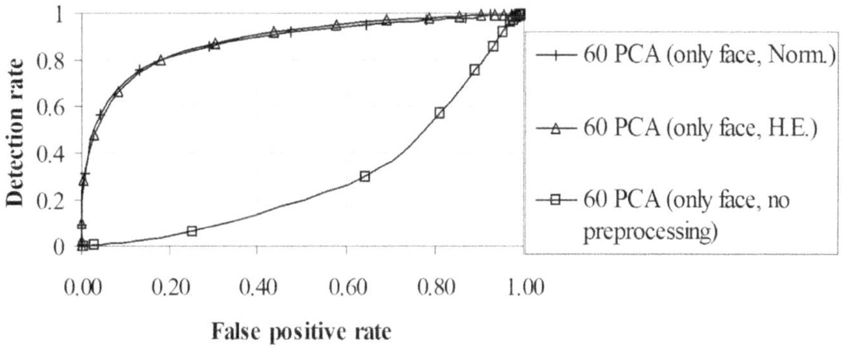

FIGURE 6.24 ROC curves for different pre-processing methods: normalization, histogram equalization, and without pre-processing. Only face training samples are used to calculate PCA values.

positive and negative training samples. Figure 6.24 shows the performance of using normalization or histogram equalization, and both obtain better results than that without pre-processing. In the following experiments, we will use normalization and positive training samples to calculate PCA values.

By using the normalized 2,429 face and 15,228 non-face training samples and taking all 283 gray values as input to train the 2nd-degree SVM, we obtain 252 and 514 support vectors for face and non-face classes, respectively. Using these support vectors and equation (6.42), we can obtain $F(k)$, where $k = 1, 2, \ldots, 283$. The performances of using 100 features selected by this ranking method, all 283 gray values, and 100 PCA features are shown in Figure 6.25. We also experiment on Fisher's scores using equations (6.37) and (6.38), and their performance is shown in Figure 6.25. We also compare the performance of using 100 features selected using

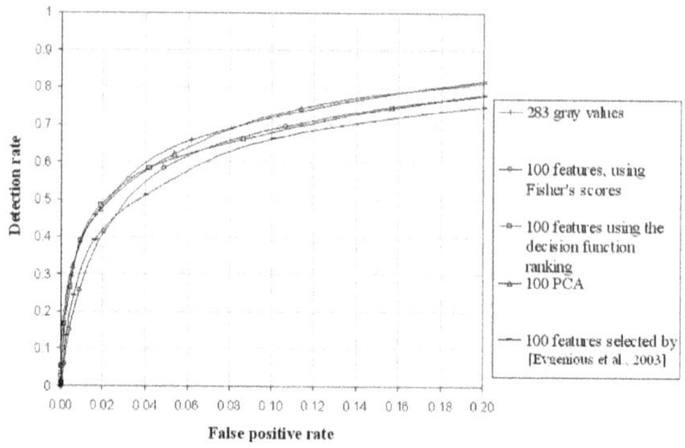

FIGURE 6.25 ROC curves for different features in input space. The pre-processing method is normalization and PCA values are calculated using only face training samples.

Evgenious et al. (2003). Figure 6.25 shows that at low false positive rates, using 100 features selected by decision function ranking performs similarly as using all 283 features and using 100 PCA values; however, it is significantly better than that using 100 features by Fisher's scores and by (Evgenious et al., 2003). This justifies the efficiency of our proposed feature reduction method.

6.7.2 Feature Reduction in the Feature Space

In feature space, the decision function *f*(**x**) of SVM can be calculated as

$$f(\boldsymbol{x}) = \sum_{i=1}^{s} \alpha_i y_i \left(\Phi(\boldsymbol{x}) \cdot \Phi(\boldsymbol{x}_i) \right) + b = \mathbf{w} \cdot \Phi(\boldsymbol{x}) + b, \tag{6.43}$$

where **w** is the support vector. For the 2nd-degree polynomial SVM with input space of dimension N and kernel $K(\mathbf{x},\mathbf{y}) = (1+\mathbf{x}\cdot\mathbf{y})^2$ the feature space is given by

$$\Phi(\boldsymbol{x}) = \left(\sqrt{2}x_1, \ldots, \sqrt{2}x_N, x_1^2, \ldots, x_N^2, \sqrt{2}x_1x_2, \ldots, \sqrt{2}x_{N-1}x_N \right), \tag{6.44}$$

of dimension $P = N(N+3)/2$. Our goal is to select a subset of q features without significantly decreasing the performance.

Suppose that we train a 2nd-degree SVM using face and non-face samples to obtain s support vectors. The support vector in the feature space can be represented as

$$\boldsymbol{w} = \sum_{i=1}^{s} \alpha_i y_i \Phi(\boldsymbol{x}_i) = (w_1, w_2, \ldots, w_P). \tag{6.45}$$

One simple method to select a subset of features is to rank according to $|w_k|$, for $k = 1,2,\ldots,P$. However, we propose a better ranking method on $\left|w_k \int Vx_k^* \left|dp\left(x_k^*\right)\right|\right|$, where x_k^* denotes the kth dimension of **x** in the feature space. Since the distribution function $dp(x_k^*)$ is unknown, we use

$$R(k) = \left| w_k \sum_{i=1}^{S} \left| x_{i,k}^* \right| \right|, \tag{6.46}$$

where $x_{i,k}^*$ denotes the k-th dimension of $\mathbf{x}_i$ in the feature space. The decision function of q features is calculated as

$$f(\boldsymbol{x},q) = \boldsymbol{w}(q) \cdot \Phi(\boldsymbol{x},q) + b, \tag{6.47}$$

where **w**(q) is the selected q features of **w**, and $\Phi(\mathbf{x},q)$ is the corresponding q features of **x**.

For a pattern **x**, we calculate the difference of two decision values using all the features and using the subset of q features as

$$\Delta f_q(\boldsymbol{x}) = \left| f(\boldsymbol{x}) - f(\boldsymbol{x},q) \right|. \tag{6.48}$$

The quantity of the sum of difference values over all the support vectors is

$$\Delta F_q = \sum_{i=1}^{S} \Delta f_q\left(\boldsymbol{x}_i\right). \tag{6.49}$$

It can be used as a criterion to evaluate the efficiency of the methods.

In the experiments, we trained a 2nd-degree polynomial SVM using 60 PCA values in the input space. The training samples are the same as before. The preprocessing method is normalization and the 60 PCA values are calculated from the transformation matrix obtained form the face training samples. We obtain 289 and 412 support vectors for face and non-face classes, respectively. The 1,890 features in the feature space can be generated by equation (6.44). The support vector in feature space $\mathbf{w} = (w_1, w_2, \ldots, w_{1890})$ can be calculated by equation (6.45). The values of ΔF_q for different q can be calculated by equation (6.49). Figure 6.26 shows the variance of ΔF_q by varying the number of feature subsets of the two ranking methods by $|w_k|$ and by our proposed method. We observe that our method leads to a faster decrease in difference; it means more efficient.

The ranking values of kth dimension in the feature space can be calculated by equation (6.46). Given a fixed number of q, for a pattern **x**, the decision value of using the selected subset of q features can be calculated by equation (6.47). We take examples of $q = 300, 500, 1000$, and the results are shown in Figure 6.27. We observe that using 1,000 features and 500 features selected by the ranking method can achieve almost the same performance as using all the 1,890 features in the feature space. On the other hand, 300 features are insufficient to achieve good performance. Heisele

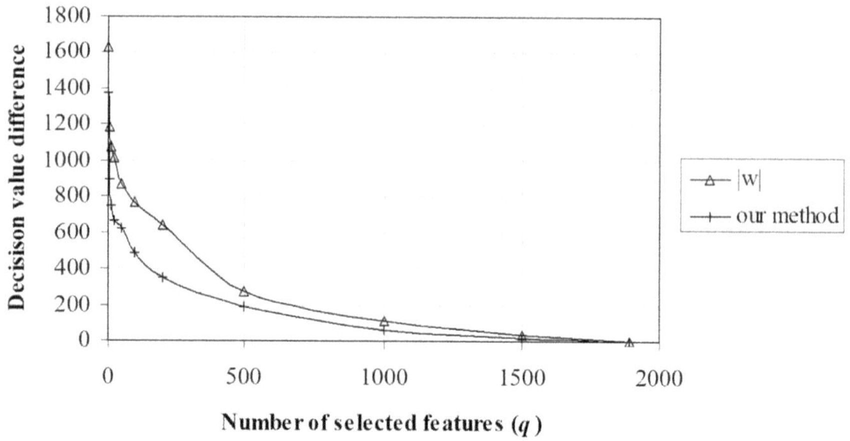

FIGURE 6.26 Decision value differences between using all the features and using a subset of q features in the feature space, with comparisons of two ranking methods.

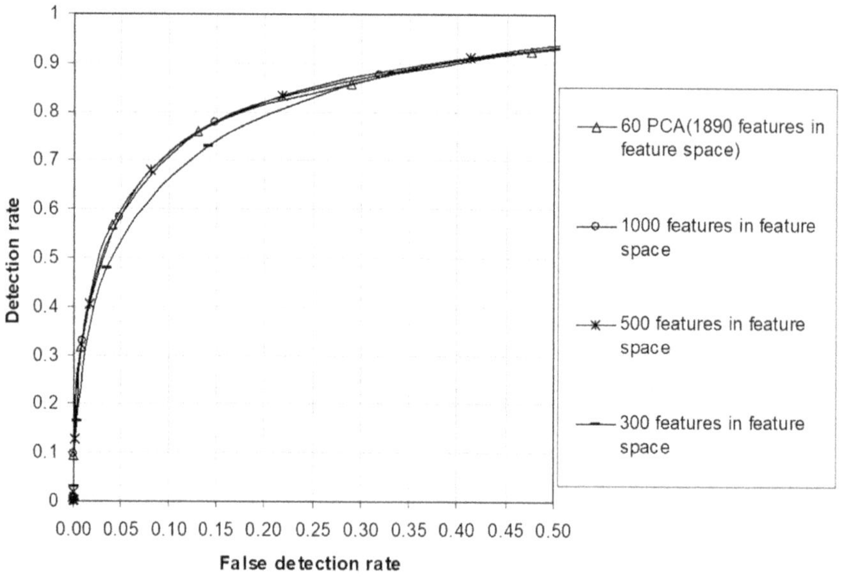

FIGURE 6.27 ROC curves for different numbers of features in the feature space. The preprocessing method is normalization and PCA values are calculated using only face training samples.

et al. (2000) used the feature ranking according to $| w_k \sum_{i=1}^{s} y_i x_{i,k}^* |$. They used 1,000 features and obtained similar performance as using all the features. When it is reduced to 500 features, the result is significantly worse.

6.7.3 Combination of Input and Feature Spaces

In this section, the performance of combining the feature reduction methods in input and feature space is evaluated. First, m features are chosen from the N input space and the 2nd-degree polynomial SVM is trained. Next, q features are selected from $P = m(m+3)/2$ features in the feature space and the decision value is calculate.

There are two methods to compute the 2nd-degree polynomial SVM, one by equation (6.39) in the input space and the other by equation (6.43) in the feature space. In equation (6.39), the number of multiplications required to calculate the decision function is $(N+1)s$. In equations (6.43) and (6.44), the number of multiplications required to calculate the decision function is $\frac{(N+3)N}{2} + \frac{(N+3)N}{2} = (N+3)N$. If $(N+1)\, s > (N+3)\, N$, it is more efficient to implement the 2nd-degree polynomial SVM in the feature space. This is always true in our experiments because the number of support vectors is more than 700, that is much larger than N. Note that $N = 283$, 60, or 100 reflects all the gray value features, 60 PCA values, or 100 features, respectively.

We train SVM using the selected 100 features to obtain 222 and 532 support vectors respectively for face and non-face classes. The 5,150 features in the feature space can be generated by equation (6.44). Figure 6.28 shows the results of using different subset of features in the feature space. Using 3,500 features selected by our method can achieve similar performance as using all 5,150 features. Figure 6.29 shows comparisons of using our combination method, 60 PCA values, and all 283 gray values. We observe that at low false positive rates, our method outperforms 60 PCA values.

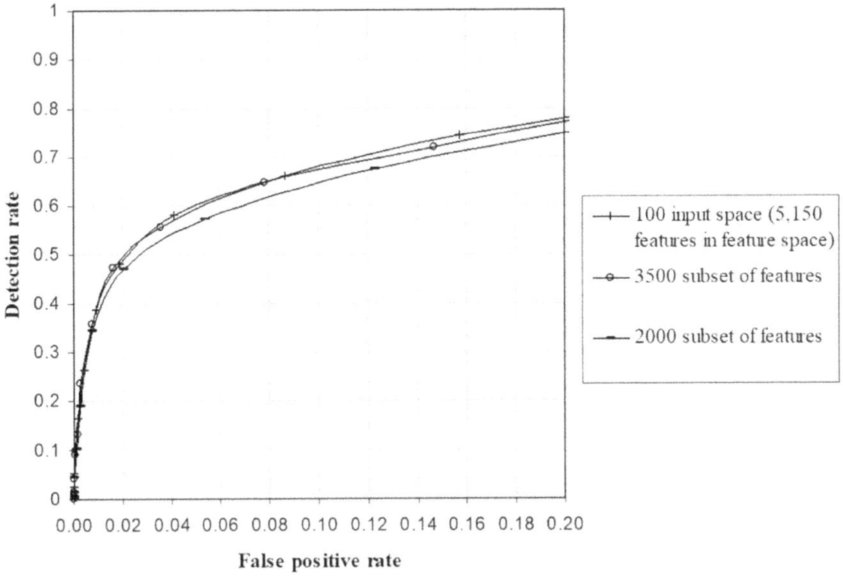

FIGURE 6.28 ROC curves of using different numbers of features in feature space. The 100 features in the input space are selected using the proposed ranking method.

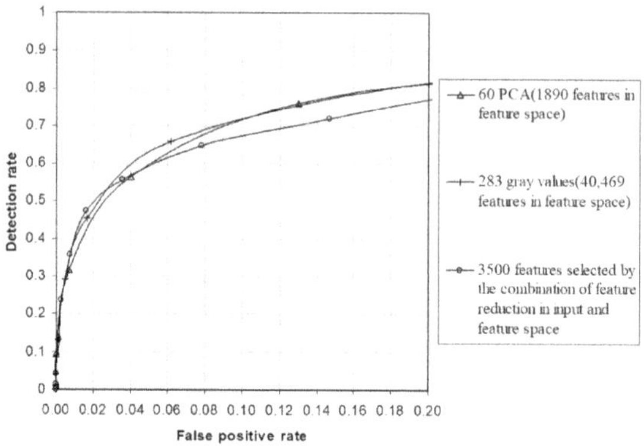

FIGURE 6.29 ROC curves of using the proposed feature reduction method, 60 PCA, and all the 283 gray values. The pre-processing method is normalization and PCA values are calculated using only face training samples.

TABLE 6.2
Numbers of Features Used in Input and Feature Space and the Numbers of Multiplication Required to Calculate the Decision Value for Different Methods

Methods	Number of Features in Input Space	Number of Features in Feature Space	Number of Multiplications
All gray values	283	40,469	80,938
PCA	60	1,890	20,760
Our method	100	3,500	8,650

We analyze computational complexity as follows. Using all 283 gray values in the input space, we need 80,938 multiplications to calculate the decision function for a pattern. Using 60 PCA values, we need 3,780 multiplications to calculate the decision function and $m \times N = 60 \times 283 = 16,980$ multiplications to obtain 60 PCA values, that gives a total of 20,760 multiplications. If applying the feature selection method in feature space and 1,000 features are used, totally $1890 + 1,000 + 16,980 = 19,870$ multiplications are required. If applying the proposed feature selection method in input space and $m = 100$ features are used, totally $(m+3)m = 103 \times 100 = 10,300$ multiplications are required. If the subset of $q = 3,500$ features in the feature space selected by the feature reduction method are used, totally $(m+3)m/2 + q = (100+3)100/2 + 3,500 = 8,650$ multiplications are needed. Table 6.2 summarizes the numbers of features used in input and feature space and the numbers of multiplication required to calculate the decision value for different methods.

Therefore, by using PCA for feature reduction in the input space and the feature reduction method in the feature space, a speed-up factor of 4.08 can be achieved. By combining the feature reduction methods in the input and feature space, a speed-up factor of 9.36 can be achieved. In fact, once the features in the feature space are determined, we do not need to project the input space to the whole feature space, but to the selected q features. This can further speed up the computation, such that only $3,500 + 3,500 = 7,000$ multiplications are required. We have shown that using the proposed feature reduction approaches, a fast non-linear SVM classifier for face detection can be developed.

REFERENCES

Agam, G. and Dinstein, I., "Regulated morphological operations," *Pattern Recognition*, vol. 32, no. 6, pp. 947–971, Jun. 1999.

Aggarwal, N. and Karl, W. C., "Line detection in images through regularized Hough transform," *IEEE Transactions on Image Processing*, vol. 15, no. 3, pp. 582–591, Mar. 2006.

Ando, S., "Image field categorization and edge/corner detection from gradient covariance," *IEEE Transactions on Pattern Analysis and Machine Intelligence*, vol. 22, no. 2, pp. 179–190, Feb. 2000.

Arnow, T. L. and Bovik, A. C., "Foveated visual search for corners," *IEEE Transactions on Image Processing*, vol. 16, no. 3, pp. 813–823, Mar. 2007.

Basak, J. and Mahata, D., "A connectionist model for corner detection in binary and gray images," *IEEE Transactions on Neural Networks*, vol. 11, no. 5, pp. 1124–1132, Sep. 2000.

Belkasim, S. O., Shridhar, M., and Ahmadi, M., "Pattern recognition with moment invariants: A comparative study and new results," *Pattern Recognition*, vol. 24, no. 12, pp. 1117–1138, Dec. 1991.

Bober, M., "MPEG-7 visual shape descriptors," *IEEE Transactions on Circuits and Systems for Video Technology*, vol. 11, no. 6, pp. 716–719, Jun. 2001.

Bonci, A., Leo, T., and Longhi, S., "A Bayesian approach to the Hough transform for line detection," *IEEE Transactions on Systems, Man, and Cybernetics, Part A*, vol. 35, no. 6, pp. 945–955, Nov. 2005.

Chatterjee, C., Kang, Z., and Roychowdhury, V. P., "Algorithms for accelerated convergence of adaptive PCA," *IEEE Transactions on Neural Networks*, vol. 11, no. 2, pp. 338–355, Mar. 2000.

Chen, C.-C., "Improved moment invariants for shape discrimination," *Pattern Recognition*, vol. 26, no. 5, pp. 683–686, May 1993.

Chuang, G. C.-H. and Kuo, C.-C. J., "Wavelet descriptor of planar curves: Theory and applications," *IEEE Transactions on Image Processing*, vol. 5, no. 1, pp. 56–70, Jan. 1996.

Cottrell, G. W. and Fleming, M. K., "Face recognition using unsupervised feature extraction," *Proceedings of International Conference on Neural Networks*, Paris, pp. 322–325, Jul. 1990.

Danielsson, P. E., "A new shape factor," *Computer Graphics and Image Processing*, vol. 7, no. 2, pp. 292–299, Apr. 1978.

Dudani, S., Breeding, K., and McGhee, R., "Aircraft identification by moment invariants," *IEEE Transactions on Computers*, vol. 26, no. 1, pp. 39–45, Jan. 1977.

Evgenious, T., Pontil, M., Papageorgiou, C., and Poggio, T., "Image representations and feature selection for multimedia database search," *IEEE Transactions on Knowledge and Data Engineering*, vol. 15, no. 4, pp. 911–920, Jul. 2003.

Flusser, J., Boldys, J., and Zitova, B., "Moment forms invariant to rotation and blur in arbitrary number of dimensions," *IEEE Transactions on Pattern Analysis and Machine Intelligence*, vol. 25, no. 2, pp. 234–246, Feb. 2003.

Flusser, J. and Suk, T., "Rotation moment invariants for recognition of symmetric objects," *IEEE Transactions on Image Processing*, vol. 15, no. 12, pp. 3784–3790, Dec. 2006.

Gao, X., Sattar, F., and Venkateswarlu, R., "Multiscale corner detection of gray level images based on log-Gabor wavelet transform," *IEEE Transactions on Circuits and Systems for Video Technology*, vol. 17, no. 7, pp. 868–875, Jul. 2007.

Harris, C. and Stephens, M., "A combined corner and edge detector," *Proceedings of the 4th Alvey Vision Conference*, Manchester, pp. 147–151, 1988.

Heisele, B., Poggio, T., Pontil, M., "Face detection in still gray images," A.I. Memo 1687, *Center for Biological and Computational Learning*, MIT, 2000.

Hu, M.-K., "Visual pattern recognition by moment invariants," *IRE Transactions on Information Theory*, vol. 8, no. 2, pp. 179–187, Feb. 1962.

Hung, K.-C., "The generalized uniqueness wavelet descriptor for planar closed curves," *IEEE Transactions on Image Processing*, vol. 9, no. 5, pp. 834–845, May 2000.

Jolliffe, I. T., *Principal Component Analysis*, Springer-Verlag, New York, 1986.

Kim, K. I., Jung, K., and Kim, H. J., "Face recognition using kernel principal component analysis," *IEEE Signal Processing Letters*, vol. 9, no. 2, pp. 40–42, Feb. 2002.

Kim, K. I., Park, S. H., and Kim, H. J., "Kernel principal component analysis for texture classification," *IEEE Signal Processing Letters*, vol. 8, no. 2, pp. 39–41, Feb. 2001.

Kiranyaz, S., Ferreira, M., and Gabbouj, M., "A generic shape/texture descriptor over multiscale edge field: 2-D walking ant histogram," *IEEE Transactions on Image Processing*, vol. 17, no. 3, pp. 377–391, Mar. 2008.

Kitchen, L. and Rosenfeld, A., "Gray-level corner detection," *Pattern Recognition Letters*, vol. 1, no. 2, pp. 95–102, Dec. 1982.

Koshimizu, H., Murakami, K., and Numada, M., "Global feature extraction using efficient Hough transform," *Proceedings on International Workshop on Industrial Applications of Machine Intelligence and Vision*, Tokyo, pp. 298–303, 1989.

Kosugi, M., "Human-face search and location in a scene by multi-pyramid architecture for personal identification," *Systems and Computers in Japan*, vol. 26, no. 6, pp. 27–38, 1995.

Krawtchouk, M., "On interpolation by means of orthogonal polynomials," *Memoirs Agricultural Institute of Kyiv*, vol. 4, pp. 21–28, 1929.

Laganiere, R., "A morphological operator for corner detection," *Pattern Recognition,* vol. 31, no. 11, pp. 1643–1652, Nov. 1998.

Lee, K.-J. and Bien, Z., "A gray-level corner detector using fuzzy logic," *Pattern Recognition Letters*, vol. 17, no. 9, pp. 939–950, Aug. 1996.

Li, M. and Yuan, B., "2D-LDA: A statistical linear discriminant analysis for image matrix," *Pattern Recognition Letters*, vol. 26, no. 5, pp. 527–532, Apr. 2005.

Liao, S. X. and Pawlak, M., "On image analysis by moments," *IEEE Transactions on Pattern Analysis and Machine Intelligence*, vol. 18, no. 3, pp. 254–266, Mar. 1996.

Lin, R.-S., Chu, C.-H., and Hsueh, Y.-C., "A modified morphological corner detector," *Pattern Recognition Letters*, vol. 19, no. 3, pp. 279–286, Mar. 1998.

Liu, C., "Gabor-based kernel PCA with fractional power polynomial models for face recognition," *IEEE Transactions on Pattern Analysis and Machine Intelligence*, vol. 26, no. 5, pp. 572–581, May 2004.

Mamistvalov, A. G., "N-dimensional moment invariants and conceptual mathematical theory of recognition n-dimensional solids," *IEEE Transactions on Pattern Analysis and Machine Intelligence*, vol. 20, no. 8, pp. 819–831, Aug. 1998.

Manjunath, B. S., Ohm, J. R., Vasudevan, V. V., and Yamada, A., "Color and texture descriptors," *IEEE Transactions on Circuits and Systems for Video Technology*, vol. 11, no. 6, pp. 716–719, Jun. 2001.

Martinez, A. M. and Kak, A. C., "PCA versus LDA," *IEEE Transactions on Pattern Analysis and Machine Intelligence*, vol. 23, no. 2, pp. 228–233, Feb. 2001.

Mikolajczyk, K. and Schmid, C., "A performance evaluation of local descriptors," *IEEE Transactions on Pattern Analysis and Machine Intelligence*, vol. 27, no. 10, pp. 1615–1630, Oct. 2005.

Nishino, K., Nayar, S. K., and Jebara, T., "Clustered blockwise PCA for representing visual data," *IEEE Transactions on Pattern Analysis and Machine Intelligence*, vol. 27, no. 10, pp. 1675–1679, Oct. 2005.

Paler, K., Föglein, J., Illingworth, J., and Kittler, J., "Local ordered greylevels as an aid to corner detection," *Pattern Recognition*, vol. 17, no. 5, pp. 535–543, 1984.

Pang, Y., Tao, D., Yuan, Y., and Li, X., "Binary two-dimensional PCA," *IEEE Transactions on Systems, Man, and Cybernetics, Part B*, vol. 38, no. 4, pp. 1176–1180, Aug. 2008.

Pawlak, M., "On the reconstruction aspect of moment descriptors," *IEEE Transactions on Information Theory*, vol. 38, no. 6, pp. 1698–1708, Nov. 1992.

Persoon, E. and Fu, K. S., "Shape discrimination using Fourier descriptors," *IEEE Transactions on Systems, Man, and Cybernetics*, vol. 7, no. 2, pp. 170–179, Mar. 1977.

Rafiei, D. and Mendelzon, A. O., "Efficient retrieval of similar shapes," *The International Journal on Very Large Data Bases*, vol. 11, no. 1, pp. 17–27, Aug. 2002.

Rockett, P. I., "Performance assessment of feature detection algorithms: A methodology and case study on corner detectors," *IEEE Transactions on Image Processing*, vol. 12, no. 12, pp. 1668–1676, Dec. 2003.

Rothe, I., Susse, H., and Voss, K., "The method of normalization to determine invariants," *IEEE Transactions on Pattern Analysis and Machine Intelligence*, vol. 18, no. 4, pp. 366–376, Apr. 1996.

Satzoda, R. K., Suchitra, S., and Srikanthan, T., "Parallelizing the Hough transform computation," *IEEE Transactions on Signal Processing Letters*, vol. 15, no. 1, pp. 297–300, Jan. 2008.

Schlemmer, M., Heringer, M., Morr, F., Hotz, I., Bertram, M.-H., Garth, C., Kollmann, W., Hamann, B., and Hagen, H., "Moment invariants for the analysis of 2D flow fields," *IEEE Transactions on Visualization and Computer Graphics*, vol. 13, no. 6, pp. 1743–1750, Nov. 2007.

Shen, D. and Ip, H. S., "Discriminative wavelet shape descriptors for recognition of 2-D patterns," *Pattern Recognition*, vol. 32, no. 2, pp. 151–165, Feb. 1998.

Shih, F. Y., Chuang, C., and Gaddipati, V., "A modified regulated morphological corner detector," *Pattern Recognition Letters*, vol. 26, no. 7, pp. 931–937, Jun. 2005.

Singh, A. and Shneier, M., "Gray level corner detection a generalization and a robust real time implementation," *Computer Vision, Graphics and Image Processing*, vol. 51, no. 1, pp. 54–69, Jul. 1990.

Sirovich, L. and Kirby, M., "Low dimensional procedure for the characterization of human faces," *Journal of the Optical Society of America*, vol. 4, no. 3, pp. 519–524, 1987.

Stork, D. G., Duda, R. O., and Hart, P. E., *Pattern Classification*, Wiley, John & Sons, 2000.

Sun, Y., Liu, W., and Wang, Y., "United moment invariants for shape discrimination," *Proceedings of IEEE International Conference on Robotics, Intelligent Systems and Signal Processing*, Changsha, pp. 88–93, 2003.

Sung, K. K. and Poggio, T., "Example-based learning for view-based human face detection," *IEEE Transactions on Pattern Analysis and Machine Intelligence*, vol. 20, no. 1, pp. 39–51, Jan. 1998.

Teh, C.-H. and Chin, R. T., "On image analysis by the method of moments," *IEEE Transactions on Pattern Analysis and Machine Intelligence*, vol. 10, no. 4, pp. 496–513, Apr. 1998.

Tian, X.-H., Li, Q.-H., and Yan, S.-W., "Regulated morphological operations with weighted structuring element," *Proceedings of 1st International Conference on Machine Learning and Cybernetics*, Beijing, vol. 2, pp. 768–771, Nov. 2002.

Tsai, D.-M., Hou, H.-T., and Su, H.-J., "Boundary-based corner detection using eigenvalues of covariance matrices," *Pattern Recognition Letters*, vol. 20, no. 1, pp. 31–40, Jan. 1999.

Turk, M. A. and Pentland, A. P., "Eigenfaces for recognition," *Journal of Cognitive Neuroscience*, vol. 3, no. 1, pp. 71–86, 1991.

Vaswani, N. and Chellappa, R., "Principal components null space analysis for image and video classification," *IEEE Transactions on Image Processing*, vol. 15, no. 7, pp. 1816–1830, Jul. 2006.

Wang, L., Wang, X., and Feng, J., "On image matrix based feature extraction algorithms," *IEEE Transactions on Systems, Man, and Cybernetics, Part B*, vol. 36, no. 1, pp. 194–197, Feb. 2006.

Weingessel, A. and Hornik, K., "Local PCA algorithms," *IEEE Transactions on Neural Networks*, vol. 11, no. 6, pp. 1242–1250, Nov. 2000.

Wong, W., Shih, F. Y., and Liu, J., "Shape-based image retrieval using Fourier descriptor and support vector machine," *Information Sciences*, vol. 177, no. 8, pp. 1878–1891, Apr. 2007.

Xie, X. and Lam, K.-M., "Gabor-based kernel PCA with doubly nonlinear mapping for face recognition with a single face image," *IEEE Transactions on Image Processing*, vol. 15, no. 9, pp. 2481–2492, Sep. 2006.

Yang, J., Zhang, D., Frangi, A. F., and Yang, J.-Y., "Two-dimensional PCA: A new approach to appearance-based face representation and recognition," *IEEE Transactions on Pattern Analysis and Machine Intelligence*, vol. 26, no. 1, pp. 131–137, Jan. 2004.

Yap, P.-T., Paramesran, R., and Ong, S.-H., "Image analysis by Krawtchouk moments," *IEEE Transactions on Image Processing*, vol. 12, no. 11, pp. 1367–1377, Nov. 2003.

Ye, J. and Li, Q., "Two-stage linear discriminant analysis via QR-decomposition," *IEEE Transactions on Pattern Analysis and Machine Intelligence*, vol. 27, no. 6, pp. 929–941, Jun. 2005.

Zhang, D. S. and Lu, G., "A comparative study of Fourier descriptors for shape representation and retrieval," *Proceedings of 5th Asian Conference on Computer Vision*, Melbourne, pp. 646–651, 2002.

Zhang, X. and Zhao, D., "A morphological algorithm for detecting dominant points on digital curves," *SPIE Proceedings, Nonlinear Image Processing*, vol. 2424, pp. 372–383, 1995.

Zhao, H., Yuen, P. C., and Kwok, J. T., "A novel incremental principal component analysis and its application for face recognition," *IEEE Transactions on Systems, Man, and Cybernetics, Part B*, vol. 36, no. 4, pp. 873–886, Aug. 2006.

Zhao, W., Chellappa, R., and Krishnaswamy, A., "Discriminant analysis of principal components for face recognition," *Proceedings of Third IEEE International Conference on Automatic Face and Gesture Recognition*, Nara, pp. 336–341, Apr. 1998.

Zheng, Z., Wang, H., and Teoh, E., "Analysis of gray level corner detection," *Pattern Recognition Letters*, vol. 20, no. 2, pp. 149–162, Feb. 1999.

Part II

Fundamentals of AI Deep Learning

7 Pattern Recognition

What is pattern recognition? An English dictionary defines a "pattern" as an example or model – something which can be copied. A pattern is also an imitation of a model. A pattern, which describes various types of objects in the physical and abstract worlds, is any distinguishable interrelation of data (analog or digital), events and/or concepts, for example, the shape of face, a table, the order of musical notes in a piece of music, the theme of a poem or a symphony, the tracks made on photographic plates by particles.

We can recognize a chair or a table from any direction, at any orientation of any color, and at a long distance. Furthermore, patterns can be perceived even when they are grossly distorted, ambiguous, incomplete or severely affected by "noise." As for an example, if a letter "A" is distorted, it could be an "A" or an "H" or probably a new letter. But if the letter appears in the word "AOT," then it is most likely an "H," and if it appears in "CAT," then it is most likely an "A." Therefore, context is of crucial importance in pattern recognition.

The pattern recognition process can be viewed as a twofold task, namely, developing decision rules based on human knowledge (learning) and using them for decision making regarding an unknown pattern (classification). The problem of pattern recognition is divided into two parts. The first part is concerned with the study of recognition mechanism of patterns by human and other living organism. This part is related to the disciplines like physiology, psychology, biology, etc. The second part deals with the development of theory and techniques for designing a device which can perform the recognition task automatically.

Applications of pattern recognition may be grouped as follows:

1. Man-machine communication: (a) automatic speech recognition, (b) speaker identification, (c) optical character recognition (OCR) systems, (d) cursive script recognition, (e) speech understanding, and (f) image understanding.
2. Biomedical applications: (a) electrocardiogram (ECG), electroencephalogram (EEG), electromyography (EMG) analysis, (b) cytological, histological, and other stereological applications, (c) X-ray analysis, and (d) diagnostics.
3. Applications in physics: (a) high-energy physics, (b) bubble chamber, and (c) other forms of track analysis.
4. Crime and criminal detection: (a) fingerprint, (b) handwriting, (c) speech sound, and (d) photographs.
5. Natural resources study and estimation: (a) agriculture, (b) hydrology, (c) forestry, (d) geology, (e) environment, (f) cloud pattern, and (g) urban quality.

DOI: 10.1201/9781003474395-9

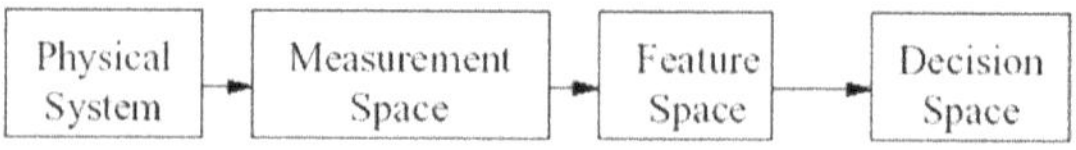

FIGURE 7.1 Operating stages in a recognition system.

6. Stereological applications: (a) metal processing, (b) mineral processing, and (c) biology.
7. Military applications: These include the six areas of applications plus (a) detection of nuclear explosions, (b) missile guidance and detection, (c) radar and sonar signal detection, (d) target identification, (e) naval submarine detection, and (f) reconnaissance.
8. Industrial applications: (a) computer-assisted design and manufacture, (b) computer graphics simulation in product testing and assembly, (c) automatic inspection and quality control in factories, (d) non-destructive testing, and (e) information systems for the handicapped.
9. Robotics and artificial intelligence: (a) intelligent sensor technology and (b) natural language processing.

The operating stages necessary in developing and implementing the decision rule in a practical pattern recognition system are indicated in blocks of Figure 7.1. In this chapter, we will introduce the unsupervised clustering algorithm, support vector machine, neural networks, the adaptive-resonance-theory (ART) network, fuzzy sets, and image analysis.

This chapter aims to present an overview of selected often-used pattern recognition topics in image processing and analysis. The approaches of syntactic or statistical pattern recognition are not covered. This chapter is organized as follows. The unsupervised clustering algorithm is introduced in Section 7.1. The Bayes classifier is presented in Section 7.2. The support vector machine is described in Section 7.3. The neural network classifier is presented in Section 7.4. The adaptive-resonance-theory network is presented in Section 7.5. The fuzzy sets in image analysis are presented in Section 7.6.

7.1 THE UNSUPERVISED CLUSTERING ALGORITHM

The *unsupervised clustering algorithm* (UCA) is based on the *two-pass mode clustering algorithm* (Jenson, 1986). The UCA only requires processing the registered multispectral image twice. In the first pass, the cluster's mean vectors are generated. In the second pass, each pixel is assigned to a cluster which represents a single type in order to produce the hard partition map. The notations are introduced here:

B: the total number of bands used. The number defines the dimensionality in spectral space. For instance, if three bands are used, a 3D spectral space is constructed.

C_{max} : the maximum number of clusters.

$r(P, k)$: the distance between the gray-value vector of a pixel P and the mean vector of the current cluster k. It can be expressed by

$$r(P,k) = \left(\sum_{i=1}^{B} \left(MEAN_i(k) - P_i \right)^2 \right)^{1/2}, \tag{7.1}$$

where $1 \le k \le C_{\max}$, $MEAN_i$ denotes the mean value of cluster k in band i, and P_i denotes the gray value in band i of the pixel P.

R: the constant radius in spectral space used to decide whether a new cluster's mean vector is needed. If r is greater than R, then a new cluster is created.

$d(k_1, k_2)$: the distance between the mean vectors of two distinct clusters k_1 and k_2. It can be expressed by

$$d(k_1, k_2) = \left(\sum_{i=1}^{B} \left(\mathrm{MEAN}_i(k_1) - \mathrm{MEAN}_i(k_2) \right)^2 \right)^{1/2} \tag{7.2}$$

where $1 \le k_1, k_2 \le C_{\max}$ and $k_1 \neq k_2$.

D: the constant radius in spectral space used to determine whether two distinct clusters should be merged. If d is less than or equal to D, the two clusters are merged.

N: the constant representing the total number of pixels to be evaluated before the cluster merging.

$n(k)$: the total number of pixels accumulated in cluster k.

7.1.1 Pass 1: Cluster's Mean Vector Establishment

First, the gray-value vector of the first pixel is placed as the initial cluster's mean vector of cluster one. The gray-value vector of the second pixel is then input to calculate its distance r from the mean vector of the first cluster. If r is greater than R, a new cluster is created with its mean vector equal to the gray-value vector. If r is less than or equal to R, the cluster's mean vector is adjusted using

$$\mathrm{MEAN}_i(k)_{\mathrm{new}} = \frac{\mathrm{MEAN}_i(k)_{old} \times n(k) + P_i}{n(k)+1} \tag{7.3}$$

If equation (7.3) is used, then $n(k)$ and the total number of pixels n_{total} in evaluation are incremented by one. The procedure is repeated when a new input is placed. If n_{total} is greater than N or k is equal to $C_{\max}$, the process being repeated is terminated. Then the cluster merging process is activated to remove unnecessary clusters. Equation (7.2) will be used to calculate the distance between two mean vectors. The selected decision radius is dependent upon the condition that occurs to activate the cluster merging process. If the activation process is due to n_{total} being greater than N, then D is chosen as the decision radius; otherwise, R is the decision radius. If d is less than or equal to the decision radius, the two clusters are merged. The mean vector of the new cluster is adjusted by

$$\text{MEAN}_i(k) = \frac{MEAN_i(k_1) \times n(k_1) + MEAN_i(k_2) \times n(k_2)}{n(k_1) + n(k_2)} \tag{7.4}$$

where $1 \le k, k_1, k_2 \le C_{\max}$ and $k_1 \ne k_2$. The total number of pixels in the new cluster, $n(k)$, is the sum of $n(k_1)$ and $n(k_2)$. After the cluster merging process is completed, if the condition results in n_{total} being greater than N, then n_{total} is set to zero; otherwise, n_{total} is left unchanged. The process of evaluating individual pixels and accumulating the clusters is repeated until n_{total} is greater than N or k is equal to $C_{\max}$

Finally, a number of clusters are built after all the individual pixels are evaluated. In the second pass, these clusters' mean vectors will be used as the feature vectors to determine the attribute of each pixel.

7.1.2 Pass 2: Pixel Classification

Pass 2 is used to classify each pixel into one of these clusters built in Pass 1. One can use the minimum distance to means method is adopted. When each pixel is evaluated, equation (7.1) is used to compute the distance in spectral space between the gray-value vector of the pixel and each cluster's mean vector. After all the distances are determined, the cluster with the minimum distance is chosen; i.e., the type of that cluster is assigned to the given pixel. By using the minimum distance to means method, the pixels with similar spectral features will be ensured to merge together.

Based on various theories, numerous clustering algorithms have been developed, and new clustering algorithms continue to appear in the literature. Yu (2005) proposed a unifying generative framework for partitional clustering algorithms, called a general c-means clustering model. Xu and Wunsch (2005) surveyed clustering algorithms for datasets appearing in statistics, computer science, and machine learning, and illustrate their applications in some benchmark datasets, the traveling salesman problem, and bioinformatics, a new field attracting intensive efforts. Lian et al. (2007) proposed a framework for evaluating the performance of cluster algorithms for hierarchical networks. Fred and Jain (2005) explored the idea of evidence accumulation for combining the results of multiple clusterings. Das et al. (2008) proposed automatic clustering using an improved differential evolution algorithm.

As combining with fuzzy logic, numerous fuzzy clustering algorithms have been proposed. Corsini et al. (2004) proposed a fuzzy relational clustering algorithm based on a dissimilarity measure extracted from data. Liu et al. (2005) presented the fuzzy clustering analysis based on the axiomatic fuzzy sets theory. Yang et al. (2008) proposed the alpha-cut implemented fuzzy clustering algorithms, which allow the data points being able to completely belong to one cluster. To deal with data patterns with linguistic ambiguity and with probabilistic uncertainty in a single framework, Lee et al. (2008) constructed an interpretable probabilistic fuzzy rule-based system that requires less human intervention and less prior knowledge than other state of the art methods. Celikyilmaz and Turksen (2008) proposed enhanced fuzzy system models with improved fuzzy clustering algorithm.

Traditional clustering algorithms (e.g., the K-means algorithm and its variants) are used only for a fixed number of clusters. However, in many clustering applications, the actual number of clusters is unknown beforehand. To easily and effectively determine the optimal number of clusters and, at the same time, construct the clusters with good validity, Pan and Cheng (2007) proposed a framework of automatic clustering algorithms that do not require users to give each possible value of required parameters (including the number of clusters).

Nock and Nielsen (2006) presented weighting clustering, which relies on the local variations of the expected complete log-likelihoods. Veenman et al. (2002) presented a partitional cluster algorithm that minimizes the sum-of-squared-error criterion while imposing a hard constraint on the cluster variance. By modifying a simple matching dissimilarity measure for categorical objects, a heuristic approach was developed in (He et al., 2005; San et al., 2004) which allows the use of the k-modes paradigm to obtain a cluster with strong intrasimilarity and to efficiently cluster large categorical datasets. Ng et al. (2007) derived the updating formula of the k-modes clustering algorithm with the new dissimilarity measure and the convergence of the algorithm under the optimization framework. Camastra and Verri (2005) presented a kernel method for clustering inspired by the classical k-means algorithm in which each cluster is iteratively refined using a one-class support vector machine.

7.2 BAYES CLASSIFIER

Consider a pattern classification problem where a sample x belongs to one of two classes, denoted as ω_1, ω_2. Assume the priori probabilities $P(\omega_1), P(\omega_2)$ are known. The density function $P(\omega_i \mid x)$ is obtained by

$$P(\omega_i \mid x) = \frac{p(x \mid \omega_i) P(\omega_i)}{p(x)} \tag{7.5}$$

According to Bayes (or Bayesian) theory, the probability of the classification error can be minimized by the following rule:

$$\begin{cases} \text{If } P(\omega_1 \mid x) > P(\omega_2 \mid x), & x \text{ is classified to } \omega_1 \\ \text{If } P(\omega_1 \mid x) < P(\omega_2 \mid x), & x \text{ is classified to } \omega_2 \end{cases} \tag{7.6}$$

If a pattern x in fact belongs to ω_1 but was classified into ω_2, it causes a loss, denoted as q_{12}; oppositely, it is denoted as q_{21}. The overall loss is

$$\varepsilon = P(\omega_1) q_{21} p(x \mid \omega_1) + P(\omega_2) q_{12} p(x \mid \omega_2) \tag{7.7}$$

The Bayes classifier intends to minimize the overall loss.

In the following, we will elaborate the binary class problem to multiple classes and extend Bayes theory to combine the results from multiple classifiers. Consider a pattern classification problem where a sample belongs to one of m classes, denoted as $\omega_1, \omega_2, \ldots, \omega_m$. According to Bayes theory, given all the decisions denoted as $j_1, j_2, \ldots, j_R$ from R classifiers for a sample, we have the combining decision rule as

$$i = \arg\max_i P(\omega_i \mid j_1 j_2 \cdots j_R), \tag{7.8}$$

where P denotes the posteriori probability and i denotes the joint decision.

In order to estimate posteriori probabilities $P(\omega_i \mid j_1 j_2 \cdots j_R)$, we build a confusion matrix by testing a total of N samples using R classifiers as

$$A\underbrace{[[1\ldots m][1\ldots m]\ldots[1\ldots m]}_{R+1}]. \tag{7.9}$$

Let $A[i, j_1, \ldots, j_R]$ denote the number of samples in class i to be given the decisions $(j_1, j_2, \ldots, j_R)$ by R classifiers. The posteriori probabilities can be calculated as

$$P(i \mid j_1 \cdots j_R) = \frac{A[i, j_1, \ldots, j_R]}{\sum_{t=1}^{m} A[t, j_1, \ldots, j_R]}. \tag{7.10}$$

Note that the sum of the posteriori probabilities of all m classes is equal to one.

$$\sum_{i=1}^{m} P(i \mid j_1, \ldots, j_R) = 1. \tag{7.11}$$

By using the decision rule in equation (7.8), we can calculate the combined minimum error $P_{\text{comb}_\text{m}\text{in}_\text{e}\text{rr}}$ as

$$P_{\text{comb}_\text{m}\text{in}_\text{e}\text{rr}} = 1 - \frac{\sum_{j_1}\sum_{j_2}\cdots\sum_{j_R} \max_i A[i, j_1, \ldots, j_R]}{N}. \tag{7.12}$$

For a single classifier s, its minimum error $P_{\text{min\_err}}(s)$ can be calculated as

$$P_{\text{min}_\text{e}\text{rr}}(s) = 1 - \frac{\sum_{j_s}\left(\sum_{j_1}\cdots\sum_{j_R} A[j_s, j_1, \ldots, j_R]\right)}{N}. \tag{7.13}$$

We can prove that the following inequality holds for any classifier s:

$$P_{\text{comb}_\text{m}\text{in}_\text{e}\text{rr}} \le P_{\text{min}_\text{e}\text{rr}}(s). \tag{7.14}$$

Proof:

$$1 - P_{\min_e \mathrm{rr}}(s) = \frac{\sum_{j_s} \left(\sum_{j_1} \cdots \sum_{j_R} A[j_s, j_1, \ldots, j_R] \right)}{N}$$

$$\le \frac{\sum_{j_s} \max_i \left(\sum_{j_1} \cdots \sum_{j_R} A[i, j_1, \ldots, j_R] \right)}{N} \le \frac{\sum_{j_s} \sum_{j_1} \max_i \left(\sum_{j_2} \cdots \sum_{j_R} A[i, j_1, \ldots, j_R] \right)}{N}$$

$$\le \frac{\sum_{j_s} \sum_{j_1} \sum_{j_2} \max_i \left(\sum_{j_3} \cdots \sum_{j_R} A[i, j_1, \ldots, j_R] \right)}{N} \le \ldots \le \frac{\sum_{j_s} \sum_{j_1} \cdots \sum_{j_R} \max_i A[i, j_1, \ldots, j_R]}{N}$$

$$= \frac{\sum_{j_1} \cdots \sum_{j_R} \max_i A[i, j_1, \ldots, j_R]}{N} = 1 - P_{\mathrm{comb\_min\_err}} \Rightarrow P_{\mathrm{comb\_min\_err}} \le P_{\mathrm{min\_err}}(s).$$

Therefore, the combined minimum error of multiple classifiers will be not greater than the minimum error of any single classifier. We illustrate the aforementioned property of Bayes combining method as follows. Given two classifiers and three classes, we assume that the confusion matrix of $A[1...3][1...3][1...3]$ is given in the following. Note that the element $A[i, j, k]$ denotes the number of samples in class i to be assigned to class j by the first classifier and to class k by the second classifier. The total number of samples, $N = 400$.

$$\begin{bmatrix} 90 & 2 & 5 \\ 1 & 5 & 5 \\ 5 & 1 & 5 \end{bmatrix}, \begin{bmatrix} 2 & 15 & 5 \\ 5 & 90 & 5 \\ 5 & 5 & 3 \end{bmatrix}, \begin{bmatrix} 1 & 5 & 10 \\ 5 & 10 & 5 \\ 10 & 5 & 95 \end{bmatrix}$$

$$A[1,1\ldots3,1\ldots3], \quad A[2,1\ldots3,1\ldots3], \quad A[3,1\ldots3,1\ldots3]$$

If the decisions from the two classifiers are $j_1 = j_2 = 1$, the posteriori probabilities can be calculated as

$$P(\omega_1 \mid j_1 j_2) = \frac{90}{90+2+1} = 0.9677,$$

$$P(\omega_2 \mid j_1 j_2) = \frac{2}{90+2+1} = 0.0215,$$

$$P(\omega_3 \mid j_1 j_2) = \frac{1}{90+2+1} = 0.0108.$$

From these results, we make the final decision to be class one according to equation (7.8). The combined minimum error of two classifiers is

$$P_{\mathrm{comb_m in_e rr}} = 1 - \frac{\sum_{j_1}\sum_{j_2} \max_i A[i, j_1, j_2]}{N} = 1 - 0.8125 = 0.1875$$

The minimum errors of the first and second classifiers are, respectively,

$$P_{\mathrm{min_e rr}}(1) = 1 - \frac{\sum_{j_1}\sum_{j_2} A[j_1, j_1, j_2]}{N} = 1 - 0.7675 = 0.2325$$

$$P_{\mathrm{min_e rr}}(2) = 1 - \frac{\sum_{j_2}\sum_{j_1} A[j_2, j_1, j_2]}{N} = 1 - 0.79 = 0.21$$

Therefore, we obtain $P_{\mathrm{comb\_min\_err}} \leq P_{\mathrm{min\_err}}(1)$ and $P_{\mathrm{comb\_min\_err}} \leq P_{\mathrm{min\_err}}(2)$.

7.3 SUPPORT VECTOR MACHINE

The *Support Vector Machine* (SVM), introduced by (Vapnik, 1995; Cortes and Vapnik, 1995), has attracted growing interest in pattern classification due to its competitive performance. It is a learning system that separates a set of input pattern vectors into two classes with an optimal separating hyperplane. The set of vectors is said to be optimally separated by the hyperplane if it is separated without error and the distance between the closest vectors to the hyperplane is maximal. SVM produces the pattern classifier by applying a variety of kernel functions (linear, polynomial, radial basis function, and so on) as the possible sets of approximating functions, by optimizing the dual quadratic programming problem, and by using structural risk minimization as the inductive principle, as opposed to classical statistical algorithms that maximize the absolute value of an error or of an error squared.

Originally, SVM was designed to handle dichotomic classes. Later, research has concentrated on expanding two-class classification to multiclass classification. Different types of SVM classifiers are used depending upon the type of input patterns: a linear maximal margin classifier is used for linearly separable data, a linear soft margin classifier is used for linearly nonseparable, or overlapping, classes, and a nonlinear classifier is used for classes that are overlapped as well as separated by nonlinear hyperplanes. All three classifiers are discussed in more detail here. It should be noted, however, that the linearly separable case is rare in real world problems.

7.3.1 Linear Maximal Margin Classifier

The case where the training patterns can be linearly separated by a hyperplane, $\mathbf{w}\cdot\mathbf{x}+b=0$, is the simplest case and provides a good foundation for the other two cases. The purpose of the SVM is to find the optimal values for $\mathbf{w}$ (e.g., $\mathbf{w}_0$) and b (e.g., b_0). After finding the optimal separating hyperplane, $\mathbf{w}_0\cdot\mathbf{x}+b_0=0$, an unseen pattern, $\mathbf{x}_t$, can be classified by the decision rule $f(x)=sign(\mathbf{w}_0\cdot\mathbf{x}+b_0)$, as shown here.

Suppose there is a set of training data, $\mathbf{x}_1,\ \mathbf{x}_2,\ \ldots,\mathbf{x}_k$, where $\mathbf{x}_i\in\mathrm{R}^n$ and $i = 1, 2, \ldots, k$. Each $\mathbf{x}_i$, belonging as it does to one of two classes, has a corresponding value y_i, where $y_i\in\{-1,\ 1\}$. The goal in this case is to build the hyperplane that maximizes the minimum distance between the two classes. Because the hyperplane is $\mathbf{w}\cdot\mathbf{x}+b=0$, the training data can be divided into two classes such that

$$\begin{cases}\mathbf{w}\cdot\mathbf{x}_i+b\geq 1 & \text{if } y_i=1\\ \mathbf{w}\cdot\mathbf{x}_i+b\leq -1 & \text{if } y_i=-1\end{cases} \tag{7.15}$$

where $\mathbf{w}\in\mathrm{R}^n$ and $b\in\mathrm{R}$. Then we obtain

$$y_i(\mathbf{w}\cdot\mathbf{x}_i+b)\geq 1\quad \forall\mathbf{x}_i,\ i=1,\ 2,\ \ldots,\ k \tag{7.16}$$

The distance between a point $\mathbf{x}$ and the hyperplane is $d(\mathbf{w},b;\mathbf{x})=|\mathbf{w}\cdot\mathbf{x}+b|\div\|\mathbf{w}\|$.

According to equation (7.15), the minimum distance between one of the two classes and the hyperplane is $\frac{1}{\|\mathbf{w}\|}$. The margin, M, which is the distance between the two classes, is $\Delta=\frac{2}{\|\mathbf{w}\|}$. Figure 7.2 illustrates the model of SVM.

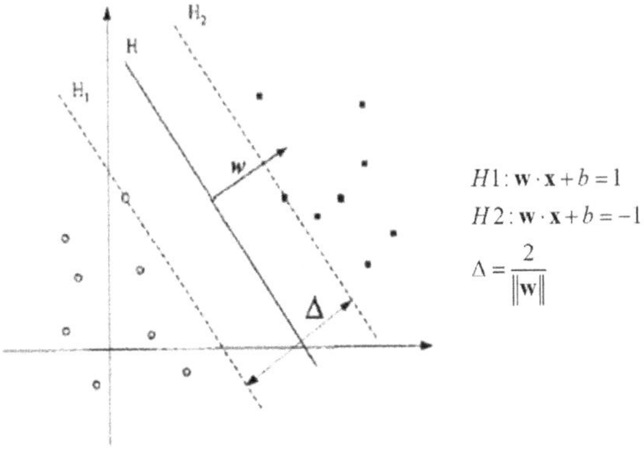

FIGURE 7.2 The model of SVM.

Finding the optimal separating hyperplane having a maximal margin requires that the following minimization problem be solved:

$$\text{Minimize:} \frac{1}{2}\mathbf{w}\cdot\mathbf{w}$$

$$\text{Subject to:} y_i\left(\mathbf{w}\cdot\mathbf{x}_i+b\right)\geq 1 \quad \forall \mathbf{x}_i, i=1,2,\ldots,k.$$

This nonlinear optimization problem with inequality constraints can be solved by the saddle point of the Lagrange function:

$$L(\mathbf{w},b,\alpha)=\frac{1}{2}\mathbf{w}\cdot\mathbf{w}-\sum_{i=1}^{K}\alpha_i(y_i(\mathbf{w}\cdot\mathbf{x}_i+b)-1) \tag{7.17}$$

where $\alpha_i \geq 0$ are the Lagrange multipliers.

By minimizing the Lagrange function with respect to **w** and b, as well as by maximizing with respect to α_i, the minimization problem can be transformed to its dual problem, called the quadratic programming problem.

$$\frac{\partial L(\mathbf{w},b,\alpha)}{\partial \mathbf{w}}\bigg|_{\mathbf{w}=\mathbf{w}_0}=\mathbf{w}_0-\sum_{i=1}^{K}\alpha_i y_i \mathbf{x}_i=0 \tag{7.18}$$

$$\frac{\partial L(\mathbf{w},b,\alpha)}{\partial b}\bigg|_{b=b_0}=\sum_{i=1}^{K}\alpha_i y_i=0 \tag{7.19}$$

By plugging equation (7.18) and equation (7.19) into equation (7.17), we obtain

$$L(\alpha)=\sum_{i=1}^{K}\alpha_i-\frac{1}{2}\sum_{i=1}^{K}\sum_{j=1}^{K}\alpha_i\alpha_j y_i y_j \mathbf{x}_i\mathbf{x}_j \tag{7.20}$$

The dual problem is then described as

$$\text{Maximize:} \sum_{i=1}^{K}\alpha_i-\frac{1}{2}\sum_{i=1}^{K}\sum_{j=1}^{K}\alpha_i\alpha_j y_i y_j \mathbf{x}_i\mathbf{x}_j$$

$$\text{Subject to:} \sum_{i=1}^{K} y_i\alpha_i=0, \quad \alpha_i\geq 0.$$

By solving the dual problem, we can determine the optimal separating hyperplane by

$$\mathbf{w}_0=\sum_{i=1}^{K}\alpha_i y_i \mathbf{x}_i \tag{7.21}$$

$$b_0 = y_i - \mathbf{w}_0 \cdot \mathbf{x}_i \tag{7.22}$$

where $\mathbf{x}_i$ belongs to support vectors and $y_i \in \{-1, 1\}$.

The unseen test data $\mathbf{x}_t$ can be classified by simply computing

$$f(x) = sign(\mathbf{w}_0 \cdot \mathbf{x}_t + b_0) \tag{7.23}$$

It can be seen that the hyperplane is determined by all the training data, $\mathbf{x}_i$, that have the corresponding attributes of $\alpha_i > 0$. We call this kind of training data as *support vectors*. Thus, the optimal separating hyperplane is not determined by the training data per se but rather by the support vectors.

7.3.2 Linear Soft Margin Classifier

As mentioned, patterns are rare to be linearly separable in real world problems. In this section, we expand SVM to handle input patterns that are overlapping or linearly nonseparable. In this case, our objective is to separate the two classes of training data with a minimal number of errors.

To accomplish this, we introduce some nonnegative slack variables ξ_i, i = 1, 2, . . . , k to the system. Thus, equations (7.15) and (7.16), in the linearly separable case presented, can be rewritten as equations (7.24) and (7.25):

$$\begin{cases} \mathbf{w} \cdot \mathbf{x}_i + b \geq 1 - \xi_i & \text{if } y_i = 1 \\ \mathbf{w} \cdot \mathbf{x}_i + b \leq -1 + \xi & \text{if } y_i = -1 \end{cases} \tag{7.24}$$

$$y_i \left(\mathbf{w} \cdot \mathbf{x}_i + b \right) \geq 1 - \xi_i \quad \forall \mathbf{x}_i, i = 1, 2, \ldots, k \tag{7.25}$$

Just as we obtained the optimal separating hyperplane in the linearly separable case, obtaining the soft margin hyperplane in the linear nonseparable case requires that the following minimization problem be solved:

$$\text{Minimize:} \frac{1}{2} \mathbf{w} \cdot \mathbf{w} + C \left(\sum_{i=1}^{K} \xi_i \right)$$

$$\text{Subject to:} \, y_i \left(\mathbf{w} \cdot \mathbf{x}_i + b \right) \geq 1 - \xi_i \text{ and } \xi_i \geq 0, i = 1, 2, \ldots, k.$$

where C is a penalty or regularization parameter.

By minimizing the Lagrange function with respect to **w**, b, and ξ_i, as well as by maximizing with respect to α_i, the minimization problem can be transformed to its dual problem as

$$\text{Maximize:} \sum_{i=1}^{K} \alpha_i - \frac{1}{2} \sum_{i=1}^{K} \sum_{j=1}^{K} \alpha_i \alpha_j y_i y_j \mathbf{x}_i \mathbf{x}_j$$

$$\text{Subject to:} \sum_{i=1}^{K} y_i \alpha_i = 0, \quad 0 \leq \alpha_i \leq C.$$

By solving the dual problem, we determine the soft margin hyperplane by

$$\mathbf{w}_0 = \sum_{i=1}^{K} \alpha_i y_i \mathbf{x}_i \tag{7.26}$$

$$b_0 = y_i - \mathbf{w}_0 \cdot \mathbf{x}_i \tag{7.27}$$

where $\mathbf{x}_i$ belongs to margin vectors, $y_i \in \{-1,\ 1\}$. It can be seen that the hyperplane is determined by all the training data, $\mathbf{x}_i$ that have the corresponding attributes of $\alpha_i > 0$. These support vectors can be divided into two categories. The first category has the attribute of $\alpha_i < C$. In this category, $\xi_i = 0$, and these support vectors lie at the distance $\frac{1}{\|\mathbf{w}\|}$ from the optimal separating hyperplane. We call these support vectors as *margin vectors*. The second category has the attributes of $\alpha_i = C$. In this category, the support vectors are correctly classified with either a distance smaller than $\frac{1}{\|\mathbf{w}\|}$ from the optimal separating hyperplane (if $0 < \xi_i \leq 1$) or they are misclassified (if $\xi_i > 1$). The support vectors in the second category are regarded as errors.

7.3.3 Nonlinear Classifier

Sometimes, the input vectors cannot be linearly separated in the input space. In this case, kernel functions, such as the polynomial or radial basis function (RBF), are used to transform the input space to a feature space of higher dimensionality. In the feature space, a linear separating hyperplane is obtained to separate the input vectors into two classes.

If $\mathbf{x} \in R^n$ is in the input space, we can map the input vector $\mathbf{x}$ from the n-dimension input space to a corresponding N-dimensional feature space through a function, ϕ. After the transformation, we know $\phi(\mathbf{x}) \in R^n$. Following the steps described in the case of linearly separable training patterns and the case of linearly nonseparable training patterns, the hyperplane and decision rule for nonlinear training patterns can be established.

In a manner similar to obtaining the hyperplane for the linearly separable training patterns in equations (7.21) and (7.22) and the hyperplane for the linearly nonseparable training patterns in equations (7.26) and (7.27), we can obtain the hyperplane for the nonlinear training pattern as

$$\mathbf{w}_0 \cdot \phi(\mathbf{x}) + b_0 = \left(\sum_{i=1}^{K} \alpha_i y_i \phi(\mathbf{x}_i) \right) \cdot \phi(\mathbf{x}) + b_0 \tag{7.28}$$

In equation (7.28), we see that the original dot products of input variables can be replaced by a function, ϕ. That is, kernel function $K(\mathbf{x}_i, \mathbf{x}) = \phi(\mathbf{x}_i) \cdot \phi(\mathbf{x})$. The decision rule for nonlinear training patterns can be established as

$$f(\mathbf{x}) = sign\left(\sum_{i=1}^{K} \alpha_i y_i K(\mathbf{x}_t, \mathbf{x}) + b_0\right) \tag{7.29}$$

7.3.4 SVM Networks

The support vector machine (SVM) is originally designed as a binary (or two-class) classifier. Researchers have proposed extensions to multi-class (or multi-group) classification by combining multiple SVMs, such as one-against-all, one-against-one and DAG-SVM (Platt et al., 2000). One-against-all method is to learn k binary SVMs during the training stage. For the ith SVM, we label all training samples by $y(j)_{j=i} = +1$ and $y(j)_{j \neq i} = -1$. In the testing, the decision function (Hsu and Lin, 2002) is obtained by

$$\text{Class of } \vec{\mathbf{x}} = \arg\max_{i=1,2,\ldots,k}\left(\left(w^i\right)^T \phi(\vec{\mathbf{x}}) + b^i\right), \tag{7.30}$$

where $(w^i)^T \phi(\vec{\mathbf{x}}) + b^i = 0$ represents the hyperplane for the ith SVM. It was shown that we can solve the k-group problem simultaneously (Herbrich, 2001).

One-against-one method is to construct $k(k-1)/2$ SVMs in a tree structure. Each SVM represents a distinguishable pair-wise classifier from different classes in the training stage. In the testing, each leaf SVM pops up one desirable class label, which propagates to the upper level in the tree until it processes the root SVM of the tree.

The directed acyclic graph (*DAG*) SVM (Platt et al., 2000) is to combine multiple binary one-against-one SVM classifiers into one multi-class classifier. It has $k(k-1)/2$ internal SVM nodes, but in the testing phase, it starts from the root node and moves to the left or right sub-tree depending on the output value of the binary decision function until a leaf node is reached, which indicates the recognized class. There are two main differences between one-against-one and DAG methods. One is that one-against-one needs to evaluate all $k(k-1)/2$ nodes, while DAG only needs to evaluate k nodes due to the different testing phase schema. The other is that one-against-one uses the bottom-up approach, but DAG uses the top-down approach.

The performance of the three methods depends on the quality of each individual binary SVM. There is no error tolerance that can adjust the system. If one SVM fails, an incorrect result could be produced. Each SVM is a standalone black box. Other than the conveyed class label, no further information is propagated to other SVMs. Shih and Zhang (2005) proposed a combined SVM and neural network method to overcome this problem.

The decision function of SVM provides not only the class label but also useful information, such as class density and distribution knowledge. We can change equation (7.16) to

$$f_d(\vec{\mathbf{x}}) = \vec{\mathbf{w}}_0 \cdot \vec{\mathbf{x}} + b_0 \tag{7.31}$$

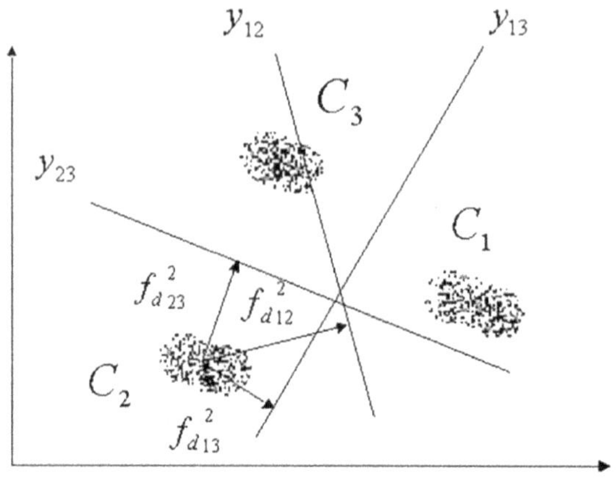

FIGURE 7.3 Decision function of a 3-class SVM.

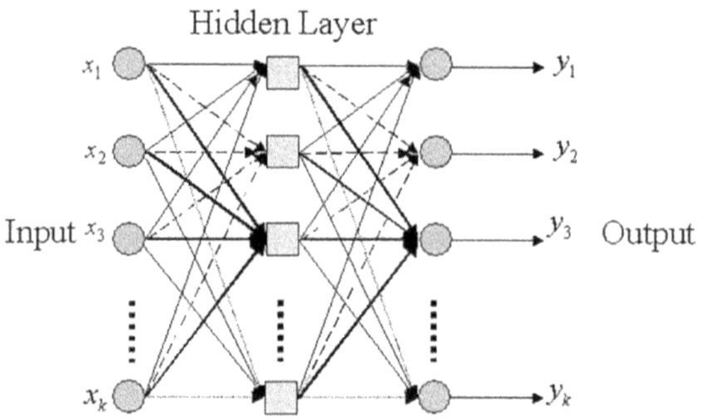

FIGURE 7.4 The architecture of our framework.

where $f_d(\vec{\mathbf{x}})$ represents the distance between the sample $\vec{\mathbf{x}}$ and the optimal hyperplane $\vec{\mathbf{w}}_0 \cdot \vec{\mathbf{x}} + b_0 = 0$. As shown in Figure 7.3, the distances f_{d12}^2, f_{d13}^2 and f_{d23}^2 compose a vector $\vec{\mathbf{f}}_d^{\,2} = \left(f_{d12}^2, f_{d13}^2, f_{d23}^2\right)^T$ to identify the position of class C_2, and y_{ij} denotes the hyperplane between classes i and j. Therefore, we can use $\vec{\mathbf{f}}_d^{\,i}$ to illustrate the distribution of class i.

We extend this idea to a layered neural network to construct a new framework as shown in Figure 7.4. For the *k-class* problem, we create k hidden neurons, each of which is a one-against-all SVM having the decision function $f_d(\vec{\mathbf{x}}) = \vec{\mathbf{w}}_0 \cdot \vec{\mathbf{x}} + b_0$ and k output nodes, which compose an output vector $\vec{\mathbf{Y}} = (y_1, y_2, ..., y_k)^T$.

For a training sample $\vec{\mathbf{x}}$ from class i, it serves all the SVMs. We obtain a new distance vector $\vec{\mathbf{f}}_d(\vec{\mathbf{x}}) = \left(f_d^{\,1}, f_d^{\,2}, ..., f_d^{\,k}\right)^T \in R^k$ to represent the sample $\vec{\mathbf{x}}$ in R^k space. Finally, the desired output vector $\vec{\mathbf{Y}}(\vec{\mathbf{x}}) = (y_1, y_2, ..., y_k)^T = \vec{\mathbf{f}}_d(\vec{\mathbf{x}}) \cdot W$, where

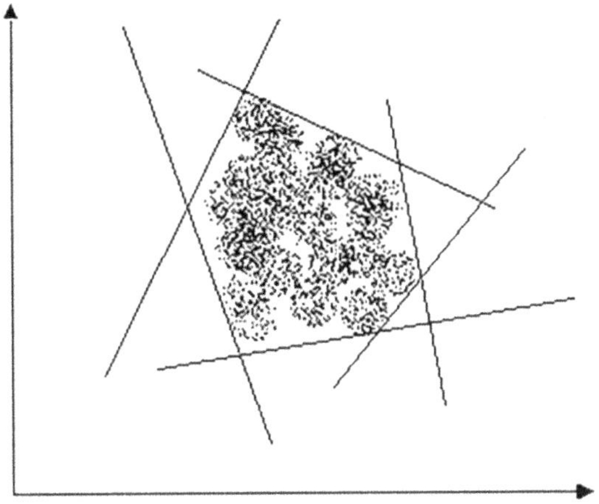

FIGURE 7.5 Hyperplanes divide the space.

W is a $k \times k$ weight matrix, and $y_i = 1$ and $y_{j \neq i} = 0$. Because $\vec{\mathbf{Y}}(\vec{\mathbf{x}})$ and $\vec{\mathbf{f}}_d(\vec{\mathbf{x}})$ are known, the weight matrix can be obtained by

$$W = F_d^{\ +} Y = \left(F_d^{\ T} F_d\right)^{-1} F_d^{\ T} Y, \tag{7.32}$$

where F_d and Y are derived by all training samples, and $F_d^{\ +}$ denotes the pseudo-inverse of F_d. Besides, we can imagine that for a k-class problem, the whole space is divided by k hyperplanes through such a framework as shown in Figure 7.5.

Traditional transformation maps the original image vectors to another high-dimensional space associated with a fixed origin point. But our proposed framework maps them to a kinetic coordinate system whose degree is the number of classes. There are three phases in our framework. The first is the constructing phase, the second is the training phase, and the third is the testing phase. The constructing phase is to build all SVMs. The dimensionality of all samples is reduced from $R^m \rightarrow R^n$ through the PCA procedure. The $\vec{\mathbf{x}} \in R^n$ will serve as the input vector into our framework. Supposedly, there are l samples for each class in the training set. For the ith SVM M_i^1, we select all training samples in class i, and any one sample from all other classes $j \neq i$ to obtain the training set T_i^1. After M_i^1 is trained, we test the complete training set. There exist some misclassified samples. We add them to T_i^1 with negative labels as a new training set T_i^2, so we can obtain a new trained SVM called M_i. If no misclassified samples are found, then $M_i^1 = M_i$. Through such a procedure, M_i can enhance its generalization. The SVM is trained by all training samples just once. After all of M_i's are established, the target outputs will be created. For a training sample $\vec{\mathbf{x}}$ from the class i, the desired output vector is

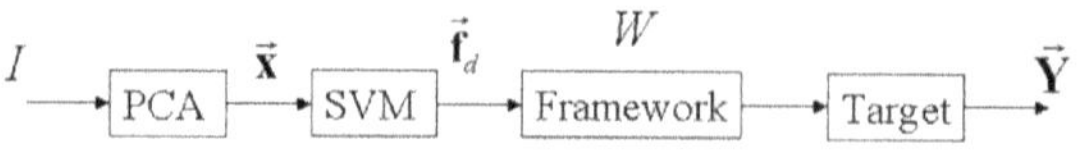

FIGURE 7.6 The overall procedures.

$\vec{\mathbf{Y}}(\vec{\mathbf{x}}) = (y_1, y_2, ..., y_k)^T$, where $y_i = 1$ and $y_{j \neq i} = 0$. The overall procedures are given in Figure 7.6.

The objective of the training phase is to compute the weight matrix. For a complete training set, the system generates $F_d W = Y$, where $F_d = \left(\vec{\mathbf{f}}_d^{\,1}, \vec{\mathbf{f}}_d^{\,2}, \ldots, \vec{\mathbf{f}}_d^{\,k \times l}\right)$ and Class of $\vec{\mathbf{x}} = \arg\max_{i=1,2,\ldots,k} y_i(\vec{\mathbf{x}})$. The output vector only contains a single "1" in the corresponding output node to identify the desired class number. We have $\vec{\mathbf{Y}}(\vec{\mathbf{x}}_{i,m}) = (y_1, y_2, ..., y_k)$, where $y_i = 1$ and $y_{j \neq i} = 0$. We expect that the output vector only contains a single "1" in the corresponding output node to identify the desired class. Once F_d and Y are known, the weight matrix can be obtained by equation (7.32). Finally, the whole system is established.

In the testing phase, for an unknown sample $\vec{\mathbf{x}}$, the system outputs a vector $\vec{\mathbf{Y}}(\vec{\mathbf{x}})$, where the location of the maximum of output nodes, as given in equation (7.33), indicates the recognized class. In other words, it is the most likely class.

$$\text{Class of } \vec{\mathbf{x}} = \arg\max_{i=1,2,\ldots,k} y_i(\vec{\mathbf{x}}) \tag{7.33}$$

Numerous improvements on SVMs and their applications have been proposed. Haasdonk (2005) presented feature space interpretation of SVMs with indefinite kernels. Lin and Lin (2003) presented a study on reduced support vector machine, which is an alternate of the standard SVM by resolving the difficulty on handling large datasets using SVM with nonlinear kernels. Xia and Wang (2004) presented a one-layer recurrent neural network for SVM learning in pattern classification and regression. Jayadeva et al. (2007) proposed twin SVM, a binary SVM classifier that determines two nonparallel planes by solving two related SVM-type problems, each of which is smaller than in a conventional SVM.

Liu and Zheng (2007) presented soft SVM and its application in video-object extraction. Li et al. (2006) developed a machine-learning technique, multitraining SVM, which combines the merits of the cotraining technique and a random sampling method in the feature space for image retrieval. Navia-Vazquez et al. (2006) proposed distributed support vector machines. Tao et al. (2008) proposed a recursive SVM, in which several orthogonal directions that best separate the data with the maximum margin are obtained. During the usual SVM biclassification learning process, the bias is chosen a posteriori as the value halfway between separating hyperplanes. Gonzalez-Abril et al. (2008) presents different approaches on the calculation of the bias when SVM is used for multiclassification.

Fei and Liu (2006) presented a new architecture named binary tree of SVM, in order to achieve high classification efficiency for multiclass problems. Anguita et al. (2003) proposed a digital architecture for SVM learning and discuss its implementation on a field programmable gate array. Anguita et al. (2006) proposed a coordinate rotation digital computer (CORDIC)-like algorithm for computing the feed-forward phase of a SVM in fixed-point arithmetic, using only shift and add operations and avoiding resource-consuming multiplications.

7.4 NEURAL NETWORKS

The neural network approach exhibits two major characteristics in the field of problem solving. First, neural network architectures are parallel and connectionist. The connections between pairs of neurons play a primary role in the function performed by the whole system. The second characteristic is that the prescriptive computer programming routine is replaced by heuristic learning processes in establishing a system to solve a particular problem. It means that a neural network is provided with representative data, also called *training data*, and some learning process which induces the data regularities. The modification of the transfer function at each neuron is derived empirically according to the training dataset in order for maximizing the goal.

Two operating phases, learning and execution, are always encountered in neural networks. During the *learning phase*, the networks are programmed to take input from training set and adjust the connection weights to achieve the desired association or classification. During the *execution phase*, the neural networks are to process input data from outside world to retrieve corresponding outputs. These two phases are different in both the processing types and the time constraints involved.

Neural networks can be considered as signal processors which are "value passing" architectures, conceptually similar to traditional analog computers. Each neuron may receive a number of inputs from the input layer or some other layers, and then produces a single output which can be as an input to some other neurons or a global network output. Many neural network models, such as Rosenblatt's Minsky's *Perceptron* (Minsky and Papert, 1969; Rosenblatt, 1962) and Fukushima's Cognitron (Fukushima, 1975), which use binary inputs and produce only binary outputs, can implement Boolean functions.

In feed-forward layered neural networks which form a simple hierarchical architecture, execution of each layer is performed in a single step. The input neurons in the first layer can be viewed as entry points for sensor information from outside world, or called sensoring neurons, and the output neurons in the last layer are considered as decision points control, or called decision neurons. The neurons in the intermediate layers, often called hidden neurons, play the role of interneurons. The network implicitly memorizes an evaluated mapping from the input vector and the output vector at the execution time, each time the network traversed. The mapping, represented as $\mathbf{X} \rightarrow \mathbf{Y}$, is just a transfer function that takes the n-dimensional input vectors $\mathbf{X} = (x_1, x_2, \cdots, x_n)$ and then outputs m-dimensional input vectors $\mathbf{Y} = (y_1, y_2, \cdots, y_m)$.

7.4.1 Programmable Logic Neural Networks

The concept of building neural network structure with dynamically programmable logic modules (DPLMs) is based on early efforts aimed at providing general design

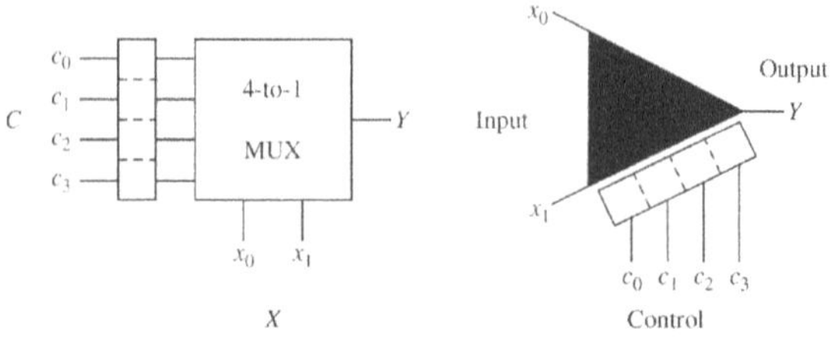

FIGURE 7.7 (a) A neuron cell consists of a 4-to-1 multiplexer and a 4-bit control register, (b) symbol for a 2-input DPLM with a 4-bit control register.

of logic circuits by using universal logic modules (Armstrong and Gecsei, 1979; Vidal, 1988). Neural networks are constructed with node modules which have only a limited number of inputs. Figure 7.7 shows a module of 2-bit input DPLM which consists of a commonly used 4-to-1 multiplexer and a 4-bit control register. In this circuit, the role of 4-bit input and 2-bit control is exchanged. The 2-bit control (**X**) is used to select any bit of the 4-bit value in the register (*C*) as the output (**Y**).

In Figure 7.7, $\mathbf{X} = (x_0, x_1)$ denotes the input vector having four combinations: {00, 01, 10, 11}. The control register *C* determines the value of the single-bit output Y from the input vector. Therefore, by loading a 4-bit control word, one of the 16 (i.e., 2^4) possible Boolean functions for two input variables can be chosen. These Boolean functions and associated control codes (i.e., the truth table) are given in Table 7.1. A 2-input DPLM with all 16 possible Boolean functions is called a *universal module*. If the 16 possible functions are not fully utilized, the size of the control register could be reduced (Vidal, 1988).

The 2-input modules are the simplest primitive units. Figure 7.8 illustrates a general-purpose 3-input DPLM composed of six 2-input modules. The desired function is selected by the corresponding control codes loaded to the six 2-input modules. For example, the AND codes loaded to all six modules acts as the AND function of the three inputs; this is called CONVERGE. Similarly, by loading the OR codes, it performs the OR function of the inputs, called DIVERGE. The accumulative combinations of primitive blocks can be used to achieve more complex architectures for various applications.

The structure of a 3-input DPLM in Figure 7.8 could be reduced into the structure in Figure 7.7. In the neighborhood operations, only three 2-input modules are needed in the pyramid structure. That is, the module in the upper level will receive inputs from the two in the lower level. Instead of loading the same control code to all modules, the new architecture can be loaded with different control codes in each level so that more functions could be performed. A number of these functions are discussed in the next section.

7.4.2 Pyramid Neural Network Structure

To illustrate the process of pyramid model, a simple example constructed with 3-input DPLMs is used. Figure 7.10 shows the structure of the network. In the network, every

TABLE 7.1
Truth Tables of 16 Possible Boolean Functions

Function	Control Code	Input (x_0 x_1) 00	01	10	11
FALSE	0000	0	0	0	0
AND	0001	0	0	0	1
LFT	0010	0	0	1	0
LFTX	0011	0	0	1	1
RIT	0100	0	1	0	0
RITX	0101	0	1	0	1
XOR	0110	0	1	1	0
OR	0111	0	1	1	1
NOR	1000	1	0	0	0
XNOR	1001	1	0	0	1
NRITX	1010	1	0	1	0
NRIT	1011	1	0	1	1
NLFTX	1100	1	1	0	0
NLFT	1101	1	1	0	1
NAND	1110	1	1	1	0
TRUE	1111	1	1	I	1

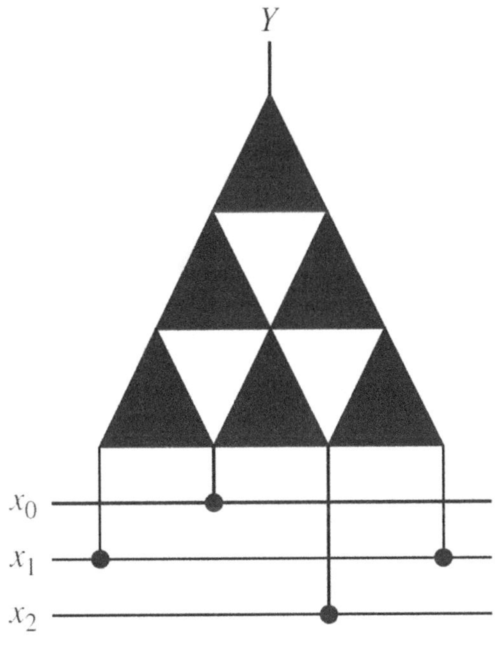

FIGURE 7.8 A 3-input primitive unit composed of six 2-input DPLMs.

FIGURE 7.9 A 3-input primitive unit composed of three 2-input DPLMs.

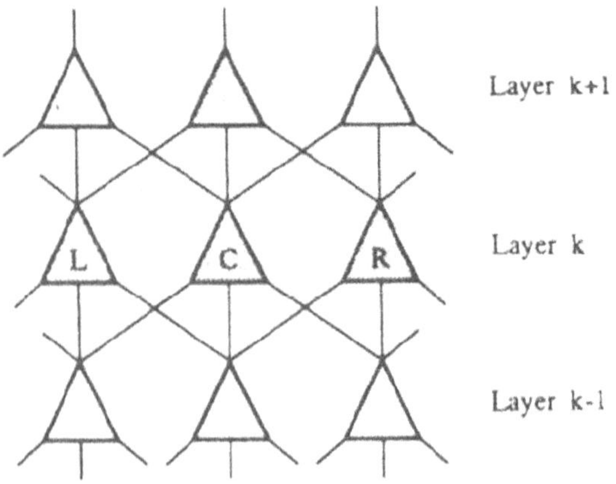

FIGURE 7.10 The structure of 2D pyramid neural network composed of 3-input DPLMs.

TABLE 7.2
Boolean Functions Performed by 3-Input DPLMs

Function Names	Equations	Control Upper	Control Lower
CONVERGE	$C_{k+1} \leftarrow L_k C_k R_k$	AND	AND
DIVERGE	$C_{k+1} \leftarrow L_k + C_k + R_k$	OR	OR
NONE	$C_{k+1} \leftarrow \bar{L}_k \bar{C}_k \bar{R}_k$	AND	NOR
CENTER-ON	$C_{k+1} \leftarrow C_k$	RITX	LFTX
		LFTX	RITX
RIGHT-ON	$C_{k+1} \leftarrow R_k$	RITX	RITX
LEFT-ON	$C_{k+1} \leftarrow L_k$	LFTX	LFTX

node (C_{k+1}) in the (k+1)th layer receives inputs from the 3 neighbors (nodes L_k, C_k, and R_k) in the kth layer and all nodes in the same layer are assigned the same function. Hence, only one control register is required for each layer. Some functions are illustrated in Table 7.2.

It has to be noted that the CONVERGE and DIVERGE operations are actually the AND and the OR Boolean functions, respectively. They are equivalent to the morphological erosion and dilation. The RIGHT-ON operation is used to detect the right neighbor and the LEFT-ON operation the left neighbor. Similarly, the CENTER-ON operation detects the center pixel.

7.4.3 Binary Morphological Operations by Logic Modules

To implement morphological operations, a new neural architecture by the use of the pyramid model is constructed. The connections among the four 3-input DPLMs shown in Figure 7.11 form a 9-input 3D pyramid module. The layers k and k + 1 are called *regular layer* and the layer k + ½ is called *intermediate layer.*

In the structure of 9-input module, each neuron $U_{k+\frac{1}{2}}(i,j)$ in the intermediate layer k + ½ receives three inputs of its column-major neighbors $(U_k(i,j-1), U_k(i,j), U_k(i,j+1))$ in the preceding layer k. The neuron $U_{k+1}(i,j)$ then collects the three outputs of its three row-major neighbors $(U_{k+\frac{1}{2}}(i-1,j), U_{k+\frac{1}{2}}(i,j), U_{k+\frac{1}{2}}(i+1,j))$ in the preceding intermediate layer k + ½. In other words, the module will extract 1D local features from each column, and then combine the three adjacent 3 × 1 features into 3 × 3 2D local feature. The overall structure of the network is shown in Figure 7.12. There is no limitation for the number of layers in a neural network; i.e., one could concatenate as many layers as desired and then assign different Boolean functions to the nodes for specific applications.

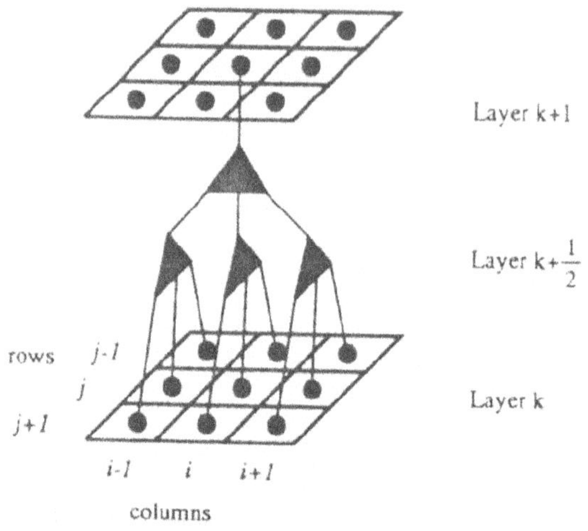

FIGURE 7.11 The 9-input pyramid module composed of four 3-input DPLMs.

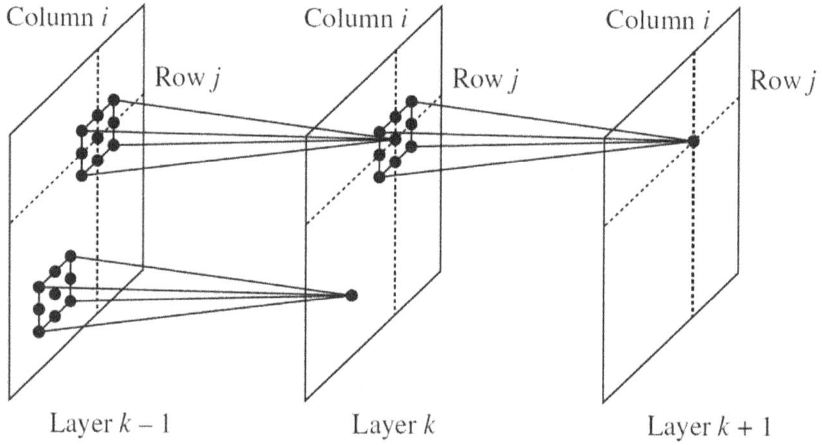

FIGURE 7.12 The overall structure of neural network formed with 3 × 3 input DPLMs for image morphological processing.

Since only one operation is applied on the whole original image at a certain time, loading different control codes to DPLMs on the same major dimension is not required. Only two control registers (one for the upper level and the other for the lower level) are connected for each column in any intermediate layer (layer k+ $\frac{1}{2}$) as the same as for each row in any regular layer (layer k) – that is, to apply a certain function with respect to a column or a row. Some examples of morphological operations performed by this neural network for extracting the local features are shown in Table 7.3.

TABLE 7.3
Morphological Operations Using Pyramid Networks

Structuring Elements	Morphological Operations	Intermediate Layers	Regular Layers
111 111 111	DILATION EROSION	DIVERGE CONVERGE	DIVERGE CONVERGE
000 111 000	DILATION EROSION	CENTER-ON CENTER-ON	DIVERGE CONVERGE
010 010 010	DILATION EROSION	NONE, CONVERGE. NONE NONE, CONVERGE. NONE	DIVERGE CONVERGE
010 111 010	DILATION EROSION	CENTER-ON, CONVERGE, CENTER-ON CENTER-ON, CONVERGE, CENTER-ON	DIVERGE CONVERGE
100 010 001	DILATION EROSION	LEFT-ON, CENTER-ON, RIGHT-ON LEFT-ON, CENTER-ON, RIGHT-ON	DIVERGE CONVERGE
001 010 100	DILATION EROSION	RIGHT-ON, CENTER-ON, LEFT-ON RIGHT-ON, CENTER-ON, LEFT-ON	DIVERGE CONVERGE

As seen in the table, for all dilations, the upper level of the modules should be loaded with the DIVERGE operation, and for all erosions, the CONVERGE operation is employed. The lower level is loaded with different operations for different structuring elements. Suppose that an erosion is applied on a binary image with the following 3×3 structuring element

1	1	1
1	1	1
1	1	1

in the layer k and pass the result to the layer $k + 1$. One may simply load CONVERGE to all 3-input DPLMs in both intermediate layer $k + ½$ and regular layer $k + 1$. The neurons in layer $k + ½$ will output a "1" if the three neighbors in column are all ones. The neurons in layer $k + 1$ will output a "1" if the three neighboring intermediate neurons in row have all ones; i.e., the nine (3×3) neighboring pixels are all ones. Similarly, for dilation with the same structuring element, DIVERGE may be loaded to all neurons.

If the structuring element is changed to

0	0	0
1	1	1
0	0	0

the erosion requires applying CENTER-ON to neurons in the layer $k + ½$ and CONVERGE to neurons in layer $k + 1$, and the dilation requires applying CENTER-ON to neurons in the layer $k + ½$ and DIVERGE to neurons in the layer $k + 1$. Another example using a different structuring element

1	0	0
0	1	0
0	0	1

is illustrated as follows. The LEFT-ON may be loaded to the first columns, CENTER-ON to the second columns, and RIGHT-ON to the third columns of each pyramid module in the layer $k + ½$. At the same time, CONVERGE is loaded in the layer $k + 1$. Therefore, any occurrence of line segments with 135° is extracted in the layer $k + 1$. Similarly, DIVERGE is also applied to the layer $k + 1$ to dilate any occurrence of "1" in the layer k to a diagonal line segment with the length 3.

A more complicated neural network could be constructed by extending DPLMs and their connections. By assigning various combinations of Boolean functions, a neural network may act in quite different behaviors. Thus, DPLM networks can be viewed as a class of *digital perceptron* which can support discrete forms of distributed parallel problem solving. They offer a new conceptual framework as well as a promising technology for developing artificial neural networks.

7.4.4 Multi-Layer Perceptron as Processing Modules

Multi-layer perceptrons are feed-forward nets with one or more layers, called *hidden layers*, of neural cells, called *hidden neurons*, between the input and output layers. The learning process of MLPs is conducted with the error back-propagation learning algorithm derived from the *generalized delta rule* (Rumelhart et al., 1986). According to (Lippmann, 1987), no more than three layers of neurons are required to form arbitrarily complex decision regions in the hyperspace spanned by the input patterns. By using the sigmoidal nonlinearities and the decision rule in selecting the largest output, the decision regions are typically bounded by smooth curves instead of line segments. If training data are sufficiently provided, an MLP can be trained to discriminate input patterns successfully. A discussion on limits of the number of hidden neurons in multi-layer perceptrons can be found in (Huang and Huang, 1991).

Most of image-processing operations, such as smoothing, enhancement, edge detection, noise removal, and morphological operations, require checking the values of neighboring pixels. Two basic morphological operations, dilation (similar to "expansion") and erosion (similar to "shrink"), are often combined in sequences for image filtering and feature extraction. The size of neighborhood may vary with applications. Eight-neighbor along with the center pixel is often used. The neighborhood processing can be implemented by look up tables to associate the relationship between input and output values. The input-output association can be realized by neural network models with nine input neurons (assuming that the 8-neighbor area is used) and one output neuron with one or more layers of hidden neurons in between. By iteratively providing input vectors to compute output from the look up table and comparing with the desired output, the error will be used to adjust the weights of connections. Therefore, MLPs can gradually and adaptively learn the input-output association. If the training MLP converges with a set of weights, it can be used to perform the same transformation on any image.

Figure 7.13 shows the MLP modules designed as the general-purpose image processing modules. The more hidden layers are used and the more complicated discriminant regions it forms in the domain space spanned with input vectors. There is a tradeoff between the complexity of MLP connections and the converge time of the MLP. Since the general-purpose MLP module is concerned, reasonably short training time to adapting the MLP and big capacity of connection weights are expected.

The error back-propagation training algorithm was given in (Rumelhart et al., 1986). A sigmoidal logistic nonlinearity

$$f(\alpha) = \frac{1}{1+e^{-(\alpha-\theta)}} \tag{7.34}$$

is used as the output function of neurons, where θ is a bias (or a threshold). Let $x_i^{(n)}$ denote the output of neuron i in layer n. Also, let w_{ij} denote the connection weight from neuron i in a lower layer to neuron j in the immediately higher layer. The activation values are calculated by

$$x_j^{(n+1)} = f\left(\sum_i w_{ij} x_i^{(n)} - \theta_j\right) \tag{7.35}$$

The major part of the algorithm is to adjust the connection weights according to the difference between the actual output of equation (7.35) and the desired output provided in the training patterns.

The equation for weights adjustment is

$$w_{ij}(t+1) = w_{ij}(t) + \eta \delta_j x_i^{(n)} \tag{7.36}$$

where η is the gain term (or called learning rate), δ_j is the error term for neuron j, and x_i is either the output of neuron i or an input when $n = 1$. The error term δ_j is

$$\delta_j = \begin{cases} x_j^{(n+1)}(1-x_j^{(n+1)})(d_j - x_j^{(n+1)}) & \text{if neuron } j \text{ is in the output layer} \\ x_j^{(n+1)}(1-x_j^{(n+1)})\sum_k \delta_k w_{jk} & \text{if neuron } j \text{ is in a hidden layer} \end{cases} \tag{7.37}$$

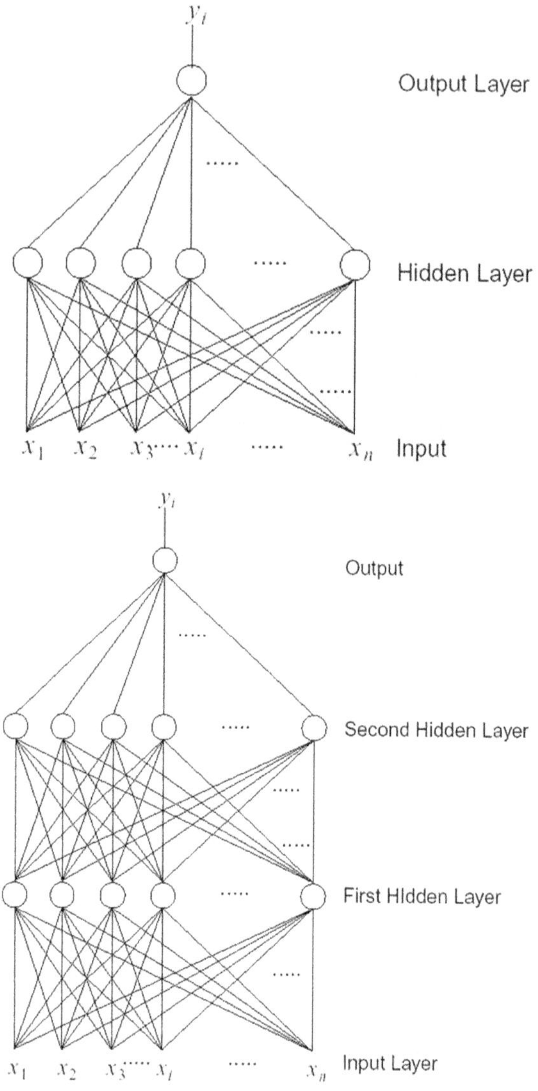

FIGURE 7.13 Multi-layer perceptron modules for the general-purpose image processing.

where k is with respect to all the neurons in the layer where neuron j is located.

The training patterns can be generated by all the variations in a local window of the specified low-level image operations. For example, in a 3×3 neighboring area, there are up to $2^9 = 512$ different training patterns, with a vector of 9 binary values and one desired output. For instance, a morphological erosion with the structuring element shown in Figure 7.14 will respond output 1 at the central pixel for the 16 input patterns in Figure 7.15 and output 0 for other input patterns.

0	1	0
1	1	1
0	1	0

FIGURE 7.14 A morphological structuring element.

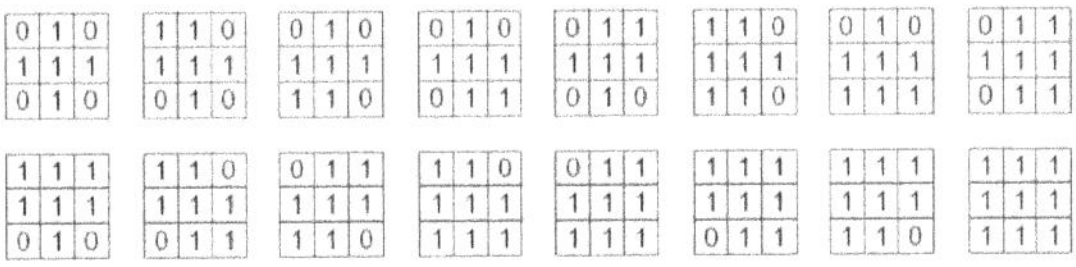

FIGURE 7.15 Patterns having desired output 1 with erosion by the structuring element in Figure 7.14.

Figure 7.16 shows the overall architecture consisting of MLP modules for low-level image operations. There is one MLP module corresponding to each pixel in the input image. Every MLP module applies the operation, which has been trained to the pixel and its $m \times m$ neighboring pixels, where m means the window size of neighborhood. The window size can be changed depending upon the desired operations for training.

Since the same operation is simultaneously applied to all the modules for all the pixels in the input image, the set of connection weights are considered the same for all MLP modules. That is, all MLP modules in the same stage can share the same set of trained connection weights.

This will significantly reduce the required number of local memories for connection weights in MLP module. In Figure 7.16(a), the pyramid, called the training module, is used for training task and provides shared memory for connection weights. Thus, no local memory is required for individual MLP modules in between the input and output images to hold the connection weights. The training phase is first turned

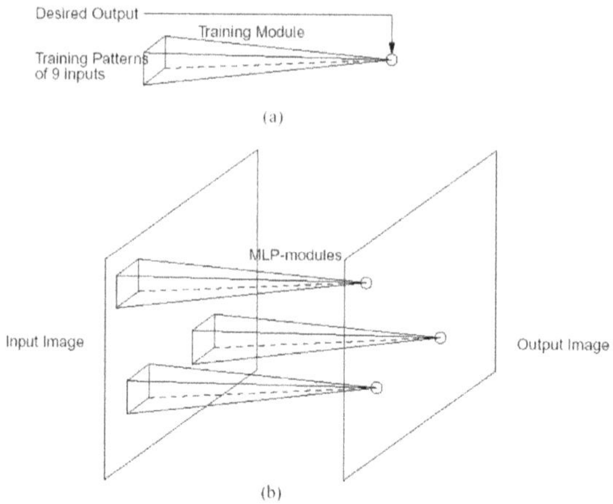

FIGURE 7.16 Overall system architecture with MLP modules for low-level image operations. (a) The pyramid is used for training tasks and storage of the trained weights. (b) Each pyramid represents an MLP module and receives m^2 inputs from its $m \times m$ neighboring pixels.

up at the training module. After the weight vector in the training module converges, the connection weights are frozen, shared, and retrieved by all the MLP modules to determine the activation value of each pixel after the operation is applied.

With the introduction of the error back-propagation learning algorithm, the MLP modules are capable of adapting themselves to the expected operations. It is also more flexible than those designed for dilation and erosion operations only (Krishnapuram and Chen, 1993).

During the experiments with different image operations, the training sets for some operations, such as dilation, erosion, and noise removal, have a small number of patterns with the desired output 1 or 0. For example, an erosion with the structuring element in Figure 7.14 has only 16 patterns with output 1, and an erosion with the structuring element shown in Figure 7.17 expects an output 1 only if all nine inputs are 1s.

By checking the training patterns, the values of some m^2 inputs do not even affect the output. Without considering a certain value of the central pixel, the training pattern set can be reduced. The same effect can be achieved by connecting the output neuron directly with the central input neuron and the MLP module operation is bypassed when the central pixel equals 0 or 1. For instance, applying an erosion to the central pixel with value 0 will never output 1. Also, applying a dilation to the central pixel with value 1 will always output 1. Such bypassing connections can be specified with the desired output as weights shared in all the MLP modules. The bypassing connections are intended to perform an exclusive OR between the central pixel and the unchanged input which is defined with respect to operations. The operation of an MLP is disable if the exclusive OR gets output 1. It means that the central pixel has the same value as specified in the shared module. Figure 7.18

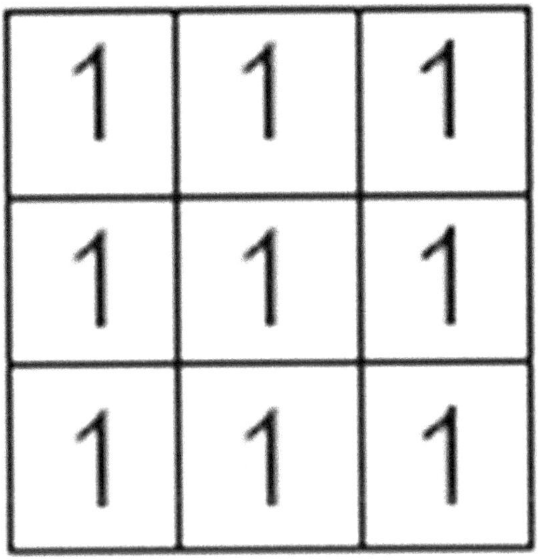

FIGURE 7.17 The structuring element with all 1s.

illustrates the bypassing connection and expresses how it affects the operation of an MLP module. The bypassing control is to enable/disable the activation of neurons in the modules and is determined by

$$E = x_c \ \textbf{XOR} \ U \tag{7.38}$$

where E means the enable/disable control to neurons, x_c is the value of the central pixel, and U is the expected input. For instance, $U = 1$ is set for dilation and $U = 0$ for erosion. The dashed lines with arrows in Figure 7.18 show the enable/disable controls to neurons, while the dotted lines indicate the bypassing connection for passing the value of the central pixel directly to the output. The output is defined by

$$\text{Output} = y \cdot E + x_c \cdot \bar{E} \tag{7.39}$$

where y is the output of the MLP module.

The proposed architecture can be used as a basis to organize multistage or recurrent networks. One may stack up with more than one stage as shown in Figure 7.19. By training the stages with a different set of patterns, one can provide and apply a sequence of image operations to the input image. The outputs of the last stage may be connected to the inputs of the first stage to form a recurrent architecture so that the same operations in sequence can be applied to the image for more than one iteration.

A layered neural network is a network of neurons organized in the form of layers. The first layer is the input layer. In the middle, there are one or more hidden layers. The last one is the output layer. The architecture of a layered neural network is

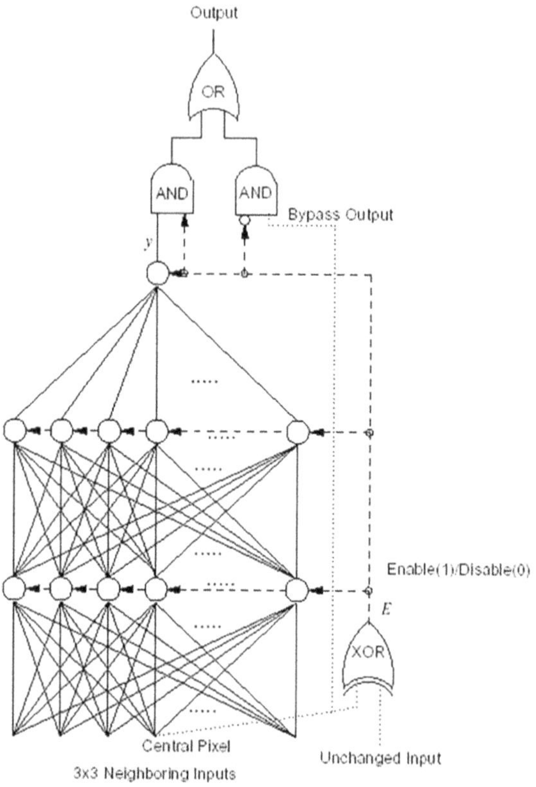

FIGURE 7.18 MLP module with a bypass control to enable/disable the activation of neurons in the modules. The dashed lines with arrows show the enable/disable controls to neurons, while the dotted lines indicate the bypassed connection for passing the value of the center pixel to the output.

shown in Figure 7.20. The function of the hidden neurons is to intervene between the external input and the network output. The output vector $\vec{\mathbf{Y}}(\vec{\mathbf{x}})$ is

$$\vec{\mathbf{Y}}(\vec{\mathbf{x}}) = \sum_{i=1}^{l} f_i(\vec{\mathbf{x}}) w_{ik} = \vec{\mathbf{F}} \cdot W, \quad (7.40)$$

where $\vec{\mathbf{F}}$ is the hidden layer's output vector corresponding to a sample $\vec{\mathbf{x}}$, W is the weight matrix with dimensions of $l \times k$, l is the number of hidden neurons, and k is the number of output nodes. In such a structure, all neurons cooperate together to generate a single output vector. In other words, the output depends on not only each individual neuron, but also all of them.

Numerous neural network algorithms and architectures have been proposed. Hou et al. (2007) presented a recurrent neural network for hierarchical control of interconnected dynamic systems. Guo and Huang (2006) employed Lyapunov functions

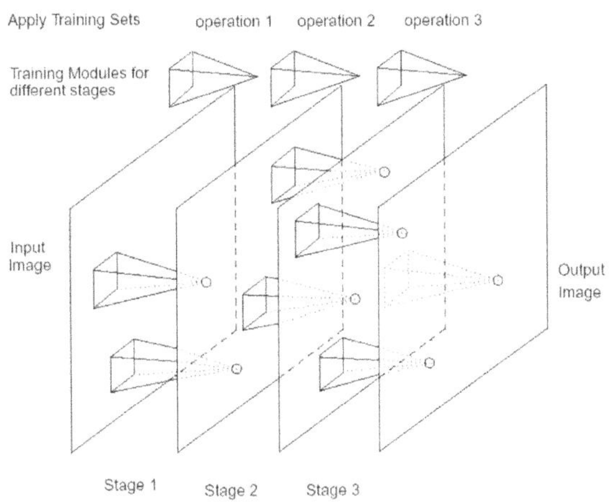

FIGURE 7.19 Stacked MLP-based architecture for multiple image operations. Only some MLP-modules are shown to simplify the figure. The separated pyramids at the top of the figure are the training modules for different stages. There may be more stages stacked on the architecture.

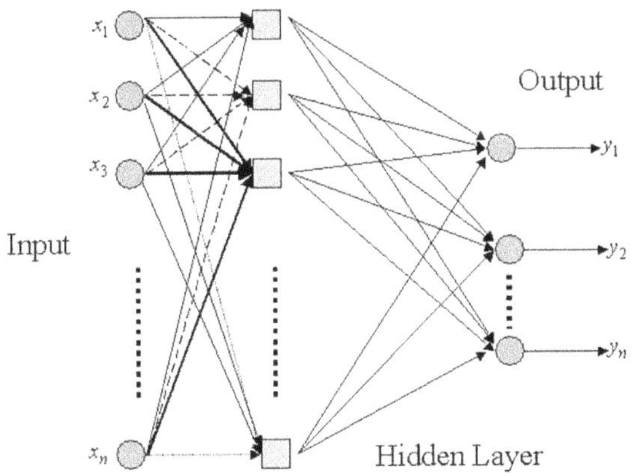

FIGURE 7.20 Architecture of a layered neural network.

to establish some sufficient conditions ensuring existence, uniqueness, global asymptotic stability, and even global exponential stability of equilibria for the Cohen and Grossberg (1983) neural networks with and without delays. Yang (2006) presented a radial basis function neural network for discriminant analysis. Jin and Liu (2008) developed the wavelet basis function neural networks. Wang and Wang (2008) used neural networks for password authentication to overcome the shortcomings of traditional approaches.

Aizenberg et al. (2008) presented blur identification by multi-layer neural network based on multivalued neurons. Wang and Peterson (2008) proposed constrained least absolute deviation neural networks. Nandedkar and Biswas (2007) proposed a fuzzy min-max neural network classifier with compensatory neurons, which uses hyperbox fuzzy sets to represent the pattern classes. Goh et al. (2008) presented a geometrical measure based on the singular value decomposition to estimate the necessary number of neurons to be used in training a single-hidden-layer feedforward neural network. Phung and Bouzerdoum (2007) proposed a neural architecture for classification of visual patterns that is motivated by the two concepts of image pyramids and local receptive fields. Savran (2007) proposed multifeedback-layer neural network. Palmes et al. (2005) proposed mutation-based genetic neural network.

7.5 THE ADAPTIVE-RESONANCE-THEORY (ART) NETWORK

In this section, we are mainly concerned with the model of adaptive resonance theory, or ART, which was developed by Carpenter and Grossberg (1987a, 1987b, 1990). This model, which forms clusters and is trained without a supervisor, can achieve the self-organization of input codes for pattern classification in response to random sequences of input patterns. With the concept of such an architecture, the process of adaptive pattern recognition is a special case of the more general cognitive process of hypothesis discovery, testing, searching, classification, and learning. This property makes it feasible to apply in more general problems of adaptively processing a big set of information sources and databases.

The ART model grew out of the analysis of a simple type of adaptive pattern recognition network, often called the *competitive learning paradigm* (Carpenter and Grossberg, 1988). Such a combination of adaptive filtering and competition is shared by many other models of pattern recognition and associative learning. There are three classes of ART architectures: ART1 (Carpenter and Grossberg, 1987a) for classifying binary patterns, ART2 (Carpenter and Grossberg, 1987b) for analog patterns, and ART3 (Carpenter and Grossberg, 1990) for parallel search of learned pattern recognition codes. Hwang and Kim (2006) developed a fast and efficient method for computing ART.

7.5.1 The ART1 Model and Learning Process

Figure 7.21 shows the typical architecture of an ART1-based neural network (Stork, 1989). The network consists of two subsystems: the *attentional subsystem* and the *orienting subsystem*, which is also called *novelty detector*. In the attentional subsystem, there are two layers of neurons, F and E, shown in Figure 7.21, which are called *feature representation field* and *exemplar (category) representation field*, respectively. Neurons in these two fields are fully connected by adaptive weights analogous to synapses in animals' nervous system. The stimuli in F pass through the connections $B[F_i : E_j]$ to the neurons in E and are multiplied by their corresponding weights. Each neuron in E sums up all of these weighted activities and responses accordingly. That is, the output μ_j of the neuron j is equal to

$$\mu_t[j] = \sum_i B_t[F_i : E_j]x_i \tag{7.41}$$

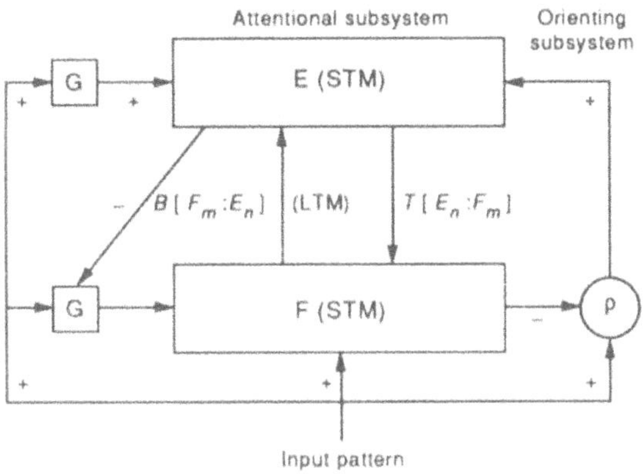

FIGURE 7.21 Typical ART1 architecture. Rectangles represent the attentional subsystem containing two fields where short-term memory (STM) patterns are stored. The weights of connections between neurons in two layers determine the long-term memory (LTM) traces. Squares with the label "G" indicate the gain control devices. The orienting subsystem with vigilance parameter ρ is represented by the circle.

where $B_t[F_i : E_j]$ denotes weights of bottom-up connections from neuron F_i in layer F to the neuron E_j in layer E at time t, and x_i is the ith component of input pattern. The layer E is capable of choosing the neuron which receives the largest total input (i.e., "winner-take-all"). The MAXNET (Lippmann, 1987), for example, with heavy literal inhibitory may choose the maximum from the input signals.

A set of interconnection weights acts as a filter, or *mask*, which is a term commonly used in the field of image processing. It transforms the activity of all neurons in layer F into a set of input values to E. Each neuron in E represents a different category. The winning neuron in E then triggers its own associative pattern learning within the vector of LTM traces that sends its inputs through the adaptive filter to E. This is very useful for applications with categorical perception – that is, classifying each input pattern as being in one and only one category.

During learning process, the first pattern is classified into the first category represented by the first neuron in the output layer, and the succeeding patterns are classified into either a newly created category or a previously created category. In addition to bottom-up connections, for computing response (i.e., sum of weighted inputs), the network is also designed to have a set of feedback (also called top-down) connections for verifying the distance between the input pattern and the exemplar related to the winning neuron. The set of weights of connections from any neuron in layer E to all neurons in layer F represents a template (or a set of critical features) for its related category.

A matching threshold (ρ) called *vigilance parameter* (ranged between 0.0 and 1.0) was introduced to determine how close a new input pattern must be to a stored exemplar to be considered similar. (Conversely, the value $1-\rho$ determines how large a mismatch can be tolerated.) An orienting system, represented by a circle with

ρ in Figure 7.21, acts as a novelty detector and handles the vigilance testing tasks. It receives input signals from both the input pattern with excitatory connections and the overall activity in layer F with inhibitory connections.

The vigilance test performed by this subnetwork can be expressed by

$$\frac{\sum_i T[E_j : F_i]x_i}{\|\mathbf{X}\|} > \rho \tag{7.42}$$

where $T[E_j : F_i]$ means the weight of the connection from neuron E_j in E to neuron F_i in F, x_i the activation value of neuron i in layer F, and $\|\mathbf{X}\|$ for the norm (length) of the input pattern vector. For ART1, $\|\mathbf{X}\| = \sum_i |x_i|$. If the input pattern satisfies the inequality, then it is classified into the category corresponding to the winning neuron. Otherwise, the subsystem sends a reset signal through the excitatory connection to the winning neuron in order to temporarily disable its output so that the system may choose another winning neuron. Such a task will be repeated until either an exemplar similar to the input pattern being chosen or no more neuron to choose, then a new neuron is created in E and assigned to represent the new category.

A top-down connection linking two active neurons approaches a strength of 1.0 because of the following dynamic equations:

$$T_0[E_j : F_i] = 1 \text{ and } T_{t+1}[E_j : F_i] = T_t[E_j : F_i]x_i, \tag{7.43}$$

where $T_0[E_j : F_i]$ is the initial weight and $T_t[E_j : F_i]$ means the weight of top-down connection from neuron E_j in E to neuron F_i in F at time t. It will be unchanged if both neurons are inactive. Suppose that a new pattern $\mathbf{X}$, whose ith component x_i is 0, is classified into a category represented by neuron E_j. If the original exemplar of the category kept in the top-down connections makes the strength of $T[E_j : F_i]$ equal to 1.0, then $T[E_j : F_i]$ decays toward 0.0 after adaptation. Therefore, the exemplar of each category will keep only the common features of all the input patterns classified into the same category.

The learning rule for the bottom-up connections is similar to top-down except that the increasing value for a connection linking two active neurons is not 1.0. Instead, the value depends on the number of active neurons. The following equations give the initial bottom-up weight and the adaptation value during learning process for the bottom-up connection from neuron F_i in F to neuron E_j in E (suppose $q \times q$ images are used as inputs):

$$\begin{cases} B_0\left[E_j : F_i\right] = \dfrac{1}{1+q^2} \\ \\ B_{t+1}\left[E_j : F_i\right] = \dfrac{T_t\left[E_j : F_i\right]x_i}{0.5 + \sum_i T_t\left[E_j : F_i\right]x_i} \end{cases} \tag{7.44}$$

Therefore, we may see that for each training pattern, the final strength of bottom-up connections between two active neurons will equally divide the maximum output (i.e., 1.0) of a neuron approximately. That is, at any time, the total strength of bottom-up connections to any neuron in E will be close to 1.0. This ensures that the response value of the neurons in E is between 0.0 and 1.0.

7.5.2 The ART2 Model

The adaptive-resonance-theory 2 (ART2) family of neural network architectures proposed by Carpenter and Grossberg (1987b) has been proven to be proficient in recognizing both binary- and analog-valued patterns. A typical ART2 neural architecture is composed of two layers of fully interconnected neurons. Each neuron has many connections emanating from and terminating on it. The adaptive connections between neurons store the long-term memory (LTM) traces of the network. The LTM represents the information that the network has learned, i.e., the structure of the categories, into which it has segregated the input patterns. The connections between neurons may be either excitatory which increase a neuron's activation, or inhibitory which decrease a neuron's activation.

The two layers (or fields) of neurons in the ART2 architecture form the attentional subsystem. The first field is named the feature representation field or F_1. Each neuron in F_1 contains six processing elements (PEs) that form three intra-PE sublayers. The main function of the feature representation field is to enhance the current input pattern's salient features while suppressing noise (Atal, 1976). This is achieved through pattern normalization and thresholding. Normalization allows an equitable comparison to be made between the input pattern and the patterns stored in the network's LTM traces (Shih et al., 1992). Thresholding maps the infinite domain of the input patterns to a prescribed range. Both normalization and thresholding are required for the processing of analog patterns. They are absent from the ART1 model since it is only capable of processing binary-valued patterns.

The second layer in the attentional subsystem is called the category representation field or F_2. Each neuron in this field represents a category (or class) that has been learned by the network. The connections emanating from a particular F_2 neuron store the exemplar of the category it represents. The exemplar represents the essential features of the category, i.e., those characteristics exhibited by all the patterns belonging to this class.

7.5.2.1 Learning in the ART2 Model

ART2 utilizes an unsupervised competitive learning technique, in which patterns are represented by points in an N-dimensional feature space, where N denotes the number of attributes in the patterns. Let $(a_1, a_2, \cdots, a_N)$ and $(b_1, b_2, \cdots, b_N)$ represent two patterns. Pattern similarity is assessed on the basis of a distance measurement in the feature space. The most common distance measurement is the Euclidean distance (d_e), which is computed as

$$d_e\left[(a_1, a_2, \cdots, a_N), (b_1, b_2, \cdots, b_N)\right] = \sqrt{(a_1 - b_1)^2 + (a_2 - b_2)^2 + \cdots + (a_N - b_N)^2} \quad (7.45)$$

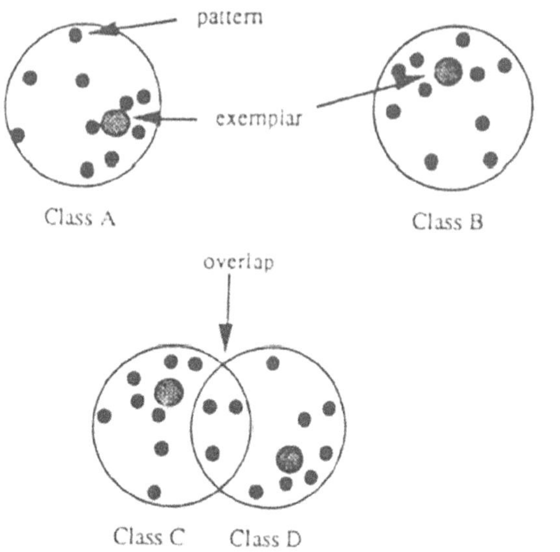

FIGURE 7.22 Illustration of the class separation problem.

Patterns that are deemed to be sufficiently close to one another are placed in the same category. The N-dimensional location of the centroid of a class represents the class's exemplar. An unsupervised learning procedure attempts to automatically discover the distributions and centroids of the categories for the patterns presented. Optimally, ART2 will discover pattern classes with minimum radii and maximum inter-centroid distances; that is, intraclass distance is minimized and interclass distance is maximized. However, this goal is not always attainable when two classes overlap with each other as illustrated in Figure 7.22.

The learning technique is considered to be competitive since each F_2 neuron competes with the others for including the current input pattern in its category code. ART2 utilizes a "winner-take-all" classification strategy, such as MAXNET (Haykin, 1992), which operates in the following manner:

(1) An input pattern is presented to the feature representation field where is normalized and thresholded.
(2) The resultant signal, which is called the short-term memory (STM), is passed through the bottom-up connections to the category representation field.
(3) Each established class in F_2 responds to the signal with an activation level, which it sends to itself through excitatory connections, and to all its neighbors through inhibitory connections.
(4) Eventually, the F_2 neuron with the highest activation will manage to inhibit the others sufficiently so that they will become disabled. The sole remaining active F_2 neuron is considered to be the winner; that is, it is the category whose exemplar most resembles the current input pattern.

> Having selected the winner in this manner, the second subsystem of ART2, named the *orienting subsystem*, is activated. The orienting subsystem determines whether the exemplar that is stored in the winning neuron's LTM traces sufficiently resembles the circulating STM pattern to be considered a match. The degree of match between two patterns is related to the cosine of the angle between them in feature space. Patterns, which are very similar are nearly parallel to each other while greatly dissimilar patterns, are orthogonal to each other.

A matching threshold called the vigilance parameter (which ranges from 0.0 to 1.0) determines how similar the input pattern must be to the exemplar to be considered a match. If the degree of match computed by the orienting subsystem exceeds the vigilance parameter, a state of resonance is attained and the STM pattern at F_1 is merged onto the winning neuron's LTM traces. Otherwise, the orienting subsystem sends a reset signal through the excitatory connections to the winning neuron to inhibit it from competing again for the current input pattern. This searching process is repeated until either an F_2 neuron passes the vigilance test or all established F_2 neurons have failed the test. In the latter case, a new category is established in the next available F_2 neuron.

The stability of pattern code learning in ART2 is achieved via read-out of the winning neuron's learned category exemplar. The F_2 neuron that wins the competition for the current input pattern sends an STM signal representing the category exemplar through the top-down connections to F_1. The structure of the F_1 neurons allows the bottom-up (input) and top-down (exemplar) STM signals to be combined. This mixing process is essential for achieving stability since without confusion between useful pattern differences and spurious baseline fluctuations could easily upset the matching process and cause spurious reset events to occur.

The actual learning process, whereby the current input pattern is encoded into the network's memory, involves modification of the bottom-up and top-down LTM traces that join the winning F_2 neuron to the feature representation field. Learning either refines the code of a previously established class based on any new information that is contained in the input pattern, or initiates code learning in a previously uncommitted F_2 neuron. In either case, learning only occurs when the system is in a resonant state. This property ensures that an input pattern does not obliterate information that has been previously stored in an established class.

7.5.2.2 Functional Link Net Pre-Processor

Performance of the ART2 model is measured by its ability to separate pattern clusters in the feature space. Unless a considerable degree of pattern separation is attained, there is likely to be much overlap of these clusters. Pao (1989) describes an approach toward increasing the separability of patterns by enhancing them in a linearly independent manner. In essence, the input patterns are represented in a space of increased dimensions; this facilitates ART2's ability to learn hyperplanes for separation.

The functional link net enhances the input patterns without actually introducing any new information. Rather, it expands each attribute in the input pattern by

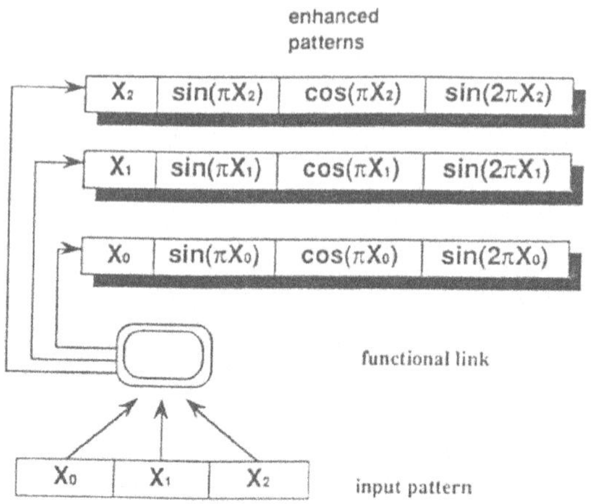

FIGURE 7.23 Functional-link net: functional expansion model.

evaluating a set of linearly independent functions with the attribute as the argument. There are two variations of the functional link net and each produces a different effect. In the functional expansion model shown in Figure 7.23, the functional link evaluates a set of three functions that span a portion of feature space for each attribute in the input pattern. The functional expansion model was demonstrated by Pao (1989) to be effective in training a network to learn a function from a limited set of sample points. In this implementation each attribute X_i in the input pattern is mapped into the following set of ten functions:

$$\left\{X_i, X_i^2, X_i^3, \sin(\pi X_i), \cos(\pi X_i), \sin(2\pi X_i), \cos(2\pi X_i), \sin(3\pi X_i), \cos(3\pi X_i), \sin(4\pi X_i)\right\}$$

In the tensor or outer-product model shown in Figure 7.24, each attribute of the input pattern multiplies the remainder of that pattern. To limit the size of these derived patterns, attribute X_i only multiplies attribute X_j if $j \geq i$. The outer-product model introduces higher-order terms in the enhanced representation in the sense that many of the terms represent joint activations (Pao, 1989).

We incorporate the functional link net into the ART2 model as an input pattern pre-processor. The functional link net operates directly on the original input pattern and passes the enhanced pattern representation to the input layer of the network. An additional benefit of the functional link net is that it allows both supervised and unsupervised learning to be performed by the same network. Furthermore, the degree of pattern enhancement that can be attained through the net eliminates the need for network architectures that contain hidden layers.

7.5.3 Improvement of ART Model

In implementing the algorithm of Carpenter and Grossberg by software simulation dealing with optical character recognition, we obtained the same result of category

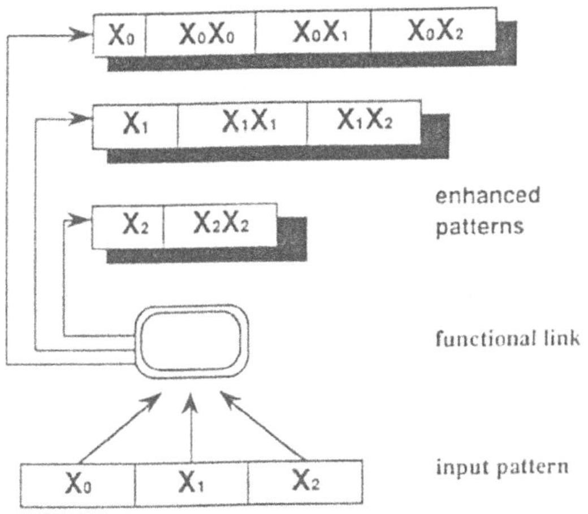

FIGURE 7.24 Functional-link net: outer-product expansion model.

classification as described in (Carpenter and Grossberg, 1987a; Carpenter, 1989; Stork, 1989). However, we have observed that the input patterns are classified with variance to their training sequences. For instance, if the sequence is {**E**, **F**, **L**}, "**F**" is classified as the same category with the exemplar "**E**" and replaces "**E**"' as a new exemplar, and then "**L**" is classified as another new category. If the sequence is {**E**, **L**, **F**}, "**L**" belongs to the category with the exemplar "**E**" and replaces it as a new exemplar, and then "**F**" is classified as another new category. If the sequence is {**F**, **L**, **E**}, then all three characters are in different categories. Even though the vigilance parameter is adjusted, the variation of categories still exists.

In this section, we will first give the problem analysis and then propose an improved model and illustrate the experimental results.

7.5.3.1 Problem Analysis

In the ART model, each neuron in the higher layer E sums up all the weighted activities from neurons in the lower layer F. The activation is calculated by

$$\mu_{E_j}^{(t)} = \sum_i B_{F_i,E_j}^{(t)} \cdot x_i \tag{7.46}$$

where $B_{F_i,E_j}^{(t)}$ denotes weights of bottom-up connections from neuron F_i to neuron E_j at time t and x_i is the ith component of the input pattern. When a winning neuron E_{j*} is chosen based on the maximum activation, the orienting subsystem activates for the vigilance test, which can be expressed as

$$\frac{\sum_i T_{E_{j*},F_i}^{(t)} \cdot x_i}{\|\mathbf{X}\|} \geq \rho \tag{7.47}$$

where $T^{(t)}_{E_{j*},F_i}$ denotes weights of top-down connections from the winning neuron E_{j*} to neuron F_i at time t, and $\|\mathbf{X}\|$ is the norm (or magnitude) of the input pattern vector, i.e., $\|\mathbf{X}\| = \sum_i |x_i|$. In equation (7.47) for vigilance testing, the numerator picks up all common pixels between the input pattern and the exemplar it's comparing with, and the denominator is the norm of the input vector. This ratio represents the percentage of the pixels of the input pattern existing in the exemplar. Since "**F**" is a subset of "**E**," the ratio is 1.0; i.e., all the pixels of "**F**" are included in "**E**." But, how can we take into account the pixels of "**E**," which are not in "**F**" (the line at the bottom)? The orienting subsystem is supposed to be able to count not only the pixels in "**F**" and not in "**E**" but also the pixels in "**E**" and not in "**F**" in order to give a fair judgment of the similarity.

Another issue is that for each new training pattern, the network repeats choosing the winning neuron and performing vigilance testing until a category is chosen. However, if none of the existing neurons in E satisfies the vigilance test, a new neuron is created to represent the new category using the pattern as an exemplar. This procedure is time-consuming and the performance is the worst when a new input pattern is quite different from all the existing exemplars. To reduce the searching time of the model, a minimum activation response may be set up so that the model need not go through the neurons with the lower response.

7.5.3.2 An Improved ART Model for Pattern Classification

The task of vigilance testing is now modified. We use a bidirectional testing to determine the similarity between an input pattern and the exemplar of a chosen category. In addition to equation (7.47), another inequality is introduced as follows to determine how similar an exemplar is to the input pattern. That is,

$$\frac{\sum_i T^{(t)}_{E_{j*},F_i} \cdot x_i}{\|\mathbf{T}\|} \geq \rho \tag{7.48}$$

where $\|\mathbf{T}\|$ is norm (or magnitude) of the exemplar vector, i.e., sum of top-down weights from the winning neuron E_{j*} to all the neurons in F. The difference between equation (7.47) and equation (7.48) is the denominator; one is the norm of the input vector and the other is the norm of the exemplar vector.

The ratio represents the percentage of the pixels of the exemplar existing in the input pattern. Only if these two equations are satisfied, the new pattern is associated and combined with the exemplar; otherwise, the next exemplar is chosen to compare with. It has to be noted that the vigilance parameters in equations (7.47) and (7.48) could be selected differently. Therefore, an improved ART model shown in Figure 7.25 includes a set of new devices which accumulates the total weights of top-down connections of an active neuron E_j and sends them to the orienting subsystem.

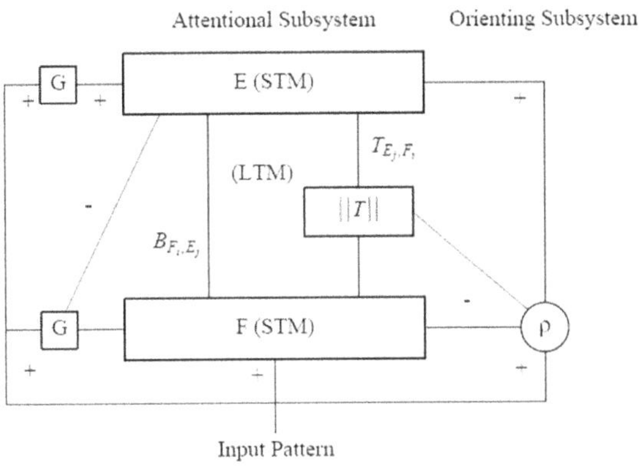

FIGURE 7.25 An improved ART model. The activation in layers F and E forms a short-term memory (STM) and the connections including bottom-up and top-down construct a long-term memory (LTM). Squares with the label "**G**" indicate the gain control devices and the orienting subsystem with vigilance parameter ρ is represented by a circle.

The algorithm of the improved ART model is described as follows:

Step 1. Initialize weights.

$$T_{E_j,F_i}^{(0)} = 1.0 \tag{7.49}$$

$$B_{F_i,E_j}^{(0)} = \frac{1}{1+N} \tag{7.50}$$

where $1 \le i \le N$ and $1 \le j \le M$ (N: number of elements in the input vector; M: number of existing exemplars). Also, the vigilance parameter ρ, where $0.0 \le \rho \le 1.0$, is selected.

Step 2. Read input pattern **X**.
Step 3. Compute neurons' responses.
The output of the neuron E_j is computed by

$$\mu_{E_j}^{(t)} = \sum_{i=1}^{N} B_{F_i,E_j}^{(t)} \cdot x_i \tag{7.51}$$

where $B_{F_i,E_j}^{(t)}$ denotes weights of bottom-up connections from F_i to E_j at time t, and x_i is the ith component of the input pattern vector **X** which can be either 0 or 1 (binary).

Step 4. Select a winning neuron with the highest response.
Let C be the total number of currently used exemplar neurons.

$$\mu^{(t)}_{E_{j*}} = \max\{\mu^{(t)}_{E_j} \mid 1 \le j \le C \text{ and } j = C+1\} \tag{7.52}$$

where the plus 1 is the first available unused neuron and E_{j*} is the winning neuron.
Step 5. Bidirectional vigilance.

Let $\|\mathbf{X}\| = \sum_i |x_i|, \|\mathbf{T}\| = \sum_i T_{E_{j^*},F_i}$, and $\|\mathbf{T}\backslash\mathbf{X}\| = \sum_i T_{E_{j^*},F_i} \cdot x_i$. If $\frac{\|\mathbf{T}\backslash\mathbf{X}\|}{\|\mathbf{X}\|} \ge \rho$ and $\frac{\|\mathbf{TX}\|}{\|\mathbf{T}\|} \ge \rho$, go to step 7; otherwise, go to step 6.
Step 6. Disable unmatched category.

Disable temporarily the output of the winning neuron chosen, and go to step 4 for choosing the next winning neuron.

Step 7. Adjust weights of the winning neuron.

$$T^{(t)}_{E^{(t+1)}_{j*,F_i}} T^{(t)}_{E_{j*,F_i}} \cdot x_i \tag{7.53}$$

$$B^{(t+1)}_{E_{j^*},F_i} = \frac{T^{(t)}_{E_{j^*},F_i} \cdot x_i}{0.5 + \sum_{i=1}^{N-1} T^{(t)}_{E^*,F_i} \cdot x_i} \tag{7.54}$$

Step 8. Enable all neurons and repeat the whole procedure by going to step 2.

The intention of adjusting the weights in the top-down connections is to store the information with respect to the exemplar of each category. Equation (7.53) shows that the new weight is derived from the intersection of the existing exemplar and the input pattern. Since the top-down weights are initialized to be all ones, the first classified exemplar is always the same as the input pattern; the updated exemplar will be the structure of the common pixels of all the patterns in each category. This step will ensure that the classified categories are independent of the input sequences.

Equation (7.50) initializes the weights of all bottom-up connections to be $1/(1+N)$, where N is the number of elements in $\mathbf{X}$. Thus, when an $\mathbf{X}$ is presented, the magnitude of any unused neuron's response will be $\|\mathbf{X}\|/(1+N)$. Referring to equations (7.47) and (7.48), a perfectly matched pattern will cause the neuron to respond 1.0, and a pattern with totally unmatched pixels will respond with the value 0.0. The higher the response value is, the more matched pixels the pattern has. Therefore, the searching procedure may skip checking the neurons with responses lower than those of unused neurons. In step 4 of the algorithm, the searching always chooses the maximum of all active E neurons plus the first available unused neuron. If all active neurons with higher response than the unused neuron could not pass the

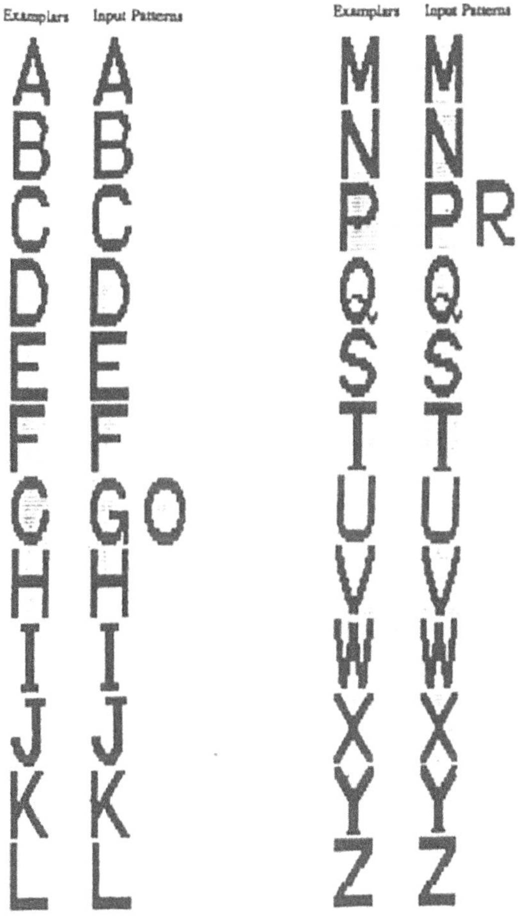

FIGURE 7.26 Result of character classification and the exemplar in each category.

vigilance testing, the unused neuron is assigned to represent the new created category. This can save significant time in searching most of the unmatched categories.

7.5.3.3 Experimental Results of the Improved Model

The input patterns used in this experiment are 16×16 optical characters. Therefore, a 256-neuron layer is designed for F and a 26-neuron layer for E. The vigilance parameter is set to be 0.8. Figure 7.26 shows the result of character classification and the exemplar in each category.

While the character "**F**" is trained, the existing exemplars already have "**A**," "**B**," "**C**," "**D**," and "**E**," and the exemplar "**E**" responds with the highest value to be the winning neuron. Thus, the vigilance testing is

$$\frac{\|\mathbf{T} \bullet \mathbf{X}\|}{\|\mathbf{X}\|} = 1.00 > \rho = 0.8$$

and

$$\frac{\|\mathbf{T} \bullet \mathbf{X}\|}{\|\mathbf{T}\|} = 0.72 < \rho = 0.8$$

Since the pattern "**F**" is a subset of the pattern "**E**," the first ratio having value 1.00 signifies that the testing passes. But on the other hand, "**E**" has more pixels than "**F**" has. This means that the second testing fails. The second highest responding neuron, i.e., the category "**B**," is next chosen and the vigilance testing still fails. Thus, the system creates a new category to represent "**F**."

With $\rho = 0.8$, the two pairs of characters {"**G**," "**O**"} and {"**P**," "**R**"} are classified into the same category, respectively. To achieve 100% classification rate, the higher vigilance parameter, e.g., 0.9, is assigned. In addition, certain noisy patterns are experimented to test the system performance. The weights of bottom-up and top-down connections are fixed. As shown in Figure 7.27, these noisy patterns can be recognized and classified into their proper categories.

Numerous ART algorithms and architectures have been proposed. Baraldi and Alpaydin (2002) proposed constructive feedforward ART clustering networks. Vasilic and Kezunovic (2005) presented an advanced pattern recognition algorithm for classifying the transmission line faults, based on supervised ART neural network with fuzzy decision rule. Araujo (2006) presented prune-able fuzzy ART neural architecture for robot map learning and navigation in dynamic environments. Chen et al. (2005) presented incremental communication for ART networks. Projective adaptive resonance theory (PART) neural network developed by Cao and Wu (2002) recently has been shown to be very effective in clustering datasets in high-dimensional spaces. Cao and Wu (2004) further provided a rigorous proof of these regular dynamics of the PART model when the signal functions are special step functions.

7.6 FUZZY SETS IN IMAGE ANALYSIS

Ever since Zadeh introduced the concept of fuzzy set theory in 1965, it has found many applications in a variety of fields. The fuzzy logic has been developed in order to capture the uncertainties associated with human cognitive processes, such as in

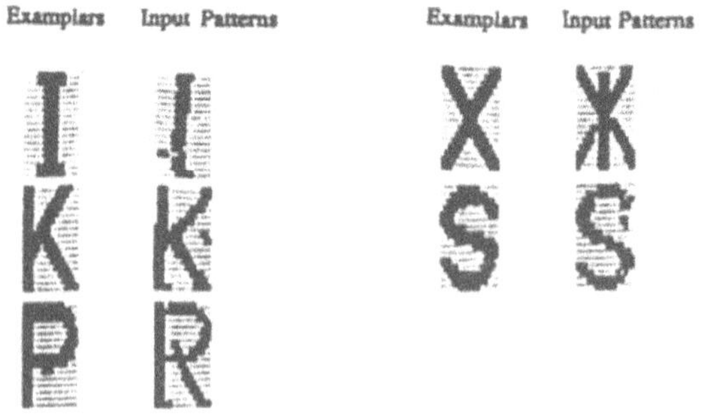

FIGURE 7.27 Testing patterns with noise and variances are also properly recognized and classified.

thinking, reasoning, perception, etc. A conventional set is defined as a collection of elements which have some common properties. The sets are known as *crisp sets* and are defined by a characteristic function as

$$f(x) = \begin{cases} 1 & \text{if } x \in A \\ 0 & \text{if } x \notin A \end{cases} \tag{7.55}$$

where A is any finite set. However, the object classes generally encountered in the real world are not so "precisely" or "crisply" defined. In most cases, several ambiguities arise in the determination of whether a particular element belongs to a set or not. A good example mentioned by Zadeh is a class of animals. This class clearly includes dogs, cats, tigers, etc. and excludes rocks, plants, houses, etc. However, an ambiguity arises in the context of objects such as bacteria and starfish with respect to the class of animals.

Unlike a crisp set in which each object is assigned a value of 0 or 1, each object in a fuzzy set is given a certain "degree of membership" which denotes the degree of belongingness of the object to the set. Therefore, a fuzzy set can be considered as a class of objects with a continuum of membership grades. The membership function assigns to each object a degree of membership value ranging from 0 to 1. The closer the membership value is to 1, the more the object belongs to the set, and vice versa.

7.6.1 Role of Fuzzy Geometry in Image Analysis

Image processing inherently bears some fuzziness in nature. This is due to the fact that the regions in the image are not always crisply defined. The approach to the analysis or interpretation of an image requires traditionally to segment the image into meaningfull regions, extract the features of each region, and finally construct the relationships among the regions. However, due to the fuzzy behavior of the images, it is convenient to regard the regions of the image as fuzzy subsets of the image. Rosenfeld (1984) generalized the standard geometrical properties to the properties of fuzzy image geometry, such as topological connectedness, adjacency, surroundedness, area, perimeter, etc. based on which many significant algorithms utilizing the concepts of fuzzy geometry in the field of image analysis have been developed.

7.6.2 Definitions of Fuzzy Sets

A fuzzy set of a grayscale image S is a mapping m from S into the values ranging [0, 1]. For any pixel $p \in S$, $\mu(p)$ is called *the degree of membership* of p. The mapping function μ is any function which follows the properties of symmetry and ambiguity around the crossover point. From the definition, one may assume that the fuzzy set is identical to the probability function. However, there are significant differences between these two. For example, the summation of all the values of the probability function should be equal to 1, which is not the case in fuzzy sets. The summation of all the values of the degree of membership in a fuzzy set need not be equal to 1.

Several definitions in fuzzy sets are given here:

(a) *Empty set*: A fuzzy set μ is said to be empty, denoted by $A = \phi$, iff $\mu(x) = 0$, $\forall x \in S$.
(b) *Equality*: Two fuzzy sets μ and ν are said to be equal, denoted by $\mu = \nu$, iff $\mu(x) = \nu(x)$, $\forall x \in S$.
(c) *Support*: The support of a fuzzy set $\propto$ is an ordinary subset of *S*, i.e., $\text{Supp}(\mu) = \{x \in S \mid \mu(x) > 0\}$.
(d) *Cross-over points*: The elements $x \in S$ such that μ (*x*) = 1/2 are called cross-over points of μ.
(e) *Normalized fuzzy set*: $\propto$ is said to be normalized iff $\exists x \in S$, $\mu(x) = 1$.
(f) *Height of* μ: The height of a fuzzy set $\propto$ is defined as hgt(μ) = sup { μ (*x*)}, i.e., the height of μ is defined to be the least upper bound of $\propto$(*x*). It is therefore obvious that for a normalized fuzzy set m, hgt(μ) = 1.
(g) *Cardinality of a fuzzy set*: When *X* is a finite set, the cardinality $|A|$ of a fuzzy set *A* is defined as $|A| = \sum \mu(x)$, which is called the *power* of *A*.
(h) *Convexity of a fuzzy set*: There are several alternative definitions to the convexity property of fuzzy sets. Listed here are three of major ones.

Definition 7.1. A fuzzy set *m* is said to be convex if for every *p*, *q* belonging to *S*, and all *r* on the line segment *pq*, it is true that $\mu(r) \geq \min[\mu(p), \mu(q)]$. This implies that μ is fuzzily convex, if it is convex in the ordinary sense.
Definition 7.2. A fuzzy set m of *X* is convex iff the sets *T* defined by $T_\alpha = \{x \in S \mid \mu(x) \geq \alpha\}$ are convex for all in the interval (0,1]. The sets T_α are called the "level sets" of μ.
Definition 7.3. A direct definition of convexity is the following: $\propto$ is convex iff

$$\mu[\lambda x_1 + (1-\lambda)x_2] \geq \min[\mu(x_1), \mu(x_2)]$$

for every x_1, x_2 belonging to *S* and in [0,1].
(i) *Cross-section*: The cross-section of m by a line *l* is defined as the restriction of μ to *l*.
(j) *Star-shapedness*: μ is star-shaped from *p* if its cross-sections by lines through *p* are all convex.

7.6.3 Set Theoretic Operations

(a) *Union*: The union of two fuzzy sets *A* and *B* with their respective membership functions $f_A(x)$ and $f_B(x)$ is a fuzzy set *C*, denoted as $C = A \cup B$, whose membership function is

$$f_C(x) = \max[f_A(x), f_B(x)], \forall x \in X$$

The union of two fuzzy sets *A* and *B* can be interpreted as the smallest fuzzy set containing both *A* and *B*. Let *D* be any fuzzy set containing both *A* and *B*. Since $f_D \geq f_A$ and $f_D \geq f_B$, $f_D \geq \max[f_A, f_B] = f_C$.

(b) *Intersection*: The intersection of two fuzzy sets *A* and *B* with their respective membership functions f_A and f_B is a fuzzy set *C*, denoted as $C = A \cap B$, whose membership function is $f_C(x) = \min[f_A(x), f_B(x)], x \in X$. The intersection of two fuzzy sets can also be interpreted as the largest fuzzy set which is contained in both.
Corollary: *A* and *B* are disjoint if their intersection *C* is empty, i.e., $f_C(x) = 0$.
(c) *Complement*: The complement of a fuzzy set $\propto$ is denoted by $\bar{\propto}$ and is defined as $\bar{\propto}(x) = 1 - \propto(x), \forall x \in S$.
(d) *Containment*: A fuzzy set $\propto$ is said to be contained in another fuzzy set ν iff $\mu(x) \leq \nu(x), \forall x \in S$.

Properties:

(a) *Commutativity*: $A \cup B = B \cup A$; $A \cap B = B \cap A$.
(b) *Associativity*: $(A \cup B) \cup C = A \cup (B \cup C)$; $(A \cap B) \cap C = A \cap (B \cap C)$.
Note: the two properties follow immediately for union and intersection because their corresponding operators of Max and Min are associative and commutative.
(c) *Idempotency*: $A \cup A = A$; $A \cap A = A$.
(d) *Distributivity*: $A \cup (B \cap C) = (A \cup B) \cap (A \cup C)$; $A \cap (B \cup C) = (A \cap B) \cup (A \cap C)$.
(e) $A \cap \phi = \phi$; $A \cup X = X$.
(f) *Identity*: The identity element with respect to union is φ and with respect to intersection is *X*.
(g) *Absorption*: $A \cup (A \cap B) = A$; $A \cap (A \cup B) = A$.
(h) *De Morgans laws*: $\overline{(A \cup B)} = \bar{A} \cap \bar{B}$; $\overline{(A \cap B)} = \bar{A} \cup \bar{B}$.
(i) *Involution*: A = A.
(j) *Equivalence formula*: $(A \cup B) \cap (A \cup C) = (A \cap B) \cup (A \cap C)$.
(k) *Symmetrical difference formula*: $(A \cap B) \cup (A \cap C) = (A \cup B) \cap (A \cup C)$.

Alternative Operators

(a) *Probabilistic operators*:
1. *Intersection*: For every $x \in X$, $\mu_{A \bullet B}(x) = \mu_A(x) \bullet \mu_B(x)$.
2. *Union*: For every $x \in X$, $\mu_{A+B}(x) = \mu_A(x) + \mu_B(x) - \mu_A(x) \bullet \mu_B(x)$.

Numerous fuzzy classification methods have been proposed. Sousa et al. (2007) proposed a fuzzy classification system to perform word indexing in ancient printed documents. Hamilton-Wright et al. (2007) proposed fuzzy classification using pattern discovery. Mansoori et al. (2008) considered the automatic design of fuzzy-rule-based classification systems from labeled data, and developed a steady-state genetic algorithm for extracting fuzzy classification rules from data. Nowicki (2008) presented an approach to fuzzy classification in the case of missing features. Cloete and van Zyl (2006) introduced a new class of induction algorithms based on fuzzy set covering principles.

REFERENCES

Aizenberg, I., Paliy, D. V., Zurada, J. M., and Astola, J. T., "Blur identification by multilayer neural network based on multivalued neurons," *IEEE Transactions on Neural Networks*, vol. 19, no. 5, pp. 883–898, May 2008.

Anguita, D., Boni, A., and Ridella, S., "A digital architecture for support vector machines: Theory, algorithm, and FPGA implementation," *IEEE Transactions on Neural Networks*, vol. 14, no. 5, pp. 993–1009, Sept. 2003.

Anguita, D., Pischiutta, S., Ridella, S., and Sterpi, D., "Feed-forward support vector machine without multipliers," *IEEE Transactions on Neural Networks*, vol. 17, no. 5, pp. 1328–1331, Sept. 2006.

Araujo, "Prune-able fuzzy ART neural architecture for robot map learning and navigation in dynamic environments," *IEEE Transactions on Neural Networks*, vol. 17, no. 5, pp. 1235–1249, Sept. 2006.

Armstrong, W. W. and Gecsei, J., "Adaptation algorithms for binary tree networks," *IEEE Transactions on Systems, Man, and Cybernetics*, vol. 9, no. 5, pp. 276–285, May 1979.

Atal, B. S., "Automatic recognition of speakers from their voices," *Proceedings of the IEEE*, vol. 64, no. 4, pp. 460–475, Apr. 1976.

Baraldi, A. and Alpaydin, E., "Constructive feedforward ART clustering networks," *IEEE Transactions on Neural Networks*, vol. 13, no. 3, pp. 645–661, May 2002.

Camastra, F. and Verri, A., "A novel kernel method for clustering," *IEEE Transactions on Pattern Analysis and Machine Intelligence*, vol. 27, no. 5, pp. 801–805, May 2005.

Cao, Y. and Wu, J., "Projective ART for clustering data sets in high dimensional spaces," *Neural Networks*, vol. 15, no. 1, pp. 105–120, Jan. 2002.

Cao, Y. and Wu, J., "Dynamics of projective adaptive resonance theory model: The foundation of PART algorithm," *IEEE Transactions on Neural Networks*, vol. 15, no. 2, pp. 245–260, Mar. 2004.

Carpenter, G. A., "Neural network models for pattern recognition and associative memory," *Neural Networks*, vol. 2, no. 4, pp. 243–57, 1987.

Carpenter, G. A. and Grossberg, S., "A massively parallel architecture for a self-organizing neural pattern recognition machine," *Computer Vision, Graphics, Image Processing*, vol. 37, no1, pp. 54–115, Jan. 1987a.

Carpenter, G. A. and Grossberg, S., "ART 2: Self-organization of stable category recognition codes for analog input patterns," *Applied Optics*, vol. 26, no. 23, pp. 4919–4930, Dec. 1987b.

Carpenter, G. A. and Grossberg, S., "The ART of adaptive pattern recognition by a self-organization neural network," *IEEE Computer*, vol. 21, no. 3, pp. 77–88, Mar. 1988.

Carpenter, G. A. and Grossberg, S., "ART3: Hierarchical search using chemical transmitters in self-organizing pattern recognition architectures," *Neural Networks*, vol. 3, no. 2, pp. 129–152, 1990.

Celikyilmaz, A. and Turksen, B., "Enhanced fuzzy system models with improved fuzzy clustering algorithm," *IEEE Transactions on Fuzzy Systems*, vol. 16, no. 3, pp. 779–794, Jun. 2008.

Chen, M., Ghorbani, A. A., and Bhavsar, V. C., "Incremental communication for adaptive resonance theory networks," *IEEE Transactions on Neural Networks*, vol. 16, no. 1, pp. 132–144, Jan. 2005.

Cloete, I. and van Zyl, J., "Fuzzy rule induction in a set covering framework," *IEEE Transactions on Fuzzy Systems*, vol. 14, no. 1, pp. 93–110, Feb. 2006.

Cohen, M. A. and Grossberg, S., "Absolute stability and global pattern formation and parallel memory storage by competitive neural networks," *IEEE Transactions on Systems, Man, and Cybernetics*, vol. 13, no. 7, pp. 815–821, Jul. 1983.

Corsini, P., Lazzerini, B., and Marcelloni, F., "A fuzzy relational clustering algorithm based on a dissimilarity measure extracted from data," *IEEE Transactions on Systems, Man, and Cybernetics, Part B*, vol. 34, no. 1, pp. 775–781, Feb. 2004.

Cortes, C. and Vapnik, V., "Support-vector network," *Machine learning*, vol. 20, no. 3, pp. 273–297, Sept. 1995.

Das, S., Abraham, A., and Konar, A., "Automatic clustering using an improved differential evolution algorithm," *IEEE Transactions on Systems, Man, and Cybernetics, Part A*, vol. 38, no. 1, pp. 218–237, Jan. 2008.

Fei, B. and Liu, J., "Binary tree of SVM: A new fast multiclass training and classification algorithm," *IEEE Transactions on Neural Networks*, vol. 17, no. 3, pp. 696–704, May 2006.

Fred, A. L. N. and Jain, A. K., "Combining multiple clusterings using evidence accumulation," *IEEE Transactions on Pattern Analysis and Machine Intelligence*, vol. 27, no. 6, pp. 835–850, Jun. 2005.

Fukushima, K., "Cognitron: A self-organizing multilayered neural network," *Biological Cybernetics*, vol. 20, no. 3, pp. 121–136, Sept. 1975.

Goh, C. K., Teoh, E. J., and Tan, K. C., "Hybrid multiobjective evolutionary design for artificial neural networks," *IEEE Transactions on Neural Networks*, vol. 19, no. 9, pp. 1531–1548, Sept. 2008.

Gonzalez-Abril, L., Angulo, C., Velasco, F., and Ortega, J. A., "A note on the bias in SVMs for multiclassification," *IEEE Transactions on Neural Networks*, vol. 19, no. 4, pp. 723–725, Apr. 2008.

Guo, S. and Huang, L., "Stability analysis of Cohen-Grossberg neural networks," *IEEE Transactions on Neural Networks*, vol. 17, no. 1, pp. 106–117, Jan. 2006.

Haasdonk, B., "Feature space interpretation of SVMs with indefinite kernels," *IEEE Transactions on Pattern Analysis and Machine Intelligence*, vol. 27, no. 4, pp. 482–492, Apr. 2005.

Hamilton-Wright, A., Stashuk, D. W., and Tizhoosh, H. R., "Fuzzy classification using pattern discovery," *IEEE Transactions on Fuzzy Systems*, vol. 15, no. 5, pp. 772–783, Oct. 2007.

Haykin, S., *Neural Networks*, The IEEE Press, 1992.

He, Z., Deng, S., and Xu, X., "Improving k-modes algorithm considering frequencies of attribute values in mode," *Proceedings of the International Conference on Computational Intelligence and Security*, Xi'an, pp. 157–162, 2005.

Herbrich, R., *Learning Kernel Classifiers Theory and Algorithms*, The MIT Press, 2001.

Hou, Z.-H., Gupta, M. M., Nikiforuk, P. N., Tan, M., and Cheng, L., "A recurrent neural network for hierarchical control of interconnected dynamic systems," *IEEE Transactions on Neural Networks*, vol. 18, no. 2, pp. 466–481, Mar. 2007.

Hsu, C. W. and Lin, C. J., "A comparison of methods for multiclass support vector machines," *IEEE Transactions on Neural Networks*, vol. 13, no. 2, pp. 415–425, Mar. 2002.

Huang, S. C. and Huang, Y. F., "Bounds on the number of hidden neurons in multilayer perceptrons," *IEEE Transactions on Neural Networks*, vol. 2, no. 1, pp. 47–55, Jan. 1991.

Hwang, S.-K. and Kim, W.-Y., "Fast and efficient method for computing ART," *IEEE Transactions on Image Processing*, vol. 15, no. 1, pp. 112–117, Jan. 2006.

Jayadeva, Khemchandani, R., and Chandra, S., "Twin support vector machines for pattern classification," *IEEE Transactions on Pattern Analysis and Machine Intelligence*, vol. 29, no. 5, pp. 905–910, May 2007.

Jenson, J. R., *Introductory Digital Image Processing – A Remote Sensing Perspective*, Prentice-Hall, 1986.

Jin, N. and Liu, D., "Wavelet basis function neural networks for sequential learning," *IEEE Transactions on Neural Networks*, vol. 19, no. 3, pp. 523–528, Mar. 2008.

Krishnapuram, R. and Chen, L. F., "Implementation of parallel thinning algorithms using recurrent neural networks," *IEEE Transactions on Neural Networks*, vol. 4, no. 1, pp. 142–147, Jan. 1993.

Lee, H.-E., Park, K.-H., and Bien, Z. Z., "Iterative fuzzy clustering algorithm with supervision to construct probabilistic fuzzy rule base from numerical data," *IEEE Transactions on Fuzzy Systems*, vol. 16, no. 1, pp. 263–277, Feb. 2008.

Li, J., Allinson, N., Tao, D., and Li, X., "Multitraining support vector machine for image retrieval," *IEEE Transactions on Image Processing*, vol. 15, no. 11, pp. 3597–3601, Nov. 2006.

Lian, J., Naik, K., and Agnew, G. B., "A framework for evaluating the performance of cluster algorithms for hierarchical networks," *IEEE/ACM Transactions on Networking*, vol. 15, no. 6, 1478–1489, Dec. 2007.

Lin, K.-M. and Lin, C.-J., "A study on reduced support vector machines," *IEEE Transactions on Neural Networks*, vol. 14, no. 6, pp. 1449–1459, Nov. 2003.

Lippmann, R. P., "An introduction to computing with neural nets," *IEEE ASSP Magazine*, vol. 4, no. 2, pp. 4–22, Apr. 1987.

Liu, X., Wang, W., and Chai, T., "The fuzzy clustering analysis based on AFS theory," *IEEE Transactions on Systems, Man, and Cybernetics, Part B*, vol. 35, no. 5, pp. 1013–1027, Oct. 2005.

Liu, Y. and Zheng, Y. F., "Soft SVM and its application in video-object extraction," *IEEE Transactions on Signal Processing*, vol. 55, no. 7, pp. 3272–3282, Jul. 2007.

Mansoori, E. G., Zolghadri, M. J., and Katebi, S. D., "SGERD: A steady-state genetic algorithm for extracting fuzzy classification rules from data," *IEEE Transactions on Fuzzy Systems*, vol. 16, no. 4, pp. 1061–1071, Aug. 2008.

Minsky, M. L. and Papert, S. A., *Perceptrons*, MIT Press, 1969.

Nandedkar, A. V. and Biswas, P. K., "A fuzzy min-max neural network classifier with compensatory neuron architecture," *IEEE Transactions on Neural Networks*, vol. 18, no. 1, pp. 42–54, Jan. 2007.

Navia-Vazquez, A., Gutierrez-Gonzalez, D., Parrado-Hernandez, E., and Navarro-Abellan, J. J., "Distributed support vector machines," *IEEE Transactions on Neural Networks*, vol. 17, no. 4, pp. 1091–1097, Jul. 2006.

Ng, M. K., Li, M. J., Huang, J. Z., and He, Z., "On the impact of dissimilarity measure in k-modes clustering algorithm," *IEEE Transactions on Pattern Analysis and Machine Intelligence*, vol. 29, no. 3, pp. 503–507, Mar. 2007.

Nock, R. and Nielsen, F., "On weighting clustering," *IEEE Transactions on Pattern Analysis and Machine Intelligence*, vol. 28, no. 8, pp. 1223–1235, Aug. 2006.

Nowicki, R., "On combining neuro-fuzzy architectures with the rough set theory to solve classification problems with incomplete data," *IEEE Transactions on Knowledge and Data Engineering*, vol. 20, no. 9, pp. 1239–1253, Sept. 2008.

Palmes, P. P., Hayasaka, T., and Usui, S., "Mutation-based genetic neural network," *IEEE Transactions on Neural Networks*, vol. 16, no. 3, pp. 587–600, May 2005.

Pan, S.-M. and Cheng, K.-S., "Evolution-based Tabu search approach to automatic clustering," *IEEE Transactions on Systems, Man, and Cybernetics, Part C*, vol. 37, no. 5, pp. 827–838, Sept. 2007.

Pao, Y.-H., *Adaptive Pattern Recognition and Neural Networks*, Addison-Wesley, New York, 1989.

Phung, S. L. and Bouzerdoum, A., "A pyramidal neural network for visual pattern recognition," *IEEE Transactions on Neural Networks*, vol. 18, no. 2, pp. 329–343, Mar. 2007.

Platt, J. C., Cristianini, N., and Shawe-Taylor, J., "Large margin DAGs for multiclass classification," *Advances in Neural Information Processing Systems*, MIT Press, vol. 12, pp. 547–553, 2000.

Rosenblatt, F., *Principles of Neurodynamics*, Spartan, New York, 1962.

Rosenfeld, A., "The fuzzy geometry of image subsets," *Pattern Recognition Letters*, vol. 2, no. 5, pp. 311–317, Sept. 1984.

Rumelhart, D. E., Hinton, G. E., and Williams, R. J., "Learning representations by back-propagating errors," *Nature*, vol. 323, pp. 533–536, Oct. 1986.

San, O., Huynh, V., and Nakamori, Y., "An alternative extension of the k-means algorithm for clustering categorical data," *International Journal of Applied Mathematics and Computer Science*, vol. 14, no. 2, pp. 241–247, 2004.

Savran, A., "Multifeedback-layer neural network," *IEEE Transactions on Neural Networks*, vol. 18, no. 2, pp. 373–384, Mar. 2007.

Shih, F. Y., Moh, J., and Chang, F.-C., "A new ART-based neural architecture for pattern classification and image enhancement without prior knowledge," *Pattern Recognition*, vol. 25, no. 5, pp. 533–542, May 1992.

Shih, F. Y. and Zhang, K., "Support vector machine networks for multi-class classification," *Pattern Recognition and Artificial Intelligence*, vol. 19, no. 6, pp. 775–786, Sept. 2005.

Sousa, J. M. C., Gil, J. M., and Pinto, J. R. C., "Word indexing of ancient documents using fuzzy classification," *IEEE Transactions on Fuzzy Systems*, vol. 15, no. 5, pp. 852–862, Oct. 2007.

Stork, D. G., "Self-organization, pattern recognition, and adaptive resonance networks," *Journal of Neural Network Computing*, vol. 1, pp. 26–42, Summer 1989.

Tao, Q., Chu, D., and Wang, J., "Recursive support vector machines for dimensionality reduction," *IEEE Transactions on Neural Networks*, vol. 19, no. 1, pp. 189–193, Jan. 2008.

Vapnik, V. *The Nature of Statistical Learning Theory*, Springer-Verlag, New York, 1995.

Vasilic, S. and Kezunovic, M., "Fuzzy ART neural network algorithm for classifying the power system faults," *IEEE Transactions on Power Delivery*, vol. 20, no. 2, pp. 1306–1314, Apr. 2005.

Veenman, C. J., Reinders, M. J. T., and Backer, E., "A maximum variance cluster algorithm," *IEEE Transactions on Pattern Analysis and Machine Intelligence*, vol. 24, no. 9, pp. 1273–1280, Sept. 2002.

Vidal, J. J., "Implementing neural nets with programmable logic," *IEEE Transactions on ASSP*, vol. 36, no. 7, pp. 1180–1190, Jul. 1988.

Wang, S. and Wang, H., "Password authentication using Hopfield neural networks," *IEEE Transactions on Systems, Man, and Cybernetics, Part C*, vol. 38, no. 2, pp. 265–268, Mar. 2008.

Wang, Z. and Peterson, B. S., "Constrained least absolute deviation neural networks," *IEEE Transactions on Neural Networks*, vol. 19, no. 2, pp. 273–283, Feb. 2008.

Xia, Y. and Wang, J., "A one-layer recurrent neural network for support vector machine learning," *IEEE Transactions on Systems, Man, and Cybernetics, Part B*, vol. 34, no. 2, pp. 1261–1269, Apr. 2004.

Xu, R. and Wunsch, D., "Survey of clustering algorithms," *IEEE Transactions on Neural Networks*, vol. 16, no. 3, pp. 645–678, May 2005.

Yang, M.-S., Wu, K.-L., Hsieh, J.-N., and Yu, J., "Alpha-cut implemented fuzzy clustering algorithms and switching regressions," *IEEE Transactions on Systems, Man, and Cybernetics, Part B*, vol. 38, no. 3, pp. 588–603, Jun. 2008.

Yang, Z. R., "A novel radial basis function neural network for discriminant analysis," *IEEE Transactions on Neural Networks*, vol. 17, no. 3, pp. 604–612, May 2006.

Yu, J., "General C-means clustering model," *IEEE Transactions on Pattern Analysis and Machine Intelligence*, vol. 27, no. 8, pp. 1197–1211, Aug. 2005.

8 Deep Learning

This chapter introduces the architecture of *convolutional neural networks* (CNNs), the hyper-parameters, learning, and activation functions involved in creating these types of CNNs. We also discuss some issues that come with training CNNs, including vanishing, exploding gradients, and overfitting. For image classification, two CNN models composed of the LeNet-5 architecture and an improved version are introduced.

8.1 INTRODUCTION

Living in an era where big data becomes a standard in all areas of the industry comes with unprecedented challenges regarding interpretation and investigation. We have seen surprising improvement in computer vision applications utilizing *deep learning* (DL), such as image retrieval, image classification, and object detection and tracking. DL aims to learn high-level abstraction through its architecture (Qian et al., 2018). The architecture of DL algorithms attempts to mimic human brain, composed of huge neurons and connections that process information in multiple stages. This biological ideology directly influences the development of deep neural networks (DNN) (Liew et al., 2016).

Since DL is a subset of machine learning, there is a mathematical approach to obtain the structure of a human brain (neurons, connections, and layers). There are primarily three vital reasons for the booming of DL nowadays: the drastic expansion of chip capacities such as graphical processing units (GPU), the low cost of computer hardware, and the impressive progress within machine learning calculations and algorithms (Guo et al., 2016) Also, there are three features to characterize a DNN: the architecture of the model (the number of layers, number of neurons, type of layers), the learning function for searching for optimal weights, and the activation function that determines whether a neuron should be part of the learning process or not.

The properties of a neural network's approximation and generalization depend heavily on its architecture. It is important to note that very wide/shallow networks are very good at memorization but not generalizing. Shallow/wide networks are not practical because they will not have every possible input value in the training set. Multiple layers in a neural network are an essential piece of the NN's architecture because having various layers can learn features at different levels of abstraction, which makes the model better at generalization due to knowing intermediate features. This ideology suggests that deeper models are more powerful in capturing more information.

One of the basic units of an artificial neural network (ANN) is the perceptron. The singular perceptron structure comprises an input, bias, weight, and an activation function that is summarized mathematically. We can denote that x as the input, w is a

DOI: 10.1201/ 9781003474395-10

vector-valued, ϕ is an activation function that performs a non-linear transformation of z, and b is a scalar value. It is expressed as here (Schmidhuber, 2015).

$$y = \phi(z) = \phi\left(w^t x + b\right) \tag{8.1}$$

Figure 8.1 shows a mathematical representation of a neuron. These basic units then build a layer connected to all the neurons in the following layer. From neuron to neuron, it generates links which create parameters. Figure 8.2 shows the artificial neural network basic architecture, where there are three layers: input layer, hidden layer, and output layer. The hidden layer has more than one layer (multi-layer perceptron). We can think of the hidden layers as a distillation layer that extracts the most relevant patterns from the inputs and propagates the information to the next layer for further feature abstraction and redundant information removal.

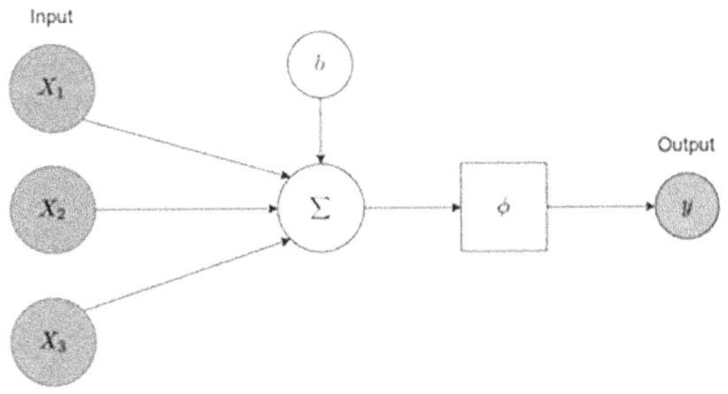

FIGURE 8.1 A mathematical representation of a neuron.

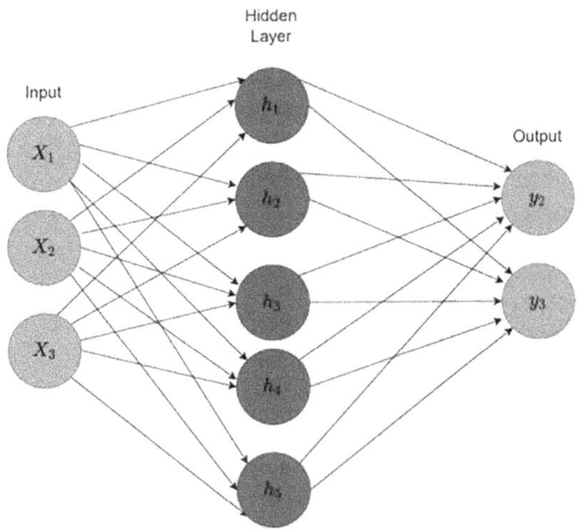

FIGURE 8.2 An artificial neural network's basic architecture.

The DNN passes the input from one neuron to the next via an activation function, and the process continues until the last layer (output layer). Through the repetition of this process, the activation function can transform a linear relationship into a non-linear relationship. Activation functions are vital in capturing relationship and features between the inputs. These functions allow the model to find optimal weights that minimize errors. Activation functions can reduce errors through a back-propagation algorithm. A gradient descent approach is used by making minor adjustments to the weights.

The feed-forward algorithm consists of connections with values that cycle forward. Their values are traveled from the input layers to hidden layers and then to the output layers. The back-propagation algorithm adjusts the neural network's weights by using the following inputs: an output vector, a target vector, and an adjusted weight vector. After obtaining the feed-forward values and errors, back-propagation is done and propagates it back to the model to make weight adjustments. Both algorithms are needed to train the neural network. During the training phase of DNNs with gradient-based back-propagation and learning, traversing the network from the final layer to the first layer yields partial derivatives. Then following chain rules, the products of the layers are computed. Note that the problem of the vanishing or exploding gradient arises when the derivatives in the layers tend to become either very small or very large. With vanishing gradients, the model becomes incapable of abstracting features due to the weights and biases of the layers not being updated effectively, which will stop the training. With exploding gradients, the model becomes unstable and unable to learn effectively (Feng et al., 2022).

DNNs are prone to overfitting when the dataset is small in volume due to the number of hyperparameters. Overfitting will lead the model to memorize the training data and perform poorly on the test set due to the lack of generalization to the unseen data. To overcome this, we increase the flexibility of DNNs to avoid overfitting through weight regularization, dilution, weight initialization, batch normalization, and learning scheduling (Khan et al., 2019).

Standard ML methods such as random forest, support vector machines (SVM), and K-nearest neighbor (KNN), while good at learning, cannot perform predictions with increasing complexity and dimensionality. On the other hand, a convolutional neural network is unique because it can extract features from images with varying complexities (Emmert-Streib et al., 2020).

8.1.1 Convolutional Layer

The convolutional layers abstract the input using a local connectivity feature extractor that calculates the hidden neurons and input. The kernel in this layer is like a window that extracts the essential features from the input. The kernel goes across the input matrix, performing convolution operations on the pixel values. The results form an activation map. That map is the output of the convolutional layer. Features are extracted and placed into the activation map and share their weights with all nearby neurons. The following parameters: kernel-size, stride, and zero-padding are hyperparameters of the convolutional layer.

8.1.2 Kernel-Size

We can consider the kernel as a window of M × N. The kernel will perform a convolutional operation on the pixels within the window, and the results from that window will be part of the activation mapping.

8.1.3 Stride

Stride is the number of pixels the kernel will move to the next position within the matrix. When the stride is small, more calculations will be performed, i.e., elemental-wise. On the other hand, fewer calculations will be performed when the stride is large because elements within the matrix are skipped.

8.1.4 Zero-Padding

This is used to preserve the dimensions of the input matrix. The zeros are padded around the border of the matrix.

8.1.5 Pool Layer

This layer comes after the convolutional layer has generated the calculations. These calculation values come together to reduce the input dimensions through the following methods: average, max, and min. At the same time, the layer retains the information and features of the input. This reduction in dimensions and feature keeping aids in improving the generalization of the model.

8.1.6 Dense Layer

The fully connected layer is a dense layer that takes the input features and applies a linear transformation through a weight's matrix. Then using an activation function, a non-linear transformation is applied to adjust the weights accordingly to determine whether the neuron should be on or off if the weight meets a threshold or not. Several hyperparameters can impact these layers, from the number of neurons to the type of activation and initial weights.

8.2 CONVOLUTIONAL NEURAL NETWORK (CNN)

The convolutional neural network (CNN) is a deep-learning algorithm that was introduced in the 1990s, but has found resurgence in recent years as technology has advanced to better utilize CNN. However, in order to understand what a convolutional neural network is, a proper understanding of deep learning and neural network algorithms is first required.

Deep learning is a field of machine learning that has grown alongside advancements in computer processing power. Machine learning has been a field of study that has dated back to the later end of the 20th century, but at that time computers did not have the computational power nor did users have the resources in sample data to make full use of some of these theoretical algorithms (Brownlee, 2019c). The next few decades have brought drastic advancements in both of these fields; until now, deep-learning algorithms are necessary to take proper advantage of the massive amount of data available.

The "deep" in deep learning is an ambiguous word, but it is meant to refer to the hierarchical learning structure that is present in algorithms, such as the convolutional neural network. These models process data through many layers until reaching a final output, and the "deep" in this case refers to the amount of layers present. All deep-learning algorithms are built off artificial neural networks, and the convolutional neural network is no exception.

Among the many different models that are prepared and trained for different purposes, there are a selection of CNN models that are trained against a large set of general images as training data. Among these models is the Google-produced GoogLeNet. Although there is slight differentiation between each of these models, the main features of any CNN are commonly shared and are defined in the following.

The convolutional neural network differentiates itself from other algorithms in its ability to preserve the multidimensionality of the input. For this reason, they have seen extensive uses in feature detection within images. It should be noted, however, that they are also used in accessing other types of data, such as documents and speech signals (Sainath et al., 2013). These networks are also classed as a hierarchical neural network, which refers to the levels of layers that are used to process data.

8.2.1 Convolutional Layers

Convolutional layers are the neurons of the network. In practice, they are filters that are applied to the image to determine where the feature exists in the image. The filter is applied across the entire input image, and the resulting output is called the *feature map*.

In a convolutional neural network, a model is trained to develop its own set of filters that are then applied to images to extract features. The resulting feature maps are then passed back into another convolutional layer so that more complex features can be identified using lower-level features. In general, the number of filters used increase as the network moves deeper into the hierarchy.

To illustrate how a filter is applied to an image to obtain a feature map, an example modified from Jason Brownlee (2019) will be used. A horizontal-line image of 8 × 8 and a 3 × 3 filter to detect horizontal lines are represented as follows:

0	0	0	0	0	0	0	0	0	0
0	0	0	0	0	0	0	0	0	0
0	0	0	0	0	0	0	0	0	0
1	1	1	1	1	1	1	1	1	1
1	1	1	1	1	1	1	1	1	1
0	0	0	0	0	0	0	0	0	0
0	0	0	0	0	0	0	0	0	0
0	0	0	0	0	0	0	0	0	0

0	0	0
1	1	1
0	0	0

In this instance, the filter would be placed along each location in the input to create a 6 × 6 feature map of horizontal lines as

0	0	0	0	0	0
0	0	0	0	0	0
3	3	3	3	3	3
3	3	3	3	3	3
0	0	0	0	0	0
0	0	0	0	0	0

This output is known as a feature graph, as it is a representation of where the horizontal lines are located within the input image. These feature graphs will be used by the pooling layers to reduce redundancy in the data.

8.2.2 Pooling Layers

Pooling layers are a tool used to reduce the dimensionality of the input feature map. This process is essential to both reduce the computational effort of the algorithm and to prevent overfitting (Brownlee, 2019). A pooling layer would take the average of a block within the feature map and return a new map of half the size. Continuing our example, our pooling layer would output this feature map as

0	0	0
3	3	3
0	0	0

This feature map would then be passed back into the convolutional layer so that a higher-level feature can be extracted from the lower-level map. It is important to note that since this algorithm is trained to develop its own filters, the features being detected may not be as simple as a horizontal or vertical line.

8.2.3 Fully Connected Layers

After many repetitions of convolutional layers and pooling layers are applied, the model will attempt to decide on what is displayed in the image. This last step is referred to as the fully connected layer. Although convolutional neural networks have originally been used for feature detection, they have since been adapted to other fields. As an example, Shi et al. (2016) used a convolutional network to reconstruct videos to higher resolutions at a faster rate than previous methods (Ng et al., 2015). In this case, this last layer is translated to make an adaptation to the image. Using a convolutional neural network to interpret videos is very possible, but it is a new challenge as it adds a new dimension to the input data.

8.3 THE LENET-5 NETWORK

LeCun et al. (1998) developed the LeNet-5 network for traffic sign recognition, facial expression recognition, Alzheimer's diseases recognition, and other fields. Figure 8.3 shows the LeNet-5 architecture, and Figure 8.4 shows its graphical representation.

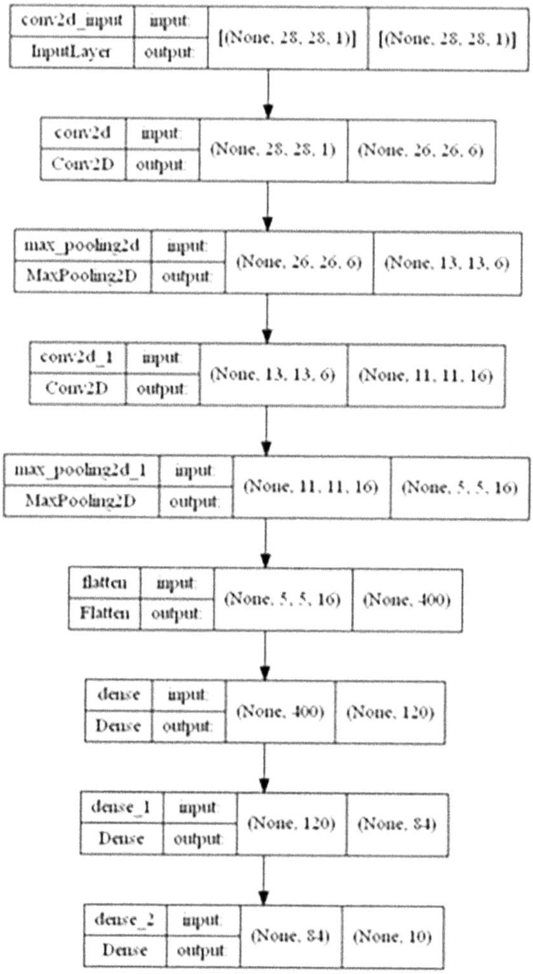

FIGURE 8.3 The architecture of the LeNet–5.

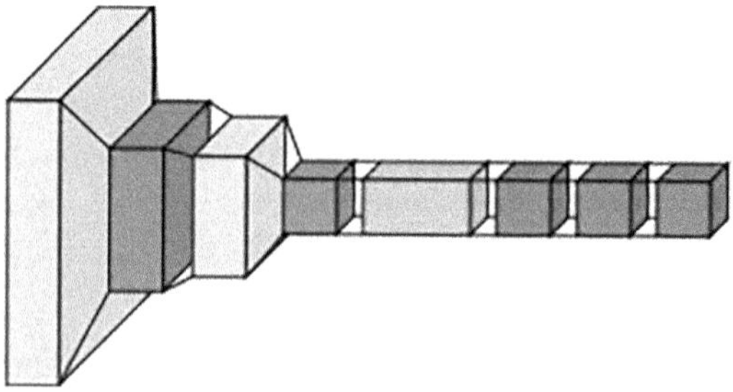

FIGURE 8.4 Graphical representation of LeNet-5 architecture.

TABLE 8.1
Hyperparameters of Layers in the LeNet-5 Architecture

Layer	Hyperparameters
InputLayer	Input_shape: 28 × 28 × 1
Conv2d	Filters:6 Kernel size: (3,3) Activation: tanh
max_pooling2d	None
conv2d_1	Filters:16 Kernel size: (3,3) Activation: tanh
max_pooling2d	None
Flatten	None
dense	Units:120 Activation: tanh
dense_1	Units:84 Activation: tanh
dense_2	Units:10 Activation: softmax

Table 8.1 lists the hyperparameters of layers of the LeNet-5 architecture. The first layer is not considered as a layer in that anything is learned. Although there are two convolutional and three dense layers, it is an assumption that after each convolutional layer, there will be some pooling that will occur.

8.4 THE IMPROVED METHODS

We improve the architecture by weight normalization, weight initialization, batch normalization, dilution, learning rate scheduling, and average pooling.

8.4.1 Weight Initialization and Normalization

The weight matrix values are random numbers from a normal uniform distribution. One of the issues arises is the vanishing and exploding gradients. We set the normal distribution variance to k/n, where k is a value dependent on the activation function, and n is the number of input nodes or output or both input and output nodes of the tensor weight. The initializer, LeCun_normal function, initializes the weights by drawing samples from a truncated normal distribution centered at 0 with the following equation $\sqrt{\frac{1}{fan_{in}}}$, where fan_{in} is the number of units in the weight tensor. This initialization of weights will be used in the SELU activation function (Li et al., 2020).

$$selu(x) = \lambda \begin{pmatrix} x & if\ x > 0 \\ \alpha e^{x} - \alpha & if\ x \leq 0 \end{pmatrix} \tag{8.2}$$

where α and λ are fixed parameters derived from inputs, and x is the input value greater than 0 or less than 0. The *selu* stands for scaled exponential linear unit. The breakthrough with this activation function is its self-normalization property, meaning that the neurons (weights, biases, and activations) will have a standard deviation of one and a mean of zero after normalization. This process avoids the exploding and vanishing gradient within DNN (Klambauer et al., 2017).

8.4.2 Batch Normalization

Batch normalization normalizes the layer's inputs by rescaling and recentering, making the neural network faster and more stable throughout the training process. In addition, this technique parametrizes the network by coordinating updates of the activation weights across several layers (Wu et al., 2019).

8.4.3 Dilution

Dilution, a dropout, is a layer that can prevent overfitting the model to the training data. The dropout layer probability is 0.0–1.0 to drop neurons in this case randomly. The probability is a hyperparameter. The dropout layer can help improve the generalization of neural networks by preventing complex co-adaptations to the training data (Hinton et al., 2012).

8.4.4 Regularization Learning Rate Scheduling

The learning rate is a piecewise approach that would be ideal where the learning rate for the optimizer can be set explicitly at every minibatch of five epochs. The adjustment is a start high then decays progressively.

8.4.5 Average Pooling

Instead of using the max-pooling layer, average pooling can retain the features of the image and other abstractions. Figure 8.5 show the architecture of the improved LeNet-5, and Figure 8. shows the graphical representation of the improved LeNet-5 architecture. Table 8.2 lists the hyperparameters of layers of the improved LeNet–5.

8.5 EXPERIMENTAL RESULTS

The MNIST dataset consists of handwritten digital images 28 × 28 pixels, composed of 10 classes (0–9), with 50,000 samples being used for training and 10,000 samples for testing. Figure 8.7 shows a sample of the MNIST dataset of handwritten digits. Figure 8.8 shows the overview comparison of loss, accuracy, validation loss,

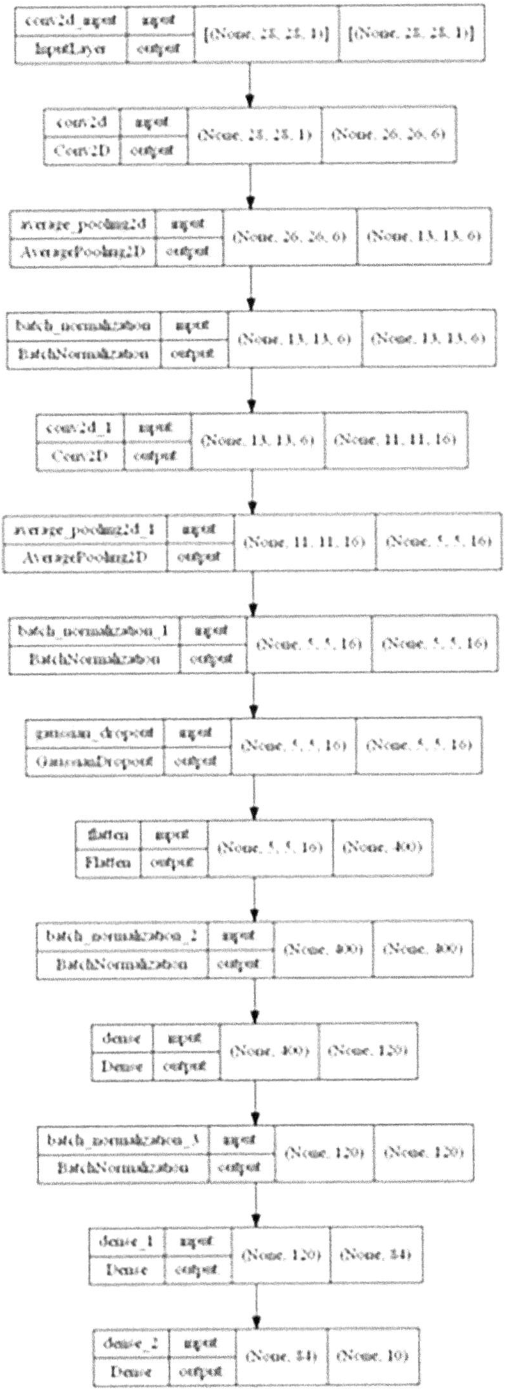

FIGURE 8.5 The architecture of the improved LeNet–5.

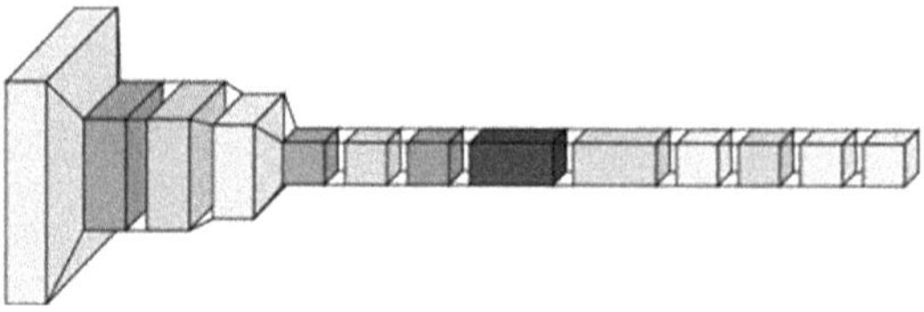

FIGURE 8.6 Graphical representation of the improved LeNet-5 architecture.

TABLE 8.2
Hyperparameters of Layers (Improved LeNet-5)

Layer	Hyperparameters
InputLayer	Input_shape: 28×28×1
Conv2d	Filters:6 Kernel size: (3,3) Kernel Initializer: LeCun normal Activation: selu
average_pooling2d	None
batch_normalization	None
conv2d_1	Filters:16 Kernel size: (3,3) Kernel Initializer: LeCun normal Activation: selu
average_pooling2d_1	None
batch_normalization	None
gaussian_dropout	0.25
Flatten	None
batch_normalization	None
dense	Units:120 Activation: selu Kernel Initializer: LeCun normal
dense_1	Units:84 Activation: selu Kernel Initializer: LeCun normal
dense_2	Units:10 Activation: softmax

validation accuracy, and learning rate. Figure 8.9 shows the cross-entropy loss at each epoch.

The LeNet-5 is experiencing divergence, meaning that it is lacking generalization. The margins that exist between both Figures 8.10 and 8.11 suggest that the LeNet-5 is overfitting slightly.

With the possibility of exploding and vanishing gradients and overfitting of data, the results suggest that the improved version of the LeNet-5 is better at generalization

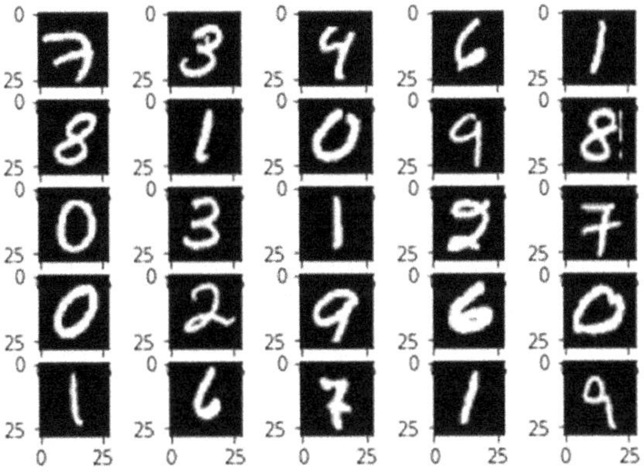

FIGURE 8.7 A sample of the MNIST dataset of handwritten digits.

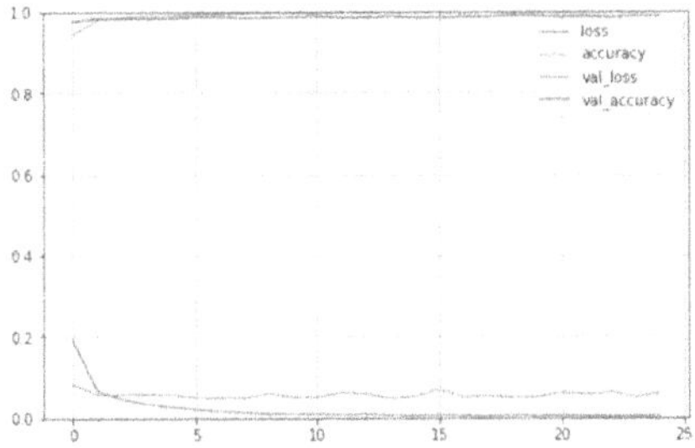

LeNet-5: Overview

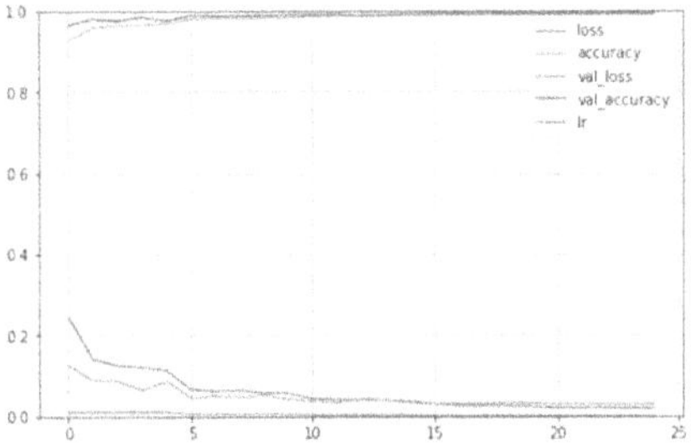

LeNet-5 Improved: Overview

FIGURE 8.8 Overview comparison.

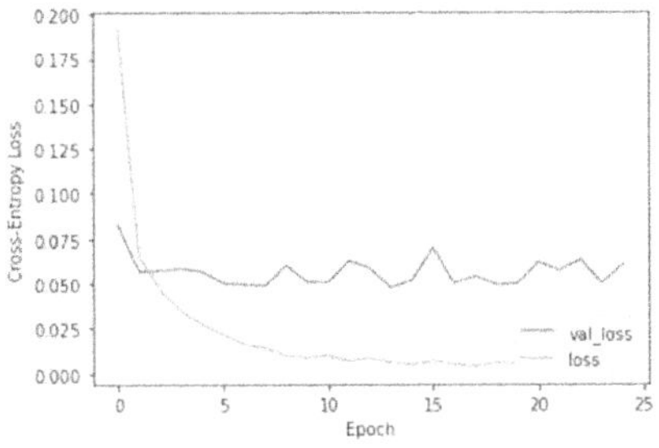

LeNet-5: Cross-Entropy loss vs Epoch

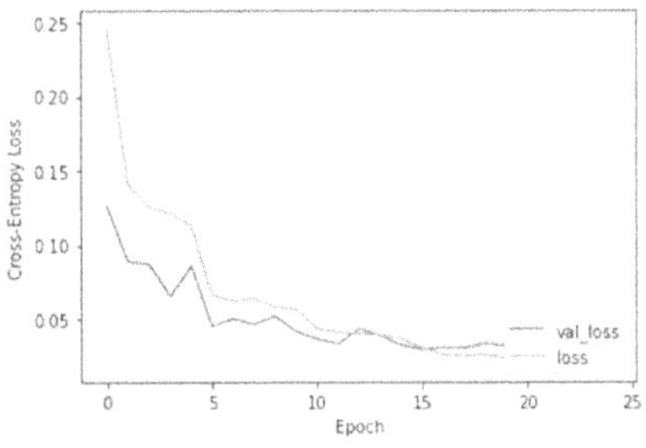

LeNet-5 Improved: Cross-Entropy loss vs Epoch

FIGURE 8.9 Cross-entropy loss at each epoch.

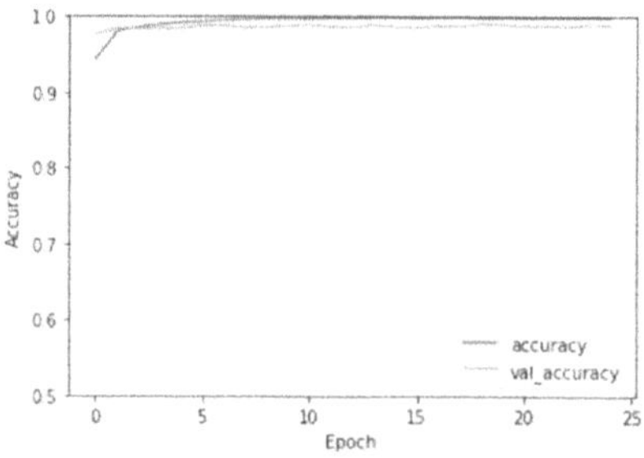

LeNet-5: Accuracy vs Epoch:

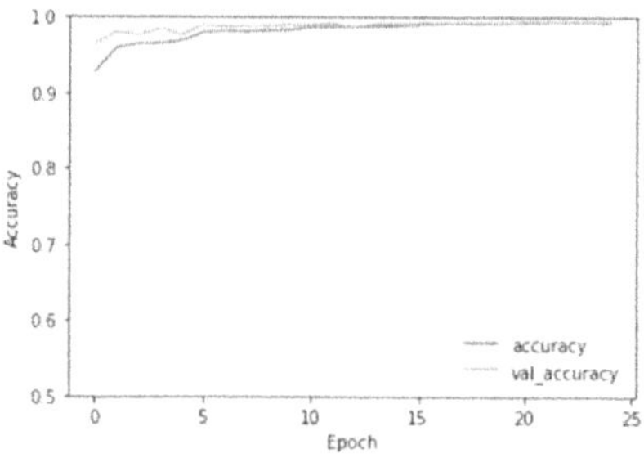

LeNet-5 Improved: Accuracy vs Epoch

FIGURE 8.10 Accuracy at each epoch.

Classification	Precision	Recall	F1-Score	Support	Jaccard Index
0	0.98	1	0.99	980	0.98
1	0.99	1	0.99	1135	0.99
2	0.99	0.99	0.99	1032	0.98
3	0.99	0.99	0.99	1010	0.98
4	0.99	0.98	0.99	982	0.98
5	0.99	0.99	0.99	892	0.97
6	0.99	0.98	0.99	958	0.98
7	0.98	0.99	0.98	1028	0.97
8	0.99	0.98	0.99	974	0.97
9	0.98	0.98	0.98	1009	0.98
Hamming Loss	**Balanced Accuracy Score**	**Matthews Correlation Coefficient**	**Accuracy**	**Mean Squared Error**	
0.013	0.987	0.986	0.988	0.065	

LeNet-5: Classification Metrics

Classification	Precision	Recall	F1-Score	Support	Jaccard Index
0	0.99	1	0.99	980	0.99
1	0.99	1	1	1135	0.99
2	1	1	1	1032	0.99
3	0.99	1	0.99	1010	0.99
4	0.99	0.99	0.99	982	0.99
5	0.99	0.99	0.99	892	0.98
6	0.99	0.99	0.99	958	0.98
7	0.99	0.99	0.99	1028	0.98
8	1	0.99	0.99	974	0.98
9	0.99	0.99	0.99	1009	0.99
Hamming Loss	**Balanced Accuracy Score**	**Matthews Correlation Coefficient**	**Accuracy**	**Mean Squared Error**	
0.007	0.993	0.993	0.993	0.022	

LeNet-5 Improved: Classification Metrics

FIGURE 8.11 Classification metrics.

and learning whether the handwritten digit is a number between 0 through 9. The LeNet-5 is still an incredible architecture to classify images; however, the hyperparameters can make the model exceptional or break.

8.6 IMAGE CLASSIFICATION WITH VISION TRANSFORMERS

Image classification is one of the age-old tasks in image processing and has been tackled by a lot of different models throughout the years. In the last decade, convolutional neural networks (CNNs) have mostly dominated this field due to their ability to quickly recognize pattern without having to scale up with the number of pixels in the image. However, attention-based transformers have taken the machine learning world by storm by becoming a valuable component in model structuring for natural language processing (NLP) due to their capacity to capture global information without depending on structural coherence expected in many traditional deep-learning structures like RNN, CNN, etc. Transformers have received success in computer vision and are performing on-par with state-of-the-art CNN models like ResNet and VGG, to name a few. In this section several popular datasets have been used to showcase the performance of these vision transformer models as opposed to transfer-learning based convolutional models. Experimental results show their dominance over convolutional-based models.

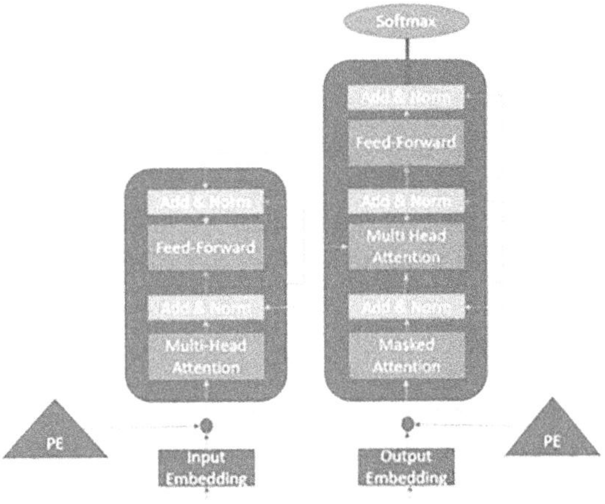

FIGURE 8.12 Original architecture of the transformer encoder-decoder used for translation.

8.6.1 Introduction

The transformer architecture was initially introduced in the paper titled "Attention Is All You Need" (Vaswani et al., 2017) in the hope of replacing recurrent neural networks and others (e.g., LSTM, GRU, etc.) in sequential data processing. Attention mechanisms (Bahdanau et al., 2014) were already prevalent in sequential modeling before transformers came to the scene. However, the vast majority of those models relied heavily on recurrent functions, which are computationally expensive and highly complex to parallelize. The transformer model becomes the first model to use solely attention mechanisms, specifically self-attention mechanisms. Figure 8.12 shows the original architecture of the transformer encoder-decoder used for translation.

Since the introduction of the transformer architecture in 2017, it has garnered huge popularity in language modelling tasks by the virtue of transfer learning methods. NLP has been greatly benefitted by such techniques and the benchmarks have been beaten several times over (Devlin et al., 2019). However, image recognition and classification are not considered strictly sequential problem. Convolutional networks have treated this problem as a kernel-based pattern recognition problem, where kernels try to figure out patterns hidden in numerous ways in a raster manner. Vision transformer treats this kernel-based problem as a sequential problem to match the input structure of vanilla transformers to gain the advantage of self-attention mechanism to process images (Dosovitskiy et al., 2021).

Vision transformer when trained with inadequate- or small-sized dataset clearly cannot outperform traditional convolutional networks. It is due to the inductive bias it lacks unlike convolutional networks, but the situation clearly changes when they take advantage of transfer learning methods. Transfer learning method usually is the pipeline of a pre-trained model built on a generic task that is unsupervised/semi-supervised (without any annotated dataset) with a flexible top layer that can be later replaced with task specific structure and fine-tune on the task-specific dataset. The first vision transformer (ViT) was developed by the researchers at Google research and was pre-trained on 14 million to 300 million images.

We take two popular datasets – the first is Ryerson Vision Lab Complex Document Information Processing (Harley et al., 2015) and the second is Bean Disease Dataset (Makerere AI Lab, 2022) – to fine tune on the vision transformer in order to test out its integrity and well-roundedness in comparison to traditional CNN. We will lay out the background behind transformer modules and different structures, and then define the vision transformer model.

8.6.2 Existing Models

8.6.2.1 Transfer-Learning Independent Methods

Convolutional neural networks are kernel-based machine-learning models that are popular for their parameter sharing techniques that help them scale up to very large models (Shih and Patel, 2019; Fang et al., 2017). In comparison to dense neural networks, CNNs are well-designed for images, where every pixel does not need to get passed through the input and carried down to the last layer.

Several methods help CNNs work the way they do. First, a kernel or a filter goes over the 2D or 3D image to extract patterns from the input image. This produces a feature map of the image that later gets scaled down by pooling methods. Pooling also utilizes a filter that represents an area of the image with fewer numbers. The convolution operation is the parameter sharing part of CNN whereas the pooling helps the models not to scale up. It takes multiple of these layers and multiple of kernels and filter sizes, representing billions of combinations of models.

8.6.2.2 Transfer Learning Dependent Methods

Transfer learning methods are widely popular in recent time due to the huge availability of data around the Internet. Transfer learning provide an opportunity to harness the knowledge gathered by unsupervised/semi-supervised models and transfer them onto task specific models by only training a little bit with task-specific datasets (Shih and Patel, 2019). The knowledge transfer is effective as to the point that transfer learning methods have become increasingly ubiquitous in all kinds of models, ranging from NLP to vision.

In computer vision, transfer learning is made popular with ResNet architectures and their variants. Keras has some pre-trained models built inside, and for our research we use some of these models to compare with our proposed methods with transformers. Table 8.3 lists an outlook of the Keras models (Team, 2016).

TABLE 8.3
An Outlook of the Keras Models (Team, 2016)

Pre-trained Model	Memory (MB)	Weights	Depth
ResNet50V2	98	25.6M	103
InceptionResNetV2	215	55.9M	449
DenseNet169	57	14.3M	338

8.6.3 The Proposed Method

8.6.3.1 Dataset

To compare different models, two widely used datasets have been selected. Both datasets are specialized for multi-class classification of images.

8.6.3.1.1 Bean Disease Dataset

This dataset contains leaf photos captured in the field by the Makerere AI Lab in partnership with the National Crops Resources Research Institute (NaCRRI), Uganda's National Agency in charge of agricultural research (Makerere AI Lab, 2022). There are three classes in this dataset:

- Healthy class
- Angular leaf spot (diseased leaf)
- Bean rust (diseased leaf)

Figure 8.13 shows a few samples of the images in the bean leaf disease dataset.

Table 8.4 gives an idea about the data distribution in the dataset.

8.6.3.1.2 Ryerson Vision Lab Complex Document Information Processing Dataset

This dataset contains extremely high caliber machine scanned images of documents of different types. These are the different document types present in the dataset (Harley et al., 2015).

1. letter
2. form
3. email

FIGURE 8.13 Data samples in the bean leaf disease dataset.

TABLE 8.4
Bean Disease Dataset Distribution

Category	Healthy	Angular Leaf Spot	Bean Rust	Total	Percentage
Training Examples	342	345	348	1035	79.89%
Testing Examples	42	43	43	128	9.89%
Validation Examples	44	44	45	131	10.01%
Total	428	432	436	1294	100%

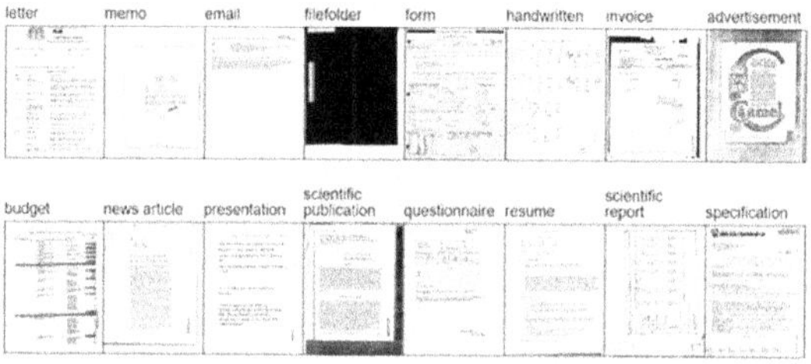

FIGURE 8.14 Samples of different classes in the RVL-CDIP dataset.

4. handwritten
5. advertisement
6. scientific report
7. scientific publication
8. specification
9. file folder
10. news article
11. budget
12. invoice
13. presentation
14. questionnaire
15. resume
16. memo

Figure 8.14 shows samples of different classes in the RVL-CDIP dataset.

There are 400K images in total, whereas 320K is for training, and the rest of 80K is divided in 40K-40K for the evaluation and validation sets, respectively.

8.6.3.2 The Proposed Method

To fully understand the mechanisms behind vision transformer, one first needs to understand the depth of the transformer model and the attention mechanism. In the following sections these components are described.

8.6.3.2.1 Self-Attention

In terms of natural language processing, the self-attention mechanism can be thought of as a way of adding more context to the word embedding of each input token.

Intuitively, the vector representation of a word *W* should change depending on the context (*C*: *c*1, *c*2, *c*3, . . . , *cn*) in a sentence, and the context is simply the companion words of *w* in that particular sentence (Vaswani, 2017). To achieve this, one could simply perform a dot product of each of the context words with the word *w*. In this way, the new representation of those context words considers the presence of the word *w*. We execute this with every word in that particular sentence taking turns and gathering some contextual information in the word vectors. Each time the word *W*, which is injecting some context into its surrounding words, can be thought of as a query, which is querying the keys – the context words and obtaining a new representation or a value with more context. Figure 8.15 shows this mechanism for word tokens. Figure 8.16 shows the self-attention mechanism for vision.

However, we also need some weights along with these word vectors that can be learned. We use some square matrices multiplied with each of these word vectors that preserve the shape of the word vectors. These matrices are essentially the parameters to be learned in the attention mechanism. Since each token in a sentence takes turns and attends to each other, this mechanism is called self-attention.

8.6.3.2.2 Scaled Dot-Product Attention

This particular attention mechanism, where we have a dot-product each time, is called a scaled dot-product attention. It has the term "scaled" in it because after performing the dot-product, some scaling is performed to keep the dot-product of larger valued vectors in check. The queries, keys, and values are also packed together in matrices *Q*, *K*, and *V* simultaneously. The following figure showcases the workflow of this scaled attention. Figure 8.17 shows the scaled dot-product self attention using database analogy.

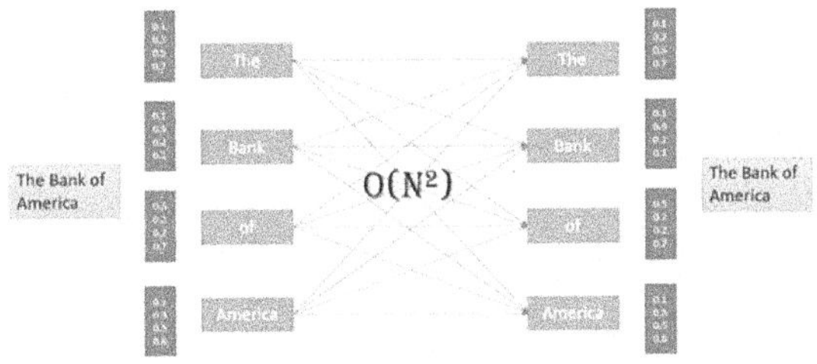

FIGURE 8.15 Self-attention mechanism original structure.

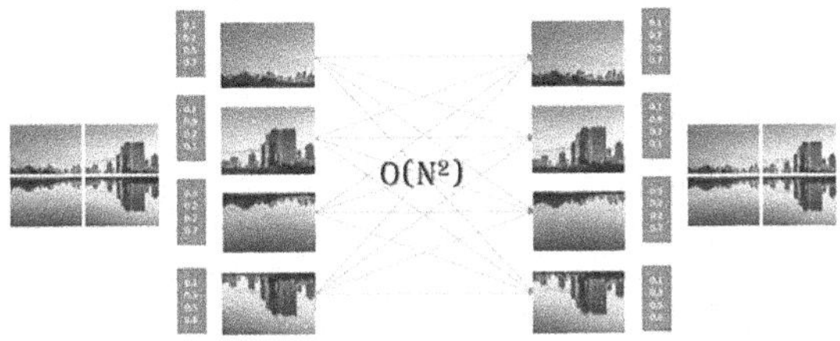

FIGURE 8.16 Self-attention mechanism for vision.

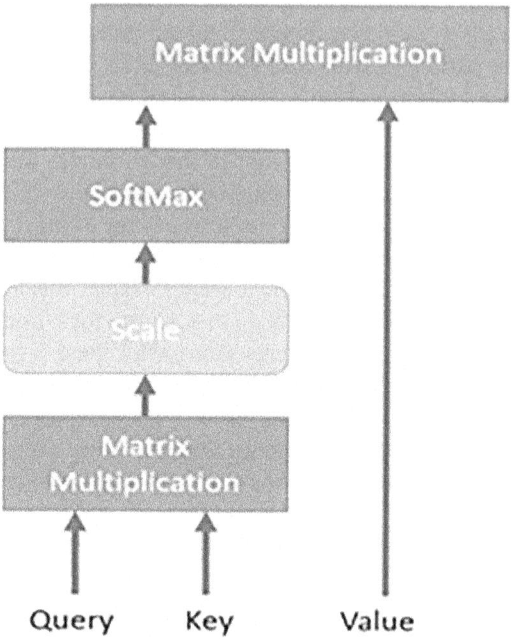

FIGURE 8.17 The scaled dot-product self-attention using database analogy.

To summarize, the queries Q query the Keys K by a matrix multiplication, get scaled by $\sqrt{dk}$, where dk is the dimension of the keys and queries, and get passed through a softmax function to obtain the weights that will add context to the keys K and result in values V, by another matrix multiplication. In case of vision, all these word tokens are replaced with image patches. Image patches are just blocks of the image from left-right top-bottom. The patches are represented with 1D vectors each and treated the same way as word tokens. Therefore, the attention mechanism for images occurs essentially the same way as for word tokens. The attention mechanism is given by

$$Attn(Q,K,V) = Softmax\left(\frac{QK^T}{\sqrt{d}}\right)V \tag{8.3}$$

8.6.3.2.2 Vision Transformer

The vision transformer simply contains the Transformer Encoder portion and changes the input structure. The image is divided into patches and later transformed into flattened representation using 1D embedding vectors. The positional embedding is added to make transformer's self-attention module aware of the position of different patches, as the self-attention module is usually unaware of the order information of its operands. The encoder contains 12 layers of the attention plus feed-forward

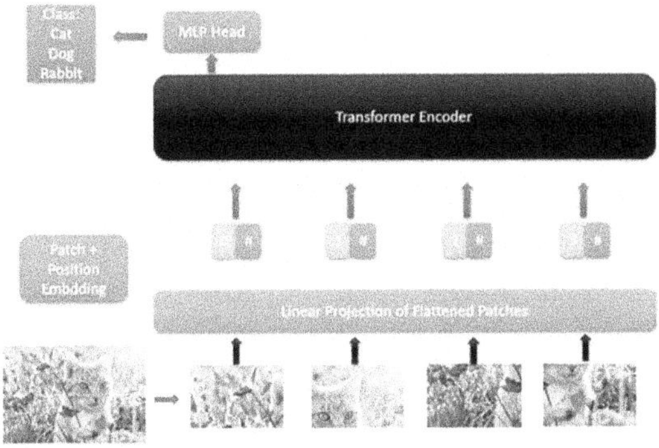

FIGURE 8.18 Vision transformer (ViT) architecture for image classification.

and normalization layers. It is passed on to a multi-layer perceptron head to do the classification. Figure 8.18 shows the workflow of this process.

8.6.4 Experimental Results

The pre-trained transformer models have a huge number of parameters, and it is impossible to train them from scratch using traditional resources. It usually requires NVIDIA V100 1024 GPUs with 1024 × 32 GB memory to train these models from scratch in more than 90 hours. We use Hugging Face Transformer's pre-trained models for vision transformers, and Keras's pre-trained ImageNet weights for convolutional networks. The experimental framework is setup in both PyTorch (HuggingFace) and Tensorflow (Keras). Some experiments are conducted in Google Colab using GPU to take advantage of Tensorflow's auto GPU capabilities. They are fine-tuned on the Beans Disease Dataset and the RVL-CDIP dataset. Following are the training hyper-parameters.

- Gradient optimizer: AdamW
- Rate of learning: 0.004
- Input size: 224×224
- Batch_size: 16 for beans and 32 for RVL-CDIP

Table 8.5 lists the accuracies and losses in both datasets using the ViT for training and evaluation. The training stage takes approximately 40 minutes for the Beans Dataset on Colab pro's environment and 5 hours, 47 minutes for the document dataset on NJIT HPC server. It clearly showcases the high-performance stats for both datasets using vision transformers. To experiment the model's integrity till saturation, a test is done to see at what iteration the performance completely saturates. As it is massively time and resource-consuming, we only execute the smaller Beans Dataset.

Table 8.6 shows the performance of ViT metrics until saturation for 6,670 epochs.

TABLE 8.5
Vision Transformer Experimental Loss and Accuracy

Dataset	Beans Dataset		RVL-CDIP Dataset	
Statistics	Training	Testing	Training	Testing
Loss	0.028300	0.194889	0.03949	0.13456
Accuracy	98.26%	94.53%	97.97%	94.88%

TABLE 8.6
The ViT Metrics until Saturation for 6,670 Epochs

Metrics	Training	Testing
Loss	0.00045	0.0018
Accuracy	99.82%	98.69%

TABLE 8.7
Comparison of ViT with Convolutional Pre-trained Models for Beans Disease Dataset

Model	Loss		Accuracy	
	Training	Testing	Training	Testing
Proposed Model	0.0283	0.1949	98.26%	94.53%
ResNet_50 (keras)	0.0045	0.3434	99.69%	92.07%
VGG_19 (keras)	8.14e-04	0.2823	100.00%	92.69%
VGG_16 (keras)	7.24e-04	0.5434	100.00%	92.13%
Inception_V3 (keras)	0.8848	1.3040	87.46%	79.45%

8.6.5 Comparison of Proposed ViT with Existing ResNet Variants

Table 8.7 showcases the comparison of existing methods with the ViT model. The experimental paradigm is almost similar. All these methods are fine-tuning methods, and hence the pre-training is completely dependent on their respective libraries. Table 8.8 shows the comparison of ViT with convolutional pre-trained models for the RVL-CDIP dataset

Figure 8.19 shows the visual representation of the performance metrics using bar charts.

From our experiments, it is evident that ViT performs better than the state-of-the-art transfer learning models in computer vision. In summary, Transformers have revolutionized the field of natural language processing, replacing convolutional neural networks in image recognition tasks specially when it comes to transfer-learning methodology. Vision transformer's capability has been vouched for by extensive research using two challenging datasets. These results indicate toward the potential for ViT and any variant it may spawn in the future.

TABLE 8.8
Comparison of ViT with Convolutional Pre-trained Models for RVL-CDIP Dataset

Model	Loss		Accuracy	
	Training	Testing	Training	Testing
Proposed Model	0.03949	0.13456	97.97%	94.88%
ResNet_50 (keras)	0.05345	0.5666	98.43%	91.34%
VGG_19 (keras)	5.14e-03	0.3432	99.01%	92.32%
VGG_16 (keras)	2.24e-03	0.4333	99.12%	93.11%
Inception_V3 (keras)	0.9234	0.9932	85.46%	78.45%

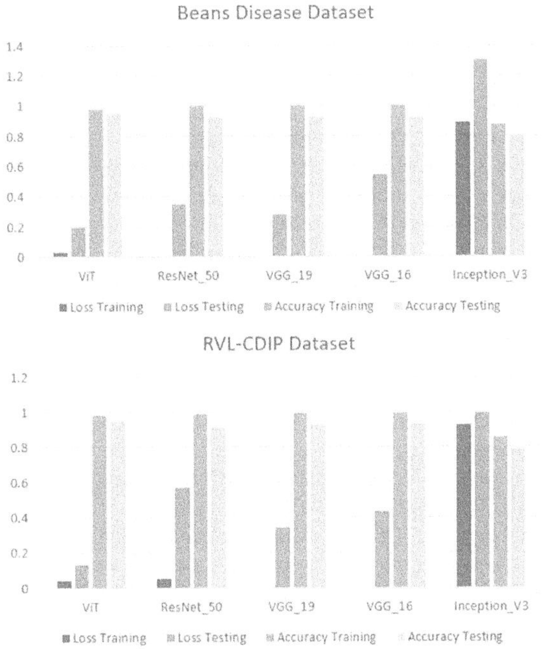

FIGURE 8.19 Visual representation of the performance metrics using bar charts.

REFERENCES

Bahdanau, D., Cho, K., and Bengio, Y., "Neural machine translation by jointly learning to align and translate," *International Conference on Learning Representations*, Sept. 2014.

Brownlee, J., "A gentle introduction to pooling layers for convolutional neural networks," *Machine Learning Mastery*, Jul. 2019a.

Brownlee, J., "Crash course in convolutional neural networks for machine learning," *Machine Learning Mastery*, Aug. 2019b.

Brownlee, J., "What is deep learning?" *Machine Learning Mastery*, Dec. 2019c.

Devlin, J., Chang, M.-W., Lee, K., and Toutanova, K., "Bert: Pre-training of deep bidirectional Transformers for language understanding," *Proceedings of the 2019 Conference of the North American Chapter of the Association for Computational Linguistics: Human Language Technologies*, vol. 1, Jun. 2019.

Dosovitskiy, A., Beyer, L., Kolesnikov, A., Weissenborn, D., Zhai, X., Unterthiner, T., Dehghani, M., Minderer, M., Heigold, G., Gelly, S., Uszkoreit, J., and Houlsby, N., "An image is worth 16x16 words: Transformers for image recognition at scale," *The Ninth International Conference on Learning Representations*, May 2021.

Emmert-Streib, F., Yang, Z., Feng, H., Tripathi, S., and Dehmer, M., "An introductory review of deep learning for prediction models with big data," *Frontier Artificial Intelligence*, vol. 3, p. 4, 2020.

Fang, L., Cunefare, D., Wang, C., Guymer, R. H., Li, S., and Farsiu, S., "Automatic segmentation of nine retinal layer boundaries in OCT images of non-exudative AMD patients using deep learning and graph search," *Biomedical Optics Express*, vol. 8, no. 5, p. 2732, 2017.

Feng, R., Zheng, K., Huang, Y., Zhao, D., Jordan, M., and Zha, Z.-J., "Rank diminishing in deep neural networks," *Proceedings of the 36th International Conference on Neural Information Processing Systems*, pp. 33054–33065, 2022.

Guo, Y., Liu, Y., Oerlemans, A., Lao, S.-Y., Wu, S., and Lew, M. S., "Deep learning for visual understanding: A review," *Neurocomputing*, vol. 187, pp. 27–48, 2016.

Harley, A. W., Ufkes, A., and Derpanis, K. G., "Evaluation of deep convolutional nets for document image classification and retrieval," *International Conference on Document Analysis and Recognition* (ICDAR), 2015.

Hinton, G. E., Srivastava, N., Krizhevsky, A., Sutskever, I., and Salakhutdinov, R. R., "*Improving Neural Networks by Preventing Co-adaptation of Feature Detectors*," arXiv: 1207.0580. https://doi.org/10.48550/arXiv.1207.0580, 2012.

Khan, S. H., Hayat, M., and Porikli, F., "Regularization of deep neural networks with spectral dropout," *Neural Networks*, vol. 110, pp. 82–90, 2019.

Klambauer, G., Unterthiner, T., Mayr, A., and Hochreiter, S., "Self-normalizing neural networks," *Proceedings of the 31st International Conference on Neural Information Processing Systems*, 2017.

Lecun, Y., Bottou, L., Bengio, Y., and Haffner, P., "Gradient-based learning applied to document recognition," *Proceedings of the IEEE*, vol. 86, no. 11, pp. 2278–2324, 1998.

Li, H., Krček, M., and Perin, G., "A comparison of weight initializers in deep learning-based side-channel analysis. *Applied Cryptography and Network Security Workshops*, I. Zhou, et al. Eds., pp. 126–143, Springer International Publishing, 2020.

Liew, S. S., Khalil-Hani, M., and Bakhteri, R., "Bounded activation functions for enhanced training stability of deep neural networks on visual pattern recognition problems," *Neurocomputing*, vol. 216, pp. 718–734, 2016.

Makerere AI Lab, "Ai-Lab-Makerere/ibean: Data Repo for the IBEAN Project of the Air Lab.," *GitHub*. https://github.com/AI-Lab-Makerere/ibean/. Accessed 1 May 2022.

Ng, J. Y.-H., Hausknecht, M., Vijayanarasimhan, S., Vinyals, O., Monga, R., and Toderici, G., "Beyond short snippets: Deep networks for video classification," *2015 IEEE Conference on Computer Vision and Pattern Recognition (CVPR)*, 2015.

Qian, S. Liu, H., Liu, C., Wu, S., and Wong, H. S., "Adaptive activation functions in convolutional neural networks," *Neurocomputing*, vol. 272, pp. 204–212, 2018.

Sainath, T. N., Mohamed, A.-R., Kingsbury, B., and Ramabhadran, B., "Deep convolutional neural networks for LVCSR," *2013 IEEE International Conference on Acoustics, Speech and Signal Processing*, 2013.

Schmidhuber, J., "Deep learning in neural networks: An overview," *Neural Networks*, vol. 61, pp. 85–117, 2015.

Shi, W., Caballero, J., Huszár, F., Totz, J., Aitken, A., Bishop, R., Rueckert, D., and Wang, Z., "Real-time single image and video super-resolution using an efficient sub-pixel convolutional neural network," *IEEE Conference on Computer Vision and Pattern Recognition*, pp. 1874–1883, Jun. 2016.

Shih, F. Y. and Patel, H., "Deep learning classification on optical coherence tomography retina images," *International Journal of Pattern Recognition and Artificial Intelligence*, vol. 34, no. 8, p. 2052002, 2019.

Team, K., "Keras documentation: Keras Applications," keras.io. https://keras.io/api/applications/#available-models, Aug. 29, 2016.

Vaswani, A., Shazeer, N., Parmar, N., Uszkoreit, J., Jones, L., Gomez, A. N., Kaiser, L., and Polosukhin, I., "Attention is all you need," *Proceedings of the 31st International Conference on Neural Information Processing Systems* (NIPS), pp. 6000–6010, Dec. 2017.

Wu, S., Li, G., Deng, L., Liu, L., Wu, D., Xie, Y., and Shi, L., "L1-norm batch normalization for efficient training of deep neural networks," *IEEE Transactions on Neural Networks and Learning Systems*, vol. 30, no. 7, pp. 2043–2051, 2019.

9 Image Processing by Deep Learning

The relentless advancement of digital image-processing technologies significantly impacts various industries, ranging from healthcare and autonomous driving to environmental monitoring and digital media creation. As digital imaging technology becomes more ingrained in these critical sectors, the demand for more sophisticated image processing solutions escalates. These solutions are not only expected to enhance visual aesthetics but also to improve the accuracy and efficiency of automated systems that rely on image data for decision-making processes. In this chapter, we explore image-processing techniques by comparing traditional filtering methods with advanced deep-learning models.

9.1 INTRODUCTION

In the realm of healthcare, for instance, enhanced image-processing techniques directly contribute to more accurate diagnoses and better patient outcomes. In autonomous driving, the ability to process and interpret visual data with high reliability is crucial for safety and navigation. Similarly, in environmental monitoring, advanced image processing helps in better understanding and responding to global environmental changes through enhanced analysis of satellite and aerial imagery. Each of these applications requires robust, efficient, and precise image-processing techniques that can handle complex, noisy images while preserving essential details.

We aim to provide evaluation of traditional filtering techniques, such as Gaussian blur, bilateral filter, and guided filter, against a deep-learning-based approach using a CNN. By conducting a series of experiments on images with various types of noise and detail requirements, we highlight the strengths and limitations of each method. The ultimate goal is to offer guidance on selecting the most appropriate image-processing technique based on specific requirements, whether it is for real-time processing, detail preservation, or noise reduction.

Blurring is a fundamental technique in image processing used for applications ranging from noise reduction and image smoothing to feature extraction and privacy protection. The literature on blur methods is extensive, reflecting their critical role in enhancing image quality and utility across various domains. We review three primary blur methods – Gaussian blur, bilateral filter, and guided filter – highlighting their development, applications, advantages, and limitations. The Gaussian blur is a foundational filter in image processing, widely used for its simplicity and effectiveness in reducing Gaussian noise. However, its isotropic smoothing effect leads to the loss of edge information.

The bilateral filter, introduced by Tomasi and Manduchi (1998), extends the Gaussian blur by incorporating a range kernel that preserves edges by considering

DOI: 10.1201/ 9781003474395-11

the intensity difference between neighboring pixels. Despite its advantages, the bilateral filter is computationally intensive, especially for large kernel sizes. The guided filter, proposed by He et al. (2013), offers a faster alternative that preserves edges through a content-aware smoothing mechanism. Although faster and less prone to artifacts than the bilateral filter, its performance in varying conditions and against different types of noise remains an area of active research. In the subsequent sections, we delve into the technical specifics of each filtering technique, evaluate them through rigorous experimental setups, and introduce a novel filtering approach that aims to combine the strengths of the existing methods while mitigating their weaknesses.

9.1.1 Gaussian Blur

The Gaussian blur filter is a fundamental tool in image processing, applied to smooth images, reduce noise, and details. It operates by convolving an image with a Gaussian function, effectively averaging pixel values with their neighbors, weighted by their spatial proximity. This technique is widely used for general blur effects and noise reduction as a pre-processing step for edge detection. The primary advantage of Gaussian blur is its effectiveness in removing Gaussian noise and its isotropic smoothing capability. However, a significant limitation is its tendency to blur edges, which can be detrimental in applications where edge preservation is critical.

The Gaussian blur applies a Gaussian function to smooth images, defined mathematically as

$$G(x,y)=\frac{1}{2\pi\sigma^2}e^{-\frac{x^2+y^2}{2\sigma^2}} \tag{9.1}$$

where x and y are the distances from the origin in the horizontal and vertical dimensions, respectively, and σ is the standard deviation of the Gaussian distribution. The effect of this operation is to average the pixel values in the image, with a preference given to nearby pixels according to the Gaussian profile.

9.1.2 Bilateral Filter

The bilateral filter extends the concept of Gaussian blurring by incorporating not only the spatial proximity of pixels but also their intensity differences. This dual consideration allows the bilateral filter to reduce noise while preserving sharp edges within the image. It finds applications in edge-preserving smoothing, detail enhancement, and stylization of images. The key to its performance lies in its ability to selectively average pixels, maintaining edge integrity where significant intensity variations occur. Despite its advantages in preserving details and textures, the bilateral filter's computational intensity and the challenge of parameter selection (e.g., diameter of pixel neighborhood, SigmaColor, SigmaSpace) pose limitations, especially for real-time applications.

The Bilateral Filter combines spatial and intensity information, providing edge-preserving smoothing. Its equation can be given by

$$I_{\text{filtered}}(p) = \frac{1}{W_p} \sum_{q \in S} G_{\sigma_s}(\| p - q \|) \cdot G_{\sigma_r}(\| I(p) - I(q) \|) \cdot I(q) \tag{9.2}$$

where $I_{\text{filtered}}(p)$ is the filtered image intensity at pixel p, G_{σ_s} and G_{σ_r} are Gaussian functions representing spatial and range kernels, respectively, σ_s and σ_r are the standard deviations for spatial and range kernels, S is the set of pixels in the neighborhood, and W_p is a normalization factor given by

$$W_p = \sum_{q \in S} G_{\sigma_s}(\| p - q \|) \cdot G_{\sigma_r}(\| I(p) - I(q) \|) \tag{9.3}$$

This filter thus considers both the spatial closeness and the intensity similarity between the pixels, enabling noise reduction without blurring edges.

9.1.3 Guided Filter

An addition to edge-preserving filters, the guided filter, offers an efficient alternative by smoothing images with consideration to a guidance image. This guidance can be the image itself or another image, providing a versatile framework for edge-preserving smoothing, image matting, and dehazing. Its advantages include faster processing times than the bilateral filter, fewer artifacts, and a natural handling of image gradients, making it suitable for real-time applications. However, the guided filter's dependence on the choice of guidance image and the tuning of parameters like the radius of guidance and epsilon (regularization parameter) necessitates careful adjustment to achieve desired results.

The guided filter is an edge-preserving filter that bases the filtering on a guidance image. The output pixel value is computed as

$$q_i = a_k I_i + b_k, \ \forall i \in \omega_k \tag{9.4}$$

where q_i is the output pixel value, I_i is the guidance image pixel value, a_k and b_k are constants computed for each window ω_k centered at pixel k. The constants are calculated based on the assumption of a linear model between the guidance I and the filtering output q, aimed to minimize the output's variance subject to edge preservation. The model parameters a_k and b_k are determined by minimizing the following cost function for each local patch:

$$E(a_k, b_k) = \sum_{i \in \omega_k} \left((a_k I_i + b_k - p_i)^2 + \epsilon a_k^2 \right) \tag{9.5}$$

where p_i is the input image pixel value, and ϵ is a regularization parameter controlling the degree of smoothness.

9.2 DEEP-LEARNING MODELS FOR IMAGE PROCESSING

Deep learning has revolutionized the field of image processing, providing powerful tools that adapt to the complexities of real-world visual data. Among the various architectures, convolutional neural networks (CNNs) have emerged as the forefront of this revolution. This section delves into the structure and function of CNNs, explaining their relevance and superiority in image-processing tasks compared to traditional methods.

CNNs are specialized deep learning architectures known for their ability to process data with grid-like topology, such as images. A typical CNN architecture consists of several types of layers: convolutional layers, activation layers, pooling layers, and fully connected layers. Each layer of a CNN performs a specific function that contributes to the network's overall ability to detect patterns and features in images.

The convolutional layer is the core building block of a CNN. It applies a convolution operation to the input, passing the result to the next layer. The convolution emulates a filter by taking the dot product of the weights of the kernel and the region of the input image to produce a feature map. The convolution operation is given by

$$f_{i,j}^{(l)} = \sum_m \sum_n I_{i+m,j+n} K_{m,n} \tag{9.6}$$

where $f_{i,j}^{(l)}$ is the output feature map at location (i, j) in layer l, I is the input image or feature map from the previous layer, K is the kernel or filter, and m, n are indices in the kernel.

9.2.1 Activation Layer

Following convolution, an activation function is applied to introduce non-linearity into the model-a critical feature as most real-world data is non-linear. This allows CNNs to learn more complex patterns. ReLU (rectified linear unit) activation function is given by

$$R(z) = \max(0, z) \tag{9.7}$$

where z is the input to the neuron. ReLU is widely used due to its computational simplicity and its ability to reduce the likelihood of the vanishing gradient problem.

9.2.2 Pooling Layer

Pooling layers reduce the dimensions of the data by combining the outputs of neuron clusters at one layer into a single neuron in the next layer. Max-pooling, one of the most common types, reduces data dimensions and processing time while making the detection of features invariant to scale and orientation changes. Max-pooling operation is given by

$$P_{i,j} = \max_{a,b \in W_{i,j}} I_{a,b} \tag{9.8}$$

where $P_{i,j}$ is the pooled output, $W_{i,j}$ is the region of the input over which the pooling is applied, and $I_{a,b}$ is the input within the pooling region.

Each of these filtering techniques contributes uniquely to the field of image processing, with specific applications benefiting from their distinct characteristics. The Gaussian blur's simplicity and effectiveness in noise reduction, the bilateral filter's capability for detail preservation, and the guided filter's efficiency and flexibility illustrate the diversity of tools available for image enhancement and analysis. However, the search for more versatile, efficient, and high-performance filtering techniques continues, driving research toward the development of new methods that can surpass the limitations of existing approaches.

The performance of a trained CNN can significantly surpass traditional methods, especially in tasks that benefit from automated feature extraction and adaptability to new, unseen data. This makes CNNs particularly valuable in dynamic environments where conditions can vary widely, such as in real-time video processing or in variable lighting conditions for outdoor cameras.

Each of these filtering techniques contributes uniquely to the field of image processing, with specific applications benefiting from their distinct characteristics. The Gaussian blur's simplicity and effectiveness in noise reduction, the bilateral filter's capability for detail preservation, and the guided filter's efficiency and flexibility illustrate the diversity of tools available for image enhancement and analysis. However, the search for more versatile, efficient, and high-performance filtering techniques continues, driving research toward the development of new methods that can surpass the limitations of existing approaches.

9.3 EXPERIMENTAL SETUP AND RESULTS

The objective of experimental setup is to evaluate and compare the performance of traditional filters – Gaussian blur, bilateral filter, and guided filter – with a CNN model in terms of their ability to reduce noise and preserve edges in images. The experiment uses a set of images, represented by a typical landscape image shown in the following, subjective to artificial Gaussian noise to simulate common real-world scenarios, where image quality may be compromised by noise.

The following steps are undertaken for each method:

1 Noise Addition: Gaussian noise is added to the original image to create a challenging environment for each filter and the CNN model.
2 Filter Application:
   - Gaussian Blur: Applied with a 5×5 kernel.
   - Bilateral Filter: Parameters are set to a diameter of 9, and sigma values for color and space set to 75.

FIGURE 9.1 The results of applying Gaussian blur, bilateral filter, and guided filter.

- Guided Filter: Applied with a radius of 8 and an epsilon value of 0.4 squared, adjusted to basic smoothing due to limitations in the available implementation.
- CNN Model: A basic CNN architecture is implemented for noise reduction to learn from the noisy images directly in a supervised learning setup. Note that the CNN model is theoretical in this context as detailed training is not feasible within the scope of this experiment.

3 Analysis: The output from each method is analyzed both visually and quantitatively using metrics such as peak signal-to-noise ratio (PSNR) and structural similarity index measure (SSIM).

Figure 9.1 shows the results from implementing the Gaussian blur, bilateral filter, and a placeholder for the guided filter from scratch applied to a synthetic image. The leftmost two image shows the synthetic image with random noise, serving as the input for our filtering algorithms. The third image from the left demonstrates the smoothing effect of the Gaussian blur. This filter effectively reduces noise by averaging pixel values with a Gaussian-weighted kernel, though it may blur edges in the process.

The fourth image shows the result of applying bilateral filter. This method preserves edges while reducing noise by considering both the spatial distance and the intensity difference between pixels. It provides a noticeable improvement in maintaining edges compared to the Gaussian blur. The rightmost image is the result of applying the guided filter, designed to preserve edges while smoothing, so the image should retain more detail compared to the Gaussian blur. The result here is subtle due to the conservative parameters chosen, but it shows how the guided filter can smooth the image while keeping the edges relatively sharp.

The CNN is adequately trained and expressed optimally by adapting its filters to both reduce noise and preserve important details based on the learning from training data. CNNs have the potential to outperform traditional methods by learning application-specific optimizations. However, the actual performance heavily depends on the quality and quantity of the training data and the network architecture used.

9.4 EDGE ENHANCEMENT USING IMAGE SUPER-RESOLUTION USING DEEP CONVOLUTIONAL NEURAL NETWORK +

In this section, we introduce a refined Super-Resolution Convolutional Neural Network (SRCNN) model designed specifically for enhancing edges in image processing applications. This advanced model builds on the core concepts of existing SRCNN frameworks, incorporating edge-specific loss functions and channel attention mechanisms to markedly enhance the perceptual quality and definition of super-resolved images. Our innovative edge loss function, which utilizes Sobel filter techniques, enables the model to focus on and accurately restore high-frequency elements, especially along edges essential for clear visualization in super-resolution processes. Additionally, the addition of a channel attention, spatial attention layer, and sharpening filter module allows for the adaptive adjustment of channel-wise feature dynamics, improving the representation of important details. Extensive testing on various image datasets confirms that our improved SRCNN model outperforms traditional methods. These results highlight the effectiveness of incorporating edge-aware and attention-based enhancements in advancing the functionality of super-resolution technologies.

9.4.1 Introduction

In the realm of digital image processing, super-resolution (SR) techniques have garnered substantial attention for their ability to enhance the resolution and quality of images beyond the capabilities of traditional imaging sensors. Central to this domain is the *Super-Resolution Convolutional Neural Network* (SRCNN), a pioneering deep-learning model that has set the foundation for numerous advancements in SR technology. SRCNN leverages the power of *convolutional neural networks* (CNNs) to learn end-to-end mappings from low-resolution to high-resolution images, achieving remarkable improvements in image clarity and detail (Dong et al., 2016).

Despite the achievements of SRCNN models and their derivatives, they frequently encounter difficulties in accurately reconstructing high-frequency details, especially at edges. These details are vital for the perceptual quality of super-resolved images. Edges and textures convey crucial information about the structure and objects within a scene, and their precise enhancement is critical for fields such as digital forensics, high-resolution photography, and remote sensing.

To tackle this issue, we present a new edge enhancement framework embedded within the SRCNN architecture. Our approach includes a unique edge loss function that combines Sobel filters (Huang et al., 2021) with mean squared error, specifically focusing on improving edges and textures. The Sobel filter, a well-established method for edge detection in image processing, offers a gradient magnitude that effectively highlights and maintains edge details throughout the super-resolution process.

Furthermore, inspired by recent developments in attention mechanisms within deep-learning frameworks, we have integrated a channel attention mechanism into our SRCNN architecture, as detailed by (Chen et al., 2022). Channel attention enables the network to selectively emphasize the most informative feature channels, thereby boosting the model's ability to depict and restore crucial image details. Additionally, we employ spatial attention to refine the process of image upscaling, ensuring that the model concentrates on areas that significantly impact edge definition and overall image quality (Lee et al., 2023). This dual attention strategy not only prioritizes the most critical features for edge enhancement but also enhances the general fidelity of the super-resolved images.

Building on this, we incorporate sub-pixel convolution, a technique pivotal for enhancing image resolution more efficiently. Sub-pixel convolution operates by rearranging the output from the previous network layers into higher resolution images, effectively increasing the image resolution at the final output stage without the need for additional computational resources to directly produce larger images (Shao et al., 2023). This method proves essential in super-resolution tasks as it allows for a finer reconstruction of details, particularly in the upscaling of low-resolution images where maintaining textural and edge clarity is critical. The use of sub-pixel convolution in our SRCNN model thus complements our attention mechanisms by providing a sharper, more defined output that closely mimics the quality of high-resolution images.

The synergy between combined edge-specific loss functions and channel attention mechanisms marks a significant advancement in SR technology. Our modified SRCNN model not only demonstrates superior performance in enhancing edges and textures but also maintains a balance between computational efficiency and image quality enhancement. This balance is crucial for deploying SR models in real-world applications where both image quality and processing speed are paramount.

Extensive experimental evaluations on a diverse set of image datasets validate the effectiveness of our approach. Compared to traditional SRCNN models and contemporary SR techniques, our edge-enhanced SRCNN model consistently achieves superior results in terms of both objective and subjective image quality metrics like peak signal-to-noise ratio (PSNR). The improved edge fidelity and texture reproduction, significantly bolstered by the inclusion of a sharpening filter as discussed in (Huang et al., 2023), contribute to a more natural and detailed visual experience, underscoring the potential of our approach in various SR applications.

9.4.2 Existing Work

The convolutional neural network for pansharpening with spatial structure enhancement operator introduces a sophisticated method for improving *multispectral* (MS) images using a CNN combined with a spatial structure enhancement operator. This approach focuses on blending the detailed spatial information from *panchromatic* (PAN) images with MS images to produce high-resolution outputs, and uses sub pixel convolution layer for upscaling. Traditional techniques often rely on simple linear models, which do not fully account for the complex processes that degrade image quality. By using CNNs, this method overcomes these limitations by addressing the issue through a non-linear perspective.

The process starts with the Sobel operator, which is used to detect edges and extract high-frequency details from both PAN and MS images, ensuring that the spatial features are accurately captured. These details are crucial for creating images that are not only high in resolution but also rich in detail. The CNN then processes these details to generate spatial feature maps, effectively retaining the information in both the spatial and spectral dimensions. This approach has been tested through various experiments, showing remarkable improvements in the clarity and detail of the images compared to previous methods (Huang et al., 2021).

The attention mechanism feedback network for image super-resolution (Huang et al., 2021) introduces a novel approach in the field of image super-resolution (SR) through the development of a super-resolution feedback network (SRFBN). This method stands out by incorporating a feedback mechanism, inspired by the human visual system, into deep learning for image SR, an aspect that has been underutilized in previous deep learning SR techniques. The SRFBN refines basic image details by looping back high-level information to earlier stages, employing *recurrent neural network* (RNN) hidden states under specific constraints to facilitate this feedback process. A unique feedback block is designed to manage these feedback connections effectively, enabling the generation of enhanced high-level representations.

One of the key strengths of SRFBN is its robust initial reconstruction capability, which allows it to progressively produce the final high-resolution image in a step-by-step manner. Furthermore, to adapt the network for more complex scenarios, where low-resolution images suffer from various types of degradation, a curriculum learning strategy is introduced. This strategy gradually increases the difficulty level, making the network more versatile and effective in handling complicated tasks.

The super-resolution combining spatial attention and channel attention (Lee et al., 2023) explores an advanced method for improving image resolution through single-image super-resolution (SISR) techniques that utilize both spatial and channel attention mechanisms. These mechanisms are integrated into a structure known as CSBlock, which effectively emphasizes high-frequency features that are often neglected in traditional CNN-based super-resolution methods. This approach allows for more precise reconstruction of high-resolution images from low-resolution inputs, showing notable improvements in performance metrics such as *PSNR* and *SSIM* when compared against standard methods like Bicubic interpolation and other deep-learning models like *VDSR* and *EDSR*.

The sub-pixel convolutional neural network (SPCNN) for image super-resolution reconstruction (Shao et al., 2023) introduces a novel approach to enhance low-resolution images using a SPCNN. This model effectively improves image resolution by translating RGB images into YCbCr mode and processing only the Y channel, thus significantly reducing computational load. The SPCNN architecture incorporates multiple convolution layers for feature extraction and non-linear mapping, alongside a residual network to prevent gradient vanishing issues. Notably, the model employs a sub-pixel convolution layer for upscaling, optimizing both the image quality and reconstruction speed. This approach demonstrates superior performance in terms of precision and time efficiency compared to traditional methods like bicubic interpolation and other neural network-based methods, such as SRCNN, marking a significant advancement in the field of image super-resolution

9.4.3 Overview of Our Edge Enhancement SRCNN

Our Edge Enhancement Super-Resolution Convolutional Neural Network (EESRCNN+) framework is distinguished by two innovative components: the *edge enhancement mechanism* (EEM) and the *channel attention module* (CAM). To address the challenges in reconstructing high-resolution images from their lower-resolution counterparts, particularly in preserving edge details and texture information, the EEM employs a novel edge loss function that integrates Sobel filter operations to accentuate edge features. This function is critical for enhancing the sharpness and clarity of edges, a common shortfall in traditional super-resolution techniques. Concurrently, the CAM is designed to recalibrate the focus of the network toward more informative features within an image by employing a mechanism that weights the importance of different channels based on their relevance to the super-resolution task. To facilitate a comprehensive understanding of our approach, it's essential to introduce the following terminologies and components integral to our EESRCNN+ model:

To facilitate a comprehensive understanding of our approach, it's essential to introduce the following terminologies and components integral to our EESRCNN+ model:

I_{LR} : Image low resolution.

I_{HR} : Image high resolution.

I_{SR} : Super-resolution image generated by EESRCNN+.

$F_{EEM}\left(I_{LR}\right)$: The edge-enhanced features extracted from I_{LR} the EEM.

$F_{CAM}\left(F_{EEM}\right)$: The feature maps obtained after applying the CAM to F_{EEM}.

L_{edge} : The edge loss calculated from I_{LR} and I_{SR}.

L_{MSE} : The mean squared error loss calculated from I_{HR} and I_{SR}.

n_1 : Number of nodes in the first layer.

n_2 : Number of nodes in the second layer.

n_3 : Number of nodes in the third layer.

n_4 : Number of nodes in the fourth layer.

The EESRCNN+ framework initiates with the processing of I_{LR} through the EEM to enhance edge definitions, followed by the application of the CAM to prioritize and amplify significant features within the edge-enhanced maps. The resultant feature maps are then further processed to reconstruct I_{SR}, aiming to closely approximate I_{HR}. The model's effectiveness is governed by a combined loss function, $L_{total} = L_{MSE} + \lambda L_{edge}$, where λ is a weighting factor that balances the contribution of edge preservation to the overall super-resolution quality.

We adopt the Kaggle Cataract dataset, which has scanned retinal images. The images are of normal eyes, cataracts, glaucomas, and retinal diseases. The dataset has a usability score of 6.8/10. Moreover, there were 609 very high-resolution retinal scans with a total size of 4 gigabytes. All the images will be down-sampled to have low resolution.

The second dataset is from the Kaggle Chest X-ray images (pneumonia). The normal chest X-ray depicts clear lungs without any areas of abnormal opacification in the image. Bacterial pneumonia typically exhibits a focal lobar consolidation, whereas viral pneumonia manifests with a more diffuse "interstitial" pattern in both lungs. The dataset has 5,863 images with 2 GB in size. The dataset has a usability score of 7.5/10.

9.4.4 EESRCNN+ Algorithm

Figure 9.2 shows the network architecture. The pre-processing phase begins with the low-resolution input image I_{LR}, which undergoes a series of preparatory steps to ready it for the super-resolution process. This involves the load_images function that rescales images to a uniform size (128 × 128 pixels) and normalizes their pixel values to fall within the [0, 1] range. The dataset of processed images is then partitioned into training and testing subsets, with the train_test_split function ensuring a randomized selection that facilitates robust model training and evaluation.

The SRCNN model learns the transformation from the low-resolution input image I_{LR} to the high-resolution target image I_{HR}. The build_data function simulates this mapping by initially down-sampling I_{HR} to generate a corresponding I_{LR}, then up-sampling I_{LR} back to the original dimensions using bicubic interpolation. This pair of I_{LR} and I_{HR} serves as the foundation for the model to understand and replicate the intricate details lost in the resolution reduction process. Further refinement is achieved by the spatial attention layer, which assesses spatial importance across the feature map. By focusing on areas with significant information, this layer helps in maintaining and highlighting critical textural details during the upscaling process.

The *edge enhancement mechanism* (EEM) is pivotal in accentuating the edges within I_{LR} to produce edge-enhanced features $F_{EEM}\left(I_{LR}\right)$. The edge_loss function computes L_{edge}, quantifying the disparity in edge details between I_{SR} and I_{HR}. By applying Sobel filters to detect edges, this function ensures that the model prioritizes edge accuracy, which is crucial for perceived image quality. The sharpening layer directly enhances the edges in $F_{EEM}\left(I_{LR}\right)$ by applying a predefined sharpening kernel. The resultant feature maps are sharper and more defined, aiding in the reconstruction of high-quality I_{SR}.

The *channel attention module* (CAM) refines $F_{EEM}\left(I_{LR}\right)$ by applying channel-wise attention, and spatial attention, producing $F_{CAM}\left(F_{EEM}\right)$. This module assesses and recalibrates the importance of each channel in the feature maps, emphasizing those channels that are most informative for reconstructing I_{HR}. Through a combination of global pooling and dense layers, the CAM dynamically adjusts channel weights, enhancing the model's focus on salient image features.

With the edge-enhanced and attention-focused features, including those recalibrated by the CAM, the SRCNN model proceeds to the reconstruction phase, where the super-resolved output image I_{SR} is synthesized. This involves a series of convolutional operations that integrate the enhanced features, gradually building up the resolution and detail to match that of I_{HR}. Each convolutional layer is carefully calibrated to preserve the enhancements from the *edge enhancement mechanism* (EEM) and CAM.

A sub-pixel convolution layer is introduced following the final convolution layers to upscale the image resolution effectively. This layer rearranges the output from a

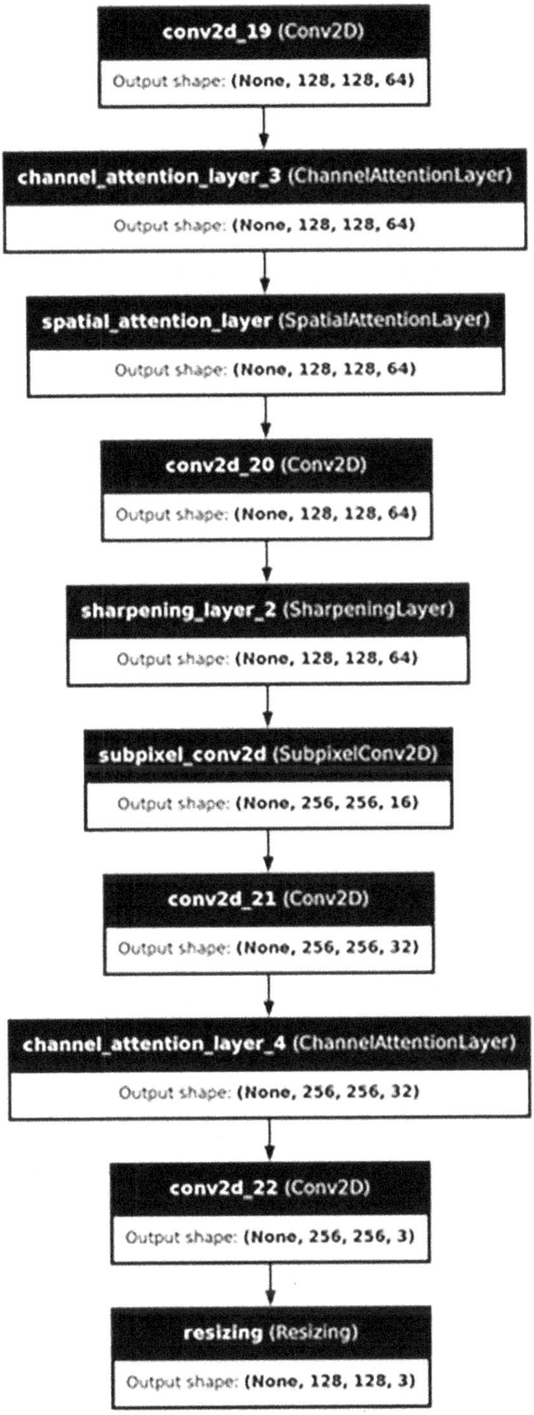

FIGURE 9.2 Network architecture.

lower-dimensional space into a higher-dimensional space, thus enhancing the resolution by a set scale factor, enhancing the sharpness and clarity of I_{SR} without introducing the artifacts typically associated with simpler upscaling methods.

The final step involves compiling the ESRCNN+ model with an optimization algorithm (Adam) and a loss function that is a composite of mean squared error L_{MSE} and edge loss L_{edge}. This combines loss function, $L_{total} = L_{MSE} + \lambda L_{edge}$, which balances the need for overall fidelity to I_{HR} with the necessity of edge precision in I_{SR}. We choose λ = 0.7 to 0.9 for edge enhancement. Lesser the value, more it will focus on structure rather than edges. The training process iteratively adjusts the model weights to minimize L_{total}, refining the model's ability to generate high-fidelity super-resolved images. Through this detailed and systematic approach, the ESRCNN+ model, bolstered by the EEM and CAM, aspires to achieve state-of-the-art performance in super-resolution, effectively bridging the gap between I_{LR} and I_{HR} while maintaining exceptional edge quality and feature salience in I_{SR}. Figure 9.3 shows an example of using EESRCNN+.

9.4.5 Experimental Results

In this section, we will commence by examining the application of the original Super-Resolution Convolutional Neural Network (SRCNN) framework to our cataract dataset. The SRCNN had $n_1 = 64$, $n_2 = 32$, and $n_3 = 3$. Utilizing the optimal 9-1-7 filter configuration within the SRCNN architecture, we attained a *peak signal-to-noise ratio* (PSNR) of 32.11 dB on average, in contrast to the 32.38 dB score of the blurred reference image. Notably, the application of SRCNN results in a significant loss of edge detail, leading to an overly generalized representation of the image. This outcome is particularly detrimental in the context of retinal images, where the preservation or enhancement of edge detail is crucial for effective analysis and interpretation.

Similarly, we apply SRCNN to the chest X-ray image, and the same configuration. We achieve the PSNR value of 30.13 dB, as compared to a low-resolution image with 29.84 dB. The color saturation looks darker, but the edges are not upscaled.

Figure 9.4 shows the SRCNN with a retina image and the SRCNN with a chest X-ray image. Note that super-resolution does not preserve the edges, making the

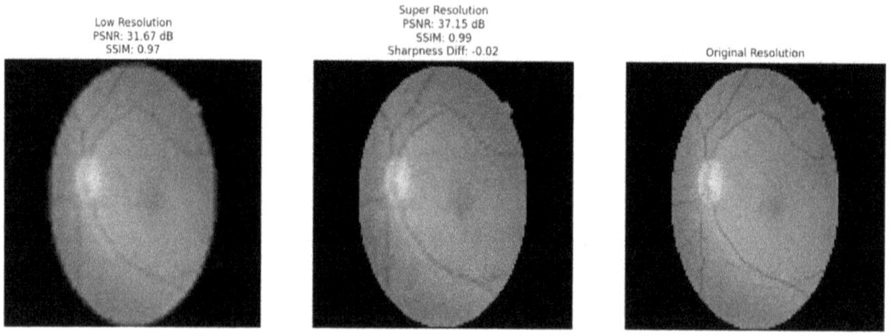

FIGURE 9.3 An example of using EESRCNN+.

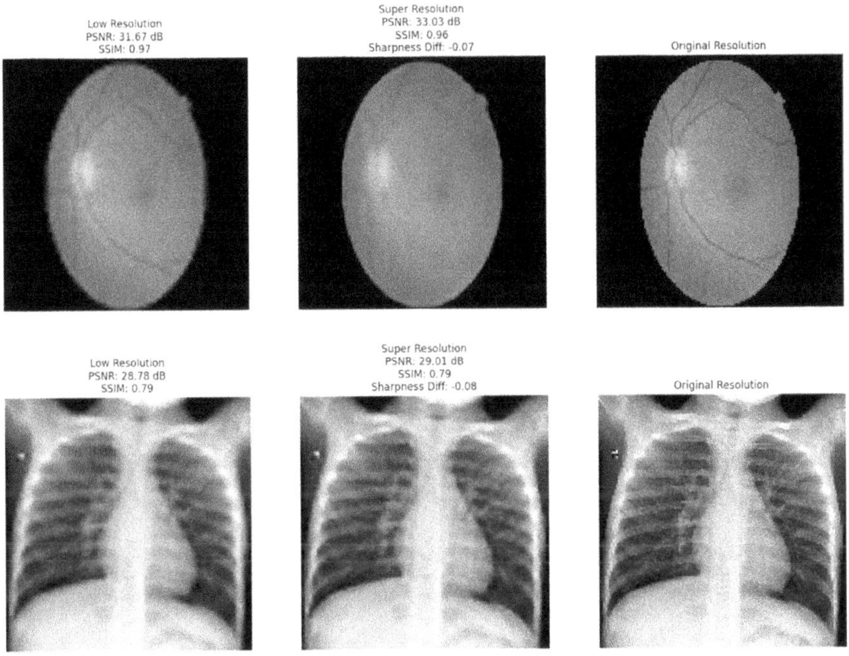

FIGURE 9.4 (a) SRCNN with a retina image, (b) SRCNN with a chest X-Ray image.

image more generalized with upscaling color saturation. The main reason that it fails as SRCNN has no attention layers and uses MSE loss function which does not enhance the edges of the images.

In the next experiment we add another layer to the SRCNN, making it a total of 4 layers. $n_1 = 64$, $n_2 = 32$, $n_3 = 32$, and $n_4 = 3$. With filter configuration of 5-7-7-3, we achieved almost the similar results. On average on the retina dataset, the super-resolution has a PSNR score of 32.35 dB which is slightly better than the original SRCNN in terms of color saturation. However, we can see that more edges are lost as compared to SRCNN. For the Chest X-ray dataset with the same configuration, we observe an average PSNR score of 29.61.

Figure 9.5 shows the Retina SRCNN with 4 layers and the Chest X-ray SRCNN with 4 layers. From the figure, we observe that SRCNN with 4 layers performs the worst. We observe the loss in edges, and for the chest X-ray images, we see almost no changes. This indicates that the increasing number of layers can be beneficial for RGB images for color saturation. However, the models still do not know where to pay attention when it comes to edges.

For the next experiment to improve the edges, we include edge loss combined with MSE loss with 4 layers configuration. Those 4 layers have configuration of $n_1 = 64$, $n_2 = 32$, $n_3 = 32$, and $n_4 = 3$. With filter configuration of 5-7-7-3. On average on the Retina dataset, the super-resolution had a PSNR score of 34.13 dB with more visual enhancement and edge detection. For the Chest X-ray dataset with the same

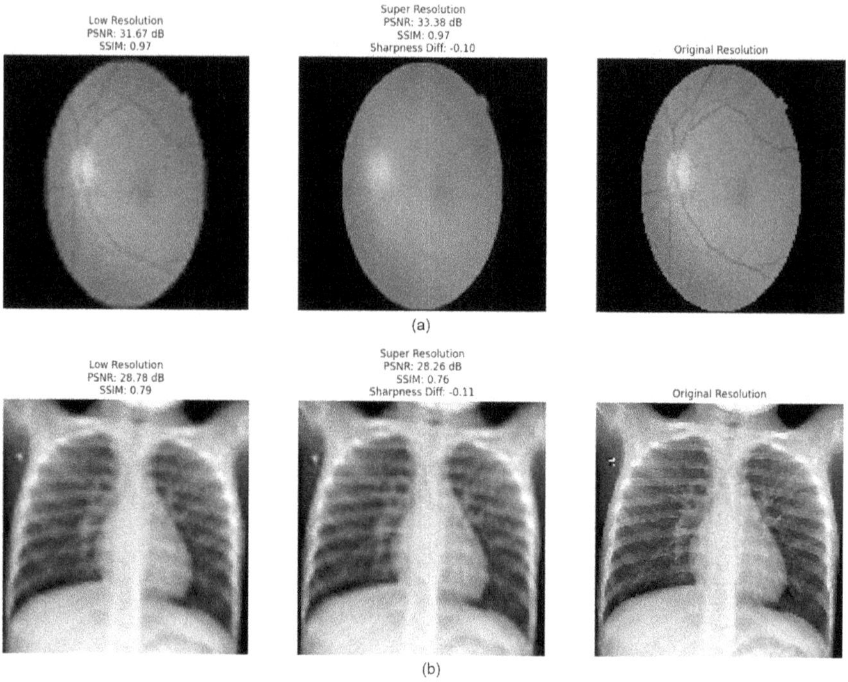

FIGURE 9.5 (a) Retina SRCNN with 4 layers, (b) chest X-ray SRCNN with 4 layers.

configuration, we observe an average PSNR score of 33.34. Figure 9.6 shows the SRCNN retina with edge loss and the Chest X-ray SRCNN with edge loss.

For the next experiment to improve the image quality, we include a sharpening filter. Those 4 layers have configuration of $n_1 = 64$, $n_2 = 32$, $n_3 = 32$, and $n_4 = 3$, with filter configuration of 5-7-7-3. On average on the Retina dataset, the super-resolution has a PSNR score of 34.64 dB with a sharper image but similar image quality. For the Chest X-ray dataset with the same configuration, we observe an average PSNR score of 33.14 dB.

Figure 9.7 shows the Retina SRCNN with sharpening and the Chest X-ray SRCNN with sharpening. We can observe that the image's edges look more refined than other models. For chest X-ray images, it is not improved much.

For the next experiment to improve the image quality, we add attention layers to our model. We set 4 layers to have configuration of $n_1 = 64$, $n_2 = 32$, $n_3 = 32$, and $n_4 = 3$, with filter configuration of 5-7-7-3. On average on the Retina dataset, the super-resolution had a PSNR score of 36.45 dB with amazing image quality. For the Chest X-ray dataset with the same configuration, we observe an average PSNR score of 33.47, which has a slight improvement.

Figure 9.8 shows the Retina EESRCNN and the Chest X-ray EESRCNN. As we can observe, adding the attention mechanism results in better image quality for retina images. However, for Chest X-ray, the image stays almost the same. The white color has more popped out resulting in better image quality.

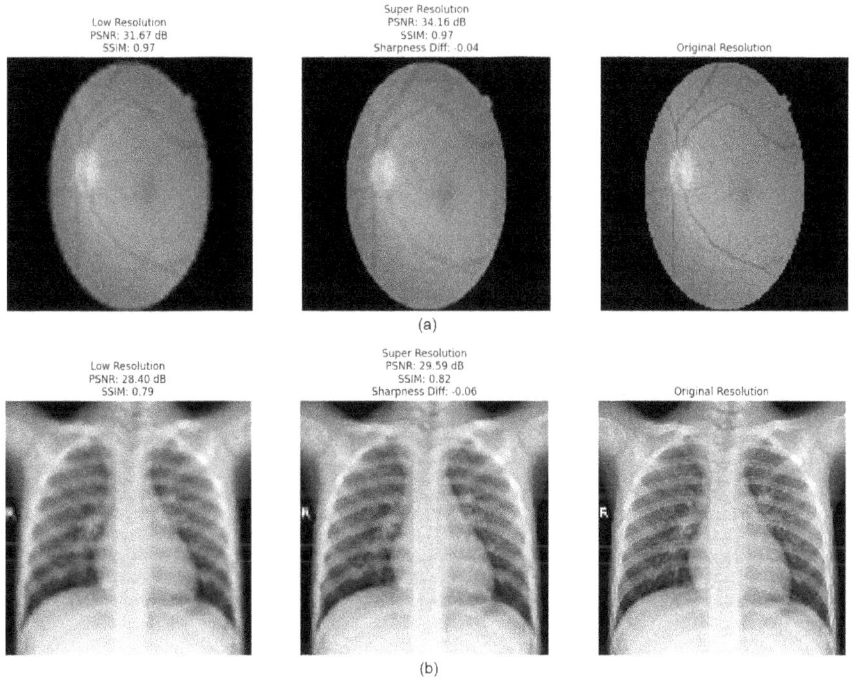

FIGURE 9.6 (a) SRCNN retina with edge loss, (b) Chest X ray SRCNN with edge loss.

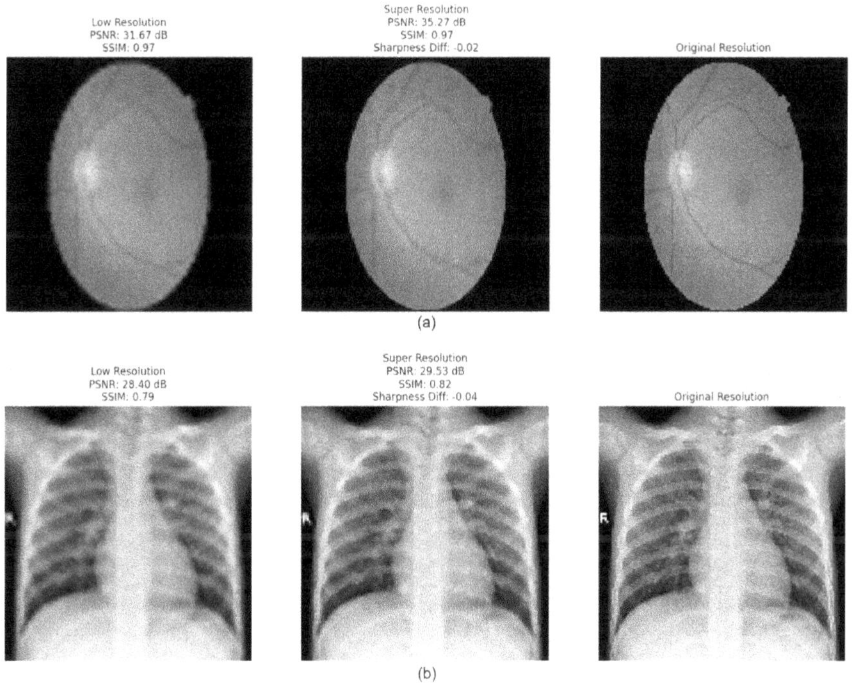

FIGURE 9.7 (a) Retina SRCNN with sharpening, (b) Chest X-ray SRCNN with sharpening.

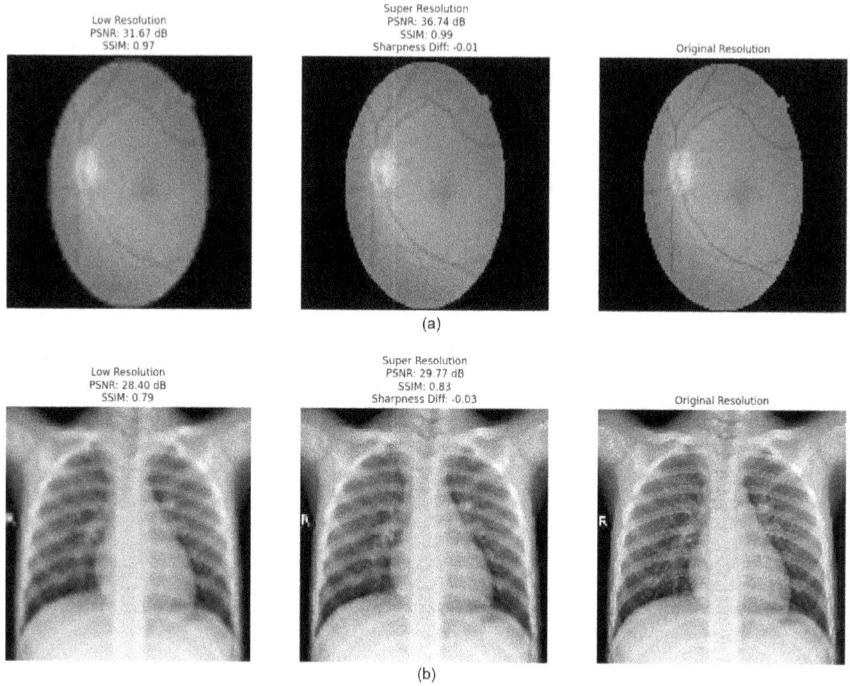

FIGURE 9.8 (a) Retina EESRCNN, (b) Chest X-ray EESRCNN.

For the experiment to improve the image quality, we add spatial attention layers to our model along with sub-pixel layer for upscaling. The 4 layers have configuration of $n_1 = 64$, $n_2 = 32$, $n_3 = 32$, and $n_4 = 3$, with filter configuration of 5-7-7-3. On average on the Retina dataset, the super-resolution had a PSNR score of 38.11 dB with the best image quality. For the Chest X-ray dataset with the same configuration, we observe an average PSNR score of 33.55 dB. That is a slight improvement compared with the last model we saw.

Figure 9.9 shows the Retina EESRCNN+ and the Chest X-ray EESRCNN+. This is the best performance model. With spatial attention and sub-pixel layer, retina images resemble more toward the original high-resolution image. For chest x-ray, the image has resulted in better image contrast.

We employ a comprehensive evaluation framework to assess the performance of our model, incorporating three pivotal metrics: *peak signal-to-noise ratio* (PSNR), *structural similarity index measure* (SSIM), and *sharpness*. PSNR offers a quantitative measure of the reconstruction quality by comparing the similarity between the original and the super-resolved images, with higher values indicating better reconstruction. SSIM, on the other hand, evaluates the perceptual quality of the super-resolved images by considering changes in structural information, luminance, and contrast, thereby providing insights into the visual impact of the reconstruction. Lastly, the sharpness metric is utilized to assess the clarity and detail enhancement in the super-resolved images, which is particularly important in maintaining the

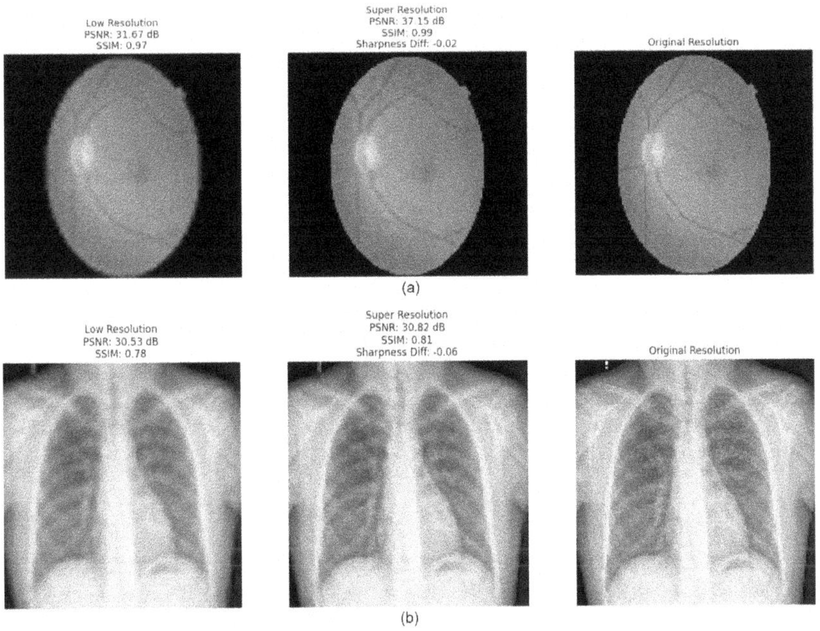

FIGURE 9.9 (a) Retina EESRCNN+, (b) Chest X-ray EESRCNN+.

integrity of fine details and edges. Together, these metrics provide a robust framework for evaluating the effectiveness of our model in enhancing image quality while preserving critical structural and perceptual features.

Our observations reveal a discernible trend where the PSNR values exhibit an upward trajectory with the incremental addition of layers and the integration of novel techniques into the SRCNN framework. This enhancement in PSNR indicates an improvement in the quality of the super-resolved images, suggesting that the model becomes more effective in reconstructing higher-fidelity images from low-resolution inputs as it grows in complexity.

Concurrently, it is noteworthy that the augmentation of the SRCNN model's complexity also leads to an increase in runtime. This escalation in execution time underscores the computational cost associated with the model's enhanced capability. Apparently, there is a tradeoff between image quality and the algorithm's execution time. This tradeoff highlights the need for careful consideration in balancing the desire for superior image reconstruction against the practical constraints of computational efficiency and resource utilization in real-world applications.

9.5 A DEEP-LEARNING APPROACH FOR IMAGE ENHANCEMENT USING GANs AND SINGLE NEURAL NETWORKS

In this chapter, we present an efficient algorithm using the deep learning method to deal with the image enhancement. Different from traditional mathematical method, the neural networks can be more easily adapted for various practical scenarios. We introduce neural models to solve the problem, which embeds several state-of-art techniques, such as SENet (squeeze and excitation networks), mish (an activation function which have achieved success in object detection models such as YOLO v4), and residual network to capture image features and enhance the lighting, blurring problem of the images. This improved one-way GAN model will process an unpaired training dataset to gain knowledge of what a good-quality image should be, and predict an enhancement when facing a new poor-quality image. In the experiment, we can see that this approach can do a better job in improving PSNR and visual effects.

9.5.1 Introduction

Images or pictures have taken an important place in our daily lives. When one is enjoying beautiful view through Internet or reminisce good old times from photos, the images can arise the pleasant feelings. Nowadays, with the development of digital devices such as cameras, smart phones, a huge number of images have been generated every day. However, these devices have limitations albeit more and more advanced they are becoming. Given the job picking pictures from a set of pictures from a scene taken by a camera and forging these pictures into a high-quality image, many state-of-the-art vision systems or devices can not complete the task very well. The reason is complicated. Natural environment, such as illumination, will incur noise and poor contrast. Moreover, the camera sensors often have a linear response to the natural light, but human vision system can deliver a more sophisticated non-linear mapping. The situation will be more severe when using a mobile phone camera, for their lens are usually small and the limited computation resources for adjusting.

Therefore, many software, such as Photoshop, emerged for this reason, to retouch the images and produce fine-quality images. It is all about algorithm. The problem prompted by imperfection of electronic devices has long been a key task in computer vision, including blurring, denoising, low-resolution, poor color, etc. Image enhancement of computer vision tends to tackle these problems by adjusting colors and improve the sharpness of the image. Classic methods include histogram equalization, sharpening, contrast adjustment color mapping, and so on. However, the result highly depends on empirical experience and skill and the operator's personal aesthetic judgment. Therefore, there is literally no standard to follow, which leads to another problem in this field, it is difficult to judge some output from certain algorithm. In addition, since the methods mentioned require many rounds of adjustment, sometimes both globally and locally, it is a little bit inefficient to implement these methods under some circumstance. Finally, they are all based on some hypothesis or specific condition, so it is very impractical to generalize one model from one domain to another.

We intend to overcome some of the weakness in aforementioned methods with the tool of deep learning. By constructing a deep model, we can learn features from the good-quality images to process the bad ones. The images for training can be chosen randomly from an image database or downloaded from Internet, which make the model benefit from the diversity of source, and we could use less human interaction in this process: let the computer do the work. After supervised training, the model can predict an enhancement through characteristic it acquired. The model is fast and easy to train but efficient. The backbone is from U-Net, which prevails in medical image detection. Some layers and structure of the original model are modified to gain more advantages. There are also some new model structures such as SENet, residual net has been adopted in the model. The data augmentation methods are implemented, such as random crop, complex mixed image, and rotation.

9.5.2 Existing Methods

Research in image enhancement has a long history, many methods and algorithms have shown their ability to address the problem. The conventional ideas, such as bilinear, bicubic, and Lancroz, can often generate a smooth image albeit seems a little rough sometimes. To improve these methods, here comes some new ideas such as neighborhood learning.

The reviving of deep learning comes with the explosive data and computing ability, which solve two key problems in deep learning before: massive training material and resource to process the back-propagation. In the computer vision or image processing field, the rise of CNN (convolutional neural networks) has played an important role. CNNs in stacked or deep structure have an excellent ability to learn the essential features of images, which make vision tasks such as object recognition, image segmentation, and image enhancement achieve some astounding success. Yan et al. (2016) proposed a very first deep-learning algorithm in the field of image enhancement by using a simple sequential CNN. Dong (2014, 2016) developed the SRCNN method that can map low-quality images to high-quality end to end.

Recently, GAN and its variants take the lead in image enhancement as well as standalone models. GAN (Generative Adversarial Network), first prompted by Ian Goodfellow et al. (2014) in his famous 2014 paper, GAN is composed of two networks: generator and discriminator. GAN and its variants are good at image generation. By using an athlete-judge model, they can study the features of given domain and then create amazing images. Therefore, GAN and its variants have been widely implemented in computer vision.

There are mainly two ways of GAN's variants that have a good performance in image enhancement: one-way GAN and two-way GAN. One-way GAN's base structure is just like the original GAN, which only has one generator and one discriminator, but with fancier generator and discriminator. Nowadays, for the sake to gain both swift and efficient networks, in the generator, a smaller CNN network usually has been deployed, and one of the most famous ones is U-Net. U-Net, brought up by Ronneberger et al. (2015), is a dumbbell-like network which include 9 convolutional blocks. It not only behaves well in medical domain, but also in areas where higher-level features are required other than just edges and contour. Like other classic CNN models, U-Net has both down-sample and up-sample mechanism. Down-sampling is carried out by max-pooling, and up-sampling is finished by bilinear interpolation. The output is a segmentation map of the original image. The output segmentation map is very useful to represent the original map and we can adopt it for many scenarios, such as image classification, calculation of differences between images, and also my topic here image enhancement base on the segmentation difference between original images and the desired domains.

VGG-16 is also well known for its power of getting segmentation map, many researchers often take the 5-th max-pooling layer's output as the input's segmentation map, which has been concurred by RichardWebster et al. (2018) in which the author take the VGG-16 trained on Imagenet as its constrain between extracted feature and ground truth and achieve a fine result. Many one-way GAN as well have employed the pre-trained VGG-16 network, not in generator but in discriminator. EnlightGAN, proposed by Jiang et al. (2021), is a one-way GAN that implements these two networks and can be trained with unpaired training images. While one-way GAN is plain to understand, the two-way GAN, which is trickier.

Figure 9.10 shows the classical GAN architecture, and a two-way GAN implemented in image enhancement is shown by Figure 9.11. The two-way GAN used to enhance image quality is usually call cycleGAN, which originated from Zhu et al. (2017). 8) implemented a cycleGAN using a U-Net variant as generator's backbone. Of course, the merits of GAN and its variants come with costs. There are two generator and two discriminators in the network. By forming a cycle, the network can cross-validate the two generator and discriminator. Although very powerful GANs and its variants are, they are notoriously hard to converge, for famous mode collapse problem which lead to experience play and instability training process originating from the dynamic model structure.

For some practical use scenarios such as mobile phones, the standalone CNN models are preferred if they have been fine-tuned. Johnson et al. (2016) proposed a method to synthesize high-resolution images from low ones. Gatys et al. (2015) also suggested an algorithm that embeds VGG-16 as its backbone to enhance the texture

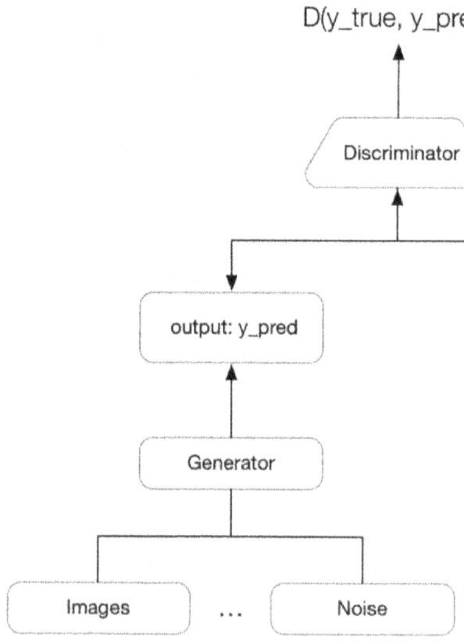

FIGURE 9.10 The classical GAN architecture.

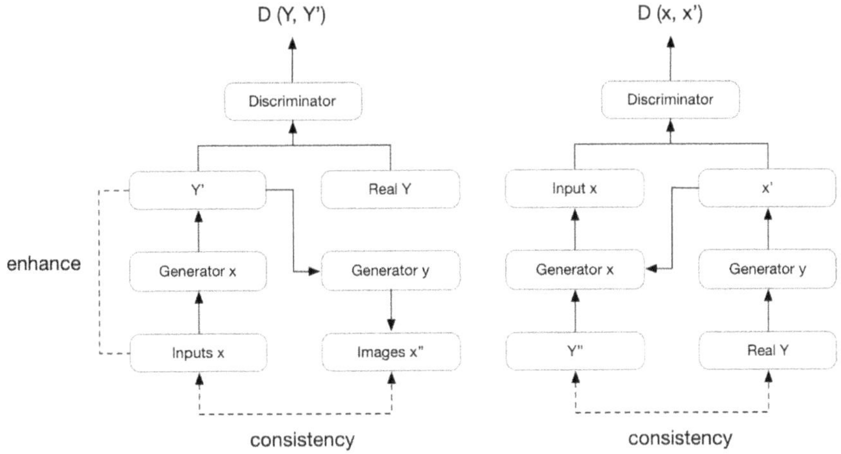

FIGURE 9.11 The cycleGAN (two-way GAN).

look of rocks. The standalone models can be trained more easily than GANs and could be relatively efficient by fine-tuning.

A problem rises that the standalone model requires paired images that are hard to gather. Some researchers use ideal images to produce the training image set artificially, for instance, blurring or adding random noise to the ideal images, that will lead to another question: the model can only acquire the features or the functions that are different from pattern existing in real poor images. Vu et al. (2018) has done a good work with this kind of model, which outperforms models such as SRCNN and has a good enhancement PSNR. Their model structure shows the power of residual convolutional network block, which are also heavily adopted in famous ResNet. However, it is also hard to be generalized to other domain, like animal images from Internet.

Based on all the knowledge presented, we implement a one-way GAN to conduct this research. The reasons are as follows:

1. One-way GAN has inherited all power GAN has in producing good-quality images.
2. One-way GAN also can work in unpaired mode when training the model.
3. Although maybe lacking some representative ability compared to two-way GAN (cycleGAN), it can save plenty of time consumed in hard coding and achieve a comparable performance.
4. Better at generalization than plain CNN models.

As mentioned before, a typical generator a one-way GAN used to enhance image quality is as Figure 9.12 (from enlightening GAN described), while its discriminator can be shown as Figure 9.13, which is a deep CNN network.

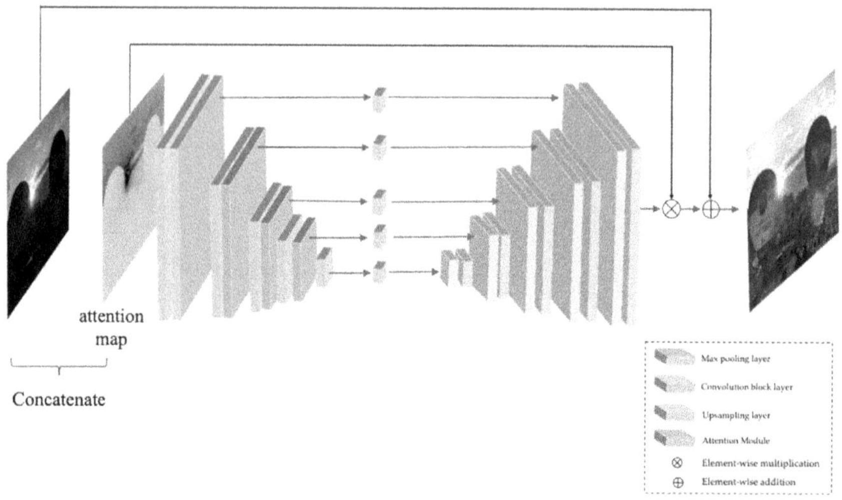

FIGURE 9.12 A typical generator of GAN.

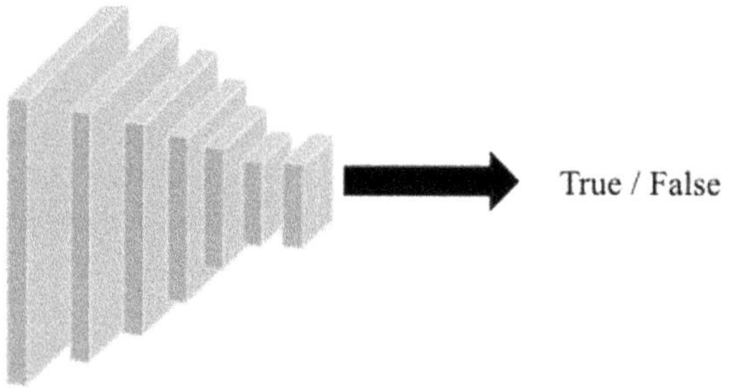

FIGURE 9.13 A typical discriminator of GAN.

9.5.3 Improved Methods

9.5.3.1 A New Activation Function

A new activation function, called mish, introduced by Misra (2020), has been used as an alternative to ReLU family and swish. Mish has been embedded to the state-of-art object recognition and localization system YOLO (you only need to look once) v4, and upstaged activation functions such as ReLU, LeakyRELU in benchmarks like Imagenet 1k and MS-COCO, using CSP-DarkNet-53, ResNet-50 as the backbone. It has the math expression like this

$$f(x) = x\tanh\left(softplus(x)\right) \tag{9.9}$$

It is smoother that ReLU family and overcomes the "negative vanish" problem as its variants.

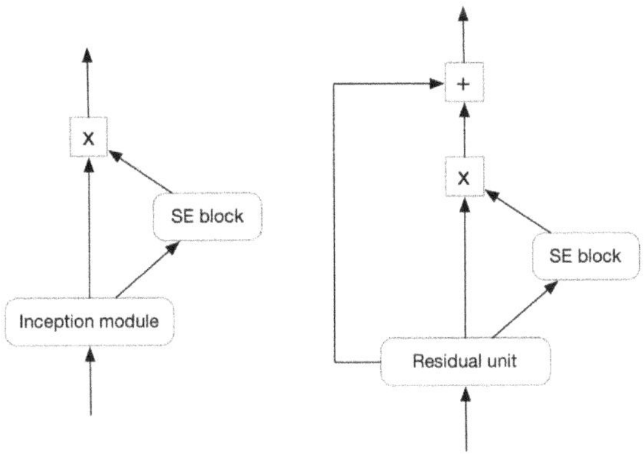

FIGURE 9.14 Two patterns of SENet.

9.5.3.2 SENet: Squeeze and Excitation Network

Proposed by Hu et al. (2020), SENet have achieved the best title of ILSVRC 2017 challenge. This structure is an extension of existing architectures such as Inception network from Google and ResNet, and boosts their performance. SENet won the title by an amazing 2.25% top-five error rate. There are two ways to implement the SENet: with Inception module or residual unit. In this paper, the Residual units are adopted with SENet. Both architectures are shown by Figure 9.14.

Its mechanism is a little bit like adaptive bit map algorithm. SENet can emphasize the important feature maps of image and give them a larger weight factor compared to the less important ones. The factor of the weight is between 0 and 1, and can be back-propagated.

9.5.3.3 Transposed Convolution

Unlike the usually implementation of bilinear interpolation for up-sampling to generate image, this paper instead uses transposed convolution technique. Transposed convolution is equivalent to first stretching the image by inserting zero value rows and columns, then play a regular convolution on the stretched area. In practice, it often achieves better performance than bilinear interpolation and faster. It is the core up-sampling technique in super-resolution.

9.5.3.4 Revised Enlighting GAN Architecture

9.5.3.4.1 Generator

In the generator design, we adopt U-Net as its backbone. U-Net is a concise and powerful architecture in today's mainstream usage; for instance, many cycleGANs have implemented it. We modify the network for a better representative ability of images, embedding the SENet, boost the up-sampling, by using transposed convolution, and replace leakyRelu for mish. The revised or improving generator network is shown in Figure 9.15, where there are six residual blocks working just as in ResNet and SENet to capture feature maps and calculate their factors. The model's input is poor-quality image sets, and the output is an enhanced image. The optimizer for back-propagation is RMSprop with its default setting.

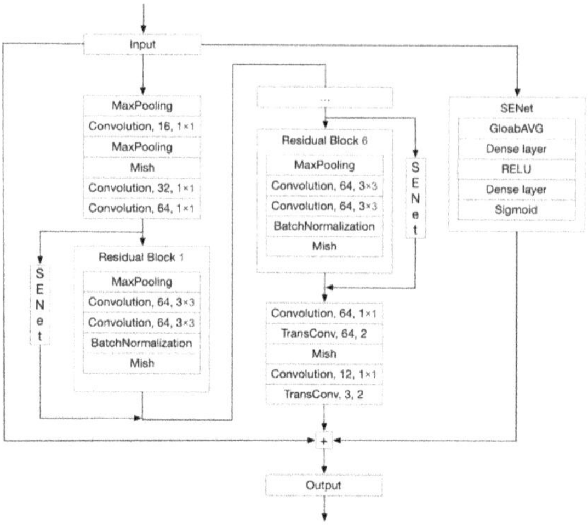

FIGURE 9.15 The improved generator model.

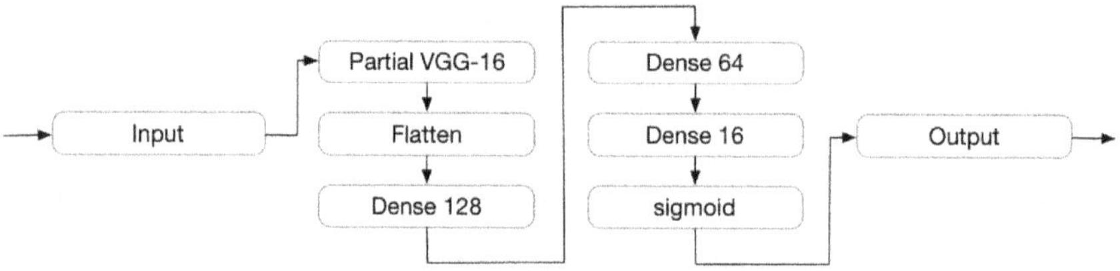

FIGURE 9.16 The discriminator structure.

9.5.3.4.2 Discriminator

Discriminator is composed of two parts: the VGG-16 model and dense network to compute prediction. Figure 9.16 demonstrates the structure of discriminator.

The partial VGG-16 is a submodel abstracted from original VGG-16. Since we just care about the segmentation information, the original model is curtailed to eliminate the last three dense layers which are for the purpose of classification. Moreover, the weights are trained on Imagenet 1000, for the well pre-trained model have a good detection ability.

In the training progress of discriminator, input is a combination of output of the generator and ground truth images.

9.5.3.4.3 Training

The training progress follows the classic training protocols of GANs. The training loop is split into two phases: phase 1 of train discriminator and phase 2 of train generator. In phase 1, first, we use generator to produce adjusted images, then combine these outputs with our ground truth images to form a test input. At last, we train discriminator with its training configurations. In phase 2, we set the discriminator to untrainable and then train the GAN model together by filling the generator poor image input and compile it with the training configurations. The workflow is shown in Figure 9.17.

Loss functions for training generator and discriminator is a little bit tricky. We reuse the loss function in Jiang et al. (2019).

9.5.4 Experiment

We choose the DIV2K dataset, in which there are 800 2K high-quality images (ground truth) and 3,300 poor-quality images of several types. They are arranged into training dataset with 3,200 images and test dataset with 100 images. The original resolution of these images is 2K; however, they are resized into size of $[224, 224, 3]$. Table 9.1 lists some image samples.

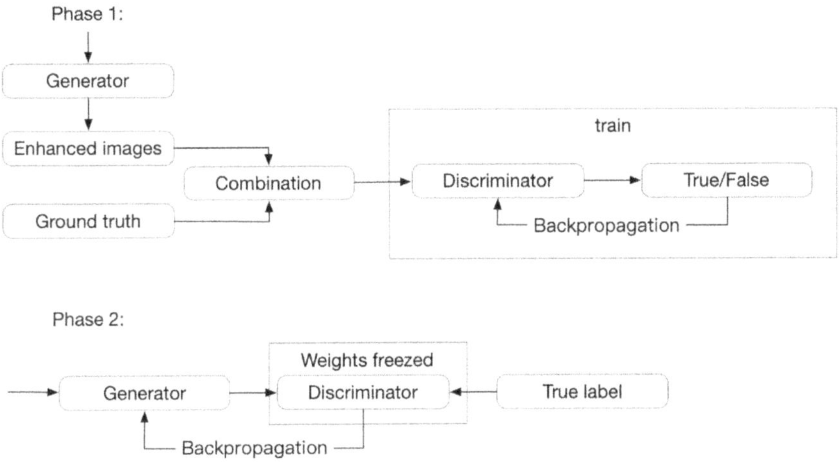

FIGURE 9.17 Training progress of one-way GAN.

TABLE 9.1
Image Samples

ground truth image sample	poor image sample 1	poor image sample 2
	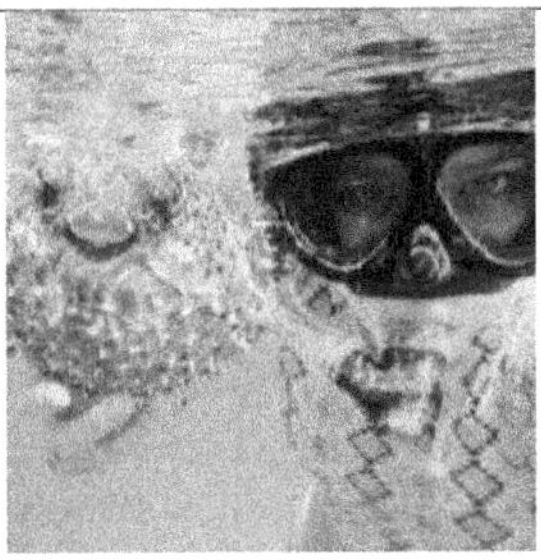	

TABLE 9.2
Training Configuration

Training Dataset Size	Ideal Image Data Size	Training Epoch	Batch Size	Test Data Size
3200	800	50	32	400

TABLE 9.3
PSNR Test Result

Number of Images	Smallest PSNR	Largest PSNR	Average PSNR
200	11.4	26.4	18.3

When the model training is over, both the generator model and its trained weights are saved on disk. Then in the training step, the model and weights are loaded to predict the enhancement. Table 9.2 lists the training configuration.

The metric to evaluate the effectiveness of the algorithm is to compare the model output of the test dataset with its ground truth image, calculate their average PSNR (peak signal-to-noise ratio), and compare their visual effect manually. The PSNR result is listed in Table 9.3. From the result, we can see that this model achieves a good PSNR over test images.

Different from PSNR test, whether the given image is good or poor is more or less a subjective question, in term of this view, some results are shown in Figure 9.18 to illustrate how well the improved model is working.

9.5.5 Comparisons with the Existing Method

The one-way GAN model is a model based on lots of knowledge we have gained in the field of deep learning. Different from classic methods such as histogram equalization, this kind of model could have the ability to solve more complex problems, such as illumination problem, while the classic methods are also very important and useful for some scenarios. Moreover, the classical methods are very fast, whereas the deep-learning models require much more time spent on data cleaning, model construction, and training.

Compared with classic standalone CNN model, the one-way GAN model could be trained without paired data but with comparable training time. The standalone CNN model often has a poor performance when facing the real-world images, such as photos generated by our mobiles, because the standalone models require paired images and could not make good use of the huge unpaired image datasets.

The cycleGAN dose has some advantages over one-way GAN; the two pairs of generators and discriminators can verify each other's result better and do not need to assign a label manually. The experiment has demonstrated that, albeit trickier and harder to converge, cycleGAN and its variants can do a job better than others so far.

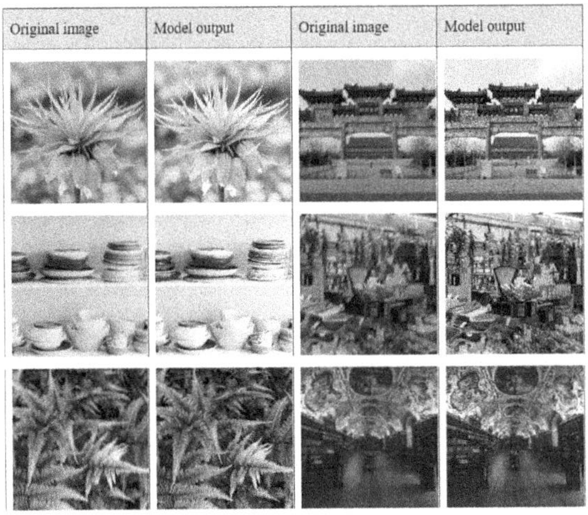

FIGURE 9.18 Some examples of the original images and the model output images.

FIGURE 9.19 Comparisons between our model, standalone model, and cycleGAN.

Nonetheless, it also requires a professional coding skill and insight into generative adversarial networks. Figure 9.19 shows the comparisons between our model, standalone model, and cycleGAN.

In summary, the one-way GAN model proposed to enhance the quality of images has a relatively good performance and is less tricky to implement. It shows its powerful adjusting ability by outperforming some classic and mainstream tools. However, image enhancement is still an active area for research, especially in some specific fields, such as astronomy, where researchers usually are stuck by the confusing images obtained from observation devices. There are still many problems waiting to be fixed. Besides, compared to fine-tuned two-way GANs, the one-way GAN lacks some cross-validation when it is generating the enhanced images.

REFERENCES

Chen, X., Jing, R., and Sun, C., "Attention mechanism feedback network for image super-resolution," *Journal of Electronic Imaging*, vol. 31, no. 4, Jul. 2022.

Chen, Y.-S., Wang, Y.-C., Kao, M.-H., and Chuang, Y.-Y., "Deep photo enhancer: Unpaired learning for image enhancement from photographs with GANs," *IEEE Conference on CVPR*, pp. 6306–6314, 2018.

Dong, C., Loy, C. C., He, K., and Tang, X., "Learning a deep convolutional network for image super-resolution," *European Conference on Computer Vision*, Springer, pp. 184–199, 2014.

Dong, C., Loy, C. C., He, K., and Tang, X., "Image super-resolution using deep convolutional networks," *IEEE Transactions on Pattern Analysis and Machine Intelligence*, vol. 38, no. 2, pp. 295–307, 2016.

Gatys, L. A., Ecker, A. S., and Bethge, M., "Texture synthesis using convolutional neural networks," *NIPS'15: Proceedings of the 29th International Conference on Neural Information Processing Systems*, vol. 1, pp. 262–270, Dec. 2015.

Goodfellow, I., Pouget-Abadie, J., Mirza, M., Xu, B., Warde-Farley, D., Ozair, S., Courville, A., and Bengio, Y., "Generative adversarial nets," *NIPS'14: Proceedings of the 28th International Conference on Neural Information Processing Systems – Volume 2*, pp. 2672–2680, Dec. 2014.

He, K., Sun, J., and Tang, X., "Guided image filtering," *IEEE Transactions on Pattern Analysis and Machine Intelligence*, vol. 35, no. 6, pp. 1397–1409, Jun. 2013.

Hu, J., Shen, L., Albanie, S., Sun, G., and Wu, E., "Squeeze-and-excitation networks," *IEEE Transactions on Pattern Analysis and Machine Intelligence*, vol. 42, no. 8, pp. 2011–2023, Aug. 2020.

Huang, W., Zhang, Y., Zhang, J., and Zheng, Y., "Convolutional neural network for pansharpening with spatial structure enhancement operator," *Remote Sensing*, vol. 13, no. 20, p. 4062, Jan. 2021.

Huang, A., Chen, J., and Jiang, N., "Reinforcement learning in low-rank mdps with density features," *Proceeding of International Conference on Machine Learning*, pp. 13710–13752, 2023.

Jiang, Y., Gong, X., Liu, D., Cheng, Y., Fang, C., Shen, X., Yang, J., Zhou, P., and Wang, Z., "EnlightenGAN: Deep light enhancement without paired supervision," *IEEE Transaction on Image Processing*, vol. 30, pp. 2340–2349, 2021.

Johnson, J., Alahi, A., and Fei-Fei, L., "Perceptual losses for real-time style transfer and super-resolution," *The European Conference on Computer Vision (ECCV)*, Springer, 2016.

Lee, D., Jang, K., Cho, S. Y., Lee, S., and Son, K., "A study on the super resolution combining spatial attention and channel attention," *Applied Science*, vol. 13, no. 6, p. 3408, 2023.

Misra, D., "Mish: A self regularized non-monotonic activation function," *The 31st British Machine Vision Virtual Conference*, 2020.

RichardWebster, B., Anthony, S., and Scheirer, W., "Psyphy: A psychophysics driven evaluation framework for visual recognition," *IEEE Transactions on Pattern Analysis and Machine Intelligence*, vol. 41, no. 9, pp. 2280–2286, 2018.

Ronneberger, O., Fischer, P., and Brox, T., "U-Net: Convolutional networks for biomedical image segmentation," *Proceeding of Conference on Medical Image Computing and Computer-Assisted Intervention (MICCAI)*, pp. 234–241, 2015.

Shao, G., Sun, Q., Gao, Y., Zhu, Q., Gao, F., and Zhang, J., "Sub-pixel convolutional neural network for image super-resolution reconstruction," *Electronics*, vol. 12, no. 17, p. 3572, 2023.

Tomasi, C., and Manduchi, R., "Bilateral filtering for gray and color images," *Sixth International Conference on Computer Vision*, pp. 839–846, Jan. 1998.

Vu, T., Nguyen, C. V., Pham, T. X., Luu, T. M., and Yoo, C. D., "Fast and efficient image quality enhancement via desubpixel convolutional neural networks," *Department of Electrical Engineering, Korea Advanced Institute of Science and Technology (KAIST)*, 2018.

Yan, Z., Zhang, H., Wang, B., Paris, S., and Yu, Y., "Automatic photo adjustment using deep neural networks," *ACM Transactions on Graphics*, vol. 35, no. 2, 2016.

Zhu, J., Park, T., Isola, P., and Efros, A. A., "Unpaired image-to-image translation using cycle-consistent adversarial networks," *Berkeley AI Research (BAIR) Laboratory*, UC Berkeley, 2017.

10 Development of Deep-Learning Framework for Mathematical Morphology

10.1 INTRODUCTION

Recent advances in computer vision are driven by deep learning, especially convolutional neural network (CNN). With the development of computer hardware, CNN has become popular during recent years, although it was introduced over 20 years ago (LeCun et al., 1989). The development of computer hardware makes the deep-learning framework become deeper and deeper. The original LeNet (LeCun et al., 1998) consists of 5 layers (i.e., 2 convolutional layers and 3 fully connected layers). In 2012 the 8-layer AlexNet (Krizhevsky et al., 2012) was able to achieve 16.4% top 5 error on ImageNet (Deng et al., 2009). Later the 19-layer VGG (Simonyan and Zisserman, 2014) and residual networks (He et al., 2016) respectively reach 7.5% top 5 error and 3.57% top 5 error on ImageNet. The CNN is not only successful in the classification tasks, but also in many other image-processing tasks, such as object detection and semantic segmentation. The R-CNN (Girshick et al., 2014), fast R-CNN (Girshick, 2015), faster R-CNN (Ren et al., 2015), and mask R-CNN (He et al., 2017) have offered fast training and flexibility on object detection problems using deep learning. The FCN (Long et al., 2015) applies the CNN without pooling and fully connected layers for solving semantic segmentation tasks.

Mathematical morphology has been applied in the areas ranging from automated vision inspection and detection to object recognition and image analysis. It is efficient in extracting topological and geometrical features, such as shape, size, connectivity, and distance. The convolution layers in deep learning play an important role as an automatic feature extraction process. However, the convolution, which is a weighted summation within the sliding window, merely captures the linear features, resulting in losing nonlinear information regarding image content, such as object position, meaningful region, shape, and size.

In mathematical morphology, dilation and erosion are two basic operations (Shih and Mitchell, 1989). Dilation is to select the maximum pixels in each sliding window, whereas erosion is to select the minimum pixels in each sliding window. Let dilation be $\phi(x)$ and erosion be $\psi(x)$. Morphological opening is defined as $\phi(\psi(x))$, whereas morphological closing is defined as $\psi(\phi(x))$. They serve as a tool for extracting representation of region shape, such as edges, skeleton, and convex hull

DOI: 10.1201/ 9781003474395-12

(Gonzalez and Woods, 2008). In industrial applications, mathematical morphology has been used for object segmentation (Zana and Klein, 2001), teeth segmentation (Said et al., 2006), texture analysis (Xia et al., 2006), and defect inspection. However, it is difficult to determine the proper pipelines of morphological operations, especially in the situation that needs sequences of morphological operations and structuring elements (Shih, 2009).

Masci et al. (2013) applied counter-harmonic mean (Angulo, 2010) to learn morphological operations in pipelines. In this paper we propose a more accurate and succinct approach to learn the mathematical morphology pipelines and the corresponding structuring elements. In addition, it provides a new nonlinear feature extractor in the deep-learning framework. Inspired by the histogram layer proposed by Sedighi and Fridrich (2017), we approximate the erosion/dilation operation by constructing a soft maximum/minimum function to create a morphological layer, which is constant and differentiable, so the morphological layer is back-propagable.

The remainder of this chapter is organized as follows. Section 10.2 introduces the convolutional autoencoder for mathematical morphology. Section 10.3 proposes the morphological layer along with the MorphNet. Section 10.4 demonstrates experimental results of the proposed models. Section 10.5 presents the multi-layer MorphNet. Finally, conclusions are drawn in Section 10.6.

10.2 CONVOLUTIONAL AUTOENCODER FOR MATHEMATICAL MORPHOLOGY

In this section we illustrate the approach of learning the mapping from an image to its morphological operated (i.e., dilated or eroded) image.

10.2.1 Convolutional Autoencoder

Autoencoder is a network for data compression and dimension reduction. It can learn a representation (code) of images and reconstruct the original image by the representation vector to achieve data compression. Convolutional autoencoder (Turchenko et al., 2017) extracts the representation of images by convolutional layers and pooling layers instead of dense layers, and reconstructs the images by deconvolutions. Learning the representation with convolutional layers can keep internal features of original images and store the information in a column vector to achieve data compression.

Figure 10.1 shows a general convolutional autoencoder with two convolutional layers and two deconvolutional layers. Since pooling layers are not weights layers, pooling is only demonstrated as increasing or decreasing of the size of rectangles.

10.2.2 Our Model

Inspired by the U-Net (Ronneberger et al., 2015), we construct an autoencoder network with skip connections to learn the morphology-operated images. The skip connection is a method that skips some layers and directly links the mirror layers.

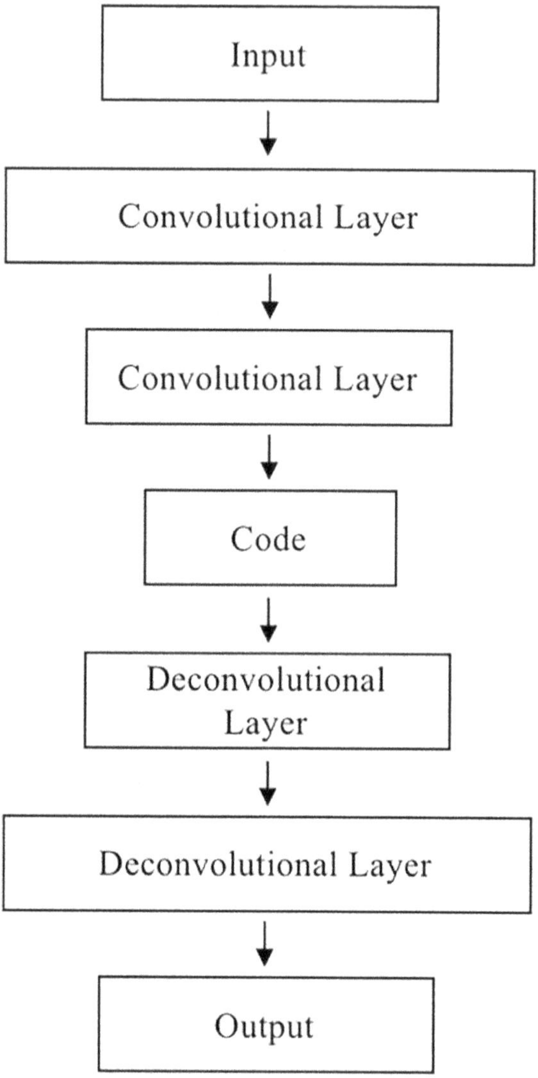

FIGURE 10.1 Structure of convolutional autoencoder.

Figure 10.2 shows the proposed network to learn the morphological operations. The original convolutional autoencoder can only learn the linear representation of original images so that we add skip connections between mirror layers to keep the structuring features of original images. Besides, convolution may lose some features in lower layers, and the skip connection can combine the lower-level features with the higher-level features, especially the structure information. The skip connection between mirror layers makes the network keep the whole generic and local specific features.

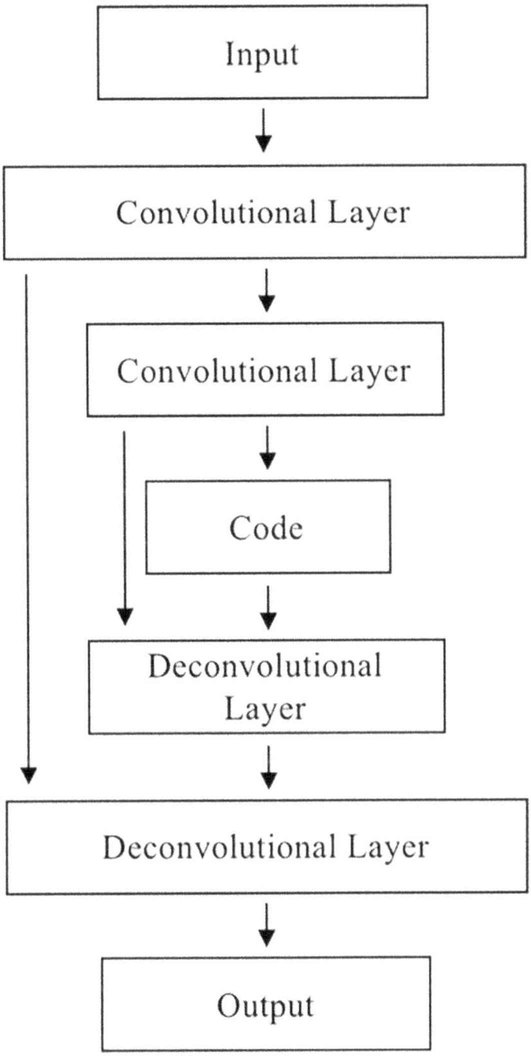

FIGURE 10.2 Convolutional autoencoder network with skip connections.

We adopt filters of size 3×3 in all the convolutional and deconvolutional layers. The first and second convolutional layers have 32 feature maps each, and the following deconvolutional layers keep the 32 channels. There is a batch normalization layer after every weight layer to help the feature maps in the same scale of weights. The network keeps the features and the weights in a similar range to avoid happening of outliers. This model maps the original images to the target images, which are the dilated/eroded results, and therefore, the distance between the target and the output images is minimized. The correctness of mapping will decide whether such a convolutional network can simulate the morphological operations. By this approach, we can match the output of the convolutional network to the dilated/eroded image. After all, it learns the target of dilation/erosion with the convolutional network.

10.3 THE PROPOSED MORPHNET

In this section, we propose the MorphNet, which consists of morphological layers, to simulate the dilation and erosion.

10.3.1 Previous Work

Masci et al. (2013) represented the dilation and erosion in deep-learning framework using counter-harmonic mean. For a grayscale image $f(x)$ and a filter $\omega(x)$, the PConv layer performs the following operation:

$$PConv(f;\omega,P)=\frac{(f^{P+1}\times\omega)(x)}{(f^{P}\times\omega)(x)}=\frac{\int_{y\in W(x)} f^{P+1}(y)\omega(x-y)dy}{\int_{y\in W(x)} f^{P}(y)\omega(x-y)dy} \tag{10.1}$$

where "*" denotes the convolution operation, P is a scalar, f^P is an image, and $W(y)$ is the support window of filter ω centered at point y. If $P<0$, it is pseudo-erosion; if $P>0$, it is pseudo-dilation; if $P=0$, it is a linear convolution. Since P cannot be infinity, this equation can only represent the pseudo-erosion and pseudo-dilation.

10.3.2 Approximation of Dilation and Erosion

Dilation and erosion are the maximum and minimum functions, respectively. Since the maximum function and the minimum function in each sliding window are non-differentiable, they cannot be directly placed into the deep neural network. We adopt the soft maximum to alternate the maximum function as

$$F(x_1,x_2,\ldots,x_n)=\ln k\left(e^{x_1}+e^{x_2}+\ldots+e^{x_n}\right)/k \tag{10.2}$$

where k is a positive constant. The soft maximum approximates the hard maximum but rounds off the corners. Figure 10.3 show the graphs of hard maximum and soft maximum, in which soft maximum can approximate hard maximum very well. The soft maximum function is smooth in the whole coordinates but has no sudden changes in direction. Besides, the soft maximum function is differentiable, making it work for the optimization algorithms. Figures 10.3(b) and 10.3(c) indicate that soft maximum approaches hard maximum when k increases.

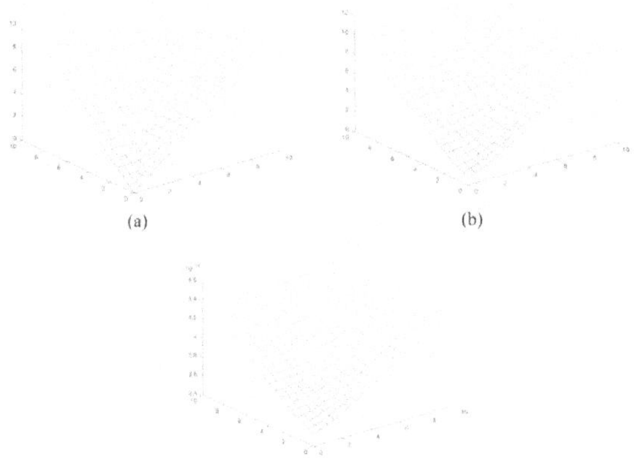

(a) (b) (c)

FIGURE 10.3 (a) The graph of hard maximum function, (b) the graph of soft maximum function when $k=1$, (c) the graph of soft maximum function when $k=10^{15}$.

Similar to soft maximum, the soft minimum can be denoted as

$$F(x_1, x_2, \ldots, x_n) = -\ln k\left(e^{-x_1} + e^{-x_2} + \ldots + e^{-x_n}\right)/k \tag{10.3}$$

10.3.3 Morphological Layer

With the differential approximation of dilation and erosion, we can construct the morphological layer. We introduce the weights ω in the dilation layer, so the expression of dilation layer is

$$Dil = \ln\left(\sum_{i=1}^{n} e^{\omega_i x_i}\right) \tag{10.4}$$

where x_i is the i -th pixel in the sliding window.

Similarly, the expression of erosion layer is

$$Ero = -\ln\left(\sum_{i=1}^{n} e^{-\omega_i x_i}\right) \tag{10.5}$$

where x_i is the i -th pixel in the sliding window. The weight is the structuring element in mathematical morphology, acting as the filter in convolution. Typically, we use structuring elements of sizes 3×3 and 5×5. The gradient of such a layer is computed by back-propagation (Werbos, 1974) with chain rule.

The objective function is denoted as $J(\omega;\hat{y},y)$, where ω is the weight, $\hat{y}$ is the output of the network, and y is the target of the network. The following is the gradient of l-th layer:

$$\frac{\partial J(\omega;\hat{y},y)}{\partial \omega_i^{(l)}} = \frac{\partial J(\omega;\hat{y},y)}{\partial z^{(l)}}\frac{\partial z^{(l)}}{\partial \omega_i} = \frac{\partial J(\omega;\hat{y},y)}{\partial z^{(l)}}\frac{e^{\omega_i x_i}x_i}{\sum_{i=1}^{n} e^{w_i x_i}} \tag{10.6}$$

where x_i is the i-th pixel in the sliding window, ω_i is the i-th pixel of weight ω, and $z^{(l)}$ is the error of l-th layer.

In contrast to the method in (Masci et al., 2013), which gave a rough approximation of dilation and erosion, we simulate dilation and erosion more accurately. In equation (10.1), P cannot be infinity so that it can only represent pseudo-dilation and pseudo-erosion. In our proposed method, we approach dilation and erosion by approximating maximum and minimum functions in a much more straightforward and efficient way.

10.3.4 Structure of the Model

To learn the structuring element from the target morphology-operated image, we construct different models with morphological layers. Figure 10.4 shows the structure of a single-layer MorphNet. We minimize the distance between the output and the target morphology-operated images by back-propagation.

The input of MorphNet is the original image, and the dilated/eroded image from the training set is the target image. The output would be the image after a single approximated morphological operation. We intend to minimize the distance between the output and the target images to learn the structuring element.

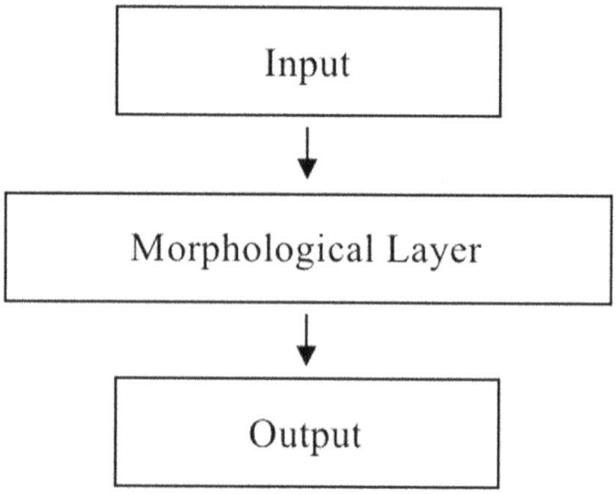

FIGURE 10.4 Structure of a single-layer MorphNet.

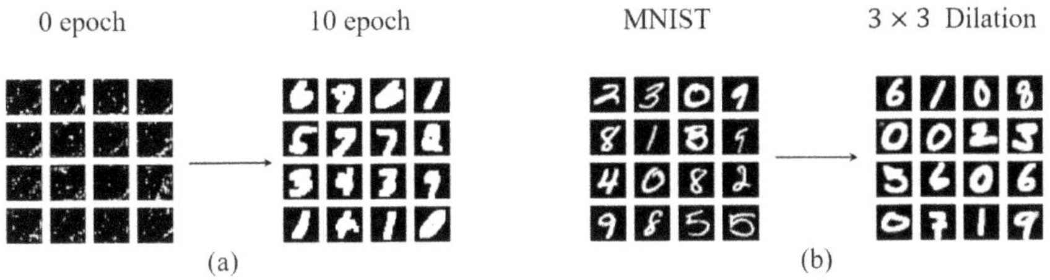

FIGURE 10.5 (a) The result of convolutional network, (b) the original and target images.

10.4 EXPERIMENTAL RESULTS

The MNIST dataset (LeCun, 1998) is trained on Nvidia Titan X in our experiment. We construct the convolutional network with Keras and minimize the square-root between the output and the target images. Let $\hat{y}_i$ denote the predicted image and y_i denote the target image, where $i = 1, 2, \ldots, n$. The square-root loss between the predicted and the target images is $loss = \frac{1}{n}\sum_{i=1}^{n}\left(\hat{y}_i - y_i\right)^2$. The learning algorithm is the stochastic gradient descent (Bottou, 2010). The training set consists of 5,000 images, which are randomly selected from the MNIST dataset. The target images are determined by a dilation from the training set.

Figure 10.5 shows that the convolutional network can learn the target from original images after 10 epochs. The convolutional network successfully learns the features of target images. After 10 epochs, the convolutional network enlarges the edge of digits. However, it cannot learn the erosion, and takes several hours to finish the whole training process. Most importantly, it can match the output image to the target image; however, the morphological operations remain unknown with structuring elements.

10.4.1 Learning Dilation and Erosion by MorphNet

10.4.1.1 The Correctness of Morphological Layer

We verify the correctness of morphological layer by an approximated morphological operation with a structuring element. Figure 10.6 shows the comparison of resulting images and the dilated/eroded images from MNIST. The first row shows the original images, the second row shows the target images, and the third row shows the results after the dilation/erosion layer. From rows 2 and 3, we observe that the morphological layer has almost the same influence on original images, whereas the dilation enlarges the digits and the erosion shrinks them. The output images of the network are almost identical to the target images by human vision.

In this paper, we test the proposed MorphNet on MNIST dataset. Since the content in the MNIST images is digit, our goal is to compare the semantic information of the target images and predictions of MorphNet. We adopt the cosine distance (Zhang and Lu, 2003) to measure the similarity between the target and output images to verify the correctness of morphological layer. Given two images S and K, the similarity is

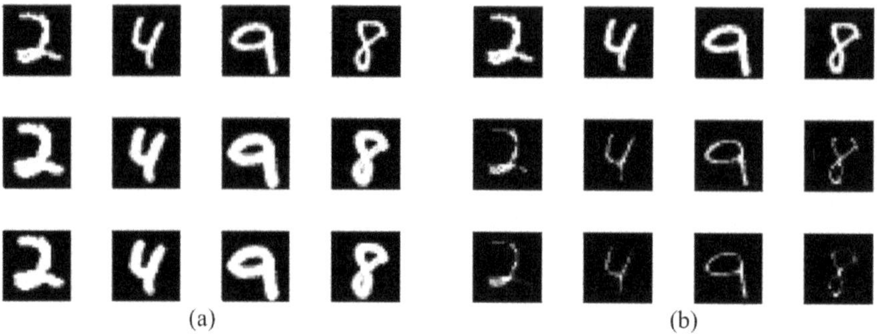

FIGURE 10.6 (a) The dilation layer, (b) the erosion layer. The first row shows the original images, the second row shows the target images, and the third row shows the output of the network.

TABLE 10.1
The Average Similarity of the Selected Network Output and the Target Images

	Dilation Layer	Erosion Layer
Average similarity	100%	99%

$$s = \frac{\sum_{i,j=1}^{n} \left(S_{ij} \times K_{ij}\right)^2}{\sum_{i,j=1}^{n} S_{ij}^2 \times \sum_{i,j=1}^{n} K_{ij}^2} \tag{10.7}$$

where S_{ij} is the pixel at i-th row, j-th column of S, and K_{ij} is the pixel at i-th row, j-th column of K. The bigger the s is, the two images will be more similar to each other. When s is 1, it indicates that two images are 100% identical to each other. Table 10.1 shows the average similarity of the selected network output and the target images. We feed 100 images from MNIST dataset to calculate the similarity. The results show that the morphological layer simulates the dilation at 100% similarity and the erosion at 99% similarity.

10.4.1.2 Learning the Structuring Element with MorphNet

We evaluate the morphological layer by constructing MorphNet and learning the structuring element from the target morphology-operated images. We learn the structuring element by setting different target morphology-operated images and diagonal, horizontal, and vertical structuring elements. The performance of MorphNet is evaluated by testing whether the MorphNet can learn the similar filters after several training epochs. The single MorphNet can successfully learn the selected structuring element when the target images are created. The training set contains 5,000 images. After 10 epochs, MorphNet can successfully converge and learn the correct

0	0	0		1	0	0		0	1	0		0	1	0
1	1	1		0	1	0		0	1	0		1	1	1
0	0	0		0	0	1		0	1	0		0	1	0

(a)

0	0	0		1	0	0		0	1	0		0	1	0
1	1	1		0	1	0		0	1	0		1	0	1
0	0	0		0	0	1		0	1	0		0	1	0

(b)

0	0	0	0	1								1	0	0	0	1
0	0	0	1	0								0	1	0	1	0
0	0	1	0	0		1	1	1	1	1		0	0	1	0	0
0	1	0	0	0								0	1	0	1	0
1	0	0	0	0								1	0	0	0	1

(c)

0	0	0	0	1								1	0	0	0	1
0	0	0	1	0								0	0	0	1	0
0	0	1	0	0		1	0	1	1	1		1	0	1	0	0
0	1	0	0	0								0	1	0	0	0
1	0	0	0	0								1	1	0	0	1

(d)

FIGURE 10.7 The original structuring elements and the structuring elements learned by the dilation layer MorphNet. (a) The 3×3 structuring elements are applied to the input images when creating target images, including horizontal, diagonal, vertical, and diamond structuring elements; (b) the corresponding structuring elements learned by the dilation layer MorphNet; (c) the original 5×5 structuring elements are applied to the input images when creating target images, including 45°, horizontal line, and crossing structuring elements; (d) the corresponding structuring elements learned by the dilation layer MorphNet.

weights as the structuring elements. The entire weights are updated by gradient, and the learning algorithm is Adadelta (Zeiler, 2012). Figure 10.7 shows the structuring elements learned by a single dilation layer MorphNet. It is observed that the dilation layer MorphNet can learn almost the same structuring elements as the original ones, although there are some biases between them.

Figures 10.8 (a) and (b) show an example of the single-layer MorphNet accurate prediction of the target images. Furthermore, MorphNet takes only a few minutes to finish the training, while the convolutional mapping needs hours to obtain the result.

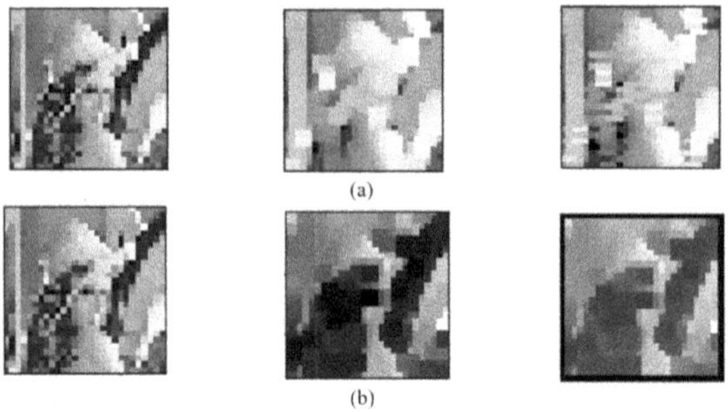

FIGURE 10.8 (a) The original Lena image, the dilated Lena image, and the prediction of MorphNet; (b) the original Lena image, the eroded Lena image, and the prediction of MorphNet.

Therefore, it is straightforward to adopt the morphological layer to simulate dilation and erosion. Besides, convolutional network cannot learn sequence of mathematical morphology and corresponding structuring elements.

10.5 MULTI-LAYER MORPHNET

In industrial applications, morphological operation pipelines usually perform better than a single morphological operation. In this section, multi-layer MorphNet is proposed to learn the proper pipeline of output images.

10.5.1 Learning the Target of Multiple Morphological Operations

Learning the target of morphological operations is difficult by traditional methods since there are multiple morphological operations in sequence as well as their corresponding structuring elements. With the help of morphological layer, we can easily learn the target of multiple morphological operations by multi-layer MorphNet. For instance, we apply three dilation operations on the grayscale images in the MNIST dataset as target images, and construct a three-layer MorphNet to learn the target images. Figure 10.9 shows the three-layer MorphNet. Let the input image be denoted as f and the three structuring elements be g_1, g_2, and g_3. Let $\oplus$ denote dilation or erosion, so the target image by three operations can be denoted as $((f \oplus g_1) \oplus g_2) \oplus g_3$. The weights of morphological layer can be considered as structuring elements, so the output of MorphNet would be

$$\text{Output} = f \oplus (g_1 \oplus g_2 \oplus g_3) \tag{10.8}$$

according to the property: $((f \oplus g_1) \oplus g_2) \oplus g_3 = f \oplus (g_1 \oplus g_2 \oplus g_3)$. If the output is equal to $((f \oplus g_1) \oplus g_2) \oplus g_3$, we can verify that MorphNet has the same property.

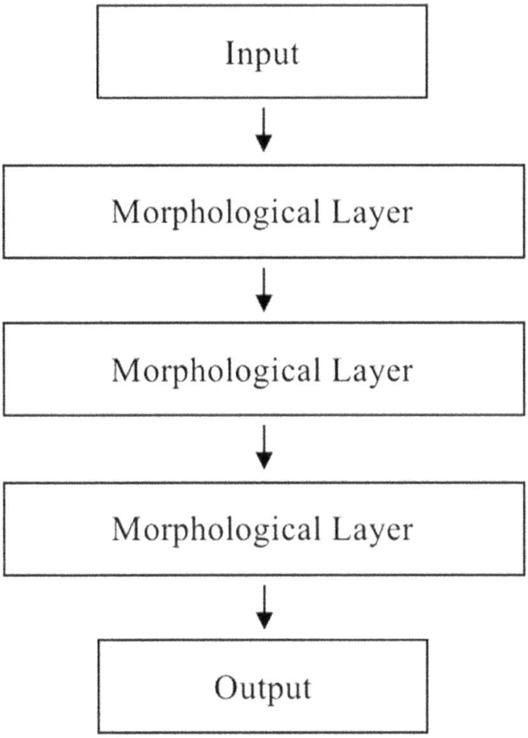

FIGURE 10.9 The three-layer MorphNet.

10.5.2 Results on Learning Multiple Morphological Operations

Based on the proposed model, we apply the multi-layer MorphNet on the MNIST dataset. The training set consists of 5,000 images, which are randomly selected from the MNIST dataset. All the images are grayscale with size 28×28.

Figure 10.10 shows the comparison of target images and network outputs. The target images are applied by three dilations with the diagonal structuring element of size 3 × 3. It is observed that the network can learn the target images as well as the structuring elements correctly. We use equation (10.7) to calculate the similarity of the target images and the predictions of network for evaluating the performance of multi-layer MorphNet. The average similarity of the selected pair of dilation multi-layer MorphNet is 100%.

Figure 10.11 shows the comparison of the target and the network output images, where the target images are applied by three dilations with three different structuring elements. The three structuring elements are diagonal, vertical, and horizontal of size 3 × 3. We compute the average similarity between the target images and the predictions of dilation multi-layer MorphNet and obtain 100%. The success of three-layer MorphNet proves that multi-layer MorphNet can learn multiple morphological operations by structuring elements correctly.

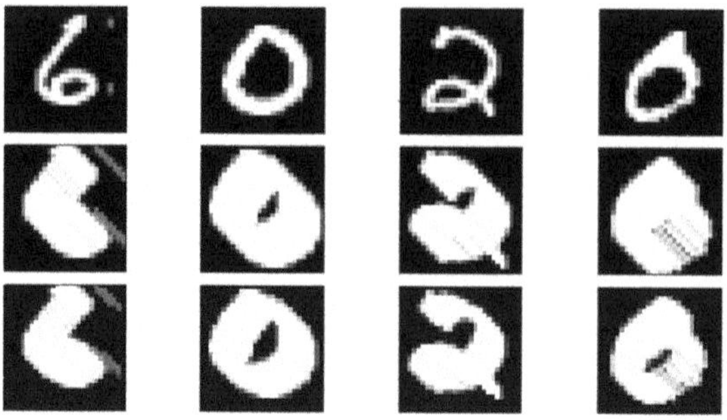

FIGURE 10.10 The comparison of the target images and the prediction of network when the target images are made by applying three dilation operations with the diagonal structuring elements of size 3 × 3. The first row shows the original images, the second row shows the target images, and the third row shows the output images.

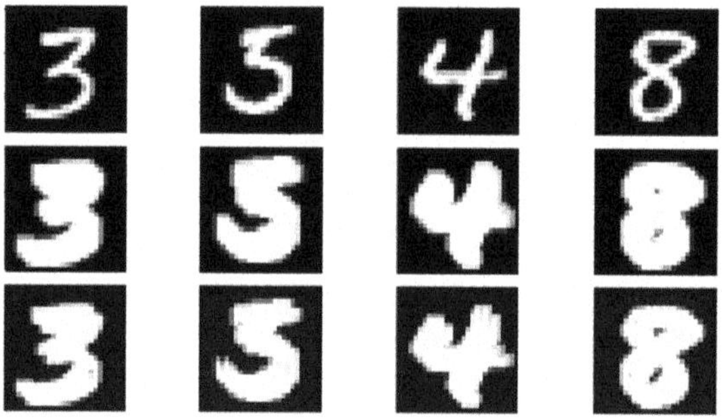

FIGURE 10.11 The comparison of target images and the prediction of network when the target images are made by three dilations with three different structuring elements. The first row shows the original images, the second row shows the target images, and the third row shows the output images.

10.6 CONCLUSIONS

We have presented the approach of deep-learning framework for morphological operations. Determining the proper pipeline of morphological operations and the corresponding structuring elements are cumbersome and time-consuming tasks in industrial applications. We develop the morphological layer, which makes morphological operations back-propagable in deep-learning networks. With the single morphological layer MorphNet, we can easily determine the structuring element and the morphological operations. The multi-layer MorphNet can learn multiple morphological operations and structuring elements successfully. We add the nonlinear MorphNet for nonlinear feature extraction in the deep-learning network. It can replace the role of convolutional layer in CNN and offset the limitation of linear feature extractor.

REFERENCES

Angulo, J., "Pseudo-morphological image diffusion using the counter-harmonic paradigm," *International Conference on Advanced Concepts for Intelligent Vision Systems*, Berlin, Heidelberg, pp. 426–437, Springer, 2010.

Bottou, L., "Large-scale machine learning with stochastic gradient descent," *Proceedings of COMPSTAT'2010*, Paris, pp. 177–186, Physica-Verlag HD, 2010.

Deng, J., Dong, W., Socher, R., Li, L. J., Li, K., and Fei-Fei, L., "Imagenet: A large-scale hierarchical image database," *Proceedings of IEEE Conference on Computer Vision and Pattern Recognition*, pp. 248–255, 2009.

Girshick, R., "Fast r-cnn," arXiv:1504.08083, 2015.

Girshick, R., Donahue, J., Darrell, T., and Malik, J., "Rich feature hierarchies for accurate object detection and semantic segmentation," *Proceedings of IEEE Conference on Computer Vision and Pattern Recognition*, pp. 580–587, 2014.

Gonzalez, R. and Woods, R., *Digital Image Processing*, 3rd edition, Prentice Hall, 2008.

He, K., Gkioxari, G., Dollár, P., and Girshick, R., "Mask r-cnn," *Proceedings of IEEE International Conference on Computer Vision*, pp. 2980–2988, 2017.

He, K., Zhang, X., Ren, S., and Sun, J., "Deep residual learning for image recognition," *Proceedings of IEEE Conference on Computer Vision and Pattern Recognition*, pp. 770–778, 2016.

Krizhevsky, A., Sutskever, I., and Hinton, G. E., "Imagenet classification with deep convolutional neural networks," *Advances in Neural Information Processing Systems*, pp. 1097–1105, 2012.

LeCun, Y., *The MNIST Database of Handwritten Digits*. http://yann.lecun.com/exdb/mnist/, 1998.

LeCun, Y., Boser, B., Denker, J. S., Henderson, D., Howard, R. E., Hubbard, W., and Jackel, L. D., "Backpropagation applied to handwritten zip code recognition," *Neural Computation*, vol. 1, no. 4, pp. 541–551, 1989.

LeCun, Y., Bottou, L., Bengio, Y., and Haffner, P., "Gradient-based learning applied to document recognition," *Proceedings of the IEEE*, vol. 86, no. 11, pp. 2278–2324, 1998.

Long, J., Shelhamer, E., and Darrell, T., "Fully convolutional networks for semantic segmentation," *Proceedings of the IEEE Conference on Computer Vision and Pattern Recognition*, pp. 3431–3440, 2015.

Masci, J., Angulo, J., and Schmidhuber, J., "A learning framework for morphological operators using counter-harmonic mean," *Proceedings of International Symposium on Mathematical Morphology and Its Applications to Signal and Image Processing*, Springer, pp. 329–340, 2013.

Ren, S., He, K., Girshick, R., and Sum, J., "Faster r-cnn: Towards real-time object detection with region proposal networks," *Advances in Neural Information Processing Systems*, pp. 91–99, 2015.

Ronneberger, O., Fischer, P., and Brox, T., "U-net: Convolutional networks for biomedical image segmentation," *Proceedings of International Conference on Medical Image Computing and Computer-Assisted Intervention*, Springer, Cham, pp. 234–241, 2015.

Said, E. H., Fahmy, D. E. M., and Ammar, H. H., "Teeth segmentation in digitized dental X-ray films using mathematical morphology," *IEEE Transactions on Information Forensics and Security*, vol. 1, no. 2, pp. 178–189, 2006.

Sedighi, V. and Fridrich, J., "Histogram layer, moving convolutional neural networks towards feature-based steganalysis," *Proceedings of IS&T, Electronic Imaging, Media Watermarking, Security, and Forensics*, San Francisco, CA, pp. 50–55, Jan. 29–Feb. 2, 2017.

Shih, F. Y., *Image Processing and Mathematical Morphology: Fundamentals and Applications*, CRC Press, 2009.

Shih, F. Y. and Mitchell, O. R., "Threshold decomposition of gray-scale morphology into binary morphology," *IEEE Transactions on Pattern Analysis and Machine Intelligence*, vol. 11, no. 1, pp. 31–42, 1989.
Simonyan, K. and Zisserman, A., "Very deep convolutional networks for large-scale image recognition," arXiv:1409.1556, 2014.
Turchenko, V., Chalmers, E., and Luczak, A., "A deep convolutional autoencoder with pooling-unpooling layers in Caffe," arXiv:1701.04949, 2017.
Werbos, P., "Beyond regression: New fools for prediction and analysis in the behavioral sciences," Doctoral Dissertation, Harvard University, 1974.
Xia, Y., Feng, D., and Zhao, R., "Morphology-based multifractal estimation for texture segmentation," *IEEE Transaction on Image Processing*, vol. 15, no. 3, pp. 614–623, 2006.
Zana, F. and Klein, J. C., "Segmentation of vessel-like patterns using mathematical morphology and curvature evaluation," *IEEE Transaction on Image Processing*, vol. 10, no. 7, pp. 1010–1019, 2001.
Zeiler, M. D., "ADADELTA: An adaptive learning rate method," arXiv:1212.5701, 2012.
Zhang, D. and Lu, G., "Evaluation of similarity measurement for image retrieval," *Proceedings IEEE International on Conference of Neural Networks and Signal Processing*, Nanjing, IEEE, Dec. 14–17, 2003.

11 Deep Morphological Neural Networks

11.1 INTRODUCTION

Mathematical morphology represents a branch of biology that deals with the forms and structures of animals and plants. It analyzes the shapes and forms of objects. In computer vision, it is used as a tool to extract image components that are useful in the representation and description of object shape (Shih and Mitchell, 1989). It is mathematical in the sense that the analysis is based on set theory, topology, lattice algebra, function, and so on. There are two elementary operations, called dilation and erosion, to enlarge and shrink the objects respectively by applying a *structuring element* (SE) as a sliding window to identify the feature on digital images. It can be applied in image analysis to extract features, such as shapes, regions, edges, skeleton, and convex hull (Shih, 2009). Therefore, it facilitates the tasks of defect extraction (Chen et al., 2015), edge detection (Zhang et al., 2015), and image segmentation (Coliban et al., 2016; Shih and Liu, 2016).

Deep learning has become an efficient tool in computer vision since the development of computer hardware brings an increased computational capability. Many deep-learning structures have been proposed based on *convolutional neural network* (CNN) (He et al., 2016). For example, LeNet (LeCun et al., 1998) was proposed for document recognition. Nowadays, deepening the CNN structures can enable many computer vision applications, especially for object recognition (Simonyan and Zisserman, 2015).

In image processing by mathematical morphology, it is pretty time-consuming and cumbersome to determine the appropriate morphological operations and SE when given a desired target. From the perspective of neural networks and deep learning, researchers have proposed the concept of morphological layers. Different from the convolutional layers that compute the linear weighted summation in each kernel on an image, we propose the dilation and erosion layers in *morphological neural network* (MNN) to approximate local maxima and minima, therefore providing nonlinear feature extractors. In contrast to the often-used CNN that trains the weights in convolution kernels, the proposed MNN intends to learn the weights in SE automatically. Besides, the morphological layer can choose either dilation or erosion automatically to operate.

Since the maximum and minimum operations in mathematical morphology are not differentiable, we need to change them into smooth and approximate functions for neural networks. Ritter and Sussner (1996) presented an attempt of MNN formulated by image algebra (Ritter and Wilson, 1996). Masci et al. (2013) approximated the dilation and erosion in deep-learning framework using counter-harmonic mean. However, only the pseudo dilation and erosion can be achieved since the formulation requires infinite

DOI: 10.1201/9781003474395-13

integers. Shih et al. (2019) developed a deep-learning framework to tackle this issue and introduced smooth local maximum and minimum layers as the feature extractors, where only approximation of dilation and erosion by the binary SE is handled. Liu et al. (2021) presented the classification of chest X-ray images using MNN.

We significantly improve the shortcoming of approximate results by learning the non-flat SE and the morphological operations to achieve accurate results of dilation and erosion. Considering the strength of morphology in analyzing shape features on images, a residual MNN pipeline and its applications are presented to validate the practicality of the proposed layer. The MNN residual structure is compared against CNN of the same structure on several datasets related to shape features. Experimental results have validated the superior of the proposed MNN in these tasks.

The remainder of this chapter is organized as follows. Section 11.2 introduces the morphology layers and the residual morphological neural network for shape classifications. Section 11.3 presents the adaptive morphological layer for determining the proper morphological operations from the original images and the desired result images. Section 11.4 shows the experimental results and performance comparisons. Finally, conclusions are drawn in Section 11.5.

11.2 DEEP MORPHOLOGICAL NEURAL NETWORK

11.2.1 The Related Work

Morphological dilation and erosion are approximated using counter-harmonic mean in (Masci et al., 2013). For a grayscale image $f(x)$ and a kernel $\omega(x)$, the core is to define a PConv layer as

$$PConv(f;\omega,P)(x)=\frac{\left(f^{P+1}*\omega\right)(x)}{\left(f^{P}*\omega\right)(x)}=\left(f*_{P}\omega\right)(x) \tag{11.1}$$

where "*" denotes the convolution, and P is a scalar controlling the choice of operation. Note that $P<0$ is pseudo-erosion, $P>0$ is pseudo-dilation, and $P=0$ is standard convolution. Since it requires P to be an unachievable infinite number in performing a real dilation or erosion, the equation can be highly inaccurate in implementation.

Shih et al. (2019) represented the dilation and erosion using the soft maximum and the soft minimum functions. Although the dilation and erosion are approximated theoretically, it fails to learn the SE accurately. By training on the samples of input images and desired morphological images, a single-layer MNN always miss some elements on the SE. Figure 11.1 shows the single-layer MNN architecture. Figure 11.2 shows some SEs learned, from which we can observe some errors, which are caused by rounding.

11.2.2 The Proposed Morphological Layer

To avoid the aforementioned rounding errors, we propose to add a bias term for differential dilation and erosion, respectively, which do not round the floating points.

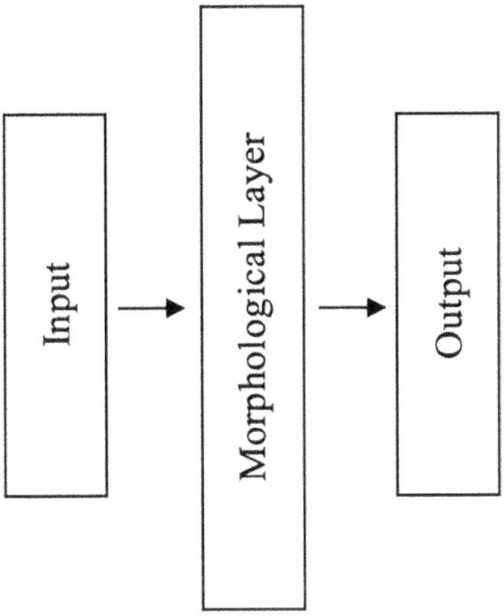

FIGURE 11.1 Architecture of the single-layer MNN.

0	0	0		1	0	0		0	1	0		0	1	0
1	1	1		0	1	0		0	1	0		1	1	1
0	0	0		0	0	1		0	1	0		0	1	0

(a)

0	0	0		1	0	0		0	1	0		0	1	0
1	1	1		0	1	0		0	1	0		1	0	1
0	0	0		0	0	1		0	1	0		0	1	0

(b)

0	0	0	0	1								1	0	0	0	1
0	0	0	1	0								0	1	0	1	0
0	0	1	0	0		1	1	1	1	1		0	0	1	0	0
0	1	0	0	0								0	1	0	1	0
1	0	0	0	0								1	0	0	0	1

(c)

0	0	0	0	1								1	0	0	0	1
0	0	0	1	0								0	0	0	1	0
0	0	1	0	0		1	0	1	1	1		1	0	1	0	0
0	1	0	0	0								0	1	0	0	0
1	0	0	0	0								1	1	0	0	1

(d)

FIGURE 11.2 (a) The horizontal, diagonal, vertical, and diamond 3×3 structuring elements applied to input images when creating target images; (b) the corresponding structuring elements learned by the single dilation layer MNN; (c) the original 45°, crossing 5×5 structuring elements, and horizontal line 1×5 structuring elements applied on the inputs images when creating target images; (d) the corresponding structuring elements learned by the single dilation layer MNN.

TABLE 11.1
Notations and Terms Used in This Paper

ω	The weight of morphological layer.
x	The input image of morphological layer.
b	The bias matrix.
$n = a \times b$	The total number of weights in a SE, a is the width and b is the height of SE.
W_i	The i-th element of the SE W.
X_i	The i-th element of the current masked window in the original image X.
$\hat{y}$	The output of the network.
y	The target of the network.

The maximum and minimum are computed in a sliding window while the neural networks intend to minimize the difference between the predicted and the target images. The notations and terms used in this paper are listed in Table 11.1.

Definition 11.1: The *differentiable binary dilation* of the j-th pixel in an output image $Y \in \mathbb{R}^n$ is defined as

$$Y_j = ln\left(\sum_{i=1}^{n} e^{W_i X_i}\right) \tag{11.2}$$

where W is a binary SE slid on the input image X, and the default stride is 1. We denote it as $W \oplus X$, where $W \in \mathbb{R}^n$ and $X \in \mathbb{R}^n$.

Definition 11.2: The *differentiable binary erosion* of the j-th pixel in an output image $Y \in \mathbb{R}^n$ is defined as

$$Y_j = -ln\left(\sum_{i=1}^{n} e^{-W_i X_i}\right) \tag{11.3}$$

where W is the binary SE slid on the input image X, and the default stride is 1. We denote it as $W \ominus X$, where $W \in \mathbb{R}^n$ and $X \in \mathbb{R}^n$.

When the binary dilation is learned,

$$ln\left(\sum_{i=1}^{n} e^{W_i X_i}\right) \geq max\left(W_i X_1, W_2 X_2, \ldots, W_n X_n\right) \tag{11.4}$$

which indicates that

$$ln\left(\sum_{i=1}^{n} e^{W_i X_i}\right) \geq X_i. \tag{11.5}$$

Therefore, we have

$$\sum_{i=1}^{n} e^{W_i X_i} \geq e^{X_i}. \tag{11.6}$$

It is clear that the equation is invalid. To tackle this issue, a slack variable ζ is added as

$$\sum_{i=1}^{n} e^{W_i X_i} \zeta \geq e^{X_i}. \tag{11.7}$$

Note that the equation is valid when $\zeta \geq \frac{e^{X_i}}{\sum_{i=1}^{n} e^{W_i X_i}}$. Similarly, a slack variable can be applied to validate the differentiable binary erosion.

We propose to apply the bias term to correct the rounding errors caused by the soft maximum and soft minimum functions. Instead of using the traditional method to apply one bias number in each filter, we define the bias as a matrix of the same size as the input image to correct the error of each point. In a binary dilation layer, the s -th feature map z of a binary dilation layer is

$$z^s = \omega \oplus x + b. \tag{11.8}$$

where $\omega \in \mathbb{R}^n$, $x \in \mathbb{R}^n$, and $b \in \mathbb{R}^n$.

The s-th feature map z in a binary erosion layer is

$$z^s = \omega \ominus x + b. \tag{11.9}$$

After adding b in equation (11.7), we obtain

$$\left(\sum_{i=1}^{n} e^{W_i X_i}\right) \cdot e^b \geq e^{X_i}. \tag{11.10}$$

Note that the equation is valid when $b \geq \ln \frac{e^{X_i}}{\sum_{i=1}^{n} e^{W_i X_i}}$. Therefore, dilation layer is correct if $b \geq \ln \frac{e^{X_i}}{\sum_{i=1}^{n} e^{W_i X_i}}$ after training. We can derive the correctness condition for the erosion layer similarly.

Definition 11.3: The *differentiable grayscale dilation* of the j-th pixel in an output image $Y \in \mathbb{R}^n$ is defined as

$$Y_j = ln\left(\sum_{i=1}^{n} e^{W_i + X_i}\right) \tag{11.11}$$

where W is the non-flat SE slid on the input image X, and the default stride is 1. We denote it as $W \oplus_g X$, where $W \in \mathbb{R}^n$, and $X \in \mathbb{R}^n$.

Definition 11.4: The *differentiable grayscale erosion* of the j-th pixel in an output image $Y \in \mathbb{R}^n$ is defined as

$$Y_j = -ln\left(\sum_{i=1}^{n} e^{-(W_i - X_i)}\right) \tag{11.12}$$

where W is the non-flat SE slid on the input image X, and the default stride is 1. We denote it as $W \ominus_g X$, where $W \in \mathbb{R}^n$, and $X \in \mathbb{R}^n$.

When learning a dilation with a non-flat SE, there should have

$$ln\left(\sum_{i=1}^{n} e^{W_i + X_i}\right) \geq max\left(W_i + X_1, \ldots, W_n + X_n\right). \tag{11.13}$$

We can obtain

$$\sum_{i=1}^{n} e^{W_i + X_i} \geq e^{W_i + X_i}. \tag{11.14}$$

It is clear that the equation is valid. Like the binary dilation layer, we apply a bias vector to correct the rounding errors caused by the soft maximum and soft minimum functions. The s-th feature map z^s of a grayscale dilation layer is

$$z^s = \omega \oplus_g x + b \tag{11.15}$$

where $\omega \in \mathbb{R}^n$, $x \in \mathbb{R}^n$, and $b \in \mathbb{R}^n$.

In grayscale erosion layer, the s-th feature map z^s is

$$z^s = \omega \ominus_g x + b \tag{11.16}$$

where $\omega \in \mathbb{R}^n$, $x \in \mathbb{R}^n$, and $b \in \mathbb{R}^n$.

The gradient of the proposed morphological layer is computed by the back-propagation (LeCun et al., 1998) following the chain rule. Denoted the objective function as $J\left(\omega, b; y, \hat{y}\right)$, the gradient $\delta^{(l)}$ of the l-th layer of the network with respect to weight E is

$$\delta^{(l)} = \frac{\partial J\left(\omega, b; y, \hat{y}\right)}{\partial \omega^{(l)}}. \tag{11.17}$$

Assuming that the learning rate is η, the weight ω of the l-th layer in t-th iteration is updated as

$$\omega_{t+1}^{(l)} = \omega_t^{(l)} - \eta\delta^{(l)}. \tag{11.18}$$

The bias b is updated as

$$b_{t+1}^{(l)} = b_t^{(l)} - \eta \frac{\partial J\left(\omega, b; y, \hat{y}\right)}{\partial b_t^{(l)}}. \tag{11.19}$$

11.2.3 Deep MNN with Stacked Morphological Layers

The target images with multiple morphological operations can be learned by stacking the morphological layers to construct a multi-layer MNN. Assuming that the l-th layer of multi-layer MNN is a dilation layer, the s-th feature map $z_s^{(l)} \in \mathbb{R}^n$ is

$$z_s^{(l)} = \omega \oplus z^{(l-1)} + b \tag{11.20}$$

where $\omega \in \mathbb{R}^n$, and $z^{(l-1)} \in \mathbb{R}^n$ is the output of $(l-1)$-th layer.

If the l-th layer of multi-layer MNN is an erosion layer, the s-th feature map of the output $z \in \mathbb{R}^n$ of current layer will become:

$$z_s^{(l)} = \omega \ominus z^{(l-1)} + b \tag{11.21}$$

where $\omega \in \mathbb{R}^n$.

Figure 11.3 illustrates the architecture of the multi-layer deep MNN. The inputs are the original images, and the outputs are the predictions of network after multiple

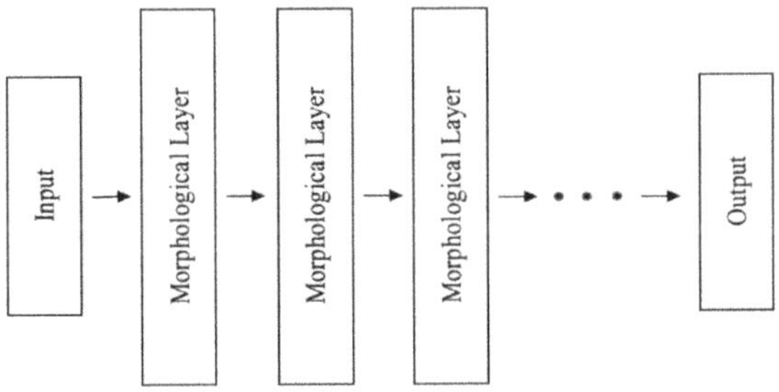

FIGURE 11.3 The multi-layer deep morphological neural network.

morphological layers. The target images are created by a sequence of morphological operations. At convergence, the deep MNN can learn the SE that minimizes the distance between the input and the target images.

The gradient of multi-layer DMNN is computed by back-propagation with chain rule. Let the objective function be denoted as $J\left(\omega, b; y, \hat{y}\right)$. The gradient $\delta^{(l)}$ of the l-th layer with respect to weight ω can be expressed as

$$\delta^{(l)} = \frac{\partial J\left(\omega, b; y, \hat{y}\right)}{\partial \omega^{(l)}} = \frac{\partial J\left(\omega, b; y, \hat{y}\right)}{\partial z^{(l)}} \frac{\partial}{\partial \omega} \sigma\left(z^{(l)}\right) \tag{11.22}$$

where $\sigma(\cdot)$ is the activation function. Assuming that the learning rate is η, the weight ω of the l-th layer in iteration t is updated by

$$\omega_{t+1}^{(l)} = \omega_t^{(l)} - \eta \delta^{(l)}. \tag{11.23}$$

11.2.4 Residual MNN

Mathematical morphology is designed to deal with shapes and structures (Haralick et al., 1987; Zana and Klein, 2001) in applications. In pattern recognition, mathematical morphology is used for pre-processing and feature extraction. In this residual MNN, we aim to apply opening with circular SE on the original image to round the corners of a shape and subtract the rounded-corner image from the original image. The morphological residuals that indicate the corners of a shape can help classification. Figure 11.4 shows an example of a morphological residual model.

We construct this residual MNN for shape classification as shown in Figure 11.5. The input of the neural network is batches of images and an erosion layer followed by a dilation layer means applying an opening on the input images. After the subtraction layer, the neural network finishes the pre-processing progress and delivers the residuals to a classifier. There are two fully connected layers to take the votes of each pixel in the residuals.

The configuration of the residual MNN is shown in Table 11.2. The number of channels m should be the same in each layer.

FIGURE 11.4 The morphological residual model. Applying an opening on the original image with a circle structuring element and then subtraction of the result image from the original image can obtain the morphological residuals.

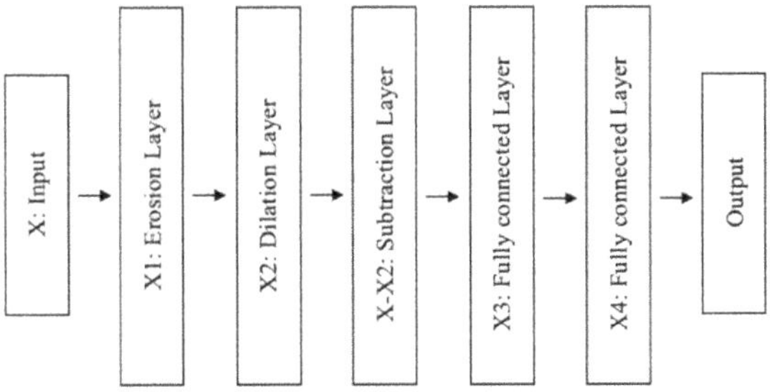

FIGURE 11.5 The architecture of residual MNN.

TABLE 11.2
The Configuration of Residual Morphological Neural Network

	Input
1	Erosion $3 \times 3 \times m$
2	Dilation $3 \times 3 \times m$
3	Subtraction m
4	FC-1024
5	FC-512
6	Soft-max

The residual MNN can be trained by back-propagation. The weights of dilation layer, erosion layer and fully connected layers are updated by equations (11.24) and (11.25). The weights are not updated in the subtraction layer; i.e., the residual MNN just transmits the gradient from the fourth layer to the second layer. Assuming that the gradient of the fourth layer is $\delta^{(4)}$, the gradient of subtraction layer is $\delta^{(3)} = \delta^{(4)}$.

11.3 ADAPTIVE MORPHOLOGICAL LAYER

Determining appropriate operation is a crucial task in the proposed MNN. Among various morphological operations, such as dilation, erosion, opening, and closing, the proposed layer aims to decide on the atomic ones including dilation and erosion.

Obviously, the difference between the differentiable binary dilation and the differentiable binary erosion is the sign before the weights. Therefore, we can apply a sign function to choose between maximum and minimum. To make the sign trainable, an extra weight is introduced in the MNN kernels. We call such a morphological layer with this extra weight as an *adaptive morphological layer*. Mathematically, the j-th pixel on the output $z \in \mathbb{R}^n$ of the dilation and erosion layers can be represented by

$$z_j = sign(a) \cdot ln\left(\sum_{i=1}^{n} e^{sign(a) \cdot \omega_i x_i}\right) + b. \quad (11.24)$$

where a is an extra trainable variable aside from ω_i and b. If $sign(a)$ is $+1$, the operation of current layer is dilation; if $sign(a)$ is -1, the operation of current layer is erosion. Since the sign function is not a continuous function and not differentiable, it cannot be used in the neural network. Therefore, we adopt a smooth sign function in the interval $[-1,1]$. The soft sign function and the hyperbolic tangent function are respectively shown here:

$$f(x)=\frac{x}{1+|x|} \tag{11.25}$$

$$g(x)=\frac{e^x-e^{-x}}{e^x+e^{-x}} \tag{11.26}$$

We replace the sign function in equation (11.24) by the hyperbolic tangent function and the soft sign function to maintain the gradient flow. Then the j-th pixel on the output $z \in \mathbb{R}^n$ of the adaptive morphological layer is computed in two ways:

$$z_j=\frac{a}{1+|a|}\cdot \ln\left(\sum_{i=1}^{n} e^{\frac{a}{1+|a|}\cdot \omega_i x_i}\right)+b \tag{11.27}$$

or

$$z_j=\frac{e^a-e^{-a}}{e^a+e^{-a}}\cdot \ln\left(\sum_{i=1}^{n} e^{\frac{e^a-e^{-a}}{e^a+e^{-a}}\cdot \omega_i x_i}\right)+b \tag{11.28}$$

where a is a trainable variable and $a \in \mathbb{R}$.

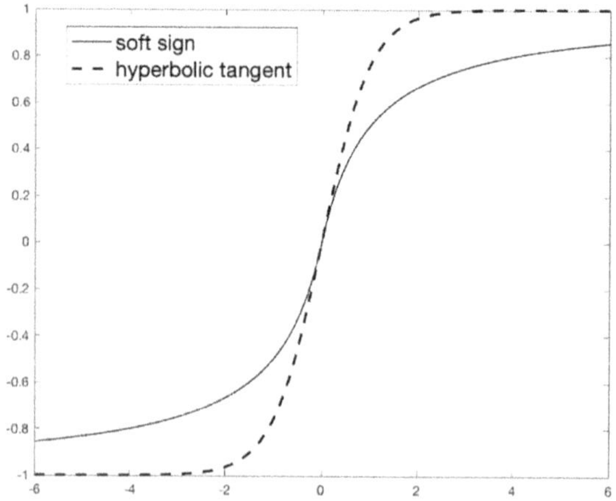

FIGURE 11.6 The soft sign function and hyperbolic tangent function.

Figure 11.6 compares the soft sign function and the hyperbolic tangent function. The hyperbolic tangent function reaches -1 and $+1$ ahead of the soft sign function in that the value of soft sign function is around -0.5 when tanh function reaches -1. Similarly, the value of the soft sign function lies around 0.5 when tanh function almost reaches $+1$. Therefore, the gradient of the soft sign is always smaller than the hyperbolic tangent function. In conclusion, the hyperbolic tangent function outperforms the soft sign function theoretically.

A single-layer MNN with the adaptive morphological layer is constructed to test performance. The input is the original images, and the target images are dilated or eroded images. The proposed adaptive morphological layer successfully learns both the target and the choice between dilation and erosion. Figure 11.7 shows the flow chart of detecting morphological operations by a single adaptive morphological layer MNN. The MNN minimizes the distance between target and output images. After converges, if the smooth sign function is $+1$, the target images are dilated images; if the smooth sign function is -1, the target images are eroded images.

In the adaptive layer, the gradients are updated by back-propagation with chain rule. The weight is updated by gradient descent for optimization. Let the objective function of such a neural network be $J\left(\omega, b, a; y, \hat{y}\right)$. The gradient $\delta^{(l)}$ of the l-th layer with respect to weight a is

$$\delta^{(l)} = \frac{\partial J\left(\omega, b, a; y, \hat{y}\right)}{\partial a^{(l)}} = \frac{\partial J\left(\omega, b, a; y, \hat{y}\right)}{\partial z^{(l)}} \frac{\partial z^{(l)}}{a^{(l)}} = \frac{\partial J\left(\omega, b, a; y, \hat{y}\right)}{\partial z^{(l)}} \phi'(a) \tag{11.29}$$

where $\phi(\cdot)$ is the soft sign or the hyperbolic tangent function.

Assuming that the learning rate is η, the weight a of the l-th layer in t-th iteration is updated by

$$a_{t+1}^{(l)} = a_t^{(l)} - \eta \delta^{(l)}. \tag{11.30}$$

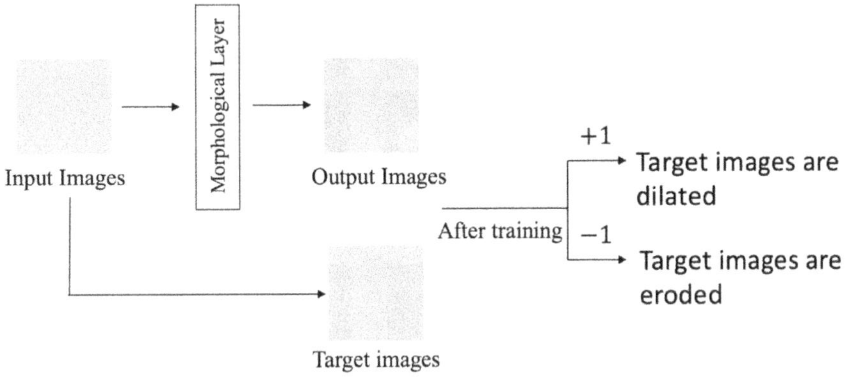

FIGURE 11.7 The flow chart of detecting morphological operations by a single adaptive morphological layer MNN.

11.4 EXPERIMENTAL RESULTS

In this section, we show the empirical studies of our proposed deep MNN to prove the correctness of our proposed morphological layers and show its performance of image classification tasks.

11.4.1 DATASETS

The experiments are performed on a 4 NVIDIA Titan X GPU system. We present our experimental results on four datasets – the MNIST dataset (LeCun et al., 1990), a self-created geometric shapes (SCGS) dataset, a German Traffic Sign Recognition Benchmark (GTSRB) dataset (Stallkamp et al., 2011), and a brain tumor dataset (Cheng, 2017) – to highlight the strength of the proposed MNN in analyzing shape features.

The MNIST (LeCun et al., 1990) dataset is a database consisting of 70,000 examples of handwritten digits 0 to 9. It has 60,000 training images and 10,000 testing images. They are all 28×28 grayscale images in 10 classes. The SCGS dataset contains 120,000 grayscale images of size 64×64 in five classes: ellipse, line, rectangle, triangle, and five-edge polygon. The images are created by randomly drawing white objects on a black background, where the size, position, and orientation are randomly initialized. There are 20,000 images in each class for training and 5,000 images used in each class for testing.

The GTSRB (Stallkamp et al., 2011) introduces a single-image, multi-class classification problem, and there are 42 classes in total. The images contain one traffic sign each, and each real-world traffic sign only occurs once. We resize all the images into 31×35, and select 31,367 images for training and 7,842 images for testing. All the images are converted to grayscale. The MRI brain tumor dataset (Cheng, 2017) contains 3,064 grayscale T1-weighted contrast-enhanced images from 233 patients with three kinds of brain tumor: meningioma (708 samples), glioma (1,426 samples), and pituitary tumor (930 samples). We resize all the images into 64×64 for classification, and 2,910 images are used for training and 154 images for testing. Figure 11.8 shows some examples from the four datasets.

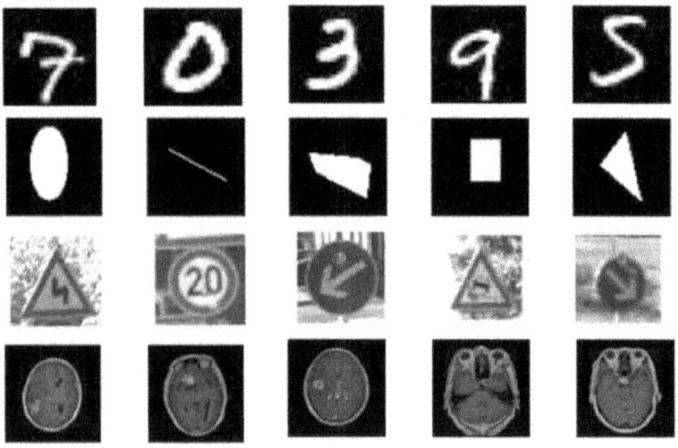

FIGURE 11.8 The examples from the four datasets in the experiments. The first row is the images from the MNIST dataset, the second row from the SCGS dataset, the third row from the GTSRB dataset, and the fourth row from the brain tumor dataset.

11.4.2 Learning the SE and Morphological Targets

We randomly select 10,000 images from MNIST dataset, and construct a single-layer MNN as shown in Figure 11.1 to learn a single binary SE. Mean squared error (MSE) is adopted to measure the distance between the target and predicted images to minimize the number outliers. The target images are created by applying dilation or erosion on the original input images. A mini-batch SGD (Li et al., 2014) with a batch size of 64 and the learning rate $\eta = 7.50$ are selected.

When learning binary SE, we adopt three examples, 3×3 diamond SE, 5×5 crossing SE, and 1×5 horizontal line SE. The experiment is repeated 100 times by randomly selecting 10,000 training images each time. The single-layer MNN has 100% accuracy on learning the 3×3 diamond SE and the 1×5 horizontal line SE, and 91% accuracy on the 5×5 crossing SE. Furthermore, if we increase the epochs from 20 to 100, the accuracy on learning the 5×5 crossing SE is increased to 97%. Figure 11.9 shows three examples, where the learned and the original structuring elements are identical.

All the settings in the grayscale morphology experiment are the same as the binary SE case, except that the targets are created by applying non-flat SE, the learning rate is $\eta = 1.0$ for learning dilated images, and $\eta = 0.5$ for learning eroded images.

We adopt the taxicab distance to measure the distance between the learned non-flat SE and the applied non-flat SE after training. Let two matrices be $\boldsymbol{A} = \left(a_{ij}\right)$ and $\boldsymbol{B} = \left(b_{ij}\right)$. The taxicab distance between $\boldsymbol{A}$ and $\boldsymbol{B}$ is computed as

$$d_1\left(\boldsymbol{A}, \boldsymbol{B}\right) = \sum\nolimits_{i=1}^{n} \sum\nolimits_{j=1}^{n} \left|a_{ij} - b_{ij}\right| \tag{11.31}$$

0	1	0
1	1	1
0	1	0

1	0	0	0	1
0	1	0	1	0
0	0	1	0	0
0	1	0	1	0
1	0	0	0	1

1	1	1	1	1

(a)

0	1	0
1	1	1
0	1	0

1	0	0	0	1
0	1	0	1	0
0	0	1	0	0
0	1	0	1	0
1	0	0	0	1

1	1	1	1	1

(b)

FIGURE 11.9 (a) The diamond 3×3 structuring element, crossing 5×5 structuring element, and 1×5 structuring element, (b) the learned structuring elements by a single dilation layer MNN after improvement.

For learning grayscale dilation and erosion, the MNN is converged in 20 epochs. For learning the morphological SE, 100 experiments are done on the same dataset to obtain the distance between the 3×3 SE and the learned SE. The average distance is 0.0706 in learning dilation and 0.0875 in learning erosion. For learning the morphological targets, the MSE is around 3.43×10^{-5} in dilation, and 7.59×10^{-5} in erosion.

Figure 11.10 show some examples of learning a non-flat SE. It is observed that the original non-flat SE and the learned SE are very close. Figure 11.11 shows the target images and the prediction of the network.

In mathematical morphology, opening and closing are also important, where the opening is an erosion followed by a dilation, and conversely, the closing is a dilation followed by an erosion. Therefore, we construct a two-layer MNN to learn the opening or closing. The corresponding targets are the opened or the closed image, and the MNN consists of two layers of dilation after erosion or erosion after dilation. The mini-batch SGD with a batch size of 64 and the learning rate $\eta = 10.0$ are selected. The loss is converged at around 0 within 10 epochs. For learning closing targets, the MSE is 3.71×10^{-2}; and the MSE is 2.90×10^{-2} when learning opening targets. Figure 11.12 shows some examples.

11.4.3 Learning the Morphological Operations

We choose 10,000 images from the MNIST dataset randomly to learn the morphological operations by the adaptive morphological layer. The MNN consists of a single adaptive morphological layer that minimizes the distance between dilated (or eroded) and the predicted images. At the convergence, the extra weight

0.2060	0.3234	0.6542
0.3551	0.5692	0.3950
0.6405	0.5834	0.5104

0.2086	0.3211	0.6521
0.3540	0.8055	0.4135
0.6261	0.5747	0.5097

(a)

0.8329	0.4865	0.9737
0.0440	0.8055	0.1752
0.6563	0.5816	0.0463

0.8361	0.4876	0.8951
0.0559	0.5054	0.1763
0.6585	0.5836	0.0573

(b)

FIGURE 11.10 (a) The top box shows the original structuring element and the bottom box shows the learned structuring elements by a single dilation layer MNN. (b) The top box shows the original structuring elements and the bottom box shows the learned structuring elements by a single erosion layer MNN. The MSE between the learned SEs and the ground truth SEs is 3.43×10^{-5} in dilation and 7.59×10^{-5} in erosion.

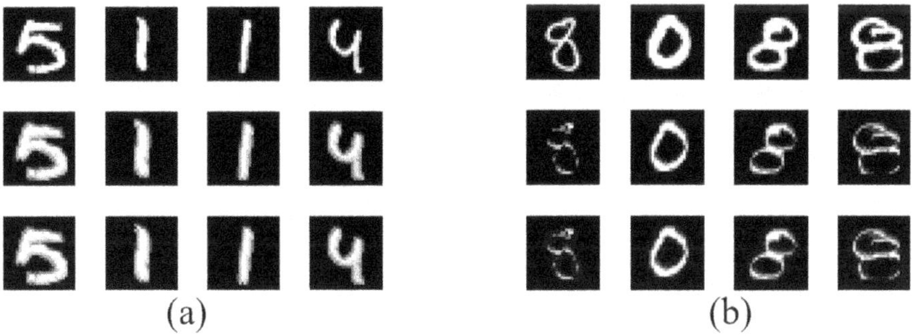

FIGURE 11.11 The results of learning grayscale (a) dilation and (b) erosion operations by MNN. The first row shows the original images, the second row shows the target images, and the third row shows the output of the network after training 20 epochs.

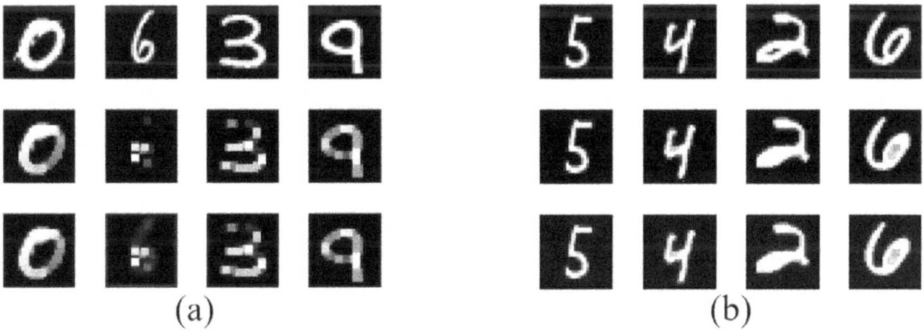

FIGURE 11.12 The results of learning (a) opening and (b) closing operations by DMNN. The first row shows the original images, the second row shows the target images, and the third row shows the output of the network after training 20 epochs. The MSE between the predictions and the ground truth is 3.71×10^{-2} when learning the closing operation and 2.90×10^{-2} when learning the opening operation.

a indicates the operation by its sign. The target images are predicted as dilated images if soft sign or hyperbolic tangent function is rounded to $+1$ and as eroded images if -1.

The mini-batch SGD is used to optimize the network. The batch size of 64 and the learning rate $\eta = 10.0$ are selected. The distance between the predicted and target images is measured by MSE loss. After 20 epochs, the single adaptive MNN converges, and the MSE loss decreases to around 3×10^{-4}. If the value of the smooth sign function larger than 0.5, we round it to 1; if the value of the smooth sign function smaller than −0.5, we round it to −1. The values of smooth sign function are in the interval $[-1,1]$. The experiment is repeated 100 times by randomly selected 10,000 images from MNIST each time. Table 11.3 shows the detection accuracy of dilation/erosion by two smooth sign functions.

TABLE 11.3
Detection Accuracy of Two Smooth Sign Functions

	Dilation	Erosion
Soft sign	100%	100%
Hyperbolic tangent	100%	100%

11.4.4 Image Classification

We adopt Adam optimizer to optimize the deep MNN in a mini-batch strategy with a batch size 64 and learning rate $\eta = 0.0001$. The residual MNN converges within 100 epochs for all the datasets. The testing accuracy of the residual MNN is 98.93% on the MNIST dataset, 98.89% on the SCGS dataset, 95.35% on the GTSRB, and 95.43% on MRI brain tumor dataset. We add a dropout layer after the second fully connected layer to prevent the overfitting when training on the GTSRB, the testing accuracy increases to 96.49%. Table 11.4 shows the configurations of the residual MNN when training on four datasets, where a indicates the number of filters applied in each layer.

To quantitatively compare the residual MNN against other methods, we add one more convolutional layer to extract more features and decrease the size of the filters from 5×5 to 3×3 in LeNet and name it as Modified LeNet (MLeNet). Table 11.5 shows the configuration of MLeNet.

Table 11.6 shows the comparisons between the residual MNN against some CNNs when $a=1$. Although the residual MNN loses on the testing accuracy as compared to some of CNNs, it has much fewer parameters. Especially in the feature extraction layers, the residual MNN has only 20 parameters in total, while the CNNs have at least thousands of parameters. We also show the comparison of the number of parameters in feature extraction layer of residual MNN with CNN in Table 11.7. From Tables 11.6 and 11.7, we conclude that residual MNN uses much fewer parameters in feature extraction layers without significantly compromising the model accuracy. The residual MNN has great a tradeoff between the computational efficiency and testing accuracy.

11.4.5 Ablation Studies

11.4.5.1 Studies on Shape Feature Extraction

To further demonstrate the advantages of the proposed morphological layers, we construct a CNN that has the same configuration with the residual MNN and compare their performance on shape related classification. Table 11.8 shows the configuration of the CNN, which is named as residual CNN, where b denotes the number of filters in each layer.

Table 11.9 shows another comparison of the residual CNN against the residual MNN on classifying the four datasets. When working on the brain tumor dataset, we remove the fourth layer shows in Table 11.8 to keep the configuration of residual CNN as same as the residual MNN in the fifth column of Table 11.4.

TABLE 11.4
Structures of Residual MNN on Four Datasets

	MNIST	SCGS	GTSRB	Brain Tumor
Erosion layer	$3\times3\times a$	$3\times3\times a$	$3\times3\times a$	$3\times3\times a$
Dilation layer	$3\times3\times a$	$3\times3\times a$	$3\times3\times a$	$3\times3\times a$
Subtraction layer	$28\times28\times a$	$64\times64\times a$	$31\times35\times a$	$64\times64\times a$
Fully connected layer	120	1024	1024	512
Fully connected layer	84	512	512	N/A
Output	10	5	43	3

TABLE 11.5
Configuration of MLeNet

	Input
1	Convolutional layer $3\times3\times16$
2	Max pooling 2×2.
3	Convolutional layer $3\times3\times32$.
4	Max pooling 2×2.
5	Convolutional layer $3\times3\times64$
6	Max pooling 2×2
7	Fully connected 2048×1
8	Fully connected 1024×1
9	Softmax

TABLE 11.6
Comparison of Residual MNN with Convolutional Neural Networks

Classifier	Dataset	Testing Accuracy	Number of Parameters
MCDNN (Ciregan, 2012)	MNIST	99.77%	2,682,470
Residual MNN	MNIST	98.93%	104,181
MLeNet	SCGS	99.50%	10,493,795
Residual MNN	SCGS	98.89%	4,721,175
MLeNet	GTSRB (Grayscale)	97.94%	4,202,339
Residual MNN	GTSRB (Grayscale)	96.49%	1,594,903
MLeNet	Brain tumor	96.10%	10,493,795
Residual MNN	Brain tumor	95.43%	4,721,175

In Table 11.9, when $a=1$ and $b=1$, the residual MNN has better testing accuracy on all these datasets than the residual CNN. When a = 16 and $b=16$, the residual MNN has better testing accuracy on GTSRB dataset. In brain tumor dataset, the residual MNN performs better when $a=16$ but loses a little bit when $a=1$. The

TABLE 11.7
Comparison of Number of Parameters in Feature Extraction Layer of Residual MNN with CNN

Model	Number of Parameters in Feature Extraction Layers
Residual MNN	20
MLeNet	2,912
MCDNN	739,900

TABLE 11.8
Configuration of Residual CNN

	Input
1	Convolutional layer $3\times3\times b$
2	Convolutional layer $3\times3\times b$
3	Subtraction layer $3\times3\times b$
4	Fully connected $2048\times b$
5	Fully connected $1024\times b$
6	Softmax

TABLE 11.9
Comparison of Residual MNN and Residual CNN

	Residual MNN ($a=1$)	Residual CNN ($b=1$)	Residual MNN ($a=16$)	Residual CNN ($b=16$)
MNIST	98.93%	97.14%	97.78%	98.18%
SCGS	98.89%	98.25%	98.90%	98.91%
GTSRB	96.49%	90.60%	97.48%	93.39%
Brain tumor	95.43%	96.10%	96.75%	94.15%

morphological layers perform better on MRI brain tumor dataset when each layer has multiple filters. Therefore, morphological layer outperforms convolutional layers if both neural networks have same structure in general. Especially on the GTSRB dataset that is closely related to shape features, morphological layer significantly improves the testing accuracy, which indicates that proposed MNN has advantages in shape feature extractions.

Therefore, with same number of parameters, the residual MNN loses a little bit on MNIST dataset but performs the best on GTSRB and brain tumor datasets which are related to shape features. In general, morphological layer works better on extracting shape features than convolutional layer when the residual MNN has same number of parameters as residual CNN.

11.4.5.2 Studies on the Efficiency of Proposed MNN

To further show the efficiency of the proposed residual MNN, we compare the inference time of residual MNN and LeNet. In the experiment, we train the models normally and compare the inference time of residual MNN with LeNet on one Titan X GPU for fair comparison. In SCGS dataset, the inference time of residual MNN is 11.58 seconds, when LeNet takes 20.13 seconds. We can see that residual MNN is twice faster than the LeNet during the model inference. Furthermore, we observe that it takes 59 seconds in average for residual MNN to finish one epoch of training, when LeNet takes 119 seconds in average. Therefore, we can easily see that residual MNN is not only efficient for training but also very efficient for model inference compared to CNN.

In summary, MNN has better accuracy in the selected datasets when it has the same number of parameters and the same structures with the CNN. In addition, MNN significantly saves parameters when it has similar accuracy with the CNN. The proposed residual MNN provides a tradeoff between model accuracy and model complexity.

11.5 CONCLUSIONS

We propose deep morphological neural networks in this paper that can learn both the operations and corresponding structuring elements in mathematical morphology. The promising performance of the proposed morphological layers, serving as an effective non-linear feature extractor, is confirmed theoretically and empirically. We also present an architecture of residual MNN for the feature extraction in shape classification tasks to validate the practicality of our morphological neural networks, which shows its superior by providing a good tradeoff between model accuracy and computational complexity.

REFERENCES

Chen, K., Zhang, Z., Chao, Y., Dai, M., and Shi, J., "Defects extraction for QFN based on mathematical morphology and modified region growing," *Proceedings of the IEEE International Conference on Mechatronics Automation*, Beijing, pp. 2426–2430, 2015.

Cheng, J., "Brain tumor dataset," *figshare*, 2, Apr. 2017.

Ciregan, D., Meier, U., and Schmidhuber, J., "Multi-column deep neural networks for image classification," *Proceedings of the IEEE Conference on Computer Vision and Pattern Recognition*, Providence, 2012, pp. 3642–3649.

Coliban, R., Ivanovici, M., and Richard, N., "Improved probabilistic pseudo-morphology for noise reduction in colour images," *IET Image Processing*, vol. 10, no. 7, pp. 505–514, Jul. 2016.

Haralick, R. M., Sternberg, S. R., and Zhuang, X., "Image analysis using mathematical morphology," *IEEE Transactions on Pattern Analysis and Machine Intelligence*, vol. PAMI-9, no. 4, pp. 532–550, Jul. 1987.

He, K., Zhang, X., Ren, S., and Sun, J., "Deep residual learning for image recognition," *Proceedings of the IEEE Conference on Computer Vision and Pattern Recognition*, Las Vegas, NV, pp. 770–778, 2016.

LeCun, Y., Boser, B. E., Denker, J. S., Henderson, D., Howard, R. E., Hubbard, W. E., and Jackel, L. D., "Handwritten digit recognition with a back-propagation network," *Proceedings of the Advances Neural Information Processing Systems*, pp. 396–404, MIT Press, 1990.

LeCun, Y., Bottou, L., Bengio, Y., and Haffner, P., "Gradient-based learning applied to document recognition," *Proceedings of the IEEE*, vol. 86, no. 11, pp. 2278–2324, Nov. 1998.

Li, M., Zhang, T., Chen, Y., and Smola, A. J., "Efficient mini-batch training for stochastic optimization," *Proceedings of the 20th ACM SIGKDD International Conference on Knowledge Discovery and Data Mining*, New York, NY, pp. 661–670, Aug. 24–27, 2014.

Liu, S., Shih, F. Y., and Zhong, X. "Classification of chest X-ray images using novel adaptive morphological neural networks," *Pattern Recognition and Artificial Intelligence*, vol. 35, no. 10, p. 2157006 (14 pages), Oct. 2021.

Masci, J., Angulo, J., and Schmidhuber, J., "A learning framework for morphological operators using counter-harmonic mean," *Proceedings of the 11th International Symposium on Mathematical Morphology Its Applications to Signal Image Processing*, Springer, Berlin, Heidelberg, pp. 329–340, 2013.

Ritter, G. X. and Sussner, P., "An introduction to morphological neural networks," *Proceedings of the 13th International Conference on Pattern Recognition*, Vienna, vol. 4, pp. 709–717, 1996.

Ritter, G. X. and Wilson, J. N., *Handbook of Computer Vision Algorithms in Image Algebra*, CRC Press, 1996.

Shih, F. Y., *Image Processing and Mathematical Morphology: Fundamentals and Applications*, CRC Press, 2009.

Shih, F. Y. and Mitchell, O. R., "Threshold decomposition of gray-scale morphology into binary morphology," *IEEE Transactions on Pattern Analysis and Machine Intelligence*, vol. 11, no. 1, pp. 31–42, Jan. 1989.

Shih, F. Y., Shen, Y., and Zhong, X., "Development of deep learning framework for mathematical morphology," *Pattern Recognition and Artificial Intelligence*, vol. 33, no. 6, p. 1954024 (16 pages), Jun. 2019.

Shih, H. and Liu, E., "Automatic reference color selection for adaptive mathematical morphology and application in image segmentation," *IEEE Transactions on Image Processing*, vol. 25, no. 10, pp. 4665–4676, Oct. 2016.

Simonyan, K. and Zisserman, A., "Very deep convolutional networks for large-scale image recognition," *Proceedings of the International Conference on Learning Representation*, San Diego, CA, May 7–9, 2015.

Stallkamp, J., Schlipsing, M., Salmen, J., and Igel, C., "The German traffic sign recognition benchmark: A multi-class classification competition," *Proceedings of the International Joint Conference on Neural Networks*, San Jose, CA, pp. 1453–1460, 2011.

Zana, F. and Klein, J. C., "Segmentation of vessel-like patterns using mathematical morphology and curve evaluation," *IEEE Transaction on Image Processing*, vol. 10, no. 7, pp. 1010–1019, Jul. 2001.

Zhang, W., Shi, D., and Yang, X., "An improved edge detection algorithm based on mathematical morphology and directional wavelet transform," *Proceedings of the International Congress on Image Signal Processing*, Shenyang, pp. 335–339, 2015.

Part III

Practical Applications

12 A Robust and Blind Image Watermarking System Based on Deep Neural Networks

12.1 INTRODUCTION

Digital image watermarking refers to the process of hiding certain messages into cover images. The transport image, called *marked-image* or *stego-image*, conveys the hidden messages (i.e., the watermark) while appears visibly similar to the cover-image. Therefore, image watermarking enables various applications including authentication, copyright protection, and covert communication (Berghel and O'Gorman, 1996). The watermark data can be encrypted (Caldelli et al., 2010; Gunjal and Manthalkar, 2010) by different purposes, such as increasing the perceivable randomness for additional security and decreasing the impact of noise for watermark integrity under attacks. In a watermarking scheme, fidelity, capacity, and robustness are considered as crucial factors, where fidelity measures the similarity between the cover- and marked-images, capacity measures the maximum amount of watermark that can be embedded, and robustness concerns the watermark extraction under variant attacks on the marked image. It is often a tradeoff between each two of the factors; for example, a higher capacity usually implies more modification on the cover-image and thus lower the fidelity, and a robust watermarking system often applies redundancy and hence lower the capacity.

Traditional image watermarking schemes place the watermark on the least significant bits (LSB) via substitutions or some mathematical operations (Tamimi et al., 2013; Shih, 2017). Although the trivial replacement enables the invisibility, LSB-based methods can be easily revealed by statistical analysis. Advanced watermarking schemes aim at the improvements from different perspectives. For example, Cox et al. (1997) embedded the watermark on the frequency spectrum for high fidelity and high security. Shih and Zhong (2016) increased the frequency domain capacity while preserving the fidelity. Pevný et al. (2010) enhanced the security by an embedding scheme that maintains the cover image statistics. Zong et al. (2015) improved the robustness by embedding the watermark into image histogram.

Incorporating deep neural networks with image watermarking has attracted increasing attentions during recent years. Compared to the significant achievements in stegananalysis (Qian et al., 2015; Pibre et al., 2016), less attempts of applying deep learning in the processes of watermark embedding and extraction are reported. Instead of manual determination of the LSB, some methods in (Sabah and Haitham,

DOI: 10.1201/9781003474395-15

2015; Alexandre and David, 2016; Robert et al., 2015) use neural networks to assign the significance of the bits for each pixel. Tang et al. (2017) proposed a variant of a generative adversarial network to determine the embedding position and the strength on cover images. Kandi et al. (2017) used two deep auto-encoders for non-blind binary watermark extraction, where in the marked image, the pixels produced by the first auto-encoder represent bit zero and the pixels produced by the second auto-encoder indicate bit one. However, the neural networks used in all aforementioned methods do not learn the rule of watermark embedding and extraction. To fully apply the fitting ability of deep learning systems to image watermarking, Baluja (2017) proposed a variant of auto-encoders to form a blind scheme aiming at high fidelity as well as high capacity. Li et al. (2019) embedded the watermark into discrete cosine domain and used convolutional neural networks to cooperate the extraction. Either the embedding or extraction rule is generalized by neural networks; however, due to fragility of neural networks (Papernot et al., 2016), the robustness issue remains a challenge since inputting a modified marked image to a pre-trained deep learning system can cause failure in watermark extraction. Like the idea in adversarial networks, Mun et al. (2017) proposed to resolve this issue by including attack simulation in the training. However, the attacks included in the training set show much higher robustness than those did not. This significantly limits the applications because of the difficulties in enumerating possible attacks in practice.

Different from aforementioned methods, we introduce a robust and blind image watermarking scheme using deep convolutional neural networks to generalize the rule of watermark embedding and extraction. The advantages of the proposed model can be summarized into threefold. First, the proposed system achieves robustness without any prior knowledge of possible attacks and distortions. Second, under the novel construction of the network structure as well as the computation of the loss, the proposed system increases the watermarking capacity as comparing against other robust image watermarking systems. Last, experimental results confirm that the proposed system has an enhanced tolerance range toward common attacks.

The remainder of this paper is organized as follows. Section 12.2 describes the proposed system. Section 12.3 provides both analytical and quantitative validations to confirm the performance, as well as an application of watermark extraction on camera-captured image to demonstrate the practicality of the proposed system. Finally, conclusion is drawn in Section 12.4.

12.2 THE PROPOSED SYSTEM

The proposed system is compatible with the deep auto-encoders (Hinton and Salakhutdinov, 2006), in which a latent space is learned through a bottleneck to compress the dimensionality of an image. It is structured as two nested auto-encoders, where the outer watermark encoder-decoder network learns a latent space of the binary watermark as the watermark code, and the inner embedder-extractor network controls the visual appearance of the watermark code by referencing the cover image. As a result, an intermediate latent space of the watermark code is obtained as the marked image that appears visually similar to the cover image, while simultaneously contains some information of the watermark. Instead of training for dimensionality

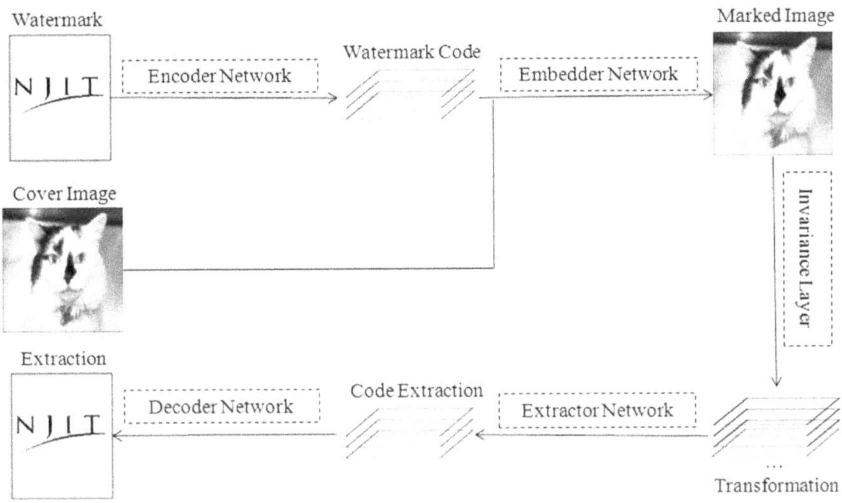

FIGURE 12.1 The overall architecture of the proposed deep watermarking system.

compression, the proposed system learns over-complete representation to secure accurate extraction as well as to achieve robustness.

The overall architecture of the proposed image watermarking system is presented in Figure 12.1. First, the encoder network prepares the binary watermark for enhanced robustness and added security. Note that the watermark can be color, grayscale, or binary images, which are converted into binary codes. The marked image is then generated by the embedder network considering both the watermark code and the cover image. An invariance layer provides invariant representation of the marked image so that its extraction of the watermark tolerates noises and distortions and therefore provides robustness. Finally, the extractor network restores the watermark code, and the decoder network reconstructs the binary watermark. The proposed system is considered as blind since the watermark extraction only takes from the marked image. The entire system is trained as a single deep network, while we separately describe each component in detail. For illustration purposes, we present the embedding of a 1,024-bit (32 × 32) watermark into a 128 × 128 × 3 color cover-image. One can adapt the sizes by slightly modifying the structure, for example, adding a fully connected layer with 1,024 neurons to receive an arbitrarily sized binary watermark.

12.2.1 The Watermark Encoder and Decoder Networks

The encoder and decoder network learns an over-complete latent space of binary watermark as watermark code. A 32 × 32 watermark is input into the encoder network and successively increased into 24 channels and 48 channels by convolutional blocks. The 48-channel feature map is the watermark code to be used in the embedder network. On the other hand, the decoder network receives a 32 × 32 × 48 watermark code and restores it to a 32 × 32 watermark. Figure 12.2 shows the structure of the encoder and decoder networks.

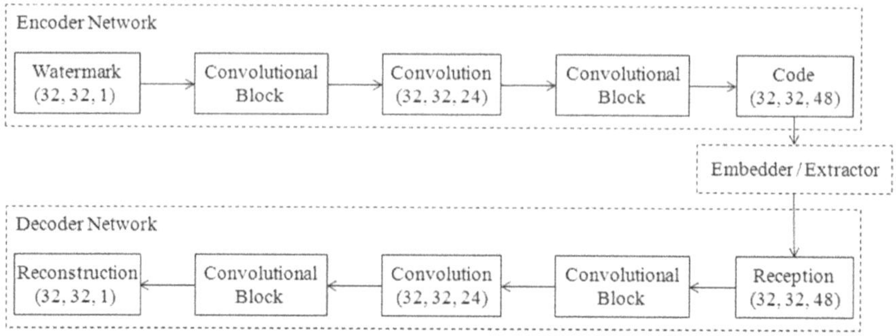

FIGURE 12.2 Structure of the encoder and decoder networks.

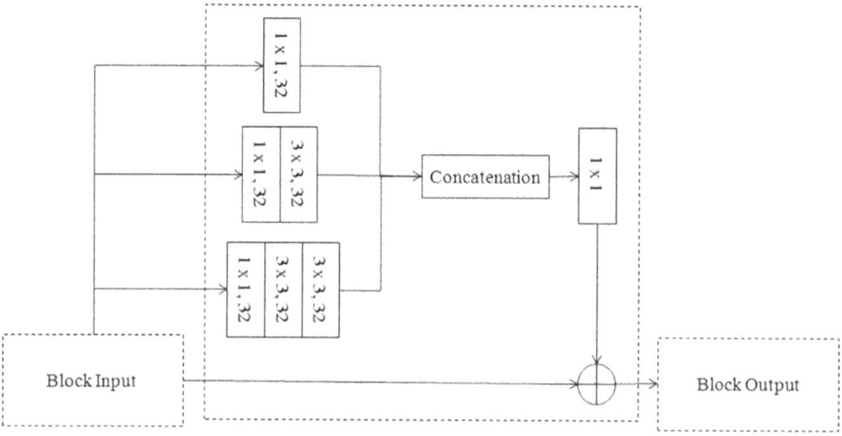

FIGURE 12.3 The convolutional block.

To increase the channels, the inception residual block (Szegedy et al., 2017) is adopted to extract crucial features as well as to preserve gradient flow. The block consists of a 1×1, a 3×3, and a 5×5 convolution and a residual connection that sums up the feature maps and the input itself so that various perception fields are applied. Each convolution has 32 filters, and the 5×5 convolution is replaced by two 3×3 convolutions for efficiency. Applying the convolutions will partition the patterns in the binary watermark into different channels, where important features can be duplicated. The convolutional block is shown in Figure 12.3.

Encoding the single-channel watermark into 48 channels produces a latent space with redundancy, decomposition, and encryption for adding security and enhancing robustness to the proposed system. As a result, this channel-increased latent space along with the invariance layer accounts for high tolerance range of robustness so that cropping 65% of the marked image yields only 8% errors in the extraction. A few 32×32 binary watermarks and their corresponding $32 \times 32 \times 48$ watermark codes (reshaped to $128 \times 128 \times 3$ for display) are shown in Figure 12.4, from which we observe the perceivable randomness.

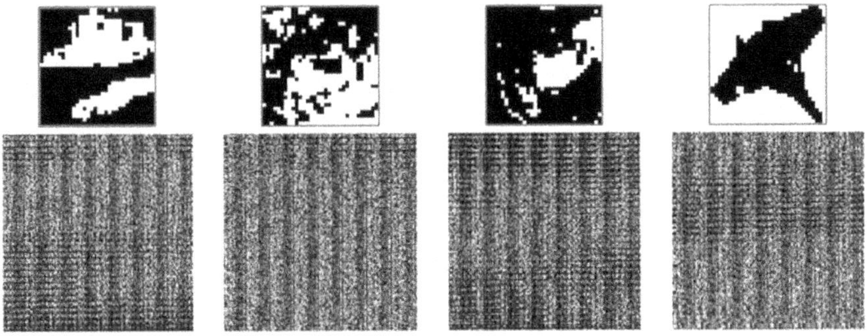

FIGURE 12.4 Some examples of the watermark code. First row: sample binary watermark. Second row: the corresponding watermark code.

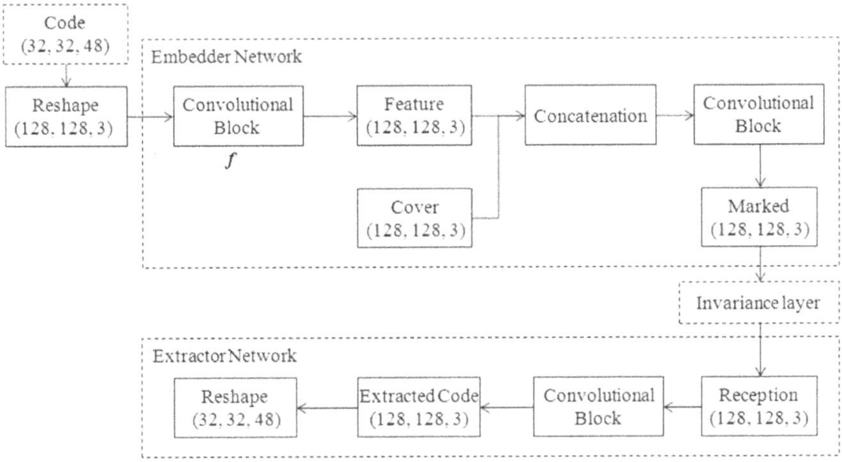

FIGURE 12.5 Detailed structure of the embedder and extractor networks.

12.2.2 The Embedder and Extractor Networks

Having the watermark code along with the cover image, the embedder and extractor network learns a latent space of the watermark code as the marked image, in which its visual appearance must be similar to the cover image and its feature must correlate with the watermark code. The detailed structure of the embedder and extractor networks is shown in Figure 12.5. A reshaped watermark code of size $128 \times 128 \times 3$ is input into the embedder network. A convolutional block is first used to extract features from the watermark code so that the code details are included in the marked image to facilitate later explanation of the loss. Then, the second convolutional block creates the $128 \times 128 \times 3$ marked image by depth concatenation of the transformed code feature and RGB channels of the cover image as considering

FIGURE 12.6 Some examples of the embedding. First row: the cover images. Second row: the marked image. Third row: the embedded and extracted watermark code.

different perception fields. The extractor network reversely restores the watermark code from the marked image.

By inputting two images, the embedder and extractor network encodes the watermark code into the least noticeable components of the cover image. Some examples of the embedding are shown in Figure 12.6. Human vision can hardly tell the differences between marked and cover images in spatial domain, while the convolutional blocks in the extractor are able to find the watermark code in feature maps.

12.2.3 The Invariance Layer

For robustness, the transformer layer learns invariant representation of the marked image. As shown in Figure 12.7, it converts the 3-channel marked image into M -channel over-complete representation with a fully connected layer, where M is the redundant parameter. Note that the higher M is, the higher robustness it can achieve.

Referring to the contractive auto-encoder (Rifai et al., 2011), this layer enables the robustness of learned representation of the marked image by a regularization term that is obtained by the fully connected layer's Frobenius norm of Jacobian matrix with regards to the training input. Mathematically, the penalty term P is given as the partial derivative of the layer to its inputs

$$P = \sum_{i,j} \left(\frac{\partial h_j(X)}{\partial X_i} \right)^2 \tag{12.1}$$

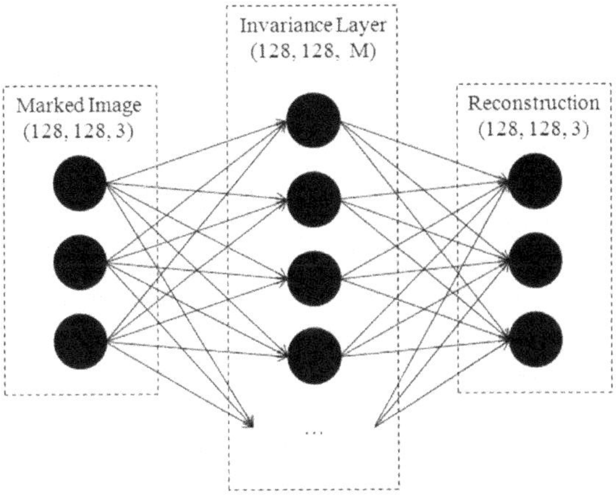

FIGURE 12.7 The invariance layer.

where X_i ($i = 1, 2, 3, \ldots$) denotes the i-th input and h_j denotes the output of the j-th hidden unit. Similar to the common gradient computation in neural networks, the partial derivative can be written as

$$\frac{\partial h_j(X)}{\partial X_i} = \frac{\partial \alpha\left(W_{ji}X_i\right)}{\partial W_{ji}X_i} W_{ji} \tag{12.2}$$

where α is an activation function and W is the weight of the layer.

In practice, we treat each channel as a single neuron for high computational efficiency, structure preservation, and enhanced robustness. Concretely, treating one pixel as an input neuron means 49,152 inputs for a $128 \times 128 \times 3$ marked image. Thus, having 147,456 units in the fully connected layer requires at least 7,247,757,312 parameters if we only set the redundant parameter M as 3, which is not practical in most of current graphical computation units and significantly lowers the efficiency. On the other hand, treating one channel as an input unit considers only 3 input units for the RGB marked image, which enables faster computation as well as a much larger M as hundreds for an enhanced robustness.

We propose to apply hyperbolic tangent as the non-linear activation function of the invariance layer for strong gradients as well as bias avoidance (LeCun et al., 1998). With α assigned as the hyperbolic tangent, P can be defined as

$$P = \sum_j \left(1 - h_j^2\right)^2 \sum_i \left(W_{ji}^T\right)^2 \tag{12.3}$$

Minimizing term P is essentially rendering the weights in the hidden layer unchangeable toward all the inputs X. However, placing it as a penalty in the total loss function enables the layer to preserve only useful information while rejecting all other noises and irrelevant information to achieve the invariance.

12.2.4 The Loss and Error Propagation

The proposed system is trained as a single and deep network by minimizing the loss function L

$$L = \mu_1 \left\| w, e \right\| + \mu_2 \left\| m, c \right\| + \mu_3 \left\| \theta(m, w) \right\| + \mu_4 P \tag{12.4}$$

where w, e, m, and c denote the watermark, the watermark extraction, the marked image, and the cover image, respectively; θ denotes the correlation computation function; P is the regularization term as in equation (12.3); and $\mu_i, i = (1,2,3,4)$ is the weight controlling the contribution of each term.

The error propagation of each term is presented in Figure 12.9. Note that $\left\| w, e \right\|$ is the cyclic term that ensures the similarity between the extraction and the original watermarks. All the components in the system apply this error term during their weights update. The $\left\| m, c \right\|$ guarantees the visual similarity between cover and marked images by comparing their contents. To explain the error term $\left\| \theta(m, w) \right\|$, we annotate f_1 and f_2 (shown in Figure 12.8) for the convolutional block f of Figure 12.5. Under this annotation, $\mu_3 \left\| \theta(m, w) \right\|$ can be computed as

$$\mu_3 \left\| \theta(m, w) \right\| = \frac{\mu_3}{2} (\| g(f_1(cw)), g(f_1(m)) \| + \| g(f_2(cw)), g(f_2(m)) \|) \tag{12.5}$$

where cw is the watermark code and g denotes the gram matrix. Besides the watermark code, the convolutional block f also extracts features on the marked image. The feature maps of the marked image must correlate with those of the watermark code. The correlation is maximized by minimizing the distance between the gram matrices. Remarkably, $\left\| m, c \right\|$ and $\left\| \theta(m, w) \right\|$ are only applied to the weights of encoder and embedder networks. All the components contain the information of the watermark under this error propagation.

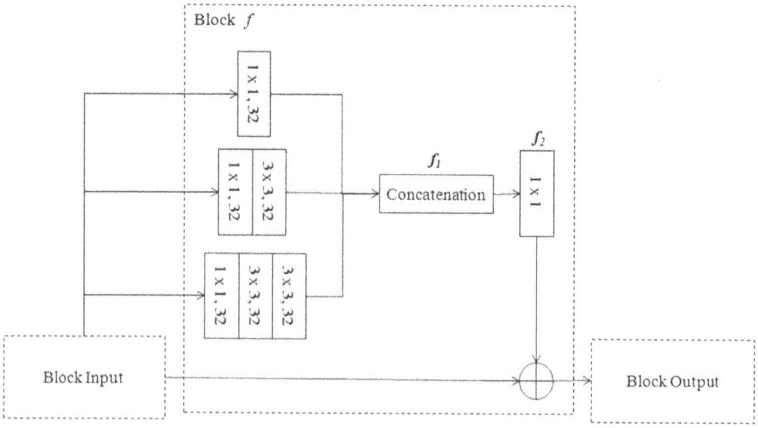

FIGURE 12.8 f_1 and f_2 in the convolutional block f.

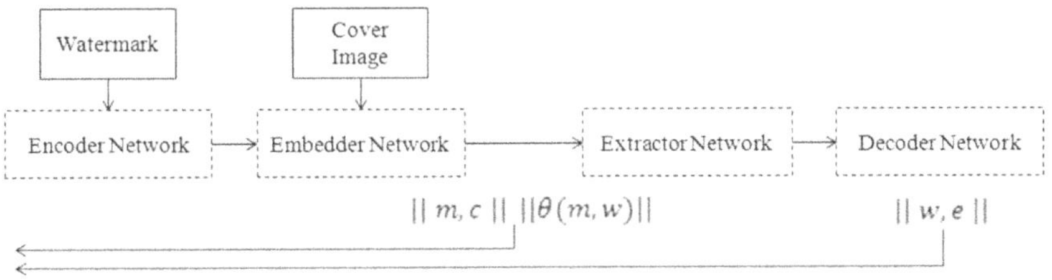

FIGURE 12.9 The error propagation of the proposed system.

12.3 EXPERIMENTAL RESULTS

12.3.1 Training and Testing

This paper presents a robust image watermarking system based on deep neural networks, with a fixed watermark payload of 1,024 bits. The proposed system is trained using ImageNet (Russakovsky et al., 2015) (rescaled to 128 × 128) as the cover image and the binary version of CIFAR (Krizhevsky and Hinton, 2009) (32 × 32) as the watermark. Both datasets include more than millions of images to introduce a large scope to the system. The ADADELTA (Zeiler, 2012) optimizer that applies a moving window in gradient updates is adopted for its ability of continuous learning after large epochs. Figure 12.10 shows the value of loss L during 200 epochs. The loss L drops smoothly and converges below 1%. The distance measures in L are set as the mean absolute error to highlight the overall performance over a few outliers. The $\mu_i, i = (1,2,3,4)$ is set to be one for equal highlight of each error term. All the layers in the system apply the rectified linear unit (ReLU) as the activation function except that the marked image and watermark extraction use sigmoid to limit the output range into (0, 1) and the invariance layer uses hyperbolic tangent.

The testing is performed on 10,000 images of Microsoft COCO dataset (Lin et al., 2014) as the cover image, and 10,000 images of the testing division of the binary CIFAR as the watermark. Both the testing cover images and testing watermarks are not used in the training. It demonstrates that the proposed system generalizes the watermarking rules without over-fitting to the training samples. The peak signal-to-noise ratio (PSNR) and bit error rate (BER) are respectively used to quantitatively evaluate the fidelity of the marked image and the quality of the watermark extraction. The PSNR is defined as

$$PSNR = 10log_{10}\left(\frac{\max(c)^2}{MSE(c,m)}\right) \tag{12.6}$$

where c and m denote the cover image and the marked image, respectively, and MSE is the mean squared error. The BER is computed as the percentage of error bits on the binarized watermark extraction. In the testing, the BER is very close to 0, indicating that the original and the extracted watermarks are identical. The testing $PSNR$ is 39.72 dB, meaning a high fidelity of the marked images so that the hidden information cannot

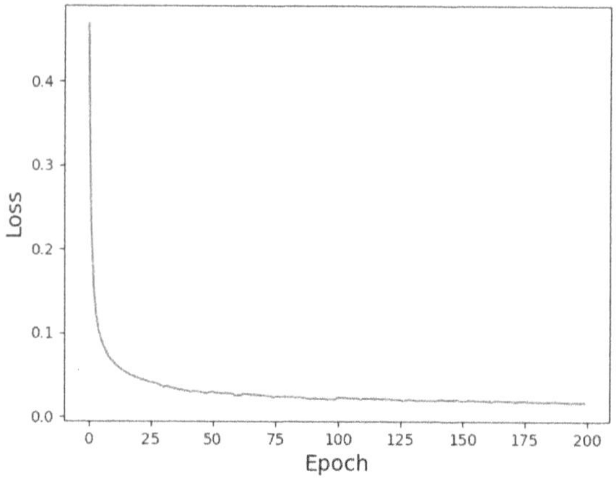

FIGURE 12.10 Training loss versus epoch.

FIGURE 12.11 A few examples. First column: the embedded and extracted watermark. Second column: the cover image. Third column: the marked image. Fourth to sixth columns: the respective residual errors of R, G, and B channels between the marked and the cover images.

be noticed by human vision. Some examples of the watermark embedding with various image content and color are shown in Figure 12.11. The residual errors showing the absolute difference in each RGB channel between the marked and the cover images are also displayed, from which we observe that the watermark is dispersed over the marked image. It provides added security to the marked image even if the cover image has leaked. By subtracting it from the marked image does not reveal the watermark information. Ranging the pixel values between 0 and 255, we compute the mean of residual errors for each RGB channel, averaging along the testing results of 2.57, 2.10, and 1.63, respectively. Similarly, the maxima of residual errors are 14.11, 24.79, and 17.08. These numbers indicate that there are some relatively spiky modifications for the extraction, but on average the watermark insertion does not alter channels a lot.

12.3.2 On Synthetic Images

We explore the proposed system to some extreme cases using synthetic images for further estimation. In particular, the synthetic situations that are not included during the training process are analyzed, and the results involving blank and random-noise images are presented here.

Figure 12.12 shows the results of embedding real watermarks into some blank cover images of black, white, red, green, and blue, where the residuals are amplified 10 times. Although the blank cover images are not included in the training, the proposed system provides acceptable results in the cases. The residual errors display more green color and the blank green marked image displays relatively more noticeable noises than those in other colors, implying that the proposed system modifies the green color slightly more.

Figure 12.13 presents the results of embedding a random binary image into a natural cover image, as well as embedding a real watermark into a random color-spotted cover. Applying a random binary image as the watermark displays good results.

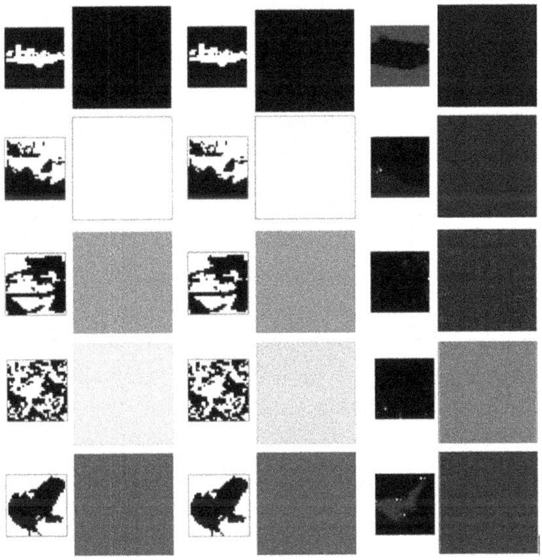

FIGURE 12.12 Embedding watermarks into blank covers. First column: the watermark. Second column: the blank cover image. Third column: the extracted watermark. Fourth column: the marked image. Fifth and sixth columns: the residual errors.

FIGURE 12.13 Embedding involving noise images. First column: the watermark. Second column: the cover image. Third column: the extracted watermark. Fourth column: the marked image. Fifth and sixth columns: the residual errors.

Although the general shape of extraction is recognizable, there are obvious distortions on the extraction when it comes to embedding a watermark into random noises. In practice, hiding a watermark into random noises indicates that the appearance of the marked media is noisy and meaningless, so the encryption methods mapping a watermark into random patterns could be used.

12.3.3 Robustness

The robustness of the proposed system against different distortions on the marked image is evaluated by analyzing the tolerance range toward the attacks. Figure 12.14 shows visual comparisons between the marked images and their distortions as well as between the original watermarks and the extractions from the distorted marked images.

Quantitatively, swept-over distortion parameters that control the strength of the attacks are applied on the testing datasets, and the average *BER* are recorded. Since the geometric distortions, such as translation, rotation, and scaling, are rectified, we focus on the responses of the proposed system against image processing attacks. Figure 12.16 presents the results of some common but challenging situations. The extracted watermarks from the proposed system respectively have 11%, 8.1%, 31%, 8.2%, 42%, and 5.1% bit errors when the distortions are a Gaussian blur with mean 0 and variance 85%, a cropping discarding 65% of the marked image, a Gaussian additive noise mean 0 and variance 20%, a JPEG compression with quality factor 10, a 20% random noise, and a 30% salt-and-pepper noise. The proposed system shows a high tolerance range on these challenges, especially for cropping, salt-and-pepper noise, JPEG compression, and the noises that randomly fluctuate the pixel values through image channels showing higher *BER*, such as Gaussian additive noise and random modificative noise. However, a 20% Gaussian noise or a 20% random noise destroys almost content of the marked image as shown in Figure 12.15. The proposed system responds acceptable performances given a decent distortion parameter of these attacks, such as 16% *BER* on 10% Gaussian noise.

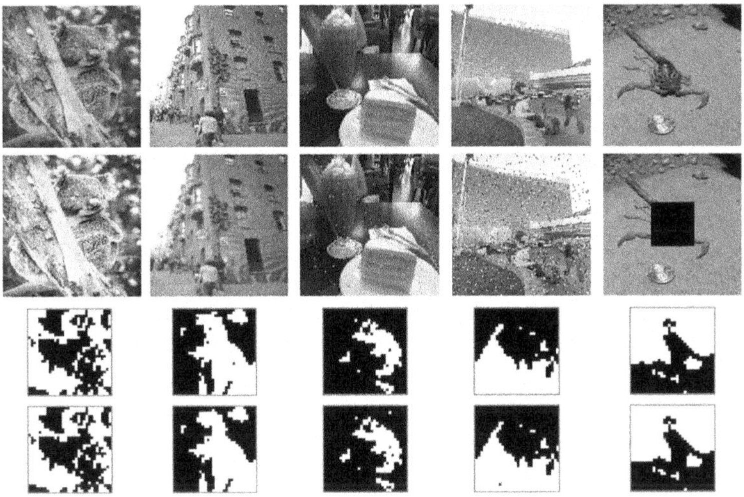

FIGURE 12.14 First row: the marked image. Second row: the distorted marked image, from left to right the operations are histogram equalization, Gaussian blur, random noise, salt-and-pepper noise, and cropping. Third row: the original watermark. Fourth row: the extraction from the distorted marked image.

FIGURE 12.15 Left: the original marked image. Middle: Gaussian additive noise with variance 20%. Right: 20% random modificative noise.

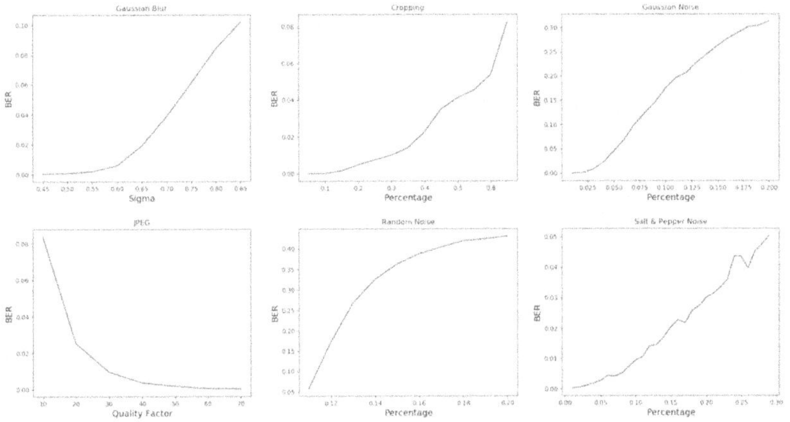

FIGURE 12.16 Distortion parameters versus *BER*.

12.3.4 Comparative Study

First, the proposed method is analytically compared against several state-of-the-art image watermarking methods that incorporate deep neural networks as shown in Table 12.1. To the best of our knowledge, Kandi et al. (2017) is the first image watermarking scheme utilizing convolutional neural networks. However, it is a non-blind scheme for achieving robustness. Embedded by increasingly changing an image block to represent a watermark bit, the system in (Mun et al., 2017) is trained to extract the watermark bits from their corresponding blocks with attack simulation and achieves both blindness and robustness. However, it requires the distortions to be included in the training phase for robustness, it is difficult to predict and enumerate all encountering attacks in practice. On the other hand, the proposed system not only achieves blindness and robustness simultaneously, but also is trained without the requirement of any prior knowledge of attack information, and hence has a wider range of applications.

Second, the proposed system is compared against several blind and robust competitors quantitatively. The selection of the competitors mainly considers variation. Mun et al. (2017) applied convolutional neural networks, and Zong et al. (2015), Zareian and Tohidypour (2013), and Ouyang et al. (2015) are classic, traditional, and robust methods with different image domains in recent years, such as histogram domain adopting statistical image features, frequency domain, and log-polar domain with summarized image features. For a fair comparison, the testing is performed on the same image sets reported in the references. The crucial results are presented in Table 12.2, where "/" denotes not applicable, S&P is the salt-and-pepper noise, and GF is Gaussian filtering. The proposed method shows clear advantages by covering more general distortion categories as well as lower *BER* under the same distortion parameters. For instance, the traditional methods such as manipulating the image histogram cannot tolerate the histogram equalization attack. In addition, the proposed method has a higher tolerance range; for example, Mun et al. (2017) and Ouyang et al. (2015) can only extract the watermark with high JPEG quality

TABLE 12.1
Comparison of the Proposed System against State-of-the-Art Image Watermarking Methods

Method	Function of the Deep Neural Networks	Blindness	Robustness	Concentration
(Tang et al., 2017)	determine the embedding position and probability	no	no	undetectability
(Kandi et al., 2017)	embedding and extraction	no	yes	robustness
(Baluja, 2017)	embedding and extraction	yes	no	capacity
(Li et al., 2018)	extraction	yes	no	undetectability
(Mun et al., 2017)	extraction	yes	yes	robustness
Ours	embedding and extraction	yes	yes	robustness

TABLE 12.2
Quantitative Comparison between the Proposal and Some Blind and Robust Competitors

Method	HE	JPEG 10	Cropping 20%	S&P 5%	GF 10%	*PSNR* (dB)	Capacity
	BER (%)						
(Mun et al., 2017)	/	/	6.61	7.98	4.81	38.01	1 bit per block
(Zong et al., 2015)	/	17.50	7.06	3.51	6.33	46.63	25 bits
(Zareian et al., 2013)	/	2.15	/	4.94	0.21	41.00	256 bits
(Ouyang et al., 2015)	/	/	7.51	9.41	27.91	36.77	24 bits
Ours	0.43	8.16	0	0.97	0	39.93	1,024 bits

of 80 to 90, while the proposed method covers as low as 10 although the method in (Zareian and Tohidypour, 2013) focuses on the JPEG having a higher performance. Remarkably, the competitors tolerate cropping 20% to 30%, while the BER is as high as 8.2% if 65% of the marked image is cropped. Finally, under a close PSNR, the proposed method outperforms the existing methods by simultaneously achieving the highest robustness and the highest capacity.

12.3.5 A Practical Application

We present one of the core applications using the proposed system to extract watermark from a camera-captured image. Success in this problem has the potential of many useful applications, such as connecting the virtual world and the real world to serve as the low-level interface for the Internet of things. Watermark detection and extraction on a camera resampled image remains a challenge. Its difficulty is mainly caused by the comprehensive combination of noises (Pramila et al., 2007), including geometric distortions, optical tilt, quality degradation, compression, and lens distortions. Researchers and engineers have been trying to address this issue from various perfectives. For example, Pramila et al. (2018) proposed to extract the watermark from a phone-camera capture of an image printed on blank paper. However, it has restricted applications since the presence of the cover image is required.

We apply the proposed blind system toward this issue. Instead of using printing, a phone camera is used to capture a marked image displayed on a laptop screen. The distortions are brought by the camera, the resolution, brightness, refresh rate, and frame rate, etc. and could be challenging to robustness issue. The scenario is related to the widely used quick-response (QR) code, where the users are directed to online resource via a scan. What dissimilar to the traditional QR code in our system is that the end users just need to scan the content image for more information and the code/watermark becomes completely invisible. As shown in Figure 12.17, the provider distributes the marked image obtained with the cover image and the watermark information via our marking app containing the trained encoder and embedder networks. The user installs our scanning app containing the trained decoder and extractor networks, and scans the image displayed on a screen for the hidden information.

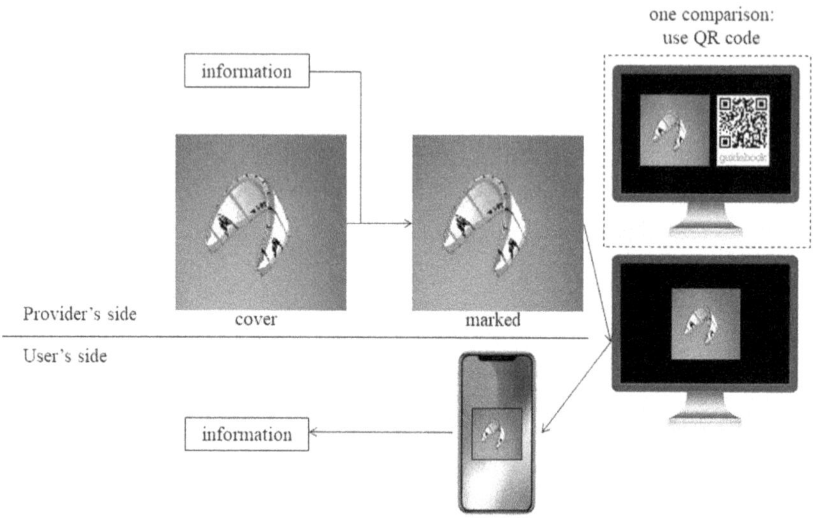

FIGURE 12.17 The process of the application.

In practice, error correction coding can be used for further protection to the watermark. For instance, a classic Reed Solomon (RS) code (Reed and Solomon, 1960) can correct up to 30% error in the extraction. In this paper, we present the raw results for visual comparison as well as for simplicity. To test the system under realistic situation, a protype is developed and a raw 32× 32 binary watermark (see Figure 12.18) is used for its clear structure. Five volunteers are asked to take a few pictures of some marked images displayed 425px × 425px on a 2,560 × 1,440 screen, with the camera of a mobile phone. Two rules are told to the users. First, as shown in the user's interface, the entire image should be placed as large as possible inside the region of interest (ROI). As a prototype of demonstrating purpose, this rule facilitates our segmentation that the largest contour inside the ROI is the marked image so that we can focus on the test of the proposed system instead of some complicated segmentation algorithms. In addition, placing the image largely in the ROI helps us to capture desired details and features for the extraction. Second, the camera should be kept as still as possible. Although the proposed system tolerates some blurring effects, it is not designed to extract watermark in a high-speed motion.

The prototype only analyzes the ROIs, and hand-taken pictures can hardly be completely parallel to the screen. Therefore, there exist some geometrical, affine, and perspective distortions. As the proposed system does not include the robustness toward these attacks, the image registration technique in (Brown, 1992; Zokai and Wolberg, 2005) is adopted. To simplify the prototype as in Figure 12.19, four corners of the largest contour inside the ROI are used as the reference points. The contoured content is mapped on the bird view plane, and the watermark is extracted from the rectified image.

Figure 12.20 presents a few extractions and their corresponding ROIs. The BERs from left to right are 3.71%, 4.98%, 1.07%, 4.30%, and 8.45%. We observe that the closer the picture is taken, the lower the error is. The more parallel between the camera

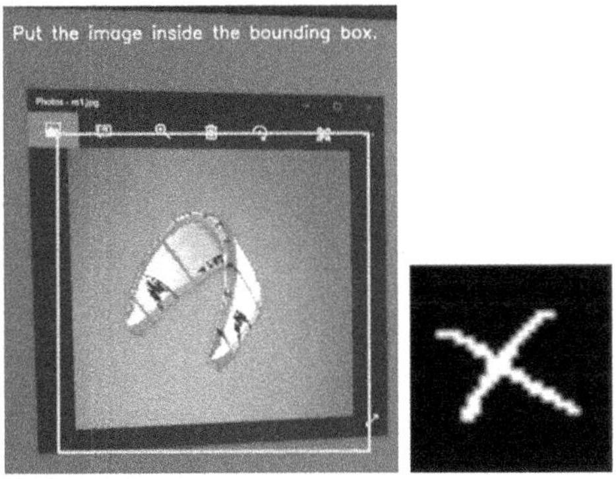

FIGURE 12.18 Left: the appearance of the prototype, Right: the sample binary watermark.

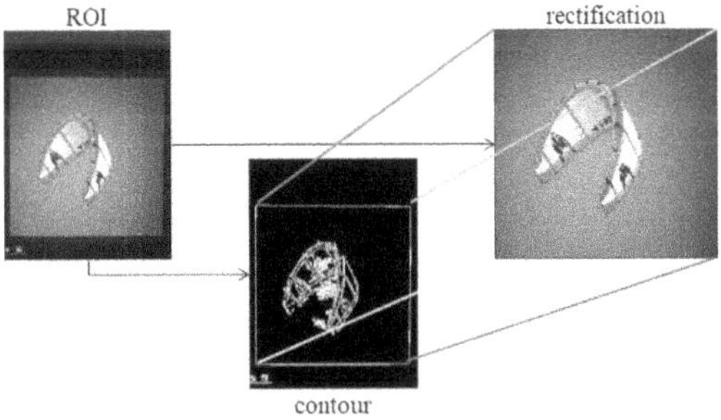

FIGURE 12.19 A marked image rectification inside an ROI.

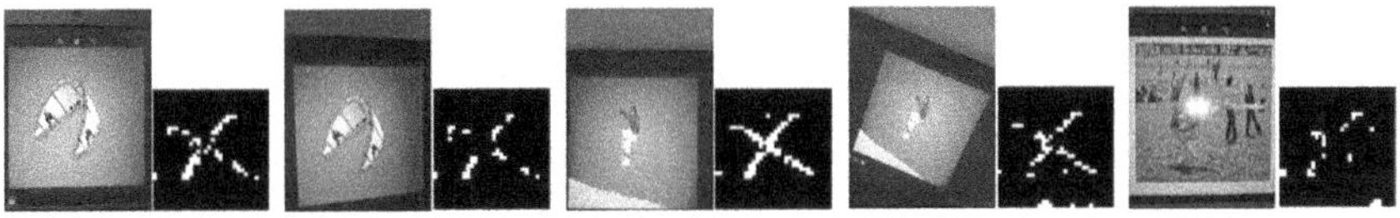

FIGURE 12.20 A few extractions and the ROIs.

and the screen, the lower the error is. Note that the tolerance limit is around 30° in this test. Also, the flashlight brings more errors since it may over- and underexpose some image areas. We may turn off the flashlights in this case since the screen has backlit. In total from the user test, we have 20 images, and the average BER is 5.13%. This is the raw result without the error correction code and can be considered as acceptable since the RS code can correct 30% errors theoretically. Moreover, the scanning app extracts the watermark within one second as it only applies the pre-trained weights in the extractor and decoder networks to the marked image rectification.

In summary, experimental results confirm that the proposed system extracts the watermark well on a rectified marked image with the registration techniques. Hence, it can be used in solving the challenging problem of watermark extraction on camera-captured images.

12.4 CONCLUSION

This paper introduces a robust and blind image watermarking system. A novel structure of applying deep convolutional neural networks is proposed to learn the watermark embedding and extraction rules with the constraint of loss function. The robustness is achieved without any prior knowledge of possible attacks and distortions. Comprehensive evaluations are presented to confirm the superiority and a challenging application is introduced to validate the practicality of the proposed system. In the future work, statistical features such as the probability density functions could be merged into the deep neural networks so that the embedding and extraction can consider the image distribution for a higher robustness.

REFERENCES

Alexandre, S. B. and David, C. J., "Artificial neural networks applied to image steganography," *IEEE Latin America Transactions*, vol. 14, no. 3, pp. 1361–1366, 2016.

Baluja, S., "Hiding images in plain sight: Deep steganography," *Proc. Intl. Conf. Advances in Neural Information Processing Systems*, pp. 2066–2076, 2017.

Berghel, H. and O'Gorman, L., "Protecting ownership rights through digital watermarking," *Computer*, vol. 29, no. 7, pp. 101–103, 1996.

Brown, L. G., "A survey of image registration techniques." *ACM Computing Surveys (CSUR)*, vol. 24, no. 4, pp. 325–376, 1992.

Caldelli, R., Francesco, F., and Becarelli, R., "Reversible watermarking techniques: An overview and a classification," *EURASIP Journal on Information Security*, no. 134546, 2010.

Cox, I. J., Kilian, J., Leighton, F. T., and Shamoon, T., "Secure spread spectrum watermarking for multimedia," *IEEE Transactions on Image Processing*, vol. 6, no. 12, pp. 1673–1687, 1997.

Gunjal, B. and Manthalkar, R. R., "An overview of transform domain robust digital image watermarking algorithms," *Journal of Emerging Trends in Computing and Information Sciences*, vol. 2, no. 1, pp. 37–42, 2010.

Hinton, G. E. and Salakhutdinov, R. R., "Reducing the dimensionality of data with neural networks," *Science*, 313, no. 5786, pp. 504–507, 2006.

Kandi, H., Mishra, D., and Gorthi, S. R. S., "Exploring the learning capabilities of convolutional neural networks for robust image watermarking," *Computers & Security*, vol. 65, pp. 247–268, 2017.

Krizhevsky, A. and Hinton, G., "Learning multiple layers of features from tiny images," *Technical Report*, University of Toronto, vol. 1, no. 4, p. 7, 2009.

LeCun, Y., Bottou, L., Orr, G. B., and Müller, K. R., "Efficient backprop," *Neural Networks: Tricks of the Trade*, pp. 9–50. Springer, Berlin, Heidelberg, 1998.

Li, D., Deng, L., Gupta, B. B., Wang, H., and Choi, C., "A novel CNN based security guaranteed image watermarking generation scenario for smart city applications," *Information Science*, vol. 479, pp. 432–447, 2019.

Lin, T. Y., Maire, M., Belongie, S., Hays, J., Perona, P., Ramanan, D., Dollár, P., and Zitnick, C. L., "Microsoft coco: Common objects in context," *Proceedings of the European Conference on Computer Vision*, pp. 740–755. Springer, Cham, 2014.

Mun, S. M., Nam, S. H., Jang, H. U., Kim, D., and Lee, H. K., "A robust blind watermarking using convolutional neural network," arXiv:1704.03248, 2017.

Ouyang, J., Coatrieux, G., Chen, B., and Shu, H., "Color image watermarking based on quaternion Fourier transform and improved uniform log-polar mapping," *Computers & Electrical Engineering*, vol. 46, pp. 419–432, 2015.

Papernot, N., McDaniel, P., Jha, S., Fredrikson, M., Celik, Z. B., and Swami, A., "The limitations of deep learning in adversarial settings," *IEEE European Symposium on Security and Privacy*, pp. 372–387, 2016.

Pevný, T., Filler, T., and Bas, P., "Using high-dimensional image models to perform highly undetectable steganography," *International Workshop on Information Hiding*, Springer, Berlin, Heidelberg, pp. 161–177, 2010.

Pibre, L., Pasquet, J., Ienco, D., and Chaumont, M., "Deep learning is a good steganalysis tool when embedding key is reused for different images, even if there is a cover source mismatch," *Electronic Imaging*, vol. 1, no. 11, 2016.

Pramila, A., Keskinarkaus, A., and Seppänen, T., "Camera based watermark extraction-problems and examples," *Proceedings of the Finnish Signal Processing Symposium*, 2007.

Pramila, A., Keskinarkaus, A., and Seppänen, T., "Increasing the capturing angle in print-cam robust watermarking," *Journal of Systems and Software*, vol. 135, pp. 205–215, 2018.

Qian, Y., Dong, J., Wang, W., and Tan, T., "Deep learning for steganalysis via convolutional neural networks," *Media Watermarking, Security, and Forensics*, vol. 9409, International Society for Optics and Photonics, 2015.

Reed, I. S. and Solomon, G., "Polynomial codes over certain finite fields," *Journal of the Society for Industrial and Applied Mathematics*, vol. 8, no. 2, pp. 300–304, 1960.

Rifai, S., Vincent, P., Muller, X., Glorot, X., and Bengio, Y., "Contractive auto-encoders: Explicit invariance during feature extraction," *Proceedings of the 28th International Conference on Machine Learning*, pp. 833–840, Omnipress, 2011.

Robert, J., Eva, V., and Martin, K., "Neural network approach to image steganography techniques," *Mendel*, pp. 317–327. Springer, 2015.

Russakovsky, O., Deng, J., Su, H., Krause, J., Satheesh, S., Ma, S., Huang, Z., Karpathy, A., Khosla, A., Bernstein, M., and Berg, A. C., "Imagenet large scale visual recognition challenge," *International Journal of Computer Vision*, vol. 115, no. 3, pp. 211–252, 2015.

Sabah, H. and Haitham, B., "Artificial neural network for steganography," *Neural Computing and Applications*, vol. 26, no. 1, pp. 111–116, 2015.

Shih, F. Y., *Digital Watermarking and Steganography: Fundamentals and Techniques*, 2nd edition, CRC Press, Boca Raton, FL, Apr. 2017.

Shih, F. Y. and Zhong, X., "Intelligent watermarking for high-capacity low-distortion data embedding," *International Journal of Pattern Recognition and Artificial Intelligence*, vol. 30, no. 5, p. 1654003, 2016.

Szegedy, C., Ioffe, S., Vanhoucke, V., and Alemi, A. A., "Inception-v4, inception-resnet and the impact of residual connections on learning," in *AAAI*, vol. 4, p. 12, 2017.

Tamimi, A. A., Abdalla, A. M., and Al-Allaf, O., "Hiding an image inside another image using variable-rate steganography," *International Journal of Advanced Computer Science and Applications*, vol. 4, no. 10, 2013.

Tang, W., Tan, S., Li, B., and Huang, J., "Automatic steganographic distortion learning using a generative adversarial network," *IEEE Signal Processing Letters*, vol. 24, no. 10, pp. 1547–1551, 2017.

Zareian, M. and Tohidypour, H. R., "Robust quantisation index modulation-based approach for image watermarking," *IEEE Transactions on Image Processing*, vol. 7, no. 5, pp. 432–441, 2013.

Zeiler, M. D., "ADADELTA: An adaptive learning rate method," arXiv:1212.5701, 2012.

Zokai, S. and Wolberg, G., "Image registration using log-polar mappings for recovery of large-scale similarity and projective transformations," *IEEE Transactions on Image Processing*, vol. 14, no. 10, pp. 1422–1434, 2005.

Zong, T., Xiang, Y., Natgunanathan, I., Guo, S., Zhou, W., and Beliakov, G., "Robust histogram shape-based method for image watermarking," *IEEE Transactions on Circuits and Systems for Video Technology*, vol. 25, no. 5, pp. 717–729, 2015.

13 Deep Learning Classification on Optical Coherence Tomography Retina Images

13.1 INTRODUCTION

Image classification intends to categorize all the pixels in an image into one of the predefined classes. It forms the basis for computer vision tasks, such as localization, detection, and segmentation. Although the classification seems natural to human beings, it is much more challenging for an automated system. This task can be conducted using machine-learning algorithms, which contain supervised and unsupervised learning approaches. Supervised learning is learning a function that maps an input to an output based on training dataset (Russell and Norvig, 2010). An example is neural networks, which are a biologically inspired programming paradigm enabling a computer to learn from observational data (Fukushima, 1975; Fukushima, 1980; Marblestone et al., 2016). Unsupervised learning is learning to produce inferences from datasets consisting of input data without labeled responses. An example is cluster analysis, which groups a set of objects in such a way that objects in the same group are more similar to each other than to those in different groups.

Deep learning is a technique for learning features in neural networks. During recent years, deep-learning models have been used in nonlinear information processing for feature extraction and transformation, as well as for pattern analysis and classification (Rawat and Wang, 2017; Bengio et al., 2013). One of the deep-learning frameworks is a convolutional neural network (CNN), which is applied in image recognition, classification, and detection tasks.

Multi-layer neural networks compromise various convolutional layers followed by fully connected layers, which can learn to perform classification without being programmed by certain task-specific rules (Nielsen, 2015). The CNN can learn to identify the images containing dogs from analyzing those training images of labeling as a dog or non-dog. It can achieve this classification without any prior knowledge regarding different features of a dog, such as facial structure, eye size, color, etc. However, the main problem with CNN is the requirement of more data (i.e., there might be few categories for which we do not file enough data to train CNN [Huang et al., 2017; Kermany et al., 2018]) and processing power, that everybody may not have. Additionally, the image classifiers built from CNN can classify only the given images into the categories, in which it trains. If intending to apply that classifier on other categories, we must retrain it from scratch which may take many hours or

DOI: 10.1201/9781003474395-16

days. To overcome this problem, we adopt the transfer learning technique, which uses a pre-trained model by someone else for a similar task instead of building a model from scratch. We replace the first few layers and tweak the model according to our usage (Tajbakhsh et al., 2016), e.g., VGG-16 (Szegedy et al., 2015), AlexNet (Krizhevsky et al., 2012), GoogLeNet (Szegedy et al., 2015), and Inception-V3 (Szegedy et al., 2016).

The optical coherence tomography (OCT) (Huang et al., 1991) is an imaging technique that applies coherent light to capture micrometer-resolution, two- and three-dimensional images from within optical scattering media. It is a non-invasive imaging test, which uses light waves to take cross-section pictures of the retina. The images are evaluated by ophthalmologists to see each distinct layer of the retina. It allows the ophthalmologists to map and measure their thickness. These measurements can provide treatment guidance for glaucoma and diseases of the retina. The retinal diseases include age-related macular degeneration (AMD) and diabetic eye disease. Useful information extracted with the combination of OCT scans, image processing, and segmentation techniques can provide a detailed overview of different retinal layers and associated diseases (Kafieh et al., 2019). The growing need for OCT in retinal disease analysis necessitates investigation into a fully automated approach (Baum et al., 2007; Yonetsu et al., 2013; Halupka et al., 2018; Lu et al., 2019).

In this chapter, we develop a deep learning classifier for classifying optical coherence tomography (OCT) retina images among the given four classes (NORMAL, CNV, DME, and DRUSE). Data augmentation is used to help neural networks learn/see the images differently. The transfer learning technique needs a minimal amount of data to train an image classification model. The remainder of this chapter is organized as follows. Section 13.2 introduces the OCT retina diseases training data. Section 13.3 presents the data augmentation technique. Section 13.4 describes the network architecture. Section 13.5 develops the transfer learning technique. Section 13.6 shows the experimental results. Conclusions are drawn in Section 13.7.

13.2 THE OCT RETINA DISEASES TRAINING DATA

We will create an image classifier to train on the following three most common diseases in the retina (Yorston, 2003).

1. Choroidal neovascularization (CNV): It involves the growth of new blood vessels that originate from the choroid through a break in the Bruch membrane into the sub-retinal pigment epithelium (sub-RPE) or subretinal space. The CNV is a significant cause of visual loss as shown in Figure 13.1.
2. Diabetic macular edema (DME): It is an accumulation of fluid in the macula, part of the retina that controls our most specific vision abilities due to leaking blood vessels, which is shown in Figure 13.2. In order to develop DME, one must first have diabetic retinopathy that damages the blood vessels in the retina, resulting in vision impairment.
3. Drusen: They are yellow deposits under the retina as shown in Figure 13.3. Drusen are made up of lipids, a fatty protein and likely increase a person's risk in developing age-related macular degeneration (AMD).

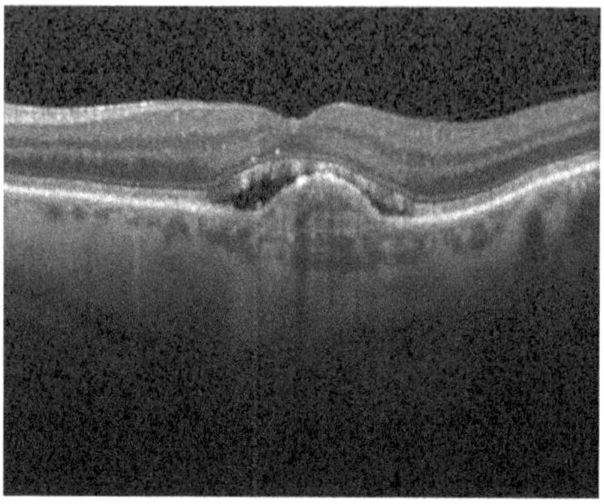

FIGURE 13.1 OCT scan of CNV.

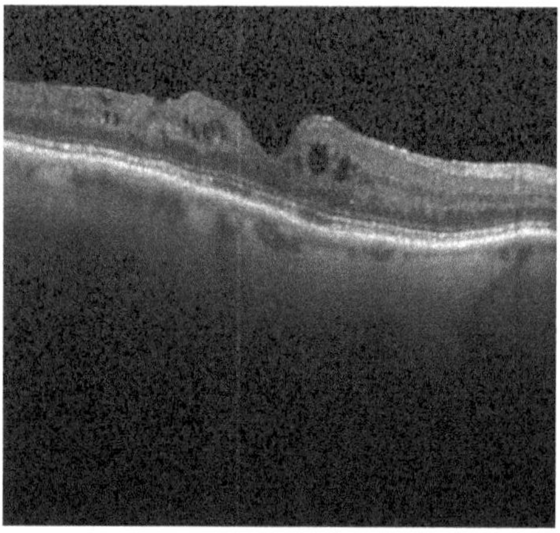

FIGURE 13.2 OCT scan of DME.

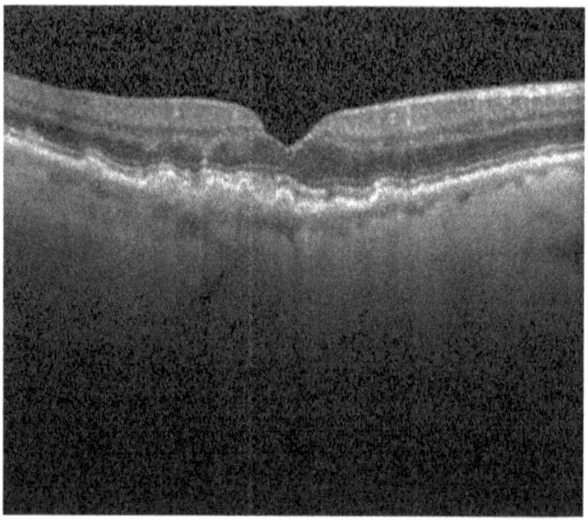

FIGURE 13.3 OCT scan of drusen.

In addition to the three diseases in the retina, Figure 13.4. shows the image of the normal retinal OCT. From Figures 13.1–13.4, we can observe that there is a minute difference between the four classes. Therefore, making the image classifier to be able to classify these diseases is a challenging task.

The dataset (Kermany, 2017) containing spectral-domain optical coherence tomography (SD-OCT) (Forooghian et al., 2008) images and spectralis OCT (Ctori and Huntjens, 2015), Heidelberg Engineering, Germany, was selected from retrospective cohorts of adult patients from the Shiley Eye Institute of University of California, San Diego, the California Retinal Research Foundation, Medical Center Ophthalmology Associates, the Shanghai First People's Hospital, and Beijing Tongren Eye Center between July 1, 2013, and March 1, 2017. The dataset is organized into two folders (train and test) and contains subfolders for each image category (NORMAL, CNV, DME, and DRUSEN). There are 84,452 OCT images (JPEG) of 512 × 496 pixels, as shown in Table 13.1.

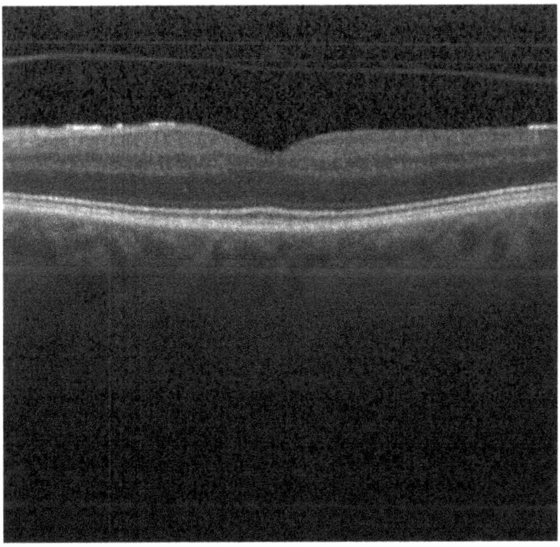

FIGURE 13.4 OCT scan of normal retina.

TABLE 13.1
The Dataset Containing 84,452 OCT Retina Images Labeled with Four Classes

Class	Training Images	Test Images
CNV	37,205	242
DME	11,348	242
DRUSEN	8,616	242
NORMAL	26,315	242

13.3 THE DATA AUGMENTATION AND GRAYSCALE TO RGB MAPPING

13.3.1 Data Augmentation

Data augmentation helps prevent the network from overfitting and memorizing the exact details of the training images. It solves the problem of overfitting as we will have more data points by applying the operations like rotation, changing lighting conditions, cropping, and translating. In this chapter, we intend to use the following two techniques.

1. Random reflection: In this technique, we randomly choose the image and reflect the selected image horizontally. When function RandReflection returns true (1), each image is reflected horizontally with 50% probability. When RandReflection is false (0), no images are reflected.
2. Random translation: Range of horizontal/vertical translation is applied to the input image. Translation distance is measured in pixels. An image is shifted from its original position to some random amounts in x- and y-axes, which is in the range from –30 to 30 pixels. Figure 13.5 shows the result of random translation.

13.3.2 Grayscale Mapping

The images in the dataset are grayscale (1 channel). To train the image classification model using a pre-trained network, we convert them into three channel images, RGB images. Figure 13.6 shows an image of a standard OCT scan with predefined color map.

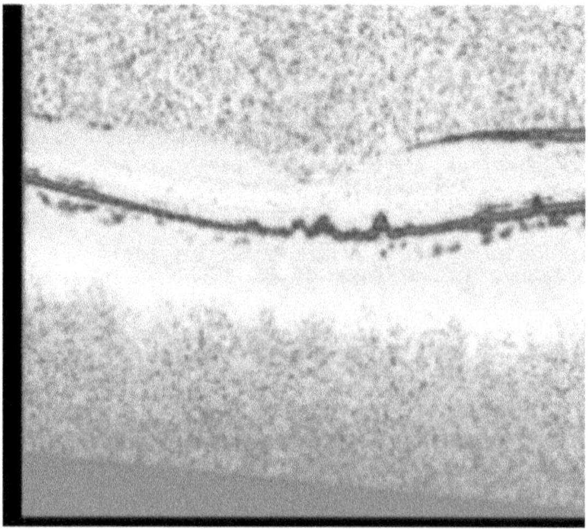

FIGURE 13.5 Random translation on x- and y-axes.

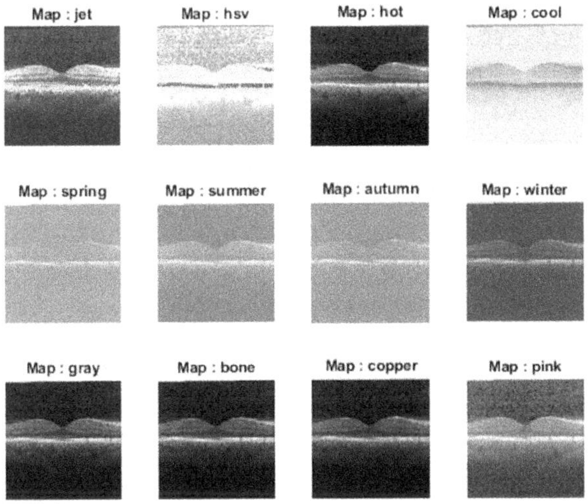

FIGURE 13.6 An image of a normal OCT scan with predefined color map.

13.4 THE NETWORK ARCHITECTURE

Layers are made up of stacks of neurons, and the collection of layers is organized into three main parts: input layers, hidden layers, and output layers. Convolution is the first building block of CNN in the convolutional layer to produce a feature map. The pooling layer is used after the convolutional layer to reduce the size. This layer reduces the number of parameters and thus reduces the computation. The most common pooling technique is max-pooling, where we take a filter of $N \times N$ and apply a maximum operation over $N \times N$–sized part of the image. The fully connected layer combines the features learned by the previous convolution and pooling layer and feeds it to the next fully connected layer. The last fully connected layer combines all the learned features to classify the images. Therefore, the last layer will have the same number of neurons as the number of classes in classification.

Batch normalization layer normalizes the activations and gradients propagating through a neural network to speed up network training and reduce the sensitivity to network initialization. The softmax activation function normalizes the output of the fully connected layer to be positive numbers or probabilities that sum to one. The final classification layer uses the probabilities to assign the input to one of the mutually exclusive classes and compute the loss. All the layers of the CNN model are listed in Table 13.2, and its flow chart is shown in Figure 13.7.

There are a few parameters that need to set up for training the neural network. We use stochastic gradient descent with momentum (SGDM) for optimizing neural networks. The learning rate is 0.01, and the number of epochs is 6. We shuffle the training data at every epoch.

TABLE 13.2
All the Layers of the CNN Model

Layer Number	Layer Variable Name	Layer Name	Comments
1	imageinput'	Image Input	227×227×3 images with "zerocenter" normalization
2	conv_1'	Convolution	8 3×3×3 convolutions with stride [1 1] and padding "same"
3	batchnorm_1'	Batch Normalization	Batch normalization with 8 channels
4	relu_1'	ReLU	ReLU
5	maxpool_1'	Max Pooling	2×2 max pooling with stride [2 2] and padding [0 0 0 0]
6	conv_2'	Convolution	8 3×3×8 convolutions with stride [1 1] and padding "same"
7	batchnorm_2'	Batch Normalization	Batch normalization with 8 channels
8	relu_2'	ReLU	ReLU
9	maxpool_2'	Max Pooling	2×2 max pooling with stride [2 2] and padding [0 0 0 0]
10	conv_3'	Convolution	16 3×3×8 convolutions with stride [1 1] and padding "same"
11	batchnorm_3'	Batch Normalization	Batch normalization with 16 channels
12	relu_3'	ReLU	ReLU
13	maxpool_3'	Max Pooling	2×2 max pooling with stride [2 2] and padding [0 0 0 0]
14	conv_4'	Convolution	32 3×3×16 convolutions with stride [1 1] and padding "same"
15	batchnorm_4'	Batch Normalization	Batch normalization with 32 channels
16	relu_4'	ReLU	ReLU
17	fc'	Fully Connected	4 fully connected layer
18	softmax'	Softmax	softmax
19	classoutput'	Classification Output	crossentropyex with "NORMAL" and 3 other classes

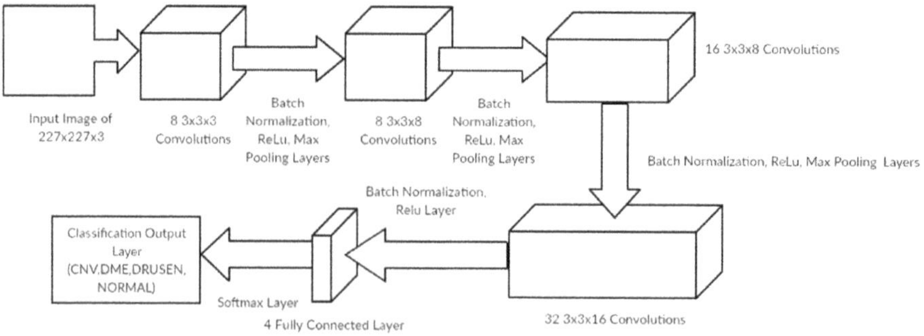

FIGURE 13.7 Flow chart of the CNN model.

13.5 THE TRANSFER LEARNING TECHNIQUE

Transfer learning reuses the knowledge gained while solving one problem and applying to a different but related problem. It enables a deep neural network to train faster with much fewer data, and this quality of transfer learning is beneficial for real-world problems since sometimes we do not have many data labels to train a complex model. Therefore, we apply an already-trained model, which trained on a large dataset, and then fine-tune some parameters for our predictive model. In this chapter, an image classifier of OCT retinal images is developed based on VGG-16, which is a second least dense pre-trained neural network. It was trained with 1,000 different classes, and we intend to build an image classifier for four classes. The proposed architecture of the VGG-16 model is listed in Table 13.3, and its flow chart is shown in Figure 13.8.

TABLE 13.3
All the Layers of the Classifier Based on VGG-16

Layer Number	Layer Variable Name	Layer Name	Comments
1	input'	Image Input	224×224×3 images with "zerocenter" normalization
2	conv1_1'	Convolution	64 3×3×3 convolutions with stride [1 1] and padding [1 1 1 1]
3	relu1_1'	ReLU	ReLU
4	conv1_2'	Convolution	64 3×3×64 convolutions with stride [1 1] and padding [1 1 1 1]
5	relu1_2'	ReLU	ReLU
6	pool1'	Max Pooling	2×2 max pooling with stride [2 2] and padding [0 0 0 0]
7	conv2_1'	Convolution	128 3×3×64 convolutions with stride [1 1] and padding [1 1 1 1]
8	relu2_1'	ReLU	ReLU
9	conv2_2'	Convolution	128 3×3×128 convolutions with stride [1 1] and padding [1 1 1 1]
10	relu2_2'	ReLU	ReLU
11	pool2'	Max Pooling	2×2 max pooling with stride [2 2] and padding [0 0 0 0]
12	conv3_1'	Convolution	256 3×3×128 convolutions with stride [1 1] and padding [1 1 1 1]
13	relu3_1'	ReLU	ReLU
14	conv3_2'	Convolution	256 3×3×256 convolutions with stride [1 1] and padding [1 1 1 1]
15	relu3_2'	ReLU	ReLU
16	conv3_3'	Convolution	256 3×3×256 convolutions with stride [1 1] and padding [1 1 1 1]
17	relu3_3'	ReLU	ReLU
18	pool3'	Max Pooling	2×2 max pooling with stride [2 2] and padding [0 0 0 0]
19	conv4_1'	Convolution	512 3×3×256 convolutions with stride [1 1] and padding [1 1 1 1]
20	relu4_1'	ReLU	ReLU
21	conv4_2'	Convolution	512 3×3×512 convolutions with stride [1 1] and padding [1 1 1 1]
22	relu4_2'	ReLU	ReLU
23	conv4_3'	Convolution	512 3×3×512 convolutions with stride [1 1] and padding [1 1 1 1]
24	relu4_3'	ReLU	ReLU
25	pool4'	Max Pooling	2×2 max pooling with stride [2 2] and padding [0 0 0 0]
26	conv5_1'	Convolution	512 3×3×512 convolutions with stride [1 1] and padding [1 1 1 1]
27	relu5_1'	ReLU	ReLU

(*Continued*)

TABLE 13.3
Continued

Layer Number	Layer Variable Name	Layer Name	Comments
28	conv5_2'	Convolution	512 $3\times3\times512$ convolutions with stride [1 1] and padding [1 1 1 1]
29	relu5_2'	ReLU	ReLU
30	conv5_3'	Convolution	512 $3\times3\times512$ convolutions with stride [1 1] and padding [1 1 1 1]
31	relu5_3'	ReLU	ReLU
32	pool5'	Max Pooling	2×2 max pooling with stride [2 2] and padding [0 0 0 0]
33	fc6'	Fully Connected	4096 fully connected layer
34	relu6'	ReLU	ReLU
35	drop6'	Dropout	50% dropout
36	fc7'	Fully Connected	4096 fully connected layer
37	relu7'	ReLU	ReLU
38	drop7'	Dropout	50% dropout
39	‘	Fully Connected	4 fully connected layer
40	‘	Softmax	Softmax
41	‘	Classification Output	Crossentropyex

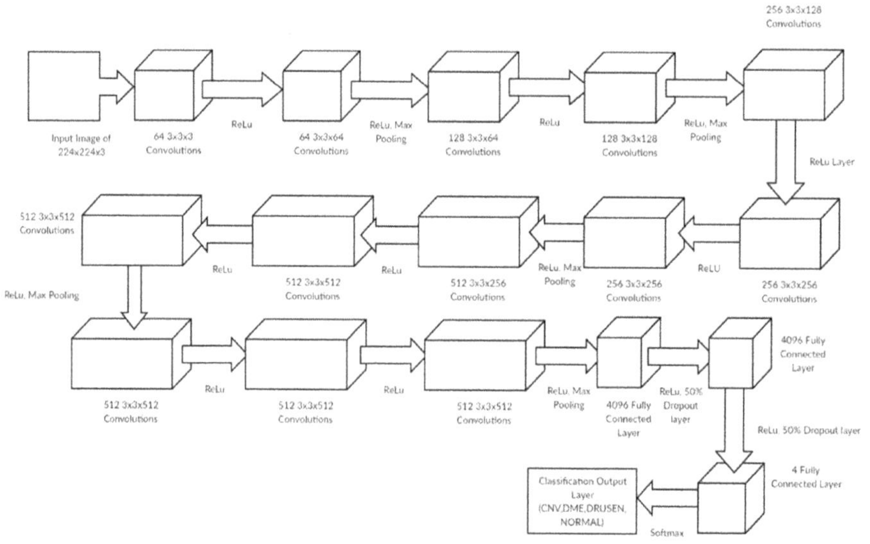

FIGURE 13.8 Flow chart for the VGG-16 model.

We use stochastic gradient descent with momentum (SGDM) for optimizing neural networks. The learning rate is set to be 0.0001, which is small enough to slow down the learning in the transferred layer. However, the learning speed is increased in the fully connected layer. This combination leads to faster learning in the newer layer and slower in other layers. The number of epochs is 6, which is low because we do not want to change the weights of our pre-trained neural network.

13.6 EXPERIMENTAL RESULTS

Figure 13.9 shows the training accuracy versus the number of iterations for the fine-tuned VGG-16 model. Figure 13.10 shows the training loss versus the number of iterations for the fine-tuned VGG-16 model. Figure 13.11 shows the validation accuracy versus the number of iterations plot for the fine-tuned VGG-16 model. Figure 13.12 shows the validation loss versus the number of iterations plot for the fine-tuned VGG-16 model.

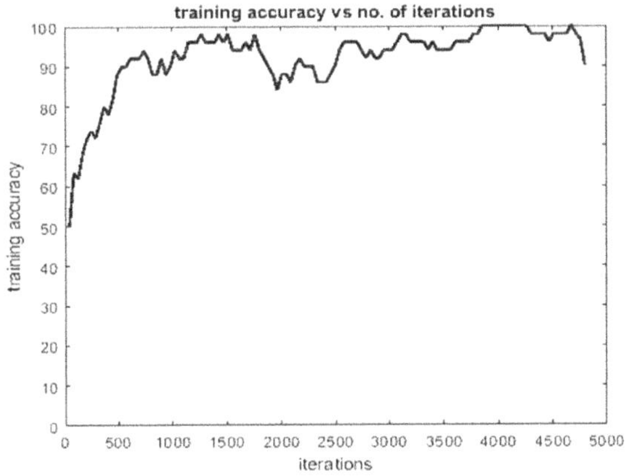

FIGURE 13.9 Training accuracy versus the number of iterations for the fine-tuned model.

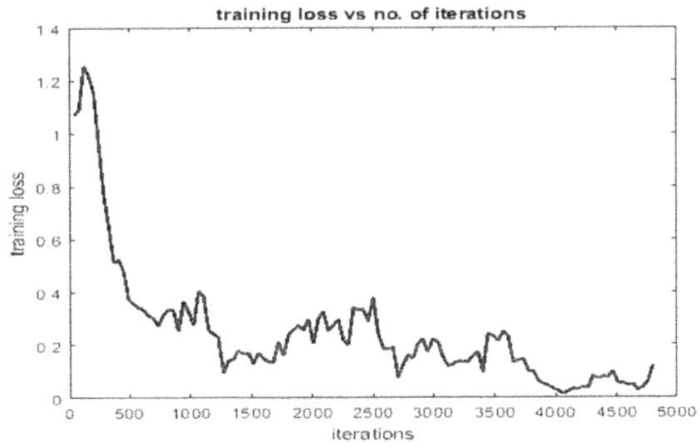

FIGURE 13.10 Training loss versus the number of iterations for the fine-tuned model.

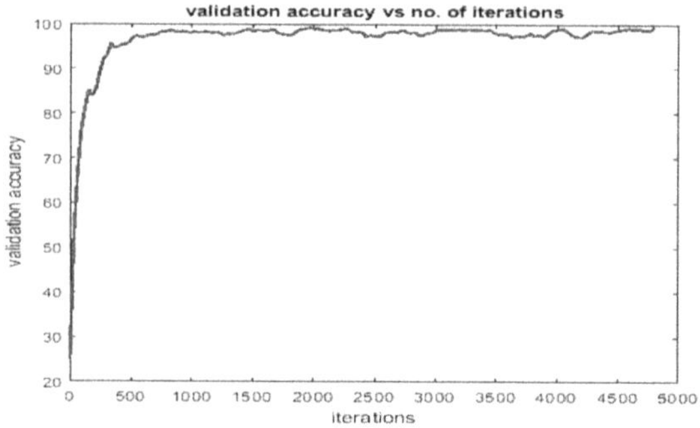

FIGURE 13.11 Testing accuracy versus the number of iterations plot for the fine-tuned model.

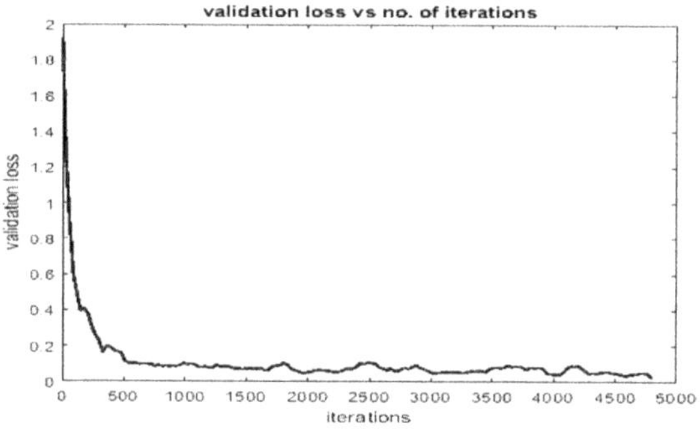

FIGURE 13.12 Validation loss versus the number of iterations plot for the fine-tuned model.

Table 13.4 shows the confusion matrix of the CNN model, where the horizontal blue row at the bottom is the precision, and the vertical blue column is the recall. This model can learn features of **DRUSEN** class more accurately (99.1% highest true positive rate) than other classes in the model. The model faces difficulty in learning the features of **CNV** (the highest error rate of 8.4% false positive rate).

Table 13.5 shows the confusion matrix of the CNN model with data augmentation, where the horizontal blue row at the bottom is the precision, and the vertical blue column is the recall. This model can learn features of **DRUSEN** class more accurately (98.9% the highest true positive rate) than other classes in the model. The model faces difficulty in learning the features of **CNV** (the highest error rate of 21.4% false positive rate).

Table 13.6 shows the confusion matrix of the CNN model with color-mapped RGB images, where the horizontal blue row at the bottom is the precision and the vertical blue column is the recall. This model can learn features of **DME** class more accurately (95.1% the highest true positive rate) than other classes in the model. The model faces difficulty in learning the features of **CNV** (the highest error rate of 48.4% false positive rate).

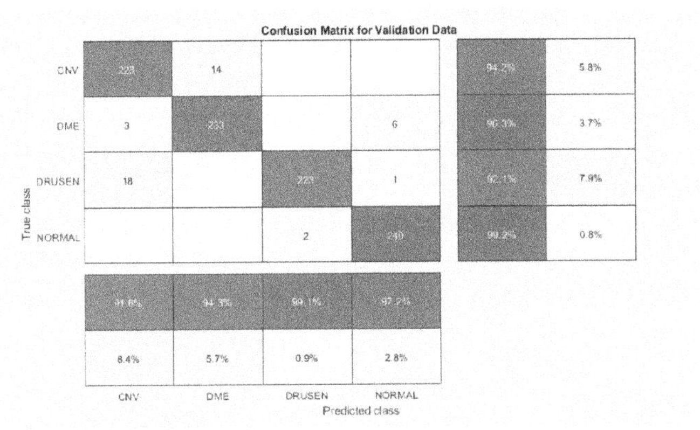

TABLE 13.4 Confusion Matrix of the CNN Model

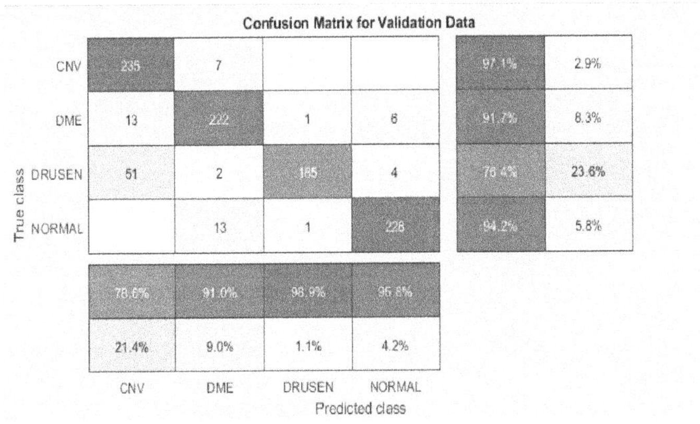

TABLE 13.5 Confusion Matrix of the CNN Model with Data Augmentation

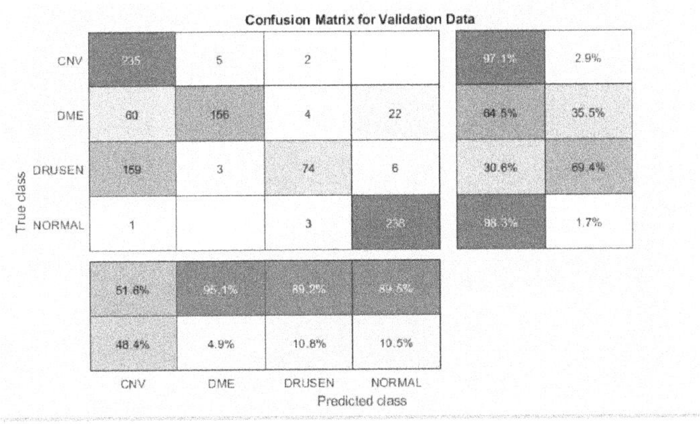

TABLE 13.6 Confusion Matrix of the CNN Model with Color-Mapped RGB Images

Table 13.7 shows the confusion matrix of the CNN model with color-mapped RGB images and data augmentation, where the horizontal blue row at the bottom is the precision, and the vertical blue column is the recall. This model can learn features of **DME** class more accurately (94.2% the highest true positive rate) than other classes in the model. The model faces difficulty in learning the features of **CNV** (the highest error rate of 47.7% false positive rate).

Table 13.8 shows the confusion matrix of the fine-tuned model with color-mapped RGB images, where the horizontal blue row at the bottom is the precision, and the vertical blue column is the recall. There are 968 images in the testing set. This model

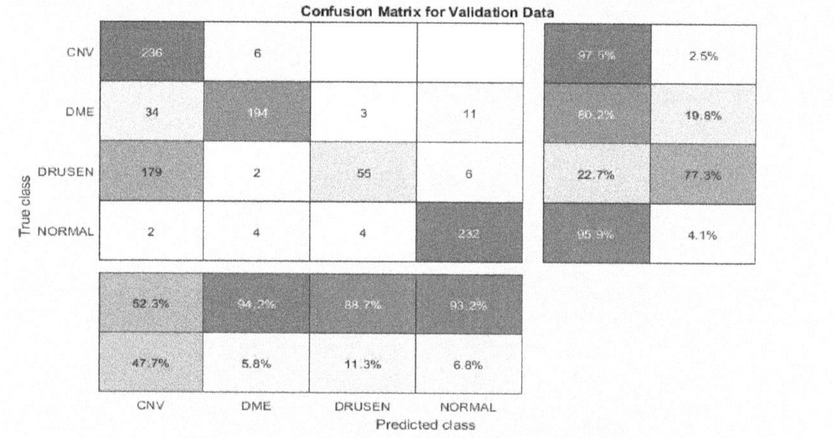

Confusion Matrix for Validation Data

True class \ Predicted class	CNV	DME	DRUSEN	NORMAL		
CNV	236	6			97.5%	2.5%
DME	34	194	3	11	80.2%	19.8%
DRUSEN	179	2	55	6	22.7%	77.3%
NORMAL	2	4	4	232	95.9%	4.1%
	52.3%	94.2%	88.7%	93.2%		
	47.7%	5.8%	11.3%	6.8%		

TABLE 13.7 Confusion Matrix of the CNN Model with RGB-Mapped Images and Data Augmentation

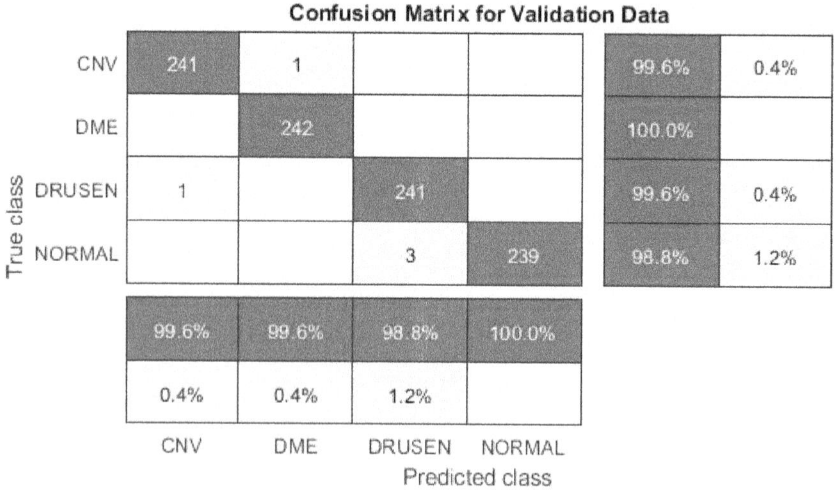

Confusion Matrix for Validation Data

True class \ Predicted class	CNV	DME	DRUSEN	NORMAL		
CNV	241	1			99.6%	0.4%
DME		242			100.0%	
DRUSEN	1		241		99.6%	0.4%
NORMAL			3	239	98.8%	1.2%
	99.6%	99.6%	98.8%	100.0%		
	0.4%	0.4%	1.2%			

TABLE 13.8 Confusion Matrix of the Fine-Tuned Model with Color-Mapped RGB Images

can learn the **NORMAL** category more accurately (the highest true positive rate of 100.0%). The model faces difficulty in learning the features of **DRUSEN** (the highest error rate of 1.2% false positive rate).

Table 13.9 shows the confusion matrix of the fine-tuned model with color-mapped RGB images and data augmentation, where the horizontal blue row at the bottom is the precision and the vertical blue column is the recall. This model can learn **NORMAL** category more accurately (the highest true positive rate of 100%). The model faces difficulty in learning the features of **CNV** (the highest error rate of 6.6% false positive rate).

Table 13.10 shows the comparisons among sixteen different models. The experimental results show that transfer learning is an optimization, a shortcut to saving time, and getting better performance.

Table 13.10 shows that the fine-tuned models perform better than CNNs though they are trained on fewer data. Converting grayscale into RGB color space using HSV color map helps the neural network to learn more features about the given images. Karri et al. (2017) obtained an accuracy of 94%, whereas we achieve an average accuracy of 99.48% on four classes. The average precisions for DME and NORMAL category are 86% and 99%, respectively, in Karri et al. (2017). However, we achieve precisions of 99.6% and 100% on DME and NORMAL, respectively.

The experiments show that VGG-16 model produces the best result among all the models. We have used HSV color map to convert one channel grayscale image to the 3-channel RGB image. To extend our experiments, we have applied a few other color maps on the VGG-16 model trained with 8,000 images (2,000 images from

Confusion Matrix for Validation Data

True class \ Predicted class	CNV	DME	DRUSEN	NORMAL		
CNV	241	1			99.6%	0.4%
DME	2	240			99.2%	0.8%
DRUSEN	15		227		93.8%	6.2%
NORMAL		5	1	236	97.5%	2.5%
	93.4%	97.6%	99.6%	100.0%		
	6.6%	2.4%	0.4%			

TABLE 13.9 Confusion Matrix of the Fine-Tuned Model with RGB-Mapped Images and Data Augmentation

TABLE 13.10
Comparisons among 16 Different Models

Model Name	Test Accuracy	Training Accuracy	Total Training Images	Image Size	Learning Rate	Number of Epochs	Testing Set
CNN	95.45	97.34	83,484	227×227×1	0.01	6	968
CNN with data augmentation	89.88	91.62	83,484	227×227×1	0.01	6	968
CNN with HSV mapped images	72.62	81.87	83,484	227×227×3	0.01	6	968
CNN with HSV and data augmentation	74.07	79.43	83,484	227×227×3	0.01	6	968
AlexNet with HSV mapped images	98.14	94.04	8,000	227×227×3	0.0001	6	968
AlexNet with HSV mapped images.	98.76	94.07	83,484	227×227×3	0.0001	6	968
AlexNet with HSV mapped images and data augmentation	90.81	93.51	8,000	227×227×3	0.0001	6	968
VGG-16 with HSV mapped images	99.48	97.71	8,000	227×227×3	0.0001	6	968
VGG-16 with HSV mapped images	98.76	96.66	83,484	227×227×3	0.0001	6	968
VGG-16 with HSV mapped images and data augmentation	97.52	93.43	8,000	227×227×3	0.0001	6	968
inceptionv3 with HSV mapped images	97.42	98.96	8,000	299×299×3	0.0001	6	968
inception v3 with HSV mapped images	97.31	93.41	83,484	299×299×3	0.0001	6	968
GoogLeNet with HSV mapped images	97.83	96.71	8,000	224×224×4	0.0001	6	968
GoogLeNet with HSV mapped images	96.59	93.42	83,484	224×224×4	0.0001	6	968
ResNet18 with HSV mapped images	99.17	96.04	8,000	224×224×3	0.0001	6	968
ResNet18 with HSV mapped images	98.24	93.51	83,484	224×224×3	0.0001	6	968

each category) and 968 images for testing. Table 13.11 shows the testing and training accuracies of each color map trained on the VGG-16 model, where HSV and gray color maps deliver the best testing accuracy.

In Figure 13.13, none of the models has a decreasing slope in the test accuracy plot, which shows none is overfitting. Combining the results of Figure 13.13. and Table 13.10, we conclude that the transfer learning on pre-trained models with the

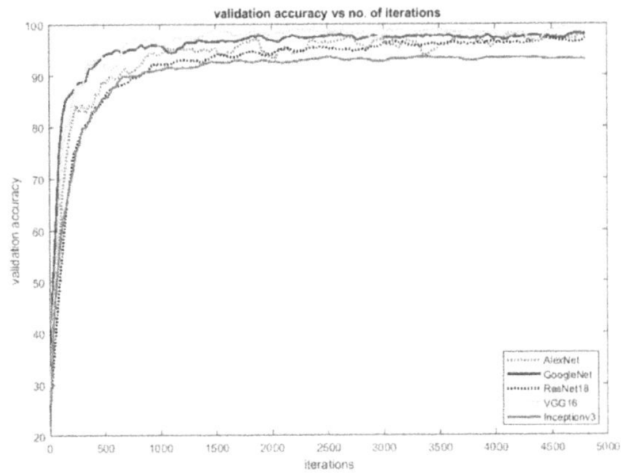

FIGURE 13.13 The plot of validation accuracy vs. iterations for five pre-trained models.

TABLE 13.11
The Testing and Training Accuracies of Each Color Map Trained on the VGG-16 Model

Color Map	Testing Accuracy	Training Accuracy
Jet	98.97	98.25
HSV	99.48	97.71
Hot	98.76	97.32
Cool	99.38	96.81
Spring	98.24	95.16
Summer	99.38	97.82
Autumn	99.17	97.25
Winter	98.66	98.06
Gray	99.48	98.86
Bone	99.38	98.70
Copper	99.07	98.10
Pink	98.04	96.34

deep neural network architecture using a small amount of training data can effectively prevent overfitting.

13.7 CONCLUSIONS

In this chapter, two approaches are used to perform image classification on OCT retinal images. The first approach is to build an image classifier from scratch, and the second approach uses transfer learning from a pre-trained model, including VGG-16, AlexNet, ResNet18, GoogLeNet, and InceptionV3. We also apply data augmentation to increase the number of data points, which solves the problem of overfitting as more data points are obtained by applying image rotation, changing lighting conditions, cropping, and translating. We map grayscale retina images to RGB color space using the HSV color map. The pre-trained models produce better performance initially. There is a confusion matrix for each method to give an insight into how different approaches affect the classification results. If one has abundant data, it is better to create own model with own parameters; otherwise, the pre-trained model is used as the starting point.

REFERENCES

Baum, K. G., Helguera, M., and Krol, A., "Fusion viewer: A new tool for fusion and visualization of multimodal medical data sets," *Journal of Digit Imaging*, vol. 21, pp. 59–68, 2007.

Bengio, Y., Courville, A., and Vincent, P., "Representation learning: A review and new perspectives," *IEEE Transactions on Pattern Analysis and Machine Intelligence*, vol. 35, pp. 1798–1828, 2013.

Ctori, I. and Huntjens, B., "Repeatability of foveal measurements using spectralis optical coherence tomography segmentation software," *PLoS ONE*, vol. 10, no. 6, 2015.

Forooghian, F., et al., "Evaluation of time domain and spectral domain optical coherence tomography in the measurement of diabetic macular edema," *Investigative Ophthalmology & Visual Science*, vol. 49, no. 10, pp. 4290–4296, 2008.

Fukushima, K., "Cognitron: A self-organizing multilayered neural network," *Biological Cybernetics*, vol. 20, pp. 121–136, 1975.

Fukushima, K., "Neocognitron: A self-organizing neural network model for a mechanism of pattern recognition unaffected by shift in position," *Biological Cybernetics*, vol. 36, pp. 193–202, 1980.

Halupka, K. J., Antony, B. J., Lee, M. H., Lucy, K. A., Rai, R. S., Ishikawa, H., Wollstein, G., Schuman, J. S., and Garnavi, R., "Retinal optical coherence tomography image enhancement via deep learning," *Biomedical Optics Express*, vol. 9, no. 12, pp. 6205–6221, 2018.

Huang, D., et al., "Optical coherence tomography," *Science*, vol. 254, pp. 1178–81, 1991.

Huang, Z., Pan, Z., and Lei, B., "Transfer learning with deep convolutional neural network for SAR target classification with limited labeled data," *Remote Sensing*, vol. 9, no. 9, p. 907, 2017.

Kafieh, R., Rabbani, H., and Kermani, S., "A review of algorithms for segmentation of optical coherence tomography from retina," *Journal of Medical Signals and Sensors*, vol. 3, pp. 45–60, 2019.

Karri, S. P., Chakraborty, D., and Chatterjee, J., "Transfer learning based classification of optical coherence tomography images with diabetic macular edema and dry age-related macular degeneration," *Biomedical Optics Express*, vol. 8, no. 2, pp. 579–592, 2017.

Kermany, D., "Labeled optical coherence tomography (OCT) for classification," *Mendeley Data*, vol. 1, 2017.

Kermany, D. S., Goldbaum, M., Cai, W., Valentim, C. S., Liang, H., Baxter, S. L., McKeown, A., Yang, G., Wu, X., and Yan, F., "Identifying medical diagnoses and treatable diseases by image-based deep learning," *Cell*, vol. 172, no. 5, pp. 1122–1131, 2018.

Krizhevsky, A., Sutskever, I., and Hinton, G. E., "ImageNet classification with deep convolutional neural networks," *Advances in Neural Information Processing Systems*, vol. 25, no. 2, 2012.

Lu, D., Heisler, M., Lee, S., Ding, G. W., Navajas, E., Sarunic, M. V., and Beg, M. F., "Deep-learning based multiclass retinal fluid segmentation and detection in optical coherence tomography images using a fully convolutional neural network," *Medical Image Analysis*, vol. 54, pp. 100–110, May 2019.

Marblestone, A. H., Wayne, G., and Kording, K. P., "Toward an integration of deep learning and neuroscience," *Frontiers in Computational Neuroscience*, vol. 10, no. 5, 2016.

Nielsen, M. A., *Neural Networks and Deep Learning*, Determination Press, 2015.

Rawat, W. and Wang, Z., "Deep convolutional neural networks for image classification: A comprehensive review," *Neural Computation*, vol. 29, pp. 1–98, 2017.

Russell, S. J. and Norvig, P., *Artificial Intelligence: A Modern Approach*, 3rd edition, Prentice Hall, 2010.

Szegedy, C., Liu, W., Jia, Y., Sermanet, P., Reed, S., and Anguelov, D., "Going deeper with convolutions," *Proceedings of the IEEE Conference on Computer Vision and Pattern Recognition*, Boston, MA, pp. 1–9, 2015.

Szegedy, C., Vanhoucke, V., Ioffe, S., Shlens, J., and Wojna, Z., "Rethinking the inception architecture for computer vision," *Proceedings of the IEEE Conference on Computer Vision and Pattern Recognition*, pp. 2818–2826, 2016.

Tajbakhsh, N., Tajbakhsh, N., Shin, J. Y., Gurudu, S. R., Hurst, R. T., Kendall, C. B., Gotway, M. B., and Liang, J., "Convolutional neural networks for medical image analysis: Full training or fine tuning?," *IEEE Transactions on Medical Imaging*, vol. 35, no. 5, pp. 1299–1312, May 2016.

Yonetsu, T., Bouma, B. E., Kato, K., Fujimoto, J. G., and Jang, I., "Optical coherence tomography," *Circulation Journal*, vol. 77, no. 8, pp. 1933–1940, 2013.

Yorston, D., "Retinal diseases and VISION 2020," *Community Eye Health*, vol. 16, pp. 19–20, 2003.

14 Classification of Ecological Data by Deep Learning

14.1 INTRODUCTION

Deep learning has been widely used in image processing, computer vision, and pattern recognition. It is a branch of machine learning based on applying a large amount of data for training a model. Different from traditional machine-learning methods, the deep-learning model can learn features automatically from vast amount of data samples and does not require domain expert's assistance in building feature extractor. It can be categorized into supervised and unsupervised learning, to be applied for various tasks on respective types of data. For instance, one can apply the convolutional neural network (CNN) for image classification or the recursive neural network (RNN) for language processing. In computer vision, CNN is an effective framework to recognize and classify different targets.

The CNN model was initially proposed by LeCun and Yoshua (1995) as LeNet in 1995. Due to the limited computing power and incomplete mathematical proof, LeNet was difficult to be accepted by most of researchers. While with recent improvements in computing capacity and speed, CNN model has achieved better performance than traditional machine-learning methods in various fields, such as object classification, object detection, natural language processing, etc.

AlexNet (Krizhevsky et al., 2012) was developed in 2012 with a more complex structure than LeNet. AlexNet contains millions of parameters. It is constructed by five convolution layers, including max-pooling layer, dropout layer and three fully connected layers. AlexNet won the championship in the 2012 ImageNet competition with the test error rate of 15.4%. In 2014, the Visual Geometry Group at University of Oxford proposed the VGG network (Simonyan and Zisserman, 2014), which constructs the network with more convolutional layers. Microsoft Research Asian proposed the ResNet (He et al., 2016), which achieved the test error rate of 3.6% in the 2015 ImageNet competition. It used a residual block to avoid the vanishing gradient problem in back-propagation. However, it took two to three weeks to finish training on an 8-GPU machine.

The Google company proposed GoogLeNet (Szegedy et al., 2015) with 22 layers. GoogLeNet achieved a promising result with error rate of 6.7% on the 2015 ImageNet competition. Its inception module uses different kernel sizes in feature extraction and reconstructs the feature maps using 1×1 convolution (Lin et al., 2013). Later, more advanced versions have been developed, including inception v2 (Ioffe and Szegedy, 2015), inception v3 (Szegedy et al., 2015), and inception v4 (Szegedy

DOI: 10.1201/9781003474395-17

et al., 2016). The inception v2 introduces batch normalization to produce a fixed distribution for the output of each layer. The inception v3 and inception v4 adopt factorizing convolution to reduce the parameters of convolutional blocks. The inception residual block (Szegedy et al., 2015) was constructed by adding a residual connection between activation functions for solving the vanishing gradient problem.

In ecology, one task is to identify bee species from bee wing images. Traditional machine-learning methods, such as random forest, artificial neural networks, support vector machines, and genetic algorithms (Goodfellow et al., 2014; Thessen, 2016; Bell, 1999; Cutler et al., 2007; Boddy and Morris, 1999; Chen et al., 2000; Silva, 2015), have already been applied on classification of ecological image data. Researchers have adopted support vector machines, artificial neural networks, Naïve Bayes (John and Langley, 1995), k-nearest neighbors (Shinn et al., 2000), and logistic classifier (Witten et al., 2011) in classifying bee wings. However, the combination of deep learning and ecology has seldom been explored. Ecologists have shown increasing interests in building more efficient species classification systems using deep-learning neural networks. In other application of deep learning and ecology, Schneider et al. (2018) used the RNN to classify different types of animals from trap camera data.

Species are various in nature and one specie usually has different kinds of subspecies. Two problems occurred in training CNN models on the ecological data are the limited image dataset and imbalanced sample numbers among different classes. Therefore, the test accuracies for the existing CNN models on the ecological dataset are low. In this paper, we utilize three methods to increase the test accuracy in classifying the ecological dataset. The first method uses data augmentation (Zhun et al., 2017; Luke and Geoff, 2017; Francisco et al., 2018; Cubuk, 2018) to enlarge the dataset by performing transformation operations. The second method adopts transfer learning (Shin et al., 2016; West et al., 2007; Weiss et al., 2016; Shao, 2015; Amir et al., 2018) by applying the parameters of a pre-trained CNN model to the proposed classification task. Transfer learning uses a highly generalized knowledge and transfers this knowledge to learn the ecological datasets. As compared with model training in random initialization, the pre-trained model converges much faster and achieves better performance. The third method combines data augmentation with transfer learning so the test accuracy in ecological datasets can be further improved.

The rest of this paper is organized as follows. Section 14.2 introduces the deep-learning framework, including LeNet-5, AlexNet, VGG, residual neural network, and inception models. Section 14.3 presents the classification in the ecological dataset. Section 14.4 describes the improvement on the test accuracy. Section 14.5 shows experimental results. Finally, conclusions are drawn in Section 14.6.

14.2 DEEP-LEARNING FRAMEWORK

Deep learning requires a large amount data to train and evaluate learning model's performance. Figure 14.1 shows the structure of LeNet-5, which is initially intended for the classification of handwritten digits. It is composed by several layers with different functions. Like other machine-learning models, LeNet-5 needs a feature representation method to compress one (i.e., a grayscale image) or three (i.e., a RGB image) 2D matrices into feature representation.

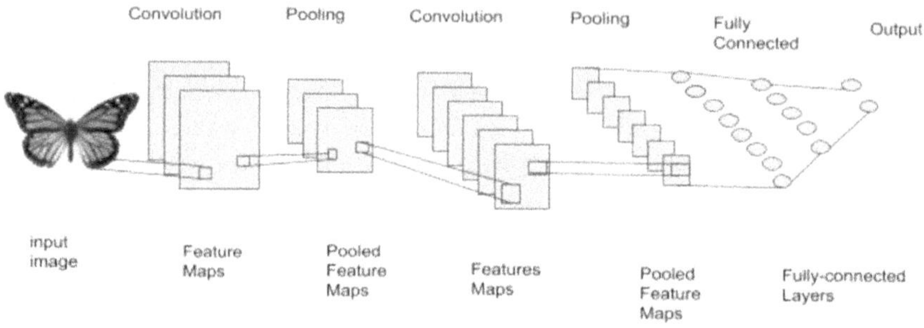

FIGURE 14.1 The structure of LeNet-5.

In LeCun's design, LeNet-5 contains an input layer, a convolution layer to extract features, and a pooling layer to reduce unnecessary data. After a second connection of a convolutional layer and a pooling layer, feature representation is fed to a fully connected artificial neural networks for classification. In the convolutional layer, the input is one or several images with one or three channels. We perform convolution several times with different filters, so there are several output images, called feature maps. The convolutional layers extract different local features using different filters to learn all the features. The convolutional layer followed by an activate function is described as

$$h^k = f(\sum\nolimits_{l \in L} x^l \otimes w^k + b^k) \tag{14.1}$$

where h^k is the latent representation of k-th feature map of the current layer, f is the activation function, x^l is the l-th feature map of group of feature maps L of the previous layers or the l-th channel of the input images with totally L channels in the case of the first layer of the network, $\otimes$ denotes the 2D convolution operation, and w^k and b^k denote the weights (filters) and biases of the k-th feature map of the current layer, respectively. A nonlinear function, called ReLU (rectified linear unit), works as the activation function f, which can be written as $f(x) = \max(0, x)$. This function will stay 0 when x is less than 0 but return to be x for any positive input. The ReLU works well for neural network models because it allows the models to compute non-linearities and interaction.

Let a softmax function be defined as

$$p_i = \frac{r^{z_i}}{\sum_{k=1}^{K} e^{z_k}}, i = 1, 2, 3, \ldots, K \tag{14.2}$$

where z_i is an element of the input tensor. Using the softmax function, we can transform an N-dimensional vector of real numbers into a vector of real numbers in the range of (0,1). The loss function is the cross-entropy, which is denoted as

$$H(y, p) = -\sum\nolimits_i y_i log(p_i) \tag{14.3}$$

where y_i is the label of i-th input image and p_i is the i-th element of the output of softmax function.

The pooling layer is designed to perform down-sampling on image data to reduce the size of feature maps. There are two typical pooling methods: average pooling and max-pooling. Average pooling is used to compute the average value in a small area and max-pooling is used to extract the maximum value. After sufficient information is acquired from convolutional layers and pooling layers, the fully connected layer is used to map the output to linearly separable space and flatten the matrix into a vector. Then, softmax is used for regression to classify the data, so the output of the last fully connected layer is the predicted label.

AlexNet (Krizhevsky et al., 2012) is the first deep convolutional neural network with five convolutional layers using ReLU as an activation function. In feature extraction, AlexNet uses large convolution kernels to extract features. In ILSVRC 2010, AlexNet obtained the top 1 and top 5 error rates of 37.5% and 17.0%, respectively.

VGG neural network (Simonyan and Zisserman, 2014) was developed by Visual Geometry Group, University of Oxford. In the 2014 ILSVRC (ImageNet Large Scale Visual Recognition Competition), VGG-16 obtained an error rate of 8.8% and VGG-19 obtained an error rate of 9.0%. In the VGG model, stacked convolution kernels with 3 by 3 are used. Note that two 3-by-3 convolution kernels equal to a 5-by-5 effective convolution area, three 3-by-3 kernels equal to a 7-by-7 effective area, and so on. The purpose of using stack convolutions is to reduce parameters in the learning process. The VGG-16 contains two 5-by-5 convolutional layers and three 7-by-7 convolutional layers and the VGG-19 contains two 5-by-5 convolutional layers and three 9-by-9 convolutional layers. However, when more convolution layers are stacked together, a vanishing gradient problem may happen. It is occurred during back-propagation when several small derivatives are multiplied together after the same activation function. The problem of a small gradient will cause the parameters not to be updated effectively.

In order to solve the vanishing gradient problem, a new convolutional block, called residual block, is introduced in residual neural network (He et al., 2016). By adding a shortcut connection between the input x to learn residual mapping $F(x)$ before the activation function, the output $x+F(x)$ is able to maintain a higher overall derivative. With residual connections, the residual neural network can add up to 152 layers. It won the competition in 2015 ILSVRC.

The inception block was introduced by GoogLeNet (Szegedy et al., 2015), which uses different kernel sizes. In inception block, 1×1 convolution, 3×3 convolution, 5×5 convolution, and 3×3 max-pooling are used at the same time using the same convolution. The 1×1 convolution with ReLu activation works as dimension reduction to reconstruct the feature maps (Lin et al., 2013). Figure 14.2 shows the inception block in GoogLeNet.

GoogLeNet (Inception v1) is the first version of inception model, continued by Inception v2 (Ioffe and Szegedy, 2015), Inception v3 (Szegedy et al., 2015), and Inception v4 (Szegedy et al., 2016). Inception v1 is the winner of 2014 ILSVRC (ImageNet Large Scale Visual Recognition Competition). Inception v2 introduced a concept termed as batch normalization to normalize the value distributions of layers'

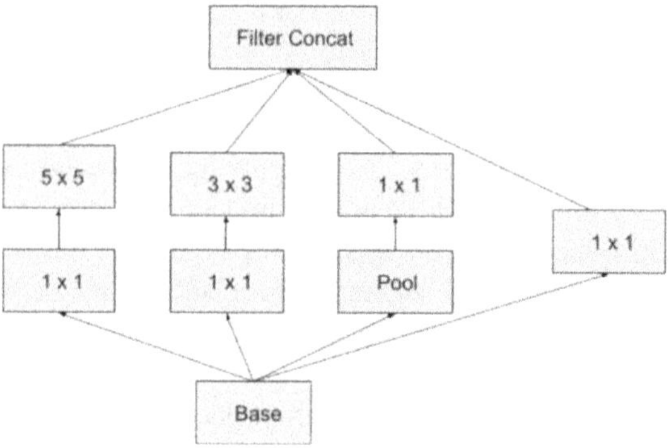

FIGURE 14.2 Inception module with dimension reduction.

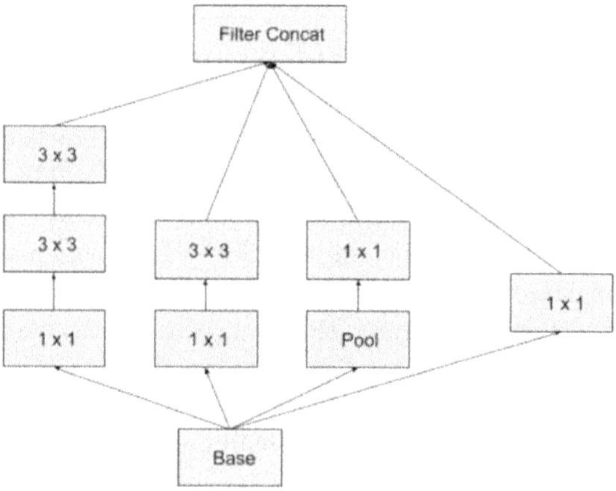

FIGURE 14.3 Factorization into a smaller convolution.

output and keep the distribution fixed. Inception v3 introduced factorizing convolution to reduce parameters. Two kinds of factorizing convolutions were introduced in (Szegedy et al., 2016), where small kernel convolutions are used to replace large convolutions and asymmetric convolutions are used to replace symmetric convolutions. Figure 14.3 shows a factorization in smaller convolutions, where the 5×5 convolution area is replaced by two 3×3 convolution areas.

Asymmetric convolution with one $n \times 1$ followed by one $1 \times n$ convolution can replace a $n \times n$ area. The purpose of using asymmetric convolution is to reduce the number of operations and maintain the network's efficiency. With asymmetric convolution, a new version of inception module is shown in Figure 14.4.

The Inception-ResNet-v1 and Inception-Resnet-v2 were introduced in Inception v4 (Szegedy et al., 2016), where a shortcut connection is added between two activation functions. Three Inception residual blocks in Inception-ResNet-v1 and Inception Resnet-v2 are shown in Figure 14.5.

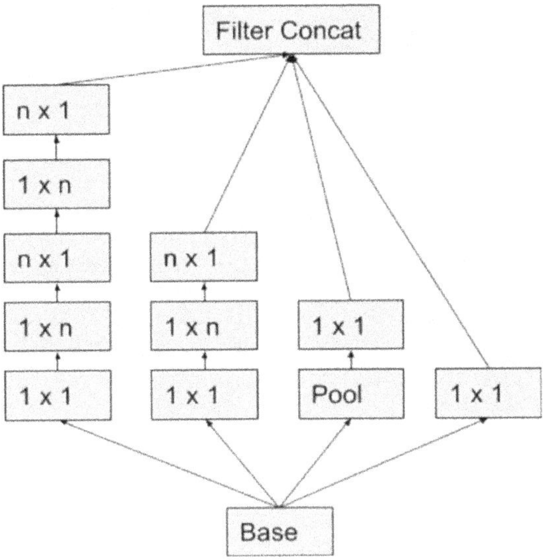

FIGURE 14.4 Factorization into asymmetric convolution.

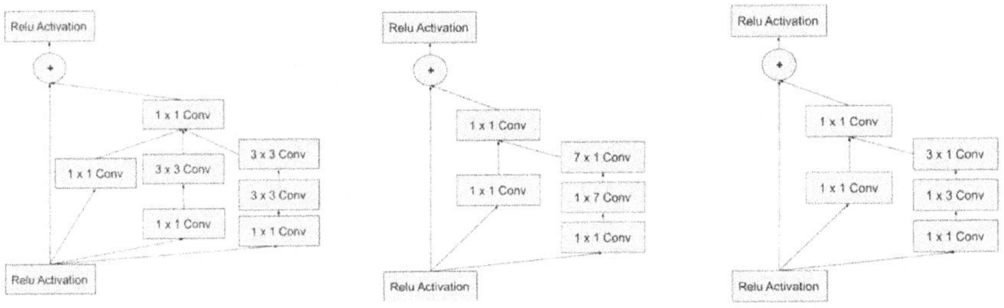

FIGURE 14.5 Inception-residual modules in inception-residual v2.

14.3 CLASSIFICATION IN ECOLOGICAL DATASET

14.3.1 Ecological Dataset

Two ecological datasets of bee wing and butterfly are used. The bee wing dataset is a relatively small and unbalanced dataset. There are 19 types of bee wing, with totally 755 samples, including 566 images for training and 189 images for testing. Within the 19 classes, 8 main classes are agapostemon, augochlora, augochlorella, augochlorella, ceratina, dialictus, halictus, and osmia. Ceratina contains three subclasses, which are ceratinacalcarata, ceratinadupla, and ceratinametallica. Dialictus contains four subclasses, which are dialictusbruneri, dialictusillinoensis, dialictusimitatus, and dialictusrohweri. Halictus contians two subclasses, which are halictusconfusus and halictusligatus. Osmia contains five subclasses, which are osmiaatriventis,

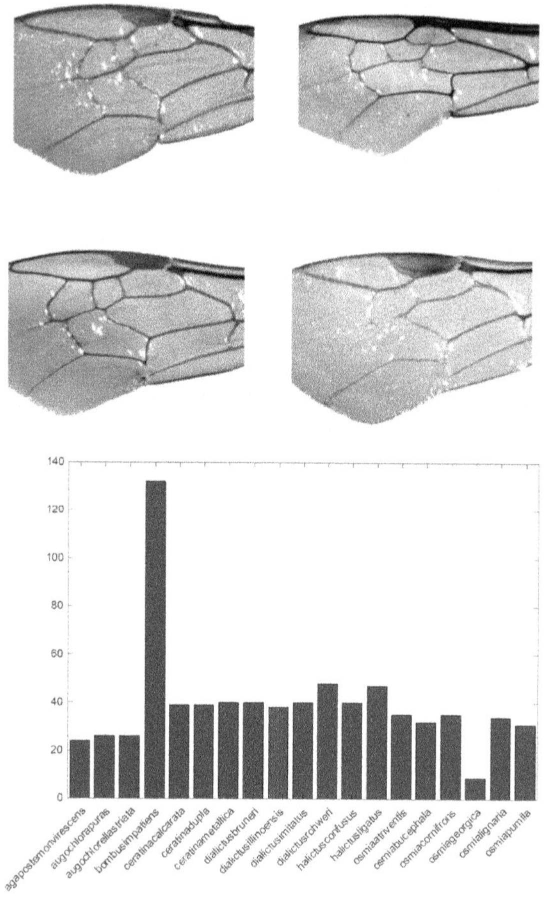

FIGURE 14.6 (a) Samples in bee wing dataset, (b) sample distribution in all classes.

osmiabucephala, osmiacornifrons, osmiageorgica, and osmialignaria. The samples among subclasses are similar but difficult to be distinguished by human beings. In Figure 14.6, (a) shows some samples in bee wing dataset and (b) shows the sample distribution in all classes.

The butterfly dataset (Wang et al., 2009) is a small and relatively balanced dataset. It contains 10 classes of RGB butterfly species, ranging from 55 to 100 images per class. The total dataset contains 832 images, where 627 images are used for training and 205 images for testing. In Figure 14.7, (a) shows some images in the butterfly dataset and (b) shows the sample distribution in all classes.

14.3.2 Classification within the Original Dataset

Deep-learning models are mainly designed for large datasets, such as ImageNet. Considering the ecological datasets are relatively small, we test and evaluate the ecological datasets using seven convolutional models, including LeNet-5, AlexNet,

FIGURE 14.7 (a) Samples in butterfly dataset, (b) sample distribution in all classes.

TABLE 14.1
Test Accuracy on the Ecological Datasets

	Bee Wing	Butterfly
LeNet-5	87.78%	70.24%
AlexNet	86.04%	79.85%
VGG-16	17.74%	12.17%
VGG-19	17.72%	12.28%
ResNet50	86.54%	75.36%
Inception v3	87.16%	78.84%
InceptionResNetV2	87.72%	79.98%

VGG-16, VGG-19, Residual Net 50, InceptionV3, and Inception Residue V2. The test results are listed in Table 14.1.

A similar test accuracy of 87% is obtained by LeNet-5, AlexNet, ResNet50, Inception v3, and InceptionResNetV2, except by VGG-16 and VGG-19. LeNet-5 is a two-layer convolutional neural network, which can extract features in the bee wing dataset. AlexNet uses five convolutional layers but obtains a lower test accuracy of 86%, indicating that a vanishing gradient problem may occur.

The VGG-16 and VGG-19 models suffer a convergence problem in training which could be caused by a small size of data samples. The ResNet50 consists of more stacked convolutional layers than the VGG models. It uses a residual connection to avoid the vanishing gradient problem. Inception v3 uses an inception block with different convolution kernel sizes to enrich the feature map. Inception residual neural network combines inception blocks with residual connection.

The test accuracy on the bee wing dataset is analyzed as follows. The test accuracy of each class is shown in Figure 14.8(a). From Figure 14.8(b), a relatively lower test accuracy is observed between subclass species. In Ceratina, the classification of Ceratinadupla subclass obtains a test accuracy of 70%, which is 17% lower than the overall accuracy. In Halictus, the classification of Halictusconfusus subclass has a test accuracy of 60%, which is 27% lower than the overall accuracy. In Osmia, the classification of Osmiageorgica subclass has a test accuracy of zero, since the two samples are classified as the Osmiageorgica, which is another subclass of Osmia. Figure 14.8(c) shows a heap map of the confusion matrix.

Using AlexNet and InceptionResV2 models, two similar test accuracies of 79% are obtained on the butterfly dataset. Using LeNet, a lower test accuracy of 70% is obtained due to its insufficient number of convolution layers. The VGG-16 and VGG-19 models face a similar convergence problem in the bee wing dataset. Using ResNet50, a test accuracy of 75% is obtained, indicating that the residual connection

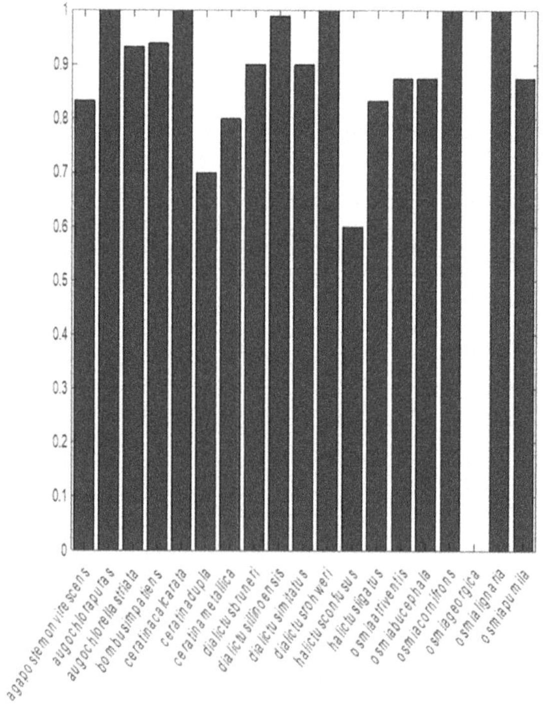

FIGURE 14.8 (a) Test accuracy in the bee wing dataset, (b) bee wing subclass classification, (c) heatmap of confusion matrix. Note that labels are from 1 to 19, respectively representing from Agapostemonvirescens to Osmiapumila.

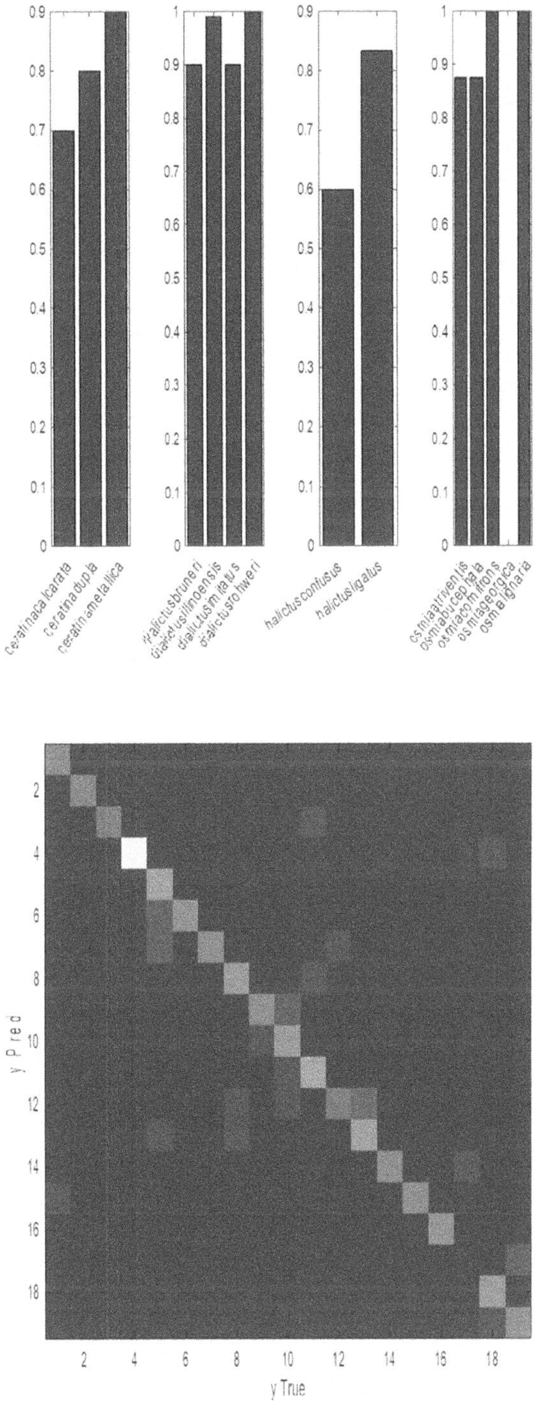

FIGURE 14.8 (Continued)

makes the model to go deeper. Note that the InceptionResV2 model obtains a higher test accuracy than InceptionV3, showing a promising feature extraction ability for the inception residual block.

In order to increase the accuracy of applying deep-learning models on small datasets, we propose to use data augmentation and transfer learning in classifying the ecological datasets.

14.4 IMPROVEMENT ON THE TEST ACCURACY

14.4.1 Data Augmentation

As compared to the dataset size often used in training a CNN model, the size of bee wing dataset is relatively small. Data augmentation is used to artificially generate new images from the original dataset by image transformations. It can improve the robustness of learning models and avoid overfitting. The data augmentor (Marcus et al., 2017) consists of three steps. The first applies image-processing functions sequentially through a pipeline. The second uses a predefined probability for each operation, which controls how likely the operation is applied. The third generates a large number of new images depending on the number of operations and the range of values used in each operation.

Perspective skewing is considered as an image transformation to view the object from different angles. Users can define a direction to perform skewing. Figure 14.9. shows the resulting images after perspective skewing functions are applied.

Elastic distortion allows users to make random distortions on the original image, while the image's aspect ratio is maintained. Figure 14.10 shows how this function works on the bee wing dataset. We observe that after elastic distortion, some of the boundaries are slightly modified.

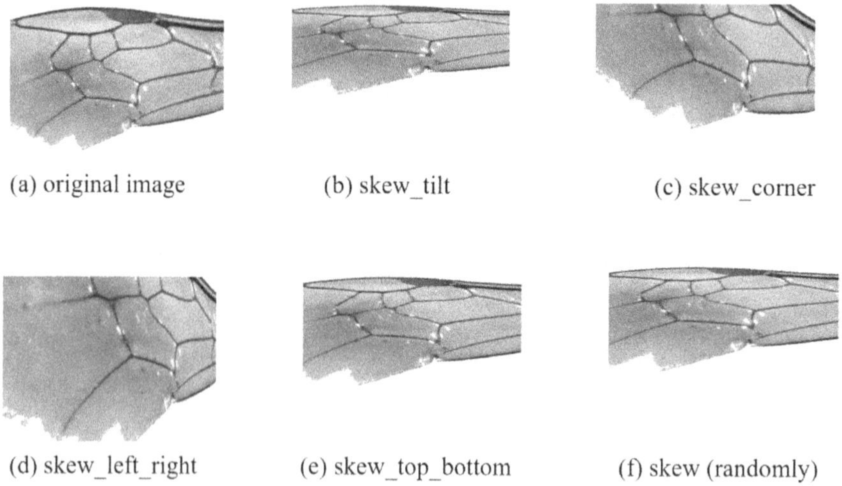

FIGURE 14.9 Perspective skewing performed on bee wing dataset. (a) An original image, (b)–(f) the images after perspective skewing by certain directions.

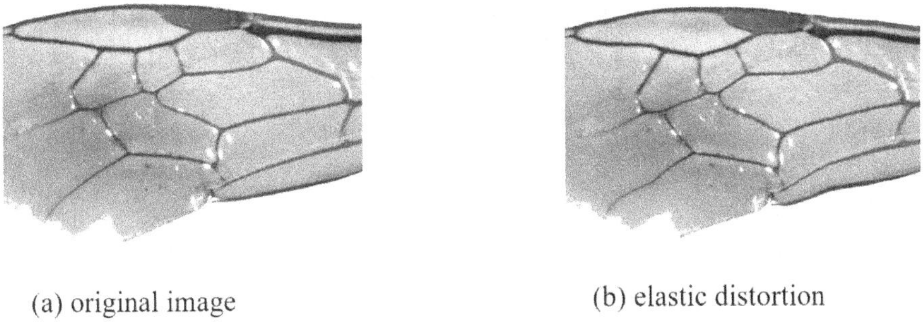

FIGURE 14.10 Elastic distortion on bee wing dataset. (a) An original image and (b) the image after elastic distortion.

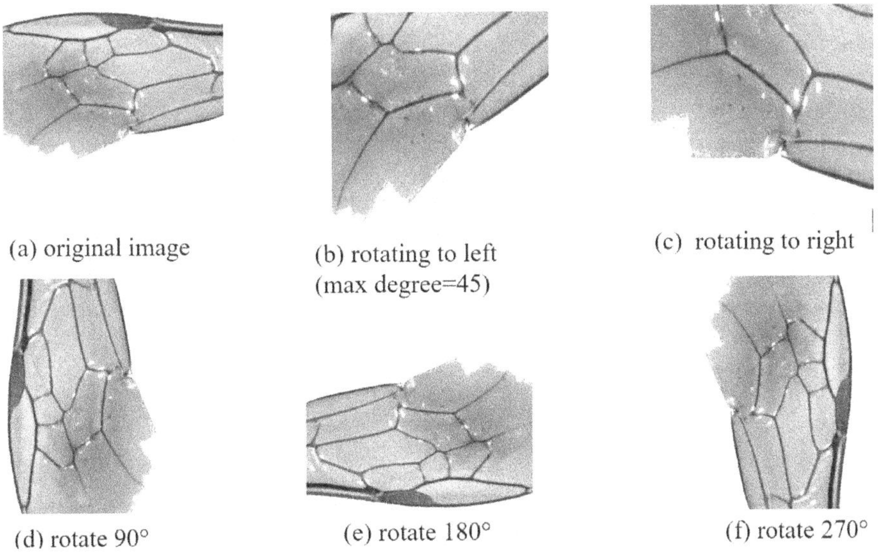

FIGURE 14.11 Rotation on bee wing dataset. (a) An original image, (b) and (c) the images rotated by two random angles (ranging from −45° to 45°) with a zoom-in effect, (d)–(e) the images rotated by 90°, 180°, or 270°, respectively.

Rotation operation intends to rotate an image in a number of ways, such as rotating 90°, 180°, or 270°. However, it could be performed by a random degree, which incorporates zoom-in or zoom-out from the original image. Figure 14.11 shows the rotation on the bee wing dataset.

Shearing intends to tilt an image along one of its sides. It can be tilted from left-to-right or right-to-left. Figure 14.12 shows a shearing on the bee wing dataset. Other functions in data augmentation include image resizing, image cropping, and mirroring.

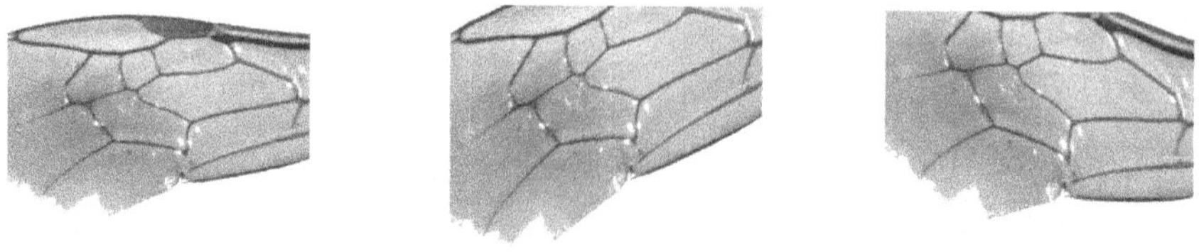

(a) original image (b) shear to left (c) shear to right

FIGURE 14.12 Shearing on bee wing dataset. (a) An original image and (b)–(c) the images after shearing by random directions (degrees from −45° to 45°).

14.4.2 Transfer Learning

Transfer learning (or inductive learning) is to gain knowledge from one task and reuse it to fulfill another task (Marcus et al., 2017). It is used when the difficulty presents to acquire a large size of datasets in real life. In this paper, six pre-trained CNN models are used to perform transfer learning on the ecological data.

In (Yosinski et al., 2014), some features are learned from convolutional neural networks that contain more common features, such as edge detectors or color blob detectors, to be used in other tasks. Note that the later layers in CNN represent progressively more specific to the details of the classes contained in the original dataset. The transfer learning is organized into three steps. First, we take a pre-trained convolutional neural network and remove its fully connected layers. Second, we add new fully connected layers at the bottom of the pre-trained model to train the ecological data. Third, we fine-tune some higher-level portion of the network.

14.4.3 Newly Designed Convolution Blocks

In the inception models, different convolutional kernel sizes are used for feature extraction. Inspired by this idea, we redesign the inception block and the inception residual block using four convolutional kernels, which are 1×1 Same Conv, 3×3 Same Conv, 5×5 same Conv, and 7×7 same Conv. The outputs are concatenated together and then passed to a 1×1 Conv. We replace the max-pooling layers by 7 x 7 same convolution to include a larger convolution kernel for detecting a wider and larger area. By combining more information in feature map, the CNN model can be more sensitive in telling the difference among different classes.

By using a different number of convolution blocks and subsampling layers in the bee wing dataset, we can compare the performance of redesigned inception block and inception residual block. Figure 14.13(a) shows the redesigned inception block and Figure 14.13(b) shows the redesigned inception residual block. The LeNet-5 has two convolutional layers to extract enough features. While AlexNet has five convolutional layers, its test accuracy faces a degradation problem. Figure 14.13(c) shows the model to compare the redesigned inception and the inception residual block.

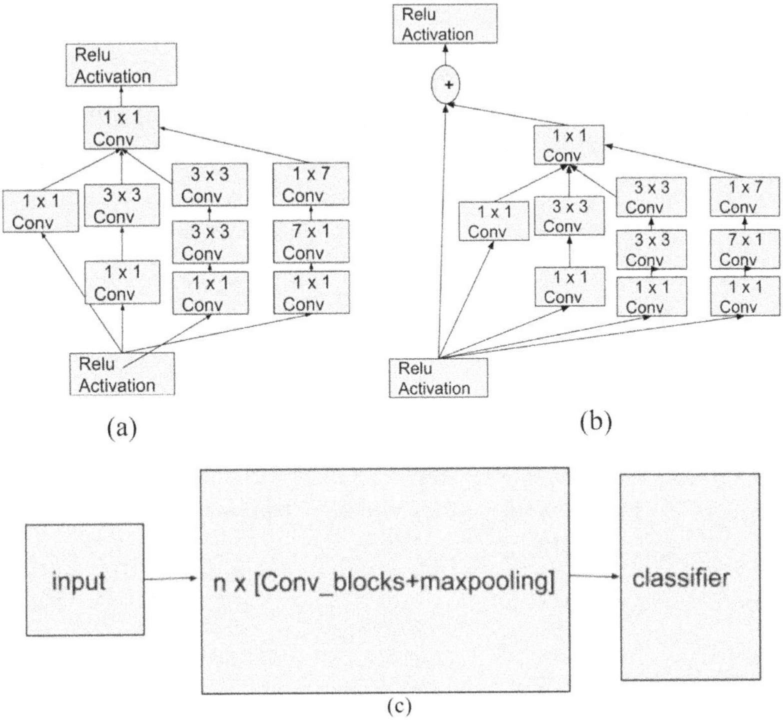

FIGURE 14.13 (a) Newly designed inception block, (b) newly designed inception residual block, (c) different $n \times$ Conv_blocks classification model.

14.5 EXPERIMENTAL RESULTS

Using data augmentation, each class is enlarged to contain 500 image samples in the training dataset. However, the number of images in the original test dataset is kept the same. In our experiments, we use a Linux workstation with 4-way Nvidia Titain GPUs. The average time to train a randomly initialized convolutional neural network on the ecology dataset can be varied from 10 minutes (LeNet-5 model) to 25 minutes (Inception ResNet V2 model). The average time to train a pre-trained deep CNN model on the ecology dataset can be varied from 10 minutes (VGG-16 model) to 20 minutes (Inception ResNet V2 model). The average time to test a CNN model can be varied from 5 seconds to 20 seconds.

14.5.1 Improvement with Data Augmentation and Transfer Learning

Table 14.2 shows the test accuracy in classifying the bee wing dataset. The test accuracy on the original dataset shows similarly at 87%. By applying data augmentation, the test accuracy obtains higher in each model. A test accuracy close to 90% is shown by using LeNet-5, AlexNet, and Inception models. Note that the convergence problem occurred by VGG-16 and VGG-19 models in classifying the original bee wing dataset is solved by using data augmentation. However, the test accuracy of applying each

TABLE 14.2
Test Accuracy on the Bee Wing Dataset

Bee Wing	Original Dataset	Data Augment	Transfer Learning	Transfer & Aug
LeNet-5	87.78%	89.97%	–	–
AlexNet	86.04%	89.8%	90.37%	91.28%
VGG-16	17.74%	88.7%	92.58%	93.41%
VGG-19	17.72%	87.34%	91.67%	93.19%
ResNet50	86.54%	89.34%	92.5%	93.12%
Inception V3	87.16%	91.46%	92.28%	93.95%
InceptionResNetV2	87.72%	90.91%	92.97%	94.40%

TABLE 14.3
Test Accuracy on the Butterfly Dataset

Butterfly	Original Dataset	Data Augment	Transfer Learning	Transfer & Aug
LeNet-5	70.24%	71.41%	–	–
AlexNet	79.85%	80.83%	89.28%	92.75%
VGG-16	12.17%	79.91%	90.65%	95.04%
VGG-19	12.28%	80.33%	90.73%	94.66%
ResNet50	75.36%	87.21%	92.60%	96.88%
Inception V3	78.84%	88.32%	93.10%	96.10%
InceptionResNetV2	79.98%	88.94%	93.67%	96.07%

model with data augmentation only improves slightly, which indicate that performing image transformations on the bee wing dataset is hard to generate diversified features.

Fine-tuning a pre-trained model is a relatively time efficiency method than training the model from a random initialization. With a highly generalized knowledge in the pre-trained model, the test accuracy in each model can be improved to 92% in the bee wing dataset. Furthermore, by combining data augmentation and transfer learning, the pre-train model shows that a much better test accuracy of 96% is achieved.

Table 14.3 shows the test accuracy on the butterfly dataset, where LeNet-5 achieves 70.24% and AlexNet reaches 79.85%. However, ResNet50, Inception V3, and InceptionResNetV2 achieve higher test accuracies of 87–89% by using data augmentation. This indicates that in the butterfly dataset, deeper convolution models can extract more detailed features. The convergence problem in VGG-16 and VGG-19 models is solved by using data augmentation. Using transfer learning, the test accuracies can be improved up to 93%. By combining data augmentation and transfer learning, a test accuracy of 96.88% is achieved using the ResNet50 model.

14.5.2 Improvement with Inception and Inception Residual Block

In order to improve the test accuracies on the bee wing dataset, different numbers of inception blocks and inception residual blocks are used to build CNN models. Table 14.4 shows the test accuracy on the original bee wing dataset.

TABLE 14.4
Test Accuracy on the Original Bee Wing Dataset

Original Dataset	Inception Block	Inception Residual Block
2× Blocks	90.04%	92.89%
3× Blocks	90.04%	92.05%
4× Blocks	89.24%	92.09%
5× Blocks	88.75%	92.90%

By using two inception blocks, a test accuracy of 90% is achieved. As compared to the test accuracy of 87% in LeNet-5, the inception blocks can detect more details using four different convolution layers. However, with more inception blocks added up, the performance has a degradation which is caused by the vanishing gradient. While checking the results in the inception residual blocks, a similar test accuracy of 92% is achieved. With the residual connection, more convolution blocks can be added up and the vanishing gradient problem can be avoided.

14.6 CONCLUSIONS

The development of convolutional neural network models on the classification of ecological data is presented in this paper. The dataset of bee wing images is very difficult for human eyes to recognize; thus, convolutional neural network models are applied due to its ability in automatic feature extraction and summarization. Seven CNN models with random initialization are trained in the original ecological datasets. However, the test accuracy needs to be improved due to the limited dataset problem. VGG models suffers a convergence problem due to a small amount of data samples.

Data augmentation and transfer learning are then used to improve the results. By applying augmented datasets and pre-trained models, the performance of using CNN models is improved. The convergence problem in VGG models is solved. Experimental results show that by combining transfer learning and data augmentation, classification of the ecological datasets can achieve the best performance.

Furthermore, redesigned inception blocks and inception residual blocks are used to improve the performance in classifying the bee wing dataset. By applying a residual connection in redesigned inception residual blocks, the models are able to overcome the vanishing gradient problem and maintain a better performance than those only using inception blocks.

REFERENCES

Amir, Z., Sax, A., Shen, W., Guibas, L., Malik, J., and Savarese, S., "Disentangling task transfer learning," *Proceedings of the IEEE Conference on Computer Vision and Pattern Recognition*, Salt Lake City, UT, pp. 3712–3722, 2018.

Bell, J., "Tree-based methods," In *Machine Learning, Methods for Ecological Applications*, A. H. Fielding, Ed., pp. 89–106, Springer, 1999.

Boddy, L. and Morris, C., "Artificial neural networks for pattern recognition," *Machine Learning, Methods for Ecological Applications*, A. H. Fielding, Ed., pp. 37–88, Springer, 1999.

Chen, D. G., Hargreaves, N. B., Ware, D. M., and Liu, Y., "A fuzzy logic model with genetic algorithm for analyzing fish stock-recruitment relationships," *Canadian Journal of Fisheries and Aquatic Sciences*, vol. 57, no. 9, pp. 1878–1887, 2000.

Cubuk, E. D., "AutoAugment: Learning augmentation policies from data," arXiv:1805.09501, 2018.

Cutler, D. R., Edwards, T. C., Beard, K. H., Cutler, A., Hess, K. T., Gibson, J., and Lawler, J. J., "Random forests for classification in ecology," *Ecology*, vol. 88, no. 11, pp. 2783–2792, 2007.

Francisco, J. M., Moreno-Barea, J., Strazzera, F., JerezUrda, D., and Franco, L., "Forward noise adjustment scheme for data augmentation," *Proceedings of the IEEE Symposium Series on Computational Intelligence*, Bangalore, pp. 728–734, 2018.

Goodfellow, I. J., Pouget-Abadie, J., Mirza, M., Xu, B., Warde-Farley, D., Ozair, S., Courville, A., and Bengio, Y., "Generative adversarial networks," *Machine Learning*, arXiv:1406.2661v1, 2014.

He, K., Zhang, X., Ren, S., and Sun, J., "Deep residual learning for image recognition," *Proceedings of the IEEE Conference on Computer Vision and Pattern Recognition*, Las Vegas, NV, pp. 770–778, 2016.

Ioffe, S. and Szegedy, C., "Batch normalization: Accelerating deep network training by reducing internal covariate shift," arXiv:1502.03167, 2015.

John, G. and Langley, P., "Estimating continuous distributions in Bayesian classifiers," *Proceedings of the Eleventh Conference on Uncertainty in Artificial Intelligence*, Montreal, Quebec, pp. 338–345, 1995.

Krizhevsky, A., Sutskever, I., and Hinton, G. E., "ImageNet classification with deep convolutional neural network," in *Advances in Neural Information Processing Systems*, vol. 25, F. Pereira, C. J. C. Burges, L. Bottou, and K. Q. Weinberger, Eds., Curran Associates, Inc., 2012, pp. 1097–1105.

LeCun, Y. and Bengio, Y., "Convolutional networks for images, speech, and time series," *The Handbook of Brain Theory and Neural Networks*, pp. 255–258, MIT Press, 1995.

Lin, M., Chen, Q., and Yan, S., "Network in network," arXiv:1312.4400, 2013.

Luke, T. and Geoff, N., "Improving deep learning using generic data augmentation," arXiv:1708.06020, 2017.

Marcus, D., Stocker, C., and Holzinger, A., "Augmentor: An image augmentation library for machine learning," arXiv:1708.04680, 2017.

Schneider, S., Taylory, G. W., and Kremer, S. C., "Deep learning object detection methods for ecological camera trap data," arXiv:1803.10842v1, 2018.

Shao, L., "Transfer learning for visual categorization: A survey," *IEEE Transactions on Neural Networks and Learning Systems*, vol. 26, no. 5, pp. 1019–1034, 2015.

Shin, H. C., Roth, H. R., Gao, M., Lu, L., Xu, Z., and Nogues, I., "Deep convolutional neural networks for computer-aided detection: CNN architectures, dataset characteristics and transfer learning," *IEEE Transactions on Medical Imaging*, vol. 35, no. 5, pp. 1285–1298, May 2016.

Shinn, A., Kay, J., and Sommerville, C., "The use of statistical classifiers for the discrimination of species of the genus gyrodactylus (monogenea) parasitizing salmonids," *Parasitology*, vol. 120, no. 3, pp. 261–269, 2000.

Silva, F., "Automated bee species identification through wing images," Thesis, University of São Paulo, 2015.

Simonyan, K. and Zisserman, A., "Very deep convolutional networks for large-scale image recognition," arXiv:1409.1556, 2014.

Szegedy, C., Li, Y., Ma, L., Bai, Z., and Chan, C.-W., "Going deeper with convolutions," *Proceedings of the IEEE Conference on Computer Vision and Pattern Recognition*, Boston, MA, Jun. 8–10, 2015.

Szegedy, C., Ioffe, S., Vanhoucke, V., and Alemi, A., "Inception-v4, Inception-ResNet and the impact of residual connections on learning," arXiv:1602.07261v2, 2016.

Szegedy, C., Vanhoucke, V., Ioffe, S., Shlens, J., and Wojna, Z., "Rethinking the inception architecture for computer vision," arXiv:1512.00567, 2015.

Thessen, A. E., "Adoption of machine learning techniques in ecology and earth science," *PeerJ PrePrints*, vol. 4, p. e1720v1, 2016. https://doi.org/10.7287/peerj.preprints.1720v1.

Wang, J., Markert, K., and Everingham, M., "Learning models for object recognition from natural language descriptions," *Proceedings of the 20th British Machine Vision Conference*, London, 2009.

Weiss, K., Khoshgoftaar, T. M., and Wang, D., "A survey of transfer learning," *Journal of Big Data*, vol. 3, no. 9, 2016. https://doi.org/10.1186/s40537-016-0043-6.

West, J., Ventura, D., and Warnick, S., *Spring Research Presentation: A Theoretical Foundation for Inductive Transfer*, Brigham Young University, College of Physical and Mathematical Sciences, 2007.

Witten, I. H., Frank, E., and Hall, M., *Data Mining: Practical Machine Learning Tools and Techniques*, Morgan Kaufmann, 2011.

Yosinski, J., Clune, J., Bengio, Y., and Lipson, H., "How transferable are features in deep neural networks?" *Advances in Neural Information Processing Systems*, vol. 27, pp. 3320–3328, Dec. 2014.

Zhun, Z., Liang, Z., Guoliang, K., Shaozi, L., and Yi, Y., "Random erasing data augmentation," arXiv e-prints, 2017.

15 Joint Learning for Pneumonia Classification and Segmentation on Medical Images

15.1 INTRODUCTION

In the past few years, pneumonia has been ranked among the top 10 causes of death in the United States of America. Detecting pneumonia at its early stage can help save a patient's life. Thus, doctors are eager to detect pneumonia with the assistance of modern computer systems. An effective automatic pneumonia identification system is able to assist doctors in finding possible pneumonia samples and localizing target areas. There are two main requirements for a pneumonia image analysis system. First, the system should identify potential pneumonia samples from healthy ones. Second, the system should be able to highlight and segment the target area accurately.

Deep learning (LeCun et al., 2015) has attracted increasing attention during recent years. It is the representation learning based on a large amount of data. In computer vision, a widely used model is convolutional neural network (CNN), which was first proposed by Yann LeCun in 1995 (LeCun and Bengio, 1995). Due to the limitations of computing capacity and incomplete mathematical proof, deep learning was difficult to be accepted by researchers in the beginning. Thanks to the expanding growth of computational ability, different structures of CNN have been developed for image classification. In AlexNet (Krizhevsky et al., 2012), VGG model (Simonyan and Zisserman, 2015), and residual model (He et al., 2016), a fixed kernel size was used in the convolution layers. GoogLeNet (Szegedy et al., 2015, 2016, 2017; Ioffe and Szegedy, 2015) consists of four different feature maps named Inception modules. DenseNet (Huang et al., 2017) was proposed to use channel-wise concatenations. ChestXNet (Rajpurkar et al., 2017) applied a 121-layer end-to-end dense convolution structure to detect 14 diseases in chest X-rays data. With more complex structures being applied, the parameter in deep-learning model increases tremendously.

Multi-objective detection (Girshick, 2015; Ren et al., 2015; He et al., 2017; Redmon et al., 2016) and semantic segmentation (Leclerc et al., 2019) are essential applications in convolutional neural networks. The semantic segmentation focuses on classifying each pixel to a given class, while the objective detection task focuses on separating the foreground from the background. The FCN (fully convolution network) (Shelhamer et al., 2015) was designed for semantic segmentation using one deconvolution operation to up-sample after feature extraction and generate a

DOI: 10.1201/9781003474395-18

pixelwise label map for prediction. Deconvolution network (Noh et al., 2015) used a VGG-16-layer as the auto-encoder for image representation and improved FCN by a gradual deconvolution and up-sampling layers. SegNet (Badrinarayanan et al., 2015) implemented an encoder and decoder structure for robust semantic pixelwise labeling on medical images. Zhou et al. (2017) showed a segmentation task using FCN in the 3D CT (computed tomography) images. However, the high ratio of the up-sampling layer makes the output label a rough representation without detail. To improve performance, Roth et al. (2018) and Wong et al. (2018) both implemented a U-Net structure (Ronneberger et al., 2015) for different segmentation tasks.

Researchers are more interested in knowing why a model makes predictions. Interpreting the reason that deep learning behaves can help make the model more trustworthy. The main idea for this research is to find essential regions that can affect the model's prediction in the testing images. Zeiler and Fergus (2014) introduced a method of using the probability map to eliminate some of the high probability pixels. However, it is difficult to select a suitable 15 kernel size. Simonyan et al. (2014) proposed a class saliency map-based explainable method for the classification model. By computing the gradient between the change of a predicted class label and pixels, some critical areas can be determined. This method successfully highlights a rough location in the testing image. Karargyris et al. (2019) applied it to bone age prediction on chest X-ray dataset. Zhou et al. (2017) modified the last fully connected layer to be an average-pooling layer and utilized feature maps to predict the regions of a given class. The gradient-weighted class activation mapping, or grad-CAM (Selvaraju et al., 2017), was applied to explain any CNN-based classifier without changing the model's structure.

Shih et al. (2019) proposed morphological neural network (MNN) as an alternative to convolutional layers for image classification. In contrast to CNN models (Yeh et al., 2020), the MNN has fewer parameters and more efficiency in shape classification and extraction. The morphological layers are used as a pre-processing module for the proposed joint-task learning model. The attention mechanism helps to increase the model representation power and guide the CNN model to focus on target areas. Chen et al. (2017a) showed the attention mechanism in the spatial domain to focus on objects and corresponding areas. Hu et al. (2018) introduced a channel-wise attention module, called squeeze-and-excitation, to weigh the feature maps in training. The CBAM (Woo et al., 2018) combines the spatial attention module and the channel-attention module.

In this chapter, we first propose a baseline model that learns image classification and segmentation simultaneously. Two algorithms of saliency map and grad-CAM for image classification model explanation are adopted. Second, an image-pre-processing module and an attention module are applied to refine the baseline model. Experimental results show these modules can separately improve the performance of the joint-task learning model. However, when the following modules are combined, the unguided MNN layers change the gradient and cause the saliency map and grad-CAM focusing on irreverent area. To overcome the problem, we apply the attention module to refine the feature maps between morphological layers in both channel-wise and spatial attention modules. The MBAM successfully helps the model to focus on the corresponding area with higher confidence.

The rest of this chapter is organized as follows. Section 15.2 introduces the proposed model design. Section 15.3 presents the image pre-processing and visual attention modules. Section 15.4 describes experimental results. Finally, conclusions are drawn in Section 15.5.

15.2 THE PROPOSED MODEL DESIGN

15.2.1 Baseline Model

In this section, the joint-task learning model for classification and segmentation is presented. The model performs binary classification that separates pneumonia samples from healthy ones. The classifier is based on VGG-16 and contains three parts: the input layer, feature extraction layers, and fully connected layers. The loss function using binary cross-entropy is defined by

$$BCE\_Loss = -\frac{1}{N}\sum\nolimits_{i=1}^{N} y_i log\ log\left(p(y_i)\right) + (1-y_i) log\ log\left(1-p(y_i)\right). \tag{15.1}$$

where y_i is the label (i.e., 1 for pneumonia pixel and 0 for healthy pixel) and $p(y_i)$ is the predicted probability of the pixel belonging to pneumonia for all N pixels. In the segmentation task, the model is required to output a pixelwise label map, where the target area is labeled as 1 while other areas as 0. The segmentation model is an encoder-decoder structure. The encoder converts an input image x into a latent-space representation h as $h = f(x)$. The decoder reconstructs the input from latent space representation h to a label map r as

$$r = g(h). \tag{15.2}$$

The autoencoder can be described as

$$r = g(f(x)). \tag{15.3}$$

By encoding the input image into latent representation and decoding it back to a label map, each pixel is assigned a label in the reconstruction process. Pixels labeled as 1 represent belonging to an opacity area, while the normal area is labeled as 0.

The segmentation model is a U-Net-like structure. The loss function in our segmentation model uses mean square error, which can be described as the summation of squared distances between ground truth map and decoded label map. Let y_i represent the ground truth for i-th pixel and Y_i represent the model's prediction for i-th pixel. The mean square error loss is computed as

$$MSE = \frac{1}{N}\sum\nolimits_{i=1}^{N}(y_i - Y_i)^2 \tag{15.4}$$

The joint-task learning model combines the classification and segmentation models with the sharing of feature extraction layers. The proposed joint-task learning

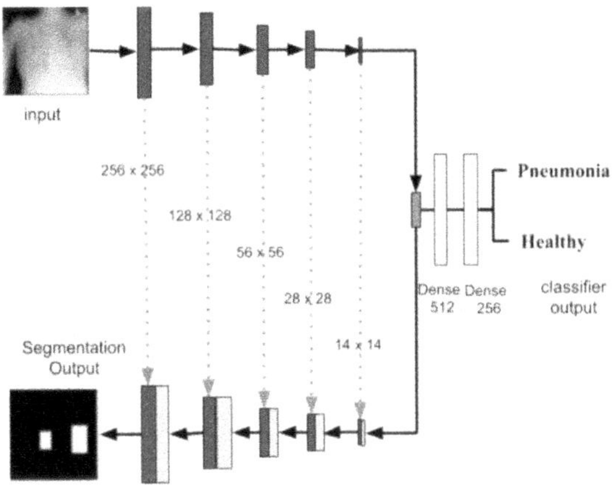

FIGURE 15.1 The proposed joint-task learning model.

model is shown in Figure 15.1. An input image is first going through convolutional layers for feature extraction. Second, the feature maps are fed into dense layers for classification and output the class types: Pneumonia or Healthy. At the same time, the feature maps are fed into the decoder for segmentation. Finally, in the segmentation model, the feature maps in the first step are concatenated with the feature maps and output the segmentation maps.

15.2.2 Class Saliency Map and Grad-CAM

When the training of the joint-task learning model is finished, a class saliency map (Chen et al., 2017) and a grad-CAM (Selvaraju et al., 2017) are used to interpret the classifier and visualize the corresponding area which has a great influence. A high class score means a relatively high influence. The class saliency maps compute the class score $S_c(I)$ from a given test image I as

$$S_c(I) = w_c^T I + b_c \quad (15.5)$$

where the label for image I is c. The class score's derivative w is computed as

$$w = \frac{\partial S_c}{\partial I} \quad (15.6)$$

By computing w in back-propagation, the pixels which have a stronger influence in determining class-score can be found. Thus, the class saliency map is determined by the classification model and class c. By visualizing the corresponding saliency map, one can understand why the classification model makes such a decision. Although the class saliency map is not a restrict segmentation tool, especially in lung CT images, it can still highlight corresponding pixels.

The grad-CAM performs a weakly supervised localization according to the image's label and the gradient of the model's last convolutional layer. For a given image and its label, the image is forward-propagated to the CNN model, and a confidence score is obtained for its corresponding label. The signal is then back-propagated to produce the feature maps. Finally, a ReLU activation function is used to combine the feature maps to show where the model is focused on when the prediction is made. Compared to CAM (Zhou et al., 2016), the grad-CAM is a generalization method and can be applied to any CNN model without modifying the model's structure. By visualizing the testing samples of using class saliency map and grad-CAM in different models, it is possible to visualize whether the model focuses on the correct area or not.

15.3 IMAGE PRE-PROCESSING AND VISUAL ATTENTION MODULES

In this section, we describe image pre-processing and visual attention modules, which can help the baseline model classifier to focus on the lung area.

15.3.1 Image-Pre-Processing Module with Morphological Layers

Mathematical morphology is a widely used approach for shape representation and image pre-processing. Two fundamental morphological operations are dilation and erosion. Let the input image be I and the structuring element be s. Dilation is denoted as $I \oplus s$, which expands the image by the structuring element. Erosion is denoted as $I \ominus s$, which shrinks the image by the structuring element.

The opening is typically used for contour smoothing, especially for breaking thin connections between components and enlarging small holes or gaps. It is defined as an erosion followed by a dilation as

$$I \circ s = \left(I \ominus s\right) \oplus s \tag{15.7}$$

Different from opening, the closing can be used for connecting narrow areas and filling in small holes or gaps. It is defined as a dilation followed by an erosion as

$$I \bullet s = \left(I \oplus s\right) \ominus s \tag{15.8}$$

Figure 15.2 shows two sample images from the Kaggle Pneumonia dataset, which are processed using dilation and erosion with a 6×6 structure element of all 1s. Figure 15.3 shows two sample images, which are processed using closing and opening with a 6×6 structure element of all 1s.

Our previous work on morphological neural network (Shih et al., 2019) is applied as pre-processing and a feature extraction layer is used for classification. Dilation can expand some of the small areas while enlarging some of the noisy areas. Erosion can clean the background by eliminating some noisy areas, but at the same time, filtering out some pixels. Opening and closing can smooth the contour, where closing tends to fill in some holes and opening tends to make them larger. Figure 15.4 shows four basic morphological operations using morphological layers.

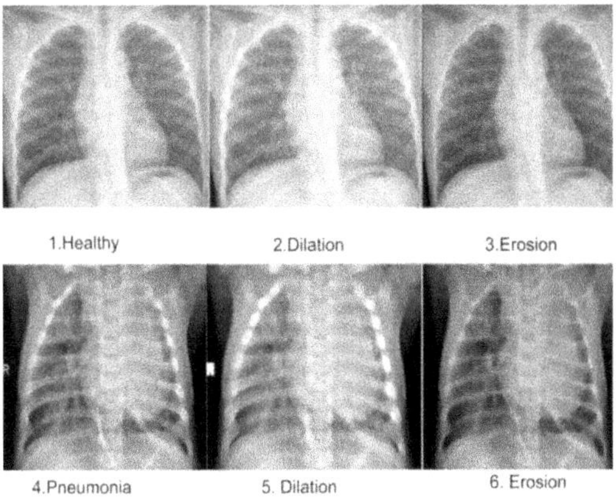

FIGURE 15.2 Sample images after morphological operations. Column 1 shows input images; column 2 shows dilation; column 3 shows erosion.

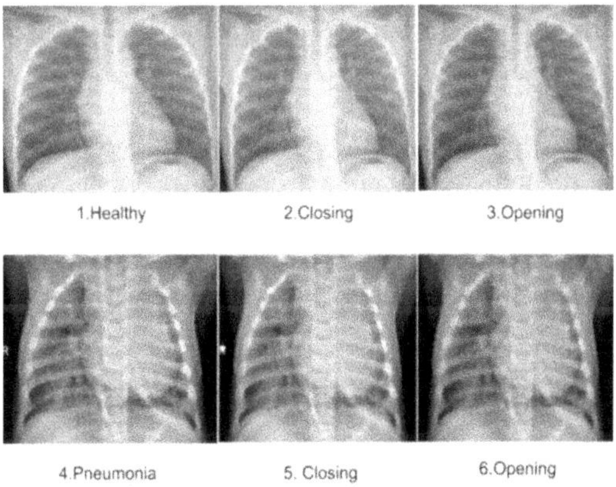

FIGURE 15.3 Sample images after morphological operations. Column 1 shows input images; column 2 shows closing; column 3 shows opening.

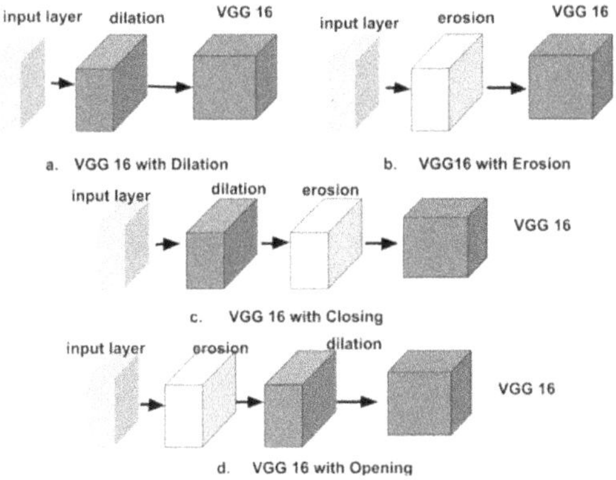

FIGURE 15.4 Morphological image-pre-processing modules with basic morphological operations.

15.3.2 Visual Attention Modules

Two attention modules, convolutional block attention module (CBAM) (Chen et al., 2017) and morphological block attention module (MBAM), are applied separately to improve the performance of the joint-task learning model, as shown in Figure 15.5. We propose MBAM to refine the feature maps between morphological layers and correctly locate a target area.

15.4 EXPERIMENTAL RESULTS

We have conducted experiments of combining different modules with the proposed joint-task learning model. In segmentation, a U-Net-like structure is used for reconstructing the masks. Considering that the ground truth is given by a bounding box instead of pixelwise label maps, performing a pixelwise segmentation may encode non-opacity regions inside a bounding box and further influence the model's prediction. The bounding box may indicate a rough area containing the lung opacity but cannot annotate each pixel. The segmentation model may not be able to preciously recognize a target area. Thus, we evaluate the performance of the joint-task learning model by showing both the segmentation model and the weakly supervised segmentation result.

The dataset from Kaggle's RSNA (Radiological Society of North America) Pneumonia Detection Challenge (Kotikalapudi et al., 2018) is used, which contains CT chest images in the DICOM format. The pixel in the opacity area is labeled as 1, indicating a potential pneumonia sample; otherwise, it is labeled as 0. In Figure 15.6, (a) shows an image which does not contain the opacity area and (b) shows an image

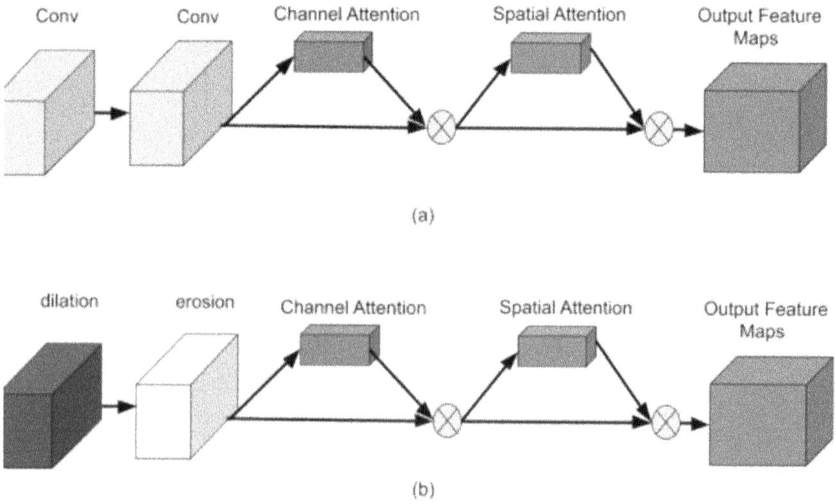

FIGURE 15.5 Visual attention modules: (a) convolutional block attention module, (b) morphological block attention module.

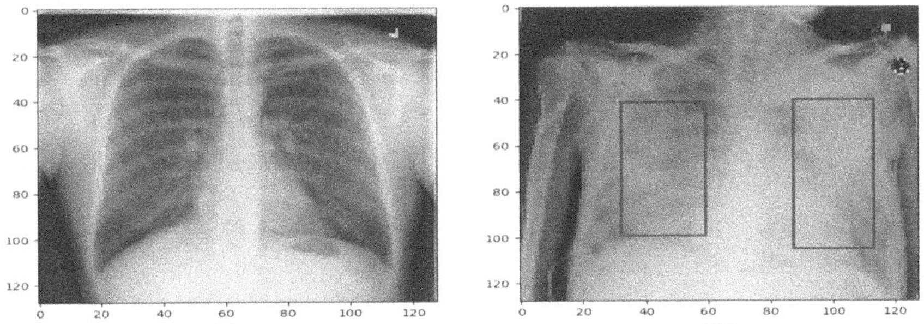

(a) (b)

FIGURE 15.6 Sample images in RSNA Pneumonia Detection Challenge. (a) Healthy body, (b) sample with lung opacity.

containing two opacity areas. The dataset contains 9,555 samples with pneumonia and 8,851 normal (healthy) samples. We randomly shuffle and divide it into three groups: training data, validation data, and testing data, which respectively have 13,804 (75%), 920 (5%), and 3,862 (20%) images.

15.4.1 Performance of the Baseline Joint-Task Learning Model

To design the proposed joint-task learning model, two main problems need to be solved. First, it is difficult for the classification and segmentation models to converge at the same time. The reason is the classification model converges much faster than the segmentation model. In the segmentation model, the decoder part has similar parameters with the encoder part, which is far more overweight than the parameters in classification model. Second, the parameter in the convolutional layers should be sufficient to extract the features and cannot be overweighed due to the limited computational capacity. Thus, the classification model uses a VGG-16 structure and the segmentation model use a U-Net structure.

The joint-task learning model is compared against different models. For classification, it is compared with ResNet-50, and for segmentation, it is compared with SegNet, FCN, and DeepLab V3. The performance comparison of these models is listed in Table 15.1.

For classification, VGG-16 and ResNet achieve a similar test accuracy. Our proposed joint-task learning model, FCN, and SegNet use a VGG-16 as feature extractor. However, in the up-sampling part our joint-task learning model uses a U-Net structure, which adds the corresponding feature maps from previous feature extractors. Compared to FCN and SegNet, our proposed joint-task learning model can directly combine previous feature maps in the feature extraction process to achieve a higher mean-average precision. As compared against the semantic segmentation model, the Deep Lab V3 (Chen et al., 2017), our joint-task learning model can achieve similar performance. Although Deep Lab V3 has fewer parameters and a slightly better performance, it cannot perform the classification task as our model can.

TABLE 15.1
Performance Comparison of the Proposed Joint-Task Model vs. Other Models

Model	Classification Accuracy	Classifier Parameter	Segmentation MAP	Total Parameter
Joint-task Model	89.27%	9.1 Million	0.5945	25 million
SegNet (Badrinarayanan et al., 2015)	/	/	0.5072	21.8 million
FCN (Shelhamer et al., 2015)	/	/	0.4368	9.1 million
ResNet-50 (Szegedy et al., 2017)	88.73%	25.6 million	/	25.6 million
Deep Lab V3 (Chen et al., 2017)	/	/	0.6012	2.5 million

15.4.2 Performance of the Different Joint-Task Learning Models

The baseline model classifier utilizes a VGG-16 structure, which is combined with different modules: morphological layers, CBAM, and MBAM. Table 15.2 shows different combinations of morphological layers as a pre-processing module with a VGG-16 classifier on the Kaggle Pneumonia dataset. The performance of CNN classifier works as a baseline model and achieves a accuracy at 89.13%. It is observed that the opening + closing + VGG-16 model achieves a relatively high test accuracy. In Figure 15.2, it is clear to find a dilation can blur the CT image, while an erosion can clear the noise. The pre-processing module using a dilation layer has a relatively weak performance than the erosion layer + CNN model. The opening and closing operations are both designed for contour smoothing. The better performance for the image-pre-processing module is through two different smoothing layers, which add more smoothing so the infected samples are easier to be recognized.

Figure 15.7 shows our proposed models, where (a) VGG-16 model, (b) the structure of morph layers + VGG-16, (c) the structure of CBAM + VGG-16, (d) the structure of morph layers + CBAM + VGG-16, and (e) the structure of MBAM + CBAM + VGG-16.

The performance of the proposed joint-task learning model is listed in Table 15.3. As compared to the baseline model, the MNN + VGG-16 model achieves a 5.13% improvement in classification and 2.32% improvement in segmentation. The reason for this improvement is caused by the image pre-processing layers using morphological layers. The MNN layers use soft minima or soft maxima function to respectively approximate dilation or erosion, which mathematically performs the morphological filtering on input images to enrich the feature maps.

The CBAM+VGG-16 model utilizes the CBAM mechanism to refine the feature maps between convolutional layers and improves the classification model by 4.58% and the segmentation model by 13.33%. The reason for this improvement is that CBAM guides the model in both spatial domain and channel-wise domain.

The MNN + CBAM + VGG-16 model combines MNN and CBAM. Even though the classification rate is increased by 1.58% and the segmentation MAP is increased by 5.4%, it is still worse than MNN + VGG-16 and CBAM + VGG-16. The reason is that MNN layers and CBAM change the gradients in original images.

TABLE 15.2
Test Accuracy for Classification Accuracy Different Morphological Layers + VGG-16

Model	Classification Accuracy
VGG-16	89.13%
Dilation + VGG-16	88.38%
Erosion + VGG-16	91.62%
Closing +VGG-16	93.02%
Opening+VGG-16	92.78%
Opening + Closing + VGG-16	94.32%
Closing + Opening+ VGG-16	94.14%

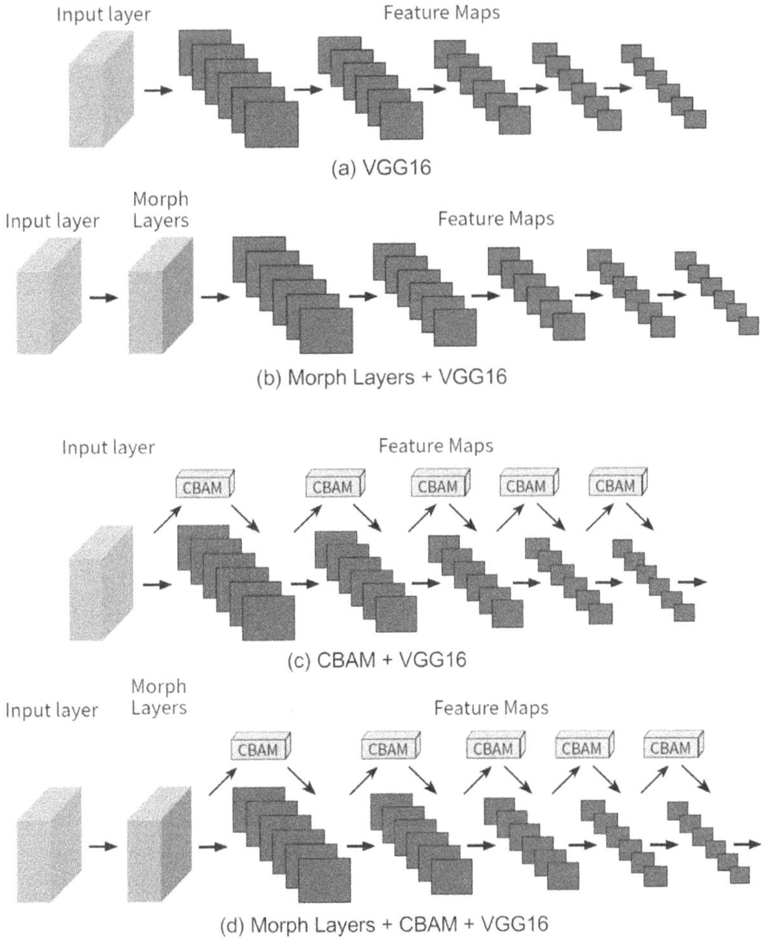

FIGURE 15.7 The proposed models.

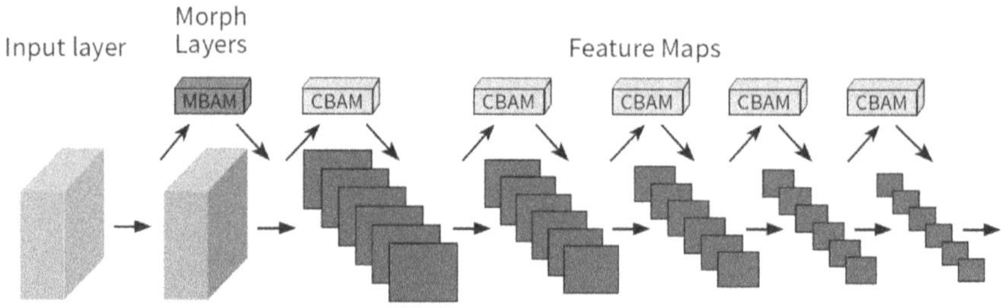

(e) MBAM + CBAM + VGG16

FIGURE 15.7 (Continued)

TABLE 15.3
Test Accuracy for Joint-Task Learning Model with Different Modules

Model	Classification Accuracy	Segmentation MAP
VGG-16	89.27%	58.45%
MNN+ VGG-16	94.14%	60.73%
CBAM + VGG-16	93.85%	71.78%
MNN+CBAM+VGG-16	90.85%	63.85%
MBAM+CBAM+VGG-16	95.73%	78.72%

The MBAM + CBAM + VGG-16 model refines the feature maps between convolutional layers and between morphological layers. Experimental results show that it improves the classification accuracy by 6.46% and the segmentation by 20.27%, as compared to the baseline model. The MBAM correctly guides the MNN layers in the training process to correct the gradient in MNN + CBAM + VGG-16, where the gradient is changed due to unorganized feature maps in morphological layers.

15.4.3 Evaluate Model Performance by Class Saliency Map and Grad-CAM

We apply class saliency maps and grad-CAM on four random samples from the test dataset to illustrate the model performance. Since the joint-task learning models have confidence ranging from 89% to 95%, it is critical to interpret whether the classifiers can detect the correct area. The class saliency map shows the corresponding influential pixels when the classifier makes its prediction. The grad-CAM shows the probability map to indicate which area has a high possibility when the classifier makes the prediction. By attaching the segmentation model's prediction with bounding boxes, we can finally decide whether this model is trusted. Figure 15.8 shows different model's performance on four pneumonia samples. The first row shows the segmentation prediction in a red bounding box, while the ground truth is displayed

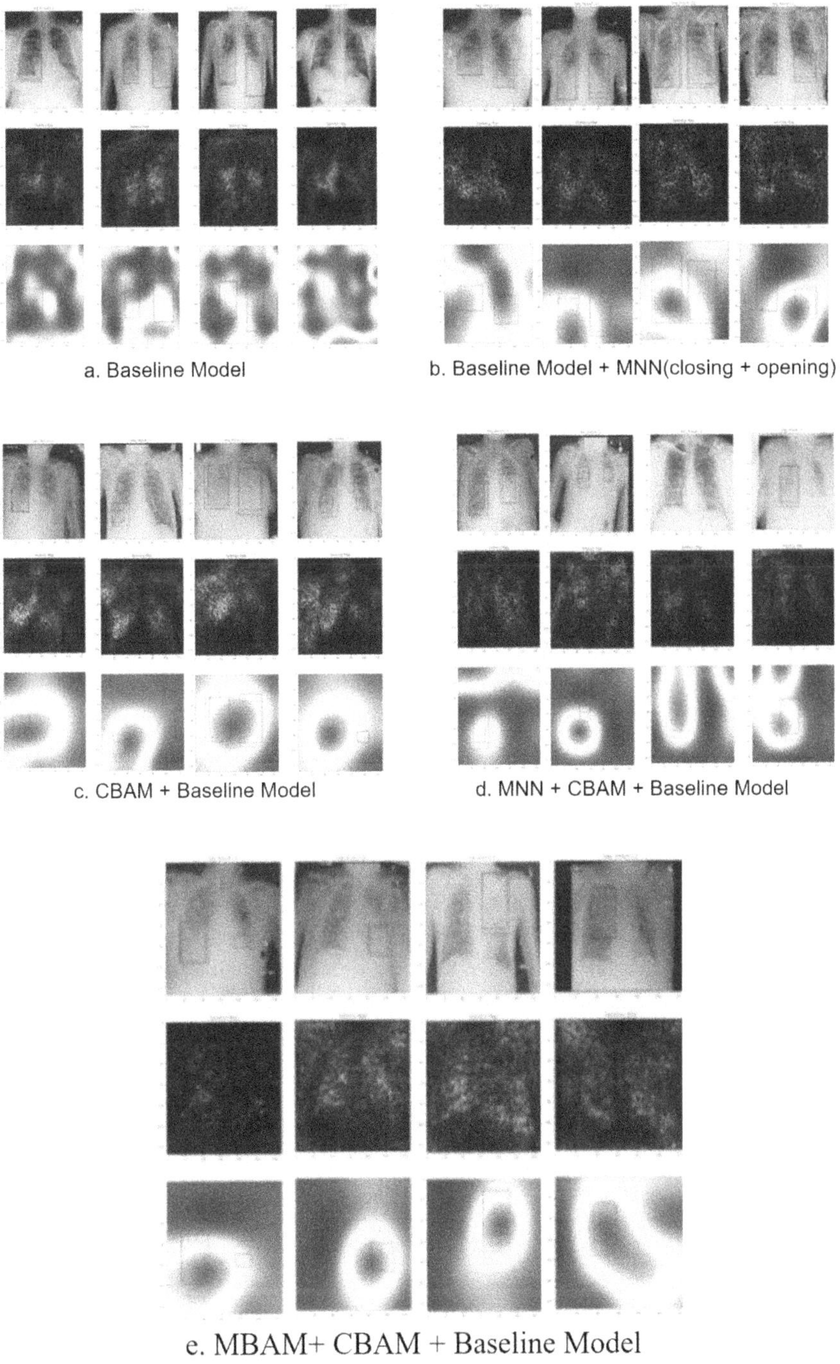

FIGURE 15.8 Class saliency map and grad-CAM for different models.

as a blue bounding box. The second row shows the class saliency map, and the third row shows the grad-CAM attention map.

Figure 15.8(a) shows that the samples are all classified as pneumonia. The class saliency map shows a weak segmentation of the lung area. The grad-CAM maps show that the baseline model is more likely to focus on the corners or bottom instead of the lung area when making its prediction. The target area has a relatively low attention probability. Thus, the baseline model has poor performance because the classifier makes its prediction based on the wrong attention area.

Figure 15.8(b) shows the baseline model with morphological layers. The class saliency map shows possible influential pixels. The morphological layers improve the model to focus on the correct attention area, so the grad-CAM can focus on the target area instead of other areas of the test images in the baseline model. Figure 15.8(c) shows the samples for the baseline model with convolutional block attention module, which successfully improves the baseline model by channel-wise attention and spatial attention modules. Compared to the baseline model, the CBAM guides the model to focus on target areas correctly.

Figure 15.8(d) shows the samples for the baseline model combined with morphological layers and CBAM. Since the morphological layers are not well guided, the image-pre-processing module misleads the model to focus on other areas. Figure 15.8(e) shows the samples for the baseline model combined with MBAM and CBAM. Compared to the grad-CAM maps in Figure 15.8(d), the morphological layers are well guided by attention modules. Thus, the model can focus on the correct target with higher confidence and solve the problems as shown in Figures 8(a) and 8(d).

15.5 CONCLUSIONS

In this chapter, a joint-task learning model is proposed for pneumonia classification and segmentation. The effectiveness of this model is proven by comparing different classification or segmentation models. From visualizing the class saliency map and grad-CAM map, we find that the baseline model's classifier focuses on other areas instead of the target area. The image pre-processing and attention modules are developed to refine the joint-task learning model. Experimental results show that the CBAM or the morphological layers can help the proposed joint-task learning model to focus on the correct area with higher confidence. Furthermore, by combining the MBAM and CBAM to the baseline model, the proposed joint-task learning model not only achieves the best classification test rate at 95.73% and the best mean-average precision of 0.7872 but also helps the classification model to focus on the correct area.

REFERENCES

Badrinarayanan, V., Handa, A., and Cipolla, R., "SegNet: A deep convolutional encoder-decoder architecture for robust semantic pixel-wise labelling," *Proceedings of the IEEE Conference on Computer Vision and Pattern Recognition*, Boston, MA, Jun. 2015.

Chen, L., Zhang, H., Xiao, J., Nie, L., Shao, J., and Chua, T. S., "SCA-CNN: Spatial and channel-wise attention in convolutional networks for image captioning," *Proceedings of the International Conference on Computer Vision and Pattern Recognition*, Honolulu, HI, Jul. 2017a.

Chen, L. C., Papandreou, G., Kokkinos, I., Murphy, K., and Yuille, A. L., "DeepLab: Semantic image segmentation with deep convolutional nets, atrous convolution, and fully connected CRFs," *IEEE Transactions on Pattern Analysis and Machine Intelligence*, vol. 40, no. 4, pp. 834–848, Apr. 2017b.

Girshick, R., "Fast R-CNN," *Proceedings of the International Conference on Computer Vision*, Santiago, Dec. 2015.

He, K., Gkioxari, G., Dollár, P., and Girshick, R., "Mask r-cnn," *Proceedings of the International Conference on Computer Vision*, Venice, Oct. 2017.

He, K., Zhang, X., Ren, S., and Sun, J., "Deep residual learning for image recognition," *Proceedings of the IEEE Conference on Computer Vision and Pattern Recognition*, Las Vegas, NV, Jun. 2016.

Hu, J., Shen, L., and Sun, G., "Squeeze-and-excitation networks," *Proceedings of the International Conference on Computer Vision and Pattern Recognition*, Salt Lake City, UT, Jun. 2018.

Huang, G., Liu, Z., van der Maaten, L., and Weinberger, K., "Densely connected convolutional networks," *Proceedings of the IEEE Conference on Computer Vision and Pattern Recognition*, Honolulu, HI, Jul. 2017.

Ioffe, S. and Szegedy, C., "Batch normalization: Accelerating deep network training by reducing internal covariate shift," *Proceedings of the 32nd International Conference on Machine Learning*, Lille, vol. 37, pp. 448–456, Jul. 2015.

Karargyris, A., Kashyap, S., Wu, J. T., Sharma, A., Moradi, M., and Syeda-Mahmood, T., "Age prediction using a large chest X-ray dataset," *Proceedings of the SPIE Medical Imaging Conference*, San Diego, CA, Jun. 2019.

Kotikalapudi, R., et al., *Keras-vis*. https://github.com/raghakot/keras-vis. Accessed 5 Jun. 2018.

Krizhevsky, A., Sutskever, I., and Hinton, G. E., "ImageNet classification with deep convolutional neural network," *Advances in Neural Information Processing Systems*, vol. 25, no. 2, Jan. 2012.

Leclerc, S., Smistad, E., Pedrosa, J., Østvik, A., Cervenansky, F., Espinosa, F., Espeland, T., Andreas Rye Berg, E., Jodoin, P., Grenier, T., Lartizien, C., D'hooge, J., Løvstakken, L., and Bernard, O., "Deep learning for segmentation using an open large-scale dataset in 2D echocardiography," *IEEE Transactions on Medical Imaging*, vol. 38, no. 9, pp. 2198–2210, 2019.

LeCun, Y., and Bengio, Y., "Convolutional networks for images, speech, and time series," *The Handbook of Brain Theory and Neural Networks*, M. A. Arbib, Ed., MIT Press, 1995.

LeCun, Y., Bengio, Y., and Hinton, G. E., "Deep learning," *Nature*, vol. 521, pp. 436–444, May 2015.

Noh, H., Hong, S., and Han, B., "Learning deconvolution network for semantic segmentation," *Proceedings of the International Conference on Computer Vision*, Santiago, Dec. 2015.

Rajpurkar, P., Irvin, J., Zhu, K., Yang, B., Mehta, H., Duan, T., Ding, D., Bagul, A., Langlotz, C., and Shpanskaya K., "CheXNet: Radiologist-level pneumonia detection on chest X-rays with deep learning," *Proceedings of the IEEE Conference on Computer Vision and Pattern Recognition*, Honolulu, HI, Jul. 2017.

Redmon, J., Divvala, S., Girshick, R., and Farhadi, A., "You only look once: Unified, real-time object detection," *Proceedings of the IEEE Conference on Computer Vision and Pattern Recognition*, Las Vegas, NV, Jun. 2016.

Ren, S., He, K., Girshick, R., and Sun, J., "Faster R-CNN: Towards real-time object detection with region proposal networks," *Proceedings of the Advances in Neural Information Processing Systems*, Montreal, Dec. 2015.

Ronneberger, O., Fischer, P., and Brox, T., "U-Net: Convolutional networks for biomedical image segmentation," *Proceedings of the International Conference on Medical Image Computing and Computer-Assisted Intervention*, Munich, pp. 234–241, Oct. 2015.

Roth, H. R., Shen, C., Oda, H., Oda, M., Hayashi, Y., Misawa, K., and Mori, K., "Deep learning and its application to medical image segmentation," *Journal of the Japanese Society of Medical Imaging Technology*, vol. 36, no. 2, pp. 63–71, 2018.

Selvaraju, R. R., Cogswell, M., Das, A., Vedantam, R., Parikh, D., and Batra, D., "Grad-CAM: Visual explanations from deep networks via gradient-based localization," *Proceedings of the International Conference on Computer Vision*, Venice, Oct. 2017.

Shelhamer, E., Long, J., and Darrell, T., "Fully convolutional networks for semantic segmentation," *Proceedings of the IEEE Conference on Computer Vision and Pattern Recognition*, Boston, MA, 2015.

Shih, F. Y., Shen, Y., and Zhong, X., "Development of deep learning framework for mathematical morphology," *International Journal Pattern Recognition and Artificial Intelligence*, vol. 33, no. 6, p. 1954024 (16 pages), Jun. 2019.

Simonyan, K., Vedaldi, A., and Zisserman, A., "Deep inside convolutional networks: Visualizing image classification models and saliency maps," *Proceedings of the International Conference on Learning Representations*, Banff, Apr. 2014.

Simonyan, K. and Zisserman, A., "Very deep convolutional networks for large-scale image recognition," *Proceedings of the International Conference on Learning Representations*, San Diego, CA, May 2015.

Szegedy, C., Ioffe, S., Vanhoucke, V., and Alemi, A., "Inception-v4, inception-resnet and the impact of residual connections on learning," *Proceedings of the 31st AAAI Conference on Artificial Intelligence*, San Francisco, CA, pp. 4278–4284, Feb. 2017.

Szegedy, C., Liu, W., Jia, Y., Sermanet, P., Reed, S., and Anguelov, D., "Going deeper with convolutions," *Proceedings of the IEEE Conference on Computer Vision and Pattern Recognition*, Boston, MA, Jun. 2015.

Szegedy, C., Vanhoucke, V., Ioffe, S., Shlens, J., and Wojna, Z., "Rethinking the inception architecture for computer vision," *Proceedings of the International Conference on Computer Vision and Pattern Recognition*, Las Vegas, NV, Jun. 2016.

Wong, K., Syeda-Mahmood, T., and Moradi, M., "Building medical image classifiers with very limited data using segmentation networks," *Medical Image Analysis*, vol. 49, pp. 105–116, 2018.

Woo, S., Park, J., Lee, J.-Y., and Kweon, I. S., "CBAM: Convolutional block attention module," *Proceedings of the European Conference on Computer Vision*, Munich, Sept. 2018.

Yeh, C.-H., Huang, C.-H., and Kang, L.-W., "Multi-scale deep residual learning-based single image haze removal via image decomposition," *IEEE Transactions on Image Processing*, vol. 29, no. 1, pp. 3153–3167, Dec. 2020.

Zeiler, M. D. and Fergus, R., "Visualizing and understanding convolutional networks," *Proceedings of the European Conference on Computer Vision*, Zurich, pp. 818–833, Sept. 2014.

Zhou, B., Khosla, A., Lapedriza, A., Oliva, A., and Torralba, A., "Learning deep features for discriminative localization," *Proceedings of the International Conference on Computer Vision and Pattern Recognition*, Las Vegas, NV, Jun. 2016.

Zhou, X., Takayama, R., Wang, S., Hara, T., and Fujita, H., "Deep learning of the sectional appearances of 3d CT images for anatomical structure segmentation based on an FCN voting method," *Medical Physics*, vol. 44, no. 10, pp. 5221–5233, Oct. 2017.

16 Classification of Chest X-Ray Images Using Novel Adaptive Morphological Neural Networks

16.1 INTRODUCTION

Deep-learning neural networks have been developed rapidly since the speed in computer hardware was significantly improved. Nowadays, many deep-learning models in computer vision are based on convolutional neural networks (CNNs). The CNN model was initially proposed by LeCun and Bengio (1995) as the LeNet in 1995. Since then, different structures based on the CNN have been developed for computer vision tasks in image classification and segmentation. At early time of these models, the CNN used a fixed kernel size in the convolution layers, such as AlexNet (Krizhevsky et al., 2012) and VGG model (Szegedy et al., 2015). Later the residual neural network model (He et al., 2016) was developed to use a fixed kernel size with a shortcut connecting the input and output layers, which makes the CNN model much deeper and accurate.

In 2015, Google first proposed their models, named GoogLeNet (Ioffe and Szegedy, 2015), which consists of 4 different feature maps called inception modules. The inception neural networks (Szegedy et al., 2016, 2017; Rajpurkar et al., 2017) are advanced versions of GoogLeNet, where the inception blocks use both different kernel sizes and residual shortcut. DenseNet (Huang et al., 2017) was proposed to use channel-wise concatenations for feature extraction. Other applications in deep learning and computer vision are multi-objective detection (Girshick, 2015; Ren et al., 2017; He et al., 2017; Redmon et al., 2016) and semantic segmentation (Leclerc et al., 2019). The semantic segmentation focuses on classifying each pixel to a given class, whereas the objective detection task focuses on separating the foreground from the background.

The FCN (fully convolution network) (Shelhamer et al., 2015) was designed for semantic segmentation using one deconvolution operation to up-sample after feature extraction and generate a pixelwise label map for prediction. The deconvolution network (Noh et al., 2015) uses a VGG-16-layer as the auto-encoder for image representation and improves the FCN by a gradual deconvolution and up-sampling layers. The SegNet (Badrinarayanan et al., 2015) implemented an encoder and decoder structure for robust semantic pixelwise labeling on medical images. Zhou et al. (2017) showed a segmentation task using the FCN in the 3D CT (computed tomography) images. However, the

DOI: 10.1201/9781003474395-19

high ratio of the up-sampling layer makes the output label a rough representation without details. To improve performance, Roth et al. (2018) and Wong et al. (2018) both implemented a U-Net structure (Ronneberger et al., 2015) for different segmentation tasks.

The CheXNet (Rajpurkar et al., 2017) applied a 121-layer end-to-end dense convolution structure to detect 14 diseases in chest X-ray data. When more complex structures are applied, the parameters are increased tremendously in the deep-learning model. Although the CNN is a dominant model in computer vision, a major problem of containing a giant number of parameters is occurred to require multi-advanced GPUs. Therefore, in this chapter we develop the novel adaptive morphological neural networks for feature extraction to reduce the number of parameters as well as to achieve higher classification rates. Recent researchers' progresses (Yeh et al., 2020a, 2020b; Lin et al., 2020) are based on CNN and put this model to a higher level.

Mathematical morphology has been used in effectively extracting object features, such as shapes, regions, edges, skeleton, and convex hull, which can improve the object representation and description (Shih and Mitchell, 1989; Haralick et al., 1987). Similar to a mask used in the convolution operation, mathematical morphology needs a structure element to perform the operation on the image. Two essential operations are dilation and erosion, and other operations are different combinations. Dilation tends to enlarge objects, while erosion tends to shrink it. Another application for mathematical morphology is image pre-processing like morphological filtering (Shih, 2009).

Shih and Moh (1992) proposed to implement morphological operations using programmable neural networks. Davidson and Hummer (1993) presented morphological neural networks (MNNs) with applications. Masci et al. (2013) proposed a method using counter harmonic mean for dilation and erosion in the deep-learning framework. Shih et al. (2019) proposed a morphological deep-learning framework using smooth local minimum and local maximum to simulate erosion and dilation, respectively.

In this chapter, we propose an adaptive MNN model for the classification of chest X-ray images. Radiologists use chest X-ray images to diagnose diseases in the lung area. However, these images are noisy and hard to analyze the diseases, such as bacteria pneumonia, virus pneumonia, or simply healthy. Moreover, we apply our model to recognize possible samples of the recent COVID-19 pandemic cases. We use different morphological layers, including dilation, erosion, opening, closing, etc., combined with convolutional neural networks. It can help convolutional neural networks to refine the feature extraction process. Furthermore, we develop adaptive morphological layers for feature extraction, which can determine a suitable morphological operation and structure elements in the training process.

The remainder of this chapter is organized as follows. Section 16.2 introduces the morphological neural networks. Section 16.3 presents the novel adaptive MNN model. Section 16.4 describes the datasets and experimental results. Finally, conclusions are drawn in Section 16.5.

16.2 THE MORPHOLOGICAL NEURAL NETWORKS

Deep learning (LeCun et al., 2015) is an essential part in machine learning, which requires a large amount data to train a model and then evaluate the model's performance on different datasets. In computer vision, the convolutional neural networks have been widely used in many areas. The LeNet-5 (LeCun and Bengio, 1995) was initially designed for the classification of handwritten digits. It contains an input layer, a feature extraction layers, and a pooling layer to reduce unnecessary data. After a second connection of a convolutional layer and a pooling layer, the feature representation is fed to a fully connected artificial neural network for classification. In the convolutional layer, the input is one or several images with one or three channels, which could be grayscale or RGB. The convolution is performed several times with different filters, so there are several output images called feature maps.

The convolutional layers extract different local features using different filters, making the whole network to learn all the features in the input images. The convolutional layer followed by an activate function is described as

$$h^k = f\left(\sum_{l\epsilon L} x^l \otimes w^k + b^k\right) \tag{16.1}$$

where h^k is the latent representation of k-th feature map of the current layer, f is the activation function, x^l is the l-th feature map of group of feature maps L of the previous layers or the l-th channel of the input images with totally L channels in the case of the first layer of the network, $\otimes$ denotes the 2D convolution operation, and w^k and b^k denote the weights (filters) and biases of the k-th feature map of the current layer, respectively.

A nonlinear function, called ReLU (rectified linear unit), works as the activation function f, which can be written as $f(x) = \max(0, x)$. This function will stay 0 when x is less than 0 but return to be x for any positive input. The ReLU works well for neural network models because it allows the models to compute non-linearities and interaction, which makes ReLU a commonly used activation function. The loss function is the cross-entropy with softmax function as

$$p_i = \frac{e^{z_i}}{\sum_{k=1}^{K} e^{z_k}}, i = 1, 2, 3, \ldots, K \tag{16.2}$$

where z_i is an element of the input tensor. With the softmax function, we can transform an N-dimensional vector of real numbers into a vector of real numbers in range of (0,1). The cross-entropy loss function is a widely used alternative of squared error. It is denoted as

$$H(y, p) = -\sum_i y_i log(p_i) \tag{16.3}$$

where y_i is the label of i-th input image and p_i is the i-th item of the output of the softmax function.

The pooling layer is designed to perform down-sampling on image data to extract useful information and reduce the size of feature maps. Typically, there are two different down-sampling methods: average pooling and max-pooling. Average pooling is used to compute the average value as feature in a small area, and max-pooling is used to extract the maximum value in a small area. After sufficient information is acquired from convolutional layers and pooling layers, the fully connected layer is used to map the output to linearly separable space and flatten the matrix into a vector. Then softmax is used for regression to classify the data, so the output of the last fully connected layer would be the predicted label.

Mathematical morphology is a widely used approach for shape representation and image pre-processing. Two fundamental morphological operations are dilation and erosion. Let the input image be I and the structuring element be s. The dilation operation is denoted as $I \oplus s$, which expands the image by the structuring element. The erosion is denoted as $I \ominus s$, which shrinks the image by the structuring element.

The opening is typically used for contour smoothing, especially for breaking thin connections between components and enlarging small holes or gaps. It is defined as an erosion followed by a dilation as

$$I \circ s = \left(I \ominus s\right) \oplus s \tag{16.4}$$

Different from opening, the closing can be used for connecting narrow areas and filling in small holes or gaps. It is defined as a dilation followed by an erosion as

$$I \bullet s = \left(I \oplus s\right) \ominus s \tag{16.5}$$

Figure 16.1 shows two sample images for chest X-ray images, which are processed using dilation and erosion with a 6×6 structure element of all 1s. Figure 16.2 shows two sample images, which are processed using closing and opening with a 6×6 structure element of all 1s.

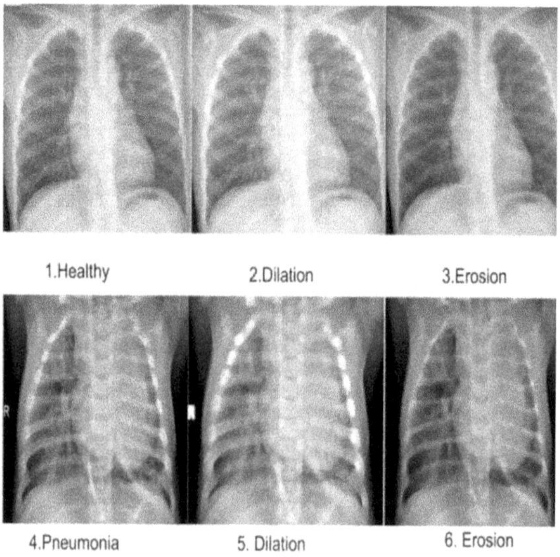

FIGURE 16.1 Sample images after morphological operations. Column 1 shows input images; column 2 shows dilation; column 3 shows erosion.

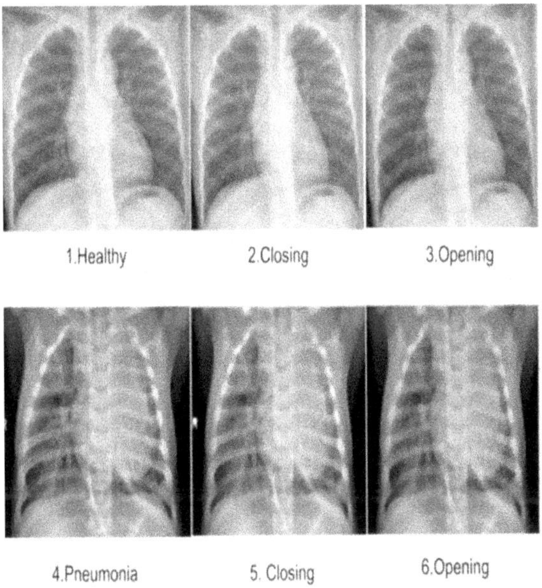

FIGURE 16.2 Sample images after morphological operations. Column 1 shows input images; column 2 shows closing; column 3 shows opening.

For the X-ray images, the dilation operation can expand some of the small areas while enlarging some of the noisy areas. The erosion can clean the background by eliminating some noisy areas, but at the same time, filtering out some pixels. Opening and closing can smooth the contour, where closing tends to fill in some holes and opening tends to make them larger. Other morphological operations include the top-hat transformation and the bottom-hat transformation. The top-hat transformation is denoted as $I - I \circ s$, and the bottom hat transformation is denoted as $I \cdot s - I$.

The morphological neural network (MNN) is another type of deep-learning framework. Similar to the convolutional layers in CNN, the morphological layers work as a feature extraction tool. With the differential approximation of dilation and erosion, we can construct the morphological layer. For the j-th pixel in an output image Y, the dilation layer is defined as

$$Y_j = \ln\left(\sum_{i=1}^{n} e^{W_i X_i}\right) \tag{16.6}$$

Where W represents the corresponding structure element and X represents the input image. For the j-th pixel in an output image Y, the erosion layer is defined as

$$Y_j = -\ln\left(\sum\nolimits_{i=1}^{n} e^{-W_i X_i}\right) \tag{16.7}$$

The weight is the structuring element in mathematical morphology, acting as a filter in convolution. Similar to the convolution, the weights can be in different shapes.

For simplicity, we adopt 3 × 3 and 5 × 5 in this chapter. The gradient of such a layer is computed by back-propagation with chain rule. With such approximation, we can easily obtain the gradient.

16.3 THE ADAPTIVE MORPHOLOGICAL NEURAL NETWORKS

In this section, we develop different deep-learning models for the classification of chest X-ray images. Several mathematical morphological operations, such as dilation, erosion, closing, opening, top-hat, and bottom-hat, are combined. However, the operation types are required to specify in these models before the deep neural networks are trained. We develop adaptive morphological neural networks for pneumonia classification using the adaptive layers, which do not require to specify the morphological operation type in each layer. The basic morphological neural networks using morphological layers are shown in Figure 16.3, where (a) is the structure

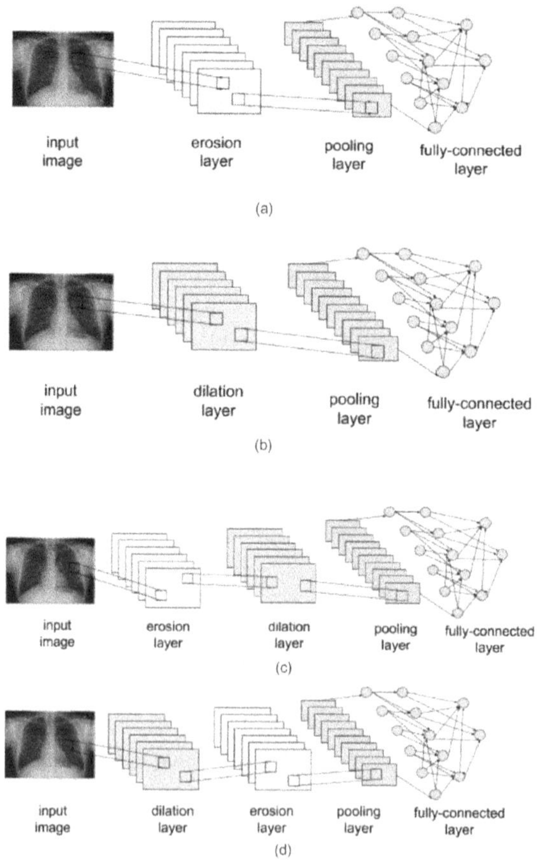

FIGURE 16.3 Morphological neural network structures for various mathematic morphological operations. (a) Erosion classifier for pneumonia chest X-ray images, (b) dilation classifier, (c) opening classifier, (d) closing classifier, (e) top-hat classifier, and (f) bottom-hat classifier.

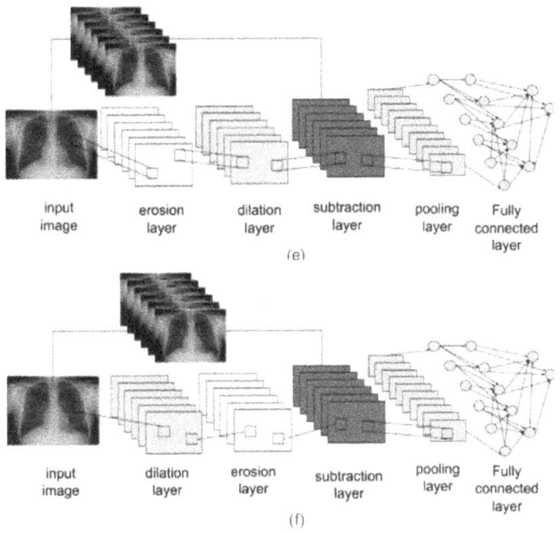

FIGURE 16.3 (Continued)

of the MNN model performing an erosion operation, (b) is the structure performing a dilation operation, (c) and (d) are the structures performing an opening and a closing, respectively, and (e) and (f) are the structures performing the top-hat and the bottom-hat operations, respectively.

Morphological operations can be various due to different combinations of dilations and erosions. Observing from equations (16.6) and (16.7), the only difference between dilation and erosion layers is the sign before the weights. Therefore, a trainable weight for sign function is used to decide the morphological operation types (dilation or erosion). The proposed adaptive morphological layer is defined as

$$z_j = sign(a) * \ln\left(\sum_{i=1}^{n} e^{sign(a)*\omega_i x_i}\right) + b \tag{16.8}$$

where a is an extra trainable variable aside with ω_i and b. If $sign(a)$ is $+1$, the adaptive morphological layer carries out a dilation operation layer; however, if $sign(a)$ is -1, the adaptive morphological layer carries out an erosion operation. Note that the sign function cannot be used in a deep neural network since it is not continuous which makes equation (16.8) undifferentiable.

To solve the undifferentiability problem, an improved sign function in the interval $[-1, +1]$ is applied for the adaptive morphological layer. The proposed morphological adaptive layer is defined as

$$Z_j = \frac{e^a - e^{-a}}{e^a + e^{-a}} \cdot \ln\left(\sum_{i=1}^{n} e^{\frac{e^a - e^{-a}}{e^a + e^{-a}} \omega_i x_i}\right) + b. \tag{16.9}$$

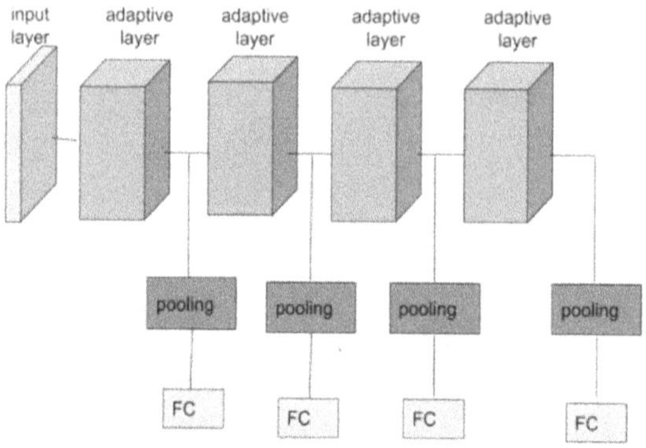

FIGURE 16.4 Stacked adaptive morphological deep leaning model.

With the proposed sign function, the adaptive morphological layers can self-learn a morphological type: dilation or erosion. A novel structure is proposed to decide the most suitable depth of the adaptive layer for pneumonia classification. Figure 16.4 shows the structure of the proposed stacked adaptive morphological deep-learning model. The activation functions are added before each pooling layer. After the pooling layer, the feature maps are processed by a fullyconnected layer and output the class predictions. The design is intended to decide the best depth for the stacked adaptive layers.

16.4 DATASETS AND EXPERIMENTAL RESULTS

Two datasets of the chest X-ray images are used to evaluate the performance of the developed models. We compare our experimental results against seven currently available state-of-the-art CNN models, which include LeNet, VGG-16, ResNet-50, DenseNet, SqueezeNet, MobileNet, and Inception v4.

16.4.1 The Chest X-Ray Datasets

We use two datasets to evaluate the performance of our developed models: the chest X-Ray dataset (Rahman et al., 2020) and the COVID-19 dataset (Iandola et al., 2016). The chest X-ray dataset was from Kaggle competition, which contains two categories of pneumonia and normal in 5,863 X-ray images. We split those images into three subsets: 4,398 images are used for training, 1,375 images are used for testing, and 93 images are used for validation. To balance the training sample, we apply data augmentation in the training process.

The COVID-19 dataset contains 219 positive cases and 1,341 normal cases, where 165 positive cases and 1,005 normal cases are randomly selected in the training process. For the test dataset, 43 positive samples and 43 normal samples are used. The validation dataset contains 11 positive samples and 68 normal samples. To balance the cases in the training process, each category is augmented to 10,000 new images using image augmentation techniques. In the experiment, all the images are resized to 256×256.

16.4.2 Experimental Results

Table 16.1 and Table 16.2 show the experimental results of the basic morphological neural networks in two datasets. The erosion and the dilation classifiers use only one layer for feature extraction. In comparison, the erosion classifier achieves a 95.27% accuracy rate on the chest X-ray dataset, whereas the dilation classifier achieves a test accuracy rate of 98.10%. The reason is that the erosion classifier tends to shrink the images. The performances for opening and closing are similar to each other since both operations tend to eliminate the noise. The recall, precision, accuracy, and F1 score rates are computed by the following equations:

$$Recall = \frac{True\,Positive}{True\,Positive + False\,Negative} = \frac{True\,Positive}{Total\,Active\,Positive} \tag{16.10}$$

$$Precision = \frac{True\,Positive}{True\,Positive + False\,Positive} = \frac{True\,Positive}{Total\,Predicted\,Positive} \tag{16.11}$$

$$Accuracy = \frac{True\,Positive + True\,Negative}{Total\,Testing\,samples} = \frac{Correct\,Prediction}{Total\,Testing\,samples} \tag{16.12}$$

$$F1 = \frac{2 * Precision * Recall}{Precision + Recall} \tag{16.13}$$

Table 16.3 shows the test accuracy of the stacked adaptive morphological neural network model. We observe that the best performance for the stacked adaptive morphological neural network is achieved at six layers. An obvious overfitting is occurred when the seventh adaptive layer is stacked. For the chest X-ray dataset, the best performance is 98.75%, and for the COVID-19 dataset, the best performance is 97.33%.

Table 16.4 shows the comparison of our proposed models against seven existing CNN models, including LeNet, VGG-16, ResNet-50, DenseNet, SqueezeNet,

TABLE 16.1
Test Accuracy for Basic Morphological Neural Networks in Chest X-Ray Dataset

Chest X-Ray Dataset	Recall	Precision	Accuracy	F1 Score	Total Parameter
Erosion	95.7%	96.06%	95.27%	0.9588	0.81 million
Dilation	98.21%	98.47%	98.10%	0.9834	0.81 million
Closing	98.85%	98.35%	98.41%	0.9860	0.82 million
Opening	98.60%	98.09%	98.10%	0.9834	0.82 million
Top-hat	98.22%	98.01%	97.89%	0.9811	0.83 million
Bottom-hat	97.21%	96.60%	96.45%	0.9690	0.83 million

TABLE 16.2
Test Accuracy for Basic Morphological Neural Networks in the COVID-19 Dataset

COVID-19 Dataset	Recall	Precision	Accuracy	F1 Score	Total Parameter
Erosion	95.23%	93.02%	94.71%	0.9411	0.81 million
Dilation	95.35%	95.35%	96.26%	0.9535	0.81 million
Closing	95.45%	97.67%	96.57%	0.9655	0.82 million
Opening	93.33%	97.67%	95.97%	0.9545	0.82 million
Top-hat	93.18%	95.34%	95.15%	0.9430	0.83 million
Bottom-hat	95.23%	93.02%	94.79%	0.9411	0.83 million

TABLE 16.3
Test Accuracy Stacked Adaptive Morphological Neural Network Model

Stacked Numbers	Chest X-Ray Dataset	COVID-19 Dataset	Total Parameter
1	75.13%	75.43%	0.81 million
2	80.35%	84.66%	0.81 million
3	89.41%	91.19%	0.82 million
4	93.02%	94.97%	0.82 million
5	97.39%	95.97%	0.83 million
6	**98.75%**	**97.33%**	**0.84 million**
7	96.10%	95.10%	0.85 million
8	93.16%	92.15%	0.88 million
9	90.33%	90.26%	0.9 million

TABLE 16.4
Comparison to CNN Models

Model	Chest X-Ray Dataset	COVID-19 Dataset	Total Parameter
The proposed stacked adaptive MNN	98.75%	97.33%	0.84 million
LeNet (LeCun and Bengio, 1995)	85.92%	79.68%	1.4 million
VGG-16 (Rajpurkar et al., 2017)	95.77%	93.27%	9.1 million
ResNet (Huang et al., 2017)	98.69%	96.78%	25.6 million
DenseNet (Leclerc et al., 2019)	98.91%	97.44%	30.2 million
SqueezeNet (Iandola et al., 2016)	90.53%	90.26%	0.49 million
MobileNet (Howard et al., 2017)	91.02%	92.21%	4.2 million
Inception v4 (He et al., 2017)	99.04%	97.77%	65 million

MobileNet, and Inception v4. It is observed that the proposed MNN models can achieve similar or even better performance than those seven CNN models. Comparing to the best performance of the Inception v4 model, the proposed model seems to achieve a slightly lower rate. However, the total of parameters in the proposed model is much significantly lower by 98.7% than the parameters in the Inception v4 model. Even comparing to the CNN model of SqueezeNet having the least parameters, our proposed model can achieve a much higher performance.

16.5 CONCLUSIONS

In this chapter, the adaptive morphological neural networks are developed for the classification tasks in chest X-ray images. Traditional deep-learning models require a giant number of parameters to achieve high classification rates. The proposed models can successfully reduce the number of parameters in the CNN models. Experimental results show that the MNN models can reach the better performance with much fewer parameters in chest X-ray datasets. This advantage makes the MNN models more competitive than the CNN models to deploy in the website or other platforms. Considering the effectiveness for the MNN models in the classification task, our future work is to apply such a model to other computer vision tasks, such as image segmentation and objective detection.

REFERENCES

Badrinarayanan, V., Handa, A., and Cipolla, R., "SegNet: A deep convolutional encoder-decoder architecture for robust semantic pixel-wise labelling," *Proceedings of the IEEE International Conference on Computer Vision*, Santiago, vol. 39, pp. 1–14, Dec. 2015.

Davidson, J. L. and Hummer, F., "Morphology neural networks: An introduction with applications," *Circuits Systems and Signal Process*, vol. 12, pp. 177–210, Jun. 1993.

Girshick, R., "Fast r-CNN," *Proceedings of the IEEE International Conference on Computer Vision*, Santiago, pp. 1440–1448, Dec. 2015.

Haralick, R. M., Sternberg, S. R., and Zhuang, X., "Image analysis using mathematical morphology," *IEEE Transactions on Pattern Analysis and Machine Intelligence*, vol. 9, no. 4, pp. 532–550, Jul. 1987.

He, K., Gkioxari, G., Dollár, P., and Girshick, R., "Mask r-CNN," *Proceedings of the IEEE Conference on Computer Vision and Pattern Recognition*, Honolulu, HI, pp. 2961–2969, Jul. 2017.

He, K., Zhang, X., Ren, S., and Sun, J., "Deep residual learning for image recognition," *Proceedings of the IEEE Conference on Computer Vision and Pattern Recognition*, Las Vegas, NV, pp. 770–778, Jun. 2016.

Howard, A. G., Zhu, M., Chen, B., Kalenichenko, D., Wang, W., Weyand, T., Andreetto, M., and Adam, H., "MobileNets: Efficient convolutional neural networks for mobile vision applications," arXiv:1704.04861 [cs.CV], 2017.

Huang, G., Liu, Z., Maaten, L., and Weinberger, K. Q., "Densely connected convolutional networks," *Proceedings of the IEEE Conference on Computer Vision and Pattern Recognition*, Honolulu, HI, pp. 4700–4708, Jul. 2017.

Iandola, F. N., Han, S., Moskewicz, M. W., Ashraf, K., Dally, W. J., and Keutzer, K., "SqueezeNet: AlexNet-level accuracy with 50x fewer parameters and <0.5MB model size," arXiv:1602.07360v4 [cs.CV], 2016.

Ioffe, S. and Szegedy, C., "Batch normalization: Accelerating deep network training by reducing internal covariate shift," *Proceedings of the 32nd International Conference on Machine Learning*, Lille, vol. 37, pp. 448–456, Jul. 2015.

Kermany, D., Zhang, K., and Goldbaum, M., "Labeled optical coherence tomography and chest X-ray images for classification," *Mendeley Data*, vol. 2, 2018.

Krizhevsky, A., Sutskever, I., and Hinton, G. E., "ImageNet classification with deep convolutional neural network," *Advances in Neural Information Processing Systems*, vol. 25, no. 2, pp. 1097–1105, Jan. 2012.

Leclerc, S., Smistad, E., Pedrosa, J., Østvik, A., Cervenansky, F., Espinosa, F., Espeland, T., Andreas Rye Berg, E., Jodoin, P., Grenier, T., Lartizien, C., D'hooge, J., Løvstakken, L., and Bernard, O., "Deep learning for segmentation using an open large-scale dataset in 2D echocardiography," *IEEE Transactions on Medical Imaging*, vol. 38, no. 9, pp. 2198–2210, 2019.

LeCun, Y. and Bengio, Y., "Convolutional networks for images, speech, and time series," *The Handbook of Brain Theory and Neural Networks*, pp. 255–258, MIT Press, 1995.

LeCun, Y., Bengio, Y., and Hinton, G. E., "Deep learning," *Nature*, vol. 521, pp. 436–444, May 2015.

Lin, C.-Y., Tao, Z., Xu, A.-S., Kang, L.-W., and Akhyar, F., "Sequential dual attention network for rain streak removal in a single image," *IEEE Transactions on Image Processing*, vol. 29, pp. 9250–9265, Sept. 2020.

Masci, J., Angulo, J., and Schmidhuber, J., "A learning framework for morphological operators using counter-harmonic mean," *Proceedings of the 11th International Symposium on Mathematical Morphology: Its Applications Signal Image Processing*, Berlin, Heidelberg, pp. 329–340, Springer, 2013.

Noh, H., Hong, S., and Han, B., "Learning deconvolution network for semantic segmentation," *Proceedings of the IEEE International Conference on Computer Vision*, Santiago, pp. 1520–1528, Dec. 2015.

Rahman, T., Chowdhury, M., and Khandakar, A., *COVID-19 Radiography Database*, vol. 1, 2020. https://www.kaggle.com/tawsifurrahman/covid19-radiography-database?

Rajpurkar, P., Irvin, J., Zhu, K., Yang, B., Mehta, H., Duan, T., Ding, D., Bagul, A., Langlotz, C., Shpanskaya, K., Lungren, M. P., and Ng, A. Y., "CheXNet: Radiologist-level pneumonia detection on chest X-rays with deep learning," arXiv:1711.05225 [cs.CV], 2017.

Redmon, J., Divvala, S., Girshick, R., and Farhadi, A., "You only look once: Unified, real-time object detection," *Proceedings of the IEEE Conference on Computer Vision and Pattern Recognition*, Las Vegas, NV, pp. 779–788, Jun. 2016.

Ren, S., He, K., Girshick, R., and Sun, J., "Faster R-CNN: Towards real-time object detection with region proposal networks," *IEEE Transactions on Pattern Analysis and Machine Intelligence*, vol. 39, no. 6, pp. 1137–1149, Jun. 2017.

Ronneberger, O., Fischer, P., and Brox, T., "U-Net: Convolutional networks for biomedical image segmentation," *Proceedings of the International Conference on Medical Image Computing and Computer-Assisted Intervention*, Munich, pp. 234–241, Oct. 2015.

Roth, H. R., Shen, C., Oda, H., Oda, M., Hayashi, Y., Misawa, K., and Mori, K., "Deep learning and its application to medical image segmentation," *Journal of the Japanese Society of Medical Imaging Technology*, vol. 36, no. 2, pp. 63–71, 2018.

Shelhamer, E., Long, J., and Darrell, T., "Fully convolutional networks for semantic segmentation," *Proceedings of the IEEE Conference on Computer Vision and Pattern Recognition*, Boston, MA, pp. 3431–3440, Jun. 2015.

Shih, F. Y., *Image Processing and Mathematical Morphology: Fundamentals and Applications*, Taylor & Francis Group, CRC Press, 2009.

Shih, F. Y. and Mitchell, O. R., "Threshold decomposition of grayscale morphology into binary morphology," *IEEE Transactions on Pattern Analysis and Machine Intelligence*, vol. 11, no. 1, pp. 31–42, Jan. 1989.

Shih, F. Y. and Moh, J., "Implementing morphological operations using programmable neural networks," *Pattern Recognition*, vol. 25, no. 1, pp. 89–99, Jan. 1992.

Shih, F. Y., Shen, Y., and Zhong, X., "Development of deep learning framework for mathematical morphology," *International Journal of Pattern Recognition and Artificial Intelligence*, vol. 33, no. 6, p. 1954024, Jun. 2019.

Szegedy, C., Ioffe, S., Vanhoucke, V., and Alemi, A., "Inception-v4, inception-resnet and the impact of residual connections on learning," *Proceedings of the 31st AAAI Conference on Artificial Intelligence*, San Francisco, CA, pp. 4278–4284, Feb. 2017.

Szegedy, C., Liu, W., Jia, Y., Sermanet, P., Reed, S., Anguelov, D., Erhan, D., Vanhoucke, V., and Rabinovich, A., "Going deeper with convolutions," *Proceedings of the IEEE Conference on Computer Vision and Pattern Recognition*, Boston, MA, pp. 1–9, Jun. 2015.

Szegedy, C., Vanhoucke, V., Ioffe, S., Shlens, J., and Wojna, Z., "Rethinking the inception architecture for computer vision," *Proceedings of the International Conference on Computer Vision and Pattern Recognition*, Las Vegas, NV, pp. 2818–2826, Jun. 2016.

Wong, K., Syeda-Mahmood, T., and Moradi, M., "Building medical image classifiers with very limited data using segmentation networks," *Medical Image Analysis*, vol. 49, pp. 105–116, 2018.

Yeh, C.-H., Huang, C.-H., and Kang, L.-W., "Multi-scale deep residual learning-based single image haze removal via image decomposition," *IEEE Transactions on Image Processing*, vol. 29, pp. 3153–3167, 2020a.

Yeh, C.-H., Lin, M.-H., Chang, P.-C., and Kang, L.-W., "Enhanced visual attention-guided deep neural networks for image classification," *IEEE Access*, vol. 8, pp. 163447–163457, 2020b.

Zhou, X., Takayama, R., Wang, S., Hara, T., and Fujita, H., "Deep learning of the sectional appearances of 3d CT images for anatomical structure segmentation based on an FCN voting method," *Medical Physics*, vol. 44, no. 10, pp. 5221–5233, Aug. 2017.

17 Land-Cover Image Segmentation Based on Individual Class Binary Masks

17.1 INTRODUCTION

Remote-sensing techniques have been developed over the past decades to acquire data without being in contact of the target object or data source. Their application on land-cover image segmentation has attracted significant attention during recent years. Image segmentation is the process of partitioning an image into multiple segments based on region and class of the segments. This procedure is intended to better understand and process the image for further analysis. It lays the foundation for future tasks and helps in eliminating outliers in the image data. Without segmentation, a huge amount of image data needs to be manually cropped or cleaned, which is a very time-consuming process.

Of all the many tasks involved in image processing and cleaning of visual data, image segmentation has always been one of the most arduous tasks. Labeling and segmenting the different images of thousands of images in a dataset is a herculean task, but is an essential part of a lot of computer vision applications. It has always been a precursor to multiple higher-level tasks, such as object detection and recognition. Image segmentation lays the groundwork of bounding boxes, that helps other algorithms perform better and remove noise from the images. However, shoddy segmentation results could cause outliers and problems in the future classification processes, and therefore, it is one of the most important tasks (Shi and Malik, 2000).

Image segmentation intends to separate coherent regions from one another. Thresholding is used to extract foreground from background based on pixel intensity or color. This technique can be applied to text and document images and other images with similar clear-cut difference between the two (Ingle and Kaur, 2017; Antonacopoulos and Ritchings, 1995). Researchers also use texture and add other factors for a better split across the regions. A more complex version can range from based on color to using more complex domains to sort out regions. Many filters, like the Gabor filter can be used to group texture of a region of pixels (Sabri and Fieguth, 2004; Liu and Zhou, 2004). The clustering itself can be done with k-means or histogram methods. The k-means clustering algorithm clusters all the color regions within groups of k. Hence, the regions near to a center chosen by the algorithm are represented by the same color (Kalam and Manikandan, 2011; Akhtar et al., 2014; Surlakar et al., 2016).

DOI: 10.1201/9781003474395-20

A better way to represent and group data is semantic image segmentation or grouping regions based on their classes. This makes a lot more sense in real world especially, where images are filled with noise, and a simple addition in the background or outlier can cause large variations in the output classes and trained data. Semantic image segmentation classifies and partitions different classes of an image, with an output mask of different values denoting different regions. This method has been extensively employed with the advent of convolutional neural networks. A general layout adopted from the VGG-16-layer net, and further deconvolution and unpooling has led to pixel-wise masks that segment regions effectively. This method has largely been successful in many industries and various types of images (Noh et al., 2015). The papers from Minaee et al. (2022) and Ghosh et al. (2019) compile such developments in image segmentation.

Primarily in the medical image segmentation with U-Net and further into V-net, medical image segmentation has never had such higher accuracies with traditional methods of segmentation and classification (Milletari et al., 2016). Medical imaging and segmentation have always gone hand in hand. Grouping and organization for images have always been a requirement in many fields. But it is more pronounced in the medical field, where such details equate presence of a condition and abnormality. It has been used extensively in the segmentation of those images, such as brain, cornea, skin lesions, and individual organs. This has also been predominantly used in the segmentation and analysis of cancer and disease detection (Zhang et al., 2001; Fadhel et al., 2016; Yuan et al., 2017).

In real life, image segmentation is used to extract more information out of images by segmenting them into classes, especially in the field of automated cars and security, where image segmentation can identify different objects and living things in an image (Tseng and Jan, 2018; Ren et al., 2017). Another important use of image segmentation is in the field of land-cover segmentation. Here, we segment terrain to classify them into different types of land and water bodies using images from satellites. This is prominently used in many applications like planning and construction, where gauging a terrain is of tantamount importance. Even in Google Maps, the terrain is mapped according to its type to indicate the importance of terrain segmentation.

Land-cover image segmentation has been one of the most difficult types of segmentation due to the difference in quality of images, noises in different images, and high number of different features even within the same class of images. Although the segmented class is not required to be of high accuracy or precision, this leads to a more challenging case of segmentation (Kuo et al., 2018; Ghosh et al., 2018).

Artificial neural networks have come a long way since its inception in the '90s. With advancements in both computer hardware and techniques, we are able to process more data, extract more features, and create more complex networks. In this regard, convolutional neural networks have been a milestone in the world of computer vision. Unlike traditional neural networks, they are able to correlate spatial differences in images and extract features that make a lot more sense. Basically, as the name implies, convolutional neural networks use the convolution operation to extract features, an old-age technique used for the generation of image processing.

Originally, convolution shows how one signal is modified when passing through another. In image processing and computer vision, convolution is based on filters to extract features from images. There have been many neural network models built on this principle. With large image sizes, artificial neural networks need to be sufficiently large to accommodate the data, but with convolutional neural nets, this problem is reduced exponentially. Since with each layer the data size becomes smaller and smaller, more features are extracted with lesser weights to be stored. Some of the recent developments in image processing can be attributed to these types of networks (Sun et al., 2021; Yeh et al., 2021). These models have extensively been used in object detection and classification models, such as VGG-16 and ResNet families (He et al., 2016; Simonyan and Zisserman, 2014). This method serves as the backbone of image segmentation.

Fully convoluted network (FCN) is the first milestone in semantic segmentation. It produces class-wise pixel by pixel classification or segmentation masks using convolutional neural networks. Traditionally, convolutional neural networks produce a single output based on class or detection of objects. However, in FCN we can extrapolate this data to produce masks that are based on the class or detected object and represent the regions of the image (Long et al., 2015).

In this chapter, we present a novel model that augments segmentation using smaller networks to segment individual classes. The combined network is trained on the same data but with the combined and trained masks using categorical cross entropy. The remainder of this chapter is organized as follows. Section 17.2 introduces the convolution-deconvolution–based networks. We propose the novel advanced model in Section 17.3. The experimental results and performance comparisons are provided in Section 17.4. Finally, conclusions are drawn in Section 17.5.

17.2 CONVOLUTION-DECONVOLUTION–BASED NETWORKS

With the advent of FCNs, new possibilities are open to the world of semantic segmentation. The features extracted could not only be output for image classification, but now it can also be used as input for image segmentation tasks. With this, many networks were built to profit from the features extracted by doing so. As the name implies, these networks have two parts to them – the backbone and classifier. The backbone is a general convolutional neural network, which is usually a VGG-16 or a ResNet model, that acts to extract features using the conventional convolution, down-sampling, and max-pooling layers to extract spatial features for identifying and classifying objects in the image. But what is different from traditional convolutional neural networks is that it is fed forward to a decoder-deconvolution network. The deconvolution network does the exact opposite of the backbone. It up-samples the data and deconvolutes to produce more pixels. Finally, with the help of a softmax layer, the mask that contains the segmentation is then produced.

We propose a different model for image segmentation. First, the segmentation mask of separating individual classes by individual decoder networks is conducted, and then the outputs are combined and post-processed to produce the required output mask. The basic outline of the model is given in Figure 17.1. We adopt U-Net in the individual class convolution network. The U-Net, introduced by Ronneberger et al.

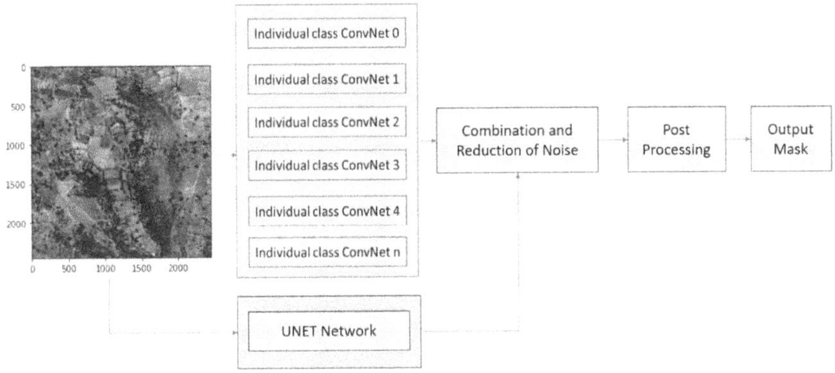

FIGURE 17.1 The basic outline of the proposed model.

(2015), was originally invented to be used in brain image segmentation. It has been very robust in its usage and has produced exceptional results in all types of image segmentation. The U-Net consists of two parts – down-sampling and up-sampling – and at each down-sample the feature map is convoluted by 2×2, which halves the feature channels followed by a ReLU. They are up-convoluted along the up-sampling phase by 2×2. U-Net is also a popular model, adopted into many recent developments in image processing applications (Ghosh et al., 2018; Yeh et al., 2020).

17.3 THE PROPOSED ADVANCED MODEL

It is noted that the precision of the convolutional net with only a single class is far above compared to metrics of models which segment all the classes together. We use this property to our advantage in creating a model that incorporates individual segmented masks into a whole to produce a mask based on individual masks. Thus, we can obtain an array of individual convolutional nets for separate classes. Each class is fed into the output from the combined class model to augment and incorporate with better results.

Ensemble has already been tried and tested as a methodology to improve metrics of different networks. There have been many models that use the process of popular voting or other mechanisms to combine multiple models. Some of the popular models in image segmentation can be referred to the literature, including Noh et al. (2015), Choudhury et al. (2020), and Ali et al. (2019). Those models use multiple networks to perfect the same mask. In contrast, we propose a novel model of building the mask with the multiple separate neural networks, and combine the multiple masks using a combination engine to generate the segmented mask. The proposed advanced model is illustrated in Figure 17.2.

In our model, the large land-cover image is split into sub-images of size 512×512, which are then simultaneously passed through the multiple individual networks and the combined network. Each of them outputs a mask, which is then processed for blob detection. The regions that do not overlap are removed. By the ordering of the

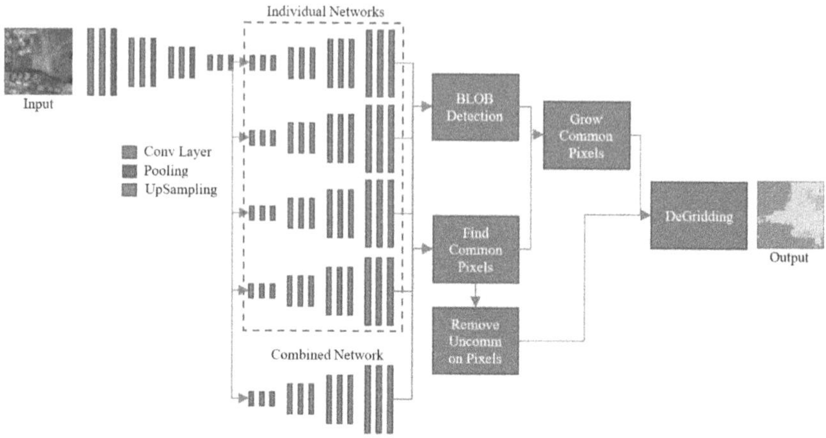

FIGURE 17.2 The proposed advanced model.

class with the largest region in the base mask, the regions are grown using the previously detected blobs to remove any regions which are not segmented. Finally, it is postprocessed to produce the output mask.

17.3.1 Individual Networks

The individual networks are a set of U-Nets, trained to produce masks that are single classes, corresponding to 1 for the class that is being trained and 0 otherwise. The input to the networks is the same as given to the entire model. The output is the single class mask of 1s and 0s. Each of these U-Nets is based on the MobileNetV2 architecture (Sinha and El-Sharkawy, 2019; Sandler et al., 2018). The individual networks are trained by separating the specific class from the ground truth image to produce the binary ground truth masks for each of the classes in the training dataset. Additionally, we remove the class 0 since if all the networks equal to 0 then it would indicate class 0.

17.3.2 The Combined Network

Similar to the individual networks, the combined network is a U-Net with a MobileNetV2 as the backbone. Instead of training on a binary mask, the network is trained on the complete ground truth mask. Thus, it produces the complete prediction mask, containing all the classes of the dataset. The reason of adopting the U-Net is because it empirically produces better results as compared against other similar models of SegNet and DeepLab, the results of which are shown in the following sections.

17.3.3 Blob Detection

In image processing, blobs are defined as a region of connected pixels with the same color or intensity. In our case, each blob is a group of pixels that are 8-connected, meaning the pixels in the region can be horizontally, vertically, or diagonally connected to each other. The blobs are detected based on binary masks for the outputs from the individual networks. Each blob is labeled, and a map of the blobs and their labels are forwarded to the region growing block.

In the case of the combined network, a separate set of grayscale images for each of the classes are prepared to represent the corresponding classes. It is then similarly processed for blob detection, producing the set of label images. This set of images is then passed to the next block.

17.3.4 Uncommon Pixels Removal

The pixels from the combined network mask that do not overlap with their corresponding pixels in the class image masks will be removed. We create a grayscale mask for each class in the base image, using which we sequentially compare with the masks of the class images produced from the individual networks. If a pixel is not set to 1 in both the grayscale images, the pixel is set to 0. In this way we can isolate the common pixels from the images, producing an image with only the common regions from both the masks.

17.3.5 Region Growing

We select the order of class pixels to grow by finding the number of pixels congruent with a class, and then choose the class that has the highest number of pixels in the image. For each pixel in the base image, we select the blob corresponding to both the individual mask and the combined mask to produce the possible pixel growth location for that pixel. The pixels, which have the value 0 in the base mask and are not originally classified as class 0, are chosen. The empty location with the value of the class is filled if the pixel is in one of the possible growth locations. This process is repeated for each of the different classes to produce the final augmented mask.

17.3.6 Post-Processing

Since U-Nets can produce the issue of gridding artifacts, we take the mean of segmented pixels across an 8×8 block to get rid of these artifacts. It is iterated across the entire image to produce solid regions of pixels. This entire process produces a 512×512 output image, which is combined with other cropped parts of the whole image to obtain the final output image.

17.4 EXPERIMENTAL RESULTS

17.4.1 DeepGlobe Dataset

Image segmentation datasets are difficult to create because of the sheer manpower required to label and classify each segment. Many datasets are updated over years and require a lot of work hours. Each segmentation mask can classify the semantic components correctly. However, each of these masks must be done by hand. There are various semi-automatic tools that take user's input only for correcting wrongly segmented masks.

The DeepGlobe dataset (Demir et al., 2018) is a highly detailed satellite imagery dataset that has semantic segmentation from building, road extraction, and terrain classification. It contains 803 images of high-resolution satellite images throughout the globe, each with a semantic segmentation mask that details the type of terrain. There are six types of terrain – urban land, agricultural land, range land, forest land, water, barren land, and unknown. This dataset is a benchmark for remote-sensing models and segmentation tasks. Figure 17.3 shows two images in the DeepGlobe dataset, where (a) and (c) are the original images and (b) and (d) are the ground-truth segmentation masks. For the purpose of testing evaluation, we use 80% in the training set of the DeepGlobe dataset as the training set for our model and the remaining 20% as the testing set for performance comparison.

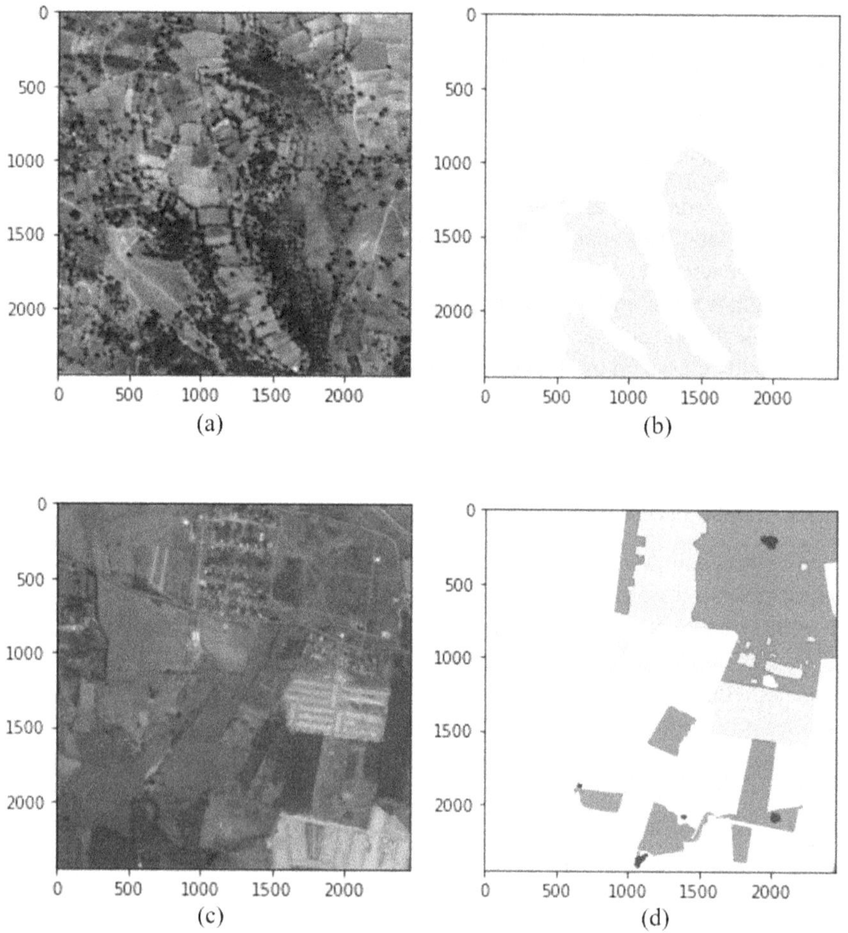

FIGURE 17.3 DeepGlobe dataset. (a) and (c) are the original images, and (b) and (d) are the ground-truth segmentation masks.

17.4.2 PARAMETERS

We use cross entropy as the loss function in semantic segmentation, which is defined as

$$H(p,q) = -\sum y_{truth} \log y_{pred} \tag{17.1}$$

where y_{truth} is the ground truth classification of each pixel and y_{pred} is the predicted classification. The accuracy is a simple metric defined as the number of correctly classified pixels with respect to the total number of pixels averaged across all the pixels across all the images in the testing dataset.

The mean intersection over union (mIoU) is another metric to measure the quality of the segmentation mask with respect to the ground truth. It is given as the mean across the classes of intersection of true mask and predicted mask divided by their union. The IoU is defined as

$$IoU = \frac{True\,Positive}{True\,Positive + False\,Positive + False\,Negative} \tag{17.2}$$

17.4.3 RESULTS

Our experiments are conducted on a machine with a nRTX 2070 mobile GPU, which has 2,304 CUDA cores, i7–9750H processor, and 16 GB of RAM. The images from the DeepGlobe dataset are separated into two sets: training and testing datasets, with the testing dataset having 20%. To accommodate the model in the GPU, the images of resolution 2048 × 2048 are cropped into 16 sub-images of 512 × 512 and then fed to the proposed model.

The individual models are plagued with the problem of overfitting. Hence, we have to introduce noise to the training dataset. The training data are augmented randomly by flipping and adding Gaussian noise. We also augment the U-Net model for binary classification to a smaller and lightweight model based on MobileNetV2. With this we are able to obtain the testing results similar to the training results. The training results of the augmented segmentation are provided in Table 17.1 and the testing results are listed in Table 17.2.

The results of degridding are shown in Table 17.3, where degridding the base image can produce better results as compared to those without the process.

TABLE 17.1
The Training Result of the Augmented Segmentation

Class	1	2	3	4	5	6
Mean IoU	0.7975	0.7023	0.7505	0.6529	0.6477	0.6427
Accuracy	0.9293	0.8572	0.8768	0.8358	0.8527	0.7846

TABLE 17.2
The Testing Result of the Augmented Segmentation

Class	1	2	3	4	5	6
Mean IoU	0.8603	0.6729	0.7820	0.5803	0.6698	0.6451
Accuracy	0.9523	0.8442	0.8897	0.7926	0.8602	0.7920

TABLE 17.3
The segmentation results for the models

Model	Mean IoU
Model	52.9 %
Model + Post-Processing	53.0 %
Model + Base Degridding	53.3 %

For performance comparisons, we evaluate the existing state-of-the-art models: U-Net, SegNet, and Deep Lab. The SegNet, developed by Badrinarayanan et al. (2017), uses the convolution to carry out on the original image. Unlike other approaches, the images are resized for lower processing time since convolutional networks require the full range of features from the original image. It consists of two parts of convolution/downscaling and deconvolution/upscaling. During the downscaling, the images are convoluted using a 2 × 2 max-pooling. For the upscaling, softmax is used to classify each pixel.

The MaskLab, developed by Chen et al. (2018), builds on R-CNN of 5 × 5 and 64 filters-based model and gathers regional logits to separate different segments and classes of an image. The DeepLab, developed by Chen et al. (2017), works on a similar principle, using DCNN based model for up-sampling to extract dense features. And instead of deconvolution, DeepLab uses Atrous convolution of features to extract dense features, but without an increase in parameters. The Deep Aggregation model from Kuo et al. (2018) uses the DeepLab network as the base and adds extra networks to make the model. The stacked U-Net model from Ghosh et al. (2018) stacks a series of U-Nets to improve segmentation results. Other networks have also been developed, which use a larger crop size, like the model from Shan and Wang (2022), but they require a larger number of CUDA cores and more processing.

Table 17.4 shows the comparisons of the segmentation results among the proposed and other models. As we can see from the results, our proposed model has the highest mean IoU as compared to similar models. The ensemble model consists of popular voting of SegNet, DeepLab, based on Xception and U-Net, and based on MobileNetV2 for reference purposes. Figure 17.4 shows the segmentation results of applying the proposed model.

TABLE 17.4
The Comparisons of the Segmentation Results among Different Models

Model	Mean IoU
The Proposed Model	**53.3 %**
UNet (Ronneberger et al., 2015)	49.1 %
SegNet (Badrinarayanan et al., 2017)	47.5 %
DeepLab (Chen et al., 2017)	39.6 %
Ensemble	50.7 %
Deep Aggregation Net (Kuo et al., 2018)	52.7 %
Stacked UNet (Ghosh et al., 2018)	50.6 %

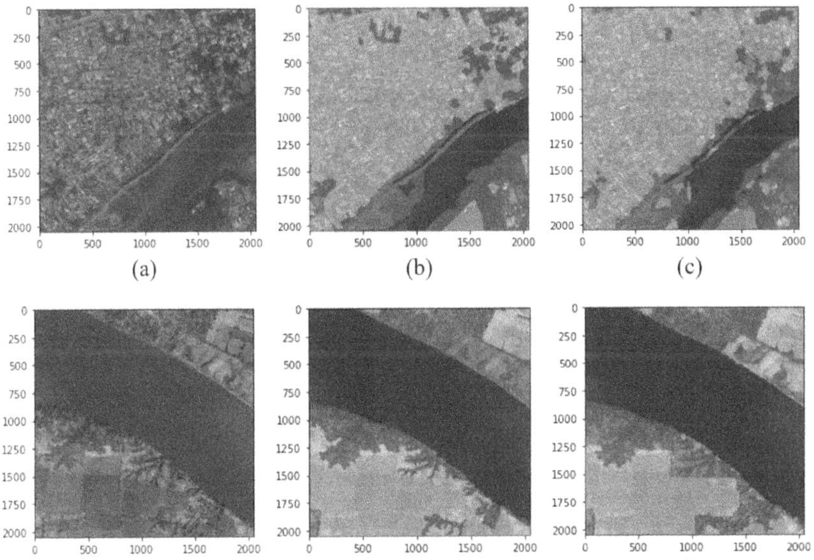

(a) (b) (c)

(d) (e) (f)

FIGURE 17.4 The segmentation results of applying the proposed model. (a) and (d) are the original images, (b) and (e) are the ground-truth images superimposed on the original, and (c) and (f) are the segmented images superimposed on the original.

17.5 CONCLUSIONS

In this chapter, we have proposed a novel method and a unique viewpoint on multi-class segmentation. Instead of looking at the problem together and parsing through a neural network, we develop a network, consisting of separate single class networks working to augment the segmentation results. Experimental results show that the proposed method produces higher metrics than several existing state-of-the-art models. This new method paves a new way to build and improve image segmentation models and problems. Our future work is to expand the results of our network in different datasets and computer vision tasks.

REFERENCES

Akhtar, N., Agarwal, N., and Burjwal, A., "K-mean algorithm for image segmentation using neutrosophy," *Proceedings of the International Conference on Advances in Computing, Communications and Informatics*, Delhi, pp. 2417–2421, 2014.

Ali, R., Hardie, R. C., Narayanan, B., and De Silva, S., "Deep learning ensemble methods for skin lesion analysis towards melanoma detection," *Proceedings of the IEEE National Aerospace and Electronics Conference*, Dayton, OH, pp. 311–316, 2019.

Antonacopoulos, A., and Ritchings, R. T., "Segmentation and classification of document images," *Proceedings of the IEE Colloquium on Document Image Processing and Multimedia Environments*, London, pp. 16/1–16/7, 1995.

Badrinarayanan, V., Kendall, A., and Cipolla, R., "Segnet: A deep convolutional encoder-decoder architecture for image segmentation," *IEEE Transactions on Pattern Analysis and Machine Intelligence*, vol. 39, no. 12, pp. 2481–2495, 2017.

Ben Fadhel, Y., Ktata, S., and Kraiem, T., "Cardiac scintigraphic images segmentation techniques," *Proceedings of the 2nd International Conference on Advanced Technologies for Signal and Image Processing*, Monastir, pp. 364–369, 2016.

Chen, L.-C., Hermans, A., Papandreou, G., Schroff, F., Wang, P., and Adam, H., "Masklab: Instance segmentation by refining object detection with semantic and direction features," *Proceedings of the IEEE Conference on Computer Vision and Pattern Recognition*, pp. 4013–4022, 2018.

Chen, L.-C., Papandreou, G., Kokkinos, I., Murphy, K., and Yuille, A. L., "Deeplab: Semantic image segmentation with deep convolutional nets, atrous convolution, and fully connected CRFS," *IEEE Transactions on Pattern Analysis and Machine Intelligence*, vol. 40, no. 4, pp. 834–848, 2017.

Choudhury, A. R., Parajuli, B., and Kumar, P., "QuadRoad: An ensemble of CNNs for road segmentation," *Procedia Computer Science*, vol. 176, pp. 138–147, 2020.

Demir, I., Krzysztof, K., David, L., Pang, G., Huang, J., Basu, S., Hughes, F., Tuia, D., and Raskar, R., "DeepGlobe 2018: A challenge to parse the earth through satellite images," *Proceedings of the IEEE/CVF Conference on Computer Vision and Pattern Recognition Workshops*, pp. 172–181, 2018.

Ghosh, A., Ehrlich, M., Shah, S., Davis, L., and Chellappa, R., "Stacked U-Nets for ground material segmentation in remote sensing imagery," *Proceedings of the IEEE/CVF Conference on Computer Vision and Pattern Recognition Workshops*, Salt Lake City, UT, pp. 257–261, 2018.

Ghosh, S., Das, N., Das, I., and Maulik, U., "Understanding deep learning techniques for image segmentation," *ACM Computing Surveys*, vol. 52, pp. 1–35, Sept. 2019.

He, K., Zhang, X., Ren, S., and Sun, J., "Deep residual learning for image recognition," *Proceedings of the IEEE Conference on Computer Vision and Pattern Recognition*, Las Vegas, NV, pp. 770–778, 2016.

Ingle, P. D. and Kaur, P., "Adaptive thresholding to robust image binarization for degraded document images," *Proceedings of the 1st International Conference on Intelligent Systems and Information Management*, Aurangabad, pp. 189–193, 2017.

Kalam, R. and Manikandan, K., "Enhancing k-means algorithm for image segmentation," *Proceedings of the International Conference on Process Automation, Control and Computing*, Coimbatore, pp. 1–4, 2011.

Kuo, T., Tseng, K., Yan, J., Liu, Y., and Wang, Y. F., "Deep aggregation net for land cover classification," *Proceedings of the IEEE/CVF Conference on Computer Vision and Pattern Recognition Workshops*, Salt Lake City, UT, pp. 252–256, 2018.

Liu, Y. and Zhou, X., "Automatic texture segmentation for texture-based image retrieval," *Proceedings of the 10th International Multimedia Modelling Conference*, Brisbane, Queensland, pp. 285–290, 2004.

Long, J., Shelhamer, E., and Darrell, T., "Fully convolutional networks for semantic segmentation," *Proceedings of the IEEE Conference on Computer Vision and Pattern Recognition*, pp. 3431–3440, 2015.

Milletari, F., Navab, N., and Ahmadi, S., "V-Net: Fully convolutional neural networks for volumetric medical image segmentation," *Proceedings of the Fourth International Conference on 3D Vision*, Stanford, CA, pp. 565–571, 2016.

Minaee, S., Boykov, Y. Y., Porikli, F., Plaza, A. J., Kehtarnavaz, N., and Terzopoulos, D., "Image segmentation using deep learning: A survey," *IEEE Transactions on Pattern Analysis and Machine Intelligence*, vol. 44, no. 7, Jul. 2022.

Noh, H., Hong, S., and Han, B., "Learning deconvolution network for semantic segmentation," *Proceedings of the IEEE International Conference on Computer Vision*, Santiago, pp. 1520–1528, 2015.

Ren, W., Pan, J., Cao, X., and Yang, M., "Video deblurring via semantic segmentation and pixel-wise non-linear kernel," *Proceedings of the IEEE International Conference on Computer Vision*, Venice, pp. 1086–1094, 2017.

Ronneberger, O., Fischer, P., and Brox, T., "U-Net: Convolutional networks for biomedical image segmentation," *Lecture Notes in Computer Science*, vol. 9351, pp. 234–241, 2015.

Sabri, M. and Fieguth, P., "A new Gabor filter based kernel for texture classification with SVM," *Lecture Notes in Computer Science*, vol. 3212, pp. 314–322, 2004.

Sandler, M., Howard, A., Zhu, M., Zhmoginov, A., and Chen, L., "MobileNetV2: Inverted residuals and linear bottlenecks," *Proceedings of the IEEE/CVF Conference on Computer Vision and Pattern Recognition*, Salt Lake City, UT, pp. 4510–4520, 2018.

Shan, L. and Wang, W., "DenseNet-based land cover classification network with deep fusion," *IEEE Geoscience and Remote Sensing Letters*, 2022.

Shi, J. and Malik, J., "Normalized cuts and image segmentation," *IEEE Transactions on Pattern Analysis and Machine Intelligence*, vol. 22, no. 8, pp. 888–905, Aug. 2000.

Simonyan, K. and Zisserman, A., "Very deep convolutional networks for large-scale image recognition," arXiv 1409.1556, 2014.

Sinha, D. and El-Sharkawy, M., "Thin MobileNet: An enhanced mobilenet architecture," *Proceedings of the IEEE 10th Annual Ubiquitous Computing, Electronics & Mobile Communication Conference*, New York, NY, pp. 280–285, 2019.

Sun, H., Zheng, X., and Lu, X., "A supervised segmentation network for hyperspectral image classification," *IEEE Transactions on Image Processing*, vol. 30, pp. 2810–2825, 2021.

Surlakar, P., Araujo, S., and Sundaram, K. M., "Comparative analysis of k-means and k-nearest neighbor image segmentation techniques," *Proceedings of the IEEE 6th International Conference on Advanced Computing*, Bhimavaram, pp. 96–100, 2016.

Tseng, Y. and Jan, S., "Combination of computer vision detection and segmentation for autonomous driving," *Proceedings of the IEEE/ION Position, Location and Navigation Symposium*, Monterey, CA, pp. 1047–1052, 2018.

Yeh, C.-H., Huang, C.-H., and Kang, L.-W., "Multi-scale deep residual learning-based single image haze removal via image decomposition," *IEEE Transactions on Image Processing*, vol. 29, pp. 3153–3167, 2020.

Yeh, C.-H., Lin, C.-H., Kang, L.-W., Huang, C.-H., Lin, M.-H., Chang, C.-Y., and Wang, C.-C., "Lightweight deep neural network for joint learning of underwater object detection and color conversion," *IEEE Transactions on Neural Networks and Learning Systems*, vol. 33, no. 11, pp. 6129–6143, 2021.

Yuan, Y., Chao, M., and Lo, Y., "Automatic skin lesion segmentation using deep fully convolutional networks with Jaccard distance," *IEEE Transactions on Medical Imaging*, vol. 36, no. 9, pp. 1876–1886, Sept. 2017.

Zhang, Y., Brady, M., and Smith, S., "Segmentation of brain MR images through a hidden Markov random field model and the expectation-maximization algorithm," *IEEE Transactions on Medical Imaging*, vol. 20, no. 1, pp. 45–57, Jan. 2001.

18 FPA-Net

Frequency-Guided Position-Based Attention Network for Land-Cover Image Segmentation

18.1 INTRODUCTION

Earth's land cover is a crucial source of information that describes both natural ecosystems and human activities on the planet surface. It has been utilized by researchers across a range of fields, including landscape ecology, urban planning, environmental modeling, and disaster management. Remote-sensing technology intends to collect surface data over a large area at regular intervals, facilitating efficient land cover classification. In remote sensing land cover classification, pixels and objects that share similar spectral characteristics are classified and grouped into available categories like forests, urban lands, water, barren lands, agriculture lands, range lands, and so on.

Various techniques have been developed for land-cover segmentation and classification, including standard statistical algorithms as well as machine-learning methods like random forests and support vector machines. Ulmas and Liiv (2020) proposed a U-Net based method to classify land-cover images. However, their method struggled to demonstrate better performance on all the categories of land-cover images. To overcome the imbalance performance, Peng et al. (2019) used UNet++ for end-to-end change detection in high-resolution satellite images and achieved better detection performance. Somasunder and Shih (2021) used the U-Net like architecture based on individual class binary masks and combined masks for land-cover image segmentation. The U-Net like architectures were also used for other segmentation applications, such as MA-Net for liver and tumor segmentation (Fan et al., 2020), and can produce comparable outcomes. This method incorporates both spatial and channel dependencies using attention mechanism. However, MA-Net has higher computational and memory complexities. Furthermore, its channel dependency is not of much help in dealing with land-cover segmentation since the feature patterns in land-cover images are less heterogeneous, unlike the patterns in liver and tumor datasets.

To solve the problems of higher computational complexity and redundancy while keeping the performance better in terms of different metrics, we propose a new segmentation method based on the U-Net shaped encoder-decoder architecture with

DOI: 10.1201/9781003474395-21

frequency-guided attention mechanism added. The key contributions of this chapter are summarized as follows.

(1) We propose a novel segmentation method based on the U-Net shaped encoder-decoder architecture for land-cover image segmentation.
(2) We utilize the position-based attention mechanism to capture the spatial dependency among features maps.
(3) We extend our network with frequency-guided component to provide additional support with high-frequency features that effectively improve the performance of segmentation.
(4) The experimental results demonstrate the better performance of our proposed method as compared against existing methods while maintaining less computational complexity.

The rest of this chapter is organized as follows. Section 18.2 describes the background and related work. Section 18.3 proposes the new land-cover image segmentation model. Section 18.4 presents experimental results and analysis. Finally, conclusions are drawn in Section 18.5.

18.2 BACKGROUND AND RELATED WORK

18.2.1 Encoder-Decoder-Based

The encoder-decoder-based approaches are popular for segmentation tasks among recent researchers because of their strong capabilities of capturing complex features in images. Many researchers have been working on land-cover segmentation by encoder-decoder-based methods, such that U-Net convolutional neural networks (CNNs) have advantages for automated land cover classification compared to traditional machine-learning approaches (Boston et al., 2022; González-Vélez et al., 2021; Fan et al., 2022; Son et al., 2022). Multi-Scale Convolutional Block (MSCB) (Li et al., 2022) can capture both local and global features of land-cover images. U-Net, UNet++, MA-Net, and other variations of U-Net-shaped methods, such as Deep UNet (Singh and Nongmeikapam, 2023), DensePPMUNet-a (Xiang et al., 2023), DeepLab3 (Rao et al., 2022), and DeepLabv3+ (Sertel et al., 2022), can provide outstanding performance in segmentation.

U-Net along with residual networks and transfer learning (Safarov et al., 2022) can provide better segmentation results. UNet++ based architectures with feature fusion and up-sampling of small samples can extract better semantic features (Wang et al., 2022). A lightweight complex-valued DeepLabv3+ (L-CV-DeepLabv3+) (Yu et al., 2022) was proposed for polarimetric synthetic aperture radar (PolSAR) image segmentation. To combine the attention mechanisms, extract and aggregate the local geometric features effectively, Li et al. (2022a) proposed a novel convolution operator called AGGM, which combines the attention mechanisms and graph geometric moments convolution, while only using raw multispectral LiDAR point clouds as the input for land-cover classification. A pseudo-Siamese network (Chen and Bruzzone, 2022) was utilized to minimize the contrastive loss between the learned features of a ResUNet branch and a gumble-softmax vector quantizer branch. FPNPSADLV3+ network (Yuan et al., 2022) was proposed to capture fine edge and small objects information in remote sensing images. CHeGCN (Liu et al., 2022) was developed to integrate a graph convolutional network (GCN) with a convolutional neural network (CNN) that utilizes topological relations among land-cover objects.

Multi-scale feature aggregation (Shen et al., 2022; Chen et al., 2023; Khan et al., 2023) was used to address the issues of classic convolutional neural networks' poor generalization capacity. DKDFN (Li et al., 2022) used a multi-head encoder and a knowledge-guided multibranch decoder to guide the segmentation process. DPPNet (Sravya et al., 2022) used a depth-wise pyramid pooling (DPP) block and a dense block with multi-dilated depth-wise residual connections for accurate localization and context information in the segmentation map. In addition to the usual deep-learning model, interactive segmentation modules were used to finely segment and refine the objects manually (Xu et al., 2022a).

Wang et al. (2022a) introduced region merging with local scale, over- and under-segmentation recognition and significantly improved the segmentation results. MKANet (Zhang et al., 2022) addressed the issues of down-sampling and cropping images before sending them to deep neural network by incorporating shared kernels along with parallel and shallow architectures that can be used for larger image patches. MSNet (Weng et al., 2022) used the three-way parallel feature fusion network to identify intra-class and inter-class information for better segmentation effect. Attention mechanism has been popular to focus on important features (Wang, 2022; Chu et al., 2022). For instance, RAANet (Liu et al., 2022) used attention module and residual structure to enhance important features and suppress unnecessary features.

18.2.2 Transformer-Based

Recently, transformer-based deep-learning models have become popular for computer vision tasks including segmentation and classifications. Scheibenreif et al. (2022) combined vision transformers with pre-trained self-supervised model for the segmentation task. Xu et al. (2023) developed RSSFormer, which used Adaptive Transformer Fusion to suppress the background noise and enhance the foreground saliency. Shi et al. (2022) used multimodal fusion, attention mechanisms, and spectral and spatial representation learning (SSRL) to explore the spatial and channel dependencies.

UAVformer (2023) used composite transformer network with combination of multi-level features and adaptive feature fusion modules to gain fruitful feature fusion. Class-Guided Swin Transformer (Meng et al., 2022) was proposed for remote sensing image segmentation, where Swin Transformer was used in encoder and Class-Guided Transformer was used in decoder. In (Feng et al., 2022), Gabor filter was used along with Swin Transformer to enhance the segmentation edges for segmenting the occlusion and shadow parts in boundary. In general, these transformer-based architectures are of higher computational complexity.

18.2.3 GAN-Based

A generative adversarial network (GAN) is widely used in image segmentation and classification. Several variations of GAN-based models have been used for land-cover segmentation such as in (Ji et al., 2020; Desai and Ghose, 2022; Li et al., 2020; Xie et al., 2021; Joseph, 2021). Recently, Kwak and Kim (2023) used semi-supervised CycleGAN and EfficientNet to overcome the problem of high volumes of dataset requirement. Abdollahi et al. (2020) used the GAN and SegNet models with bidirectional convolutional LSTM to solve the problem.

Encoder network was incorporated in GAN to reduce the requirement of large volumes of data (Ansith and Bini, 2022). GAN-FuzzyNN (Gite and Gupta, 2023) combined GAN and fuzzy neural network (FNN) to tackle the issue of land cover change detection. Desai and Ghose (2022) used the GAN-based semi-supervised network with the help of pool-based active learning sampling strategy to select labeled sample images. However, these methods struggle to provide computational efficiency.

18.3 THE PROPOSED LAND-COVER IMAGE SEGMENTATION MODEL

In this section, we propose a novel method consisting of res block, down-sampling, position-based attention, and up-sampling. Figure 18.1 shows the overview of our proposed Frequency-Guided Position-Based Attention Network (FPA-Net) architecture. In the encoding part we find the important features helpful for segmentation, and in the decoding part we generate the corresponding mask for a land-cover image to recognize the regions of different categories. Res block is used to capture the high-dimensional features and send them via skip connection to the decoder to help generate more plausible segmentation masks. Down-sampling reduces the size of the spatial dimension, while generating a greater number of feature channels. Inspired by MA-Net, we add the attention mechanism for land-cover segmentation to find the spatial dependency among feature maps. In the decoding process, the nearest interpolation is used to increase the spatial dimensions, while reducing the number of feature channels with the additional features coming through the skip connections. Furthermore, we extend our network with frequency-guided component to support the network with high-frequency features.

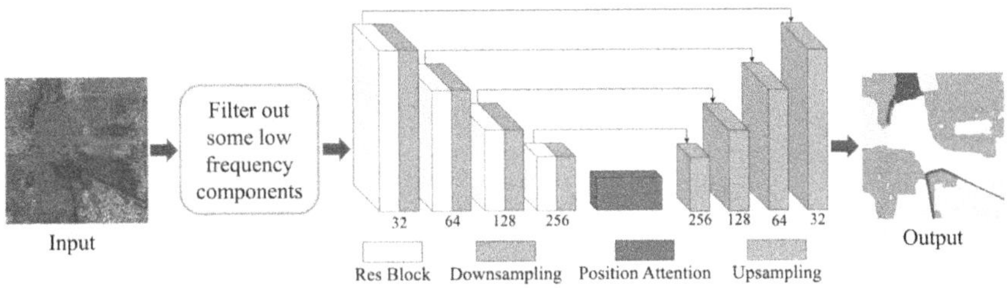

FIGURE 18.1 Overview of our proposed FPA-Net architecture.

18.3.1 Res Block

Figure 18.2 shows the res block, which captures the important features by three units of 3 × 3 convolution followed by batch normalization and ReLU activation function. In addition, we add the 1 × 1 convolution followed by batch normalization and ReLU activation function to control the number of channels.

18.3.2 Position-Based Attention

We adapt the MA-Net with the self-attention mechanism to search for spatial dependency among feature maps. Figure 18.3 shows the position-based attention block, which is crucial for land-cover image segmentation as the same features may be available in different spatial regions. In the attention block, we incorporate the key-query-value mechanism as in the vision transformer (Vaswani et al., 2017). Since there are three branches – top, center, and bottom – we match them with key, query, and value of attention mechanism. The result of convolution in top branch is multiplied with that of center branch to represent the "key × query" operation. Then it is multiplied with the result in bottom branch to represent the operation of "key × query × value" to find the attention score of the features.

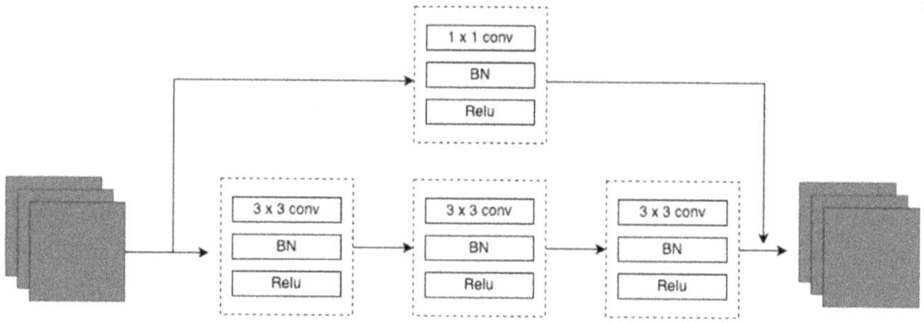

FIGURE 18.2 Res block.

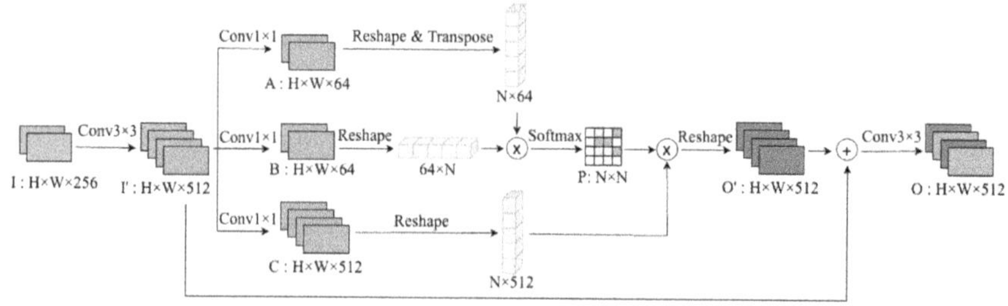

FIGURE 18.3 The position-based attention block.

18.3.3 Low-Frequency Component Filtering

We extend our network with an additional component that converts the image into frequency domain and filter out some low-frequency content. Then we convert the filtered frequency image back to spatial domain that is sent to the encoder.

18.4 EXPERIMENTAL RESULTS AND ANALYSIS

18.4.1 Dataset

To evaluate the effectiveness of our proposed method FPA-Net, we conduct the experiments on DeepGlobe dataset (Demir et al., 2018), Gaofen Image Dataset (GID-15) (Tong et al., 2020), and Land Cover AI dataset (Boguszewski et al., 2021). The DeepGlobe dataset contains 1,949 land-cover images along with the corresponding masks, indicating the position of different categories of land cover as shown in Figure 18.4.

The DeepGlobe dataset contains seven classes: urban, agriculture, rangeland, forest, water, barren, and unknown, which are denoted by different colors as shown in Table 18.1.

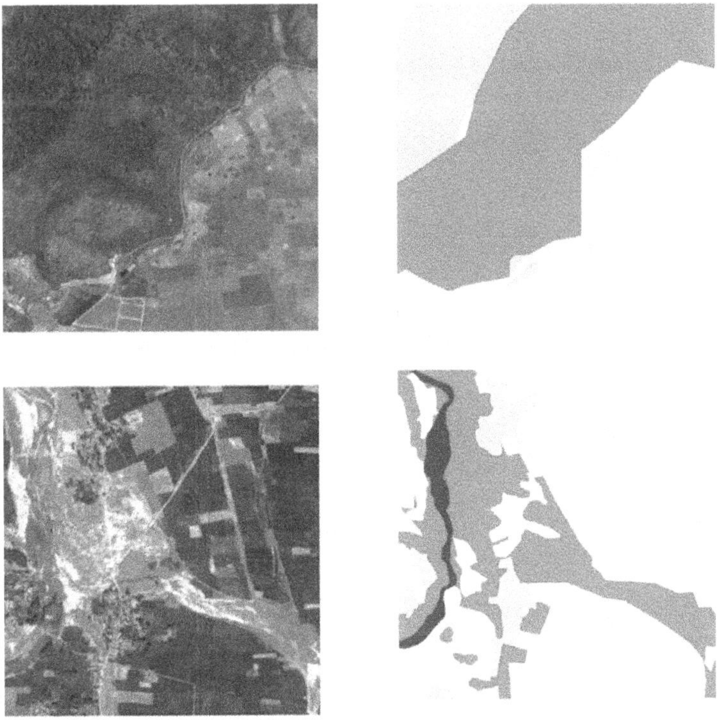

FIGURE 18.4 Sample land-cover images (left) and the corresponding masks (right) from the DeepGlobe dataset.

TABLE 18.1
DeepGlobe Dataset Categories with RGB Color Codes

Index	Name	R	G	B	Color
0	urban land	0	255	255	
1	agriculture land	255	255	0	
2	range land	255	0	255	
3	forest land	0	255	0	
4	water	0	0	255	
5	barren land	255	255	255	
6	unknown	0	0	0	

FIGURE 18.5 A sample land-cover image (left) and the corresponding mask (right) from GID–15.

The Gaofen Image Dataset (GID-15) consists of 150 high-resolution images. We pre-process the dataset into 3,000 smaller images by cropping. Figure 18.5 shows a sample image and its corresponding ground-truth mask from the GID-15 dataset.

The GID-15 dataset contains 16 classes: industrial land, urban residential, rural residential, traffic land, paddy field, irrigated land, dry cropland, garden plot, arbor woodland, shrub land, natural grassland, artificial grassland, river, lake, pond, and unknown background, which are denoted by different colors as shown in Table 18.2.

TABLE 18.2
GID-15 Dataset Categories with RGB Color Codes

Index	Name	R	G	B	Color
0	unknown	0	0	0	
1	industrial land	200	0	0	
2	urban residential	250	0	150	
3	rural residential	200	150	150	
4	traffic land	250	150	150	
5	paddy field	0	200	0	
6	irrigated land	150	250	0	
7	dry cropland	150	200	150	
8	garden plot	200	0	200	
9	arbor woodland	150	0	250	
10	shrub land	150	150	250	
11	natural grassland	250	200	0	
12	artificial grassland	200	200	0	
13	river	0	0	200	
14	lake	0	150	200	
15	pond	0	200	250	

The Land Cover AI dataset consists of 41 high-resolution images. We pre-process the dataset into 3,015 smaller images by cropping. Figure 18.6 shows a sample image and its corresponding segmentation mask from the Land Cover AI dataset.

The Land Cover AI dataset contains 5 classes: building, woodland, water, road, and unknown, which are denoted by different colors as shown in Table 18.3.

18.4.2 Evaluation Metrics

We evaluate the performance of the proposed method by several metrics, including accuracy, IoU, precision, recall, and F1 score. They are represented as follows:

$$Accuracy = \frac{Correct\ predictions}{Total\ Predictions} = \frac{TP + TN}{TP + TN + FP + FN} \tag{18.1}$$

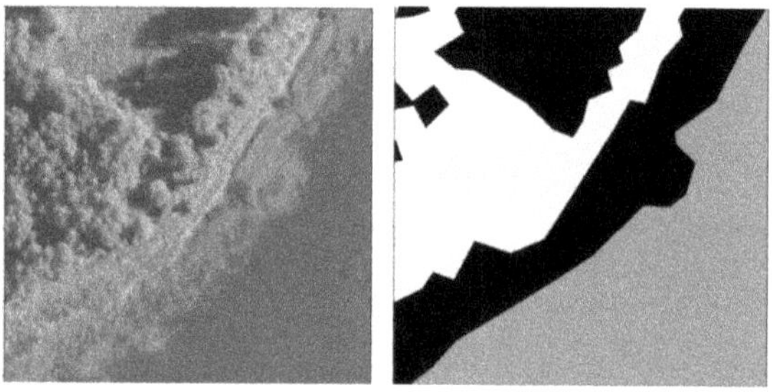

FIGURE 18.6 A sample land-cover image (left) and the corresponding masks (right) from the Land Cover AI dataset.

TABLE 18.3
The Land Cover AI Dataset Categories with RGB Color Codes

Index	Name	R	G	B	Color
0	unknown	0	0	0	
1	building	0	255	255	
2	woodland	255	255	0	
3	water	255	0	255	
4	road	0	255	0	

$$IoU = \frac{Intersection\,of\ prediction\,and\ ground\ truth}{Union\,of\ prediction\,and\ ground\ truth} = \frac{TP}{TP + FP + FN} \tag{18.2}$$

$$Precision = \frac{FP}{TP + FP} \tag{18.3}$$

$$Recall = \frac{TP}{TP + FN} \tag{18.4}$$

$$F1\,Score = \frac{2 * Precision * Recall}{Precision + Recall} \tag{18.5}$$

where *TP* denotes true positive, *TN* denotes true negative, *FP* denotes false positive, and *FN* denotes false negative.

18.4.3 Experimental Results and Comparisons

We use Pytorch Lightning framework and run our experiments on Google Colab with GPU support. The performance of the proposed method FPA-Net is compared against the existing state-of-the-art techniques, such as U-Net (Ulmas and Liiv, 2020), UNet++ (Peng et al., 2019), MA-Net (Fan et al., 2020), MV2 DeepLabv3+ (Lu et al., 2022), and STU-Net (Huang et al., 2023). Experimental results show that our proposed FPA-Net performs the best in terms of accuracy, IoU, precision, recall, and F1 score on DeepGlobe, GID-15, and Land Cover AI datasets. In addition, our proposed FPA-Net possesses low computational complexity and low memory requirement. Tables 18.4, 18.5, and 18.6 show the performance comparisons of the proposed FPA-Net and other techniques on the DeepGlobe, GID-15, and Land Cover AI datasets, respectively.

Table 18.4 shows that the proposed FPA-Net outperforms all the mentioned techniques as it achieves the highest of 96.2% accuracy and 78.4% mean IoU. Furthermore, the precision, recall, and F1 score also verify the dominance of our proposed method. The loss, IoU, accuracy, precision, recall, and F1 score curves obtained on the training and validation data from the DeepGlobe dataset are shown in Figures 18.7 and 18.8, respectively.

The network training was stopped when the validation loss did not improve relative to the previous five consecutive runs. The training and validation losses decrease rapidly in the beginning and then gradually became relatively stable in the end. The

TABLE 18.4
Evaluation of the Methods on the DeepGlobe Dataset

Metric	U-Net	UNet++	MA-Net	DeepLabv3+	STU-Net	FPA-Net (Proposed)
Accuracy	0.951	0.956	0.948	0.954	0.908	**0.962**
IoU	0.731	0.752	0.720	0.748	0.555	**0.784**
Precision	0.830	0.846	0.821	0.842	0.681	**0.869**
Recall	0.830	0.846	0.821	0.842	0.681	**0.869**
F1 score	0.830	0.846	0.821	0.842	0.681	**0.869**

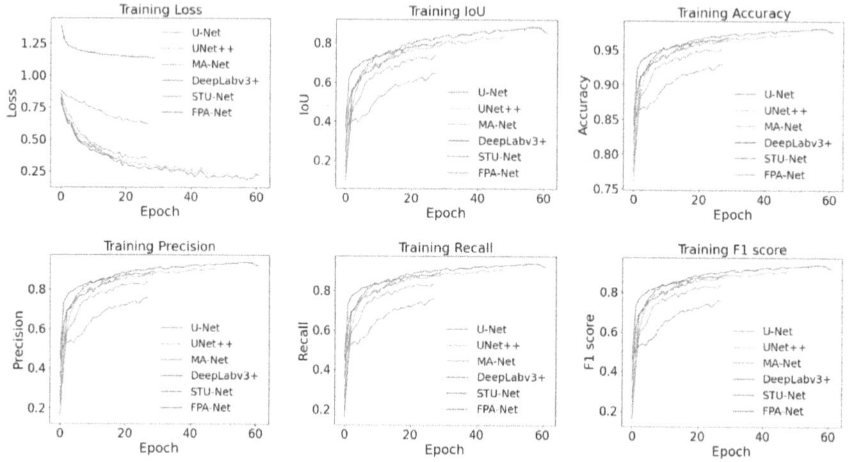

FIGURE 18.7 Performance on the training data of DeepGlobe dataset.

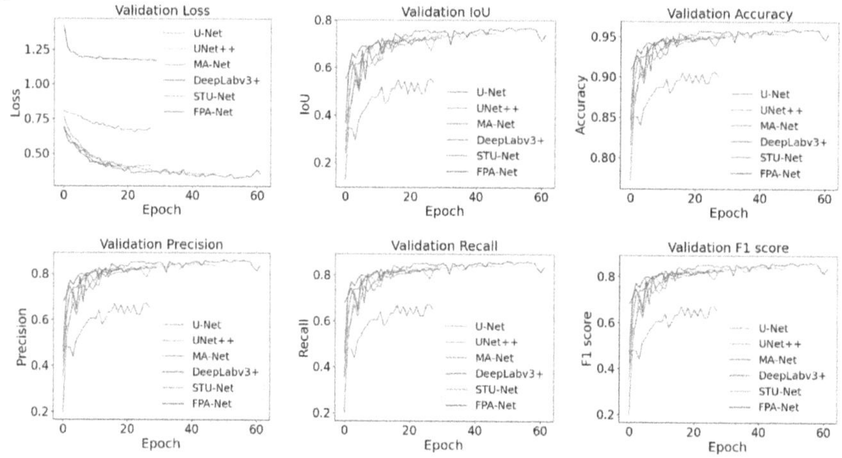

FIGURE 18.8 Performance on the validation data of DeepGlobe dataset.

MA-Net, DeepLabv3+, and STU-Net are stopped in 30 epochs, while other methods kept running for more epochs. The validation loss of FPA-Net is less than other methods all the way to the last epoch. We observe that the other metrics, such as mean IoU, accuracy, precision, recall, and F1 score, show the superiority of FPA-Net at every epoch.

Table 18.5 demonstrates the performance of the proposed method along with other methods on the GID-15 dataset. It contains 16 classes, which is more complex than the DeepGlobe dataset. The results show that our method of FPA-Net still outperforms the existing methods by achieving accuracy of 97.2% and mean IoU of 67.1%. Furthermore, the F1 score along with precision and recall are also the best.

Figures 18.9 and 18.10 show the loss, IoU, accuracy, precision, recall, and F1 score curves by different methods on the training and validation data from the GID-15

TABLE 18.5
Evaluation of the Methods on GID-15 Dataset

Metric	U-Net	UNet++	MA-Net	DeepLabv3+	STU-Net	FPA-Net (Proposed)
Accuracy	0.968	0.968	0.964	0.968	0.937	**0.972**
IoU	0.626	0.624	0.600	0.631	0.379	**0.671**
Precision	0.746	0.745	0.719	0.747	0.499	**0.780**
Recall	0.746	0.745	0.719	0.747	0.499	**0.780**
F1 score	0.746	0.745	0.719	0.747	0.499	**0.780**

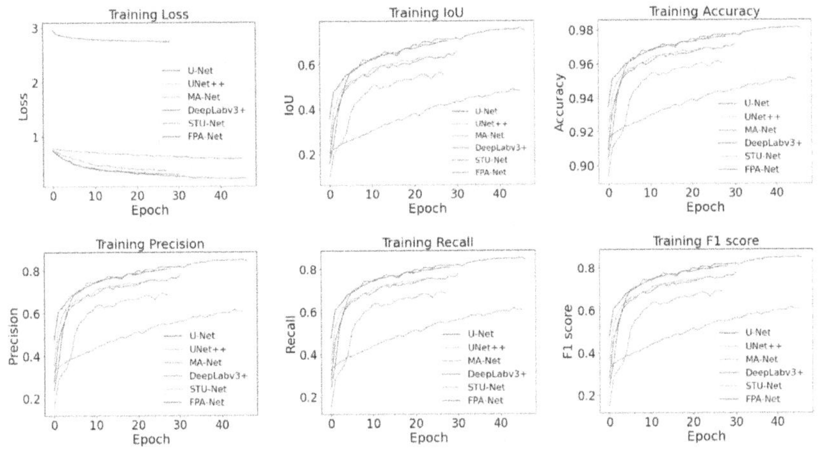

FIGURE 18.9 Performance of different methods on training data of the GID-15 dataset.

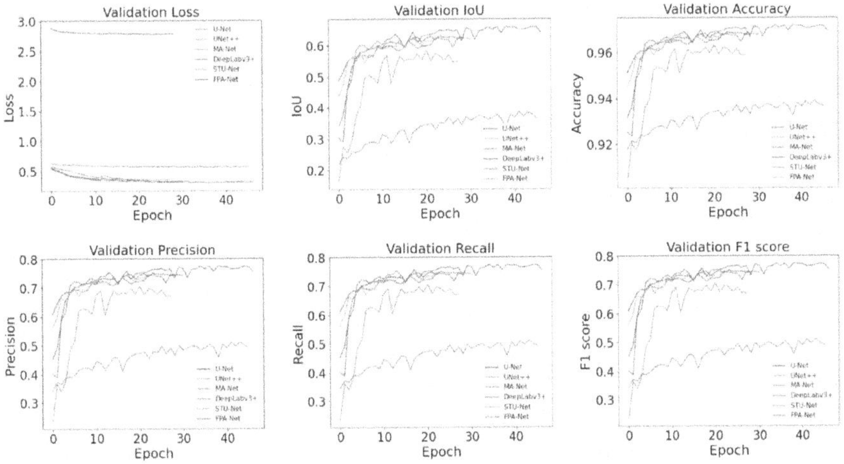

FIGURE 18.10 Performance of different methods on validation data of the GID-15 dataset.

TABLE 18.6
Evaluation of the Methods on the Land Cover AI Dataset

Metric	U-Net	UNet++	MA-Net	DeepLabv3+	STU-Net	FPA-Net (Proposed)
Accuracy	0.978	0.978	0.979	0.978	0.957	**0.980**
IoU	0.904	0.903	0.910	0.904	0.828	**0.912**
Precision	0.945	0.945	0.949	0.946	0.892	**0.950**
Recall	0.945	0.945	0.949	0.946	0.892	**0.950**
F1 score	0.945	0.945	0.949	0.946	0.892	**0.950**

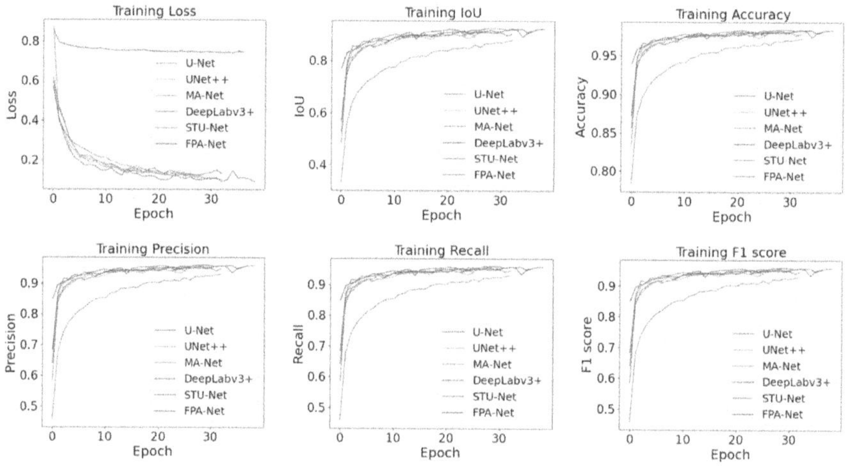

FIGURE 18.11 Performance of different methods on training data of the Land Cover AI dataset.

dataset, respectively. We observe that both training and validation losses decrease and tend to become stable with the increase in epochs. Furthermore, the mean IoU, accuracy, precision, recall, and F1 score gradually increases with the increase in epochs. Since the training and validation losses are close to each other, it indicates that the training procedure has not been overfitted to training data and the generalization capability of the trained models are good enough. These experimental curves clearly verify the dominance of our proposed FPA-Net.

Table 18.6 provides performance comparisons on the Land Cover AI dataset. We observe that FPA-Net achieves accuracy of 98% and mean IoU of 91.2%, which is the best among all of them. Also, the other metrics such as precision, recall and F1 score verify the superiority of our proposed FPA-Net.

The loss, IoU, accuracy, precision, recall, and F1 score curves on the Land Cover AI dataset for the training and validation data are shown in Figures 8.11 and 8.12, respectively. The proposed FPA-Net achieves relatively high accuracy after just a few epochs (5–10) with gradual improvement in later epochs. In general, the training and validation accuracies, mean IoU, precision, recall, and F1 score of all the models reach relatively high values all the way to the last epoch with limited noise. Experiments with all the methods are stopped when their training does not improve for consecutive five epochs. All the models with the lowest validation loss are selected to avoid potential overfitting.

Figure 18.13 depicts the predictions of our proposed FPA-Net along with other models which helps to visualize the effectiveness of our method as compared against other existing methods. We observe that the proposed method can closely predict the ground-truth more than other methods. Even in case of GID-15 sample images which are more complex than other datasets, our method can generate better segmentation masks, where other methods struggle to address the issue of identifying the complex shape in the image.

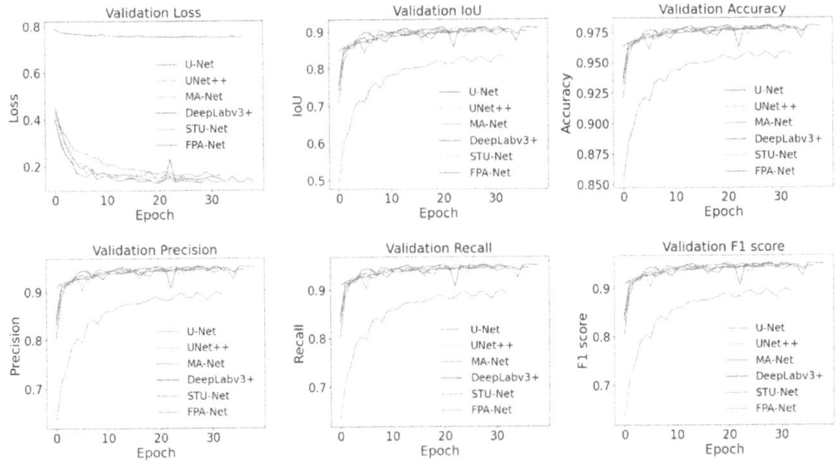

FIGURE 18.12 Performance of different methods on validation data of the Land Cover AI dataset.

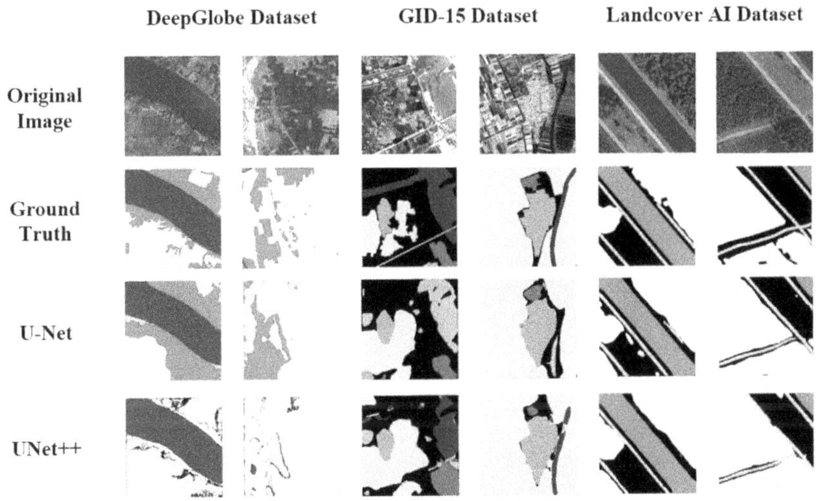

FIGURE 18.13 Prediction results of different models on the DeepGlobe, GID-15, and Land Cover AI datasets.

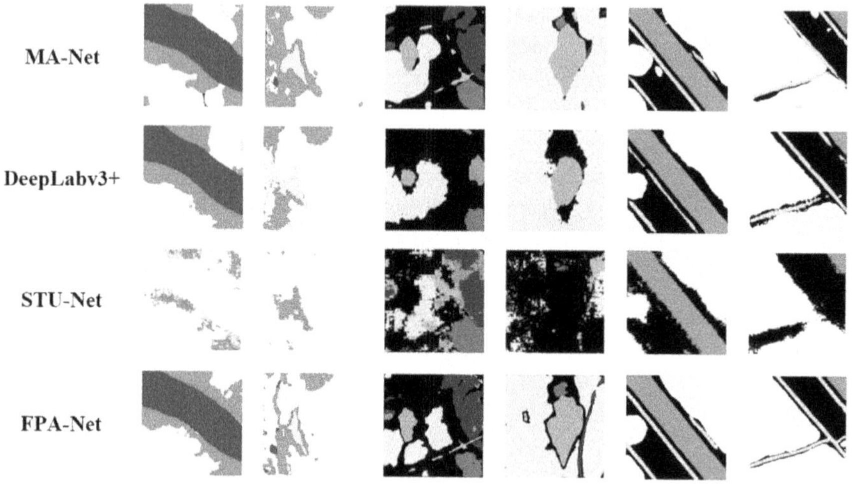

FIGURE 18.13 (Continued)

18.5 CONCLUSIONS

In this chapter, we design a novel FPA-Net for land-cover image segmentation. This encoder-decoder-based method includes frequency-guided component, residual blocks, position-wise attention, and skip connections. We incorporate the idea of attention mechanism to utilize the spatial dependency among feature maps. Also, we extend our network with frequency-guided component to incorporate high-frequency features that contributes to improve the performance of our method on land-cover image segmentation. Our experiments demonstrate that the proposed method can provide outstanding performance in terms of several quantitative evaluation metrics and qualitative measurement, as compared to existing segmentation methods. However, there is more scope to improve our model for segmentation in other related applications, for example, the images containing more complex patterns and the classes being more heterogeneous in nature.

REFERENCES

Abdollahi, A., Pradhan, B., Gite, S., and Alamri, A., "Building footprint extraction from high resolution aerial images using generative adversarial network (GAN) architecture," *IEEE Access*, vol. 8, pp. 209517–209527, 2020.

Ansith, S. and Bini, A. A., "Land use classification of high resolution remote sensing images using an encoder based modified GAN architecture," *Displays*, vol. 74, 2022, p. 102229.

Boguszewski, A., Batorski, D., Ziemba-Jankowska, N., Dziedzic, T., and Zambrzycka, A., "LandCover. ai: Dataset for automatic mapping of buildings, woodlands, water and roads from aerial imagery," *Proceedings of the IEEE/CVF Conference on Computer Vision and Pattern Recognition*, Nashville, TN, Jun. 20 to Jun. 25, 2021.

Boston, T., Van Dijk, A., Larraondo, P. R., and Thackway, R., "Comparing CNNs and random forests for landsat image segmentation trained on a large proxy land cover dataset," *Remote Sensing*, vol. 14, no. 14, p. 3396, 2022.

Chen, J., Xia, M., Wang, D., and Lin, H., "Double branch parallel network for segmentation of buildings and waters in remote sensing images," *Remote Sensing*, vol. 15, no. 6, p. 1536, 2023.

Chen, Y. and Bruzzone, L., "An approach based on contrastive learning and vector quantization to the unsupervised land-cover segmentation of multimodal images," *IGARSS 2022–2022 IEEE International Geoscience and Remote Sensing Symposium*, pp. 4811–4814, IEEE, 2022.

Chu, B., Chen, J., Chen, J., and Pei, X., "SDCAFNet: A deep convolutional neural network for land-cover semantic segmentation with the fusion of PolSAR and optical images," *IEEE Journal of Selected Topics in Applied Earth Observations and Remote Sensing*, vol. 15, pp. 8928–8942, 2022.

Demir, I., Krzysztof, K., David, L., Pang, G., Huang, J., Basu, S., Hughes, F., Tuia, D., and Raskar, R., DeepGlobe 2018: A challenge to parse the earth through satellite images," *Proceedings of the IEEE/CVF Conference on Computer Vision and Pattern Recognition Workshops*, Salt Lake City, UT, pp. 172–181, Jun. 18 to Jun. 22, 2018.

Desai, S. and Ghose, D., "Active learning for improved semi-supervised semantic segmentation in satellite images," *Proceedings of the IEEE/CVF Winter Conference on Applications of Computer Vision*, Waikoloa, HI, Jan. 3 to Jan. 8, 2022.

Fan, T., Wang, G., Li, Y., and Wang, H., "Ma-net: A multi-scale attention network for liver and tumor segmentation," *IEEE Access*, vol. 8, pp. 179656–179665, 2020.

Fan, Z., Zhan, T., Gao, Z., and Li, R., "Land cover classification of resources survey remote sensing images based on segmentation model," *IEEE Access*, vol. 10, pp. 56267–56281, 2022.

Feng, D., Zhang, Z., and Yan, K., "A semantic segmentation method for remote sensing images based on the Swin transformer fusion Gabor filter," *IEEE Access*, vol. 10, pp. 77432–77451, 2022.

Gite, K. R. and Gupta, P., "GAN-FuzzyNN: Optimization based generative adversarial network and Fuzzy neural network classification for change detection in satellite images," *Sensing and Imaging*, vol. 24, no. 1, pp. 1–26, 2023.

González-Vélez, J. C., Martinez-Vargas, J. D., and Torres-Madronero, M. C., "Land cover classification using CNN and semantic segmentation: A case of study in Antioquia, Colombia," *Smart Technologies, Systems and Applications: Second International Conference, SmartTech-IC 2021*, Quito, Dec. 1–3, 2021, Revised Selected Papers, Springer International Publishing, Cham, 2022.

Huang, Z., Wang, H., Deng, Z., Ye, J., Su, Y., Sun, H., He, J., Gu, Y., Gu, L., Zhang, S., and Qiao, Y., "STU-Net: Scalable and transferable medical image segmentation models empowered by large-scale supervised pre-training," arXiv:2304.06716 [cs.CV], Apr. 2023.

Ji, S., Wang, D., and Luo, M., "Generative adversarial network-based full-space domain adaptation for land cover classification from multiple-source remote sensing images," *IEEE Transactions on Geoscience and Remote Sensing*, vol. 59, no. 5, pp. 3816–3828, 2020.

Joseph, S., "Accurate segmentation for low resolution satellite images by discriminative generative adversarial network for identifying agriculture fields," *Journal of Innovative Image Processing*, vol. 3, no. 4, pp. 298–310, 2021.

Khan, S. D., Alarabi, L., and Basalamah, S., "DSMSA-Net: Deep spatial and multi-scale attention network for road extraction in high spatial resolution satellite images," *Arabian Journal for Science and Engineering*, vol. 48, no. 2, pp. 1907–1920, 2023.

Kwak, T. and Kim, Y., "Semi-supervised land cover classification of remote sensing imagery using CycleGAN and EfficientNet," *KSCE Journal of Civil Engineering*, pp. 1–14, 2023.

Li, D., Shen, X., Guan, H., and Yu, Y., "AGFP-Net: Attentive geometric feature pyramid network for land cover classification using airborne multispectral LiDAR data," *International Journal of Applied Earth Observation and Geoinformation*, vol. 108, pp. 102723, 2022a.

Li, R., Zheng, S., Duan, C., Wang, L., and Zhang, C., "Land cover classification from remote sensing images based on multi-scale fully convolutional network," *Geo-Spatial Information Science*, vol. 25, no. 2, pp. 278–294, 2022b.

Li, X., Luo, M., Ji, S., and Zhang, L., "Evaluating generative adversarial networks based image-level domain transfer for multi-source remote sensing image segmentation and object detection," *International Journal of Remote Sensing*, vol. 41, no. 19, pp. 7343–7367, 2020.

Li, Y., Zhou, Y., Zhang, Y., and Zhong, L., "DKDFN: Domain knowledge-guided deep collaborative fusion network for multimodal unitemporal remote sensing land cover classification," *ISPRS Journal of Photogrammetry and Remote Sensing*, vol. 186, pp. 170–189, 2022c.

Liu, R., Tao, F., Liu, X., Na, J., Leng, H., Wu, J., and Zhou, T., "RAANet: A residual ASPP with attention framework for semantic segmentation of high-resolution remote sensing images," *Remote Sensing*, vol. 14, no. 13, p. 3109, 2022.

Liu, Z., Tang, P., Zhang, W., and Zhang, Z., "CNN-enhanced heterogeneous graph convolutional network: Inferring land use from land cover with a case study of park segmentation," *Remote Sensing*, vol. 14, no. 19, p. 5027, 2022.

Lu, J., Zhou, B., Wang, B., and Zhao, Q., "Land cover classification of remote sensing images based on improved DeeplabV3+ network," *Journal of Physics: Conference Series*, vol. 2400, no. 1, p. 012035, 2022.

Meng, X., Yang, Y., Wang, L., Wang, T., Li, R., and Zhang, C., "Class-guided Swin transformer for semantic segmentation of remote sensing imagery," *IEEE Geoscience and Remote Sensing Letters*, vol. 19, pp. 1–5, 2022.

Peng, D., Zhang, Y., and Guan, H., "End-to-end change detection for high resolution satellite images using improved UNet++," *Remote Sensing*, vol. 11, no. 11, p. 1382, 2019.

Rao, D. R., Noorjahan, S., and Fathima, S. A., "Classification of land cover usage from satellite images using deep learning algorithms," *2022 International Conference on Electronics and Renewable Systems (ICEARS), IEEE*, 2022.

Safarov, F., Temurbek, K., Jamoljon, D., Temur, O., Chamberlain, J., Akmalbek, A., Abdusalomov, B., and Cho, Y., "Improved agricultural field segmentation in satellite imagery using TL-ResUNet architecture," *Sensors*, vol. 22, no. 24, p. 9784, 2022.

Scheibenreif, L., Hanna, J., Mommert, M., and Borth, D., "Self-supervised vision transformers for land-cover segmentation and classification," *Proceedings of the IEEE/CVF Conference on Computer Vision and Pattern Recognition*, New Orleans, LA, Jun. 18 to Jun. 24, 2022.

Sertel, E., Ekim, B., Ettehadi, P., and Kabadayi, M. E., "Land use and land cover mapping using deep learning based segmentation approaches and VHR worldview-3 images," *Remote Sensing*, vol. 14, no. 18, p. 4558, 2022.

Shen, X., Weng, L., Xia, M., and Lin, H., "Multi-scale feature aggregation network for semantic segmentation of land cover," *Remote Sensing*, vol. 14, no. 23, p. 6156, 2022.

Shi, W., Qin, W., and Chen, A., "Towards robust semantic segmentation of land covers in foggy conditions," *Remote Sensing*, vol. 14, no. 18, p. 4551, 2022.

Singh, N. J. and Nongmeikapam, K., "Semantic segmentation of satellite images using deep-unet," *Arabian Journal for Science and Engineering*, vol. 48, no. 2, pp. 1193–1205, 2023.

Somasunder, S. and Shih, F. Y., "Land cover image segmentation based on individual class binary masks," *International Journal of Pattern Recognition and Artificial Intelligence*, vol. 35, no. 16, p. 2154034, 2021.

Son, S., Lee, S., Bae, J., Ryu, M., Lee, D., Park, S., Seo, D., and Kim, J., "Land-cover-change detection with aerial orthoimagery using SegNet-based semantic segmentation in Namyangju City, South Korea," *Sustainability*, vol. 14, no. 19, p. 12321, 2022.

Sravya, N., Goyal, P., Lal, S., and Nalini, J., "DPPNet: An efficient and robust deep learning network for land cover segmentation from high-resolution satellite images," *IEEE Transactions on Emerging Topics in Computational Intelligence*, vol. 7, pp. 128–139, 2022.

Tong, X.-Y., Xia, G.-S., Lu, Q., Shen, H., Li, S., You, S., and Zhang, L., "Land-cover classification with high-resolution remote sensing images using transferable deep models," *Remote Sensing of Environment*, vol. 237, p. 111322, 2020.

Ulmas, P. and Liiv, I., "Segmentation of satellite imagery using u-net models for land cover classification," arXiv:2003.02899, 2020.

Vaswani, A., Shazeer, N., Parmar, N., Uszkoreit, J., Jones, L., Gomez, A. N., Kaiser, Ł., and Polosukhin, I., "Attention is all you need," *Advances in Neural Information Processing Systems*, vol. 30, 2017.

Wang, H., "Remote sensing image segmentation model based on attention mechanism," *2022 5th International Conference on Advanced Electronic Materials, Computers and Software Engineering (AEMCSE), IEEE*, 2022.

Wang, L., Wang, J., Zhang, X., and Wang, L., "Deep segmentation and classification of complex crops using multi-feature satellite imagery," *Computers and Electronics in Agriculture*, vol. 200, p. 107249, 2022a.

Wang, Y., Wu, L., Qi, Q., and Wang, J., "Local scale-guided hierarchical region merging and further over-and under-segmentation processing for hybrid remote sensing image segmentation," *IEEE Access*, vol. 10, pp. 81492–81505, 2022b.

Weng, L., Gao, J., Xia, M., and Lin, H., "MSNet: Multifunctional feature-sharing network for land-cover segmentation," *Remote Sensing*, vol. 14, no. 20, p. 5209, 2022.

Xiang, D., Zhang, X., Wu, W., and Liu, H., "DensePPMUNet-a: A robust deep learning network for segmenting water bodies from aerial images," *IEEE Transactions on Geoscience and Remote Sensing*, vol. 61, pp. 1–11, 2023.

Xie, J., Fang, L., and Li, S., "Super resolution guided deep network for land cover classification from remote sensing images," *IEEE Transactions on Geoscience and Remote Sensing*, vol. 60, pp. 1–12, 2021.

Xu, L., Liu, Y., Shi, S., Zhang, H., and Wang, D., "Land-cover classification with high-resolution remote sensing images using interactive segmentation," *IEEE Access*, vol. 11, pp. 6735–6747, 2023a.

Xu, R., Wang, C., Zhang, J., Xu, S., Meng, W., and Zhang, X., "RSSFormer: Foreground saliency enhancement for remote sensing land-cover segmentation," *IEEE Transactions on Image Processing*, vol. 32, pp. 1052–1064, 2023b.

Yu, L., Zeng, Z., Liu, A., and Xie, X., "A lightweight complex-valued DeepLabv3+ for semantic segmentation of PolSAR image," *IEEE Journal of Selected Topics in Applied Earth Observations and Remote Sensing*, vol. 15, pp. 930–943, 2022.

Yuan, P., Zhao, Q., Zheng, Y., Wang, X., and Hu, B., "Capturing small objects and edges information for cross-sensor and cross-region land cover semantic segmentation in Arid areas," *IEEE Journal of Selected Topics in Applied Earth Observations and Remote Sensing*, vol. 16, pp. 983–997, 2022.

Zhang, Z., Lu, W., Cao, J., and Xie, G., "MKANet: An efficient network with Sobel boundary loss for land-cover classification of satellite remote sensing imagery," *Remote Sensing*, vol. 14, no. 18, p. 4514, 2022.

19 Defense against Adversarial Attacks Based on Stochastic Descent Sign Activation Networks on Medical Images

19.1 INTRODUCTION

Medical images, such as magnetic resonance imaging (MRI), computational tomography (CT), and histopathology, provide detailed information for diagnosing various diseases. With more accurate and efficient classification systems on medical images being deployed, the demands on robust medical machine-learning systems have increased. The systems can help the experts diagnose diseases and accelerate treatment processes.

The classification approaches of human experts and machine-learning systems are different. The human experts search for abnormal areas which are distinguishable from normal portions using their knowledge. While the machine-learning techniques use a function to map the healthy and unhealthy as the specific labels. They learn the information in a supervised way. Therefore, the quality and resolution of medical image datasets could impact the performance of machine learning.

Machine-learning algorithms have been proven to achieve high accuracy in the classification tasks and more new modules have been proposed to enhance accuracy (Yeh et al., 2020); however, they could misclassify by minor perturbations in such data known as adversarial attacks (Papernot et al., 2016a, 2016b; Carlini and Wagner, 2017; Panda et al., 2019). Adversarial examples have been shown to transfer across models, making it possible to perform transfer-based (substitute model) black-box attacks. Transfer adversarial attacks and boundary attacks are the most lethal as they can be performed effectively without access to the model's parameters (Brendel et al., 2017).

The attackers can fool machine-learning systems with adversarial images, which are often imperceptible to human eyes. In other words, the models could make mistakes by these adversarial inputs, which are intentionally crafted. As a result, machine-learning systems would generate false results, misdiagnosis, or even insurance fraud.

Researchers have investigated adversarial attacks on medical images and mainly focused on testing the robustness of deep-learning models designed for medical

DOI: 10.1201/9781003474395-22

image analysis (Finlayson et al., 2019; Wetstein et al., 2020). Paschali et al. (2018) showed that classification accuracy drops from above 87% on the regular medical images to almost 0% on the adversarial examples. Hirano et al. (2021) demonstrated UAPs achieved over 80% success rates on DNNs model. Many defense methods have been proposed to defend against adversarial attacks, in which adversarial training is most prevalent (Goodfellow et al., 2015; Tramèr et al., 2018). However, this tends to lower accuracy on clean test data. To overcome this problem, the transfer-based methods were developed (Dong et al., 2019; Chen et al., 2020), but they are still vulnerable. Thus, adversarial robustness is still an open problem in machine learning.

Gradient-free trained sign activation networks have been proven to be able to defend against adversarial attacks with a higher possibility (Xie et al., 2019; Xue et al., 2020a). These networks are trained with a stochastic coordinate descent algorithm (Xue et al., 2020b; Yang et al., 2020), and their higher minimum distortions indicate that an image must comply with a more distinct modification to fool a model. In this chapter we adopt the gradient-free stochastic coordinate descent algorithm for training sign activation networks on medical image datasets, including MRI, chest X-ray, and histopathology. The rest of this chapter is organized as follows. Section 19.2 presents the proposed sign activation networks. Section 19.3 describes the datasets and experimental results. Conclusions are drawn in Section 19.4.

19.2 THE PROPOSED SIGN ACTIVATION NETWORKS

We propose to train the sign activation networks with a gradient-free stochastic coordinate descent algorithm, which is named as the *stochastic coordinate descent* (SCD).

19.2.1 The Stochastic Coordinate Descent (SCD)

We denote a given binary class data $x_i \in R^d$ and $y_i \in \{-1, +1\}$, for i = 0, 1, . . ., n−1. A linear classifier $w \in R^d$, $w_0 \in R$ minimizes the empirical risk for a given loss function defined as

$$L_{scd} = \sum_i L(w, w_0, x_i, y_i) \tag{19.1}$$

We start with a random solution $w_i \in N(0, 1)$, $w_0 \in N(0, 1)$, for $i = 0, 1, \ldots, d-1$ and iteratively make incremental changes that improve the risk. In each iteration, we select a random set of features (coordinates) from w called F. For each feature $w_i \in F$, we add/subtract a learning rate η and then determine w_0 that optimizes the risk. We compute all possible values of w_0 defined as

$$w_0 = \frac{w_i^T x_i + w_{i+1}^T x_{i+1}}{2} \tag{19.2}$$

for $i = 0, 1, \ldots, n-2$ and select the one that minimizes the loss L_{scd}. A random sample of the training data in each iteration is generated to avoid local minima. To train a single hidden layer network, we apply SCD to the final node and then to a randomly selected hidden node in each iteration of the algorithm. We apply parallelism and several heuristics in practice to speed up the run time.

19.2.2 Network Implementation

We train the following three types of sign activation networks using the proposed algorithm:

(1) SCD01: A 01-loss in the final node
(2) SCDCE: Cross-entropy loss in the final node
(3) SCDCEBNN: Cross-entropy in the final node with binary weights throughout the model

The basic architecture of SCD models is shown in Figure 19.1. The training procedure is implemented in Python, Numpy, and Pytorch (Paszke et al., 2019). Since sign activation is non-convex, our training process starts from a different random initialization. We run it 100 times and output the majority vote.

To illustrate the run time and clean test accuracies, we compare our models with the convolutional networks LeNet (LeCun et al., 1998), ResNet18 (He et al., 2016),

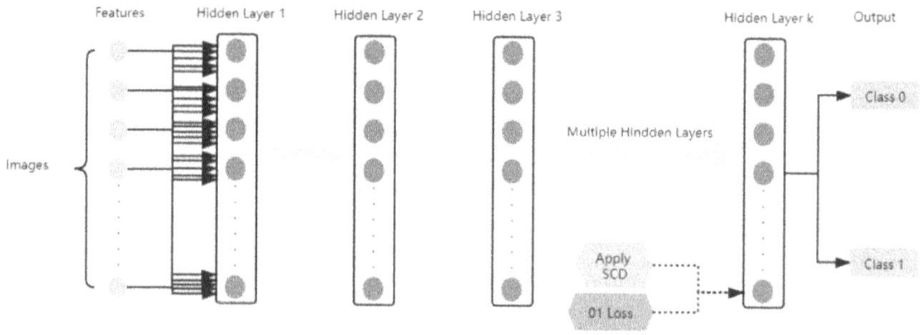

(a)

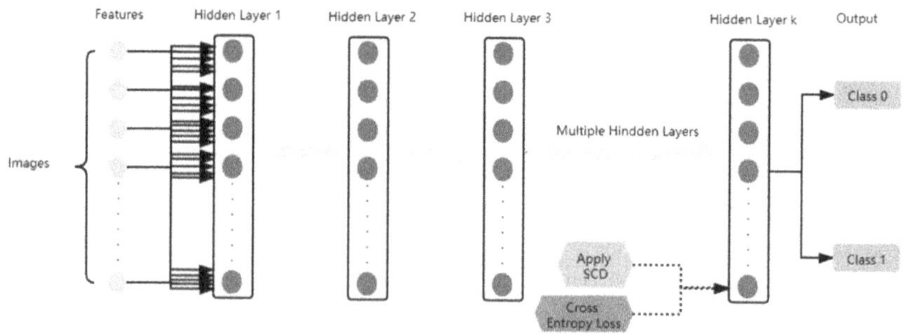

(b)

FIGURE 19.1 The architecture of SCD models. (a) The sign activation networks with our algorithm and 01-loss in the final node, (b) the sign activation networks with our algorithm and cross-entropy loss in the final node.

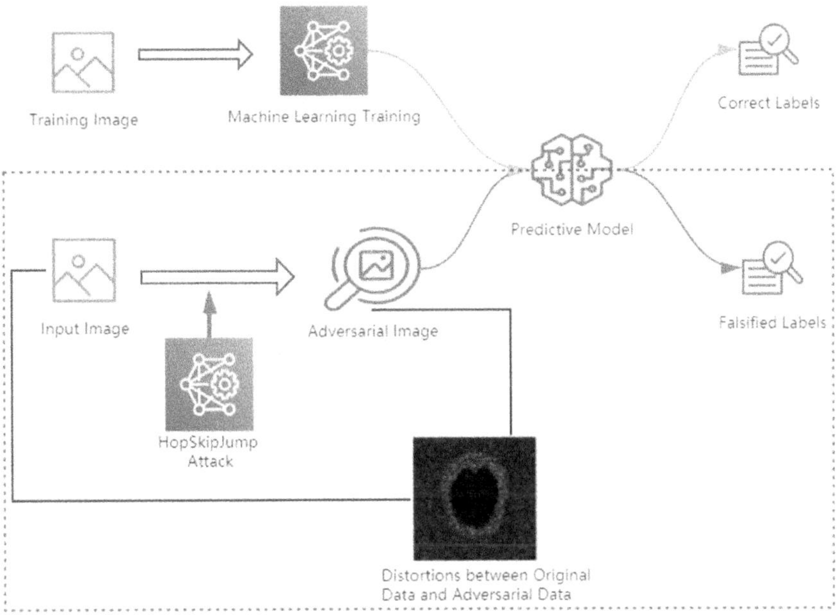

FIGURE 19.2 The procedure of attacking the models with HopSkipJump.

and a single hidden layer of 20 nodes to the equivalent network with sigmoid activation and logistic loss function (denoted as MLP). The MLP classifier in scikit-learn is used to implement MLP and the Larq library with the approximation to the sign activation. In addition, we use the HopSkipJump implementation in the IBM Adversarial Robustness Toolkit (Chen et al., 2020). It is a family of algorithms and includes both untargeted and targeted attacks optimized for L2 and L∞ similarity metrics respectively. The model is developed based on a novel estimate of the gradient direction using binary information at the decision boundary. Theoretical analysis and experiments show HopSkipJump requires significantly fewer model parameters than several state-of-the-art decision-based adversarial attacks. It also achieves competitive performance in attacking several widely used defense mechanisms.

In Figure 19.2, a predictive model is attacked by HopSkipJump to generate an adversarial image, which would fool the model. To obtain an as accurate of estimation as possible, we run HopSkipJump 10 times. In each time, we use an initial pool size of 1,000 random data points and maximum iterations of 100 to report the minimum value. For a single data point, this typically takes several hours to finish. Thus, we can report the distortion of only five random points.

19.3 EXPERIMENTAL RESULTS

19.3.1 DATASETS

We use three popular medical imaging datasets: BraTs18, Chest X-rays, and Colorectal Histopathology, to evaluate the classification accuracy.

19.3.1.1 BraTs18

The BratS18 dataset contains 210 high-grade glioma (HGG) and 75 low-grade glioma (LGG) MRI images with binary masks for the tumor. Each 3D MRI image contains 155 slices of size 240 × 240. We use the FLAIR modality images for all the experiments because the entire tumor is represented well by this modality. In total, we have 17,100 abnormal and 18,500 normal images for training. For testing, we have 1,800 abnormal and 1,900 normal images. We show more experimental results on other modalities, where ANT-GAN presents impressive synthesis quality. A more detailed medical description of the data is available in (Bakes et al., 2018). We down-sample two classes to be a balanced dataset, and each class contains 1,462 images, which are resized to 96 × 96. We split training and testing datasets by a ratio of 80:20.

19.3.1.2 Chest X-Rays

The chest X-ray images (anterior-posterior) are selected from retrospective cohorts of pediatric patients of one to five years of age from Guangzhou Women and Children's Medical Center, Guangzhou, China. All chest X-ray imaging was performed as part of patients' routine clinical care. The dataset is organized into two folders (train and test) and contains subfolders for each image category (pneumonia/normal). There are 5,863 X-ray images and 2 categories (pneumonia/normal). All chest radiographs are initially screened for quality control by removing all low quality or unreadable scans. Two expert physicians then grade the diagnoses for the images before being cleared for training the AI system. To account for any grading errors, the evaluation set is checked by a third expert. We resize the images to 96 × 96 and down-sample to 1,584 for each category as the balanced dataset. We have 3,168 images in total, which are split into training and testing sets by a ratio of 80:20.

19.3.1.3 Colorectal Histopathology

This dataset represents a collection of textures in histological images of human colorectal cancer (Kather et al., 2016). Ten anonymized H&E stained CRC tissue slides are obtained from the pathology archive at the University Medical Center Mannheim, Heidelberg University, Mannheim, Germany. The low-grade and high-grade tumors are included in this set, and no further selection is applied. The slides are first digitized, and then the contiguous tissue areas are manually annotated and tessellated to create 625 non-overlapping tissue tiles of size 150×150 (19.74 μm×74 μm). Thus, the texture features of different scales are included, ranging from individual cells (approximate 10 μm) to larger structures such as mucosal glands (> 50 μm).

The following eight types of tissue are selected for analysis: tumor epithelium, simple stroma, complex stroma, immune cells, debris, normal mucosal gland, adipose tissue, and background (no tissue). Together, the resulting 5,000 images represent the training and testing sets. We randomly pick 2 classes, immune cells, and normal mucosal glands, resize them to 96 × 96, and split train and test sets with a ratio of 80:20. Aside from the difference in imaging tissue and modality of these three datasets, the images are shown in Figure 19.3.

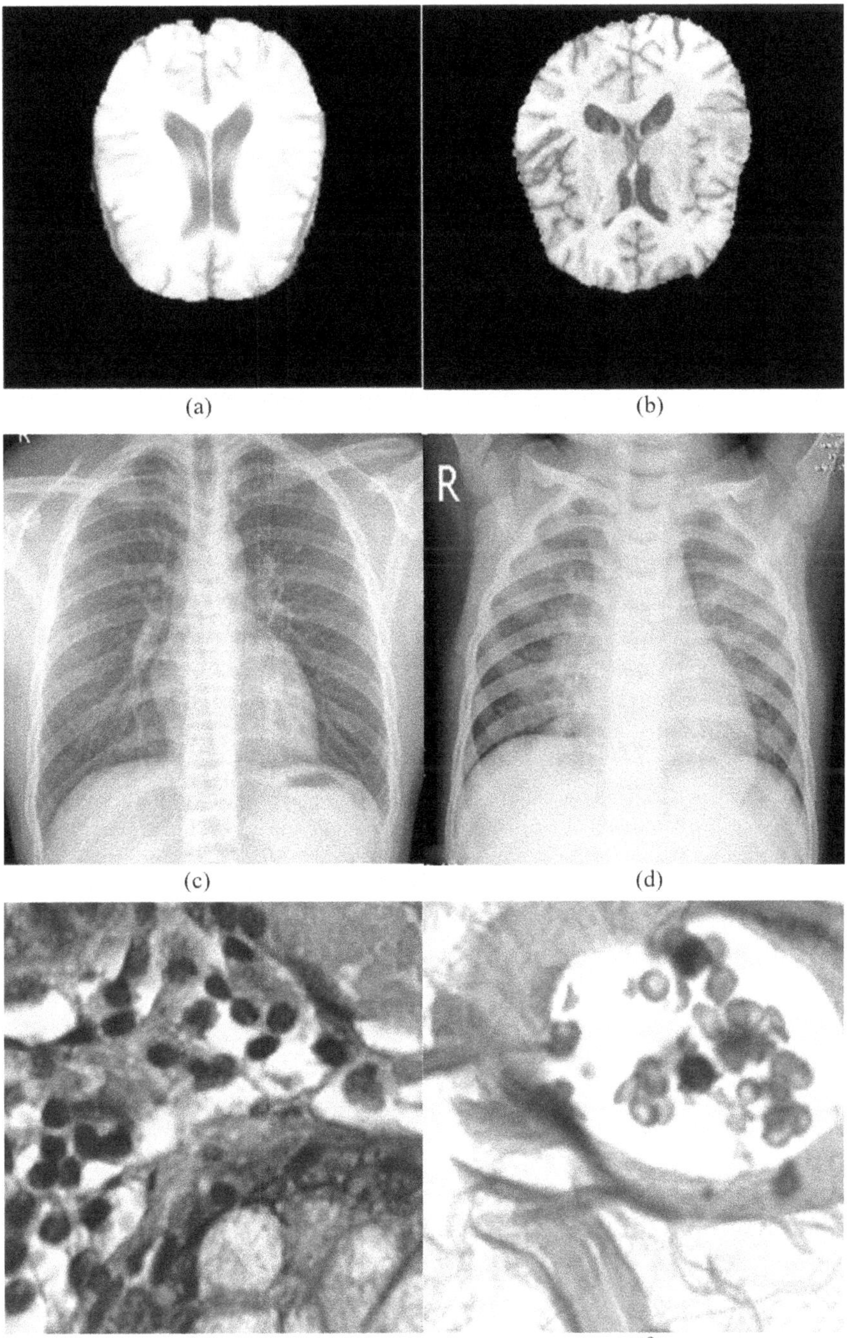

(a) (b) (c) (d) (e) (f)

FIGURE 19.3 The sample images from three datasets. (a) Normal brain MRI, (b) abnormal brain MRI, (c) healthy chest X-ray, (d) pneumonia-affected chest X-ray, (e) and (f) two different classes of human colorectal cancer, normal mucosal glands, and immune cells.

19.4 QUALITATIVE ANALYSIS

19.4.1 Evaluation of the Test Accuracy

We first conduct experiments to compare the clean test accuracies of all seven models on chest X-ray, histopathology, and BraTs18. The results are listed in Table 19.1. On the chest X-ray dataset, the convolutional networks LeNet (LeCun et al., 1998) and ResNet18 (He et al., 2016) have higher accuracies since they have the advantage of convolutions. On the histopathology, the MLP and random forest (Breiman, 2001) have higher accuracies. On BraTs18, the ResNet18, LeNet and random forest have higher accuracies, but other models are not too far behind.

19.4.2 Evaluation of the Defense Ability by L2 Distance

We compare the minimum distortion required to make an adversarial image on different models to evaluate the defense ability of adversarial attacks. The larger the value, the more robust the model since a significant distortion is likely to be detected in advance. Finding the exact minimum distortion is an NP-hard problem evaluated in ReLU-activated neural networks (Katz et al., 2017; Sinha et al., 2017) and tree ensemble classifiers (Kantchelian et al., 2016). Even the approximation of the minimum distortion in ReLU-activated neural networks is NP-hard (Weng et al., 2018).

The distortions reported by HopSkipJump have been shown to be lower (i.e., tighter and more accurate) than other boundary attack methods (Nicolae, et al., 2018). Therefore, we run the HopSkipJump boundary-based black-box attack (Chen et al., 2020) to determine the adversarial distortion of randomly selected images from the BraTs18, chest X-ray, and the colorectal cancer histopathology validation datasets.

The HopSkipJump is run ten times on each image to report the minimum value.

As shown in Figure 19.4, we observe that after 90 iterations, the distortions are minimum and become stable. Therefore, considering the best results and the computational ability, we pick 100 as the maximum iteration.

We quantitatively measure the robustness of defense by measuring the distance between normal and abnormal samples under the L2 metric as did for most attacks (Tramer and Boneh, 2019). The Lp distance is the difference between original examples and adversarial examples, defined as $\|d\|_p = (\sum_{i=0}^{n} |vi|^p)^{1/p}$. Common

TABLE 19.1
Average Accuracy of Validation Data on BraTs18, Chest X-ray, and Histopathology Image Datasets

	SCD01	SCDCE	SCDCEBNN	MLP	LeNet	ResNet18	Random Forest
BraTs18	98.38%	98.92%	95.31%	98.76%	99.1%	99.64%	99.07%
Chest X-ray	90.69%	91.32%	89.12%	88.72%	92.59%	94.32%	89.12%
Histopathology	99.2%	99.6%	99.6%	100%	99.6%	99.6%	100%

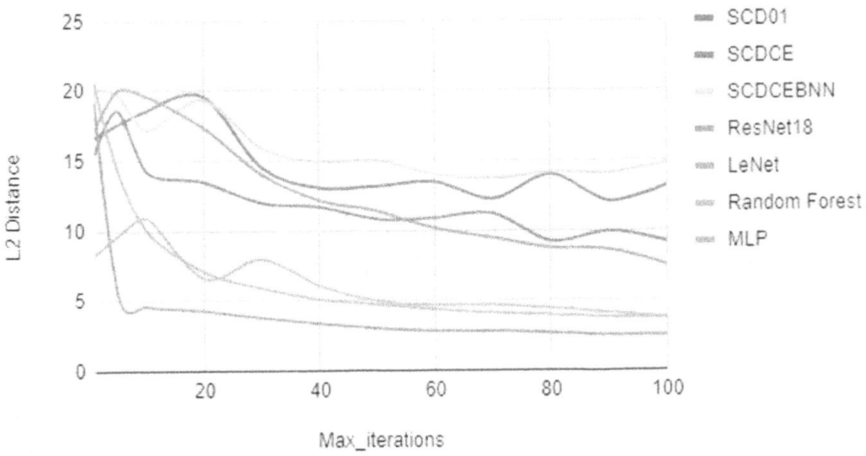

FIGURE 19.4 Variations of L2 distance on one image with different maximum iterations during HopSkipJump attack on different models.

TABLE 19.2
Average Minimum Estimated L2 Adversarial Distortion of on BraTs18 Datasets as Given by HopSkipJump When Attacking Different Models

	SCD01	SCDCE	SCDCEBNN	MLP	LeNet	ResNet18	Random Forest
Image 1	14.61	19.13	23.47	8.95	12.28	2.00	3.44
Image 2	10.55	13.44	16.18	4.32	9.06	1.95	4.03
Image 3	8.17	12.05	15.13	2.75	7.47	1.82	2.12
Image 4	7.49	13.00	3.33	3.67	7.50	2.50	3.33
Image 5	8.75	11.66	2.87	3.99	8.75	2.12	3.99
Average	9.06	12.23	13.7	4.38	8.27	2	2.78

choices of p include L0, a measure of the number of pixels changed; L2, the standard Euclidean norm; and L∞, a measure of the maximum absolute value change to any pixel. If the distortion under any of these three distance metrics is small, the images will likely appear visually similar.

Table 19.2 shows the average adversarial distortions of random test images from the BraTs18. The gradient-free trained sign networks have the higher distortions than other state-of-the-art models, and the SCDCEBNN has the highest distortion.

We plot the original and adversarial images of "Image 1" from BtaTs18 dataset in Figure 19.5 to get a visual feel for the distortions. The first six adversarial images have a high distortion, where (b)–(d) are the adversarial images from SCD models. Note that they have higher distortions than the currently available state-of-the-art models. Clearly, there are more noises than the original, while it is hard to observe the difference by human eyes in the other images.

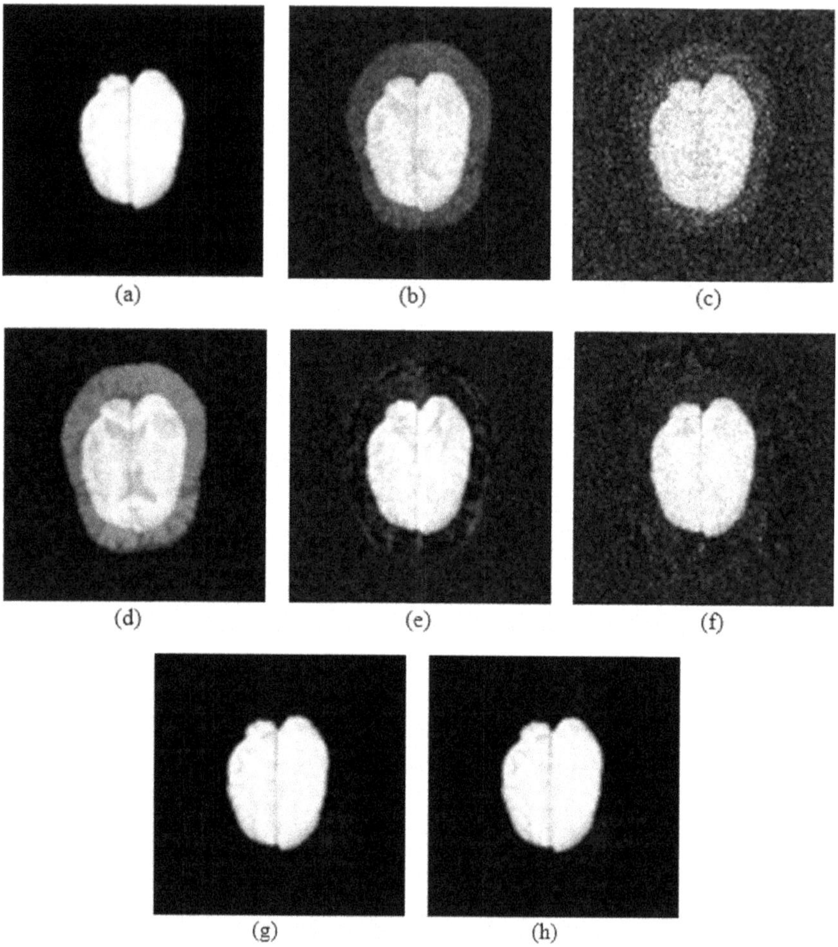

FIGURE 19.5 Visualization of original images and adversarial images among different networks from BraTs18 dataset. (a) The original image, (b) the adversarial example that will fool SCD01, (c) the adversarial example that will fool SCDCE, (d) the adversarial example that will fool SCDCEBNN, (e) the adversarial example that will fool MLP, (f) the adversarial example that will fool LeNet, (g) the adversarial example that will fool ResNet18, and (h) the adversarial example that will fool Random Forest.

Table 19.3 lists the average adversarial distortions of random test images from the Chest X-ray dataset, where MLP is the second best after SCDCE.

To get a visual feel for the distortions, Figure 19.6 shows the original and adversarial images of "Image 1" from Chest X-ray dataset. They all have higher distortions, among which SCDCE has the highest.

Table 19.4 lists the average adversarial distortions of random test images from the colorectal dataset. The average distortion of SCDCEBNN is highest.

Figure 19.7 shows the original image and adversarial images of human colorectal histopathology dataset which shows a visual feel for the distortions. All three SCD

TABLE 19.3
Average Minimum Estimated L2 Adversarial Distortion of on Chest X-ray Datasets as Given by HopSkipJump When Attacking Different Models

	SCD01	SCDCE	SCDCEBNN	MLP	LeNet	ResNet18	Random Forest
Image 1	10.59	18.40	17.50	14.78	4.28	1.03	18.53
Image 2	10.48	16.39	12.64	15.08	2.86	0.45	11.28
Image 3	9.00	17.55	10.50	14.49	4.19	0.64	9.18
Image 4	9.26	7.68	11.68	10.71	0.34	0.07	12.01
Image 5	7.24	14.02	10.16	12.91	4.10	0.43	2.39
Average	9.31	14.81	12.49	13.60	3.15	0.52	10.68

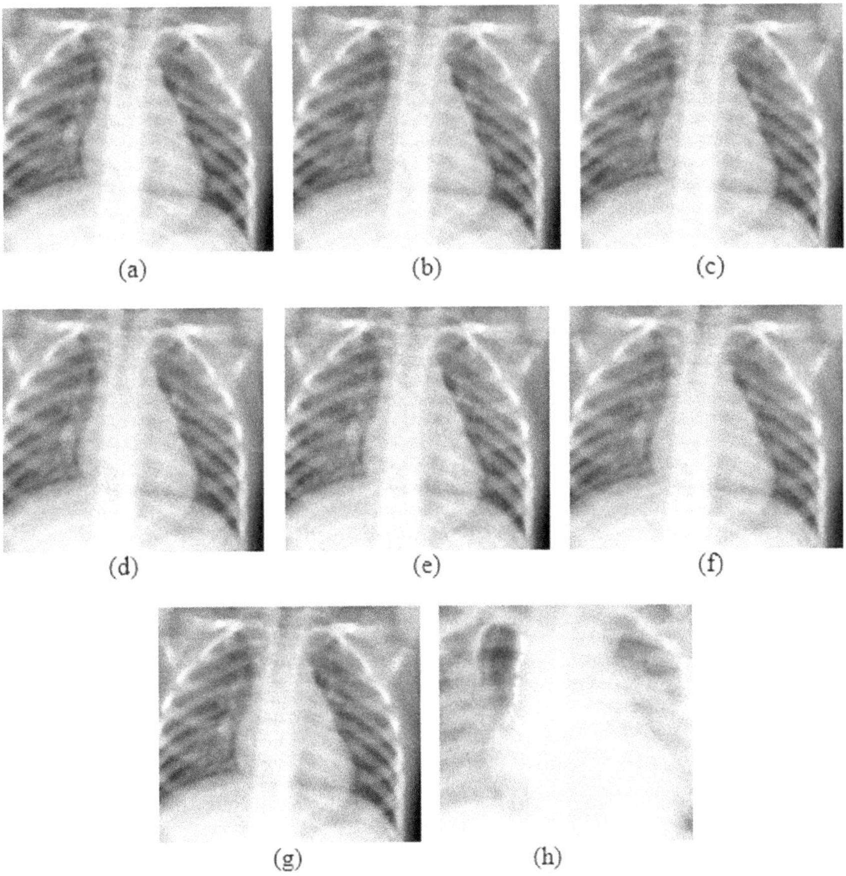

FIGURE 19.6 Visualizations of original images and adversarial images among different networks from Chest X-ray dataset. (a) The original image, (b) the adversarial example that will fool SCD01, (c) the adversarial example that will fool SCDCE, (d) the adversarial example that will fool SCDCEBNN, (e) the adversarial example that will fool MLP, (f) the adversarial example that will fool LeNet, (g) the adversarial example that will fool Resnet18, and (h) the adversarial example that will fool Random Forest.

TABLE 19.4
Average Minimum Estimated L2 Adversarial Distortion of on Colorectal Cancer Histopathology Datasets as Given by HopSkipJump When Attacking Different Models

	SCD01	SCDCE	SCDCEBNN	MLP	LeNet	ResNet18	Random Forest
Image 1	28.3	41	41.32	9.9	29	31.6	19.9
Image 2	4.4	6.3	9.2	2.8	7	6.2	3.9
Image 3	35.8	36.1	44.71	9.9	36.8	39.8	30.4
Image 4	30	38.6	43.02	12	24.1	19.1	28.7
Image 5	17.2	26.5	28.97	7.7	17.1	19	13.4
Average	24.1	29.7	33.44	8.5	22.8	23.1	19.2

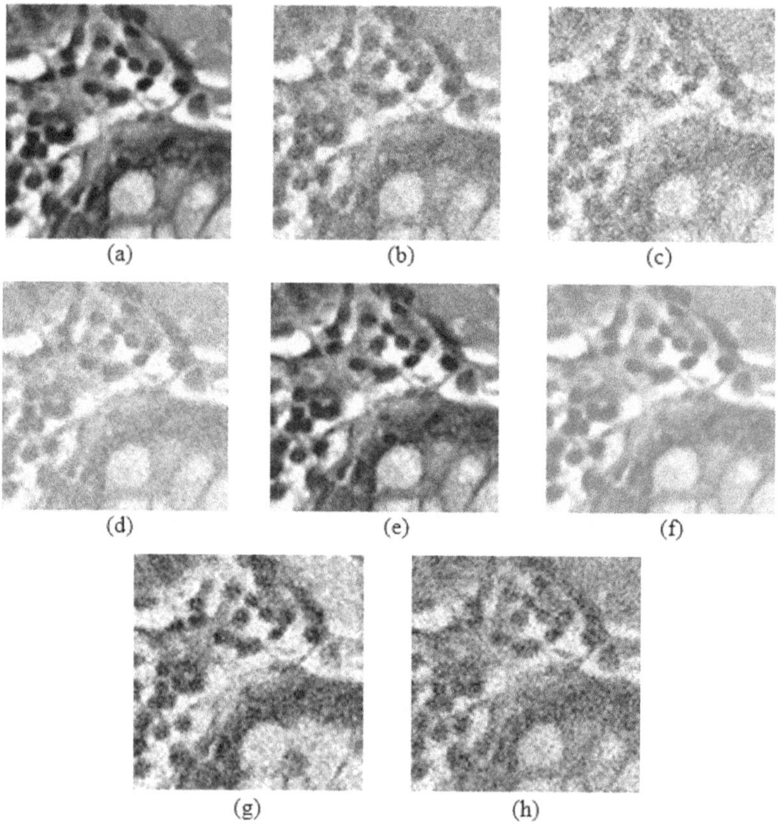

FIGURE 19.7 Visualizations of original images and adversarial images among different networks from colorectal histopathology dataset. (a) The original image, (b) the adversarial example that will fool SCD01, (c) the adversarial example that will fool SCDCE, (d) the adversarial example that will fool SCDCEBNN, (e) the adversarial example that will fool MLP, (c) the adversarial example that will fool LeNet, (e) the adversarial example that will fool ResNet18, (f) the adversarial example that will fool Random Forest.

models have higher distortions than other models. Compared with other adversarial images, SCDCE adversarial image is full of more colorful ditties. It is hard to identify whether the morphology is potentially abnormal.

Table 19.5 lists the average minimum estimated L2 adversarial distortion on all three datasets. The distortions of the SCD models are even higher with SCDCEBNN taking the lead and twice better than all other models.

19.4.3 Evaluation of Defense Ability by Transferability

Another evaluation is to make use of the transferability property (Hirano et al., 2021). Given two models, $F(\cdot)$ and $G(\cdot)$, an adversarial example trained on F will be an adversarial example on G, even if they are trained in completely different manners or on different datasets. There has been a significant amount of available methods to construct adversarial examples (Carlini and Wagner, 2017; Biggio et al., 2013; Moosavi-Dezfooli et al., 2016; Papernot et al., 2016; Szegedy et al., 2013) and to make networks robust against adversarial examples (Papernot et al., 2016; Rozsa et al., 2016; Shaham et al., 2018; Zheng et al., 2016). No defenses have been able to classify adversarial examples correctly. Thus, correctly classifying adversarial examples is difficult.

In the previous section, we showed how the attackers generated adversarial samples on different models; here, we test all these adversarial examples on all models. If a model G can detect the adversarial examples from another model F and classify them correctly, the model G is more robust against adversarial attack. Table 19.6

TABLE 19.5
Average Minimum Estimated L2 Adversarial Distortion on All Three Datasets

	SCD01	SCDCE	SCDCEBNN	MLP	LeNet	ResNet18	Random Forest
Average	13.67	18.26	19.12	8.59	10.96	8.43	10.74

TABLE 19.6
Results for Classifying One Random Image and All Adversarial Examples

	SCD01	SCDCE	SCDCEBNN	MLP	LeNet	ResNet18	Random Forest
Original Test Image	Y	Y	Y	Y	Y	Y	Y
Adversarial Image from SCD01	–	Y	Y	Y	Y	N	Y
Adversarial Image from SCDCE	Y	–	Y	N	Y	N	Y
Adversarial Image from SCDCEBNN	Y	Y	–	Y	N	N	N
Adversarial Image from MLP	Y	Y	Y	–	Y	N	Y
Adversarial Image from LeNet	Y	Y	Y	Y	–	N	Y
Adversarial Image from ResNet18	Y	Y	Y	Y	Y	–	Y
Adversarial Image from Random Forest	Y	Y	Y	Y	Y	N	–

TABLE 19.7
Average Accuracy of All Models When Classifying the Adversarial Examples

	SCD01	SCDCE	SCDCEBNN	MLP	LeNet	ResNet18	Random Forest
Average Accuracy	88.89%	88.89%	85.19%	81.48%	81.48%	38.89%	57.14%

shows the results for classifying one random image and all adversarial examples. We can see that a random image can be classified by all models, which is marked as "Y." The targeted adversarial samples are misclassified by their targeting models, respectively. If the model can identify the adversarial example correctly, it is marked as "Y"; otherwise, it is marked as "N." Our models SCD01, SCDCE, and SCDCEBNN can detect all adversarial examples and classify them correctly.

Table 19.7 shows the average accuracy of all models when classifying the adversarial examples. We can see that our proposed models have higher accuracies, which are 88.89% and 85.19%. They can identify fake examples and are hard to be fooled by adversarial attacks. In other models like MLP and LeNet are the best, but their accuracies are still lower than our proposed models.

19.5 CONCLUSIONS

In this chapter, we present a model that is robust to adversarial attacks in MRI images, chest X-ray, and histopathology images. We show that higher distortions are required when adversarial attacking is applied on the gradient-free trained sign networks with SCD compared with state-of-the-art models. Experimental results on classifying the adversarial samples show that our models' accuracy is more competitive, and thus, the adversarial attack can easily be detected on our models. To develop a robust medical machine-learning models which can deter attack in advance, more research is required to verify the results on a larger cohort and show the results on the different adversarial attack, such as white box. We plan to develop a medical AI imaging system that can detect and deter adversarial attack in advance in the future work.

REFERENCES

Bakes, S., Reyes, M., Jakab, A., Bauer, S., Rempfler, M., and Crimi, A., "Identifying the best machine learning algorithms for brain tumor segmentation, progression assessment, and overall survival prediction in the BRATS challenge," arXiv:1811.02629, 2018.

Biggio, B., Corona, I., Maiorca, D., Nelson, B., Šrndić, N., Laskov, P., Giacinto, G., and Roli, F., "Evasion attacks against machine learning at test time," *Proceedings of the Joint European Conference on Machine Learning and Knowledge Discovery in Databases*, Würzburg, pp. 387–402, Sept. 2013.

Breiman, L., "Random forests," *Machine Learning*, vol. 45, pp. 5–32, 2001.

Brendel, W., Rauber, J., and Bethge, M., "Decision-based adversarial attacks: Reliable attacks against black-box machine learning models," arXiv:1712.04248, 2017.

Carlini, N. and Wagner, D., "Towards evaluating the robustness of neural networks," *Proceedings of the IEEE Symposium on Security and Privacy*, San Jose, CA, pp. 39–57, Mar. 2017.

Chen, J., Jordan, M. I., and Wainwright, M. J., "Hopskipjump attack: A query-efficient decision-based attack," *Proceedings of the IEEE Symposium on Security and Privacy*, San Francisco, CA, pp. 1277–1294, May 2020.

Dong, Y., Pang, T., Su, H., and Zhu, J., "Evading defenses to transferable adversarial examples by translation-invariant attacks," *Proceedings of the IEEE Conference on Computer Vision and Pattern Recognition*, Long Beach, CA, pp. 4312–4321, Jun. 2019.

Finlayson, S. G., Bowers, J. D., Ito, J., Zittrain, J. L., Beam, A. L., and Kohane, I. S., "Adversarial attacks on medical machine learning," *Science*, vol. 363, no. 6433, pp. 1287–1289, Mar. 2019.

Goodfellow, I. J., Shlens, J., and Szegedy, C., "Explaining and harnessing adversarial examples," *Proceedings of the 3rd International Conference on Learning Representations*, San Diego, CA, ICLR, May 2015.

He, K., Zhang, X., Ren, S., and Sun, J., "Deep residual learning for image recognition," *Proceedings of the IEEE Conference on Computer Vision and Pattern Recognition*, Las Vegas, NV, pp. 770–778, Jun. 2016.

Hirano, H., Minagi, A., Soudry, D., and Takemoto, K., "Universal adversarial attacks on deep neural networks for medical image classification," *BMC Medical Imaging*, pp. 1–13, Jan. 2021.

Kantchelian, A., Tygar, J. D., and Joseph, A., "Evasion and hardening of tree ensemble classifiers," *Proceedings of the International Conference on Machine Learning*, New York City, vol. 48, pp. 2387–2396, Jun. 2016.

Kather, J. N., Weis, C.-A., Bianconi, F., Melchers, S. M., Schad, L. R., Gaiser, T., Marx A., and Zöllner, F. G., "Multi-class texture analysis in colorectal cancer histology," *Scientific Reports*, Rep. 6, p. 27988, Mar. 2016.

Katz, G., Barrett, C., Dill, D. L., Julian, K., and Kochenderfer, J., "Reluplex: An efficient SMT solver for verifying deep neural networks," *Proceedings of the International Conference on Computer Aided Verification*, Heidelberg, pp. 97–117, Jul. 2017.

LeCun, Y., Bottou, L., Bengio, Y., and Haffner, P., "Gradient-based learning applied to document recognition," *Proceedings of the IEEE*, vol. 86, no.11, pp. 2278–2324, 1998.

Moosavi-Dezfooli, S. M., Fawzi, A., and Frossard, P., "Deepfool: A simple and accurate method to fool deep neural networks," *Proceedings of the IEEE Conference on Computer Vision and Pattern Recognition*, Las Vegas, NV, pp. 2574–2582, Jun. 2016.

Nicolae, M., Sinn, M., Tran, M., Buesser, B., Rawat, A., Wistuba, M., Zantedeschi, V., Baracaldo, Chen, N. B., Ludwig, H., Molloy, I. M., and Edwards, B., "Adversarial robustness toolbox," v1.0.0. arXiv:1807.01069, 2018.

Panda, P., Chakraborty, I., and Roy, K., "Discretization based solutions for secure machine learning against adversarial attacks," *IEEE Access*, vol. 7, pp. 70157–70168, 2019.

Papernot, N., McDaniel, P., and Goodfellow, I., "Transferability in machine learning: From phenomena to black-box attacks using adversarial samples," arXiv:1605.07277, 2016a.

Papernot, N., McDaniel, P., Jha, S., Fredrikson, M., Celik, Z. B., and Swami, A., "The limitations of deep learning in adversarial settings," *Proceedings of the IEEE European Symposium on Security and Privacy*, Saarbrücken, pp. 372–387, Mar. 2016b.

Papernot, N., McDaniel, P., Wu, X., Jha, S., and Swami, A., "Distillation as a defense to adversarial perturbations against deep neural networks," *Proceedings of the IEEE Symposium on Security and Privacy*, San Jose, CA, pp. 582–597, May 2016c.

Paschali, M., Conjeti, S., Navarro, F., and Navab, N., "Generalizability vs. robustness: Investigating medical imaging networks using adversarial examples," *Proceedings of the Conference on Medical Image Computing and Computer Assisted Intervention*, Granada, pp. 493–501, Sept. 2018.

Paszke, A., Gross, S., Massa, F., Lerer, A., Bradbury, J., Chanan, G., and Killeen, T., "PyTorch: An imperative style, high-performance deep learning library," *Proceedings of the Advances in Neural Information Processing Systems*, Vancouver, pp. 8024–8035, Dec. 2019.

Rozsa, A., Rudd, E. M., and Boult, T. E., "Adversarial diversity and hard positive generation," *Proceedings of the IEEE Conference on Computer Vision and Pattern Recognition*, Las Vegas, NV, pp. 410–417, Jun. 2016.

Shaham, U., Yamada, Y., and Negahban, S., "Understanding adversarial training: Increasing local stability of supervised models through robust optimization," *Neurocomputing*, vol. 307, pp. 195–204, Sept. 2018.

Sinha, A., Namkoong, H., and Duchi, J., "Certifiable distributional robustness with principled adversarial training," arXiv:1710.10571, 2017.

Szegedy, C., Zaremba, W., Sutskever, I., Bruna, J., Erhan, D., Goodfellow, I., and Fergus, R., "Intriguing properties of neural networks," arXiv:1312.6199, 2013.

Tramer, F. and Boneh, D., "Adversarial training and robustness for multiple perturbations," *Proceedings of the Advances in Neural Information Processing Systems*, Vancouver, pp. 5858–5868, Dec. 2019.

Tramèr, F., Kurakin, A., Papernot, N., Goodfellow, I., Boneh, D., and McDaniel, P., "Ensemble adversarial training: Attacks and defenses," *Proceedings of the 6th International Conference on Learning Representations*, Vancouver, BC, May 2018.

Weng, T., Zhang, H., Chen, H., Song, Z., Hsieh, C., Boning, D., Dhillon, I. S., and Daniel, L., "Towards fast computation of certified robustness for ReLU networks," arXiv:1804.09699, 2018.

Wetstein, S. C., González-Gonzalo, C., and Bortsova, G, "Adversarial attack vulnerability of medical image analysis systems: Unexplored factors," arXiv:2006.06356, 2020.

Xie, M., Xue, Y., and Roshan, U., "Stochastic coordinate descent for 0/1 loss and its sensitivity to adversarial attacks," *Proceedings of the 18th IEEE International Conference on Machine Learning and Applications*, pp. 299–304, 2019.

Xue, Y., Xie, M., and Roshan, U., "On the transferability of adversarial examples between convex and 01 loss models," *Proceedings of the IEEE International Conference on Machine Learning and Applications*, Miami, FL, 2020a.

Xue, Y., Xie, M., and Roshan, U., "Towards adversarial robustness with 01 loss neural networks," *Proceedings of the 19th IEEE International Conference on Machine Learning and Applications*, Miami, FL, 2020b.

Yang, Z., Yang, Y., Xue, Y., Shih, F. Y., Ady, J., and Roshan, U., "Accurate and adversarially robust classification of medical images and ECG time-series with gradient-free trained sign activation neural networks," *Proceedings of the IEEE International Conference on Bioinformatics and Biomedicine*, Seoul, pp. 2456–2460, 2020.

Yeh, C., Lin, M., Chang, P., and Kang, L., "Enhanced visual attention-guided deep neural networks for image classification," *IEEE Access*, vol. 8, pp. 163447–163457, 2020.

Zheng, S., Song, Y., Leung, T., and Goodfellow, I., "Improving the robustness of deep neural networks via stability training," *Proceedings of the IEEE Conference on Computer Vision and Pattern Recognition*, Las Vegas, NV, pp. 4480–4488, Jun. 2016.

20 Adaptive Image Reconstruction for Defense against Adversarial Attacks

20.1 INTRODUCTION

Deep neural networks (DNNs) have been adopted to achieve high accuracies in many tasks (Yeh et al., 2020), even better than human beings. Unfortunately, recent researches have been shown that adversarial attacks can force DNN networks to make a wrong decision by adding slight perturbations to the input data (Papernot et al., 2016b, 2016c; Carlini and Wagner, 2017; Panda et al., 2019). The success of adversarial attacks leads to security threats; thus, the robustness of deep-learning algorithms needs to be reevaluated before deploying them in the real world. It is crucial to ensure that neural networks can detect abnormal inputs in a safer and more secure way.

Recent studies (Goodfellow et al., 2015; Moosavi-Dezfooli et al., 2016; Szegedy et al., 2014) have demonstrated that the DNN-based image classifiers can misclassify the adversarial examples well-crafted by adversarial attacks. For instance, Goodfellow et al. (2015) showed that an image with perturbations added can be misclassified with very high confidence by GoogLeNet (Szegedy et al., 2015). However, a human observer can still correctly classify it without noticing the existence of the introduced perturbations. Papernot et al. (2017) showed that a crafted stop sign is incorrectly classified as a yield sign. All of these studies indicate that the adversaries can potentially use the crafted images to inflict severe damages.

To defend against adversarial attacks, some researchers proposed the techniques of improving model robustness and detecting malicious behaviors. Most of them require modifying the target model. One of the most straightforward methods is called *adversarial training*, which uses as many adversarial examples as possible to retrain the network and improve classification accuracy (Wang et al., 2019; Shi and Han, 2018; Papernot et al., 2016a). Papernot et al. (2016b) introduced another defense technique, named *defensive distillation*, to train two networks as a distillation. The first network produces probability vectors to label the original dataset, while the newly labeled dataset is used to train the other network.

Recent studies have been focused on detecting adversarial examples directly. Xu et al. (2018) proposed training the detection modules using a set of adversarial examples and the corresponding benign ones. However, this technique requires modifying the model and a sufficient number of adversarial examples.

DOI: 10.1201/9781003474395-23

Although the researchers have shown certain achievements in defending against adversarial attacks, the cost of training time and the computational complexity to retrain the models and modify the network architectures are impractical. Generating appropriate adversarial examples for training and statistical testing is quite expensive and depends on a comprehensive prior knowledge of various potential adversarial techniques. Even worse, attackers can compose adversarial examples by the revamped attack algorithms, usually unknown to the defender in advance. In this way, there is a good chance for the adversarial examples to evade the classifier. Moreover, most of the existing defense techniques are model-specific. Rebuilding or retraining a classifier would consume extra significant time. Therefore, attackers can easily craft efficacious fake examples.

Other methods to defend against adversarial attacks intend to reconstruct the input images. Liao et al. (2018) proposed a pixel denoiser method to remove the noise on high-level representation. Xie et al. (2019) developed a feature denoising architecture to smooth input images by applying non-local means or other filters directly on the feature level. Zhang et al. (2021) proposed an image reconstruction network based on residual block structures to suppress the noise. Zhao et al. (2021) divided pixels into low-sensitive and high-sensitive groups based on the contributions to image classification, and then developed a structure-preserving low-rank image completion method on high-sensitive points to remove noise.

However, there exist a problem with these techniques. Reconstruction and denoise methods apply a smoothing filter on the input images to reduce noise, but inevitably removing some details on the object's interest. Different adversarial attackers magnify different scales of perturbations even on the same images. The perturbations on a compelling adversarial image from the semantic and grayscale datasets are very different for attackers. In other words, using the same parameters on an adversarial attack to generate an adversarial sample on ImageNet images cannot generate a successful one from MNIST database (Modified National Institute of Standards and Technology database). As a result, using the same reconstructed strategy may overly reduce the noise and lead to a new misclassification.

To address the aforementioned challenges, we develop a novel scheme that reconstructs the input image effectively and detects the adversarial samples in advance by an adaptive method to avoid excessive noise reduction. The framework is lightweight, so it can be assembled into any network, and only a few arguments need to be set up. The rest of this chapter is organized as follows. Section 20.2 describes the adversarial perturbations and removal. Section 20.3 presents the proposed method. Section 20.4 shows our experimental results. Finally, conclusions are drawn in Section 20.5.

20.2 THE ADVERSARIAL PERTURBATIONS AND REMOVAL

20.2.1 The Perturbations from Adversarial Attacks

We observe that the difference between the original and the crafted adversarial images looks like noise in visualization. In Figure 20.1, the first column shows the clean images, the middle column shows the adversarial images generated by the

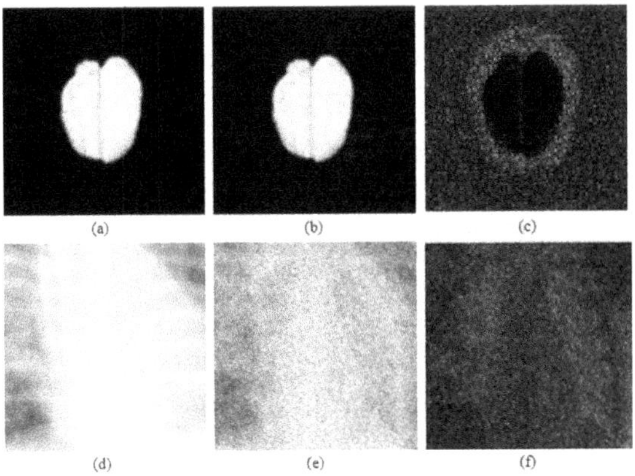

FIGURE 20.1 The perturbations on the image. (a) The clean image from BraTs18, (b) the adversarial image crafted by HopSkipJump black-box attack on the ResNet18 model, (c) the noise added by the attacker, (d) the clean image from a chest X-ray, (e) the adversarial image crafted by HopSkipJump black-box attack on the ResNet18 model, and (f) the noise added by the attacker.

HopSkipHump black-box attack (Yang et al., 2022) on the ResNet18 (He et al., 2016) model, and the last column shows the noise added by the attacker. It suggests that adversarial perturbations are difficult to detect by human eyes at the pixel level but lead to substantial noise at feature levels.

The perturbations from the adversarial attack are added layer by layer until the classifier makes a wrong decision. Different datasets require different scale perturbations even trained by the same model and attacker. For the same images trained on the same model, the effective fake samples are different with various attackers. We can observe that the noise added on the two different datasets in Figures 20.1(c) and 20.1(f) are distinct, even though they are both crafted by the HopSkipJump attack on the ResNet18 model.

20.2.2 Perturbation Removal

Let G denote an adversarial attacker who generates a fake sample $I'(x, y)$ to fool the model F by adding perturbation $\varepsilon(x, y)$ on the original image $I(x, y)$ as

$$I'(x, y) = I(x, y) + \varepsilon(x, y) \tag{20.1}$$

where x and y denote the spatial coordinates. Note that the perturbation $\varepsilon(x, y)$ varies largely among attacker G, classifier F, and the clean input I. Assume that the model F can correctly classify any clean instance. By applying the reconstruction techniques to remove $\varepsilon(x, y)$, we can generate a reconstructed image $I''(x, y)$, whose classified result is identical to the clean image I as

$$F(I) = F(I'') \tag{20.2}$$

Note that it is impossible to apply a single technique to remove all different perturbations. Image reconstruction techniques can reduce some noise, but over denoising happens since the noise are different among variant cases. Therefore, the reconstructed image $I''(x, y)$ is a higher-quality image than the adversarial one $I'(x, y)$ but a lower quality than the original clean image $I(x, y)$. Our goal is to minimize the difference $d(x, y)$ as

$$d(x, y) = I(x, y) - I''(x, y) \tag{20.3}$$

In this way, the model F has a higher chance to make the correct decision. Therefore, the challenge is to remove as many perturbations $\varepsilon(x, y)$ as possible when an unknown input is given with added unknown noise.

20.2.3 Entropy Values

To analyze the internal information characteristics, we use Shannon entropy (Tsai et al., 2008; Cooper and Miller, 2000) to measure the uncertain distributions and complexity features in images. If the entropy is high, it means the image includes more information. Let the gray level k have a probability P_k. The entropy H is calculated as

$$H = -\sum_k P_k * log_2(P_k) \tag{20.4}$$

Let

$$k = \sum_{(x,y)} k(x, y) \tag{20.5}$$

At the pixel $k(x, y)$, the probability of this pixel is $P_{xy} = k(x, y)/k$. Therefore, we have

$$H(x, y) = log_2(k) - \frac{1}{k}\sum_{(x,y)} k(x, y) * log_2(k(x, y)) \tag{20.6}$$

The minimum entropy value of zero is occurred when the image pixel value is constant at any location. The maximum entropy value is related to the total number of grayscales. For instance, the maximum entropy of an image with 256 gray levels is $log_2(256) = 8$. Figure 20.2 shows the entropy values among different images, where (a) has a smaller entropy than (c), indicating it contains less interesting information. Note that the difference between the clean and adversarial images of (a) and (b) is larger than that of (c) and (d). We can conclude that the simpler images need larger perturbations to produce successfully adversarial samples. Because the entropy values of simple images are usually small, the greater noise have more possibilities to impact the interest of the images, so it will increase the differences of entropy values between original images and fake ones. In contrast, color images have lager entropy values, and therefore, they usually request tiny noise to craft the successful adversarial images.

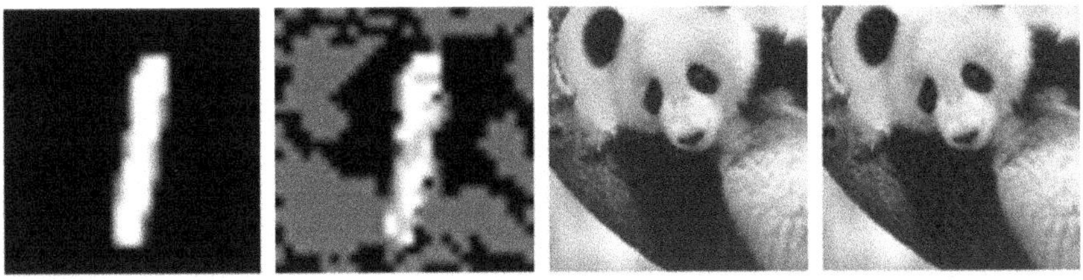

(a) (b) (c) (d)

FIGURE 20.2 (a) The clean image with entropy of 1.5222, (b) the adversarial image crafted by FGSM attacker with entropy of 5.4241, (c) the clean image from ImageNet with entropy of 4.9742, and (d) the adversarial image crafted by FGSM attacker with entropy of 7.5302.

20.2.4 Adaptive Smoothing

Image smoothing intends to reduce and suppress image noise. We use the average smoothing filter as shown in Figure 20.3, where (a) is a square mask of 3 × 3 and (b) is a plus-shape mask of 5 × 5.

The Gaussian template is another method to reduce the blur in the smoothing process and obtain a more natural smoothing effect. The average smoothing treats the same weight to all the pixels in the neighborhood. Therefore, it is natural to think about increasing the weight of the neighbors close to center and reducing the weight of distant neighbors. Figure 20.4 shows a 3 × 3 Gaussian template.

Increasing filter size to reduce the noise is an approach to remove as many noises as possible, but this method also causes the image over blurring. We use the adaptive Wiener filter to reconstruct data by removing noise without significantly blurring the structures in the image. The neighborhood $a(n1, n2)$ is used to estimate local mean μ and variance σ^2 as

$$\mu = \frac{1}{n \times m} \sum_{n1,n2 \in x} a(n1, n2) \tag{20.7}$$

$$\sigma^2 = \frac{1}{n \times m} \sum_{n1,n2 \in x} (a(n1, n2) - \mu)^2 \tag{20.8}$$

where x is the $n \times m$ local neighborhood of each pixel. The pixelwise filter to suppress noise is represented as

$$b(n1, n2) = \mu + \frac{\sigma^2 - \gamma^2}{\sigma^2} \left(a(n1, n2) - \mu\right) \tag{20.9}$$

where γ^2 is the noise variance, which is default as the average of all local variance in equation (20.8).

1/9 *

1	1	1
1	1	1
1	1	1

(a)

1/9 *

0	0	1	0	0
0	0	1	0	0
1	1	1	1	1
0	0	1	0	0
0	0	1	0	0

(b)

FIGURE 20.3 Different average smoothing filter masks. (a) A square mask of 3 × 3 and (b) a plus-shape mask of size 5 × 5.

1/16 *

1	2	1
2	4	2
1	2	1

FIGURE 20.4 The 3 × 3 Gaussian mask template.

20.3 THE PROPOSED METHOD

20.3.1 The Architecture

Figure 20.5 shows the proposed framework. If the classified results of the original image $X(x, y)$ and the related reconstructed image $X''(x, y)$ are the same, we say $X(x, y)$ is a clean image; otherwise, it is an adversarial sample.

Two factors decide the reliability of this detection scheme. One is the robustness of model F. We previously presented in (Yang et al., 2020, 2022) that the ability to defend against adversarial attacks differs vastly on different models. For example, LeNet has a higher tolerant ability than ResNet18 (Zhang et al., 2021), but is similar to MLP classifier. This means the reconstructed image $X''(x, y)$ cannot be classified by ResNet18, but may work on LeNet. The other is the reconstructed technique, such that the perturbations added on the clean images are random, and all the removed noise made by pre-processing methods are different in different scenarios.

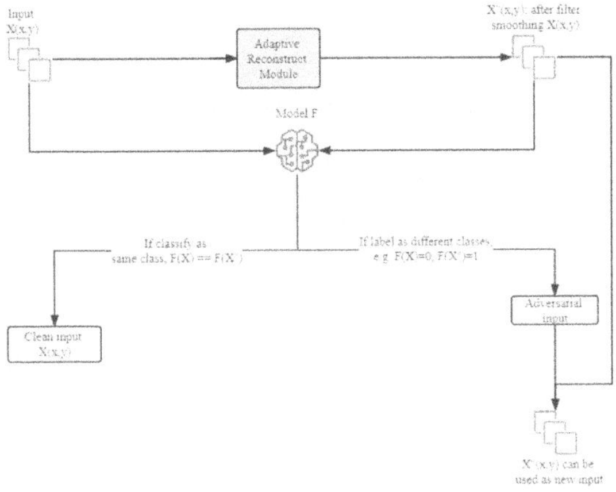

FIGURE 20.5 Detect adversarial samples by comparing the original classified result and the reconstructed one.

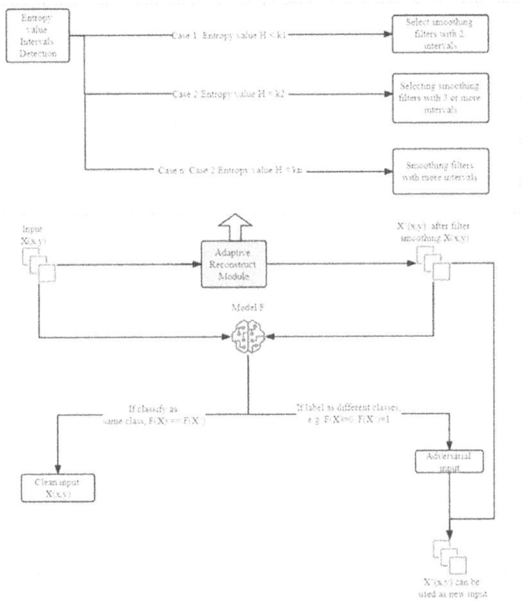

FIGURE 20.6 The architecture of adaptive reconstruction module. Computing and training the entropy values are divided into different intervals. In different cases, the module will apply different filters to reconstruct the images.

We propose a novel reconstruction module with an adaptive process as shown in Figure 20.6, which applies variant smoothing filters on different inputs respectively. The images of high entropy values require few distortions, and those of low entropy values need large perturbations. Based on different entropy values, we can decide if the smoothing filters should be applied or what kind of filters can be applied.

20.3.2 Parameters Setting

As discussed before, the filters may excessively smooth while reconstructing the images, and the models may return a new misclassification. To solve this problem, we adopt the uniform method to divide the entropy value into intervals evenly, and each interval shares the same size. Entropy values decide the quantities of intervals. The following adaptive filter selection algorithm is developed to select a filter for the image with different entropy values.

Adaptive Filter Selection Algorithm

Input: H: entropy value; k: intervals; F: initiate spatial filter; $I(x, y)$: original image input; F_i: another filter; i: the quantities of filters predefined; n: entropy value; $m1$: interval number; $m2$: interval number
Output: $I'(x, y)$: the smoothed image

```
if H < n and k == m1,
    return I(x, y)
if H < n+1 and k == m2,
    return I(x, y)
for all i filters:
    if abs(I(x, y) - F(I(x, y))) <= abs(I(x, y) - F_i(I (x, y))),
        return I'(x, y) = F(I(x, y))
    otherwise, return I'(x, y) = F_i(I (x, y))
end
```

The proposed algorithm is tested on the MINIST and ImageNet-subset (i.e., two classes) images to set the interval size threshold values. Then, we use FGSM (Fast Gradient Sign Method) (Goodfellow et al., 2015) and PGD (Projected Gradient Descent) (Madry et al., 2018) to attack ResNet18 and craft adversarial images. The PGD is a well-known attack method for adversarial training. The FGSM has been proven that it has higher training efficiency and better performance than the PGD. After applying the standard convolution spatial filer of size 5 × 5, the entropy values of the MNIST dataset are smaller than 4, and those of the ImageNet are larger than 6.

We test different interval numbers and use the MNIST and ImageNet-subset as the experimental baseline. Since the entropy values of MNIST are all smaller than 4, and those of ImageNet are typical more extensive entropy value set; i.e., all are greater than 6. As shown in Table 20.1, the accuracy of MNIST is continuously high. Although it is improved with increasing intervals but not over 1%, we can conclude that two intervals are good enough. The accuracy of ImageNet-subset is enhanced when the number of intervals starts from six.

TABLE 20.1
The Accuracy of MNIST and ImageNet-Subset under Different Numbers of Intervals

	The Number of Intervals						
Dataset	**2**	**3**	**4**	**5**	**6**	**7**	**8**
MNIST	97.14%	97.28%	97.42%	97.57%	97.64%	97.65%	97.71%
ImageNet-subset (2 classes)	54.64%	78.05%	81.96%	85.86%	92.10%	92.88%	92.96%

20.4 EXPERIMENTAL RESULTS

In this section, we present our experiments on the ImageNet-subset classification dataset and three popular medical imaging datasets – BraTs18, Chest X-rays, and Colorectal Histopathology – to evaluate the proposed algorithm.

20.4.1 Datasets

20.4.1.1 BraTs18

The BratS18 dataset is 210 high-grade glioma (HGG) and 75 low-grade glioma (LGG) MRI with binary masks for the tumor. Each 3D MRI contains 155 slices of size 240 × 240. We use the FLAIR modality images for all the experiments because the entire tumor is represented well by this modality. In total, we have 17,100 abnormal and 18,500 normal images for training. For testing, we have 1,800 abnormal and 1,900 normal images. We show more experimental results on other modalities, where ANT-GAN presents impressive synthesis quality. A more detailed medical description of the data can be referred to (Yang et al., 2022). We down-sample two classes to be a balanced dataset, and each class contains 1,462 resized images of 96 × 96. The training and testing datasets are split into a ratio of 80:20.

20.4.1.2 Chest X-ray

The chest X-ray images (anterior-posterior) are selected from retrospective cohorts of pediatric patients of one to five years old from Guangzhou Women and Children's Medical Center, Guangzhou, China. All chest X-ray imaging was performed as part of patients' routine clinical care. The dataset is organized into two folders (training and testing) and contains subfolders for each image category (pneumonia/normal). There are 5,863 X-ray images and 2 categories (pneumonia/normal). All chest radiographs are initially screened for quality control by removing all low-quality or unreadable scans. Two expert physicians then grade the diagnoses for the images before being cleared for training the AI system. A third expert checks the evaluation set to account for any grading errors. We resize the images to 96 × 96 and down-sample to 1,584 for each category as the balanced dataset. We have 3,168 images in total, which are split into training and testing sets by a ratio of 80:20.

20.4.1.3 Colorectal Histopathology

This dataset represents a collection of textures in histological images of human colorectal cancer. Ten anonymized H&E stained CRC tissue slides are obtained from the pathology archive at the University Medical Center Mannheim, Heidelberg University, Mannheim, Germany. The low-grade and high-grade tumors are included in this set, and no further selection is applied. The slides are first digitized, and then the contiguous tissue areas are manually annotated and tessellated to create 625 non-overlapping tissue tiles of size $150 * 150$ ($74\,\mu m * 74\,\mu m$). Thus, the texture features of different scales are included, ranging from individual cells (approximate ten μm) to larger structures such as mucosal glands ($> 50\,\mu m$).

The following eight tissue types are selected for analysis: tumor epithelium, simple stroma, complex stroma, immune cells, debris, normal mucosal gland, adipose tissue, and background (no tissue). Together, the resulting 5,000 images represent the training and testing sets. We randomly pick two classes, immune cells and normal mucosal glands, resize them to 96 × 96, and split training and testing sets with a ratio of 80:20.

20.4.2 Qualitative Analysis

Two networks, ResNet18 and SCD01MLP, are used to evaluate the defense performance. We previously showed in Yang et al. (2022) that ResNet18 has the worst defense ability, while SCD01MLP is the best.

20.4.2.1 Evaluation against White-Box Attacks

We train both networks on the clean datasets, and the classification accuracies are all higher than 90%. Because generating the PGD adversarial examples does not work on non-convex SCD models (Yang et al., 2022), we will not consider the detecting samples from SCD (stochastic coordinate descent) models. Table 20.2 evaluates the accuracy of ResNet18 on detecting white-box PGD attack examples. We apply different filters and obtain that the adaptive smoothing filter works better. Table 20.3 shows the accuracy of SCD01MLP on detecting white-box PGD attack examples. The adaptive filters work better on BraTS18 datasets, but the Gaussian smoothing filters perform better on human colorectal histopathology images.

TABLE 20.2
Performance of ResNet18 on Detecting PGD Adversarial Examples with Different Smoothing Filters

Accuracy	Clean Accuracy on ResNet18	3×3 Spatial Average Smoothing	5×5 Spatial Average Smoothing	3×3 Gaussian Smoothing	Adaptive Smoothing with Mean
Colorectal Histopathology	99.60%	25%	25%	54.16%	54.16%
BraTs18	99.64%	29.16%	29.16%	66.67%	70.83%
ChestXray	94.32%	16%	25%	41.66%	41.66%

TABLE 20.3
Performance of SCD01MLP on Detecting PGD Adversarial Examples with Different Smoothing Filters

Accuracy	Clean Accuracy on SCD01MLP	3×3 Spatial Average Smoothing	5×5 Spatial Average Smoothing	3×3 Gaussian Smoothing	Adaptive Smoothing with Mean
Colorectal Histopathology	99.20%	29.16%	54.16%	66.67%	54.16%
BraTs18	98.38%	29.16%	66.67%	79.16%	79.16%
ChestXray	90.69%	25%	41.66%	41.66%	41.66%

20.4.2.2 Evaluation against Black-Box Attacks

In Tables 20.4 and 20.5, we list the accuracies of detecting black-box HopSkipJump implementation in the IBM attack examples from Yang et al. (2022). We set up four different filters and train both networks on a clean dataset to achieve comparative classification accuracies. The adversarial examples are classified on the pertained models without defense methods. The experimental results show that the adaptive and Gaussian filters work better on three datasets, but the adaptive filters perform better in removing black-box perturbations.

20.4.2.3 Evaluation against Non-Adversarial Samples

To evaluate whether the filters can excessively remove the information from images, we evaluate the accuracies on clean images. Table 20.6 shows that the filters decrease the accuracy, but not over 1%. The filter improves the classification results or promises the baseline accuracy.

20.4.2.4 Evaluation with Transfer Adversarial Samples

Transferability is another vital property of a machine-learning system. A model has the robust defense ability if it can correctly classify the adversarial sample from other unknown models. We show the classification rates on transfer examples in Tables 20.7 and 20.8. The SCD01MLP can correctly classify most adversarial examples from

TABLE 20.4
Performance of ResNet18 on Detecting HopSkipJump Adversarial Examples with Different Smoothing Filters

Accuracy	Clean Accuracy on ResNet18	3×3 Spatial Average Smoothing	5×5 Spatial Average Smoothing	3×3 Gaussian Smoothing	Adaptive Smoothing with Mean
Colorectal Histopathology	99.60%	41.66%	45.83%	58.33%	58.33%
BraTs18	99.64%	41.66%	50%	62.50%	62.50%
ChestXray	94.32%	37.50%	37.50%	37.50%	45.83%

TABLE 20.5
Performance of SCD01MLP on Detecting HopSkipJump Adversarial Examples with Different Smoothing Filters

Accuracy	Clean Accuracy on SCD01MLP	3×3 Spatial Average Smoothing	5×5 Spatial Average Smoothing	3×3 Gaussian Smoothing	Adaptive Smoothing with Mean
Colorectal Histopathology	99.20%	45.83%	58.33%	58.33%	62.50%
BraTs18	98.38%	50%	62.50%	62.50%	62.50%
ChestXray	90.69%	45.83%	37.50%	37.50%	58.33%

TABLE 20.6
Performance of Detecting Clean Examples with Different Smoothing Filters

Accuracy	Clean Accuracy on SCD01MLP	3×3 Spatial Average Smoothing	5×5 Spatial Average Smoothing	3×3 Gaussian Smoothing	Adaptive Smoothing with Mean
Colorectal Histopathology	99.20%	99.80%	98.69%	99.80%	99.20%
BraTs18	98.38%	99.41%	98.38%	99.41%	99.67%
Chest X-ray	90.69%	92.23%	90.69%	93.78%	93.78%

TABLE 20.7
Performance of SCD01MLP on Detecting Transferred Adversarial Examples

Accuracy	Clean Accuracy on SCD01MLP	No Defense	3×3 Spatial Average Smoothing	5×5 Spatial Average Smoothing	3×3 Gaussian Smoothing	Adaptive Smoothing with Mean	Best-Case Improvement
Colorectal Histopathology	99.20%	85.19%	85.83%	83.33%	86.89%	86.89%	1.70%
BraTs18	98.38%	88.89%	89%	88.41%	89.00%	91.17%	2.28%
ChestXray	90.69%	82.13%	83.00%	83.00%	85.83%	83.00%	3.70%

TABLE 20.8
Performance of ResNet18 on Detecting Transferred Adversarial Examples

Accuracy	Clean Accuracy on ResNet18	No Defense	3×3 Spatial Average Smoothing	5×5 Spatial Average Smoothing	3×3 Gaussian Smoothing	Adaptive Smoothing with Mean	Best case Improvement
Colorectal Histopathology	99.60%	26.60%	31.66%	25.83%	38.33%	38.33%	11.73%
BraTs18	99.64%	38.89%	45.83%	43.92%	52.50%	52.50%	13.61%
ChestXray	94.32%	21.40%	31.66%	31.66%	37.50%	37.50%	16.10%

ResNet18 attacked by the black-box attack. The best accuracy is 91.19%, and the worst is 83%. The classification rates of ResNet18 are improved up to 16.1%.

20.4.2.5 Evaluation by Other Detection Methods

We detect the PGD adversarial examples generated by ResNet18 models in MNIST dataset to keep the fair scenario. Table 20.9 shows our approach is slightly lower than the adversarial training, but higher than others. The classification performance on adversarial training is difficult to defeat. However, the adversarial training takes a longer time and requires a sufficient number of adversarial examples to retrain the networks.

TABLE 20.9
Performance Comparisons with Other Methods

Method	Performance
Feature denoising for improving adversarial robustness (Xie et al., 2019)	55.70%
Defense against adversarial attacks by reconstructing images (Zhang et al., 2021)	48–98%
Low-rank completion of high-sensitivity points (Zhao et al., 2021)	62.20%
Adversarial training (Shi and Han, 2018)	98.00%
Ours	**97.14%**

Furthermore, when a new attack technique is developed, we have to retrain the models again to improve the performance. Compared with that, our method has a distinct advantage in consideration of training time and computation consumption.

20.5 CONCLUSIONS

In this chapter, we propose a novel adaptive framework that can reconstruct the images and detect the adversarial attacks in advance. By comparing the classification results of the original images and the reconstructed ones, the proposed method can detect the adversarial samples successfully. An adaptive reconstruction process is developed based on the entropy values to avoid excessive noise reduction during reconstructing images. The significant contributions of the proposed method are that it does not need to retrain the model and can be assembled into any network.

We evaluate the performances in both white-box and black-box attack scenarios. To show the universal feature, we test our method on different target models. Experimental results show that it achieves significant improvement on correctly classifying the adversarial samples and provides a higher detection accuracy than the existing techniques. Our future work will focus on extending more adversarial attacks and target models, including larger cohort datasets.

REFERENCES

Carlini, N. and Wagner, D., "Towards evaluating the robustness of neural networks," *Proceeding of the IEEE Symposium on Security and Privacy*, San Jose, CA, pp. 39–57, Mar. 2017.

Cooper, H. L. and Miller, M. I., "Information measures for object recognition accommodating signature variability," *IEEE Transactions on Information Theory*, vol. 46, no. 5, pp. 1896–1907, 2000.

Goodfellow, I., Shlens, J. and Szegedy, C., "Explaining and harnessing adversarial examples," *Proceedings of the 2015 International Conference on Learning Representations*, San Diego, CA, May 7–9, 2015.

He, K., Zhang, X., Ren, S., and Sun, J., "Deep residual learning for image recognition," *Proceedings of the IEEE Conference on Computer Vision and Pattern Recognition*, Las Vegas, NV, pp. 770–778, Jun. 2016.

Liao, F., Liang, M., Dong, Y., and Pang, T., "Defense against adversarial attacks using high-level representation guided denoiser," *Proceedings of the IEEE/CVF Conference on Computer Vision and Pattern Recognition*, Salt Lake City, UT, 2018.

Madry, A., Makelov, A., Schmidt, L., Tsipras, D., and Vladu, A., "Towards deep learning models resistant to adversarial attacks," *6th International Conference on Learning Representations*, Vancouver, BC, Apr. 2018.

Moosavi-Dezfooli, S.-M., Fawzi, A., and Frossard, P., "Deepfool: A simple and accurate method to fool deep neural networks," *Proceedings of the IEEE Conference on Computer Vision and Pattern Recognition*, pp. 2574–2582, 2016.

Panda, P., Chakraborty, I., and Roy, K., "Discretization based solutions for secure machine learning against adversarial attacks," *IEEE Access*, vol. 7, pp. 70157–70168, 2019.

Papernot, N., McDaniel, P., and Goodfellow, I., "Transferability in machine learning: From phenomena to black-box attacks using adversarial samples," arXiv:1605.07277, 2016b.

Papernot, N., McDaniel, P., Goodfellow, I., Jha, S., Celik, Z. B., and Swami, A., "Practical black-box attacks against machine learning," *Proceedings of the 2017 ACM on Asia Conference on Computer and Communications Security*, pp. 506–519, 2017.

Papernot, N., McDaniel, P., Jha, S., Fredrikson, M., Celik, Z. B., and Swami, A., "The limitations of deep learning in adversarial settings," *Proceeding of the IEEE European Symposium on Security and Privacy*, Saarbrücken, pp. 372–387, Mar. 2016a.

Papernot, N., McDaniel, P., Wu, X., Jha, S., and Swami, A., "Distillation as a defense to adversarial perturbations against deep neural networks," *Proceedings of the 2016 IEEE Symposium on Security and Privacy*, pp. 582–597, 2016c.

Shi, Y. and Han, Y., "Schmidt: Image augmentation for black-box adversarial attack," *2018 IEEE International Conference on Multimedia and Expo*, San Diego, CA, pp. 1–6, 2018.

Szegedy, C., Liu, W., Jia, Y., Sermanet, P., Reed, S., Anguelov, D., Erhan, D., Vanhoucke, V., and Rabinovich, A., "Going deeper with convolutions," *Proceedings of the IEEE Conference on Computer Vision and Pattern Recognition*, Boston, MA, pp. 1–9, Jun. 7–12, 2015.

Szegedy, C., Zaremba, W., Sutskever, I., Bruna, J., Erhan, D., Goodfellow, I., and Fergus, R., "Intriguing properties of neural networks," *Proceedings of the 2014 International Conference on Learning Representations*, Banff, Apr. 14–16, 2014.

Tsai, D. Y., Lee, Y., and Matsuyama, E., "Information entropy measure for evaluation of image quality," *Journal of Digital Imaging*, vol. 21, no. 3, pp. 338–347, 2008.

Wang, J., Zhang, T., Liu, S., Chen, P. Y., Xu, J., Fardad, M., and Li, B., "Beyond adversarial training: Min-max optimization in adversarial attack and defense," *Nuclear Physics, Section A*, vol. 986, pp. 1–22, Jun. 2019.

Xie, C., Wu, Y., Maaten, L. V. D., Yuille, A. L., and He, K., "Feature denoising for improving adversarial robustness," *Proceedings of the IEEE/CVF Conference on Computer Vision and Pattern Recognition*, Long Beach, CA, 2019.

Xu, W., Evans, D., and Qi, Y., "Feature squeezing: Detecting adversarial examples in deep neural networks," *Proceedings of the Network and Distributed System Security Symposium*, San Diego, CA, Feb. 18–21, 2018.

Yang, Y., Shih, F. Y., and Roshan, U., "Defense against adversarial attacks based on stochastic descent sign activation networks on medical images," *International Journal of Pattern Recognition and Artificial Intelligence*, pp. 2254005–2254021, 2022.

Yang, Z., Yang, Y., Xue, Y., Shih, F. Y., Ady, J., and Roshan, U., "Accurate and adversarially robust classification of medical images and ECG time-series with gradient-free trained sign activation neural networks," *Proceedings of the IEEE International Conference on Bioinformatics and Biomedicine*, Seoul, pp. 2456–2460, 2020.

Yeh, C., Lin, M., Chang, P., and Kang, L., "Enhanced visual attention-guided deep neural networks for image classification," *IEEE Access*, vol. 8, pp. 163447–163457, 2020.

Zhang, S., Gao, H., and Rao, Q., "Defense against adversarial attacks by reconstructing images," *IEEE Transactions on Image Processing*, pp. 6117–6129, 2021.

Zhao, Z., Wang, H., Sun, H., Yuan, J., Huang, Z., and He, Z., "Removing adversarial noise via low-rank completion of high-sensitivity points," *IEEE Transactions on Image Processing*, pp. 6485–6497, 2021.

21 A Novel Multi-Data-Augmentation and Multi-Deep-Learning Framework for Counting Small Vehicles and Crowds

21.1 INTRODUCTION

During recent decades, the rapidly growing development of science and technology has significantly impacted human living environment. Unmanned aerial vehicles (UAVs) especially offer a broad field of views with increased mobility and flexibility, making them applicable in numerous industrial fields, such as geographic information collection, traffic monitoring, item delivery, communication network relay stations, radar UAV detection (Ren et al., 2021), visual UAV tracking (Li et al., 2022), and urban scene segmentation of UAV images (Yi et al., 2023). UAVs possess high mobility and remote-control capabilities, allowing easy access to most areas. With the use of high-resolution cameras, detailed and comprehensive ground variations can be captured from an aerial perspective without any omissions (AI CUP, 2022).

In the field of computer vision, addressing the challenges posed by drone UAV-captured video sequences is crucial. Counting very small vehicles and crowds in UAV images or videos presents a formidable task due to several inherent difficulties:

(1) Non-Fixed Cameras: The cameras mounted on drones are not fixed, introducing variability in the captured images or videos.
(2) Diverse View Angles: Images or videos are obtained from different view angles, adding complexity to the task of counting small vehicles and crowds.
(3) Small Object Sizes: Vehicles and crowds often appear in very small sizes, demanding precision in detection and counting.
(4) Variations in Illumination and Occlusion: The presence of diverse illumination conditions and occlusion further complicates the detection process.
(5) Differing Orientations: Objects of the same class may exhibit different orientations, requiring robust orientation-aware models.
(6) Jittery Captured Images or Videos: The inherent instability of drone-captured images or videos, characterized by jitter, is added into the challenge.

DOI: 10.1201/9781003474395-24

FIGURE 21.1 The captured objects appear in different view angles from drones.

FIGURE 21.2 The detection results of applying Chen et al. method to detect cars, buses, trucks, motorcycles, and persons in Figure 21.1.

Figure 21.1 illustrates these challenges, where there are small cars, heavy-duty vehicles, motorcycles, and persons in variant view angles and small sizes.

Chen et al. (2021) proposed a parallel residual bi-fusion feature pyramid network (PRB-FPN) with a parallel design and multiple improvements to accurately detect both small and large objects in images or videos. When it is applied on Figure 21.1, the detection result is shown in Figure 21.2, where many cars, trucks, motorcycles, and persons in the far distances cannot be detected. Their model is not capable of detecting small and crowded objects in UAV images under difficult conditions of varying view angles, small object sizes, and illumination and occlusion effects. Therefore, further research is needed to develop more robust and accurate object detection models for UAV applications. Another model of YOLOv7 (Wang et al., 2022a) can achieve the highest accuracy among all known real-time object detectors. However, when it is applied to Figure 21.1, the result is shown in Figure 21.3, where many cars, buses, trucks, motorcycles, and persons in the far distances cannot be detected although showing a slight improvement.

FIGURE 21.3 The detection results of applying YOLOv7 to detect cars, buses, trucks, motorcycles, and persons in Figure 21.1.

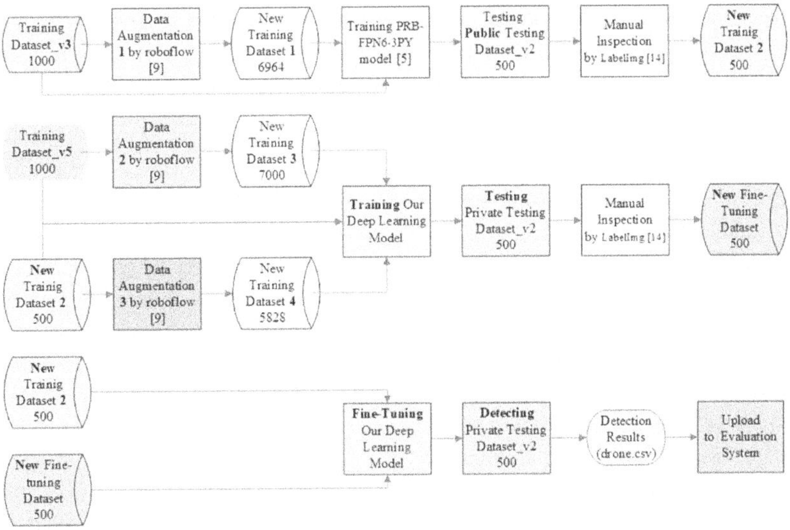

FIGURE 21.4 The general pipeline of the proposed MDAMDL framework.

In this chapter, we propose a multi-data-augmentation and multi-deep-learning (MDAMDL) framework to address the stated challenges by taking advantages of several techniques for achieving high accuracy and efficiency. Its pipeline is illustrated in Figure 21.4, where the strengths of PRB-FPN and PRB-FPN6–3PY (Chen et al., 2021), YOLOv7-E6E (Wang et al., 2022), CBAM attention (Woo et al., 2018; Mohenghui, 2022), and data augmentation (Data Augmentation, 2023) are adopted.

The PRB-FPN model detects small objects, while the PRB-FPN6-3PY model can detect both small and large objects. The YOLOv7-E6E model enables fast and accurate object detection, and the CBAM attention mechanism improves object feature representation. Data augmentation techniques are used to increase training data diversity and improve model generalization performance. The proposed framework is evaluated on the Public Testing Dataset_v2, which contains small vehicles and crowded scenes in UAV images. Experimental results show that it is a promising solution for accurately counting small vehicles and crowds in UAV images and has potential applications in various fields, such as traffic monitoring, security surveillance, and disaster response.

The rest of this chapter is organized as follows. Section 21.2 discusses the related work of using deep-learning techniques for vehicle counting, crowd counting, and small object detection. Details of the proposed method are presented in Section 21.3. Section 21.4 provides experimental results and the comparisons of the AI Cup 2022: Vehicle and Crowd Counting Competition. Finally, conclusions are drawn in Section 21.5.

21.2 RELATED WORK

In the realm of computer vision, deep-learning models have demonstrated exceptional performance. Kaur and Singh (2022) presented a concise overview of object detection, covering topics such as object detection frameworks, backbone convolutional neural networks, challenges in object detection, applications, and future research challenges in designing deep neural networks. However, in this chapter, we develop an innovative multi-data-augmentation and multi-deep-learning framework to solve the more difficult problems in counting very small vehicles and crowds.

21.2.1 Vehicle Counting

Hsieh et al. (2017) proposed the layout proposal networks (LPNs) and spatial kernels for simultaneous car counting and localization in drone-recorded videos. However, their method is limited to counting cars only at a fixed height of approximately 40 meters and assumes that the counted cars are parked in specific layouts of four parking lots. Real-world scenarios involve drones flying at different heights to capture various types of objects, including small cars, large cars (e.g., trucks), motorcycles, and persons with diverse layouts. Youssef and Elshenawy (2021) explored the application of a cascade region-based convolutional neural network (R-CNN) for automatic vehicle counting and tracking in aerial video streams. Their approach utilizes feature pyramid networks and a cascade R-CNN architecture to achieve precise vehicle detection and classification. However, their method is limited to counting and tracking only vehicles such as cars, vans, trucks, and buses, but cannot perform crowd counting.

Anitha and Arun (2021) conducted a comprehensive literature review on object detection and tracking using UAVs for various applications. Their study reviewed and summarized the findings of existing research papers in this area, and identified research gaps that need to be addressed. Kilic and Ozturk (2023) developed a single-shot detector model, named HLCNN, for detecting and counting cars in aerial images. Their method employs a heatmap of target car instances predicted by the HLCNN model. They improved the CNN architecture by adding three convolutional layers as adaptation layers instead of fully connected layers and utilized VGG-16 as a backbone CNN.

Wang's (2022) method utilized deep learning for vehicle detection in UAV video data. The video data was captured from a DJI Phantom3 Professional drone hovering 50 meters above the road, with the camera lens pointed at $-90°$ with a resolution of 3840×2160. The algorithm achieved a high vehicle detection rate of 96.49%.

However, the method is limited to vehicle detection and may not be suitable for other object detection or counting tasks.

21.2.2 Crowd Counting

Jin et al. (2021) proposed a method for detecting pedestrians in low-quality images using a playground dataset containing 5,752 images with 31,041 annotations. These images were captured from cameras positioned at a height of over 3 meters above the ground, with pedestrians averaging 87×41 in size and an image resolution of 480×640. To enhance the resolution of low-quality images, they proposed a super-resolution detection (SRD) network and a modified faster R-CNN network to relocate occluded pedestrians. However, their cameras were fixed unlike the dynamic camera position of UAV images, and the average sizes of pedestrians in their dataset are larger than those in UAV images. Therefore, their method is suitable for detecting pedestrians in low-quality images captured by fixed cameras, but cannot be directly applied to UAV images due to their distinct characteristics.

Yeh et al. (2022) proposed the lightweight deep neural network for joint learning of underwater object detection and color conversion. Ptak et al. (2022) presented a method for crowd counting in low-altitude aerial images obtained from an unmanned aerial vehicle. They evaluated several neural network architectures suitable for onboard image processing using edge computing devices, while minimizing performance loss. Their experiments showed that the image resolution significantly affects the lightweight quality and should be considered before adopting more complex neural network models for accuracy improvement. Their research solely focused on crowd counting and did not include the counting of vehicles or motorcycles.

Zhao et al. (2022) proposed a method called PDNet, which employs a multi-task learning approach combining point regression and density map regression to address challenges associated with small targets, scale variations, and complex backgrounds. The PDNet consists of a backbone for extracting multi-scale features, a dilated feature fusion (DFF) module, a density map attention (DMA) module, a density map branch, and a pointed branch. Their method was evaluated on the DroneCrowd dataset and outperformed state-of-the-art methods in crowd counting. However, their research only focused on crowd counting, but the detection and counting of other objects, such as vehicles or motorcycles, are not addressed.

21.2.3 Small Objects Detection Using Deep Learning

Ajaz et al. (2022) proposed a method for object detection in drone images by modifying the architecture of Tiny YOLOv3. Their custom architecture comprises a total of 31 layers, including 16 convolutional layers, 3 detection (YOLO) layers, and the remaining layers for max-pooling and routing. Experimental results showed that their modified architecture outperforms the original Tiny YOLOv3. Wang et al. (2021) proposed an improved algorithm for small-scale object detection based on YOLOv3. They utilized an improved k-medians clustering method to address the model instability caused by singularity, instead of the previous k-means. Furthermore, they proposed a local enhancement method to strengthen weak features for small-scale object detection by paralleling a branch on the backbone.

Chen et al. (2022) proposed an AFYOLO structure for detecting small objects. The structure includes the DenseNet module, channel attention, spatial attention, and feature pyramid network. Ablation studies on the ImageNet classification task and MS-COCO object detection task verify the effectiveness of the proposed attention module and FPN. Qu et al. (2022) proposed an enhanced model for SSD (single-shot multibox detector) by improving the shallow network's feature extraction ability through multi-scale feature fusion and an attention mechanism module. They utilized RFB (receptive Field block) to expand the object's receptive field and extract richer semantic information.

Wang et al. (2022) proposed a modified network structure to incorporate data augmentation, image pyramid, and attention mechanism to the Yolov5 network to enhance the detection of small objects. Experimental results on the Visdrone2019 dataset and DOTA dataset demonstrate a significant improvement in small object detection, defined as objects with a size below 32 × 32 pixels. Liu et al. (2022) proposed a small object detection network that utilizes multi-level feature fusion and feature augmentation. The network consists of a feature enhancement module and a multi-level feature fusion module. They also introduced a strategy that combines Soft-NMS and CIOU to address the issue of missed detection of occluded small objects.

Ye et al. (2022) introduced a global-local feature-enhanced network designed to detect dense and small objects in UVA-vision. However, their work is mainly on the detection of ships, cars, and airplanes, not suitable for cars, buses, trucks, persons, and motorcycles. Zhu et al. (2022) organized three challenge workshops in conjunction with ECCV 2018, ICCV 2019, and ECCV 2020, successfully attracting over 100 teams from all over the world. Additionally, they curated and released the VisDrone dataset, captured by drones over various urban and suburban areas in 14 different cities across China, spanning from north to south. The dataset's expansive coverage across diverse urban settings adds valuable insights to the capabilities and challenges of visual analysis in drone applications.

21.3 THE PROPOSED PIPELINE MDAMDL SYSTEM

This section describes the proposed pipeline framework for a multi-data-augmentation and multi-deep-learning (MDAMDL) system designed for vehicles and crowds counting. The system comprises four new training datasets and one new fine-tuning dataset, three data augmentation methods, two deep-learning models for training and one for fine-tuning, two testing methods, and one object detection method. Furthermore, the system integrates two manual inspections to ensure accurate labeling.

21.3.1 2022 AI Cup Competition Dataset and Data Preparation

Regarding the Vehicle and Crowd Counting Competition held in 2022 (AI CUP, 2022), the competition provided two datasets for training and testing. The Training Dataset_v3 dataset contains 1,000 UAV images with a size of 1920 × 1080, and 1,000 labels for four classes, including car, HOV, person, and motorcycle. However, the labels are not very accurate, so the competition organizers provided a new version of the training dataset, Training Dataset_v5, with more accurate labels for the same number of images. The Public Testing Dataset_v2 contains 500 UAV images without ground truth labels, which contestants use to test their trained models and submit the results to the evaluation system for scoring. The Public Leaderboard

TABLE 21.1
New Training Dataset 1

Data Augmentation 1	Descriptions	Amount	Size
Method 1	Horizontal Flip Rotation: between –45° and +45° Noise: up to 5% of pixels	1,741	640 × 360
Method 2	90° Flip: clockwise, counterclockwise Rotation: between –45° and +45°	1,741	640 × 360
Method 3	Horizontal Flip Rotation: between –30° and +30° Brightness: between –5% and +5%	1,741	640 × 360
Method 4	Horizontal Flip Rotation: between –12° and +12° Brightness: between –5% and +5% Noise: up to 5% of pixels	1,741	640 × 360

displays the scores of the submitted results. Additionally, the competition organizers provided the Private Testing Dataset_v2, which includes 500 UAV images of the same size as the training dataset for contestants to test and upload their results. The scores are displayed on the Private Leaderboard. Overall, contestants can use these datasets to train and test their models to participate in the competition.

In order to tackle the challenges of counting very small vehicles and crowds in UAV images, the initial step is to augment the size of the training dataset. We have created four novel training datasets and one additional fine-tuning dataset. The details of these five datasets are elaborated in the following.

21.3.1.1 New Training Dataset 1

To generate New Training Dataset 1, we apply data augmentation 1 as shown in Table 21.1. This approach utilizes online roboflow (Data Augmentation, 2023) to augment Training Dataset_v3 by four techniques. The resultant dataset comprises 6,964 UAV images, with each image sized at 640 × 342 pixels, and contains 1,741 images for each of the four augmentation methods.

21.3.1.2 New Training Dataset 2

To acquire New Training Dataset 2, we train the PRB-FPN6–3PY deep-learning model on Training Dataset_v3 (1,000) and New Training Dataset 1 (6,964) to identify the four classes (car, HOV, person, and motorcycle). Subsequently, we utilize this detector to evaluate the Public Testing Dataset_v2, resulting in the production of new labels. We then manually verify these new labels utilizing LabelImg (Tzutalin, 2023) to rectify any errors and develop New Training Dataset 2, containing 500 UAV images and 500 new labels.

21.3.1.3 New Training Dataset 3

To generate New Training Dataset 3, we use data augmentation 2 as shown in Table 21.2. This method uses online roboflow to augment Training Dataset_v5 with seven techniques. The resulting dataset comprises 7,000 UAV images, with each technique contributing 1,000 images. The sizes of the augmented images vary from 640 × 360 to 1920 × 1080 pixels.

21.3.1.4 New Training Dataset 4

To acquire New Training Dataset 4, we employ data augmentation 3 as depicted in Table 21.3. This approach involved utilizing online roboflow to augment New Training Dataset 2 by six techniques. The resultant dataset comprises 5,828 images, with each technique contributing a distinct number of images. The sizes of the augmented images vary from 640 × 360 to 1920 × 1080 pixels.

TABLE 21.2
New Training Dataset 3

Data augmentation	Descriptions	Amount	Size
Method 1	Horizontal Flip	1,000	1920 × 1080
Method 2	Horizontal Flip Rotation: between −45° and +45°	1,000	1920 × 1080
Method 3	Horizontal Flip Clipping: Horizontal ±15°, Vertical ±15° Noise: up to 5% of pixels	1,000	640 × 360
Method 4	Horizontal Flip Rotation: between −22° and +22° Brightness: between −20% and +20%	1,000	640 × 360
Method 5	Horizontal Flip Rotation: between −45° and +45° Hue: between −25° and +25°	1,000	640 × 360
Method 6	90° rotation: clockwise, counterclockwise Hue: between −25° and +25°	1,000	640 × 360
Method 7	Saturation: between −4% and +4% Brightness: between −2% and +2% Bounding Box: Flip: Horizontal, Vertical	1,000	640 × 360

TABLE 21.3
New Training Dataset 4

Data augmentation	Descriptions	Amount	Size
Method 1	90° rotation: clockwise, counterclockwise	1,000	640 × 360
Method 2	Horizontal Flip Noise: up to 5% of pixels	1,000	640 × 360
Method 3	90° rotation: clockwise, counterclockwise Rotation: between −22° and +22° Brightness: between −5% and +5%	1,000	640 × 360
Method 4	Bounding Box: Flip: Horizontal Brightness: between −5% and +5%	998	640 × 360
Method 5	90° rotation: clockwise, counterclockwise	830	1920 × 1080
Method 6	Clipping: horizontal ±15°, vertical ±15° Bounding Box: Noise: Up to 5% of pixels	1,000	640 × 360

21.3.1.5 New Fine-Tuning Dataset

To generate the New Fine-Tuning Dataset, we train our proposed deep-learning model on Training Dataset_v5 (1,000), New Training Dataset 3 (7,000), New Training Dataset 2 (500), and New Training Dataset 4 (5,828) to detect four classes (car, HOV, person, and motorcycle). We then utilize this detector to evaluate the Private Testing Dataset_v2, leading to the production of new labels. We subsequently manually verify these new labels utilizing LabelImg to rectify any errors and generate the New Fine-Tuning Dataset, containing 500 UAV images and 500 new labels.

21.3.2 The Proposed PRB-FPN-CBAM Model

The primary aim of the proposed MDAMDL system is to accurately count small vehicles and crowds in UAV images. To attain optimal bounding box performance in each UAV image, we integrate the cutting-edge object detection models, including the PRB-FPN, PRB-FPN6–3PY, YOLOv7-E6E, CBAM attention, and Inception Architecture (Szegedy et al., 2015, 2016; Pytorch, 2020].

Our developed deep-learning object detection model, denoted as the PRB-FPN-CBAM, is illustrated in Figure 21.5(a), providing an overview of the entire model. On the other hand, Figure 21.5(b) delves into the individual components comprising the PRB-FPN-CBAM model, offering a comprehensive understanding of its internal structure.

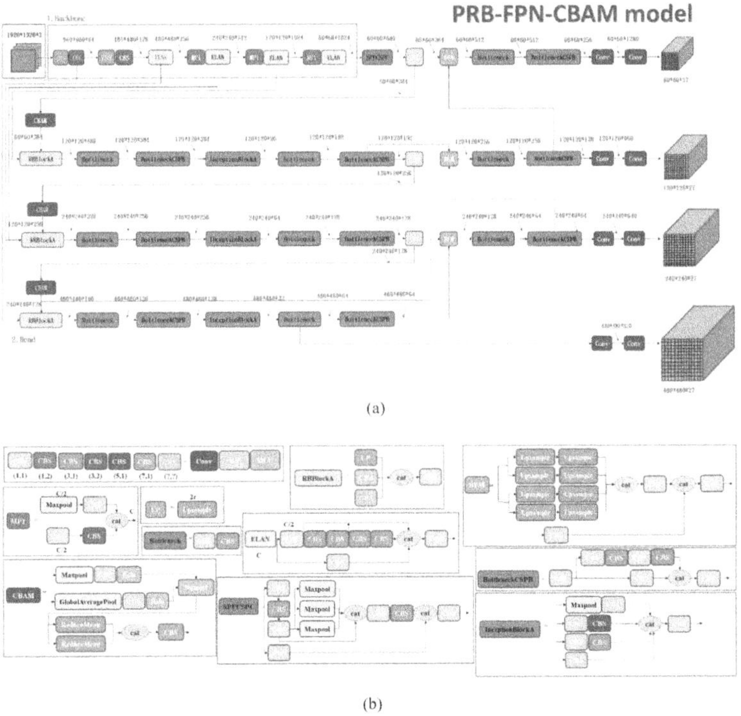

FIGURE 21.5 The diagram of the proposed model. (a) PRB-FPN-CBAM model. (b) The components in PRB-FPN-CBAM model

(1) **Backbone**: The proposed model employs the PRB-FPN model as its backbone to ensure superior detection accuracy. While many state-of-the-art object detectors utilize deep CNN structures to achieve exceptional performance in detecting medium and large-sized objects, their ability to detect smaller objects remains limited. This is mainly due to the reduction in feature map resolution that occurs after simple pooling in the feature pyramid (FP). As a result, tiny objects (less than 32×32 pixels) can become a single-pixel feature vector after five pooling layers in the last FP layer, resulting in insufficient spatial resolution for precise discrimination. To overcome this issue, we use only three pooling layers in the FP.

(2) **Head:** The head of the proposed model comprises a bi-fusion model and three bottom-up fusion modules (BFMs) adapted from the PRB-FPN6–3PY model. Additionally, we adopt the lead head and aux head from the YOLOv7-E6E model. The bi-fusion model is designed to fuse the multi-level features extracted from the backbone network and the BFM to obtain high-quality feature representations. The BFMs are responsible for fusing features from different levels in the feature pyramid. The lead head and aux head are specialized in detecting large and small objects, respectively. The integration of these components in the head enhances the detection accuracy of both small and large objects in UAV images.

(2.1) **One Bi-Fusion Model:** In the PRB-FPN6–3PY model, a parallel feature pyramid fusion design is proposed to address the challenging task of object detection, considering objects of all scales. This approach creates multiple fusion paths to capture features suitable for detecting objects of all sizes, including small and large objects. We use three pooling layers in the feature pyramid and one bi-fusion model, which combines contextual features from adjacent layers in a concurrent manner.

(2.2) **Three Bottom-Up Fusion Modules (BFMs):** Our proposed model incorporates the bottom-up fusion module (BFM) from the PRB-FPN6–3PY model to enhance the localization of small and large objects. The BFM divides the C channels of the feature map into 4C channels using a Re-Org block, which helps to preserve spatial information and generate robust semantic features through 1×1 convolution, thereby improving small object detection.

(2.3) **Lead Head and Aux Head:** The lead head and aux head in our proposed model are adapted from the YOLOv7-E6E model. The lead-head-guided and coarse-to-fine lead-head-guided label assigners are used to guide both the auxiliary head and lead head by the lead head prediction. The lead-head-guided label assigner calculates the soft label based on the prediction result of the lead head and ground truth, and generates a set of soft labels through an optimization process. This set of soft labels is used as the target training model for both the auxiliary head and lead head. The coarse-to-fine lead-head-guided label assigners also generate soft labels based on the predicted result of the lead head and the ground truth. However, this process generates two sets of soft labels: coarse and fine labels. The fine label is the same as the soft label generated by the lead-head-guided label assigner, while the coarse label relaxes the constraints of the positive sample assignment process, allowing more grids to be treated as positive targets.

(2.4) **Convolutional Block Attention Module (CBAM):** We adopt the CBAM mechanisms to enhance the representation power of CNNs by emphasizing important features and suppressing irrelevant ones. It is composed of two modules: channel and spatial attentions. The channel attention module learns to weigh the importance of each feature map channel based on its global distribution, while the spatial attention module learns to highlight informative spatial locations within each channel. The CBAM can be integrated into any CNN architecture with minimal overhead, and is end-to-end trainable along with the base CNNs. In our proposed model, the CBAM module is incorporated between the CBS and the RBBlockA modules.

(2.5) **Inception Architecture:** GoogLeNet's Inception architecture (Szegedy et al., 2015) was designed to perform efficiently under strict memory and computational constraints, making it suitable for processing large datasets. However, its complexity poses challenges for modification, and simply scaling up the architecture may lead to a loss in computational gains. To address this, Szegedy et al. (2016) proposed general Inception principles and optimization ideas that can effectively scale up convolution networks. They showed that high-quality results can be achieved with a low receptive field resolution of 79×79, which is beneficial for detecting relatively small objects. By factorizing convolutions and aggressively reducing dimensions inside the neural network, they demonstrated that networks can be built with low computational costs while maintaining high quality. In our proposed model, we integrate the InceptionBlockA (Pytorch, 2020) between the BottleneckCSPB module and the Bottleneck module.

21.3.3 Fine-Tuning the Proposed PRB-FPN-CBAM Model

To achieve higher scores on the AI Cup 2022 Private Leaderboard, we fine-tuned our model using New Training Dataset 2 and the New Fine-Tuning Dataset. The fine-tuning process produces an object detector, capable of detecting all four object classes. The resulting object detector is used to detect the Private Testing Dataset_v2 and generate detection results in the AI Cup 2022 upload format. These detection results are submitted to the evaluation system to obtain the Private Score on the Private Leaderboard. The final ranking in the competition is determined based on the scores on the Private Leaderboard.

21.4 EXPERIMENTAL RESULTS

We conduct experiments to evaluate the performance of the proposed framework pipeline on the Private Testing Dataset_v2 of the 2022 AI Cup Competition. It is implemented using an Ubuntu 20.04 LTS operating system with 8 NVIDIA Tesla V100 32 GB GPUs, Nvidia 450.119.04 driver, CUDA 11.7 driver, and Python 3.8 for training. For testing and detecting, we use an Ubuntu 18.04.6 LTS operating system with 1 NVIDIA TESLA T4 15 GB GPU, Nvidia 460.32.03 driver, CUDA 11.2 driver, and Python 3.8.15. The source code of our MDAMDL framework is made available publicly on GitHub – cmtsai2023/MDAMDL-Framework.

21.4.1 Datasets and Evaluation Metrics

The harmonic mean ($Hmean_{TIoU}$) score is used as the evaluation metric for the 2022 AI Cup Competition dataset. The formula for the $Hmean_{TIoU}$ score involves $Recall_{TIoU}$, $Precision_{TIoU}$, and $SCORE_{dis}$, which are calculated by the following steps:

(1) For each ground true box G_i, we find the detection box D_j that has the largest intersection over union (IoU) with G_i (greater than 0.5), and the predicted label is the same as the ground truth label. Calculate the $TIoU^{Recall}$ score by

$$TIoU_i^{Recall} = \frac{A(G_i \cap D_j)}{A(G_i \cup D_j)} \times \frac{A(G_i \cap D_j)}{A(G_i)} \tag{21.1}$$

where $A(G_i)$, $A(G_i \cap D_j)$, and $A(G_i \cup D_j)$ are the area of the G_i, the intersection area of the G_i and D_j, and the union area of the G_i and D_j, respectively. If no detection box D_j meets the conditions, the item is calculated with 0.

(2) For each detection box D_j, we find the ground true box G_i that has the largest IoU with D_j (greater than 0.5), and the ground truth label is the same as the predicted label. Calculate the $TIoU^{Precision}$ score by

$$TIoU_j^{Precision} = \frac{A(G_i \cap D_j)}{A(G_i \cup D_j)} \times \left(1 - \frac{A\left(U_{k \neq i}\left\{D_j \cap G_k - D_j \cap G_k \cap G_i\right\}\right)}{A(D_j)}\right) \tag{21.2}$$

where $A(D_j)$ is the area of the detection box D_j. If no ground true box G_i meets the conditions, the item is calculated with 0.

(3) Calculate the $Recall_{TIoU}$ and $Precision_{TIoU}$ by

$$Recall_{TIoU} = \frac{1}{Num_{gt}} \sum_{i=1}^{Num_{gt}} TIoU_i^{Recall} \tag{21.3}$$

$$Precision_{TIoU} = \frac{1}{Num_{dt}} \sum_{j=1}^{Num_{dt}} TIoU_{ij}^{Precision} \tag{21.4}$$

where Num_{gt} and Num_{dt} denote the total number of ground true boxes and detection boxes in all images, respectively.

(4) For each ground true box G_i, we find the detection box D_j that has the largest IoU with G_i (greater than 0.5), and the predicted label is the same as the ground truth label. Calculate the $SCORE_{dis}$ by

$$SCORE_{dis} = \frac{1}{Num_{gt}} \sum_{i=1}^{Num_{gt}} e^{\frac{-\left\|G_i(x,y) - D_j(x,y)\right\|_2^2}{c}} \tag{21.5}$$

where $G_i(x,y)$ and $D_j(x,y)$ are the coordinates of the center point of the ith ground true box and the jth detection box, Num_{gt} is the total number of ground true boxes, and C is a normalization constant.

(5) Calculate the harmonic mean ($Hmean_{TIoU}$) by

$$Hmean_{TIoU} = \frac{3 \times Recall_{TIoU} \times Precision_{TIoU} \times SCORE_{dis}}{Recall_{TIoU} \times Precision_{TIoU} + Precision_{TIoU} \times SCORE_{dis} + Recall_{TIoU} \times SCORE_{dis}} \quad (21.6)$$

It gives equal weight to $Recall_{TIoU}$, $Precision_{TIoU}$, and $SCORE_{dis}$ in calculating the final score.

21.4.2 Implementation Details

In this section, we provide implementation details of our proposed detection model in the training and testing processes. First, we use the PRB-FPN6–3PY model to test the Public Testing Dataset_v2 and produce New Training Dataset 2. The detection model is then trained using New Training Dataset 2 with an input size of 1280×1280 for 300 epochs while utilizing the default p6 parameters. For the first test, the detection model is evaluated with an input size of 1280×1280, a confidence threshold of 0.5, and an IoU threshold of 0.65.

In the second stage of training, we train our proposed model to detect objects in the Private Testing Dataset_v2 and generate the New Fine-Tuning Dataset. The detection model is trained for 300 epochs with an input size of 1920×1920, using the default p6 parameters. We test the model on the Public Testing Dataset_v2 with an input size of 1920×1920 to obtain detection results, which are uploaded to the evaluation system to obtain scores on the Public Leaderboard. After experimenting with different confidence and IoU threshold parameters, we find that the optimal values are 0.28 and 0.35, respectively. We use these parameters to test the Private Testing Dataset_v2 to generate the New Fine-Tuning Dataset. Different detection results using various parameters are also uploaded to the AI Cup 2022 evaluation system and their scores on the Public Leaderboard are shown.

In the third training, We use the detection model obtained in the second training and the best two parameters to fine-tune the New Training Dataset 2 and the New Fine-Tuning Dataset with input size 1920×1920 for 350 epochs with the default p6 parameters. In the last detection step, we use the fine-tuned detection model to detect the Private Testing Dataset_v2, upload the detection results to the evaluation system, and show the score in the Private Leaderboard.

21.4.3 Comparison with Other Teams

Our proposed system was submitted to the AI CUP (2022) Vehicle and Crowd Counting Competition for evaluation. Table 21.4 shows that our system achieved an $Hmean_{TIoU}$ score of 0.74645 on the public leaderboard, ranking it first among the 106 teams that participated in the competition. For the final competition, as shown in Table 21.5, our system achieved an $Hmean_{TIoU}$ score of 0.765476 on the private leaderboard, ranking it first among the 68 teams that submitted their results.

TABLE 21.4
Public Leaderboard in the AI CUP (2022)

Rank	Team Name	Upload Times	$Hmean_{TIoU}$ Scores
1	**TEAM_2334 (ours)**	120	**0.74645**
2	TEAM_2432	114	0.745735
3	TEAM_2082	33	0.739696
4	TEAM_2144	161	0.739406
5	TEAM_2610	44	0.739363
6	TEAM_2060	123	0.737379
7	TEAM_2484	43	0.737348
8	TEAM_2059	86	0.73687
9	TEAM_2422	46	0.736586
10	TEAM_2189	39	0.7356

TABLE 21.5
Private Leaderboard in the AI CUP (2022)

Rank	Team Name	Upload Times	$Hmean_{TIoU}$ Scores
1	**TEAM_2334 (ours)**	120	**0.765476**
2	TEAM_2432	114	0.764277
3	TEAM_2080	33	0.758381
4	TEAM_2144	161	0.756348
5	TEAM_2610	44	0.754987
6	TEAM_2060	123	0.754827
7	TEAM_2484	43	0.754777
8	TEAM_2059	86	0.754556
9	TEAM_2189	46	0.753994
10	TEAM_2449	39	0.749105

21.4.4 Comparison with YOLOv7-E6E Model

We compare the results of our proposed model with the YOLOv7-E6E model. Both models are trained on New Training Datasets 3 and 4, using an input size of 1920 × 1920 for 350 epochs with the default p6 parameters. The trained models are tested on the Public Testing Dataset_v2 to compare Precision, Recall, mAP@.5, and mAP@.5:.95. Tables 21.6 and 21.7 list the testing results of the YOLOv7-E6E model and our proposed model, respectively. Note that YOLOv7-E6E model has higher Recall than our proposed model in the all, car, and person classes.

The Precision, mAP@.5, and mAP@.5:.95 of all classes are higher in our proposed model than the YOLOv7-E6E model. We also test both models on the Private Testing Dataset_v2, and the results are listed in Tables 21.8 and 21.9. Our proposed model outperforms the YOLOv7-E6E model in terms of Precision, mAP@.5, and mAP@.5:.95 for all classes, as well as for car, HOV, person, and motorcycle classes.

TABLE 21.6
Testing Results by the YOLOv7-E6E Model on Public Testing Dataset_v2

Class	Images	Labels	Precision	Recall	mAP@.5	mAP@.5:.95
all	500	16457	0.835	**0.901**	0.888	0.598
car	500	11974	0.976	**0.973**	0.986	**0.878**
HOV	500	757	0.893	0.93	0.951	0.7
person	500	2135	0.649	**0.819**	0.733	0.373
motorcycle	500	1591	**0.823**	0.882	**0.883**	0.532

TABLE 21.7
Testing Results by the Proposed Model on Public Testing Dataset_v2

Class	Images	Labels	Precision	Recall	mAP@.5	mAP@.5:.95
all	500	16457	**0.844**	0.889	**0.893**	**0.621**
car	500	11974	**0.977**	**0.973**	**0.987**	0.804
HOV	500	757	**0.911**	**0.943**	**0.955**	**0.724**
person	500	2135	**0.699**	0.741	**0.749**	**0.398**
motorcycle	500	1591	0.788	**0.9**	0.881	**0.557**

TABLE 21.8
Testing Results by YOLOv7-E6E Model on Private Testing Dataset_v2

Class	Images	Labels	Precision	Recall	mAP@.5	mAP@.5:.95
all	500	19995	0.952	0.926	0.926	0.701
car	500	12827	0.988	0.973	0.979	0.864
HOV	500	863	0.968	0.935	0.94	0.807
person	500	3918	0.885	0.905	0.889	0.485
motorcycle	500	2387	0.966	0.891	0.896	0.648

TABLE 21.9
Testing Results by PRB-FPN-CBAM Model on Private Testing Dataset_v2

Class	Images	Labels	Precision	Recall	mAP@.5	mAP@.5:.95
all	500	19995	**0.994**	**0.942**	**0.95**	**0.812**
car	500	12827	**0.999**	**0.974**	**0.981**	**0.937**
HOV	500	863	**0.998**	**0.949**	**0.955**	**0.892**
person	500	3918	**0.987**	**0.927**	**0.941**	**0.62**
motorcycle	500	2387	**0.992**	**0.917**	**0.923**	**0.8**

21.4.5 Ablation Study

To demonstrate the effectiveness of incorporating CBAM in our proposed model, we conduct an ablation study by removing CBAM to create a less CBAM model, named PRB-FPN. We evaluate the performance of both models on the Public Testing Dataset_v2 and the Private Testing Dataset_v2, and the results are listed in Tables 21.10 and 21.11, respectively. From Tables 21.7 and 21.10, we observe that the Recall in all classes and the car class for the PRB-FPN model is higher than that of the proposed model. For all other metrics, the proposed model outperforms the PRB-FPN model. Furthermore, as shown in Tables 21.9 and 21.11, the results obtained by the proposed model are superior to those obtained by the PRB-FPN model across all classes, including car, HOV, person, and motorcycle. Therefore, experimental results show that incorporating CBAM in our proposed model leads to improved performance.

21.4.6 Qualitative Results

The detection results of cars, HOVs (buses and trucks), people, and motorcycles in Figure 21.1 by the YOLOv7-E6E model and the proposed PRB-FPN-CBAM model are shown in the following. In Figure 21.6(a), the YOLOv7-E6E model fails to detect a very small car at the top of the road (blue arrow), whereas the proposed model successfully detects it in Figure 21.6(b). However, neither model can detect the very small objects, for example, two persons riding motorcycles on the top road (indicated by orange arrow).

TABLE 21.10
Testing Results by PRB-FPN Model on Public Testing Dataset_v2

Class	Images	Labels	Precision	Recall	mAP@.5	mAP@.5:.95
all	500	16457	0.821	**0.891**	0.884	0.6
car	500	11974	0.97	**0.974**	0.984	0.782
HOV	500	757	0.9	0.93	0.948	0.692
person	500	2135	0.652	0.772	0.742	0.386
motorcycle	500	1591	0.761	0.889	0.863	0.538

TABLE 21.11
Testing Results by PRB-FPN Model on Private Testing Dataset_v2

Class	Images	Labels	Precision	Recall	mAP@.5	mAP@.5:.95
all	500	19995	0.953	0.933	0.933	0.703
car	500	12827	0.989	0.975	0.98	0.853
HOV	500	863	0.979	0.927	0.931	0.778
person	500	3918	0.89	0.919	0.907	0.508
motorcycle	500	2387	0.955	0.91	0.913	0.674

(a)

(b)

FIGURE 21.6 Detection results of Figure 21.1. (a) The results are obtained by the YOLOv7-E6E model, (b) results are obtained by the proposed model.

21.4.7 Comparison Results on VisDrone

To assess the generalization capabilities of the proposed model across various scenarios, we conduct comparative experiments using the publicly available VisDrone datasets (Zhu et al., 2022). Specifically, our focus is on the Task 1 challenge, which involves object detection in images within the VisDrone-DET dataset. This task entails the detection of objects across ten predefined categories, including pedestrians, people, bicycles, cars, vans, trucks, tricycles, awing-tricycles, buses, and trucks. These objects are captured in individual images from cameras mounted on drones.

TABLE 21.12
Comparison Results of Various Methods on the Testset-Dev in the VisDrone-DET Dataset.

Methods	Images	AP_{50}
CornerNet (Law and Deng, 2018)	1,610	41.18%
Light-RCNN (Li et al., 2017)	1,610	39.56%
FPN (Lin et al., 2017)	1,610	39.57%
Cascade R-CNN (Cai and Vasconcelos, 2018)	1,610	37.84%
DetNet (Li et al., 2018)	1,610	37.54%
RefineDet (Zhang et al., 2018)	1,610	37.27%
RetinaNet (Lin et al., 2017)	1,610	31.67%
Ours	1,610	**52.00%**

The VisDrone-DET dataset comprises 6,471 training images, 548 validation images, and 1,610 testing images. Following the training of the proposed model on the VisDrone-DET dataset for 100 epochs, we conduct a comparative analysis against existing works in the literature. The comparison involves our proposed model and the following methods: FPN (Lin et al., 2017a), RetinaNet (Lin et al., 2017), Light-RCNN (Li et al., 2017), Cascade R-CNN (Cai and Vasconcelos, 2018), RefineDet (Zhang et al., 2018), DetNet (Li et al., 2018), and CornerNet (Law and Deng, 2018). This comparative evaluation aims at providing insights into the efficacy of the proposed model by benchmarking its performance against established methods in the field of object detection on drone-captured images.

The results on the Testset-Dev within the VisDrone-DET dataset are presented in Table 21.12. Key findings from the table are as follows. CornerNet (Law and Deng, 2018) achieves the highest AP_{50} score of 41.18% by utilizing the Hourglass-104 backbone for feature extraction. FPN (Lin et al., 2017) and RetinaNet (Lin et al., 2017) employ additional stages designed to handle objects with various scales. However, DetNet (Li et al., 2018) takes a different approach by re-designing the backbone network for object detection, maintaining spatial resolution to achieve a notable AP_{50} score of 37.54%. RefineDet (Zhang et al., 2018) outperforms RetinaNet with the ResNet-101 backbone to reach an AP_{50} score of 37.27%. This improvement is attributed to RefineDet's use of an object detection module to regress the locations and sizes of objects based on coarsely adjusted anchors from the anchor refinement module.

In comparisons, our proposed model demonstrates superior performance, as achieving the highest AP_{50} score of 52.00%. This highlights the efficacy of our model in object detection tasks, outperforming established approaches in the context of the VisDrone-DET dataset's Testset-Dev.

21.4.8 Results for Different Scenes of Object Detection on VisDrone

To demonstrate the versatility of the proposed model, we employed the Testset-Dev from the VisDrone-DET dataset for detection purposes. This specific dataset subset facilitates a comprehensive evaluation of the model's object detection capabilities

FIGURE 21.7 Detection results in difference scenes. (a) Night conditions with occlusion, (b) night conditions with large objects, (c) night conditions with objects of various sizes, (d) haze weather, (f) sunlight weather, and (g) small objects.

across different scenes, providing valuable insights into its robustness and performance in varied environmental conditions. Figure 21.7 provides a comprehensive overview of the detection results achieved by the proposed model across various scenes. Each figure visually represents the model's performance in distinct environmental scenarios, underscoring its capability to handle challenges, such as low-light conditions, occlusion, diverse object sizes, weather variations, and detection of small objects.

21.5 CONCLUSIONS

In this chapter, we propose an MDAMDL framework for counting small vehicles and crowds in UAV images, which achieves the first place in the AI Cup 2022: Vehicle and Crowd Counting Competition. Our framework utilizes multiple data augmentations to increase the training dataset, and deep learning methods to generate new training data for improved performance. We also develop a novel deep-learning object detection model, named the PRB-FPN-CBAM model, which combines state-of-the-art object detection models for car, bus, truck, person, and motorcycle detection, achieving higher detection accuracy. Experimental results show that the MDAMDL framework outperforms the state-of-the-art models in detecting small vehicles and crowded scenes in UAV images. In the future, we aim to address the challenge of detecting very small objects, such as two persons riding motorcycles as shown in Figure 21.6. Additionally, we plan to further enhance the accuracy of our proposed model in object detection.

21.5.1 Acknowledgments

This work was supported by the Ministry of Science and Technology, Taiwan, under Grants MOST 108-2221-E-845-003-MY3.

REFERENCES

AI CUP, *Intelligent counting of UAV flight vehicles (vehicle and crowd counting competition).* https://tbrain.trendmicro.com.tw/Competitions/Details/25, 2022.

Ajaz, A., Salar, A., Jamal, T., and Khan, A. U., "Small object detection using deep learning," arXiv:2201.03243v1, 2022. https://doi.org/10.48550/arXiv.2201.03243.

Anitha, R., and Arun, K. S., "A review on object detection in unmanned aerial vehicle surveillance," *International Journal of Cognitive Computing in Engineering*, vol. 2, pp. 215–228, 2021. https://doi.org/10.1016/j.ijcce.2021.11.005.

Cai, Z. and Vasconcelos, N., "Cascade R-CNN: Delving into high quality object detection," *Proceedings of the IEEE Conference on Computer Vision and Pattern Recognition*, pp. 6154–6162, 2018.

Chen, J. Y., Liu, S. P., Zhao, L. A., Chen, D. F., and Zhang, W. H., "Small object detection combining attention mechanism and a novel FPN," *Journal of Intelligent & Fuzzy Systems*, vol. 42, no. 4, pp. 3691–3703, 2022.

Chen, P. Y., Chang, M. C., Hsieh, J. W., and Chen, Y. S., "Parallel residual bi-fusion feature pyramid network for accurate single-shot object detection," *IEEE Transactions on Image Processing*, vol. 30, pp. 9099–9111, 2021. https://doi.org/10.1109/TIP.2021.3118953.

Data Augmentation from Roboflow. https://app.roboflow.com/login, 2023.

Hsieh, M. R., Lin, Y. L., and Hsu, W. H., "Drone-based object counting by spatially regularized regional proposal networks," *IEEE International Conference on Computer Vision (ICCV)*, pp. 4145–4153, 2017.

Jin, Y., Zhang, Y., Cen, Y., Li, Y., Mladenovic, V., and Voronin, V., "Pedestrian detection with super-resolution reconstruction for low-quality image," *Pattern Recognition*, vol. 115, p. 107846, 2021. https://doi.org/10.1016/j.patcog.2021.107846.

Kaur, R. and Singh, S., "A comprehensive review of object detection with deep learning," *Digital Signal Processing*, vol. 132, p. 103812, 2023. https://doi.org/10.1016/j.dsp.2022.103812.

Kilic, E. and Ozturk, S., "An accurate car counting in aerial images based on convolutional neural networks," *Journal of Ambient Intelligence and Humanized Computing*, vol. 14, pp. 1259–1268, 2023. https://doi.org/10.1007/s12652-021-03377-5.

Law, H. and Deng, J., "CornerNet: Detecting objects as paired keypoints," *Proceeding of the European Conference on Computer Vision*, pp. 765–781, 2018.

Li, S., Liu, Y., Zhao, Q., and Feng, Z., "Learning residue-aware correlation filters and refining scale for real-time UAV tracking," *Pattern Recognition*, vol. 127, p. 108614, 2022. https://doi.org/10.1016/j.patcog.2022.108614.

Li, Z., Peng, C., Yu, G., Zhang, X., Deng, Y., and Sun, J., "Light-head R-CNN: In defense of two-stage object detector," arXiv:1711.07264, 2017.

Li, Z., Peng, C., Yu, G., Zhang, X., Deng, Y., and Sun, J., "DetNet: Design backbone for object detection," *Proceeding of the European Conference on Computer Vision*, pp. 339–354, 2018.

Lin, T., Goyal, P., Girshick, R. B., He, K., and Dollar, P., "Focal loss for dense object detection," *Proceeding of the IEEE International Conference on Computer Vision*, pp. 2999–3007, 2017a.

Lin, T.-Y., Dollar, P., Girshick, R., He, K., Hariharan, B., and Belongie, S. J., "Feature pyramid networks for object detection," *Proceedings of the IEEE Conference on Computer Vision and Pattern Recognition*, pp. 936–944, 2017b.

Liu, B. K., He, F. J., Du, S. Q., Li, J. C., and Liu, W. J., "An advanced YOLOv3 method for small object detection," arXiv:2212.02809v2, 2022. https://doi.org/10.48550/arXiv.2212.02809.

Mohenghui.yolov5_attention. https://github.com/mohenghui/yolov5_attention, 2022.

Ptak, B., Pieczyński, D., Piechocki, M., and Kraft, M., "On-board crowd counting and density estimation using low altitude unmanned aerial vehicles – looking beyond beating the benchmark," *Remote Sensing*, vol. 14, no. 10, p. 2288, 2022. https://doi.org/10.3390/rs14102288.

Pytorch implementation for Inception Architecture. https://github.com/pytorch/vision/blob/main/torchvision/models/inception.py, 2020.

Qu, Z., Han, T., and Yi, T., "MFFAMM: A small object detection with multi-scale feature fusion and attention mechanism module," *Applied Sciences*, vol. 12, no. 18, p. 8940, 2022. https://doi.org/10.3390/app12188940.

Ren, J. and Jiang, X., "A three-step classification framework to handle complex data distribution for radar UAV detection," *Pattern Recognition*, vol. 111, p. 107709, 2021. https://doi.org/10.1016/j.patcog.2020.107709.

Szegedy, C., Liu, W., Jia, Y., Sermanet, P., Reed, S., Anguelov, D., Erhan, D., Vanhoucke, V., and Rabinovich, A., "Going deeper with convolutions," *Proceedings of the IEEE Conference on Computer Vision and Pattern Recognition (CVPR)*, pp. 1–9, 2015. https://doi.org/10.1109/CVPR.2015.7298594.

Szegedy, C., Vanhoucke, V., Ioffe, S., Shlens, J., and Wojna, Z., "Rethinking the inception architecture for computer vision," *2016 IEEE Conference on Computer Vision and Pattern Recognition (CVPR)*, pp. 2818–2826, 2016. https://doi.org/10.1109/CVPR.2016.308.

Tzutalin. *LabelImg*. https://github.com/tzutalin/labelImg, 2023.

Wang, C. Y., Bochkovskiy, A., and Liao, H. Y. M., *YOLOv7: Trainable bag-of-freebies sets new state-of-the-art for real-time object detectors*. https://doi.org/10.48550/arXiv.2207.02696, 2022a.

Wang, K., Liu, M. Z., and Ye, Z. J., "An advanced YOLOv3 method for small-scale road object detection," *Applied Soft Computing*, vol. 112, p. 107846, 2021. https://doi.org/10.1016/j.asoc.2021.107846.

Wang, X., "Vehicle image detection method using deep learning in UAV video," *Computational Intelligence and Neuroscience*, vol. 2022, p. 8202535, 2022, 10 pages. https://doi.org/10.1155/2022/8202535.

Wang, X. B., Zhu, D. K., and Yan, Y., "Towards efficient detection for small objects via attention-guided detection network and data augmentation," *Sensors*, vol. 22, no. 19, p. 7663, 2022b. https://doi.org/10.3390/s22197663.

Woo, S., Park, J., Lee, J. Y., and Kweon, I. S., "Cbam: Convolutional block attention module," *2018 Proceedings of the European Conference on Computer Vision (ECCV)*, pp. 3–19, 2018.

Ye, T., Qin, W., Li, Y., Wang, S., Zhang, J., and Zhao, Z., "Dense and small object detection in UAV-vision based on a global-local feature enhanced network," *IEEE Transactions on Instrumentation and Measurement*, vol. 71, pp. 1–13, 2022.

Yeh, C. H., Lin, C. H., Kang, L. W., Huang, C. H., Lin, M. H., Chang, C. Y., and Wang, C. C., "Lightweight deep neural network for joint learning of underwater object detection and color conversion," *IEEE Transactions on Neural Networks and Learning Systems*, vol. 33, no. 11, pp. 6129–6143, 2022.

Yi, S., Liu, X., Li, J., and Chen, L., "UAVformer: A composite transformer network for urban scene segmentation of UAV images," *Pattern Recognition*, vol. 133, p. 109019, 2023. https://doi.org/10.1016/j.patcog.2022.109019.

Youssef, Y., and Elshenawy, M., "Automatic vehicle counting and tracking in aerial video feeds using cascade region-based convolutional neural networks and feature pyramid networks," *Transportation Research Record*, vol. 2675, no. 8, pp. 304–317, 2021. https://doi.org/10.1177/0361198121997833.

Zhang, S., Wen, L., Bian, X., Lei, Z., and Li, S. Z., "Single-shot refinement neural network for object detection," *Proceedings of the IEEE Conference on Computer Vision and Pattern Recognition*, pp. 4203–4212, 2018.

Zhao, L., Bao, Z. W., Xie, Z. J., Huang, G. Y., and Ur Rehman, Z. H., "A point and density map hybrid network for crowd counting and localization based on unmanned aerial vehicles," *Connection Science*, vol. 34, no. 1, pp. 2481–2499, 2022. https://doi.org/10.1080/09540091.2022.2130878

Zhu, P., Wen, L., Du, D., Bian, X., Fan, H., Hu, Q., and Ling, H., "Detection and tracking meet drones challenge," *IEEE Transactions on Pattern Analysis and Machine Intelligence*, vol. 44, no. 11, pp. 7380–7399, 2022. https://doi.org/10.1109/TPAMI.2021.3119563.

22 Drug Toxicity Prediction by Machine-Learning Approaches

22.1 INTRODUCTION

Over the past decades, the illicit drug market has been reshaped by the proliferation of clandestinely produced designer drugs. Synthetic drugs are increasing in popularity in the United States, who are acting to ban certain substances found in drugs. However, the new drug derivatives appear on the illicit drug market, making it difficult for legislation to keep up. Since human beings are daily exposed to millions of chemical molecules and compounds, it is critically important to provide efficient and accurate toxicity prediction. In the past, drug toxicity prediction was carried out by testing in the laboratory, which is very time-consuming. Nowadays with the improvement of molecule-based machine-learning approaches (Wang et al., 2019; Yang et al., 2019; Lu et al., 2019), toxicity prediction has been becoming increasingly efficient and reliable.

Among the machine-learning techniques, *random forest* (RF) (Ho, 1995) is appropriate to recognize the infections from patients' records (Pu et al., 2017) and can improve the molecule prediction for kinase ligands, hormones, and so forth (Cano et al., 2017). *Multi-layer perceptron* (MLP) (Rosenblatt, 1961) is a powerful machine-learning approach, used by Stokes et al. (2021), to predict the actions among different drugs. Moreover, MLP is adopted for drug design to generate variant compounds automatically by certain predefined properties (Gómez-Bombarelli et al., 2018). For instance, Walters and Barzilay (2020) proposed to apply a fully connected neural network for molecule prediction.

As a subset of machine learning, deep learning is widely used in the drug discovery problems. Khrabrov et al. (2017) discovered anticancer agents through molecule fingerprints by *recurrent neural network* (RNN). Autoencoder was introduced to deliver molecules through conducting experiments right into vermin (Zhavoronkov et al., 2019). The wide-range applications of deep learning on drug discovery have drawn researchers' interests in developing more powerful deep-learning models.

However, the input, such as a molecule, can be of arbitrary size and shape. Currently, most of machine-learning and deep-learning approaches can only handle the inputs with a fixed size, i.e., a fixed vector, matrix or tensor. This limitation constraints the applications of machine-learning approaches on drug discovery, especially toxicity prediction. Alternatively, a deep-learning model called *graph neural network* (GNN) (Hamilton et al., 2017), which take graphs as its input to perform feature extraction, was proposed to conduct graph mining, such as node classification, link prediction, graph classification, graph generation, etc. Xu et al. (2019)

DOI: 10.1201/9781003474395-25

studied its theoretical representation and proposed a *graph isomorphism network* (GIN). For example, ChemRL (Fang et al., 2022) is an effective molecule representation learning approach, which successfully facilitates molecular property prediction.

With the help of self-supervised learning and graph transformer, GROVER (Rong et al., 2020) gains significant improvements on a huge unlabeled molecule dataset with 100 million parameters on 10 million unlabeled molecules. Afterward, the growing interests in the representation learning schemes benefit the extended scope of GNNs. Among them, PretrainGNN (Hu et al., 2020) uses a strategy to avoid negative transfer and improve generalization for downstream tasks and achieves state-of-the-art performance for molecular property prediction. Furthermore, Uni-Mol (Zhou et al., 2022) significantly improves the performance of molecule property prediction on various datasets, such as Tox 21 (Mayr et al., 2016), ClinTox (Novick et al., 2013), ToxCast (Richard et al., 2016), SIDER (Kuhn et al., 2016), HIV (Mayr et al., 2016), and BACE (Subramanian et al., 2016). In general, these graph neural networks have successfully reached the state-of-the-art performance on many graph-related applications, such as link prediction, graph classification, node classification, and graph generation.

In this chapter, we aim at applying various machine-learning approaches to solve the toxicity prediction problem. To observe and compare the performance of toxicity prediction, we adopt six datasets: Tox 21, ClinTox, ToxCast, SIDER, HIV, and BACE. We select five machine-learning techniques, including random forest, multi-layer perceptron, logistic regression, graph conv (Duvenaud et al., 2015), and GIN (Xu et al., 2019), to investigate the effects of different models on different toxicity prediction datasets. The rest of this chapter is organized as follows. Section 22.2 briefly introduces the machine-learning techniques used for toxicity prediction. Section 22.3 describes the platform of experimental design. Section 22.4 presents the experimental results and quantitative analysis. Conclusions are drawn in Section 22.5.

22.2 THE MACHINE-LEARNING TECHNIQUES FOR TOXICITY PREDICTION

In this section, we briefly introduce five machine-learning techniques, including random forest, multi-layer perceptron, logistic regression, graph conv, and GIN, for toxicity prediction.

22.2.1 Random Forest

Random forest, consisting of a bunch of decision trees, is a supervised algorithm that is able to perform regression and classification (Ho, 1995). Each decision tree delivers a classification assignment, and a majority voting is formed to make the final classification prediction. By increasing the diversity of individual decision trees, we can significantly improve the generalization ability of the model. In random forest, each leaf node is assigned to the majority class of its training samples, and then the whole forest can predict the class for a given sample based on the majority vote over all trees.

Figure 22.1 shows the flowchart of a random forest. The trees in random forest run in parallel, and each tree is built separately. The decision tree offers flexibility since it can conduct either regression or classification task.

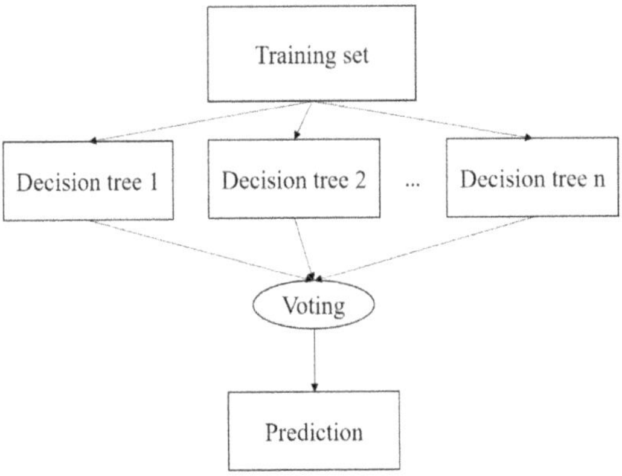

FIGURE 22.1 The flow chart of a general random forest model.

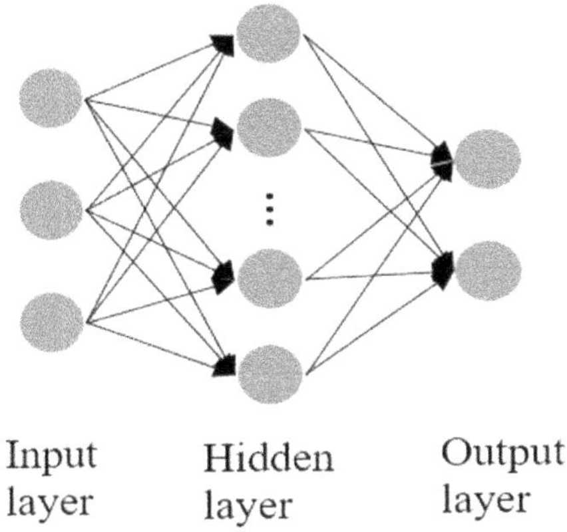

FIGURE 22.2 Illustration of a three-layer MLP. The input layer takes input data and feeds it into the neural network. The hidden layer extracts the features by its neurons, and then the output layer delivers the prediction of the network.

22.2.2 Multi-Layer Perceptron

Multi-layer perceptron consists of an input layer, an output layer, and at least one hidden layer (Xu et al., 2019). Figure 22.2 illustrates a three-layer perceptron. Given a set of data $\{x_i\}$ and weight w_i, we can formulate the weight adjustment rules as follows:

$$w_i \leftarrow w_i + \Delta w_i, \tag{22.1}$$

$$\Delta w_i = \alpha\left(y - \hat{y}\right)x_i, \tag{22.2}$$

where $\alpha \in (0,1)$ is the learning rate, y is the output of the MLP, and $\hat{y}$ is the target. If $y = \hat{y}$, i.e., the prediction is correct, the weight will not be updated; otherwise, the weight will be adjusted by the error $\left(y - \hat{y}\right)$.

The neurons of the MLP are updated by feed-forward pass manner while the weights are updated by back-propagation. Because of the large capacity of a MLP, the model is easy to overfit, so early stopping and regularization are developed to overcome the overfitting problem.

22.2.3 Logistic Regression

Logistic regression is a linear classifier extended from linear regression. A logistic function is adopted to map the output of the linear equation to [0, 1]. The standard logistic function (i.e., sigmoid function) is defined as

$$f(z) = \frac{1}{1+e^{-z}}. \tag{22.3}$$

Figure 22.3 shows the plot of the sigmoid function.

A linear equation predicts $\hat{y}$ as

$$\hat{y} = \sum_{i=1}^{n} w_i x_i + b, \tag{22.4}$$

Where w_i is the i-th element of the weight vector w, x_i is the i-th element of the input vector x, and b is the bias. Therefore, we can convert the linear equation to the logistic regression problem as

$$f(z = \hat{y}) = \frac{1}{1+e^{-\sum_{i=1}^{n} w_i x_i - b}}, \tag{22.5}$$

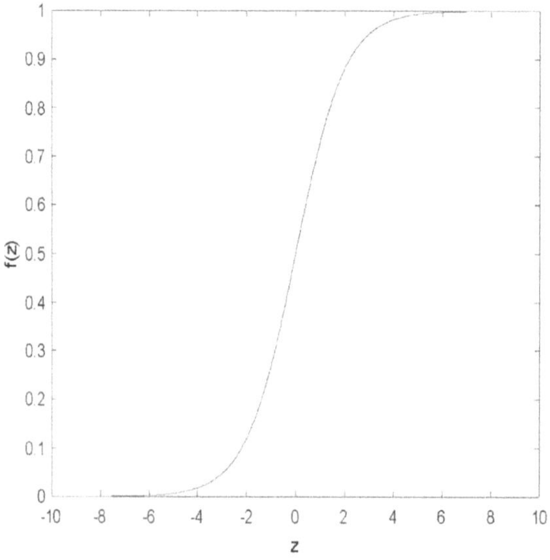

FIGURE 22.3 The plot of the sigmoid function. It maps the input $z \in R$ to the range of [0, 1].

We choose a proper threshold value to determine whether the input sample belongs to class 0 or class 1. To solve the equation, maximum likelihood estimation or gradient descent can be used.

22.2.4 Graph Conv

Graph conv (Duvenaud et al., 2015) introduces a convolutional neural network that operates on the graphs of arbitrary size and shape. It follows the general idea of convolutional neural network but with circular fingerprints. By applying a hashing function at each layer, each atom and its neighbors' structuring information is combined. An indexing operation combines all the feature vectors from the hashing function into a fingerprint of the molecule. Thus, circular fingerprints are able to generate each layer's feature, which is analogous to convolutional layers and global pooling layers in the convolutional neural network.

22.2.5 Graph Neural Networks

Graph neural networks (GNNs) are an effective framework for the representation learning of graphs and use the technique such that the representation vector of a node is calculated by recursively aggregating and transforming representation vectors of its neighboring nodes. Since learning with graph structured data, such as molecules, social, biological, and financial networks, requires effective representation of their graph structure, it is appropriate to be applied for drug property classification.

Let $G = \{V, E\}$ denote an undirected graph of a K-layer graph neural network $f(\cdot)$. The propagation of k-th layer is represented as

$$g_n^{(k)} = AGGREGATION^{(k)}\left(\left\{h_{n'}^{(k-1)} : n' \in N(n)\right\}\right) \tag{22.6}$$

$$h_n^{(k)} = COMBINE^{(k)}\left(h_n^{(k-1)}, g_n^{(k)}\right), \tag{22.7}$$

where $h_n^{(k)}$ is the embedding of the vertex v_n at the k-th layer with $h_n^{(0)} = x_n$, $N(n)$ is the set of vertices adjacent to v_n, and $AGGREGATION^{(k)}(\cdot)$ and $COMBINE^{(k)}(\cdot)$ are functions of the GNN layer. After the K-layer, a $READOUT$ function calculates the output to obtain the embeddings of the graph. An MLP is adopted to perform the downstream task as

$$f(G) = READOUT\left(\left\{h_n^{(k)} : v_n \in V, k \in K\right\}\right), z_G = MLP(f(G)). \tag{22.8}$$

Figure 22.4 shows the overview of GNN. In order to maximize its representational power, GIN (Xu et al., 2019) was developed by using an aggregator to produce different node embeddings when dealing with non-isomorphic graphs. Besides, global pooling is added to produce the graph embedding by using the mean, sum or max of every node embedding. In general, GIN is a more powerful GNN since it largely improves the representational power of the traditional GNN.

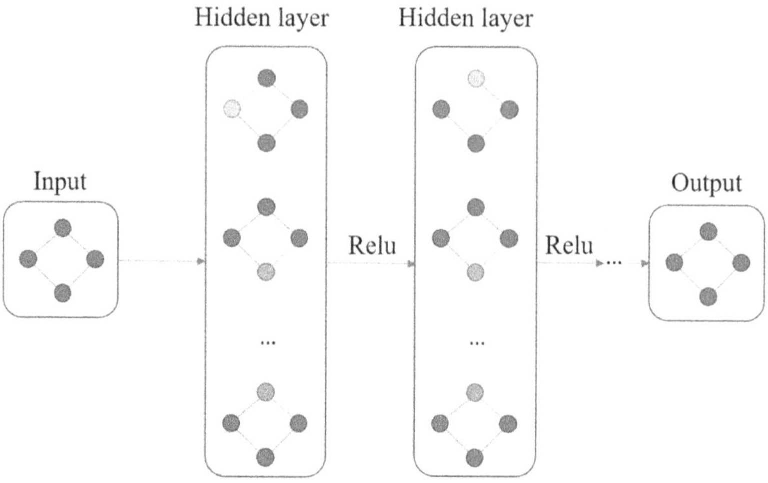

FIGURE 22.4 An overview of a general graph neural network.

22.3 THE PLATFORM OF EXPERIMENTAL DESIGN

To conduct performance comparisons of different models for toxicity prediction, we adopt six datasets: Tox 21, ClinTox, ToxCast, SIDER, HIV, and BACE, which are described in the following.

Tox 21 (Mayr et al., 2016) dataset is a public dataset created to measure the toxicity of chemical compounds. It contains 7,950 chemical compounds and labels for classifications on 12 essays of toxicity. Among them, we randomly select 6,242 for the training set, 683 for the validation set, and 784 for the testing set.

ClinTox (Novick et al., 2013), a toxicity prediction dataset, consists of 1,451 chemical compounds for toxicological study with labels of 2 toxicity tasks. Among them, we randomly select 1,182 for the training set, 148 for the validation set, and 148 for the testing set. Figure 22.5 shows some examples of molecules from ClinTox.

ToxCast (Richard et al., 2016) dataset is another public toxicity prediction data collection that provides 8,575 molecules with 617 classification tasks. Among them, we randomly select 6,860 samples for the training set, 857 samples for the validation set, and 857 for testing set.

SIDER (Kuhn et al., 2016) dataset is an open source dataset that makes efforts to aggregate the side effects of marketed drugs to human beings. It contains 1,427 approved drugs with 27 classification tasks. We use 1,141 of them for training, 143 for validation, and the rest of them for testing.

HIV (Mayr et al., 2016) dataset is a dataset created to experimentally measure abilities to inhibit HIV replication. It includes 41,127 molecules with 1 classification task. 32,901 of them are considered as training data, 4,113 of them are for testing, and another 4,113 are for the testing set.

BACE (Subramanian et al., 2016) dataset contains binary binding results for a set of inhibitors of human β-secretase. It includes 1,513 molecules with 1 classification task. The size of the training set is 1,210, the validation size is 151, and the testing size is 152.

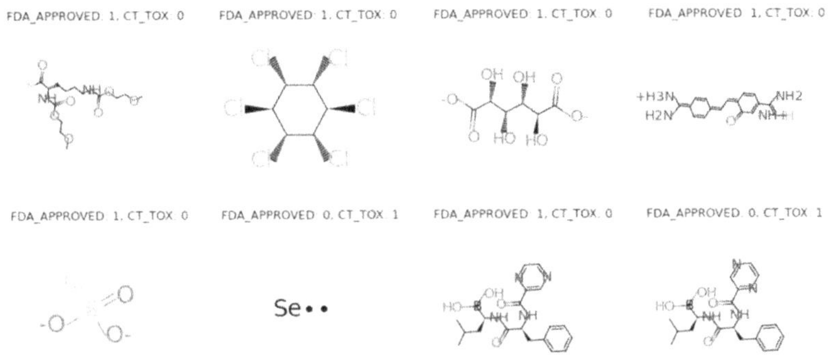

FIGURE 22.5 Some selected molecules from ClinTox dataset.

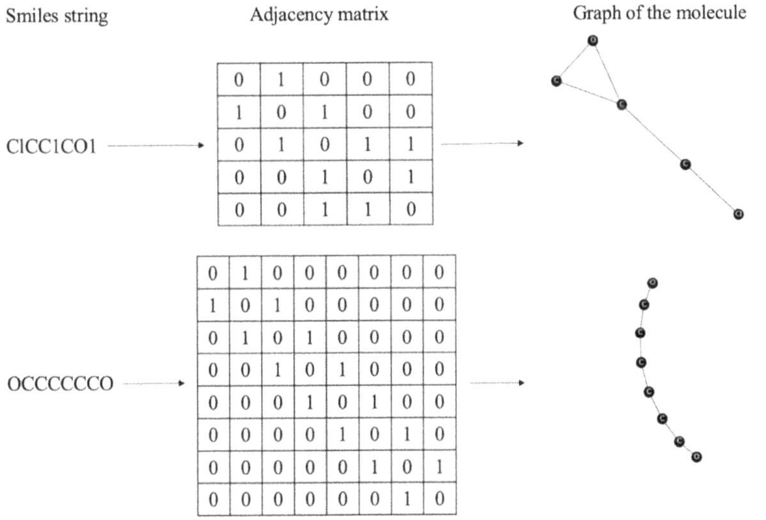

FIGURE 22.6 An example of converting a SMILES string to an adjacency matrix, and its corresponding graph visualization.

A great amount of chemical molecular representation has been developed over the years due to the increasing needs of computer processing on all structural and chemical characteristics. Chemical structures are popularly represented as the molecular graphs: 2D or 3D representation and the input strings: SMILES (Simplified Molecular Input Line Entry System), Canonical SMILES, IUPAC, and more. Depending on the representation, those molecules are stored in different formats. Once the molecule is represented in the correct format, converting a 3D representation to a 2D or an input string of SMILES becomes a simple format conversion problem. We take the SMILES string as the input to encode molecules, which is converted to graph representation by the Python library RDKit.

Figure 22.6 shows an example of a molecule's SMILES string selected from Tox 21 and its corresponding graph representation, i.e., adjacency matrix. A visualization

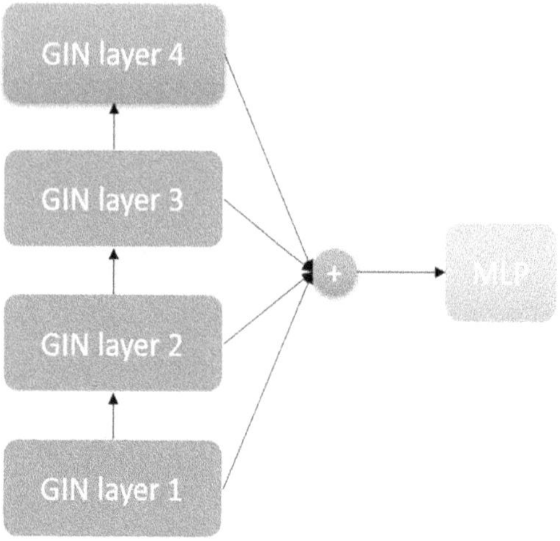

FIGURE 22.7 Illustration of our GIN model. It consists of four GIN layers.

of the selected molecule is presented at the end. By representing a molecule as a graph, we can include the topological information of a molecule, the nodes and their neighbors, and even the connectivity properties.

In this way, we can preserve as much information as possible of a molecule, which is definitely beneficial to the molecule's property prediction. In our experiments, we select five machine-learning techniques – random forest, multi-layer perceptron, logistic regression, graph conv, and GIN – for investigating their performance of toxicity prediction on different datasets. We create 500 decision trees for the random forest, and adopt Gini impurity to split up nodes to form the trees. For MLP, a three-layer perceptron is constructed, and 300 epochs are executed with a softmax classifier at the end of the output layer.

For the deep-learning models, a graph conv with four hidden layers is optimized by the Adam optimizer with a learning rate of 0.0001 in a mini-batch strategy and the batch size of 1,024. To perform a fair comparison, we adopt GIN with four hidden layers and keep the Adam optimizer with the same learning rate 10^{-4}. and the batch size 256. For both deep-learning models, we adopt the binary cross entropy loss and train them for 500 epochs. More specifically, we illustrate the architecture of our GIN model in Figure 22.7. The dimension of each hidden layer is 512, and the output dimension is 1,024. Furthermore, we adopt a batch norm layer inside each of the GIN hidden layers.

All our experiments are implemented by Pytorch (Paszke et al., 2019) using four NVIDIA Titan XP GPUs. Additionally, graph conv and GIN are implemented by TrochDrug (Zhu et al., 2022), which provides drug discovery models in Pytorch.

22.4 EXPERIMENTAL RESULTS AND QUANTITATIVE ANALYSIS

We use the ROC AUC score to measure the performance of toxicity prediction. It calculates the area under the ROC curve, which is a chart to show the tradeoff between TPR (true positive rate) and FPR (false positive rate) for every threshold. The TPR and FPR can be formulated as

$$TPR = \frac{TP}{TP + FN}, \tag{22.9}$$

$$FPR = \frac{FP}{FP + TN}, \tag{22.10}$$

where *TP* and *FN* are true positive and false negative, respectively, and *TN* is true negative in the confusion matrix. Therefore, the ROC AUC score is in the range [0, 1], where the higher the better. In other words, the ROC AUC score measures the ability of distinguishing positive and negative classes apart.

We use the accuracy to measure the classification performance. A threshold of 0.5 is selected for classification. If the prediction is greater than 0.5, it belongs to class 1; otherwise, it belongs to class 0. The accuracy is formulated as

$$Accuracy = \frac{TP + TN}{TP + TN + FP + FN}, \tag{22.11}$$

where *FP* is false positive in the confusion matrix. Intuitively, it divides the number of correct predictions by the total number of predictions.

We use the recall to show the ability of a classifier to classify the positive samples. It is actually the TPR, which is computed by equation (9). Therefore, it is in the range [0, 1], and the higher the better.

Tables 22.1, 22.2 and 22.3 show the performance of four methods on Tox 21, ClinTox, ToxCast, SIDER, HIV, and BACE. We can observe that as comparing with random forest, MLP, and logistic regression, the deep-learning approaches significantly improve the ROC AUC scores on all the datasets. The graph conv achieves 0.6720, 0.8282, 0.6190, 0.5524, 0.7517, and 0.8838 on Tox 21, ClinTox, ToxCast, SIDER, HIV, and BACE, respectively. Furthermore, we can gain larger improvements over graph conv by GIN. On Tox 21, we obtain around 0.12 improvement over graph conv, while 0.10 increase is achieved on ClinTox. We slightly increase about 0.01 ROC AUC score for the ToxCast dataset. In SIDER, the GIN significantly

TABLE 22.1
ROC AUC Score of Different Machine-Learning Approaches on Toxicity Prediction Datasets

ROC AUC Score	Tox 21	ClinTox	ToxCast	SIDER	HIV	BACE
RF	0.6536	0.6975	-	0.6205	0.6966	0.8752
MLP	0.6497	0.6708	-	0.5922	0.6878	0.7641
LR	0.6570	0.7470	-	0.5917	0.6999	0.7982
Graph Conv (Subramanian et al., 2016)	0.6720	0.8282	0.6190	0.5524	0.7517	0.8838
GIN (Xu et al., 2019)	**0.8045**	**0.9240**	**0.6330**	**0.6750**	**0.8009**	**0.8971**

TABLE 22.2
Accuracy of Different Machine-Learning Approaches on Toxicity Prediction Datasets

Accuracy (%)	Tox 21	ClinTox	ToxCast	SIDER	HIV	BACE
RF	92.93	93.58	–	**78.11**	97.11	**73.03**
MLP	91.24	**94.25**	–	73.99	**97.16**	67.76
LR	76.96	82.43	–	67.37	61.10	71.71
Graph Conv (Subramanian et al., 2016)	87.72	93.91	81.01	46.91	96.45	58.55
GIN (Xu et al., 2019)	**93.09**	49.99	**81.66**	45.13	96.08	57.24

TABLE 22.3
Recall of Different Machine-Learning Approaches on Toxicity Prediction Datasets

Recall	Tox 21	ClinTox	ToxCast	SIDER	HIV	BACE
RF	0.0525	0.5000	–	0.7138	0.0846	0.5978
MLP	0.1820	0.6464	–	0.6619	0.2000	0.5652
LR	0.4479	0.6817	–	0.6601	**0.6923**	0.6522
Graph Conv (Subramanian et al., 2016)	0.2977	0.7212	0.1789	0.6998	0.3767	0.8730
GIN (Xu et al., 2019)	**0.4948**	**0.8507**	**0.2811**	**0.7171**	0.4470	**0.8923**

improves the ROC AUC score by around 0.05 as comparing with graph conv. In HIV, GIN improves around 0.05 compared to graph conv, and it slightly gains 0.01 more on graph conv. The recall score shows that the similar pattern as the ROC AUC score. Although logistic regression performs the best on HIV dataset, GIN still has superiority performance on toxicity prediction. Therefore, the deep-learning approaches have significant improvement, so we conclude that GIN performs the best on classifying safe drugs.

When it comes to the classification accuracy, we cannot find a pattern because we have inconsistent results on different datasets. Due to the imbalanced property of the drug property prediction datasets, the classification accuracy is not a good performance metric for drug property prediction tasks. From the quantitative analysis, we observe that graph neural networks outperform other approaches, and it shows the effectiveness of GNN on drug property predictions tasks. With the arbitrary input shape and size, GNN is able to take the accurate representation of the graphs as input and extract the features to obtain the graph embedding for downstream tasks. Furthermore, the weights of the GNN are not only on the neurons, but also on the edges and nodes of the graphs, which helps it learn the graph structures as well. In this way, GNN learns features from both the nodes and the edges of the graphs

TABLE 22.4
The ROC AUC Score of Different Hidden Layer Dimension for GIN on Toxicity Prediction Datasets

Hidden dim.	Tox 21	ClinTox	ToxCast	SIDER	HIV	BACE
256	0.7707	0.7073	**0.6496**	0.5965	**0.8401**	0.8970
512	**0.8045**	0.9240	0.6330	**0.6750**	0.8009	0.8971
1024	0.7755	**0.9786**	0.6212	0.5891	0.7575	**0.9375**

and makes it more powerful on learning graph topology, as comparing with other machine-learning and deep-learning approaches.

Based on the results in Table 22.2, the classification accuracy has an inconsistent pattern with ROC AUC score and recall because of the imbalanced labels in drug property prediction datasets. In real world applications, we aim at identifying fewer toxic drugs from large amounts of safe drugs, so it is crucial to have a higher recall instead of classification accuracy. The performance metrics like ROC AUC score and recall are better than the classification accuracy.

Furthermore, we conduct an ablation study on GIN by adjusting the dimension of each hidden layer. As shown in Table 22.4, when adjusting the dimension of the hidden layer for different datasets, we can slightly improve the performance of GIN. Especially on ClinTox, SIDER, and HIV datasets, by increasing the dimension of the hidden layer, the ROC AUC score increases from 0.9240 to 0.9786, from 0.5965 to 0.6750, and from 0.8009 to 0.8401, respectively. Therefore, by adjusting some hyper-parameters, GIN still has potential to improve its performance on toxicity prediction.

22.5 CONCLUSIONS

In this chapter, we apply machine-learning approaches to drug property prediction tasks, especially toxicity prediction. We show that deep-learning approaches, especially GNN, are able to learn graph representation of chemical molecules. Furthermore, GNN takes molecules as graphs of arbitrary size and shape and aggregates the graph embedding by targeting nodes and edges, making it to achieve better results over other machine-learning approaches.

Moreover, from the experimental results on different performance metrics, we observe that the most commonly used classification accuracy in other application areas may not be suitable for drug property prediction due to that our main goal for drug property prediction is to clearly distinguish the positive samples from negative samples. Therefore, using ROC AUC score is a great way to measure the drug property prediction models.

Drug property prediction, especially toxicity prediction, is a crucial task in modern health environment. As shown in the property prediction datasets in this chapter, we observe that GIN is able to distinguish between safe and toxic drugs, and it is the best approach to perform such tasks in the real-world applications under the supervision of ROC AUC score and recall.

REFERENCES

Cano, G., Garcia-Rodriguez, J., Garcia-Garcia, A., Perez-Sanchez, H., Benediktsson, J. A., Thapa, A., and Barr, A., "Automatic selection of molecular descriptors using random forest: Application to drug discovery," *Expert Systems with Applications*, vol. 72, pp. 151–159, 2017.

Duvenaud, D. K., Maclaurin, D., Iparraguirre, J., Bombarell, R., Hirzel, T., Aspuru-Guzik, A., and Adams, R. P., "Convolutional networks on graphs for learning molecular fingerprints," *Advances in Neural Information Processing Systems*, vol. 28, 2015.

Fang, X., Liu, L., Lei, J., He, D., Zhang, S., Zhou, J., Wang, F., Wu, H., and Wang, H., "Chemrlgem: Geometry enhanced molecular representation learning for property prediction," *Nature Machine Intelligence*, vol. 4, pp. 127–134, 2022.

Gómez-Bombarelli, R., Wei, J. N., Duvenaud, D., Hernández-Lobato, J. M., Sánchez-Lengeling, B., Sheberla, D., Aguilera-Iparraguirre, J., Hirzel, T. D., Adams, R. P., and Aspuru-Guzik, A., "Automatic chemical design using a data-driven continuous representation of molecules," *ACS Central Science*, vol. 4, no. 2, pp. 268–276, 2018.

Hamilton, W. L., Ying, R., and Leskovec, J., "Representation learning on graphs: Methods and applications," *IEEE Data Engineering Bulletin*, vol. 40, no. 3, pp. 52–74, 2017.

Ho, T. K., "Random decision forests," *Proceedings of 3rd International Conference on Document Analysis and Recognition*, vol. 1, pp. 278–282, IEEE, Aug. 1995.

Hu, W., Liu, B., Gomes, J., Zitnik, M., Liang, P., Pande, V., and Leskovec, J., "Strategies for pre-training graph neural networks," *International Conference on Learning Representations*, pp. 1–22, 2020.

Khrabrov, K., Mamoshina, P., Zhavoronkov, A., Kadurin, A., Vanhaelen, Q., Kazennov, A., and Aliper, A., "The cornucopia of meaningful leads: Applying deep adversarial autoencoders for new molecule development in oncology," *Oncotarget*, vol. 8, no. 7, pp. 10883–10890, 2017.

Kuhn, M., Letunic, I., Jensen, L. J., and Bork, P., "The SIDER database of drugs and side effects," *Nucleic Acids Research*, vol. 44, no. D1, pp. D1075–D1079, 2016.

Lu, C., Liu, Q., Wang, C., Huang, Z., Lin, P., and He, L., "Molecular property prediction: A multilevel quantum interactions modeling perspective," *Proceedings of the AAAI Conference on Artificial Intelligence*, vol. 33, no. 1, pp. 1052–1060, Jul. 2019.

Mayr, A., Klambauer, G., Unterthiner, T., and Hochreiter, S., "DeepTox: Toxicity prediction using deep learning," *Frontiers in Environmental Science*, vol. 3, p. 80, 2016.

Novick, P. A., Ortiz, O. F., Poelman, J., Abdulhay, A. Y., and Pande, V. S., "SWEETLEAD: An in silico database of approved drugs, regulated chemicals, and herbal isolates for computer-aided drug discovery," *PLoS One*, vol. 8, no. 11, p. e79568, 2013.

Paszke, A., Gross, S., Massa, F., Lerer, A., Bradbury, J., Chanan, G., Killeen, T., Lin, Z., Gimelshein, N., Antiga, L., Desmaison, A., Köpf, A., Yang, E., DeVito, Z., Raison, M., Tejani, A., Chilamkurthy, S., Steiner, B., Fang, L., Bai, J., and Chintala, S., "PyTorch: An imperative style, high-performance deep learning library," *Advances in Neural Information Processing Systems*, vol. 32, 2019.

Pu, Y., Wang, W., Henao, R., Chen, L., Gan, Z., Li, C., and Carin, L., "Adversarial symmetric variational autoencoder," *Advances in Neural Information Processing Systems*, vol. 30, pp. 1–15, 2017.

Richard, A. M., Judson, R. S., Houck, K. A., Grulke, C. M., Volarath, P., Thillainadarajah, I., Yang, C., Rathman, J., Martin, M. T., Wambaugh, J. F., Knudsen, T. B., Kancherla, J., Mansouri, K., Patlewicz, G., Williams, A. J., Little, S. B., Crofton, K. M., and Thomas, R. S., "ToxCast chemical landscape: Paving the road to 21st century toxicology," *Chemical Research in Toxicology*, vol. 29, no. 8, pp. 1225–1251, 2016.

Rong, Y., Bian, Y., Xu, T., Xie, W., Wei, Y., Huang, W., and Huang, J., "Self-supervised graph transformer on large-scale molecular data," *Advances in Neural Information Processing Systems*, vol. 33, pp. 12559–12571, 2020.

Rosenblatt, F., *Principles of Neurodynamics Perceptrons and the Theory of Brain Mechanisms*, Cornell Aeronautical Lab Inc, 1961.

Stokes, A., Hum, W., and Zaslavsky, J., "A minimal-input multilayer perceptron for predicting drug-drug interactions without knowledge of drug structure," *STEM Fellowship Journal*, vol. 6, no. 1, pp. 19–23, 2021.

Subramanian, G., Ramsundar, B., Pande, V., and Denny, R. A., "Computational modeling of β-secretase 1 (BACE-1) inhibitors using ligand based approaches," *Journal of Chemical Information and Modeling*, vol. 56, no. 10, pp. 1936–1949, 2016.

Walters, W. P. and Barzilay, R., "Applications of deep learning in molecule generation and molecular property prediction," *Accounts of Chemical Research*, vol. 54, no. 2, pp. 263–270, 2020.

Wang, X., Li, Z., Jiang, M., Wang, S., Zhang, S., and Wei, Z., "Molecule property prediction based on spatial graph embedding," *Journal of Chemical Information and Modeling*, vol. 59, no. 9, pp. 3817–3828, 2019.

Xu, K., Hu, W., Leskovec, J., and Jegelka, S., "How powerful are graph neural networks?" *International Conference on Learning Representations*, New Orleans, LA, May 6–9, 2019.

Yang, K., Swanson, K., Jin, W., Coley, C., Eiden, P., Gao, H., Guzman-Perez, A., Hopper, T., Kelley, B., Mathea, M., Palmer, A., Settels, V., Jaakkola, T., Jensen, K., and Barzilay, R., "Analyzing learned molecular representations for property prediction," *Journal of Chemical Information and Modeling*, vol. 59, no. 8, pp. 3370–3388, 2019.

Zhavoronkov, A., Ivanenkov, Y. A., Aliper, A., Veselov, M. S., Aladinskiy, V. A., Aladinskaya, A. V., Terentiev, V. A., Polykovskiy, D. A., Kuznetsov, M. D., Asadulaev, A., Volkov, Y., Zholus, A., Shayakhmetov, R. R., Zhebrak, A., Minaeva, L. I., Zagribelnyy, B. A., Lee, L. H., Soll, R., Madge, D., Xing, L., Guo, T., and Aspuru-Guzik, A., "Deep learning enables rapid identification of potent DDR1 kinase inhibitors," *Nature Biotechnology*, vol. 37, no. 9, pp. 1038–1040, 2019.

Zhou, G., Gao, Z., Ding, Q., Zheng, H., Xu, H., Wei, Z., Zhang, L., and Ke, G., "Uni-Mol: A universal 3D molecular representation learning framework," *ChemRxiv*, pp. 1–31, 2022.

Zhu, Z., Shi, C., Zhang, Z., Liu, S., Xu, M., Yuan, X., and Tang, J., "Torchdrug: A powerful and flexible machine learning platform for drug discovery," arXiv:2202.08320, 2022.

23 An Efficient Detection and Recognition System for Multiple Motorcycle License Plates Based on Decision Tree

23.1 INTRODUCTION

Motorcycles are considered as small, convenient vehicles, which have been widely used in Asian as well as other tropical countries. Due to increasingly tremendous amounts of roadside parking motorcycles, it is an urgent need for the parking bureau to acquire the technology of automatic motorcycle license plate detection and recognition since manual key-in of license plates numbers is error-prone and labor-intensive. Therefore, the development of such an accurate license plates detection and recognition system is critical, having lots of applications including parking administration, toll processing, and parking fee collection. Moreover, it can help police officers visually search for stolen motorcycles by simply scanning license plates in patrols.

The *vehicle license plate recognition* (VLPR) or *automated license plate recognition* (ALPR) has been studied intensively and extensively over the last decade. There are various weather conditions when vehicles are driven in the real world, which can affect the license plate detection and recognition accuracy. These challenges include uneven illumination, skewed views, blurriness, different distances from the camera, distortion, multiple license plates in the image, night, brake lights, headlights, adverse atmospheric conditions, complex backgrounds, unclear vehicle plates, low-quality surveillance camera, and so on (Björklund et al., 2019).

Motorcycles, in contrast to automobiles, are more compact and versatile, thus their license plates appear in the image view almost arbitrarily with cluttered scenes and complex backgrounds. Therefore, the *motorcycle license plate recognition* (MLPR) is more challenging and difficult than the VLPR. When the existing VLPR systems are directly applied to the MLPR, it often leads to undesirable results. Figure 23.1 shows some examples of generating the failure results of detecting "no vehicle" when the WPOD-NET (Li et al., 2018) is used to detect motorcycle license plates.

Conventional ALPR methods include license plate detection, tilt correction, character segmentation, and character recognition. Hsu et al. (2013) proposed application-oriented license plate recognition in three scenarios: access control, law enforcement, and road patrol. However, their method may produce errors due to imprecise cropping in the detection and segmentation phases to induce misclassification in the

DOI: 10.1201/9781003474395-26

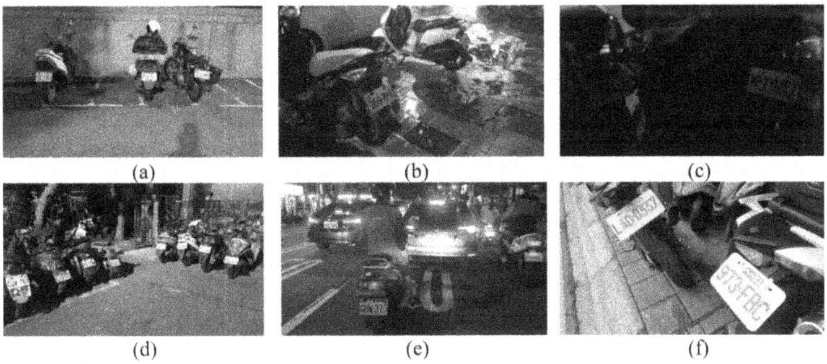

FIGURE 23.1 Examples of motorcycle license plate recognition. (a) Three motorcycles in low light conditions with a variable amount of skew in each license plate, (b) two motorcycles at a rainy night with a large amount of horizontal and vertical skews in the view, (c) three motorcycles in poor lighting condition, (d) eight motorcycles captured with camera flash lighting, (e) five license plates captured at night with blurriness, bright lights, large viewing distance, and skewed angles, and (f) two motorcycles with yellow background in large skew orientations.

recognition phase. Al-Shemarry et al. (2018) used many AdaBoost cascades and a 3-level local binary pattern classifier to detect license plate regions. However, their method is limited with one license plate per image, and no character segmentation is performed for recognition.

Khare et al. (2019) proposed to segment license plate characters using partial character reconstruction based on the characteristics of stroke width in the Laplacian and gradient domain. This approach may result in character components with incomplete shapes. Molina-Moreno et al. (2019) detected license plates based on a scale-adaptive deformable part-based model. Their method is effective in detecting license plates under various viewing distances and scales. However, their method struggles with low lighting conditions.

In this chapter, we propose an efficient decision-tree-based deep-learning framework for motorcycles, including scooters and heavy motorcycles. Our method starts with license plate detection to determine the input image orientation, and explicitly identify variant view skews of the license plates to effectively process the license plate rectification and recognition pipeline. The rest of this chapter is organized as follows. Section 23.2 briefly discusses the related work of vehicle and motorcycle license plate recognition. Our proposed algorithm is presented in Section 23.3. Section 23.4 provides experimental results and performance comparisons with currently available methods and commercial systems. Finally, conclusions are drawn in Section 23.5.

23.2 THE RELATED WORK

The vehicle license plate recognition techniques can be categorized into three classes: (1) conventional, (2) deep-learning-based, and (3) hybrid methods. Deep-learning-based methods use a single or multiple convolutional neural networks (CNNs). Hsu et al. (2017) proposed one CNN that modified Yolov1 (Redmon et al., 2015) or Yolov2 (Redmon et al., 2016) deep networks to detect license plates. However, they only focus on the detection but not the recognition of license plates. Li et al. (2019)

proposed a unified deep neural network to detect and recognize vehicle license plates in natural scene images. Their network can localize license plates and recognize the letters simultaneously in a single forward pass. This approach avoids intermediate error accumulation and accelerates the processing speed. However, it requires a detailed annotated dataset for model training.

Selmi et al. (2020) used a mask region CNN framework for license plate detection and recognition in complex scenes. Wang et al. (2020) proposed a Chinese VLPR method to directly recognize license plates through a single-CNN architecture, which consists of (i) a residual network to extract base features, (ii) a network to extract multi-scale features, (iii) a regression net to locate license plates and characters, and (iv) a classification network for recognition.

The two-CNN VLPR system of Björklund et al. (2019) consists of five steps: (1) preprocessing, (2) license plate detection CNN with localization, classification, redundancy reduction, and refinement, (3) license plate rectification, (4) character detector CNN with per-character localization, classification, redundancy reduction, and refinement, and (5) post-processing. Kessentini et al. (2019) proposed a two-CNN stages to tackle the problem of multi-norm and multilingual license plate detection and recognition. However, their method only deals with the simple case of one license plate per image.

Silva and Jung (2018) proposed a complete ALPR system based on three CNNs that can run with unconstrained scenarios, where the license plate might be considerably distorted in oblique views. A YOLOv2 network is adopted to detect the license plates, and the detected license plates are resized and fed into a Warped Planar Object Detection Network (WPOB-NET) for license plate rectification. Optical character recognition (OCR) is then performed. Li et al. (2018) proposed three different CNN classifiers to detect and recognize license plates. For license plate detection, a 37-class CNN classifier was used to detect the presence of text by generating a text saliency map. Another plate/non-plate CNN classifier is adopted to distinguish license plates from general text and reject false positives.

Laroca et al. (2018) proposed a multiple CNN ALPR system, where a CNN detects vehicles, a CNN detects license plates, a CNN segments the license plate character, and two CNNs recognize the license plate characters. Two data augmentation methods are employed. Negative images simulate characters from other vehicle categories, and characters are checked which can be flipped horizontally or vertically to create new instances. They achieve a recognition rate of 78.33% by aggregating temporal information using majority voting. Abdullah et al. (2018) proposed a YOLO-based three-staged network for Bangla license plate recognition in Dhaka metropolitan city. Their method applied a binary YOLOv3 classifier (Redmon et al., 2018; YOLOv3, 2018) to detect license plates. Another 11-class YOLOv3 classifier is used to detect the six Bangla digits, and yet another 32-class CNN is developed to classify the 32 Bangla characters.

Laroca et al. (2019) developed a three-staged CNN ALPR system: (i) a YOLOv2 network detects vehicles, (ii) a fast-YOLOv2 network detects license plates and classifies the license plate layout, and (iii) a CR-NET recognizes the license plates. Omar et al. (2020) proposed a four CNN ALPR system. They first pre-process the input image by Gaussian filtering and adaptive contrast enhancement, apply a deep semantic segmentation network to determine the three license plate regions, and apply a deep encoder-decoder network to segment the license plate regions. The determined license plate regions are fed into two separate CNN models for Arabic number recognition and city determination. Their method is efficient in both license plate detection and recognition; however, only the simple case of one license plate per image is considered.

Liu et al. (2018) proposed a hybrid ALPR method to locate, segment, and recognize Chinese license plates. They used a color edge method to detect the color edge features of Chinese license plate while eliminating unimportant edge points. Xie et al. (2018) presented a combined feature extraction model and back-propagation neural network to detect and recognize license plates in weak illumination and complicated background. The contrast of license plate is first enhanced, and Roberts edge operator and morphological operations are applied to obtain license plate candidate regions. However, most of their tested license plates are in frontal views without addressing the skewed view issues.

Motorcycle license plate detection and recognition are less addressed in the ALPR literature, despite their popularity and importance in in Southeast Asia and other countries. The ALPR system of Lee et al. (2004) extracts and recognizes license plates of vehicles and motorcycles on the highway. Their system consists of three stages: moving object detection, license plate extraction, and character recognition. A block-difference method is used to reduce computational time for moving object detection. Projection-based method is applied to extract license plates in various sizes. Statistical features including direction/crossing counts and peripheral background areas are used for license plate recognition.

Hsu et al. (2015) provided a comparative study on motorcycle license plate detection in terms of processing speed and accuracy. They found that histogram of oriented gradients (HOG) with support vector machines (SVMs) performs the best for motorcycle license plate detection. There exist only a few works on tracking of motorcycle license plates following the tracking by detection paradigm. Hsu and Chiu (2016) provided a comparative study on this regard. Puarungroj and Boonsirisumpun (2018) proposed two CNN networks to recognize Thai motorcycle license plates, which contain three lines of characters of the category, registered place, and registration numbers. The single-shot multibox detector (SSD) is used to segment the license plate image into three parts: letters, numbers, and the registered places. However, this method cannot detect license plates with large viewing angle or skews.

23.3 THE PROPOSED DETECTION AND RECOGNITION SYSTEM FOR MOTORCYCLE LICENSE PLATES

Our proposed detection and recognition system for motorcycle license plates is based on YOLOv3 deep-learning model in a decision-tree processing pipeline. The existing ALPR systems aim at automobiles under the assumption of a simple background surrounding the automobile license plate. However, our system can effectively detect and localize multiple motorcycle license plates under complex background, rectify variant view configurations, and recognize the license plates, all running on a case-by-case analysis as a decision tree pipeline. Besides, our system is simple, effective, flexible, and easy to train.

Figure 23.2 illustrates the proposed detection and recognition system. We categorize the viewing angles of motorcycle license plates into three types: frontal views, horizontally skewed views, and hybrid horizontally-and-vertically-skewed views. The type is determined and handled in a fixed order from simple to complex to avoid unnecessary

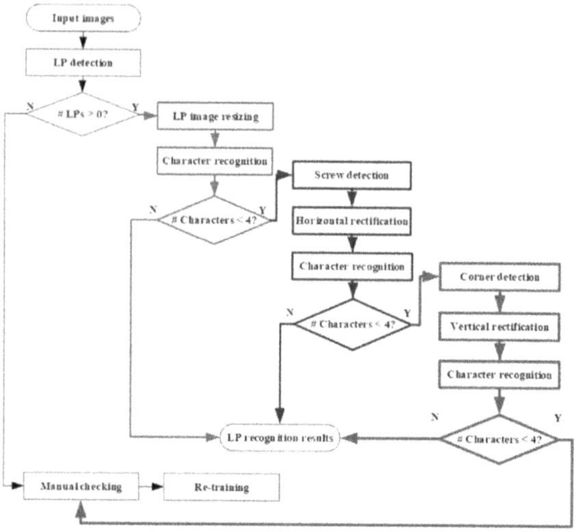

FIGURE 23.2 The proposed decision-tree-based deep-learning pipeline for motorcycles.

rectifications and extra computations. If a license plate is confidently determined to be horizontally skewed but not vertically skewed, we will only perform a horizontal rectification. This decision-tree pipeline effectively improves the amortized processing speed for multiple license plates. Note that from our experiments, there is no need to go further for rectifications with more than two parameters of hybrid horizontally-and-vertically-skewed rectification, such as the affine or projective transformations.

In general, motorcycles are typically parked in standing on a center pillar or a side pillar, which tilts the motorcycle by an angle. If the motorcycle is parked upright and captured from a frontal view, no rectification is required (as Case I pipeline in red) and the license plate character recognition can be directly performed. If the number of recognized license plate characters is less than a pre-defined threshold, the license plate is likely in a skewed view (as Case II pipeline in blue). If a sufficient number of license plate characters cannot be recognized, we will move on to Case III pipeline in purple to apply a hybrid horizontally-and-vertically-skewed rectification. If the YOLOv3 license plate detector cannot detect any license plate, the input image will be checked against ground-truth labels to decide whether a miss detection occurs. If it is, this image will be used for detector retraining or fine-tuning.

Our proposed system is efficient throughout the decision-tree pipelines and cases since only a single detector network is required. Note that YOLOv3 is used as the base detector, which can be replaced by another compatible detector without changing the pipeline. Since the YOLOv3 is not initially designed for detecting license plates nor for recognizing license plate characters, we propose a decision-tree pipeline for license plate rectification to achieve high performance and efficiency.

We train the YOLOv3 license plate detection and recognition model on the following 40 classes: the license plate, screw, corner, the dot character of the license plate, license plate digits (0 to 9), and upper-case license plate characters (A to Z), which are sufficient for the ALPR on all cases. A few hyper-parameters are adapted for training in the YOLOv3 model, and use batch size of 64, subdivisions of 16, max batches

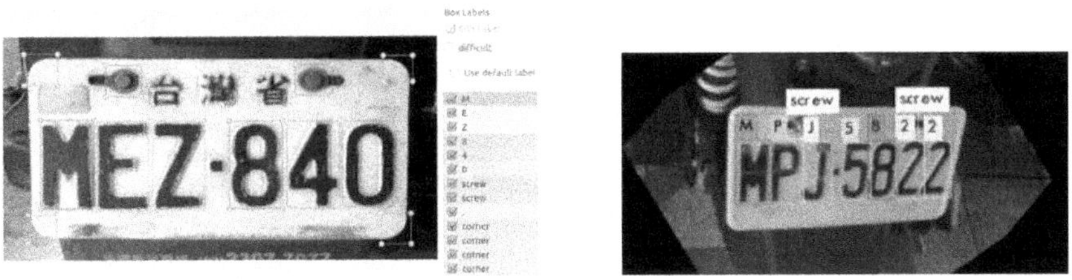

(a) (b)

FIGURE 23.3 (a) LabelImg license plate annotation example, (b) YOLOv3 license plate character and fiducial detection example.

of 50, 200, and the number of conv filters = (#classes + 1 + 4) × 3 = 135. We initialize training with ImageNet pre-trained model, and in the testing runs we lower the detection score threshold to ensure high recall of license plates since false detections can be reliably filtered out in the subsequent cases in our decision tree. We apply the LabelImg (2018) GUI tool to annotate license plates together with the screws, license plate corners, and each license plate character as shown in Figure 23.3.

CASE I: PIPELINE IN THE DECISION TREE FOR RECOGNIZING LICENSE PLATES IN FRONTAL VIEW

Given a license plate detection, we crop the license plate image and enlarge the detected image into 600 × 300 and run the YOLOv3 detector to detect individual license plate characters. This step provides a solution to go beyond the limitation of the YOLOv3 detector for being unable to detect small objects, such as the individual license plate characters. If the number of the characters in the resized license plate is greater than a threshold, this license plate is regarded as in frontal view, *i.e.*, no view rectification is required. Otherwise, the process flow will continue to the Case II pipeline for view rectification.

CASE II: PIPELINE IN THE DECISION TREE FOR RECOGNIZING LICENSE PLATE IN HORIZONTALLY SKEWED VIEW

The existing ALPR method in Jun et al. (2012) performs similar horizontal rectification by the combination of binary thresholding, Canny edge detection, and Hough transform to detect the horizontally skewed angles. However, it is sensitive to environmental factors and large view angles, and thus, post-processing, such as contrast enhancement and median filtering, is required. In our approach, we estimate the license plate horizontal skew angle by first detecting the two license plate screws as fiducials using the YOLOv3 model. In addition, if any of the license plate screws are not detected, we will have the Case III pipeline to be described next that can estimate the license plate skew. Figure 23.4 shows how the two detected screws are used to calculate the license plate horizontal skew angle.

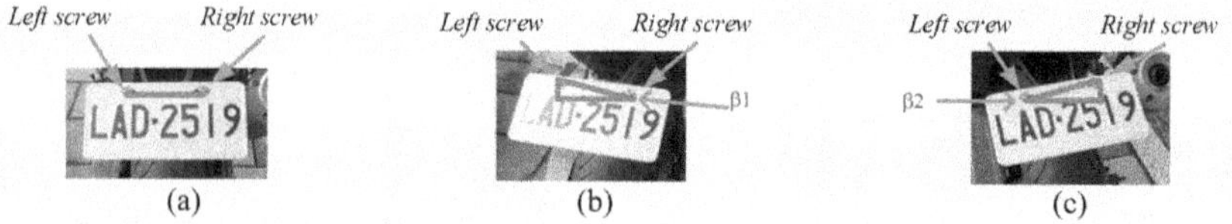

FIGURE 23.4 License plate horizontal rectification based on the two detected screws. (a) The frontal view with no horizontal skew, (b) a horizontal skew cases with a clockwise skew angle, $\beta 1$, and (c) a horizontal skew cases with a counterclockwise skew angle, $\beta 2$.

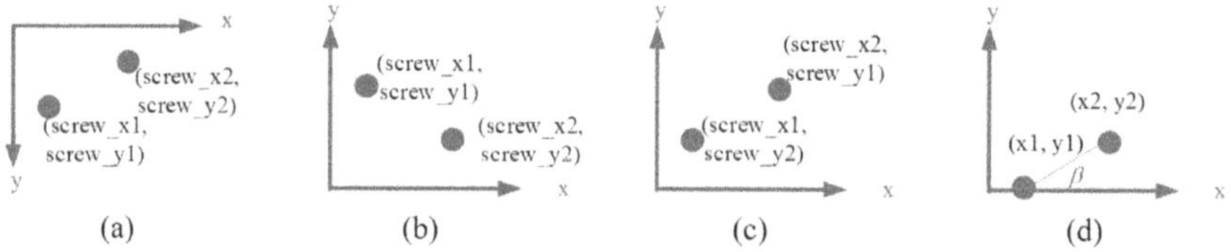

FIGURE 23.5 License plate screw coordinate transformation. (a) The YOLOv3 detection coordinate in image pixels, (b) Cartesian coordinate, (c) coordinate exchange, and (d) the standardized coordinate for skew rectification calculation.

Figure 23.5 describes the coordinate transformation used to calculate the skew angle and the horizontal rectification via exchanging of coordinates using the following equations:

$$\begin{aligned} \mathrm{x1} &= \mathrm{screw\_x1} - screw\_x1 + 1 = 1 \\ y1 &= screw\_y2 - screw\_y2 = 0 \\ x2 &= screw\_x2 - screw\_x1 + 1 \\ y2 &= screw\_y1 - screw\_y2 \end{aligned} \tag{23.1}$$

The horizontal skew angle (β) in radian is calculated by the following equation:

$$\beta = -\cos^{-1} \frac{x_1 \times x_2 + y_1 \times y_2}{\sqrt{x_1^2 + y_1^2} \times \sqrt{x_2^2 + y_2^2}} \tag{23.2}$$

Coordinate transformation for horizontal rectification is

$$\begin{cases} x' = x_0 \times \cos\beta - y_0 \times \sin\beta \\ y' = x_0 \times \sin\beta + y_0 \times \cos\beta \end{cases} \tag{23.3}$$

where (x_0, y_0) denotes image pixel positions and (x', y') denotes the transformed image positions. In some rear cases when any screw is missing, we simply perform horizontal rectification using a pre-calculated default skew parameter before moving on to the rest of the processing pipeline.

Next the YOLOv3 is performed on the rectified license plate image to detect and recognize license plate characters. If the number of recognized characters is greater than or equal to an empirically determined threshold of 4, we finalize the pipeline with an end if of a rectified license plate view. Otherwise, the pipeline continues to Case III.

CASE III: PIPELINE IN DECISION TREE FOR RECOGNIZING LICENSE PLATES IN HYBRID HORIZONTALLY-AND-VERTICALLY-SKEWED VIEW

We consider the case when the license plate is skewed both horizontally and vertically, as shown in Figure 23.6. Several existing ALPR methods require similar vertical rectification with different assumptions. For example, binary thresholding, Canny edge detection, and Hough transform are used to detect the vertical skew angle in Jun et al. (2012). However, conventional methods of this kind are sensitive to noise. In Björklund et al. (2019) and Silva et al. (2018), license plate skews are estimated by detecting the four license plate corners. In Silva et al. (2018), the locations of the four license plate corners are used to calculate a spatial transformation of the license plate image for training data augmentation.

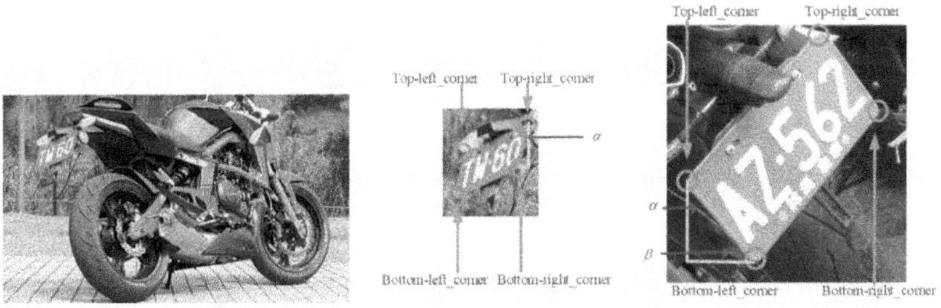

(a) (b) (c)

FIGURE 23.6 Examples of four corners in (a) a heavy motorcycle with its license plate, (b) zoom in to the cropped license plate with the small vertical skew angle α shown, and (c) another example of a cropped motorcycle license plate with a large vertical skew angle α.

We perform YOLOv3 license plate corner detection for skew rectification. The vertical skew angles α is estimated via coordinate transformation as shown in Figure 23.7. Specifically, the following equation defines coordination calculation as

$$\begin{aligned} \mathrm{x1} &= \mathrm{corner\_x1} - corner\_x1 + 1 = 1 \\ y1 &= corner\_y1 - corner\_y1 = 0 \\ x2 &= corner\_x2 - corner\_x1 + 1 \\ y2 &= corner\_y2 - corner\_y1 \end{aligned} \tag{23.4}$$

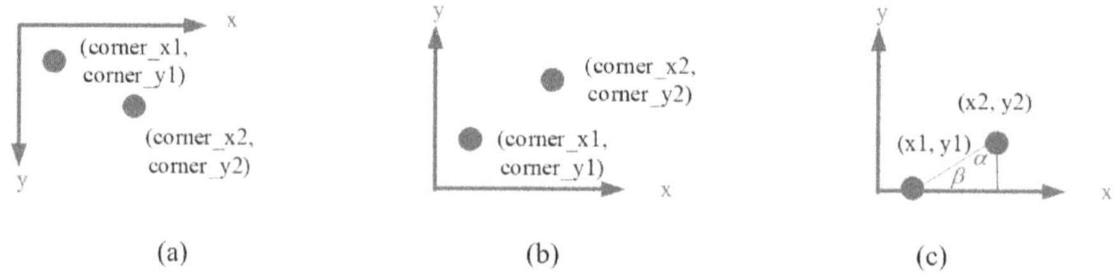

FIGURE 23.7 License plate corner coordinate transformation. (a) The YOLOv3 coordinate in image pixels, (b) Cartesian coordinate, and (c) the standardized coordinate for skew rectification calculation.

The corner angle (β) in radian is calculated as

$$\beta = \cos^{-1} \frac{x_1 \times x_2 + y_1 \times y_2}{\sqrt{x_1^2 + y_1^2} \times \sqrt{x_2^2 + y_2^2}} \tag{23.5}$$

The vertical rectification angle α is calculated as $\alpha = \pi / 2 - \beta$ as shown in Figure 23.8(c). Vertical rectification is calculated as

$$\begin{cases} x'' = y' \\ y'' = x' + y' \times \tan^{-1} \alpha + (-\alpha * 100) \end{cases} \tag{23.6}$$

where (x', y') is the coordinate of the horizontal rectified image and (x'', y'') is the coordinate of the vertical rectified image value. When vertical rectification angle (α) is used to rectify the horizontally rectified license plate, the vertical rectified license plate will still contain some distortions. That is, the license plate will be rotated toward the left or the right border of the license plate. The numbers and characters on the license plate will be unrecognizable. In order to solve this problem, the term ($-\alpha$ * 100) is used to let the numbers and characters move to the middle part of the license plate. Thus, the license plate can be completely displayed on the picture, and subsequent character recognitions can recognize them.

We denote the vertical skew angle of the left side of the license plate as α_l, and the respective one for the right side as α_2. If α_l and α_2 have the same sign, we calculate the final vertical skew angle α as $(\alpha_l + \alpha_2)/2$; otherwise, $\alpha = (\alpha_l + \alpha_2)$. Furthermore, in case if only the two left corners are detected, $\alpha = \alpha_l$; and similarly, for the right side. In case both α_l and α_2 are failed to determine, we refer to the detection of the dot character to estimate the license plate rectangular geometry. In this way, our vertical rectification can be performed more robustly against cluttered scenes and low light conditions in most cases.

Next in the pipeline, YOLOv3 character recognition is performed on the rectified license plate image. If we cannot successfully recognize four or more license plate characters, this license plate is a hard-negative sample that requires manual checking for insertion to the training set for necessary model retraining. Our pipeline continues until all detected license plates in the input image are processed.

23.4 EXPERIMENTAL RESULTS

The proposed system is programed in C and running on a Linux GPU workstation equipped with Intel(R) CPU i7–7700 3.60GHz and GeForce GTX 1080. Experimental results are conducted using the Application-Oriented License Plate (AOLP) dataset (Hsu et al., 2013) and our newly collected Taiwan Motorcycle License Plate (TMLP) dataset. We compare our method with three state-of-the-art commercial ALPR systems, namely the PLATE RECOGNIZER (Plate Recognizer, 2018), OPENALPR (2018), and Sighthound ALPR (2018).

23.4.1 Datasets

The AOLP dataset (Hsu et al., 2013) has been widely used by many state-of-the-art methods for performance comparisons. It contains 2,049 images of Taiwan license plates of vehicles, which are divided into three categories: Law Enforcement (LE) with 757 images, Access Control (AC) with 681 images, and Road Patrol (RP) with 611 images. These images have been captured at various locations and times in different weather and illumination conditions.

We have created a new Taiwan License Plate (TLP) dataset of motorcycles by ourselves, where the image size is of 1229×691 pixels, captured from real-world locations, including school motorcycle parking areas, roadsides, residential motorcycle parking areas, on the road, roadside, and in front of motorcycle shops. This dataset contains challenging real-world conditions, including view distortions, different skewed angles, poor illumination, different distances from the camera, burry, and over-saturation from vehicle headlights. The size and aspect ratio of the motorcycle license plates are varied largely due to the viewing distance and orientations.

The TLP dataset contains 4140 images in total, where 2623 images are used for training, 1105 for validation, and 412 for testing. In order to extend the proposed method to detect and recognize vehicle license plates, there are 62, 18, and 65 images in the training images which come from LE, AC, and RP subsets in the AOLP dataset, respectively. This ends up with most training images being the motorcycle license plates and a few automobile license plate images. Furthermore, each image in the training set contains at least one license plate, except that two of those images contains special customized license plates with four characters.

Our model is trained to detect the following 40 object classes: the license plate, screw, 4 corners, the dot character, number digits (0 to 9), and upper-case characters (A to Z). Table 23.1 lists the number of each object class that are annotated using LabelImg in the training images of the TLP dataset. The total numbers of the license plate, screw, corner, dot, digits, and alphabets in the training images of the TLP dataset are 1364, 2783, 4674, 1651, 5302, and 3798, respectively. A total of 19,572 samples (including all object classes) is used to train our model.

23.4.2 Evaluation Criterion

We evaluate the vehicle license plate detection and recognition using the standard precision-recall. Specifically, the true positives (TPs) are the correctly recognized license plates; false positives (FPs) are the incorrectly recognized license plates;

TABLE 23.1
Number of Annotated Objects in the 40 Classes on the TLP Training Set

class	#	class	#	class	#	class	#
LP	1364	6	585	G	163	Q	113
screw	2783	7	596	H	174	R	95
corner	4674	8	683	I	29	S	74
.	1651	9	573	J	175	T	104
0	525	A	191	K	232	U	98
1	551	B	181	L	145	V	63
2	609	C	244	M	437	W	99
3	556	D	204	N	173	X	81
4	74	E	186	O	0	Y	100
5	550	F	139	P	198	Z	100

false negatives (FNs) are the license plates that are not recognized; true negatives (TNs) are the no-detection cases where there is no license plate and is always 0 in our case. We calculate the accuracy and F1 score by

$$\begin{aligned} &Accuracy(A) = (TP + TN)/(TP + TN + FP + FN) \\ &Precision(P) = TP/(TP + FP) \\ &Recall(R) = TP/(TP + FN) \\ &F1score(F) = (2 \times PC \times RC)/(PC + RC) \end{aligned} \tag{23.7}$$

23.4.3 Performance Evaluation on the TLP Validation Set

Table 23.2 shows evaluation results of license plate detection on the TLP validation set. There are totally 1,105 validation images divided into four categories: 708 street images, 252 sideway images, 59 road images, and 86 heavy-motorcycle images. Most of the on-road images are captured at night, so there are frequently overexposures from the headlights and rear lights from nearby vehicles that affect the license plate detection performance. In this case, our method achieves perfect detection of 100% precision in all categories, where the accuracy, recall and F1 scores are pretty good. Most of the misdetection of license plates (the FN scores) are due to the small license plate size and large distance from the camera. The total accuracy, precision, recall, and F1 scores of 96.79%, 100%, 96.79%, and 98.37%, respectively, show that the proposed method is very effective at detecting license plate on the TLP validation set.

Table 23.3 shows comparisons of license plate recognition on the TLP validation set. For license plate recognition, we quantify the detection and recognition performance of each individual license plate characters. The total accuracy, precision, recall, and F1 scores of 99.72%, 99.84%, 99.88%, and 99.96%, respectively, show that the proposed method is effective on recognizing each license plate characters on the TLP validation set.

TABLE 23.2
Performance Evaluation of License Plate Detection on the TLP Validation Set

LP detection	#LP	TP	TN	FP	FN	A	P	R	F
street	1123	1111	0	0	12	98.93	100	98.93	99.46
sideway	499	487	0	0	12	97.96	100	97.60	98.78
road	221	183	0	0	38	82.81	100	82.81	90.58
heavy-motor	86	86	0	0	0	**100.0**	**100**	**100.0**	**100.0**
Total	1929	1867	0	0	62	96.79	100	96.79	98.37

TABLE 23.3
Performance Evaluation of License Plate Character Recognition on the TLP Validation Set

LP recognition	#LP characters	TP	TN	FP	FN	A	P	R	F
street	7000	6995	0	5	5	99.86	**99.93**	99.93	99.93
sideway	3080	3080	0	3	0	**99.90**	99.90	**100.0**	**99.95**
road	1132	1123	0	10	9	98.34	99.12	99.20	99.16
heavy-motor	513	513	0	1	0	99.81	99.81	100.0	99.90
Total	11725	11711	0	19	14	99.72	99.84	99.88	99.86

23.4.4 Performance Evaluation on the TLP Test Set

Table 23.4 shows evaluation results of license plate detection on the TMLP test set. There are totally 412 test images divided into three categories: 172 sideway images, 102 road images, and 138 heavy-motor images. The total accuracy, precision, recall, and F1 scores are 91.64%, 99.77%, 91.84%, and 95.64%, respectively, shows the effectiveness of the proposed method. We note that the performance of the road images is with the lowest scores due to the high misdetections (i.e., the 57 FN cases). This is because in many cases the viewing distances are too far, which results in small license plates that are hard to detect.

Table 23.5 shows evaluation results of license plate recognition on the TLP test set, where the detection and recognition of each individual license plate characters are quantified. The total accuracy, precision, recall, and F1 scores are 98.31%, 98.87%, 99.42%, and 99.15%, respectively, show that the proposed method is effective on license plate character recognition on the TLP test set.

23.4.5 Performance Evaluation for License Plate Detection on the AOLP Dataset

Table 23.6 compares the proposed license plate detection performance against six state-of-the-art methods, such as Hsu et al. (2013), Li et al. (2019), Selmi et al. (2020), Björklund et al. (2019), Li et al. (2018), and Omar et al. (2020) on the AOLP (Hsu et al., 2013) dataset. The proposed method achieved a higher accuracy over the state-of-the-art methods under the Access Control (AC) category. Our method

TABLE 23.4
Performance Evaluation Results in License Plate Detection on TLP Test Set

LP detection	#LP	TP	TN	FP	FN	A	P	R	F
sideway	447	429	0	0	18	95.97	100.0	95.97	97.95
road	342	285	0	2	57	82.85	99.30	83.33	90.62
Heavy-Motor	142	141	0	0	1	99.30	100.0	99.30	99.65
Total	931	855	0	2	76	91.64	99.77	91.84	95.64

TABLE 23.5
Performance Evaluation Results in License Plate Recognition on TLP Test Set

Recognition	Characters	TP	TN	FP	FN	A	P	R	F
sideway	2746	2734	0	24	12	98.70	99.13	99.56	99.35
road	1819	1801	0	35	18	97.14	98.09	99.01	98.55
heavy-motor	806	805	0	2	1	99.63	99.75	99.88	99.81
Total	5371	5340	0	61	31	98.31	98.87	99.42	99.15

TABLE 23.6
Comparisons of License Plate Detection Results of the Proposed Method and the Other State-of-the-Art Methods on the AOLP Dataset

Methods	AC				LE				RP			
	A	P	R	F	A	P	R	F	A	P	R	F
Hsu et al. (2013)	87.7	91.0	96.0	93.4	86.8	91.0	95.0	92.9	86.0	91.0	94	92.5
Li et al. (2019)	–	–	99.1	–	–	–	99.1	–	–	–	98.2	–
Selmi et al. (2020)	–	99.3	99.4	99.3	–	99.2	**99.2**	99.2	–	98.9	98.8	98.9
Björklund et al. (2019)	99.3	**100**	99.3	99.6	**98.8**	99.8	99.0	**99.4**	**98.8**	99.8	99.0	99.4
Li et al. (2018)	97.0	98.5	98.4	98.4	95.5	97.8	97.6	97.7	91.3	95.3	95.6	95.4
Omar et al. (2020)	–	**100**	**99.8**	**99.9**	–	**100**	96.4	98.2	–	**100**	**99.2**	**99.6**
The proposed method	**99.4**	**100**	99.4	99.7	96.3	99.8	96.5	98.1	97.8	**100**	97.8	98.9

obtains similar scores close to the highest in other categories. Björklund et al. (2019) and Omar et al. (2020) performs close to ours but with a much more complex method and architecture. As aforementioned, their methods only deal with the simple case of one license plate per image, whereas our method can handle multiple license plates in the input image.

23.4.6 Performance Evaluation for License Plate Recognition on the AOLP Dataset

Table 23.7 compares the proposed license plate recognition performance against six state-of-the-art methods (Hsu et al., 2013; Li et al., 2018, 2019; Selmi et al., 2020;

TABLE 23.7
Comparisons of License Plate Recognition Results of the Proposed Method and the Other State-of-the-Art Methods on the AOLP Dataset

Methods	AC	LE	RP	Average
Hsu et al. (2013)	88.50	86.60	85.70	86.93
Li et al. (2019)	95.59	96.43	83.80	91.94
Selmi et al. (2020)	97.80	97.40	96.30	97.17
Björklund et al. (2019)	99.40	99.80	99.60	99.60
Li et al. (2018)	94.85	94.19	88.38	92.61
Omar et al. (CRNN) (2020)	97.97	95.95	81.81	91.91
Omar et al. (YOLO) (2020)	93.75	98.91	90.42	94.36
The proposed method	**99.98**	**99.86**	**99.87**	**99.90**

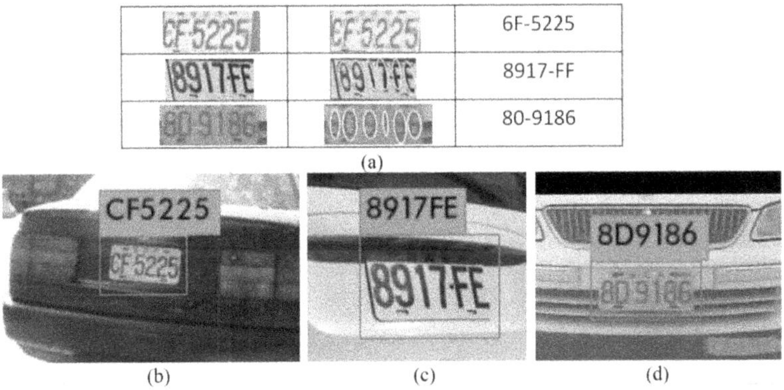

FIGURE 23.8 Visual comparison between Hsu et al. (2013) and our method on the AOLP dataset. (a) Their incorrect results, (b) and (c) are from the RP subset and (d) is from the AC subset. In all cases, our method can successfully detect and recognize the license plates.

Björklund et al., 2019; Omar et al., 2020) on the AOLP dataset. The proposed method obtains the best results of 99.98% in AC, 99.86% in LE, and 99.87% in RP.

23.4.7 Visual Comparisons of ALPR on the AOLP Dataset

We provide visual examples in comparing the proposed method against state-of-the-art ALPR methods. Figure 23.8 compares Hsu et al. (2013) and ours, where their method misinterpretes all three license plates, while our method can correctly identify these license plates. Note that they mistook the letter "c" as number "6," letter "E" as "F," and letter "D" as number "0."

Figure 23.9 compares Li et al. (2019) with our method, where they misinterpreted the letter "Y" as "X" in (b). Our system is able to correctly extract the Y. Figure 23.10

FIGURE 23.9 Visual comparison between Li et al. (2019) (top row) and our method (bottom row) on the AOLP dataset. (a, e) An AC image with correct license plate detection and recognition results; (b) an LE image with correct license plate detection but incorrect recognition result; (c, f, g) an LE image with correct license plate detection and recognition results; (d, h) an RP image with correct license plate detection and recognition results.

provides additional visual examples of Li et al. (2018) and ours in various challenging examples from the AOLP dataset.

Figure 23.11 compares Selmi et al. (2020) and our method, where they cannot recognize the license plate characters correctly, while our method can detect and recognize them correctly.

Figure 23.12 compares Björklund et al. (2019) with our method. Note that multiple CNNs are used in their method and there is only a single YOLOv3 network in our decision tree pipeline.

23.4.8 Performance Comparison with Three Commercial ALPR Systems

We compare three commercial off-the-shelf ALPR systems – namely, PLATE RECOGNIZER (2018), OPENALPR (2018), and Sighthound ALPR (2018) – with the proposed method on 100 randomly selected images (which contain 289 license plates) from our TLP dataset. Since we do not have access to the internal design or architecture of these commercial systems, we can only treat them as black boxes and evaluate the final outcome. Table 23.8 shows the license plate detection comparisons. We observe that the precision of the PLATE RECOGNIZER and OPENALPR are both 100%, while Sighthound is 99.72% and ours is 99.65%. However, our method performs better in terms of accuracy (A, 98.28%), recall (R, 98.62%), and F1 scores (F, 99.10%).

Table 23.9 shows the license plate recognition comparisons. The PLATE RECOGNIZER, OPENALPR, Sighthound, and the proposed method have 14, 13, 6, and 4 false positive (FP) recognition results, respectively. The false negative (FN) recognition results are 168, 160, 720, and 26 of the PLATE RECOGNIZER, OPENALPR, Sighthound, and the proposed method, respectively. We can observe

FIGURE 23.10 Visual examples of (a–f) Li et al. (2018) and (g–l) our method on the AOLP dataset. (a-c, e) are images from the LE set and (d, f) are from RP set, where all are with correct license plate detection and recognition outputs. (g, h) show our results on detecting both automobile and motorcycle license plates in various conditions: (g) poor lighting, (h) mixed automobile and motorcycles, (i) with a slightly tilted view, (j) with large tilted view, (k) bright license plate, and (l) poor lighting. In all cases, our method can successfully detect and recognize the license plates.

that the proposed method obtains better with 98.34%, 99.77%, 98.55%, and 99.16% in accuracy (A), precision (P) recall (R), and F1 score (F), respectively.

23.4.9 Time Complexity

Table 23.10 shows the time complexity of the proposed method on the validation set of 1105 images and the test set of 412 images. The average execution time per image

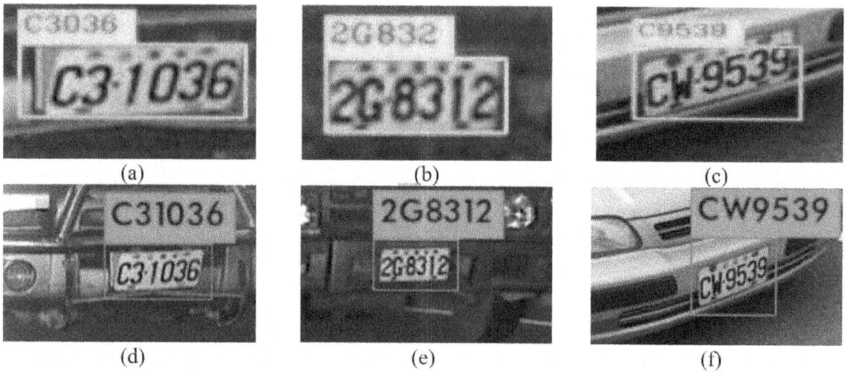

FIGURE 23.11 Visual comparison between Selmi et al. (2020) (top row) and our method (bottom row) on the AOLP dataset. Both (a, d) and (b, e) are from the LE set, and (c, f) is from the RP set. In (b) the number "1" is not successfully recognized, and in (c) the letter "W" is not successfully recognized by Selmi et al. In all cases our method can detect and recognize all license plate characters correctly.

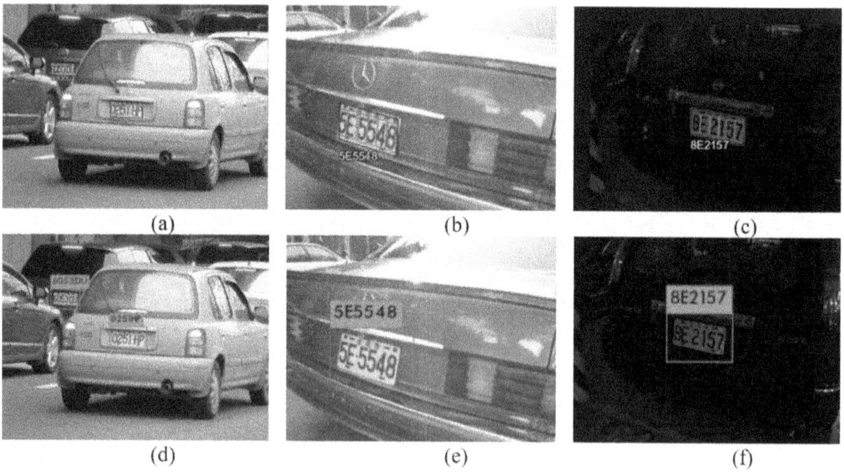

FIGURE 23.12 Visual comparison between Björklund et al. (2019) and our method on the AOLP dataset. (a, d, f) A LE image with correct license plate detection (LPD) and LPR results (b, c, e) a RP image with correct LPD and LPR results. The top three images are results from Björklund et al. The bottom three is from our proposed method.

is 97.6 ms and 125.3 ms for the validation and testing sets, respectively. Furthermore, when an image contains only a non-skew license plate, the execute time is about 48 ms. The execution time will increase as more skew license plates appear in the image.

In summary, the proposed method outperforms the three compared commercial ALPR systems by a substantial margin in terms of accuracy, precision, recall, and F1 scores. Nonetheless, there still exists plenty of space for the ALPR systems to

TABLE 23.8
Comparisons of License Plate Detection Results of the Proposed Method and Three ALPR Commercial Systems on 100 Randomly Selected Images from Our TLP Dataset

Methods	Total License Plates	LPD							
		TP	TN	FP	FN	A	P	R	F
PLATE RECO.	290	267	0	0	23	92.07	**100.0**	92.07	95.87
OPENALPR	290	266	0	0	24	91.72	**100.0**	91.72	95.68
Sighthound	290	176	0	0	114	60.69	**100.0**	60.69	75.54
The proposed method	290	**286**	0	1	4	**98.28**	99.65	**98.62**	**99.10**

TABLE 23.9
Comparisons of License Plate Recognition Results of the Proposed Method and Three ALPR Commercial Systems on 100 Randomly Selected Images from Our TLP Dataset

Methods	Total License Plates	LPR							
		TP	TN	FP	FN	A	P	R	F
PLATE RECO.	1,799	1,631	0	14	168	89.96	99.15	90.66	94.72
OPENALPR	1,799	1,639	0	13	160	90.45	99.21	91.11	94.99
Sighthound	1,799	1,079	0	6	720	59.78	99.45	59.98	74.83
The proposed method	1,799	**1,773**	0	4	26	**98.34**	**99.77**	**98.55**	**99.16**

TABLE 23.10
The Time Complexity (ms) of the Proposed Method on 1,105 Validation Set and 412 Testing Set

Datasets	Images	Resolution	LPs	Characters	Total Time	Average/Image
Validation	1,105	1229 × 691	1,929	11,725	107,848.0	97.6
Testing	412	1229 × 691	931	5,371	51,623.6	125.3

improve, especially for the motorcycles. More comparisons, license plate detection, and license plate recognition results in TLP and AOLP datasets can be found at (Tsai, 2018).

The contributions of this chapter are summarized as follows:

1. We present a decision tree based deep learning ALPR system operated based on only one YOLOv3 model that can effectively detect and recognize license plate for scooters, motorcycles, heavy motorcycles, and vehicles.
2. The proposed ALPR methods can detect and recognize multiple license plates in three major viewing categories: frontal, horizontally skewed views, and hybrid horizontally-and-vertically-skewed views.

3. The proposed ALPR methods can run in-the-wild with the real-world viewing challenges: blurry, very poor illumination, different distance from camera, distortion, multiple license plates in an image, night, brake lights, motorcycle headlights, and those problems in combinations.
4. We created a new Taiwan License Plate (TLP) dataset for ALPR with 2,623 fully annotated images (over 19,572 license plate characters) with real-world scenarios.
5. We conducted performance evaluations by comparing the proposed method against other state-of-the-art methods and three commercial ALPR systems. Experimental results demonstrated the efficiency of our method.

23.5 CONCLUSIONS

We present a method for license plate detection and recognition under a combination of real-life conditions. The proposed method combines decision tree and deep-learning methods. The decision tree is used to divide the license plate view images into three cases (frontal views, horizontally skewed views, and hybrid horizontally-and-vertically-skewed views). The deep-learning method just uses one CNN to train forty object classes. The experiments produce favorable results which are judged in terms of both the visual quality, accuracy, precision, recall, and F1 score performance. Experimental results indicate that the proposed approach not only outperforms the six state-of-the-art license plate detection methods and six state-of-the-art license plate recognition methods, but also performs considerably better than three commercial ALPR systems.

Our future work will focus on the followings: (1) optimization of the proposed program, (2) processing far license plate images that have been affected by the headlight of other motorcycles, (3) finding a way to process hard-negative license plate images, and (4) extension of the proposed method to real-world mobile and embedded applications.

REFERENCES

Abdullah, S., Mahedi Hasan, M., and Muhammad Saiful Islam, S., "YOLO-based three-stage network for Bangla license plate recognition in Dhaka Metropolitan City," *2018 International Conference on BSLP*, pp. 1–6, 2018.

Al-Shemarry, M. S., Li, Y., and Abdulla, S., "Ensemble of adaboost cascades of 3L-LBPs classifiers for license plates detection with low quality images," *Expert Systems with Applications*, vol. 92, pp. 216–235, 2018.

Björklund, T., Fiandrotti, A., Annarumma, M., Francini, G., and Magli, E., "Robust license plate recognition using neural networks trained on synthetic images," *Pattern Recognition*, vol. 93, pp. 134–146, 2019.

Hsu, G. S., Ambikapathi, A., Chung, S. L., and Su, C. P., "Robust license plate detection in the wild," *2017 14th IEEE International Conference on AVSS*, pp. 1–6, 2017.

Hsu, G. S., Chen, J. C., and Chung, Y. Z., "Application-oriented license plate recognition," *IEEE Transactions on Vehicular Technology*, vol. 62, no. 2, pp. 552–561, 2013.

Hsu, G. S. and Chiu, C., "A comparison study on real-time tracking motorcycle license plates," *2016 IEEE 12th Image, Video, and Multidimensional Signal Processing Workshop (IVMSP)*, Bordeaux, pp. 1–5, 2016.

Hsu, G. S., Zeng, S. D., Chiu, C. W., and Chung, S. L., "A comparison study on motorcycle license plate detection," *2015 IEEE ICMEW*, pp. 1–6, 2015.

Jun, F. and Shuguang, D., "A vehicle license plate location and correction method based the characteristics of license plate," *Proceedings of the 10th World Congress on ICA*, pp. 42–46, 2012.

Kessentini, Y., Besbes, M. D., Ammar, S., and Chabbouh, A., "A two-stage deep neural network for multi-norm license plate detection and recognition," *Expert Systems with Applications*, vol. 136, pp. 159–170, 2019.

Khare, V., Shivakumara, P., Chan, C. S., Lu, T., Meng, L. K., Woon, H. H., and Blumenstein, M., "A novel character segmentation-reconstruction approach for license plate recognition," *Expert Systems with Applications*, vol. 131, pp. 219–239, 2019.

LabelImg. https://github.com/tzutalin/labelImg, 2018.

Laroca, R., Severo, E., Zanlorensi, L. A., Oliveira, L. S., Goncalves, G. R., Schwartz, W. R., and Menotti, D., "A robust real-time automatic license plate recognition based on the YOLO detector," arXiv:1802.09567v6, 2018.

Laroca, R., Zanlorensi, L. A., Gonçalves, G. R., Todt, E., Schwartz, W. R., and Menotti, D., "An efficient and layout-independent automatic license plate recognition system based on the YOLO detector," arXiv:1909.01754, 2019.

Lee, H. J., Chen, S. Y., and Wang, S. Z., "Extraction and recognition of license plates of motorcycles and vehicles on highways," *Proceedings of the 17th ICPR*, pp. 356–359, 2004.

Li, H., Wang, P., and Shen, C., "Toward end-to-end car license plate detection and recognition with deep neural networks," *IEEE Transactions on Intelligent Transportation Systems*, vol. 20, no. 3, pp. 1126–1136, 2019.

Li, H., Wang, P., You, M. Y., and Shen, C. H., "Reading car license plates using deep neural networks," *Image and Vision Computing*, vol. 72, pp. 14–23, 2018.

Liu, Y. J., Huang, H., Cao, J. D., and Huang, T. W., "Convolutional neural networks-based intelligent recognition of Chinese license plates," *Soft Computing*, vol. 22, pp. 2403–2419, 2018.

Molina-Moreno, M., Gonzlez-Daz, I., and Daz-de-Mara, F., "Efficient scale-adaptive license plate detection system," *IEEE Transactions on Intelligent Transportation Systems*, vol. 20, pp. 2109–2121, 2019.

Omar, N., Sengur, A., Ganim, S., and Al-Ali, S., "Cascaded deep learning-based efficient approach for license plate detection and recognition," *Expert Systems with Applications*, vol. 149, p. 113280, 2020.

OPENALPR. https://www.openalpr.com/, 2018.

Plate Recognizer. https://platerecognizer.com/, 2018.

Puarungroj, W. and Boonsirisumpun, N., "Thai license plate recognition based on deep learning," *Procedia Computer Science*, vol. 135, pp. 214–221, 2018.

Redmon, J. and Farhadi, A., "Yolo9000: Better, faster, stronger," arXiv:1612.08242, 2016.

Redmon, J. and Farhadi, A., "Yolov3: An incremental improvement," arXiv:1804.02767, 2018. https://arxiv.org/abs/1804.02767

Redmon, J., Divvala, S., Girshick, R., and Farhadi, A. "You only look once: Unified, realtime object detection," arXiv:1506.02640, 2015.

Selmi, Z., Halima, M. B., Pal, U., and Alimi, M. A., "DELP-DAR system for license plate detection and recognition," *Pattern Recognition Letters*, vol. 129, pp. 213–223, 2020.

Sighthound ALPR. https://www.sighthound.com/products/alpr, 2018.

Silva, S. M. and Jung, C. R., "License plate detection and recognition in unconstrained scenarios," *2018 European Conference on Computer Vision (ECCV)*, Munich, pp. 580–596, 2018.

Tsai. http://cmtsai.utaipei.edu.tw/~cmtsai/MMLP/MMLP.html, 2018.

Wang, D., Tian, Y. M., Geng, W. H., Zhao, L., and Gong, C., "License plate R-Net: Recognizing Chinese license plate in complex environments," *Pattern Recognition Letters*, vol. 130, pp. 148–156, 2020.

Xie, F., Zhang, M., Zhao, J., Yang, J. Q., Liu, Y. J., and Yuan, X. Y., "A robust license plate detection and character recognition algorithm based on a combined feature extraction model and BPNN," *Journal of Advanced Transportation*, pp. 1–14, 2018.

YOLOv3. https://pjreddie.com/darknet/yolo/, 2018.

24 The Deep Hybrid Neural Network and an Application on Polyp Detection

24.1 INTRODUCTION

Colorectal cancer (CRC) accounts for 10% of overall new cases and presents higher incidence rates in developed countries (International Agency for Research on Cancer, 2018). It is considered as a "lifestyle" disease associated with a diet high in calories, animal fat, and sedentarism (Stewart and Wild, 2014). Colonoscopy is a gold-standard screening test for colorectal cancer (Lieberman et al., 2012; Laura et al., 2004; Zauber et al., 2012), one of the leading causes of cancer-related death in both the United States (Edwards et al., 2010; Siegel et al., 2013) and China (Chin et al., 2011). Colonoscopy can reduce the risk of death from colorectal cancer through the detection of tumors at an early and treatable stage, as well as through the removal of precancerous adenomas (Brenner et al., 2011). However, failure to detect adenomas may lead to the development of interval cancer. Evidence has shown that a 1% increase in adenoma detection rate (ADR) can lead to a 3% decrease in the risk of interval colorectal cancer (Corley et al., 2014). Therefore, different approaches have been developed to improve ADR and reduce the number of missed lesions. During colonoscopy, these approaches include endoscope caps, positional assisted system, and the use of imaging modalities such as narrow-band imaging (NBI), which emphasizes the capillary pattern and mucosa surface (Ishaq et al., 2017). It is a real-time automatic polyp detection system to serve as a second set of eyes to assist an endoscopist to improve ADR by making polyps more visible to the endoscopist (Aslanian et al., 2013).

Over the past decades, deep convolutional neural networks (DCNNs) (Carrinho and Falcao, 2023; Yeh et al., 2021; Deng et al., 2023) have been proved to be more powerful than traditional image-processing approaches. DCNN-based detection has been extensively applied in medical imaging in various tasks, including disease detection, segmentation, classification, and anomaly detection. However, DCNN-based detection models have some potential defects, which can impact the performance in certain scenarios. Here are a few common defects associated with DCNN-based detection models: false positives, false negatives sensitivity to image quality, and high computational complexity. Note that these defects can be mitigated or addressed to some extent through proper training techniques, data augmentation, model architecture improvements, and domain-specific fine-tuning.

Mathematical morphology (Shih and Puttagunta, 1995; Shih and Wu, 2005a, 2005b) has found extensive uses across multiple applications in image processing

DOI: 10.1201/9781003474395-27

and pattern recognition. This approach is grounded in the principles of set theory, and facilitates the extraction of object features by probing its microstructure with various forms, known as *structuring elements*. Distinct from linear convolution operators, mathematical morphological operators are nonlinear.

Recently, many researchers have been considering to replace the convolution layers by nonlinear mathematical morphological layers. Nogueira et al. (2021) showed that the DeepMorphNet can extract useful nonlinear features and address issues that traditional architectures equipped with standard convolution filters fail to resolve. Shen et al. (2022) presented the residual morphological neural networks (residual-MNN) and showed the superiority by offering an optimal balance between model accuracy and computational complexity.

In this chapter, we propose a novel concept by simultaneously adopting the linear and the nonlinear operators, i.e., to combine the convolutional and mathematical morphological layers. We generate the convolutional and morphological backbone in parallel to extract linear and nonlinear features, and then combine the features for further analysis. The remainder of this chapter is organized as follows. Section 24.2 introduces the material of datasets. Section 24.3 presents our DHNN solution to show the way of combining convolutional and mathematical morphological operators. Section 24.4 provides the experimental results. Finally, the conclusions are made in Section 24.5.

24.2 MATERIAL OF DATASET

Obtaining high-quality and efficient data for utilizing deep-learning-based solutions poses a significant challenge and is a critical concern. In medical applications, this challenge is compounded by two major issues. First, it necessitates obtaining patient's consent to address data protection concerns. Second, it involves the time-consuming yet essential task of annotating the data with the expertise of medical professionals. Therefore, in order to develop our polyp detection model, we conduct experiments using publicly available data and our collected data from Tri-Service General and Show Chwan Memorial Hospitals. We adopt nine polyp image datasets, such as Kvasir-SEG (Debesh et al., 2020), CVC-ClinicDB (Bernal et al., 2015), CVC-ColonDB (Bernal et al., 2017), CVC-300 (Vázquez et al., 2017), ETIS-LaribPolypDB (Silva et al., 2014), EndoCV2022 (Ng, 2022), EDD2020 (Ali et al., 2020), and our collected dataset for training and validation, and use PICCOLO (Sánchez-Peralta et al., 2020) for testing.

24.2.1 Training and Validation Datasets

We combine the publicly available and our collected data to form a dataset of 15,620 images. Figure 24.1 shows an overview of the data material, consisting of images and the bounding boxes. Table 24.1 shows the summary of the data material.

We list the publicly available datasets from items 1 to 7 and our collected dataset as item 8.

1. **Kvasir-SEG**: The dataset comprises 1,000 images. These images were collected and verified by experienced gastroenterologist from Vestre Viken Health Trust in Norway.

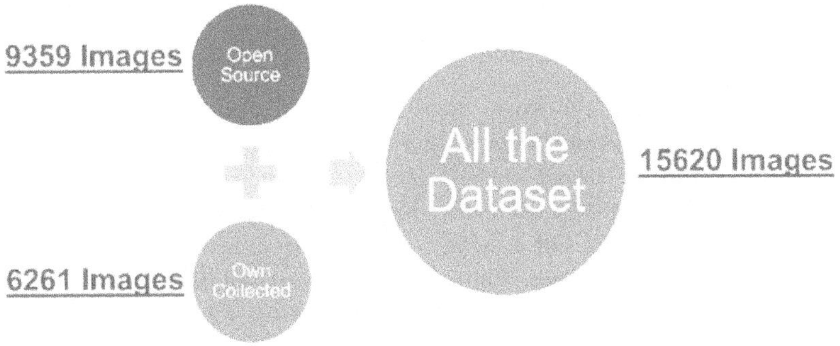

FIGURE 24.1 Overview of the data material.

TABLE 24.1
Summary of Data Material

Dataset	Category	Images	Source	Findings
Kvasir-SEG	Public	1000	WL	Polyp
CVC-ClinicDB	Public	612	WL	Polyp
CVC-ColonDB	Public	380	WL	Polyp
CVC-300	Public	60	WL	Polyp
ETI-LaribPolypDB	Public	196	WL	Polyp
EndoCV2022	Public	3292	WL NBI	Polyp
EDD2020	Public	386	WL NBI	Polyp
TSGH	Private	3317	WL NBI	Polyp
SCMH	Private	2944	WL NBI	Polyp
PICCOLO (Testing)	Public	3433	WL NBI	Adenoma Hyperplastic

2. **CVC-ClinicDB**: The dataset is a database of frames extracted from colonoscopy videos. The dataset comprises 612 standard-definition still images of 384×288 pixel resolution with 31 different polyps from 31 different sequences.
3. **CVC-ColonDB**: The dataset comprises 380 standard-definition still images of 574×500 pixel resolution. It was created by the Computer Vision Center (CVC) at the Autonomous University of Barcelona, Spain.
4. **CVC-300**: The dataset comprises 60 standard-definition still images of 574×500 pixel resolution. It was created by the Computer Vision Center (CVC) at the Autonomous University of Barcelona, Spain.
5. **ETI-LaribPolypDB**: The dataset comprises 196 standard-quality polyp images from 34 different video sequences and 196 clinicians labeled ground truth images with a size of $1,225 \times 966$ pixel resolution.
6. **EndoCV2022**: It is a challenge organized as part of the International Conference on Medical Image Computing and Computer Assisted

Intervention (MICCAI) in the field of endoscopic computer vision. The dataset comprises 3,292 standard-definition still images of 1920 × 1080 pixel resolution.

7. **EDD2020**: It is a challenge released a dataset containing five different classes with masks and bounding boxes for each image and polyp instance. The polyps are extracted from all images and stored to the relevant bounding boxes into a custom JSON file. These data contain 386 images, and the size of the images is 720 × 576.
8. **Own Collected Dataset**: Retrospective cases from Tri-Service General Hospital (TSGH) and Show Chwan Memorial Hospital (SCMH). There are 2,944 and 3,317 polyp images for TSGH and SCMH, respectively.

24.2.2 Testing Dataset

The PICCOLO dataset (Sánchez-Peralta et al., 2020) was recorded by an AVerMedia video capture and a hard drive storage. It comprises 76 lesions from 40 patients, and lesions were recorded between October 2017 and December 2019 at Hospital Universitario Basurto (Bilbao, Spain) using Olympus endoscopes (CF-H190L and-CF-HQ190L). In total, 62 out of these 76 lesions are included both white light (WL) and narrow-band imaging (NBI) frames, while the remaining 14 lesions are recorded only using WL. More than 145,000 frames were revised and out of the 80,847 frames showed a polyp, and 3,433 frames were selected for manual segmentation. Some polyp images of the PICCOLO dataset are shown in Figure 24.2.

24.3 THE DEEP HYBRID NEURAL NETWORKS

24.3.1 Background Knowledge

Mathematical morphology and convolution are two different types of operators in image processing and computer vision applications. Convolution and morphological operators are the processes of changing the shape of one function by applying another function. To clearly illustrate the key distinctions between these two operators, we adjust the respective mathematical formulations. Let f represent the grayscale image, and g denote the mask. In the context of convolution and morphological operators, g is referred to as the "kernel mask" and "structuring element," respectively.

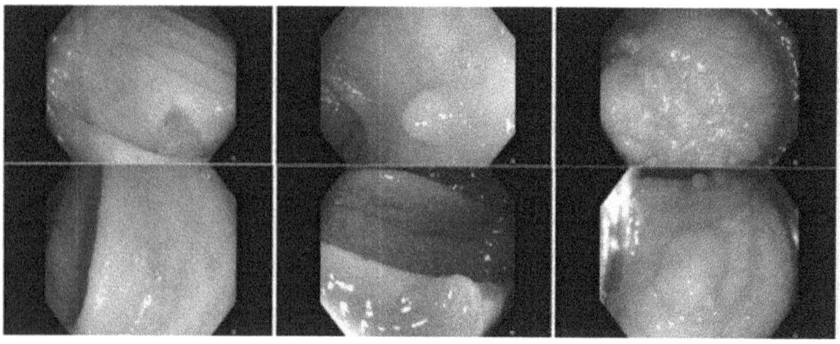

FIGURE 24.2 Some sample polyps from the PICCOLO dataset.

The convolution operator is classified as a linear operator, which is the summation of the product of the f with the mirrored and g as

$$(f*g)(x)=\sum_{x-z\in F,\ z\in G}\left[f(x-z)\cdot g(z)\right] \tag{24.1}$$

where the symbols F and G respectively represent the bounded domains for f and g.

Mathematical morphological operators are classified as nonlinear operators, which are based on set theory and topology. The operators change the shape and structure of an image based on the intersection or merge between a structuring element and the image. The morphological operator is similar to the convolution operator, except that addition and subtraction are substituted for multiplication, and maximum and minimum for summation.

The *grayscale dilation* of f by g is given by

$$(f\oplus g)(x)=\max_{x-z\in F,\ z\in G}\left[f(x-z)+g(z)\right] \tag{24.2}$$

The *grayscale erosion* of f by g is given by

$$(f\ominus g)(x)=\min_{x+z\in F,\ z\in G}\left[f(x+z)-g(z)\right] \tag{24.3}$$

The convolution operator is primarily used for smoothing images, enhancing image details, or conducting feature extraction. It can apply various convolution kernels to achieve different effects, such as Gaussian blur, edge detection, and sharpening, among others. It is commonly used in numerous image-processing fields, including image enhancement, feature extraction, pattern recognition, and medical image processing.

The morphological operator is mainly used to change the shape, size, and structure of the image. For example, dilation can expand the area, erosion can shrink the area, opening operation can eliminate small noise, and closing operation can fill small holes. It is commonly used in shape analysis, feature extraction, object detection, image segmentation, and other applications, especially suitable for processing binarized images.

24.3.2 The Proposed DHNN Architecture

As mentioned earlier, convolution and morphological operators are two distinct types of operators, so the DHNN architecture is designed to integrate the two operators as shown in Figure 24.3. Figure 24.3(a) shows the concept of traditional deep convolution neural network (DCNN) architecture, which comprises two main components: feature extraction and classification. Figure 24.3(b) presents the proposed DHNN architecture, which includes both linear and nonlinear operators with trainable parameters.

To illustrate the efficiency and effectiveness of the proposed DHNN architecture, the morphology-based feature extraction part is added to the existing DCNN module. The YOLOv5 (Zhu et al., 2021) is selected as the DCNN module since it is a state-of-the-art, real-time object detection system and has been trained on a large dataset of images. Since both linear and nonlinear features will be extracted

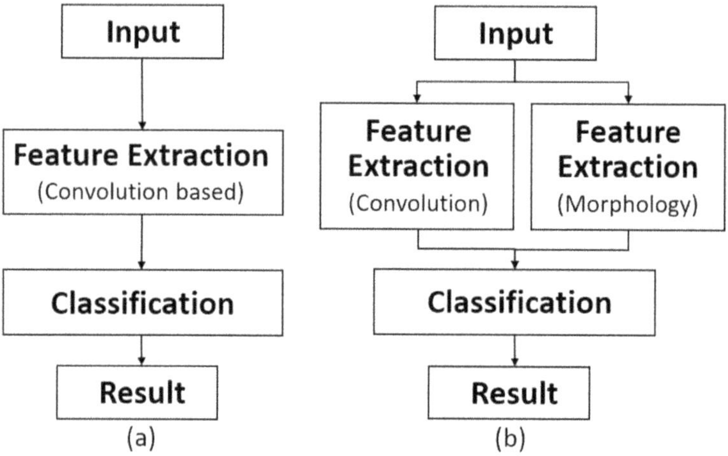

FIGURE 24.3 The comparison between DCNN and DHNN.

in our DHNN architecture, it is able to essentially enhance the ability of network to recognize and analyze different shapes and structures, and we do believe the computational complexity of the network can be reduced accordingly.

The feature extraction part (also called backbone) of the YOLOv5 can be separated into 7 stages and the DHNN architecture based on YOLOv5 called DHNN-YOLOv5 is shown in Figure 24.4, and the detailed neural network can be found in Table 24.2. Note that the convolution-based feature extraction neural network layers of the DHNN-YOLOv5 architecture are halved from the original YOLOv5, and the FLOPs and trainable parameters are respectively reduced from 290 to 205.49 and 46,625,206 to 28,962,742.

24.3.3 Modeling

The process of modeling involves designing and creating a mathematical or computational representation of a system, phenomenon, or problem. In the context of machine learning and deep learning, modeling refers to the construction of a model architecture that can learn from data and make predictions or decisions. The following are the three steps (pre-trained model, data pre-processing, and model training) involved in the modeling procedure:

(1) ***Pre-Trained Model:*** Our proposed model is pre-trained using the ImageNet dataset, which consists of over 14 million image samples with more than 1,000 different object classes. Pre-training with ImageNet allows the model to learn a wide range of image features, which enhances the accuracy and efficiency of the model. Pre-trained models using ImageNet have good portability and can be easily applied to related tasks such as image recognition and object detection. Furthermore, pre-training with ImageNet can accelerate the training process as the pre-trained model has already

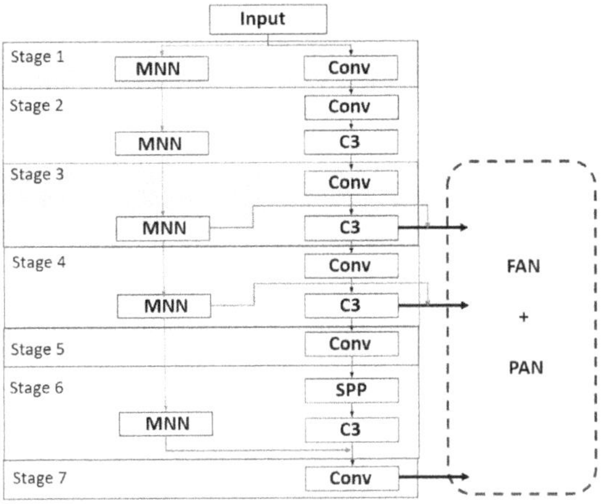

FIGURE 24.4 The feature extraction neural network layers of the DHNN-YOLOv5.

TABLE 24.2
The Comparison of Network Architecture in the Feature Extraction Part (Backbone) between DCNN and DHNN

Layers	**Output size**	**DCNN (Yolov5)**	**DHNN (Yolov5-Half + Morphology)**	
		Convolution-Based	Convolution-Based	Morphology-Based
Stage 1	256*256	1X1, 64, stride 1	1X1, 32, stride 1	Erosion 3X3, 8, stride4 Dilation 3X3, 8, stride1
Stage 2	128*128	3X3, 128, stride 2 1X1, 64, stride 1 [1X1, 32, stride 1] X3 [3X3, 64, stride 1] X3 1X1, 128, stride 1	3X3, 64, stride 2 1X1, 32, stride 1 [1X1, 16, stride 1] X3 [3X3, 32, stride 1] X3 1X1, 64, stride 1	Erosion 3X3, 8, stride1 Dilation 3X3, 8, stride1
Stage 3	64*64	3X3, 256, stride 2 1X1, 128, stride 1 [1X1, 64, stride 1] X9 [3X3, 128, stride 1] X9 1X1, 256, stride 1	3X3, 128, stride 2 1X1, 64, stride 1 [1X1, 32, stride 1] X9 [3X3, 64, stride 1] X9 1X1, 128, stride 1	Erosion 3X3, 16, stride2 Dilation 3X3, 16, stride1
Stage 4	32*32	3X3, 512, stride 2 1X1, 256, stride 1 [1X1, 128, stride 1] X9 [3X3, 256, stride 1] X9 1X1, 512, stride 1	3X3, 256, stride 2 1X1, 128, stride 1 [1X1, 64, stride 1] X9 [3X3, 128, stride 1] X9 1X1, 256, stride 1	Erosion 3X3, 32, stride2 Dilation 3X3, 32, stride1

(Continued)

TABLE 24.2
Continued

Layers	Output size	DCNN (Yolov5)	DHNN (Yolov5-Half + Morphology)	
Stage 5	10*10	3X3, 1024, stride 2	3X3, 512, stride 2	Erosion 3X3, 64, stride2
Stage 6	10*10	1X1, 512, stride 1 5X5, max pool, stride 1 9X9, max pool, stride 1 13X13, max pool, stride 1 1X1, 1024, stride 1	1X1, 256, stride 1 5X5, max pool, stride 1 9X9, max pool, stride 1 13X13, max pool, stride 1 1X1, 512, stride 1	Dilation 3X3, 64, stride1
Stage 7	10*10	1X1, 512, stride 1 [1X1, 256, stride 1] X3 [3X3, 512, stride 1] X3 1X1, 1024, stride 1 1X1, 512, stride 1	1X1, 256, stride 1 [1X1, 128, stride 1] X3 [3X3, 256, stride 1] X3 1X1, 512, stride 1 1X1, 256, stride 1	
FLOPs		290	205.49	
Parameters		46,625,206	28,962,742	

learned a significant amount of image features. It serves as a starting point for model training, reducing the training time and cost. Therefore, we believe that utilizing ImageNet for pre-training is a highly effective approach to improve the performance and efficiency of our model.

(2) ***Data Pre-Processing:*** Data pre-processing is typically performed before training a model to ensure that the data is in a suitable format and to enhance the model's performance. The following steps outline the process of data pre-processing.

24.3.3.1 Step 1: Data Augmentation

To increase the diversity and quantity of our collected samples and improve the robustness of the model, we perform data augmentation. Data augmentation is primarily carried out in the following four directions. These data augmentation techniques help introduce variations in the training data, enabling the model to generalize better and become more robust to different conditions and variations in real-world scenarios.

A. Random cropping: Images are randomly cropped into different sizes and ratios, thus increasing the diversity and quantity of training samples.
B. Flipping: Images are horizontally or vertically flipped, thereby increasing the diversity and quantity of training samples.
C. Rotation: Images are randomly rotated by a certain angle, contributing to the diversity and quantity of training samples.
D. Scaling: Images are randomly scaled to different sizes and ratios, adding to the diversity and quantity of training samples.

TABLE 24.3
Summary of Training and Validation Dataset

	WLI	NBI	Background	Total
Training	7572	2566	593	10731
Validation	842	286	66	1194

24.3.3.2 Step 2: Data Split

A total of 11,925 images were used for data training, which included images collected from open and private datasets as mentioned in Materials. These images were split into a training set and a validation set with a ratio of 9:1. The training set was used for weight updates during model training, while the validation set was used to evaluate the model's real-time performance. The training set consisted of 10,731 images, and the validation set consisted of 1,194 images as shown in Table 24.3. To ensure that the model learns features related to non-polyp samples, 659 background images (true negatives) were specifically selected and included in the training data. This helps the model distinguish between polyp and non-polyp features more effectively.

24.3.3.3 Step 3: Data Resize and Normalization

Before model training, all images are first resized to a uniform size of 480 × 480 pixel resolution. Subsequently, each channel of the image is normalized by dividing it by 255, bringing the values into the range of [0, 1]. The purpose of normalization is to aid the model in converging faster, prevent the issues of exploding or vanishing gradients, and improve the stability and accuracy of the model. The normalization process ensures that the image data is in a consistent and standardized format, allowing the model to process the inputs effectively. It helps the model better understand the relative intensities and relationships between different pixel values, promoting more efficient learning and improving the overall performance and stability of the model.

(3) *Model Training:* Our model weights are initialized through pre-training on the ImageNet dataset and then fine-tuned on our own dataset. The training process is performed using NVIDIA GeForce RTX 3090 with 12 GB of RAM. To avoid GPU resource overload during training, the training dataset is divided into smaller subsets. Each subset is processed through the model network independently, and the loss function is computed. The optimizer (Adam) then performs back-propagation to update the model weights, aiming to minimize the loss function.

Adam (Adaptive Moment Estimation) is an optimizer that utilizes adaptive learning rates. It adjusts the learning rate automatically and performs well when handling large datasets, avoiding the issue of getting trapped in local optima. After each iteration update, the model calculates the loss function on the validation set to assess whether overfitting has occurred.

We set a maximum of 300 iterations (epochs) for training. If the maximum number of iterations is reached, the training process stops. Additionally, to prevent

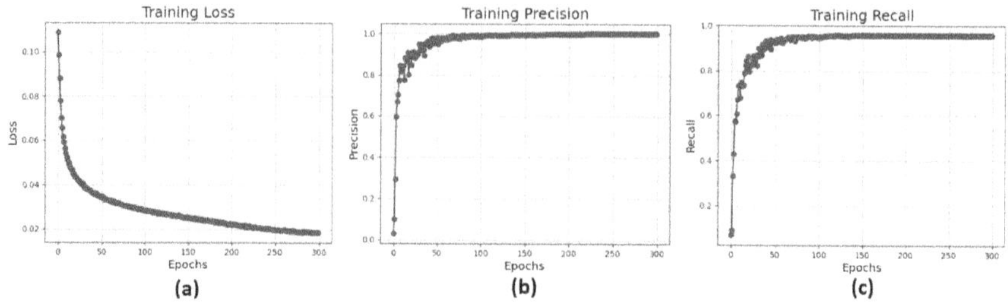

FIGURE 24.5 The loss, precision, and recall functions during training procedure.

overfitting, if the improvement in the loss function on the validation set decreases by less than 1% for five consecutive iterations, early stopping is triggered to halt the model's learning process. The loss, precision, and recall function of epoch of the training set can be found in Figure 24.5.

24.4 EXPERIMENTAL RESULTS

24.4.1 Experiment Environment

We implement the experiment environment of the detection model using Python 3.8 programming language with the open-source deep-learning framework Pytorch 1.13.1 with CUDA 11.6. The experiments are built under Ubuntu 20 operating system and ran on an Intel Core i9–11900KF CPU 2.80GHz and an RTX 3090 graphic card.

24.4.2 Evaluation Metrics

We have used standard computer vision metrics to evaluate polyp detection and localization methods on the PICCOLO dataset. Considering the problem of polyp detection, these values can be considered as follows. TP indicates that the predicted bounding box falls on the ground truth of the polyp. FP indicates that the predicted bounding box falls outside the basic truth of the polyp. FN indicates no predicted bounding box but a polyp of the frame. Finally, TN indicates that no polyps are detected in images without polyps.

Precision indicates how many of the values estimated as positive are true positive. Thus, it is an important metric for subtracting the cost of the FP number and is formulated by

$$Precision = \frac{TP}{TP + FP} \tag{24.4}$$

Recall indicates how many true positives are estimated from all positives. It is sometimes referred as sensitivity and is formulated by

$$Recall = Sensitivity = \frac{TP}{TP + FN} \tag{24.5}$$

Specificity indicates how many true negatives are estimated from all negatives. It is formulated by

$$Specificity = \frac{TN}{TN + FP} \tag{24.6}$$

The F1-score represents the harmonic mean of precision and recall metrics. In particular, it should be chosen when a balance between precision and recall is required. The F1-score is expressed by

$$F1 = \frac{2 * Precision * Recall}{Precision + Recall} \tag{24.7}$$

The area under the curve (AUC) represents the area under the ROC curve. It measures the overall performance of the binary classification model. As both TPR (True Positive Rate) and FPR (False Positive Rate) range between 0 to 1, therefore, the area will always lie between 0 and 1, and a greater value of AUC denotes better model performance.

24.4.3 Experiments

In this section, the experiment dataset are 3,433 images contained polyps and 1,003 images did not contain polyps. All polyps were confirmed histologically after biopsy. To evaluate the performance of our DHNN architecture, three backbones (DHNN, CSPDarknet53, and morphology) are made based for the YOLOv5 architecture, in which the CSPDarknet53 is the original backbone of YOLOv5.

Figure 24.6 shows the FROC (Free-Response Receiver operating characteristic curve) curves for DHNN, CSPDarknet53, and morphology backbone models, and the AUC value of the different models are 0.97, 0.94, and 0.85, respectively.

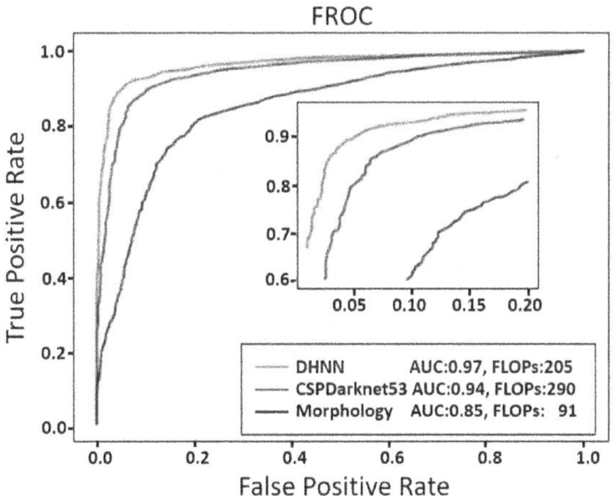

FIGURE 24.6 The AUC results of DHNN, CSPDarknet53, and morphology backbone models.

TABLE 24.4
Specificity and FP Counts Results Based on 90% Sensitivity

Method	Backbone	Specificity	FP counts
Yolov5	CSPDarknet53	89.36%	107
Yolov5	DHNN	**94.63%**	**54**
Yolov5	Morphology	46.1%	539

TABLE 24.5
Sensitivity and FN Counts Results Based on 10% FPR

Method	Backbone	Sensitivity	FN counts
Yolov5	CSPDarknet53	89.04%	376
Yolov5	DHNN	**92.92%**	**243**
Yolov5	Morphology	60.76%	1343

TABLE 24.6
F1-Score, Precision and Sensitivity Results Based on FPR Being 0.1

Method	Backbone	F1	Precision	Sensitivity
Yolov5	CSPDarknet53	0.9276	0.9680	89.04%
Yolov5	DHNN	**0.9486**	**0.9687**	**92.92%**
Yolov5	Morphology	0.7421	0.9529	60.76%

Table 24.4 and Table 24.5 show the results for the polyp detection and localization on the PICCOLO dataset, respectively. Set sensitivity to be 90%, the specificity of the DHNN, CSPDarknet53, and morphology backbone models are 94.63%, 89.3%, and 46.1%, respectively. Set the FPR to be 0.1, the sensitivity of the DHNN, CSPDarknet53, and morphology backbone models are 92.92%, 89.04%, and 60.76%, respectively.

Table 24.6 shows the results for the polyp detection and localization on the PICCOLO dataset. Set the FPR to be 0.1, the F1-score of the DHNN, CSPDarknet53, and morphology backbone models are 0.9486, 0.9276, and 0.7421, respectively. The precision of the DHNN, CSPDarknet53, and morphology backbone models are 0.9687, 0.9680, and 0.9529, respectively. The sensitivity of the DHNN, CSPDarknet53, and morphology backbone models are 92.92%, 89.04%, and 60.76%, respectively.

Table 24.7 shows the results for the polyp detection and localization on the PICCOLO dataset. Set the sensitivity to 90%, the F1-score of the DHNN, CSPDarknet53, and morphology backbone models are 0.9403, 0.9321, and 0.8762, respectively. The precision of the DHNN, CSPDarknet53, and morphology backbone models are 0.9828, 0.9665, and 0.8517, respectively. The specificity of the DHNN, CSPDarknet53, and morphology backbone models are 94.63%, 89.36%, and 90.21%, respectively.

From the experimental results, it can be found that the DHNN architecture perform better performance than the backbone models of convolution-only and

TABLE 24.7
F1-Score, Precision, and Specificity Results Based on Sensitivity Being 90%

Method	Backbone	F1	Precision	Specificity
Yolov5	CSPDarknet53	0.9321	0.9665	89.36%
Yolov5	DHNN	**0.9403**	**0.9828**	**94.63%**
Yolov5	Morphology	0.8762	0.8517	90.21%

TABLE 24.8
Performance Metrics for the Overall Image Test Dataset, for Different Subsets Attending to the Histology, Morphology, and Size of the Polyp

	Recall	Precision	F1-Score
Overall	0.929	0.969	0.949
Histology			
Adenoma	0.958	0.954	0.956
Hyperplasia	0.807	0.843	0.825
Adenocarcinoma	0.966	0.813	0.883
Morphology			
Protruded	0.977	0.926	0.951
Flat	0.881	0.934	0.906
Size			
≥5 mm	0.927	0.956	0.941
<5 mm	0.895	0.809	0.850

morphology-only. However, we want to know the further performance of proposed DHNN architecture on different subgroups. According to the pathology report, the different subsets attending to the histology, morphology, and size of the polyp. Table 24.8 shows the detailed performance results by subgroups of polyps, according to their morphology, histology, and size.

Regarding histology, images of adenoma polyps were more accurately than average predicted, achieving the highest F1 score of 0.956. Regarding morphology, there is more variation, with the F1 score obtained when testing with images of polyps of flat morphology being much lower (0.906) than the average. However, images of protruded polyps were more accurately than average predicted, achieving the highest F1 score of 0.951. Small polyps seem to be more difficult to detect, the performance was better in bigger ones than in smaller ones (F1 score of 0.941 and 0.850, respectively). The confusion matrix for polyp classifications with histology, morphology, and size are shown in Figures 24.7, 24.8 and 24.9, respectively.

We categorize the adenoma and hyperplasia polyps into two types: protruded IS lesion and flat IIa lesion. Figures 24.10 and 24.11 show the adenoma protruded IS and flat Iia lesions, respectively. Figure 24.12 shows the hyperplasia flat Iia lesion. The red bounding box represents the ground truth (GT), and the green bounding box is the detected polyp by our DHNN architecture. Note that the original

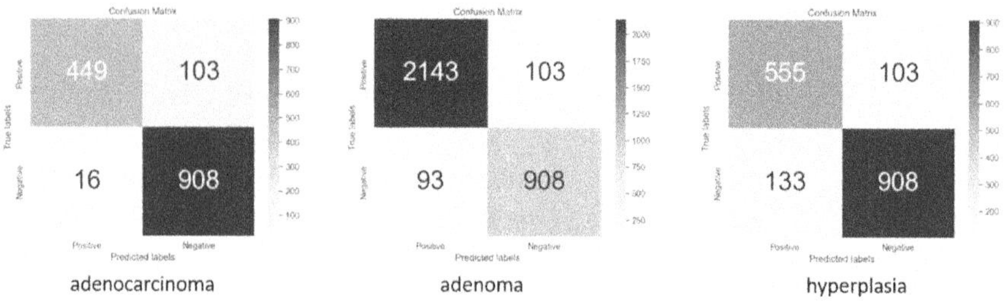

FIGURE 24.7 Confusion matrix of adenocarcinoma, adenoma, and hyperplasia classification.

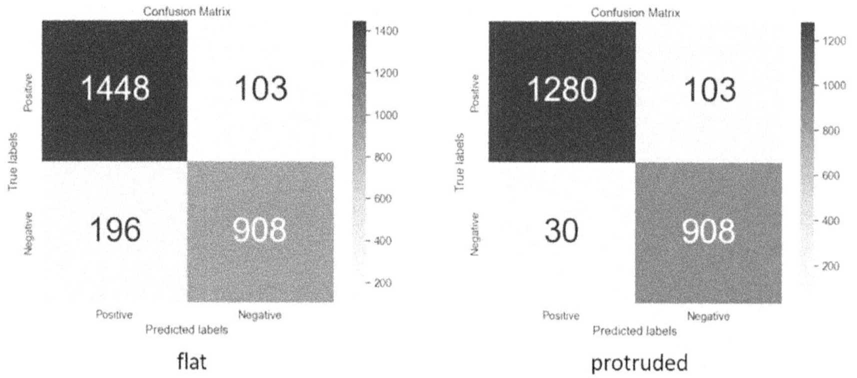

FIGURE 24.8 Confusion matrix of flat and protruded classification.

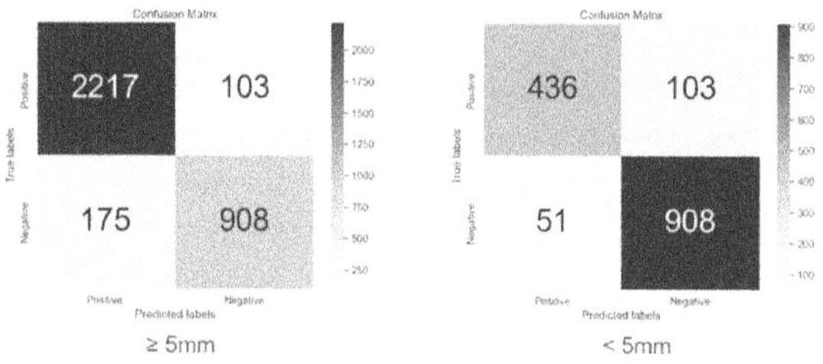

FIGURE 24.9 Confusion matrix of size smaller than five and others classification.

YOLOv5 architecture cannot detect those polyps. It can be concluded that the original YOLOv5 detection model is less likely to detect polyps when the boundaries are less distinct. However, after extracting both linear and nonlinear features, it becomes capable of detecting polyps with less conspicuous boundaries. Clinically, these polyps with unclear boundaries are also more likely to be missed. Therefore, the method proposed in this chapter can enhance the clinical polyp detection rate.

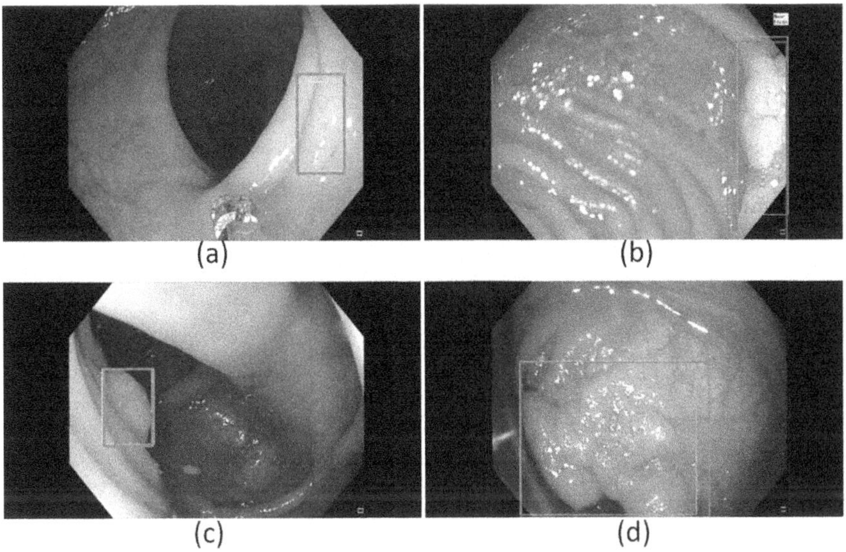

FIGURE 24.10 The adenoma protruded IS lesions detected by our DHNN architecture but fail by original YOLOv5 architecture.

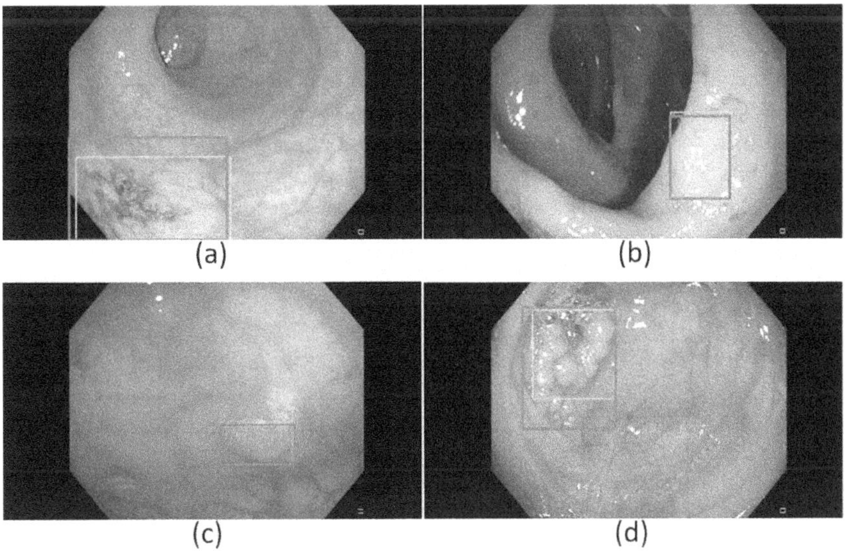

FIGURE 24.11 The adenoma protruded flat Iia lesions detected by our DHNN architecture but fail by original YOLOv5 architecture.

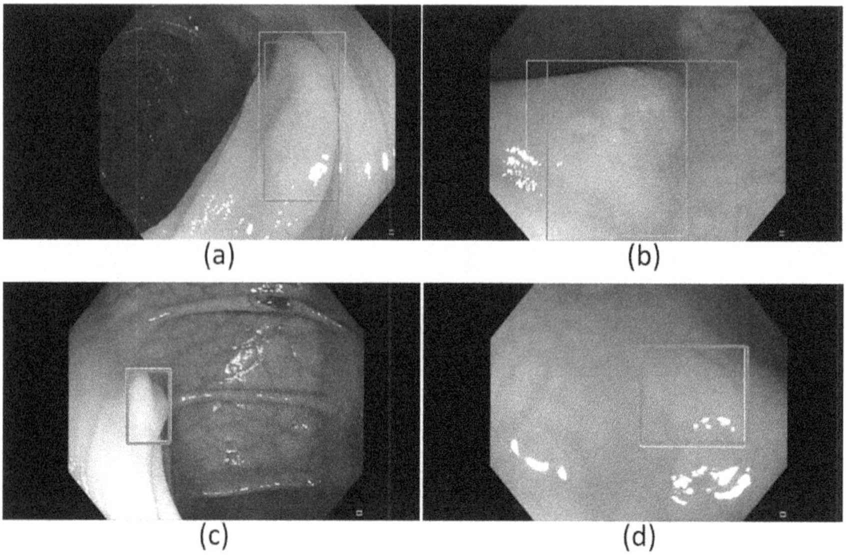

FIGURE 24.12 The hyperplasia flat Iia lesions detected by our DHNN architecture but fail by original YOLOv5 architecture.

24.5 CONCLUSIONS

In this chapter, the DHNN architecture is developed to combine the morphology (nonlinear) and convolution (linear) operators. It can enhance the ability of the network to recognize and analyze different shapes and structures in the image data. We use nine polyp image datasets, including publicly available data and our own collected data for experiments and select three backbone models for comparisons. Experimental results show the proposed DHNN architecture outperforms the best in terms of computational complexity and accurate performance. We believe our DHNN-architecture-based AI solution can be a second set of eyes to assist an endoscopist for polyp detection in the real world.

REFERENCES

Ali, S., Ghatwary, N., Braden, B., Lamarque, D., Bailey, A., Realdon, S., Cannizzaro, R., Rittscher, J., Daul, C., and East, J., "Endoscopy disease detection challenge 2020," arXiv:2003.03376, 2020.

Aslanian, H. R., Shieh, F. K., Chan, F. W., Ciarleglio, M. M., Deng, Y., Rogart, J. N., Jamidar, P. A., and Siddiqui, U. D., "Nurse observation during colonoscopy increases polyp detection: A randomized prospective study," *American Journal of Gastroenterology*, vol. 108, pp. 166–172, 2013.

Bernal, J., Sánchez, F. J., Fernández-Esparrach, G., Gil, D., Rodríguez, C., and Vilariño, F., "WM-DOVA maps for accurate polyp highlighting in colonoscopy: Validation vs. saliency maps from physicians," *Computerized Medical Imaging and Graphics*, vol. 43, pp. 99–111, 2015.

Bernal, J., Tajbakhsh, N., Sanchez, F. J., Matuszewski, B. J., Chen, H., Yu, L., Angermann, Q., Romain, O., Rustad, B., Balasingham, I., Pogorelov, K., Choi, S., Debard, Q., Maier-Hein, L., Speidel, S., Stoyanov, D., Brandao, P., Córdova, H., Sánchez-Montes, C., Gurudu, S. R., Fernández-Esparrach, G., Dray, X., Liang, J., and Histace, A., "Comparative validation of polyp detection methods in video colonoscopy: Results from the miccai 2015 endoscopic vision challenge," *IEEE Transactions on Medical Imaging*, vol. 36, pp. 1231–1249, 2017.

Brenner, H., Chang-Claude, J., Seiler, C. M., Rickert, A., and Hofmeister, M., "Protection from colorectal cancer after colonoscopy: A population-based, case-control study," *Annals of Internal Medicine*, vol. 154, pp. 22–30, 2011.

Carrinho, P. and Falcao, G., "Highly accurate and fast YOLOv4-based polyp detection," *Expert Systems with Applications*, vol. 232, Article 120834, 2023.

Chin, J., "Chinese Society of Gastroenterology. Chinese consensus: Screening, early diagnosis and treatment, comprehensive prevention of large bowel cancer," *Gastroenterology Hepatology*, vol. 20, pp. 979–995, 2011.

Corley, D. A., Jensen, C. D., Marks, A. R., Zhao, W. K., Lee, J. K., Doubeni, C. A., Zauber, A. G., de Boer, J., Fireman, B. H., Schottinger, J. E., Quinn, V. P., Ghai, N. R., Levin, T. R., and Quesenberry, C. P., "Adenoma detection rate and risk of colorectal cancer and death," *New England Journal of Medicine*, vol. 370, pp. 1298–1306, 2014.

Debesh, J., Smedsrud, H. P., Riegler, M. A., Halvorsen, P., Johansen, D., de Lange, T., and Johansen, H. D., "Kvasir-SEG: A segmented polyp dataset," *Proceedings of the trinational conference on Multimedia Modeling*, Republic of Korea, 2020.

Deng, Z.-Y., Chiang, H.-H., Kang, L.-W., and Li, H.-C., "A lightweight deep learning model for real-time face recognition," *IET Image Processing*, vol. 17, no. 13, pp. 3869–3883, 2023.

Edwards, B. K., Ward, E., Kohler, B. A., Eheman, C., Zauber, A. G., Anderson, R. N., Jemal, A., Schymura, M. J., Lansdorp-Vogelaar, I., Seeff, L. C., van Ballegooijen, M., Goede, S. L., and Ries, L. A. G., "Annual report to the nation on the status of cancer, 1975–2006, featuring colorectal cancer trends and impact of interventions (risk factors, screening, and treatment) to reduce future rates," *Cancer*, vol. 116, pp. 544–573, 2010.

International Agency for Research on Cancer, *Colorectal Cancer Factsheet*, International Agency for Research on Cancer, 2018.

Ishaq, S., Siau, K., Harrison, E., Tontini, G. E., Hoffman, A., Gross, S., Kiesslich, R., and Neumann, H., "Technological advances for improving adenoma detection rates: The changing face of colonoscopy," *Digestive and Liver Disease*, vol. 49, pp. 721–727, 2017.

Laura, C. S., Thomas, B. R., Shapiro, J. A., Marion, R. N., Diane, L. M., Leslie, S. G., Fred, B. D., Linda, D. W., and Matthew, T. M., "How many endoscopies are performed for colorectal cancer screening? Results from CDC's survey of endoscopic capacity," *Gastroenterology*, vol. 127, pp. 1670–1677, 2004.

Lieberman, D. A., Rex, D. K., Winawer, S. J., Giardiello, F. M., Johnson, D. A., and Levin, T. R., "Guidelines for colonoscopy surveillance after screening and polypectomy: A consensus update by the US Multi-Society Task Force on Colorectal Cancer," *Gastroenterology*, vol. 143, pp. 844–857, 2012.

Ng, S. A., *Endoscopic computer vision challenges 2.0.* https://endocv2022.grand-challenge.org/. Accessed 14 November 2022.

Nogueira, K., Chanussot, J., Mura, M. D., and Dos Santos, J. A., "An introduction to deep morphological networks," *IEEE Access*, vol. 9, pp. 114308–114324, 2021.

Sánchez-Peralta, L. F., Pagador, J. B., Picón, A., Calderón, A. J., Polo, F., Andraka, N., Bilbao, R., Glover, B., Saratxage, C. L., and Sánchez-Margallo, F. M., "PICCOLO white-light and narrow-band imaging colonoscopy dataset: A performance comparative of models and datasets," *Applied Sciences*, vol. 10, p. 8501, 2020.

Shen, Y., Zhong, X., and Shih, F. Y., "Deep morphological neural networks," *International Journal of Pattern Recognition and Artificial Intelligence*, vol. 36, no. 12, 2022.

Shih, F. Y. and Puttagunta, P., "Recursive soft morphological filters," *IEEE Transactions on Image Processing*, vol. 4, pp. 1027–1032, 1995.

Shih, F. Y. and Wu, Y.-T., "Decomposition of arbitrary gray-scale morphological structuring elements," *Pattern Recognition*, vol. 38, pp. 2323–2332, 2005a.

Shih, F. Y. and Wu, Y.-T., "Decomposition of binary morphological structuring elements based on genetic algorithms," *Computer Vision and Image Understanding*, vol. 99, pp. 293–302, 2005b.

Siegel, R., Naishadham, D., and Jemal, A., "Cancer statistics, 2013," *CA: A Cancer Journal for Clinicians*, vol. 63, pp. 11–30, 2013.

Silva, J., Histace, A., Romain, O., Dray, X., and Granado, B., "Toward embedded detection of polyps in WCE images for early diagnosis of colorectal cancer," *International Journal of Computer Assisted Radiology and Surgery*, vol. 9, no. 2, pp. 283–293, 2014.

Stewart, B. W. and Wild, C. P., *World Health Organization. World Cancer Report 2014*, International Agency for Research on Cancer, 2014.

Vázquez, D., Bernal, J., Sánchez, F. J., Fernández-Esparrach, G., López, A. M., Romero, A., Drozdzal, M., and Courville, A., "A benchmark for endoluminal scene segmentation of colonoscopy images," *Journal of Healthcare Engineering*, vol. 2017, pp. 1–9, Jul. 2017.

Yeh, C.-H., Lin, C.-H., Kang, L.-W., Huang, C.-H., Lin, M.-H., Chang, C.-Y., and Wang, C.-C., "Lightweight deep neural network for joint learning of underwater object detection and color conversion," *IEEE Transactions on Neural Networks and Learning Systems*, vol. 33, no. 11, pp. 6129–6143, 2021.

Zauber, A. G., Winawer, S. J., O'Brien, M. J., Lansdorp-Vogelaar, I., van Ballegooijen, M., Hankey, B. F., Shi, W., Nond, J. H., Schapiro, M., Panish, J. F., Stewart, E. T., and Waye, J. D., "Colonoscopic polypectomy and long-term prevention of colorectal-cancer deaths," *New England Journal of Medicine*, vol. 366, pp. 687–696, 2012.

Zhu, X., Lyu, S., Wang, X., and Zhao, Q., "TPH-YOLOv5: Improved YOLOv5 based on transformer prediction head for object detection on drone-captured scenarios," *Proceedings of the IEEE/CVF International Conference on Computer Vision (ICCV) Workshops*, pp. 2778–2788, 2021.

25 BFC-Cap *Background and Frequency-Guided Contextual Image Captioning*

25.1 INTRODUCTION

During recent years, image captioning has emerged as a pivotal area of research in artificial intelligence, attracting increasing attention from technological and scientific communities. Image captioning involves the automatic generation of coherent and accurate descriptions for given images, requiring a blend of computer vision and natural language processing techniques. This task is not merely about identifying objects within an image but also about understanding their attributes, actions, and the complex interactions and relationships that define, and therefore convey, the context of the scene.

The use of transformer-based techniques (Barraco et al., 2023) has greatly improved image captioning. These methods are especially good at understanding relationships in sequences, which helps in creating detailed and meaningful captions. To make the most of transformers, it is important to pull out and combine key features from images, including both details about objects and context they are in.

Earlier region-based transformer approaches (Cornia et al., 2020; Pan et al., 2020; Qin et al., 2019) leverage object-level features extracted by sophisticated object detection networks, providing detailed and specific insights into the image content, particularly regarding the objects. On the other hand, grid-based approaches (Anderson et al., 2018; Jiang et al., 2018) use local features at the grid level, which can be effective but often miss detailed information about specific objects. Moreover, using both region- and grid-based features together tends to create semantic noises during the attention process.

Region features and grid features, as shown in Figure 25.1, alone are often insufficient for representing the full range of inherent image properties. To generate captions with more background and contextual information, it is essential to develop a framework that captures key visual features during the encoding process. Therefore, we enhance these features by incorporating additional elements, such as background features and high-frequency local grid features, as shown in Figure 25.1(e) and Figure 25.1(f), respectively. The decoding phase requires a well-designed caption generator that can interpret these enhanced visual features to create accurate captions with contextually rich information.

The inclusion of complex language decoders, large language models, and vision-language models (Ramos et al., 2023a) has shown potential for more accurate image captioning. However, implementing these models is challenging due to limitations in data availability, computational power, resources, model generalization, annotation

DOI: 10.1201/9781003474395-28

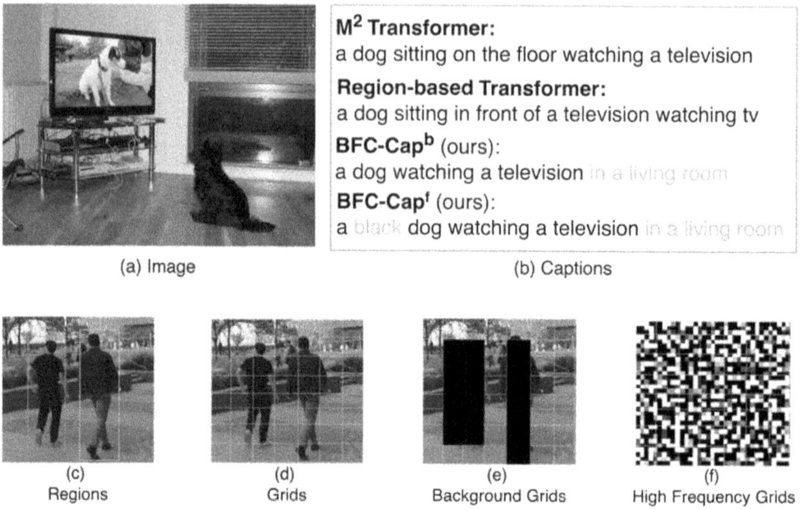

FIGURE 25.1 A comparison of our approaches (BFC-Capb and BFC-Capf) with the M2 Transformer and Region-Based Transformer (up) and an illustration of different visual features (down). Compared to the M2 Transformer and Region-Based Transformer, our approaches with background and high-frequency grid features generate captions with more background and concise contextual details.

quality, and the complex integration of vision and language components that involves aligning and fine-tuning the interplay between image and text features. As a result, there is a strong need to develop methods that can better leverage the unique features of images to deliver richer and more contextual information.

This research aims to utilize multiple visual features and multi-stream cross-attention for effective image captioning, focusing on the background and nuanced contextual understanding of visual content and its translation into natural language. In summary, our main contributions are as follows:

- We present two image captioning frameworks, BFC-Capb and its extended variant BFC-Capf, with multiple visual features including background and high-frequency features, in addition to object-level region features and their interactions.
- We propose a background feature extraction module that pays dedicated attention to background by removing grids that overlap with the bounding boxes of the object-level region features, and an object-background attention module to capture the interactions and relationships between the object features and background features.
- We propose a high-frequency grid features extraction module to enhance fine details that are critical for thorough visual understanding and enrich the overall context in captions.
- We introduce multi-stream cross-attention (MSCA) that reduces the modality gap between vision and language and captures the interaction of text

features with high-frequency grid features, objects, background, and their relationships.
- Extensive experiments on MS COCO image-captioning benchmark with both traditional and recent evaluation metrics confirm the effectiveness of our approaches in generating captions with more background and concise contextual information, in terms of both quantitative and qualitative measurements.

The rest of this chapter is organized as follows. Section 25.2 describes related work. Section 25.3 describes our proposed image captioning methods. Section 25.4 presents the experimental results and analysis. Finally, conclusions are drawn in Section 25.5.

25.2 RELATED WORK

25.2.1 General Encoder-Decoder-Based Methods

Image-captioning framework generally consists of an encoder that encodes images into different representations and a decoder that generates captions for the images. Several previous studies focus on encoder design for better representation of visual information. Complementary Shifted Transformer (CST) (Liu et al., 2023) utilizes both absolute and relative positional encoding, drop branch technique and spatial shifting of the encoding features that helps generate captions with fine grained and specific actions. However, this method struggles in specific content expressions such that a specific object may be identified as another due to lack of explicit object detection. The traditional transformers are biased to the words that have higher frequency in the datasets. CATNet (Yang et al., 2023) addresses the bias problem by incorporating global context information and capturing the multi-level inter-modal interactions. However, it learns to represent the input image and caption text in two independent representation spaces, which implicitly hinder language representation from exploiting visual information. Context-aware auxiliary guidance significantly helps in generating more descriptive captions (Song et al., 2021). Parallel cross-attention between region and grid features performs better than concatenated cross-attention and sequential cross-attention (Nguyen et al., 2022). Intermediate concept tokens generated from intermediate grid features significantly contribute to the final caption (Fang et al., 2022). DIFNet (Wu et al., 2022) incorporates segmentation features to enhance the contribution of visual information for caption prediction. Computational complexity is an important factor concerning the efficiency of an image-captioning model. One effective strategy for reducing the complexity is to exit the prediction early in internal decoding layers without passing the entire model. DeeCap (Fei et al., 2022) significantly reduces the complexity at the expense of a marginal decrease in accuracy, employing an imitation learning mechanism that predicts the deep layer activation using features from shallow layers.

25.2.2 Prompt-Learning-Based Methods

Prompt learning is a powerful technique to allow a model pre-trained on a large-scale dataset and then enhance that model for a downstream task by providing an appropriate prompting function. Embedding the prompt learning, ConCap (Wang et al., 2023) is able to simultaneously handle other domains and generate captions with different lengths and styles for each image. ViECap (Fei et al., 2023) addresses the bias issue with pre-trained models including modality bias and object hallucination and provides better generalization. However, these approaches may struggle to completely understand the rich context of images, due to their reliance on text prompts.

25.2.3 Retrieval-Augmented-Based Methods

Retrieval augmented approaches usually retrieve additional data from a data store to support the primary network. SMALLCAP (Ramos et al., 2023a) and EXTRA (Ramos et al., 2023b) generate captions given the input image and a set of captions related to images, retrieved from a datastore which facilitates textual evidence from nearest neighbor captions. However, biases in the dataset and mismatched retrieved captions may lead these models to incorrect predictions. EVCAP (Li et al., 2023), a lightweight retrieval-augmented image-captioning model, utilizes an easily expandable external visual-name memory with object names rather than full captions. However, it may not retrieve all objects that appear in the given image, leading to incomplete image descriptions, and its focus on object representations may limit other significant captioning elements such as the relationship among objects. An alternative to retrieving information from external knowledge bases can be a prototypical memory model with distribution of past keys and values through discriminative and compact prototype vectors, without the usage of external databases (Barraco et al., 2023). Unlike autoregressive approaches, which construct sentences word by word, non-autoregressive methods generate all words simultaneously. Non-autoregressive methods, particularly those enhanced with retrieval augmentation, appear to outperform the autoregressive approaches (Luo et al., 2023). Unsupervised captioning without requiring image-text pair retrieves objects related to a given text from the training set, even if these objects are not associated with the same image (Meng et al., 2022). However, unsupervised captioning struggles to exceed the performance of supervised captioning.

Our proposed image-captioning models follow the encoder-decoder paradigm with the goal of generating captions that incorporate rich background and contextual information. During the encoding phase, we utilize three types of visual features: object-level region features, background features, and high-frequency local grid features. In the decoding phase, we employ a multi-stream cross-attention mechanism to reduce the modality gap between language and visual features.

25.3 THE PROPOSED APPROACHES

In this section, we introduce two approaches, specifically BFC-Capb – a background-based image captioning, and its extended variant BFC-Capf – a frequency-guided contextual image captioning, to enhance image captions with more background and concise contextual information.

Given an input image, both approaches extract object-level region features with bounding boxes using faster R-CNN (Ren et al., 2015). A background feature extraction module exploits background features by discarding the grids that highly overlap with bounding boxes of region features. To extract high-frequency grid features, grid features are transformed into frequency domain and low-frequency components are filtered out. The high-frequency grid features enhance the fine details that are critical for thorough visual understanding and enrich the overall context in captions. Encoding layers consist of multi-head attention (MHA) (Vaswani et al., 2017) and feed-forward network (FFN) followed by addition and layer normalization. A transformer-based decoder generates a caption for the given image.

25.3.1 Preliminaries

Algorithm 1: Multi-Head Attention (MHA)

1. MHA (Q, K, V):
2. Input: Q, K, V
3. Attention: $A_i = Softmax\left(\frac{QW_i^Q\left(KW_i^K\right)^T}{\sqrt{d_k}}\right)$
4. Head: $H_i = A_i V W_i^V$
5. Multi-Head: $H = \left[H_1, H_2, H_3, \ldots, H_h\right] W^H$
6. $Y = LN\left(H + Q\right)$
7. Output: Y

As stated in Algorithm 1, MHA takes three inputs: query Q, key K, and value V. The projections are the parameter metrics $W_i^Q \in R^{d \times d_q}, W_i^K \in R^{d \times d_k}, W_i^V \in R^{d \times d_v}$, and $W_i^H \in R^{d \times d}$. Here, $d_Q = d_K = d_V = d / h$; where h is the number of heads in MHA. After linear projection of Q and K using matrix multiplication with their corresponding parameter matrix, scaling is applied by dividing the result of multiplication by the square root of dimension of K, d_k. Attention score is calculated using the softmax function after scaling. Then, the attention score is multiplied by the linear projection of V. We obtain the linear projection of V by matrix multiplication of V with its corresponding parameter matrix W_i^V. The [·] in line 5 of Algorithm 1 represents the concatenation operation, and LN is the layer normalization operation.

Algorithm 2: Multi-Layer Perceptron (MLP)

1. MLP (X)
2. Input: $X_0 = X$
3. $X_i = W_i X_{i-1} + b_i$
4. $X_i = \sigma\left(X_i\right)$
5. $Y = X_n$
6. Output: Y

Multi-Layer Perceptron (MLP) is outlined in Algorithm 2. Here, W_i is the weight matrix for the i-th layer. σ represents an activation function and in this case, it is Rectifier Linear Unit (ReLU) activation function.

Algorithm 3: Feed-Forward Network (FFN)

1. FFN (X)
2. Input: X
3. $Y = LN\left(FC\left(ReLU\left(FC\left(X\right)\right)\right) + X\right)$
4. Output: Y

Algorithm 3 states the feed-forward network. Here, LN is the layer normalization operation, FC denotes a fully connected layer, and the ReLU is the rectifier linear unit activation function.

25.3.2 The First Approach: BFC-Capb

In this section, we introduce a simple and effective approach of image captioning with background feature exploitation to enrich the generated captions with more background information. Figure 25.2 shows the overall architecture of this approach.

Given an input image, object-level region features with bounding boxes and local grid features are extracted using faster R-CNN. The background feature extraction module exploits background feature by discarding the grids that are highly overlapped with bounding boxes of region features. The object attention (OA) module takes the region features as input and finds the interactions and relationships between the region features at object level. Similarly, the attention scores of the background features from the grids, not overlapped with the bounding boxes of the region features, are obtained by the background attention (BA) module. Then, the object-background attention (OBA) module finds the interactions and relationships between the object features and background features. The outputs from the OA module and the OBA module are weighted by $\alpha \in [0,1]$ and $(1-\alpha)$ respectively, and then combined before being sent to the encoder. Encoding

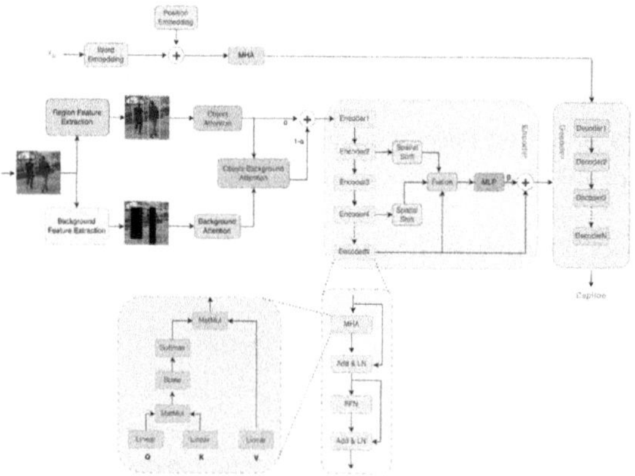

FIGURE 25.2 Overview of the proposed BFC-Capb. Given an image, the region and background feature extractors provide the object-level region features and grid-level background features, respectively. Object-background attention captures the interaction and relationship between the object features and the background features. The resulting features are combined with object features and used as input to the encoder. The decoder takes text features and visual features from the encoder as input and generates a caption for the given image.

layers consist of MHA and FFN followed by addition and layer normalization. Features from every second encoding layers are concatenated and passed through a multi-layer perceptron (MLP). The resulting output is weighted by a parameter $\beta \in [0,1]$ and then combined with the features from the last encoding layer before being sent to the decoder. Also, the partially generated caption, after being processed with word embedding, position embedding, and multi-head attention (MHA), is sent to the decoder. The transformer-based decoder (Vaswani et al., 2017) generates the caption for the given image.

25.3.2.1 Background Feature Extraction

The background feature extraction module pays dedicated attention to the background of an image so that the model generated caption contains more background information. A pre-trained feature extractor, faster R-CNN provides the object-level region features and local grid features. Leveraging the coordinates of the bounding boxes from the object detections and the coordinates of the grids, we identify the grids that are overlapped with the bounding boxes and remove them so that we have the remaining grids for the background only. We utilize the alignment graph (Luo et al., 2021) to determine if a grid is overlapped with a bounding box of region features. Thus, we obtain the background features, as depicted in Figure 25.3.

25.3.2.2 Attention Modules

The object attention (OA) module takes the object-level region features as input and applies the multi-head attention (MHA), as stated in Algorithm 1, to find the interactions and relationships between the region features. In MHA, the region features are set as query (Q), key (K), and value (V). Thus, OA returns the significant object-level region features with their interactions and relationships. Similarly, in the background attention (BA) module, MHA is used where the background features are used as query, key, and value. The BA module provides the interactions and relationships between the background features. As region features contain object-level information and background features contain background information, it is essential to find the interactions and relationships between the region features and background features,

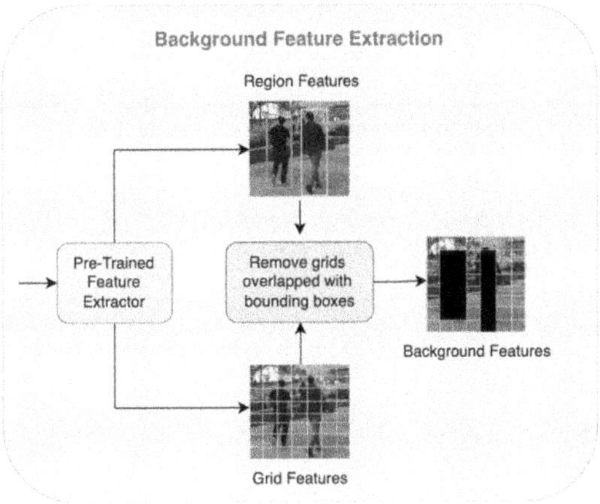

FIGURE 25.3 Background feature extraction process.

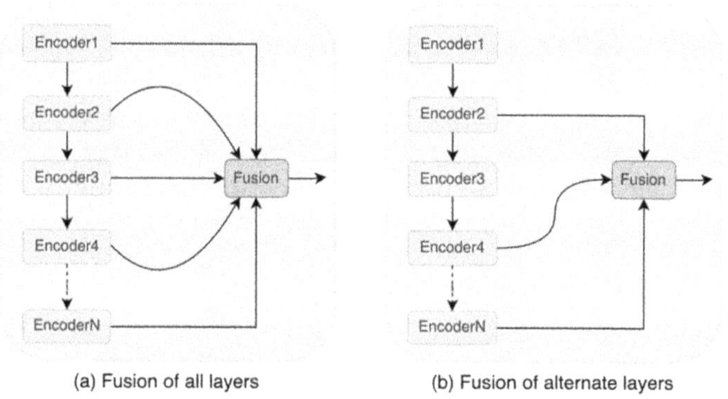

FIGURE 25.4 Two designs of fusion process.

which facilitate an additional support to generate background-aware and coherent captions. To accomplish this, we employ object-background attention (OBA), which is an MHA where background features are used as query and object-level region features are used as key and value.

25.3.2.3 Final Feature Encoding (FFE)

The encoding process is outlined in Algorithm 4. Here, we use multiple encoding layers, each of which consists of MHA and FFN. Fusion of features from the encoding layers is important as it includes both high-level and low-level features. We consider two design choices for the fusion process, as illustrated in Figure 25.4. In one design, we consider all the encoding layers in fusion. However, in the other design, as features from two consecutive encoding layers may not be significantly distinctive and may introduce redundancy in the concatenated features after the fusion, we consider concatenating the features from every second encoding layer, including the last layer. We examine the performance of these two choices through experiments, as discussed in Section 25.4.5. Before concatenation, we perform spatial shift operations to enable the features to interact with the neighbors. Multi-layer perceptron (MLP) is used after the feature fusion for mapping the concatenated features with desired dimensions.

Algorithm 4: Final Feature Encoding (FFE)

1. FFE (Z)
2. Input: Z
3. $Ei\sim = LN\left(MHA\left(Q=Z, K=Z, V=Z\right) + E_{i-1}\right)$
4. $E_i = LN\left(FFN\left(Ei\sim\right) + Ei\sim\right)$
5. $E = \left[SS\left(E_2\right), SS\left(E_4\right), ..., E_n\right]$
6. $E_M = MLP\left(E\right)$
7. $Y_Z = \beta * E_M + E_n$
8. Output: YZ

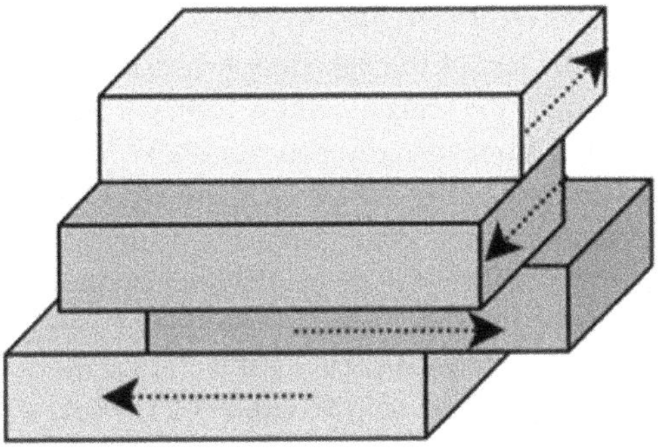

FIGURE 25.5 Spatial shift operation.

Here, Z is the weighted summation of object-level region features and the relationship between object and background features, given by

$$Z = \alpha * Y_{RR} + (1 - \alpha) * Y_{R\hat{G}} \tag{25.1}$$

where $\alpha \in [0,1]$ controls the weights of object-level region features and background features. We set $\alpha = 0.6$ which is the optimal value as discussed in Section 25.4.5. YRR and $Y_{R\hat{G}}$ are the outputs from object attention and object-background attention, respectively. In Algorithm 4, SS represents the spatial shift operation (Ma et al., 2023; Liu et al., 2023) and MLP is multi-layer perceptron, as outlined in Algorithm 2. For each channel dimension, SS operation shifts the vectors in each spatial dimension both in forward and backward directions which is illustrated in Figure 25.5 and can be represented as equations (25.2) and (25.3). The weighting factor $\beta \in [0,1]$ in Algorithm 4 is set to 0.2, inspired by Ma et al. (2023).

$$\begin{aligned} V_1[d_s : h,\ :,0 : c/4] &= V_1[0 : h-d_s,\ :,\ 0 : c/4], \\ V_1[0 : h-d_s,\ :,\ c/4 : c/2] &= V_1[d_s : h,\ :,\ c/4 : c/2], \\ V_1[:,\ d_s : w,\ c/2 : 3c/4] &= V_1[:,0 : w-d_s,\ c/2 : 3c/4], \\ V_1[:,\ 0 : w-d_s, 3c/4 : c] &= V_1[:,\ d_s : w, 3c/4 : c] \end{aligned} \tag{25.2}$$

$$\begin{aligned} V_2[:\ ,\ d_s : w, 0 : c/4] &= V_2[:,0 : w-d_s, 0 : c/4], \\ V_2[:\ ,\ 0 : w-d_s,\ c/4 : c/2] &= V_2[:,\ d_s : w,\ c/4 : c/2], \\ V_2[d_s : h,\ :,\ c/2 : 3c/4] &= V_2[0 : h-d_s,\ :,\ c/2 : 3c/4], \\ V_2[0 : h-d_s,\ :, 3c/4 : c] &= V_2[d_s : h,\ :,\ 3c/4 : c] \end{aligned} \tag{25.3}$$

where V_i is the output feature of the $2i$-th encoder layer and d_s is the shift distance of spatial shift, which determines the scope of local interaction. The output of the last encoder layer is not processed by any shift operations.

25.3.2.4 Caption Generation by BFC-Capb

Caption is generated at the end of the decoding process as outlined in Algorithm 5. We take two types of inputs: the image features from encoders and the text features. Text features are obtained from partially generated text XD using multi-head attention (MHA). Before sending XD to MHA, it is transformed into word embedding and combined with position embedding. We use a trainable word embedding layer to represent the words in a caption, where each word is mapped to a dense vector. To capture the order of words in text, we use a pre-computed sinusoidal position embedding following the method in (Vaswani et al., 2017). Text features are set as Q (query), and image features are set as K (key) and V (value) for the MHA of the decoder. This decoder has a similar architecture to the vanilla transformer (Vaswani et al., 2017), a foundational model in deep learning that introduced attention mechanisms, leading to significant advancements in various tasks including machine translation and text generation. Each decoder layer contains MHA and feed-forward network (FFN) followed by an addition and layer normalization. There is a fully connected layer (FC) and softmax function to generate a caption in an autoregressive manner.

Algorithm 5: Caption Generation (CG)

CG (XD, YZ)

Input: XD, YZ

$X_W = \text{WE}\left(X_D\right)$

$X_{WP} = X_W + \text{PE}\left(X_D\right)$

$X_C = \text{MHA}\left(Q = X_{WP}, K = X_{WP}, V = X_{WP}\right)$

$Y_{CZ} = \text{MHA}\left(Q = X_C, K = Y_Z, V = Y_Z\right)$

$Y_D = \text{LN}\left(X_C + Y_{CZ}\right)$

$Y_F = \text{FFN}\left(Y_D\right)$

$Y_L = \text{LN}\left(Y_D + Y_F\right)$

$Y_C = \text{Softmax}\left(\text{FC}\left(Y_L\right)\right)$

Output: YC

Here, X_D and Y_Z represent partially completed caption and encoding features, respectively. WE and PE denote word embedding and position embedding, respectively. Y_C represents the generated caption.

25.3.3 The Second Approach: BFC-Capf

In this section, we introduce our second image-captioning approach with high-frequency grid and background features, in addition to object-level region features and their interactions. Figure 25.6 shows the overall architecture of this approach. Given an input image, object-level region features with bounding boxes and local grid features are extracted using faster R-CNN. The background features are extracted by

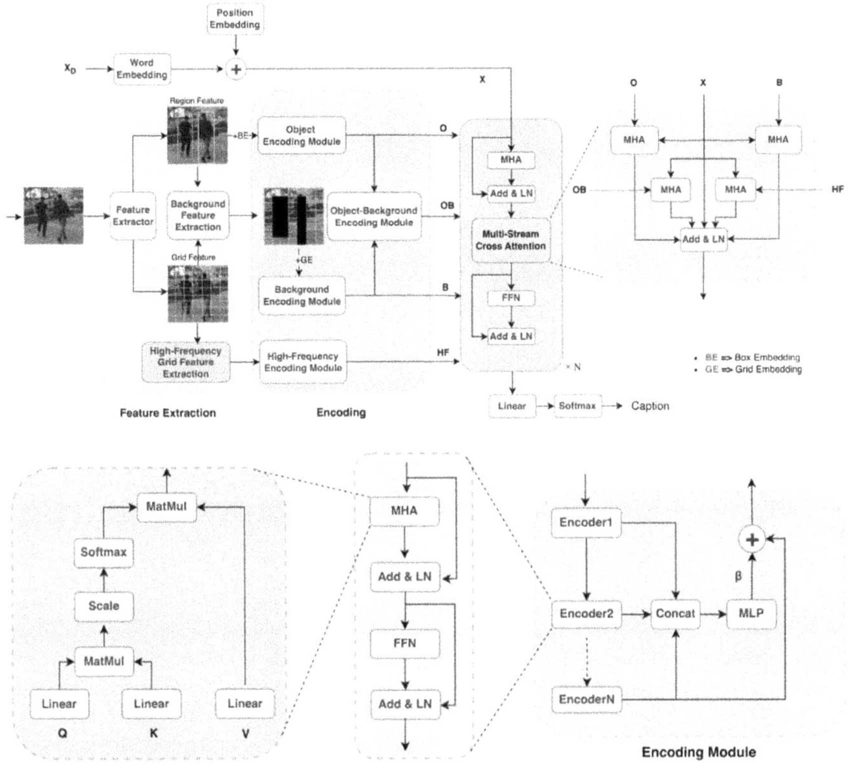

FIGURE 25.6 Overview of the proposed approach, BFC-Capf. Given an image, the feature extractor provides the object-level region features and local grid features. Additional visual features, including background and high-frequency grid features, are extracted. Four encoding modules provide effective representation of the visual features and their interactions. The resulting encoded features and text features are processed by multi-stream cross-attention to generate coherent captions with more contextual information.

removing the grids which overlap with bounding boxes of region features. High-frequency components from grid features are obtained by filtering out low-frequency components. Four encoding modules are used to find effective representations of the visual features including object-level region features, background features, object-context interactions, and high-frequency grid features. Features from all encoding layers are concatenated and weighted to reduce low-level information loss. Multi-head attention (MHA) is applied to partially completed captions with word and position embedding. The transformer-based decoder with multi-stream cross-attention generates a caption in an autoregressive manner for the given image.

25.3.3.1 High-Frequency Grid Feature Extraction

Capturing complex changes in high-dimensional feature spaces is crucial for enhancing the fine details that are critical to detailed visual understanding. This approach enables a deeper analysis of visual data, thereby contributing to image captions that

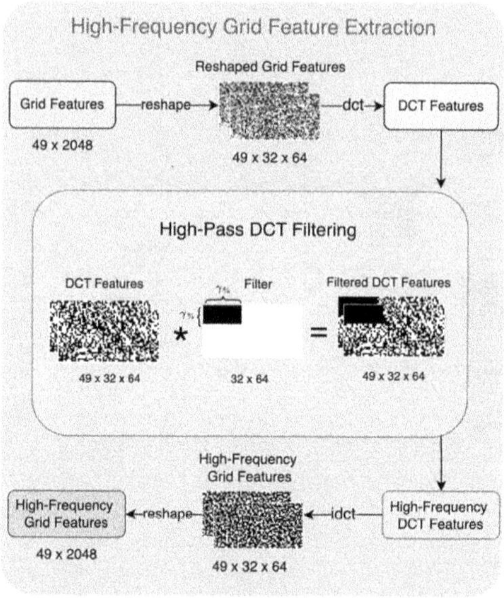

FIGURE 25.7 Frequency component filtering process. DCT transforms the grid features, extracted from faster RCNN, into DCT features in the frequency domain. High-pass DCT filter filters out low-frequency components and passes the high-frequency components. Resulting high-frequency DCT features are transformed back into the spatial domain by inverse DCT. The figures with DCT features are rescaled to a specific range for enhanced visualization.

are more contextually informed. Grid features usually represent patterns at different levels inside each grid, while high-frequency grid features can provide more distinctive patterns. Therefore, we incorporate the extraction of high-frequency grid features through frequency component filtering, as depicted in Figure 25.7. We use discrete cosine transformation (DCT) to transform grid features into DCT features in the frequency domain. The low-frequency components of DCT features are filtered out using a high pass DCT filter. This filter is designed with a cutoff ratio γ that determines how much low-frequency components will be filtered out. In this filter, the γ proportion of the rows and columns from the top-left corner are set to 0, while the remaining $(1-\gamma)$ proportion of the rows and columns are set to 1. In this case, based on the best performance from our experiment, we set 0.4 as the cutoff ratio, which implies 40% rows and columns from top-left corner containing low-frequency components are filtered out and remaining high-frequency components are used. We then convert the high-frequency features from the frequency domain back into the spatial domain. Inverse DCT is used to transform the resulting DCT image back to the spatial domain.

25.3.3.2 Encoding Modules

To get the benefit of various visual features, we use four encoding modules – object encoding, background encoding, object-background encoding, and high-frequency grid feature encoding. All the encoding modules comprise MHA and FFN. The object encoding module transforms the object-level region features into different representations with their interactions, determining how the object features are related to each other and which parts are more significant. Similarly, the background encoding module processes background features and provides transformation with their relevance and interactions. Moreover, the object-background encoding determines how object-level features are related to background features and provides different projections with the significance of their relations. Finally, the frequency feature encoding filters out the low-frequency component from the grid features and captures how much attention should be given to which part of the high-frequency features. To utilize low-level visual features, each encoding module integrates feature fusion. Feature fusion is followed by a multi-layer perceptron (MLP), which is weighted by $\beta \in [0,1]$.

25.3.3.3 Multi-Stream Cross-Attention

Multi-stream cross-attention (MSCA) bridges the modality gap between visual and language features, and captures the interaction of text features with high-frequency grid features, objects, background, and their relationships. MSCA takes four different visual encoding features along with partially completed text features XD and generates captions for images in an autoregressive manner. We use four cross-attentions followed by addition and normalization. Cross-attention between XD and object encoding features O captures the relation between object features and text features. Similarly, cross-attention between XD and background encoding features B finds out how the text features in partially completed captions are related to the features from the background encoding. It is also crucial to incorporate how the text features are related to the interactions between object and background, which is done by the cross-attention between XD and OB. Furthermore, the cross-attention between XD and high-frequency encoding features HF bridges the modality gap, and enables the generation of contextually coherent captions for images.

25.3.3.4 Caption Generation by BFC-Capf

Caption is generated at the end of the decoding process as shown in Figure 25.6. We take two types of inputs, the visual features from encoders and the text features, and apply a similar caption generation strategy as described in the first approach. Each decoding layer also contains MHA and FFN followed by an addition and layer normalization. There are a fully connected layer (FC) and softmax function to generate a caption in an autoregressive manner.

25.3.4 Objective Functions

Given a ground truth sequence $\overset{*}{y}_{1:T}$ and an image-captioning model with parameter θ, we optimize the following cross-entropy (CE) loss:

$$L_{CE} = -\sum_{1}^{T} \log p_\theta(\overset{*}{y}_T | \overset{*}{y}_{1:T-1}) \quad (25.4)$$

We further optimize the BFC-Capb model using CIDEr score and BFC-Capf using all the mentioned traditional metrics by self-critical sequence training (SCST) (Rennie et al., 2017) as follows:

$$\nabla_\theta L_{RL}(\theta) = -\frac{1}{k}\sum_{1}^{T}(r(y_{1:T}^{i}) - b)\nabla_\theta \log p_\theta\left(y_{1:T}^{i}\right) \tag{25.5}$$

where k is the beam size, r is the CIDEr score function for the BFC-Capb and average score of all the mentioned traditional metrics for the BFC-Capf, and $b = \sum_{i}(r(y_{1:T}^{i}))/k$ is the mean reward as a baseline.

25.4 EXPERIMENTAL RESULTS AND ANALYSIS

25.4.1 Implementation Details

To extract visual features, we adopt the methodology outlined in DLCT (Luo et al., 2021), leveraging the pre-trained faster R-CNN. The visual features include both region and grid features extracted using ResNeXt–101. Region features represent a collection of local image features from bounding boxes around detected objects. Faster R-CNN uses a dilated stride-1 C5 backbone and a 1×1 RoIPool with a detection head comprising two fully connected layers. For extracting grid features, the model removes dilation and employs the standard C5 layer. The grid size is 7×7, with features of 2,048 dimensions.

In our implementation, we configure the model with a dimension (dmodel) of 512 and 8 attention heads. Beam search is set to 5, β is 0.2, and both the encoder and decoder consist of 5 layers for BFC-Capb and 3 layers for BFC-Capf. During the cross-entropy (CE) training phase, the batch size is set to 50, while in the self-critical sequence training (SCST) (Rennie et al., 2017) phase, the batch size is increased to 100. We use the Adam optimizer to train our model in both phases. For BFC-Capf, we additionally employ a teacher-student (TS) training where we utilize two existing models, particularly DLCT (Luo et al., 2021) and LSTNet (Ma et al., 2023), as teacher models. In this case, our model acts as a student model and it is trained with ground truth as hard targets and teacher models' predictions as soft targets. During the training, we combine loss functions corresponding to both the hard targets and soft targets.

25.4.2 Dataset

We experimentally validate and evaluate the effectiveness of our method using the widely recognized MS COCO benchmark dataset (Lin et al., 2014) in the domain of image captioning. The MS COCO dataset includes over 120,000 images, each with five human-annotated captions. To ensure a fair comparison with existing methods, we strictly follow the dataset split known as Karpathy split and described in (Karpathy et al., 2015). This split allocates 5,000 images for validation, another 5,000 for testing, and the rest for training.

25.4.3 Evaluation Metrics

To gauge the performance of our methods, we follow the standard evaluation protocol, encompassing both traditional and modern metrics. Specifically, in addition to BLEU-1 (Papineni et al., 2002), BLEU-4 (Papineni et al., 2002), ROUGE-L (Lin et al., 2004), and CIDEr (Vedantam et al., 2015), we utilize cutting-edge image-captioning evaluation metrics, such as BERTScore (Zhang et al., 2019), CLIP-S (Hessel et al., 2021), RefCLIP-S (Hessel et al., 2021), PAC-S (Sarto et al., 2023), and RefPAC-S (Sarto et al., 2023). BERTScore computes token similarity using contextual embeddings, aligning more closely with human judgments. CLIP-S is a reference-free metric that employs the CLIP model to evaluate image-caption compatibility, mirroring human judgments without relying on references. RefCLIP-S is a reference-augmented version of CLIP-S, considering both image-caption and candidate-reference compatibility. PAC-S is a contrastive-based reference-free evaluation metric for image captioning, leveraging image-caption pairs and additional synthetically generated positives. RefPAC-S is a variant of PAC-S that also considers a set of reference captions. These recent metrics demonstrate greater robustness to challenging examples compared to traditional image-captioning evaluation metrics.

25.4.4 Performance Comparison

We conduct several experiments to verify the effectiveness of our approaches, and the performance of our approaches compared to existing relevant models on the MS COCO Karpathy test split are summarized in Table 25.1. In this table, to ensure a

TABLE 25.1
Performance Comparison with Existing Approaches on the "Karpathy" Test Split B-1, B-4, R, and C are short for BLEU-1, BLEU-4, ROUGE, and CIDEr scores, respectively. TS denotes teacher-student training. Bold values indicate the best results among compared methods.

Model	B-1	B-4	R	C
SCST (Rennie et al., 2017)	–	34.2	55.7	114.0
Up-Down (Anderson et al., 2018)	79.8	36.3	56.9	120.1
RFNet (Jiang et al., 2018)	79.1	36.5	57.3	121.9
LBPF (Qin et al., 2019)	80.5	38.3	58.4	127.6
GCN-LSTM (Yao et al., 2018)	80.5	38.2	58.3	127.6
SGAE (Yang et al., 2019)	80.8	38.4	58.6	127.8
TRRAR (Wang et al., 2023)	80.1	38.7	58.8	128.0
ORT (Herdade et al., 2019)	80.5	38.6	58.4	128.3
BCEFT (Guo et al., 2025)	76.5	35.2	56.0	128.6
DeeCap (Fei et al., 2022)	80.1	38.7	58.1	129.0
ETransCap (Mundu et al., 2024)	80.1	38.3	58.7	129.9
CDLC (Wang et al., 2024)	80.3	39.0	59.5	130.2
M2 Transformer (Cornia et al., 2020)	80.8	39.1	58.6	131.2
Latgeo (Dubey et al., 2023)	81.0	38.8	58.7	131.7
IOS-Transformer (Abdal-Hafeth and Kollias, 2024)	80.0	37.7	58.5	132.0
BFC-Capb	80.2	38.0	58.1	128.7
BFC-Capf	81.3	39.1	58.2	131.8
BFC-CapfTS	80.0	39.2	58.5	133.7

TABLE 25.2
Performance Comparison with Recent Evaluation MetricsBold values indicate the best results among compared methods.

Model	BERT-S	CLIP-S	RefCLIP-S	PAC-S	RefPAC-S
M2 Transformer	91.39	75.68	81.49	80.60	85.41
Region-Based Transformer	91.60	74.60	80.80	79.80	84.80
BFC-Capb	91.30	75.00	81.10	80.10	85.00
BFC-Capf	91.42	75.68	81.54	80.73	85.52

fair comparison, we include only those existing methods that utilize similar strategies with region and grid features and employ simple transformers in the decoding process, rather than large language or vision-language models. Although existing evaluation metrics generally assess captions without specifically evaluating the integration of background and contextual information, it is still evident from this performance comparison that our method is quantitatively competitive to the available methods, achieving a high BLEU-1, BLEU-4, and CIDEr score during self-critical sequence training. Since most of the metrics used in this table are traditional n-gram-based metrics that do not fully capture human judgment, we further evaluate our method using recent embedding-based evaluation metrics, as shown in Table 25.2. These metrics show that our method performs better than the Region-Based Transformer and M2 Transformer (Cornia et al., 2020). While embedding-based evaluation metrics are more capable of capturing human judgment compared to n-gram-based metrics, they still lack ideal human judgment. Also, both traditional and recent evaluation metrics usually do not assess the degree of background and contextual information integrated into captions. Therefore, we validate our approach, which aims to enhance contextual understanding, by analyzing the qualitative results through human perception in Section 25.4.6.

25.4.5 Ablation Study

We conduct several ablation studies to provide quantitative evidence of the effectiveness of each component of our BFC-Capb architecture. Table 25.3 shows that both the background-attention (BA) and object-background attention (OBA) contribute to the overall performance of our BFC-Capb approach. We also investigate the optimal value of the parameter α which controls the weight of the object-level region features and background features, by conducting experiments with several values of α. The value of 0.6, as shown in Table 25.4, is the optimal value as it gives the best score among other values of α. Furthermore, we conduct experiments without the spatial shift operation and without skipping any encoding layer in the fusion process. The performance results of these experiments, shown in Table 25.5, confirm the effectiveness of the spatial shift operation and our choice to use the alternate, particularly every second, encoding layers in the fusion process.

Also, for our BFC-Capf approach, we conduct several ablation studies to provide quantitative evidence of the effectiveness of each component of its architecture. The experimental results shown in Table 25.6 indicate that all encoding modules contribute

TABLE 25.3
Performance Comparison with Different Components of BFC-Capb, Particularly Object Attention (OA), Background Attention (BA), and Object-Background Attention (OBA)

OA	BA	OBA	B-1	B-4	R	C
✓	×	×	78.5	36.6	56.9	124.5
✓	✓	×	78.9	35.8	56.6	125.5
✓	✓	✓	80.2	38.0	58.1	128.7

TABLE 25.4
Performance Comparison with Different Values of the Parameter α in BFC-Capb

Parameter α	B-1	B-4	R	C
0.4	78.9	35.8	56.6	121.1
0.5	79.9	37.3	57.4	126.1
0.6	80.2	38.0	58.1	128.7
0.7	79.5	36.7	57.1	124.2

TABLE 25.5
Performance with/without Spatial Shift and Skipping Encoding Layers during Fusion in BFC-Capb

Method	B-1	B-4	R	C
BFC-Capb w/o SS	78.6	36.0	56.9	127.4
BFC-Capb w/o FS	78.9	35.9	56.9	126.2
BFC-Capb	80.2	38.0	58.1	128.7

TABLE 25.6
Performance Comparison with Different Components, Including Object Encoding (OE), Background Encoding (BE), and Object-Background Encoding (OBE), and High-Frequency (HF) Component

OE	BE	OBE	HF	B-1	B-4	R	C
✓	×	×	×	79.6	37.7	57.8	128.1
✓	✓	×	×	79.9	38.1	57.9	129.7
✓	✓	✓	×	79.9	38.4	58.1	130.3
✓	✓	✓	✓	80.0	39.2	58.5	133.7

TABLE 25.7
Performance Comparison of Different Cutoff Ratios

Cutoff ratio, γ	B-1	B-4	R	C
0.0	79.9	37.3	57.7	130.8
0.2	80.3	37.5	57.9	132.7
0.3	79.3	37.8	57.8	131.3
0.4	80.0	39.2	58.5	133.7
0.5	80.4	38.1	58.1	132.5
0.6	79.7	38.4	58.1	130.4
0.7	80.3	37.8	57.7	131.1
0.8	79.7	37.7	57.8	132.4
0.9	79.3	37.9	57.8	131.3

to the overall performance of BFC-Capf. We also investigate the optimal value of cutoff ratio, γ, which controls how much low-frequency component is filtered out while extracting high-frequency grid features. Results from several experiments with different values of γ are shown in Table 25.7 where the value of 0.4 provides the best average score. The first row in Table 25.7, with a cutoff ratio $\gamma = 0$, suggests that using complete grid features without filtering out the low-frequency components is less effective than filtering out low-frequency components and using high-frequency ones.

25.4.6 Qualitative Analysis

We demonstrate the effectiveness of our methods for image captioning with more background and contextual information, as shown by the qualitative results in Figure 25.8. The captions generated by our methods contain more background and contextual information than M2 Transformer and Region-Based Transformer. In Figure 25.8(a), our methods provide more details about the objects and background. In Figure 25.8(b), we can observe that our method provides more information about the attributes of objects such as "green vase." Additionally, the captions for all the images in Figure 25.8 provide more detailed information about the background content. Given these qualitative results, it is evident that our methods are well capable of generating captions with more background and contextual information.

25.4.7 Discussions and Limitations

In most cases, our methods are able to generate coherent captions with more background and contextual information and more accurate information regarding the objects and their attributes. However, in some cases, specifically where the salient objects highly dominate the background of the image or cover almost the entire image, our methods tend to provide incomplete captions, as shown in Figure 25.9. Therefore, constraints are observed when it encounters a significantly imbalanced object-background ratio in the image. Despite this limitation, though, our approaches' ability to generate captions with richer contextual information highlight their effectiveness.

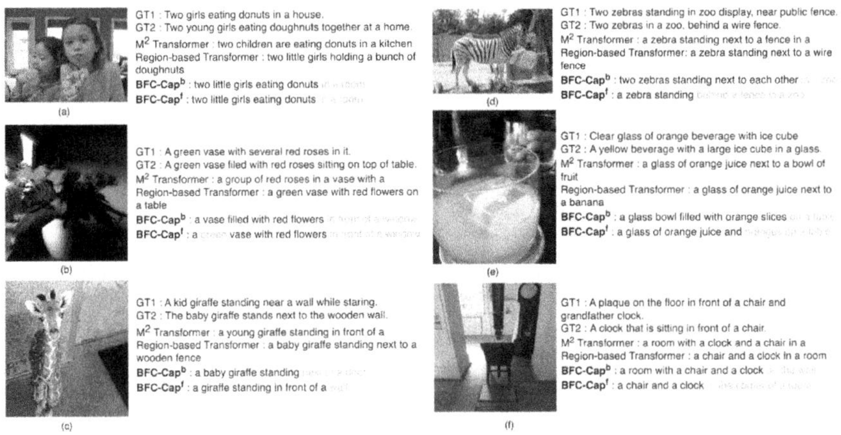

FIGURE 25.8 Examples of image-captioning qualitative results by M2 Transformer, Region-Based Transformer, and our BFC-Capb and BFC-Capf. GT1 and GT2 are two sample ground truth captions from MS COCO dataset. Our methods generate captions with more contextual information.

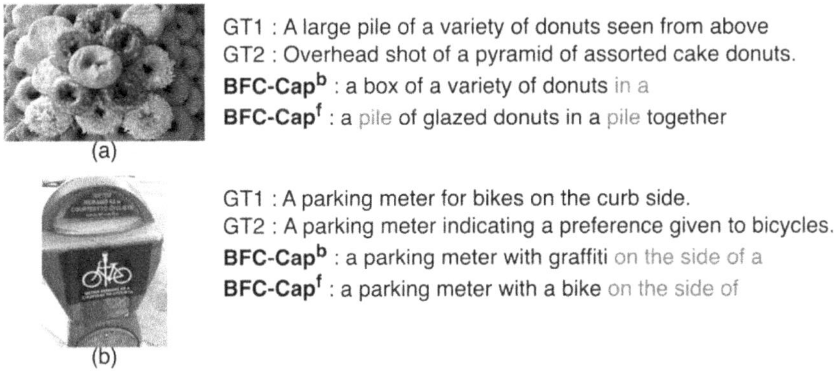

FIGURE 25.9 Limitations of our methods. Our methods struggle to generate complete captions while encountering an image with imbalanced object-background ratio or salient objects dominating the background.

25.5 CONCLUSIONS

In this chapter, we present two image-captioning frameworks with background and high-frequency visual features to generate captions with more visual understanding and contextual relevance. High-frequency features sharpen critical fine details, and background visual features add comprehensive context. In addition, we utilize object-level region features and their interactions with background features. We also introduce a multi-stream cross-attention module which bridges the modality gap between language and vision, and discovers how textual features relate to various visual features, including high-frequency local features, object-level features, contextual features, as well as their interactions. Furthermore, we integrate feature fusion in the encoding process to utilize low-level visual features. Our image-captioning methods are validated for effective image captioning with rich background and contextual information using extensive experiments.

REFERENCES

Abdal-Hafeth, D. and Kollias, S., "Insights into object semantics: Leveraging transformer networks for advanced image captioning," *Sensors*, vol. 24, no. 6, p. 1796, 2024.

Anderson, P., He, X., Buehler, C., Teney, D., Johnson, M., Gould, S., and Zhang, L., "Bottom-up and top-down attention for image captioning and visual question answering," *Proceedings of the IEEE Conference on Computer Vision and Pattern Recognition*, pp. 6077–6086, 2018.

Barraco, M., Sarto, S., Cornia, M., Baraldi, L., and Cucchiara, R., "With a little help from your own past: Prototypical memory networks for image captioning," *Proceedings of the IEEE/CVF International Conference on Computer Vision*, pp. 3021–3031, 2023.

Cornia, M., Stefanini, M., Baraldi, L., and Cucchiara, R., "Meshed-memory transformer for image captioning," *Proceedings of the IEEE/CVF Conference on Computer Vision and Pattern Recognition*, pp. 10578–10587, 2020.

Dubey, S., Olimov, F., Rafique, M. A., Kim, J., and Jeon, M., "Label-attention transformer with geometrically coherent objects for image captioning," *Information Sciences*, vol. 623, pp. 812–831, 2023.

Fang, Z., Wang, J., Hu, X., Liang, L., Gan, Z., Wang, L., Yang, Y., and Liu, Z., "Injecting semantic concepts into end-to-end image captioning," *Proceedings of the IEEE/CVF Conference on Computer Vision and Pattern Recognition*, pp. 18009–18019, 2022.

Fei, J., Wang, T., Zhang, J., He, Z., Wang, C., and Zheng, F., "Transferable decoding with visual entities for zero-shot image captioning," *Proceedings of the IEEE/CVF International Conference on Computer Vision*, pp. 3136–3146, 2023.

Fei, Z., Yan, X., Wang, S., and Tian, Q., "Deecap: Dynamic early exiting for efficient image captioning," *Proceedings of the IEEE/CVF Conference on Computer Vision and Pattern Recognition*, pp. 12216–12226, 2022.

Guo, J., Li, Y., Cheng, G., and Li, W., "Based-CLIP early fusion transformer for image caption," *Signal, Image and Video Processing*, vol. 19, no. 1, pp. 1–10, 2025.

Herdade, S., Kappeler, A., Boakye, K., and Soares, J., "Image captioning: Transforming objects into words," *Advances in Neural Information Processing Systems*, pp. 11137–11147, 2019.

Hessel, J., Holtzman, A., Forbes, M., Bras, R. L., and Choi, Y., "Clipscore: A reference-free evaluation metric for image captioning," arXiv:2104.08718, 2021.

Jiang, W., Ma, L., Jiang, Y. G., Liu, W., and Zhang, T., "Recurrent fusion network for image captioning," *Proceedings of the European Conference on Computer Vision (ECCV)*, pp. 499–515, 2018.

Karpathy, A. and Fei-Fei, L., "Deep visual-semantic alignments for generating image descriptions," *Proceedings of the IEEE Conference on Computer Vision and Pattern Recognition*, pp. 3128–3137, 2015.

Li, J., Vo, D. M., Sugimoto, A., and Nakayama, H., "EVCap: Retrieval-augmented image captioning with external visual-name memory for open-world comprehension," arXiv:2311.15879, 2023.

Lin, C.-Y., "Rouge: A package for automatic evaluation of summaries," *Text Summarization Branches Out*, pp. 74–81, 2004.

Lin, T.-Y., Maire, M., Belongie, S., Hays, J., Perona, P., Ramanan, D., Dollár, P., and Zitnick, C. L., "Microsoft coco: Common objects in context," *Proceedings of the Computer Vision–ECCV 2014: 13th European Conference*, Zurich, Part V 13, pp. 740–755, Springer International Publishing, Sept. 6–12, 2014.

Liu, Y., Yang, Y., Xiang, R., and Ma, J., "Complementary shifted transformer for image captioning," *Neural Processing Letters*, pp. 1–25, 2023.

Luo, J., Li, Y., Pan, Y., Yao, T., Feng, J., Chao, H., and Mei, T., "Semantic-conditional diffusion networks for image captioning," *Proceedings of the IEEE/CVF Conference on Computer Vision and Pattern Recognition*, pp. 23359–23368, 2023.

Luo, Y., Ji, J., Sun, X., Cao, L., Wu, Y., Huang, F., Lin, C.-W., and Ji, R., "Dual-level collaborative transformer for image captioning," *Proceedings of the AAAI Conference on Artificial Intelligence*, vol. 35, no. 3, pp. 2286–2293, 2021.

Ma, Y., Ji, J., Sun, X., Zhou, Y., and Ji, R., "Towards local visual modeling for image captioning," *Pattern Recognition*, vol. 138, p. 109420, 2023.

Meng, Z., Yang, D., Cao, X., Shah, A., and Lim, S.-N., "Object-centric unsupervised image captioning," *European Conference on Computer Vision*, pp. 219–235, Springer Nature, 2022.

Mundu, A., Singh, S. K., and Dubey, S. R., "ETransCap: Efficient transformer for image captioning," *Applied Intelligence*, vol. 54, no. 21, pp. 10748–10762, 2024.

Nguyen, V.-Q., Suganuma, M., and Okatani, T., "Grit: Faster and better image captioning transformer using dual visual features," *European Conference on Computer Vision*, pp. 167–184, Springer Nature, 2022.

Pan, Y., Yao, T., Li, Y., and Mei, T., "X-linear attention networks for image captioning," *Proceedings of the IEEE/CVF Conference on Computer Vision and Pattern Recognition*, pp. 10971–10980, 2020.

Papineni, K., Roukos, S., Ward, T., and Zhu, W.-J., "Bleu: A method for automatic evaluation of machine translation," *Proceedings of the 40th Annual Meeting of the Association for Computational Linguistics*, pp. 311–318, 2002.

Qin, Y., Du, J., Zhang, Y., and Lu, H., "Look back and predict forward in image captioning," *Proceedings of the IEEE/CVF Conference on Computer Vision and Pattern Recognition*, pp. 8367–8375, 2019.

Ramos, R., Elliott, D., and Martins, B., "Retrieval-augmented image captioning," arXiv:2302.08268, 2023b.

Ramos, R., Martins, B., Elliott, D., and Kementchedjhieva, Y., "SmallCap: Lightweight image captioning prompted with retrieval augmentation," *Proceedings of the IEEE/CVF Conference on Computer Vision and Pattern Recognition*, pp. 2840–2849, 2023a.

Ren, S., He, K., Girshick, R., and Sun, J., "Faster R-CNN: Towards real-time object detection with region proposal networks," *Advances in Neural Information Processing Systems*, vol. 28, 2015.

Rennie, S. J., Marcheret, E., Mroueh, Y., Ross, J., and Goel, V., "Self-critical sequence training for image captioning," *Proceedings of the IEEE Conference on Computer Vision and Pattern Recognition*, pp. 7008–7024, 2017.

Sarto, S., Barraco, M., Cornia, M., Baraldi, L., and Cucchiara, R., "Positive-augmented contrastive learning for image and video captioning evaluation," *Proceedings of the IEEE/CVF Conference on Computer Vision and Pattern Recognition*, pp. 6914–6924, 2023.

Song, Z., Zhou, X., Mao, Z., and Tan, J., "Image captioning with context-aware auxiliary guidance," *Proceedings of the AAAI Conference on Artificial Intelligence*, vol. 35, no. 3, pp. 2584–2592, 2021.

Vaswani, A., Shazeer, N., Parmar, N., Uszkoreit, J., Jones, L., Gomez, A. N., Kaiser, Ł., and Polosukhin, I., "Attention is all you need," *Advances in Neural Information Processing Systems*, vol. 30, 2017.

Vedantam, R., Zitnick, C. L., and Parikh, D., "Cider: Consensus-based image description evaluation," *Proceedings of the IEEE Conference on Computer Vision and Pattern Recognition*, pp. 4566–4575, 2015.

Wang, D., Hu, Z., Zhou, Y., Hong, R., and Wang, M., "A text-guided generation and refinement model for image captioning," *IEEE Transactions on Multimedia*, vol. 25, pp. 2966–2977, 2023.

Wang, N., Xie, J., Wu, J., Jia, M., and Li, L., "Controllable image captioning via prompting," *Proceedings of the AAAI Conference on Artificial Intelligence*, vol. 37, no. 2, pp. 2617–2625, 2023.

Wang, Y., Xu, J., and Sun, Y., "CLIP-driven low-cost image captioning," *2024 International Joint Conference on Neural Networks (IJCNN)*, pp. 1–8, IEEE, Jun. 2024.

Wu, M., Zhang, X., Sun, X., Zhou, Y., Chen, C., Gu, J., Sun, X., and Ji, R., "Difnet: Boosting visual information flow for image captioning," *Proceedings of the IEEE/CVF Conference on Computer Vision and Pattern Recognition*, pp. 18020–18029, 2022.

Yang, X., Tang, K., Zhang, H., and Cai, J., "Auto-encoding scene graphs for image captioning," *Proceedings of the IEEE Conference on Computer Vision and Pattern Recognition*, 2019.

Yang, X., Wang, Y., Chen, H., Li, J., and Huang, T., "Context-aware transformer for image captioning," *Neurocomputing*, p. 126440, 2023.

Yao, T., Pan, Y., Li, Y., and Mei, T., "Exploring visual relationship for image captioning," *Proceedings of the European Conference on Computer Vision*, 2018.

Zhang, T., Kishore, V., Wu, F., Weinberger, K. Q., and Artzi, Y., "Bertscore: Evaluating text generation with Bert," arXiv:1904.09675, 2019.

26 A Novel Adaptive Data Transformation for Contrastive Learning

26.1 INTRODUCTION

Contrastive learning, an approach in self-supervised learning, has its incredible ability to learn representation from unlabeled datasets. In pre-training process of using unlabeled datasets, the contrastive learning algorithms adopt Siamese encoders (Bromley et al., 1993) to take positive augmented images and minimize their distances in the embedding space (Caron et al., 2021; Coates et al., 2011; Cui et al., 2023). Evidences have shown that after pre-training, the remarkable performance can be achieved in image classification (Bachman et al., 2019), object detection, instance segmentation, key points estimation, and so on.

One challenge in contrastive learning is identifying the suitable data augmentation strategies in generating positive augmented views. Tian et al. (2020a) introduced multi-views to improve its performance. Wang and Qi (2022) provided a better data augmentation strategy for contrastive learning by retrieving stronger augmented queries from a pool of instances. However, it distorts the structure of the images and results in difficult retrieval. Another challenge is that data augmentation may accidentally create false positives. As a result, we may crop out the foreground of an original image when *random cropping* (RandomCrop) (2018) is applied to create a positive augmented image. Therefore, the generated positive augmented image is inappropriate for positive augmented views of the original image. The ContrastiveCrop (2022) generates cropped images, which could avoid false positives as occurred in RandomCrop.

It is known that multiple data augmentation applied on original images to distort the structure can improve the representation in contrastive learning. Unfortunately, there is no existing work so far to address how to determine an optimal sequence of image transformation as data augmentation to generate positive augmented views. In practice, one can use various data transformations in computer vision, such as Gaussian blur, color jittering, cropping, horizontal/vertical flip, equalization, perspective transformation, etc. It remains a question that which combination of those transformations is suitable and how to seamlessly integrate with contrastive learning. Therefore, there is a crucial demand for an effective and optimized data transformation in contrastive learning.

In this chapter, we propose an *Adaptive Data Transformation* algorithm, named **ADTrans**, in contrastive learning to adaptively select multiple transformations for maximizing the knowledge that the model can learn. To achieve this goal, an optimal data transformation is selected to make the highest objective loss, which indicates the most important area that the model needs to focus on in order to improve its performance in each training state. After one training state is finished, a number of training

DOI: 10.1201/9781003474395-29

steps are spanned until model convergence. Then an optimal transformation among all possible options based on the previously trained model is reselected for the next training state. This adaptive training process continues until there is no further transformation to improve the performance. Since multiple transformations are iteratively selected and applied, ADTrans integrates the datasets before and after applying a selected transformation to address potential distribution shifts on the positive augmented images. As a result, the performance of models trained with ADTrans is significantly improved.

Our extensive experiments on benchmark datasets show that the proposed ADTrans can significantly outperform the state-of-the-art contrastive learning algorithms, i.e., MOCO v2 (Chen et al., 2020a), SimSiam (Chen et al., 2021), BYOL (Grill et al., 2020), and SimCLR (Chen et al., 2020b), in several computer vision tasks, including object classification, object segmentation, and instance segmentation. For linear classification, our proposed ADTrans improves around 0.5% and 4% linear accuracy over the baseline approaches on ImageNet-100 with ResNet-50 backbone. Furthermore, ADTrans improves the AP (average precision) for 0.6 on VOC detection, 0.3 on COCO instance segmentation, and 0.4 on COCO detection.

The rest of this chapter is organized as follows. Section 26.2 reviews the background and related work. Section 26.3 discusses the combination of multiple data transformations. Section 26.4 proposes the Adaptive Data Transformation (ADTrans) algorithm. Experimental results are provided in Section 26.5. Finally, conclusions are drawn in Section 26.6.

26.2 BACKGROUND AND RELATED WORK

In this section, we briefly review contrastive learning and related work. Given an unlabeled training set $X = \{x_i\}_{i\in[0,n]}$, where n is the number of data points x. The information noise-contrastive estimation (named Info NCE) (Oord et al., 2018) based contrastive loss (He et al., 2020) can be formulated as follows:

$$\mathcal{L}(\theta,x) = -log\frac{\exp\left(sim\left(\theta\cdot x_i,\theta\cdot x_j\right)/\tau\right)}{\sum_{k=1}^{K}\exp\left(sim\left(\theta\cdot x_i,\theta\cdot x_k\right)/\tau\right)}, \tag{26.1}$$

where τ is a temperature coefficient, (x_i, x_j) is a positive pair of images created by applying transformation on the original image x, x_k denotes the negative images of x, and θ denotes the weights of the encoder. The product $\theta\cdot x$ is the encoded result of an image x given the model parameters θ.

In addition, $sim(\cdot)$ is a similarity function such as cosine similarity. The summation is over one positive sample and $(K-1)$ negative samples. We consider the set of the one positive sample and $(K-1)$ negative samples as a dictionary. Equation (26.1) categorizes x_i and x_j to the same class. Given the whole dataset X, the loss function is an average loss over all original images $x\in X$ as follows:

$$\mathcal{L}(\theta,X) = \frac{1}{n}\sum_{x\in X}\mathcal{L}(\theta,x). \tag{26.2}$$

26.2.1 Contrastive Learning

The *contrastive predictive coding* (CPC) (Oord et al., 2018) introduced Info NCE loss to the concept of contrastive learning by predicting future samples in time series patterns to learn the embedding from the input dataset. It is universal to image, text, and even audio data. The *contrastive multiview coding* (CMC) (Tian et al., 2020b) used an object and its different perspective views to perform contrastive representation learning. Its performance can be slightly improved with more perspective views involved.

The SimCLR (Chen et al., 2020b) proposed a very simple end-to-end contrastive learning framework for computer vision tasks. It maximizes the agreement between predictions of two encoders over two augmented views, respectively. However, a large batch size up to 8,000 is required to learn a good representation. The MoCo (He et al., 2020) and its variants (Chen et al., 2020b, 2021) largely improve limits of contrastive learning. The proposed momentum encoder provides a much more efficient way to achieve the positive and negative pairing, and improves the performance as well.

The SimSiam (Chen and He, 2021) proposed a symmetric architecture of pairing positive augmented views by a symmetrized cosine similarity loss function. Therefore, it does not need negative samples and momentum encoder, making it easy and straightforward to follow. The BYOL (Grill et al., 2020) predicted the output of one view from another view with a Siamese network, where one of the branches is a momentum encoder. Therefore, it does not need negative samples as well.

26.2.2 Data-Transformation-Based Augmentation

Data-transformation-based augmentation can improve the performance of contrastive learning (Chen et al., 2020b). There are several types of data transformation as follows: (1) geometric transformations, such as random cropping and horizontal/vertical flipping; (2) appearance transformations, such as color jittering and Gaussian blur; and (3) perspective transformations, such as getting different perspective views of the original input samples.

Wang and Qi (2022) discussed the influence of different data augmentations on contrastive learning and proposed a stronger augmentation to help improve the performance of contrastive learning. Peng et al. (2022) proposed ContrastiveCrop, which effectively generates better crops for contrastive representation learning. Zhang and Ma (2022) introduced a hierarchical data augmentation module to the general contrastive learning frameworks. Those approaches proposed stronger data augmentation by integrating transformations. However, there is no existing work that can deliver strong and optimal data augmentation for contrastive pre-training to enhance the performance.

26.3 COMBINATION OF MULTIPLE DATA TRANSFORMATIONS

In this section, we discuss how multiple data transformations can be combined to enhance the performance of contrastive learning in two perspectives: (1) Does a combination of multiple transformations improve the performance of contrastive learning? (2) Does the order of multiple transformations affect the performance of contrastive learning?

Numbers of Transformations	1	2	3
Accuracy (%)	21.54	35.95	36.87

(a)

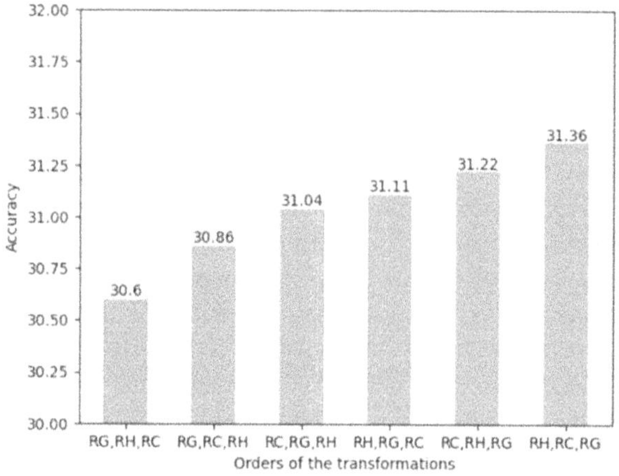

(b)

FIGURE 26.1 (a) Classification accuracy on TinyImageNet trained with linear classifier with different numbers of transformations as data augmentations, and (b) classification accuracy on TinyImageNet trained with linear classifier with different orders of transformations.

To answer these questions, we train an encoder with a ResNet-18 model as the backbone (He et al., 2016) on TinyImageNet for 800 epochs, and then transfer the embedding to linear classifier, which is trained separately for another 100 epochs given the encoder frozen. Figure 26.1(a) shows the classification accuracy on TinyImageNet trained with linear classifier after pre-training by MOCO v2 on unlabeled TinyImageNet with different numbers of transformations as data augmentations. They are pre-trained with data transformations: (1) random crop; (2) random crop, color jittering; and (3) random crop, color jittering, horizontal flip. We apply one to three transformations and observe that the image classification accuracy increases with numbers of transformations significantly. Figure 26.1(b) shows the classification accuracy on TinyImageNet trained with linear classifier after pre-training by MOCO v2 on unlabeled TinyImageNet with different orders of transformations. Note that RH, RG, and RC denote random horizontal flip, random grayscale, and random cropping, respectively. We observe that different orders of transformations vary the image classification accuracy; i.e., the lowest linear accuracy is 30.6% when the highest is 31.36%. Therefore, the numbers of and orders of data transformations applied on original images affect the performance of contrastive learning. Motivated by these results, we aim at proposing an efficient approach that is able to optimize the order and number of transformations applied on the original images in contrastive learning.

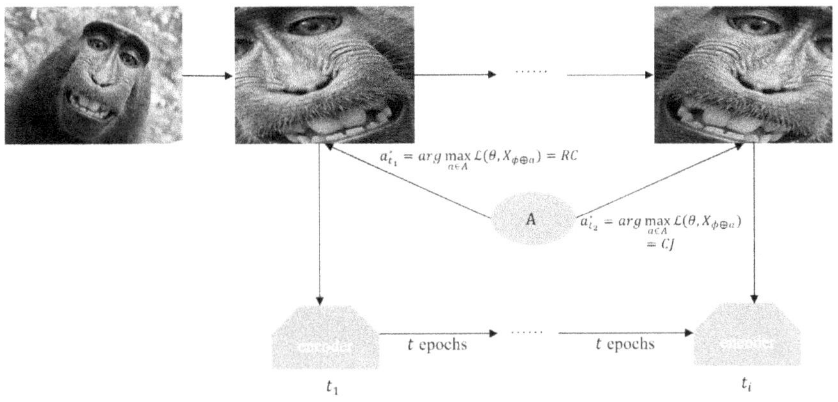

FIGURE 26.2 The proposed Adaptive Data Transformation (ADTrans).

26.4 THE PROPOSED ADAPTIVE DATA TRANSFORMATION (ADTRANS)

In this section, we propose the novel Adaptive Data Transformation (ADTrans) with problem formulation to adaptively search for optimized combinations of data transformations and address distribution shifts caused by applying those data transformations. Figure 26.2 shows its architecture, where RC and CJ denote random cropping and color jittering, respectively. Given a set of m data transformations $A = \{a_i\}_{i \in [0,m]}$, we formulate the problem of optimizing a sequential combination of data transformations as $a \in A$, denoted as ϕ, for data augmentation among a set of all possible sequential combinations ϕ.

$$\phi^*, \theta^* = arg \min_{\phi \in \Phi, \Theta} \mathcal{L}(\theta, X_\phi), \tag{26.3}$$

where X_ϕ is the augmented data obtained by applying ϕ on X. To create X_ϕ, we sequentially apply transformations in ϕ as applying the new transformation on top of the previous one. For instance, $\phi = \{\text{color jittering, random cropping, Gaussian blur}\}$. First, we apply color jittering on the original images X to obtain X_1. Second, random cropping is conducted on X_1 to obtain X_2. Finally, Gaussian blur is added on X_2 to obtain X_ϕ.

An optimal transformation that maximizes the loss function is selected from the candidate transformations set and applied on the input data iteratively every t epochs. Each optimal transformation is removed from the candidate transformations set after being selected. The selected transformations are sequentially applied on the input data to generate positive augmented views for contrastive learning models.

It is non-trivial to optimize equation (26.3) due to the large search space of ϕ^* given an unlimited number of possible sequential combinations in ϕ. The join optimization between ϕ^* and the model parameters θ^* further creates a complicated surface, which makes the problem even more challenging to find good solutions for both ϕ^* and θ^*. In fact, simply applying existing contrastive training algorithms can result in the poor local minima.

An optimal transformation that maximizes the loss function is selected from the candidate transformation set and applied on the input data iteratively every t epochs. Each optimal transformation is removed from the candidate transformation set after being selected. Those selected transformations are sequentially applied on the input data to generate positive augmented views for contrastive learning models.

An adversarial training is proposed as a mini-max surrogated objective function as follows:

$$\phi^*, \theta^* = arg \max_{\phi \in ¦} arg \min_{\theta} \mathcal{L}\left(\theta, X_{\phi}\right). \tag{26.4}$$

The optimization is to find a sequential combination of transformations ϕ that can offer the most knowledge that the model θ can learn. In other words, we maximize the loss function $\mathcal{L}\left(\theta, X_{\phi}\right)$ while training θ to minimize the loss $\mathcal{L}\left(\theta, X_{\phi}\right)$ given ϕ. This would enable us to significantly improve the representation learned by the model parameters θ for downstream tasks. The following algorithm shows the pseudo-code to optimize equation (26.4).

Algorithm: Adaptive Data Transformations (ADTrans)

Input: Training data X, stop condition ε, sampling probability p.

Initialize $\phi = \varnothing$.

Randomly initialize model parameter θ.

For $i = 0$ to $\phi - 1$

  While $\mathcal{L} > \varepsilon$ do

    Select $a \in A$ by $a = arg \max_{a \in A} \mathcal{L}\left(\theta, X_{\phi \oplus a}\right)$.

    Remove a from A by $A - \{a\}$.

    Train the model parameter θ to minimize the loss function $\mathcal{L}\left(\theta, X_{\phi \oplus a^*}\right)$.

    Mitigate distribution shift by $X_{\phi \oplus a^*} \cup p \cdot \tilde{X}_{\phi}$.

  End while

End for

Given an initially empty $\phi = \varnothing$ on line 1, ADTrans randomly initializes the model parameters θ on line 2. There are two iterative steps in our algorithm as follows:

Step 1. First, ADTrans selects the best data transformation $a \in A$ to maximize the loss function $\mathcal{L}\left(\theta, X_{\phi \oplus a^*}\right)$ given θ on line 7 as

$$a^* = arg \max_{a \in A} \mathcal{L}\left(\theta, X_{\phi \oplus a}\right), \tag{26.5}$$

where the augmented data $X_{\phi \oplus a}$ is created by applying the transformation a on X_{ϕ}.

Step 2. Given the augmented data $X_{\phi \oplus a^*}$ obtained in Step 1 until the model converges (i.e., the loss function cannot be reduced further), ADTrans trains the model parameters θ to minimize the loss function $\mathcal{L}\left(\theta, X_{\phi \oplus a^*}\right)$ as

$$\theta^* = arg \min_{\theta} \mathcal{L}\left(\theta, X_{\phi \oplus a^*}\right). \quad (26.6)$$

To minimization, we use the standard stochastic gradient descent (SGD) algorithm (Ruder, 2016).

The ADTrans updates the sequential combination of transformations $\phi = \phi \oplus a^*$ for the next training state by going back to Step 1 on line 5. The training process will stop if ADTrans has added a sufficient number of transformations in ϕ. Empirical results show that sequentially applying 3 or 4 transformations on original images can offer the best model utility.

In the adaptive contrastive training, we apply a new transformation a^* on X_{ϕ} to create a distribution shift in the input data; i.e., $X_{\phi \oplus a^*}$ is shifted from X_{ϕ}. This shift encourages the model θ to learn the new knowledge and enhance the data representation. However, a significant shift can affect the model by making it forgetting about the previously learned knowledge from X_{ϕ}. As a result, this degenerates the model utility. It can slow down the model convergence rate when ADTrans needs more training steps to converge.

To mitigate the negative impacts of data distribution shifts, ADTrans adds the previously created augmented data X_{ϕ} to the newly created augmented data $X_{\phi \oplus a^*}$ in optimizing equation (26.6) in Step 2. In fact, ADTrans randomly selects data from X_{ϕ} with a uniform probability p for each data sample. We denote this sampled dataset as $\tilde{X}_{\phi}$ as follows:

$$\theta^* = arg \min_{\theta} \mathcal{L}\left(\theta, X_{\phi \oplus a^*} \cup p \cdot \tilde{X}_{\phi}\right). \quad (26.7)$$

Therefore, ADTrans inhibits distribution shifts after applying new transformations during the training process. Note that we only use the combined augmented dataset $X_{\phi \oplus a^*} \cup p \cdot \tilde{X}_{\phi}$ to generate positive views, while the negative images are extracted from the original dataset to avoid duplicating negative images in the contrastive training.

26.5 EXPERIMENTAL RESULTS

To evaluate the effectiveness of ADTrans, we conduct extensive experiments to show its performance as compared with baseline approaches on image classification, object detection, and instance segmentation using a wide range of benchmark datasets.

26.5.1 Datasets and Model Configurations

We use 1) CIFAR 10 (Krizhevsky and Hinton, 2009), STL 10 (Caron et al., 2020), TinyImageNet, and ImageNet-100 (Russakovsky et al., 2015) for image classification, 2)

COCO (Lin et al., 2014) for object detection and instance segmentation, and 3) Pascal VOC (Everingham et al., 2010) for object detection. The CIFAR 10 consists of 60,000 color images of size 32×32 in 10 classes, where 50,000 images are used for training and 10,000 images for testing. The TinyImageNet consists of 110,000 images of size 64×64 in 200 classes, where 100,000 images are used for training set and 10,000 images for testing. The STL 10 provides 100,000 unlabeled images of size 96×96 for contrastive training, 5,000 labeled training images, and 8,000 labeled test images. The ImageNet-100 is a subset of ImageNet-1k from ImageNet Large Scale Visual Recognition Challenge 2012. It contains randomly 100 classes from ImageNet-1k. Pascal VOC has 20 classes in total. The train and validation data have 11,530 images, containing 27,450 ROI annotated objects and 6,929 segmentations. The COCO dataset contains 82,783 training images, 40,504 validation images, and 40,775 testing images. There are nearly 270K segmented people and a total of 886K segmented object instances in the train+val data alone.

We adopt MOCO v2 (Chen et al., 2020b; He et al., 2020), SimSiam (Chen et al., 2021), BYOL (Grill et al., 2020), and SimCLR (Chen et al., 2020) as the baseline contrastive learning approaches. MOCO v2 is a modification of MOCO, which provides a much more efficient way to pair positive and negative images with a momentum encoder. SimSiam is a symmetric architecture of pairing positive augmented images by a symmetrized cosine similarity. BYOL directly predicts the output of one positive view from another positive view without any negative images. SimCLR uses an end-to-end contrastive learning framework to maximize the agreement between the output of two encoders with positive images as their inputs.

We integrate ADTrans to MOCO v2 by providing optimized positive augmented images and leveraging the selection of negative samples. Therefore, we strictly follow the hyperparameters, optimizer, and loss function in MOCO v2. We set the learning rate as 0.03, the batch size as 256, and the encoder training as 800 epochs. In contrastive training of ADTrans, we use ResNet-18 and ResNet-50 (He et al., 2016) as the backbone of the encoder, which is the model θ. We constraint the number of transformations in the sequential combination ϕ to be 4 and set the probability $p = 1.0$ to mitigate the data distribution shift caused among these transformations.

For image classification, we adopt the SGD optimizer with a learning rate of 10.0 and the batch size of 256. For object detection and instance segmentation, we fine-tune and train by faster RCNN (Ren et al., 2015) with a backbone ResNet-18 on train 2012 for 40K iterations, and evaluate on val 2012 on Pascal VOC. We also train it for 100K iterations and evaluate on val 2017. For COCO detection and segmentation and Pascal VOC segmentation, we use detectron 2 for implementation. We fine-tune on the embedding learned during contrastive training, and train the softmax linear classifier 100 epochs. As for the candidate transformations set, we conduct empirical studies of MOCO v2 by applying a single transformation to the original images and evaluate the results using classification accuracy. Table 26.1 lists the top 10 most effective data transformations and their parameters in our experiments. For example, the hue of color jittering is in the range of [–0.5,0.5]; the shear operation is drawn under the range of [–0.3,0.3]; the degree of random rotation is randomly drawn from [0,20]; the kernel size is (5,5); the standard variation of Gaussian blur lies on the range of [0.1,2.0].

Since contrastive learning naturally benefits from more epochs of pre-training, we train our proposed ADTrans and all the baseline approaches for 800 epochs in our experiments. All the experiments are conducted on a server with four NVIDIA Titan XP GPUs.

TABLE 26.1
Candidate Data Transformations for Conciseness and Clarity

Transformation	Color Jittering (CJ)	Random Cropping (RC)	Vertical Flip (VF)
Map range	[−0.5, 0.5]	[0.2, 1.0]	0 and 1
Transformation	Equalization (E)	Horizontal flip (HF)	Gaussian blur (GB)
map range	0 or 1	0 or 1	(5, 5), (0.1, 2.0)
Transformation	Shear (S)	Translate (T)	Random grayscale (RG)
map range	[−0.3, 0.3]	[0, 0.3]	0 or 1
Transformation	Random rotation (RR)		
map range	[0, 20]		

26.5.2 Comparison Analysis

Table 26.2 shows the comparisons on image classification under unsupervised contrastive training. We freeze the trained model parameters, train softmax classifiers on the training set, and then test on the testing set. Using a ResNet-18 model, ADTrans significantly improves the classification accuracy on STL and TinyImageNet over the baseline approaches. In fact, ADTrans improves around 12% on STL 10 and around 9% on TinyImageNet over baseline approaches. ADTrans achieves comparable results with the baselines using CIFAR-10 dataset. Using ResNet-50 model, we test it on STL 10, TinyImageNet, and ImageNet–100. It achieves comparable results on the STL 10 dataset, while notably outperforming all baselines on TinyImageNet and ImageNet-100 datasets. It improves over 0.5% and 4% on the TinyImageNet and ImageNet-100 datasets, respectively. Therefore, our proposed ADTrans is a superior model utility on larger datasets, such as STL 10, TinyImageNet, and ImageNet–100.

The key reason for ADTrans's improvement in model utility lies on its ability to optimize the sequential combination of transformations compared with ad-hoc selection of transformations in baseline approaches. On the CIFAR 10 dataset, ADTrans introduces transformations to the data augmentation, and the order of the transformations is different from the baseline approaches. Although selecting the same set of transformations on the STL 10, TinyImageNet, and ImageNet-100 datasets, ADTrans uses a different order of these transformations resulting in a significant improvement of image classification accuracy. We observe a similar trend when different models are used for contrastive learning. Note that ADTrans selects different orders of transformation given ResNet-18 and ResNet-50 models on three different datasets.

We also evaluate ADTrans for VOC and COCO object detection. For VOC detection, our proposed ADTrans improves 0.6, 0.7, and 1.1 on AP, AP_{50}, and AP_{75}, respectively compared with MOCO v2. For COCO object detection, ADTrans improves 0.4, 0.5, and 0.9 on AP, AP_{50}, and AP_{75}, respectively more than MOCO v2. It slightly improves the performance when the feature is transferred to object detection tasks both on Pascal VOC and COCO.

For COCO instance segmentation, ADTrans outperforms 0.3, 0.3, and 0.6 over MOCO v2 on AP, AP_{50}. and AP_{75}, respectively, over MOCO v2. It achieves better performance than MOCO v2 when the feature is transferred to instance segmentation.

TABLE 26.2
Comparisons on Image Classification

	Backbone	Dataset	Transformation	Linear Acc. (%)
SimCLR	ResNet-18	CIFAR 10	RC, CJ, P, E	48.71
MOCO v2	ResNet-18	CIFAR 10	RC, CJ, P, E	47.53
SimSiam	ResNet-18	CIFAR 10	RC, CJ, P, E	**51.44**
BYOL	ResNet-18	CIFAR 10	RC, CJ, P, E	49.75
ADTrans	ResNet-18	CIFAR 10	T, CJ, RC, P	50.75
SimCLR	ResNet-18	STL 10	RC, CJ, P, E	73.60
MOCO v2	ResNet-18	STL 10	RC, CJ, P, E	65.55
SimSiam	ResNet-18	STL 10	RC, CJ, P, E	65.45
BYOL	ResNet-18	STL 10	RC, CJ, P, E	71.54
ADTrans	ResNet-18	STL 10	T, RC, CJ, E	**77.04**
MOCO v2	ResNet-18	TinyImageNet	RC, CJ, P, E	25.28
SimSiam	ResNet-18	TinyImageNet	RC, CJ, P, E	23.51
BYOL	ResNet-18	TinyImageNet	RC, CJ, P, E	26.44
ADTrans	ResNet-18	TinyImageNet	CJ, RC, P, T	**34.59**
MOCO v2	ResNet-50	STL 10	RC, CJ, P, E	81.98
SimSiam	ResNet-50	STL 10	RC, CJ, P, E	81.34
BYOL	ResNet-50	STL 10	RC, CJ, P, E	81.79
ADTrans	ResNet-50	STL 10	CJ, RC, P, E	**82.01**
MOCO v2	ResNet-50	TinyImageNet	RC, CJ, P, E	44.18
SimSiam	ResNet-50	TinyImageNet	RC, CJ, P, E	44.97
ADTrans	ResNet-50	TinyImageNet	CJ, T, RC, P	**45.51**
MOCO v2	ResNet-50	ImageNet-100	RC, CJ, P, E	80.03
BYOL	ResNet-50	ImageNet-100	RC, CJ, P, E	81.62
ADTrans	ResNet-50	ImageNet-100	RC, E, CJ, P	**85.32**

Table 26.3 shows the comparisons on object detection and instance segmentation on Pascal VOC and COCO. In general, we can see that our proposed ADTrans can achieve superior performance over MOCO v2 when the feature is transferred to downstream tasks, such as object detection and instance segmentation on large-scale datasets.

26.5.3 Ablation Studies

We conduct further experiment on evaluating the impact of the number of transformations in ϕ by applying ADTrans on TinyImageNet and STL 10 with one to five transformations for image classification. Figure 26.3 shows that the classification accuracy increases significantly from one to three transformations. For four and five transformations, the numbers of training steps for each transformation are smaller in the setting of four and five transformations compared with one to three transformations. As a result, it marginally variates or decreases when four or five transformations are applied. Our experimental results offer a guidance for end-users to select

TABLE 26.3
Comparisons on Object Detection and Instance Segmentation

	VOC Detection			COCO Segmentation			COCO Detection		
	AP	AP_{50}	AP_{75}	AP	AP_{50}	AP_{75}	AP	AP_{50}	AP_{75}
MOCO v2	13.7	28.8	10.4	26.1	43.8	27.2	28.7	46.4	29.9
ADTrans	**14.3**	**29.5**	**11.5**	**26.4**	**44.1**	**27.8**	**29.1**	**46.9**	**30.8**

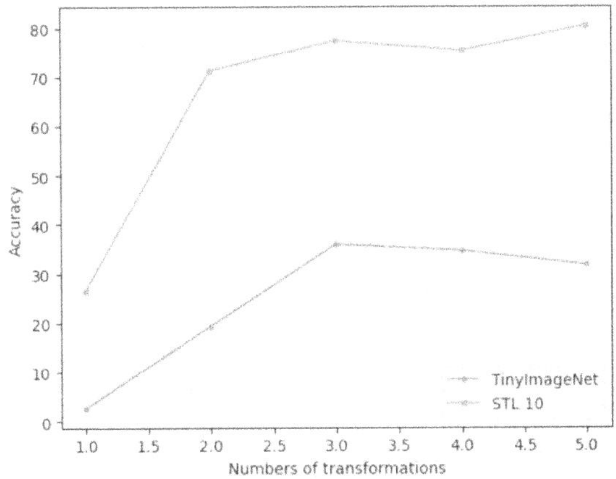

FIGURE 26.3 Image classification accuracy with different number of transformations applied on the original input data when producing positive images during contrastive training.

the best number of transformations in their datasets. In fact, end-users can stop training when the model accuracy starts to marginally variates or decreases.

Furthermore, the sampling probability p controls the proportion of X_{ϕ} to be combined with $X_{\phi \oplus a^*}$. We have investigated the impact of p on the model utility of ADTrans using TinyImageNet by varying from 0.0 to 1.0. Table 26.4 shows that without our distribution shift mitigation, i.e., $p = 0$, contrastive training does not work well with poor utility. However, the model utility significantly increases when p is increased to 0.5. Therefore, the data distribution shifts in adaptive data transformations are necessary and crucial to achieve high model utility.

Finally, we increase the dataset size when dealing with distribution shift problem for the proposed ADTrans. Contrastive learning could have better performance with more data during pre-training. We compare the performance of the proposed ADTrans with MOCO v2, which uses the same data combination strategy to show the effectiveness of ADTrans. We train ADTrans with the same configuration as aforementioned, and combine X_{ϕ} with $X_{\phi \oplus a^*}$ by setting $p = 0.5$ every 200 epochs. We also train MOCO v2 with the same combination strategy so that they have the same dataset size during the whole contrastive pre-training. A slight difference is that we apply data augmentation before contrastive pre-training for MOCO v2.

TABLE 26.4
Sampling Probability *p* and Image Classification Accuracy

Sampling Probability *P*	0.0	0.25	0.50	1.0
Image Classification Acc. (%)	2.56	29.24	34.58	34.59

TABLE 26.5
Comparison of ADTrans with MOCO v2 with the Same Dataset Size Allocation Strategy

	Dataset	Transformation	Linear Acc. (%)
MOCO v2	STL 10	RC, CJ, P, E	73.49
ADTrans	STL 10	T, RC, CJ, E	**77.04**
MOCO v2	TinyImageNet	RC, CJ, P, E	26.39
ADTrans	TinyImageNet	CJ, RC, P, T	**34.59**

Table 26.5 shows that ADTrans achieves around 3.5% over MOCO v2 on STL 10 and around 8% on TinyImageNet. Therefore, we conclude that even with same dataset size, our proposed ADTrans can still significantly improve the performance of contrastive pre-training.

26.6 CONCLUSIONS

In this chapter, we propose the Adaptive Data Transformation (ADTrans) algorithm in contrastive learning to optimize the sequential combination of data transformation. By formulating this problem as a mini-max objective function, ADTrans searches for an optimized transformation, which maximize the objective loss, to learn more knowledge in contrastive training. The sequential combination of transformations is expanded while avoiding data distribution shifts, enabling ADTrans to generate high-quality positive augmented images compared with baseline approaches. Extensive experimental results on several downstream tasks, including image classification, object detection, and instance segmentation, show a remarkable improvement of ADTrans in terms of model utility as compared against baseline approaches.

REFERENCES

Bachman, P., Hjelm, R. D., and Buchwalter, W., "Learning representations by maximizing mutual information across views," *Advances in Neural Information Processing Systems*, vol. 32, 2019.

Bromley, J., Guyon, I., LeCun, Y., Säckinger, E., and Shah, R., "Signature verification using a Siamese time delay neural network," *Advances in Neural Information Processing Systems*, vol. 6, 1993.

Caron, M., Misra, I., Mairal, J., Goyal, P., Bojanowski, P., and Joulin, A., "Unsupervised learning of visual features by contrasting cluster assignments," *Advances in Neural Information Processing Systems*, vol. 33, pp. 9912–9924, 2020.

Caron, M., Touvron, H., Misra, I., Jégou, H., Mairal, J., Bojanowski, P., and Joulin, A., "Emerging properties in self-supervised vision transformers," *Proceedings of IEEE/CVF International Conference on Computer Vision*, pp. 9650–9660, 2021.

Chen, T., Kornblith, S., Norouzi, M., and Hinton, G., "A simple framework for contrastive learning of visual representations," *International Conference on Machine Learning*, pp. 1597–1607, PMLR, Nov. 2020a.

Chen, X., Fan, H., Girshick, R., and He, K., "Improved baselines with momentum contrastive learning," arXiv:2003.04297, 2020b.

Chen, X. and He, K., "Exploring simple Siamese representation learning," *Proceedings of IEEE/CVF Conference on Computer Vision and Pattern Recognition*, pp. 15750–15758, 2021.

Chen, X., Xie, S., and He, K., "An empirical study of training self-supervised vision transformers," *Proceedings of the IEEE/CVF International Conference on Computer Vision*, pp. 9640–9649, 2021.

Coates, A., Ng, A., and Lee, H., "An analysis of single-layer networks in unsupervised feature learning," *Proceedings of the Fourteenth International Conference on Artificial Intelligence and Statistics*, pp. 215–223, JMLR Workshop and Conference Proceedings, Jun. 2011.

Cui, J., Zhong, Z., Tian, Z., Liu, S., Yu, B., and Jia, J., "Generalized parametric contrastive learning," *IEEE Transactions on Pattern Analysis and Machine Intelligence*, 2023.

Everingham, M., Van Gool, L., Williams, C. K., Winn, J., and Zisserman, A., "The pascal visual object classes (VOC) challenge," *International Journal of Computer Vision*, vol. 88, pp. 303–338, 2010.

Grill, J. B., Strub, F., Altché, F., Tallec, C., Richemond, P., Buchatskaya, E., Doersch, C., Pires, B. A., Guo, Z. D., Azar, M. G., Piot, B., Kavukcuoglu, K., Munos, R., and Valko, M., "Bootstrap your own latent-a new approach to self-supervised learning," *Advances in Neural Information Processing Systems*, vol. 33, pp. 21271–21284, 2020.

He, K., Fan, H., Wu, Y., Xie, S., and Girshick, R., "Momentum contrast for unsupervised visual representation learning," *Proceedings of IEEE/CVF Conference on Computer Vision and Pattern Recognition*, pp. 9729–9738, 2020.

He, K., Zhang, X., Ren, S., and Sun, J., "Deep residual learning for image recognition," *Proceedings of IEEE Conference on Computer Vision and Pattern Recognition*, pp. 770–778, 2016.

Krizhevsky, A. and Hinton, G., "Learning multiple layers of features from tiny images," *Technical Report*, University of Toronto, Toronto, 2009.

Lin, T. Y., Maire, M., Belongie, S., Hays, J., Perona, P., Ramanan, D., . . . and Zitnick, C. L., "Microsoft coco: Common objects in context," *Computer Vision–ECCV 2014: 13th European Conference, Zurich, Switzerland, September 6–12, 2014, Proceedings, Part V 13*, pp. 740–755, Springer International Publishing, 2014.

Oord, A. V. D., Li, Y., and Vinyals, O., "Representation learning with contrastive predictive coding," arXiv:1807.03748, 2018.

Peng, X., Wang, K., Zhu, Z., Wang, M., and You, Y., "Crafting better contrastive views for Siamese representation learning," *Proceedings of IEEE/CVF Conference on Computer Vision and Pattern Recognition*, pp. 16031–16040, 2022.

Ren, S., He, K., Girshick, R., and Sun, J., "Faster R-CNN: Towards real-time object detection with region proposal networks," *Advances in Neural Information Processing Systems*, vol. 28, 2015.

Ruder, S., "An overview of gradient descent optimization algorithms," arXiv:1609.04747, 2016.

Russakovsky, O., Deng, J., Su, H., Krause, J., Satheesh, S., Ma, S., Huang, Z., Karpathy, A., Khosla, A., Bernstein, M., Berg, A. C., and Fei-Fei, L., "ImageNet large scale visual recognition challenge," *International Journal of Computer Vision*, vol. 115, pp. 211–252, 2015.

Takahashi, R., Matsubara, T., and Uehara, K., "Ricap: Random image cropping and patching data augmentation for deep CNNs," *Asian Conference on Machine Learning*, pp. 786–798, PMLR, Nov. 2018.

Tian, Y., Krishnan, D., and Isola, P., "Contrastive multiview coding," *Computer Vision–ECCV 2020: 16th European Conference, Glasgow, August 23–28, 2020, Proceedings, Part XI 16*, pp. 776–794, Springer International Publishing, 2020a.

Tian, Y., Sun, C., Poole, B., Krishnan, D., Schmid, C., and Isola, P., "What makes for good views for contrastive learning?" *Advances in Neural Information Processing Systems*, vol. 33, pp. 6827–6839, 2020b.

Wang, X. and Qi, G. J., "Contrastive learning with stronger augmentations," *IEEE Transactions on Pattern Analysis and Machine Intelligence*, vol. 45, no. 5, pp. 5549–5560, 2022.

Zhang, J. and Ma, K., "Rethinking the augmentation module in contrastive learning: Learning hierarchical augmentation invariance with expanded views," *Proceedings of IEEE/CVF Conference on Computer Vision and Pattern Recognition*, pp. 16650–16659, 2022.

Index

I

L

M

N

O

P

Q

R

S

T

U

Z

For Product Safety Concerns and Information please contact our EU
representative GPSR@taylorandfrancis.com
Taylor & Francis Verlag GmbH, Kaufingerstraße 24, 80331 München, Germany

www.ingramcontent.com/pod-product-compliance
Ingram Content Group UK Ltd.
Pitfield, Milton Keynes, MK11 3LW, UK
UKHW021939250925
463327UK00007B/91
9781032755304